MYP Mathe

A concept-based approach

4&5
Standard

Rose Harrison • Clara Huizink
Aidan Sproat-Clements • Marlene Torres-Skoumal

OXFORD

How to use this book

Chapters

Problem-solving introduces Pólya's problem-solving steps and other strategies as a well-defined procedure for solving complex problems. All mathematics depends on problem-solving, and so problem-solving questions are included and highlighted throughout the book.

Chapters 1–3 each focus on one of the key concepts:

Form, Relationships and Logic

Chapters 4–15 each focus on one of the twelve related concepts for Mathematics:

Representation, Space, Simplification, Quantity, Measurement, Patterns, Change, Equivalence, Generalization, Justification, Models, and Systems

Creating your own units

The suggested unit structure opposite shows just one way of grouping all the topics from different chapters, from both *Standard* and *Extended*, to create units. These units have been created following the official guidelines from the IB Building Quality Curriculum. Each unit is driven by a meaningful statement of inquiry and is set within a relevant global context.

- Each chapter focuses on one of the IB's twelve related concepts for Mathematics, and each topic focuses on one of the three key concepts. Hence, these units, combining topics from different chapters, connect the key and related concepts to help students both understand and remember them.
- Stand-alone topics in each chapter teach mathematical skills and how to apply them, through inquiry into factual, conceptual, and debatable questions related to a global context. This extends students' understanding and ability to apply mathematics in a range of situations.
- You can group topics as you choose, to create units driven by a contextualized statement of inquiry.
- This book covers the MYP Standard skills framework. *MYP Mathematics 4 & 5 Extended* covers the Extended skills.

Using *MYP Mathematics 4 & 5 Standard* with an existing scheme of work

If your school has already established units, statements of inquiry, and global contexts, you can easily integrate the concept-based topics in this book into your current scheme of work. The table of contents on page vii clearly shows the topics covered in each concept-based chapter. Your scheme's units may assign a different concept to a given topic than we have. In this case, you can simply add the concept from this book to your unit plan. Most topics include a Review in context, which may differ from the global context chosen in your scheme of work. In this case, you may wish to write some of your own review questions for your global context, and use the questions in the book for practice in applying mathematics in different scenarios.

Suggested plan

The units here have been put together by the authors as just one possible way to progress through the content.

UNIT 1

Topics: Working with sets of data, working with grouped data, histograms

Global context: **Globalization and sustainability**
Key concept: **Relationships**

UNIT 2

Topics: Scatter graphs and linear regression, drawing reasonable conclusions, data inferences

Global context: **Identities and relationships**
Key concept: **Relationships**

UNIT 3

Topics: Equivalence transformations, inequalities, non-linear inequalities

Global context: **Identities and relationships**
Key concept: **Form**

UNIT 4

Topics: Rational and irrational numbers, direct and indirect proportion, fractional exponents

Global context: **Globalization and sustainability**
Key concept: **Form**

UNIT 5

Topics: Currency conversion, absolute value, converting units and reasoning quantitatively

Global context: **Globalization and sustainability**
Key concept: **Relationships**

UNIT 6

Topics: Using patterns to work backwards, quadratic functions in 2D space, equivalent forms

Global context: **Scientific and technical innovation**
Key concept: **Form**

UNIT 7

Topics: Finding patterns in sequences, making generalizations from a given pattern, arithmetic and geometric sequences

Global context: **Scientific and technical innovation**
Key concept: **Form**

UNIT 8

Topics: Simple probability, probability systems, conditional probability

Global context: **Identities and relationships**
Key concept: **Logic**

UNIT 9

Topics: Transforming functions, exponential functions

Global context: **Orientation in space and time**
Key concept: **Form**

UNIT 10

Topics: Right-angled triangles and trigonometric ratios, sine and cosine functions

Global context: **Scientific and technical innovation**
Key concept: **Relationships**

UNIT 11

Topics: Circle segments and sectors, volumes of 3D shapes, 3D orientation

Global context: **Personal and cultural expression**
Key concept: **Relationships**

UNIT 12

Topics: Using circle theorems, intersecting chords, problems involving triangles

Global context: **Personal and cultural expression**
Key concept : **Logic**

UNIT 13

Topics: Algebraic fractions, equivalent methods, rational functions

Global context: **Scientific and technical innovation**
Key concept: **Form**

Access your support website for more suggested plans for structuring units:

www.oxfordsecondary.com/myp-mathematics

Chapter openers

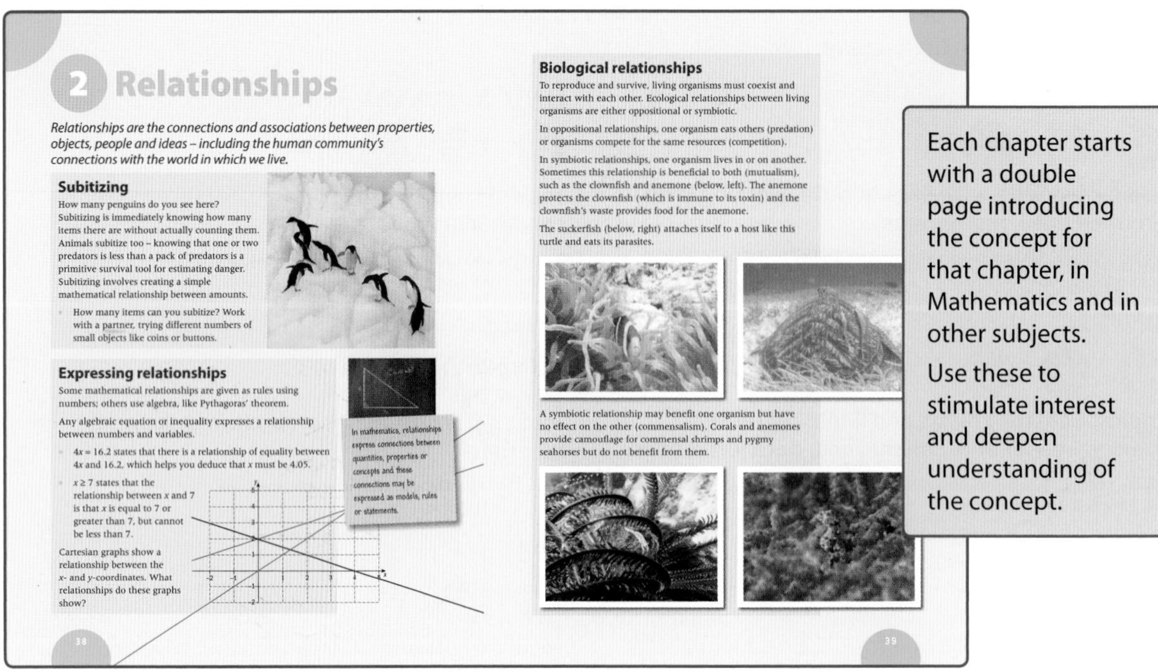

Each chapter starts with a double page introducing the concept for that chapter, in Mathematics and in other subjects.

Use these to stimulate interest and deepen understanding of the concept.

Topic opening page

Objectives – the mathematics covered in this topic.

Key concept for the topic.

ATL – the Approach to Learning taught in this topic.

Statement of Inquiry for this unit. You may wish to write your own.

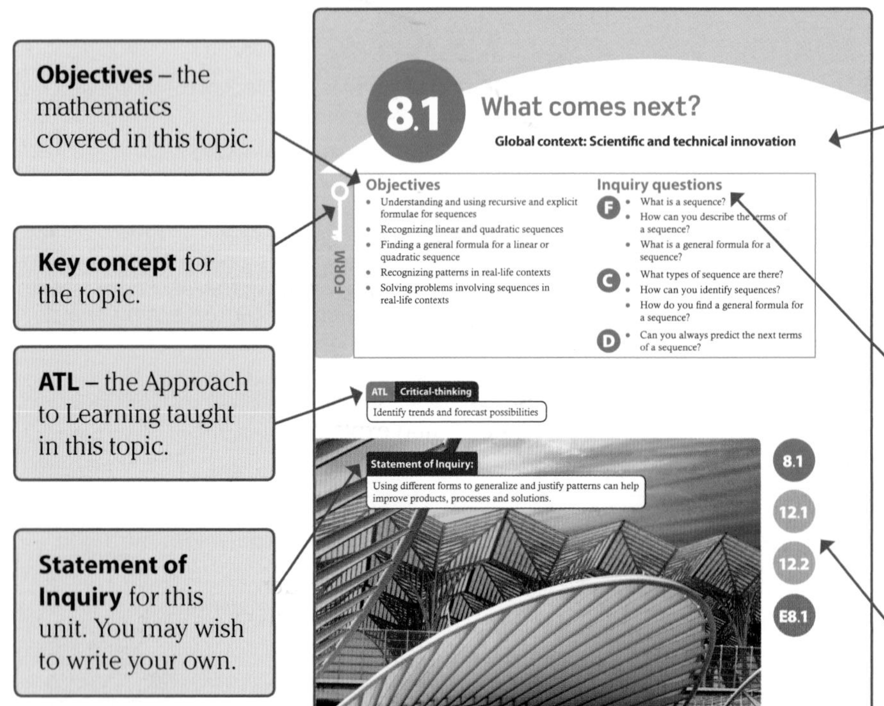

Global context – some questions in this unit are set in this context. You may wish to write your own global context to engage your students.

Inquiry questions – the factual, conceptual, and debatable questions explored in this topic.

Unit plan – shows the topics in the *Standard* and *Extended* books that you could teach together as a unit.

Learning features

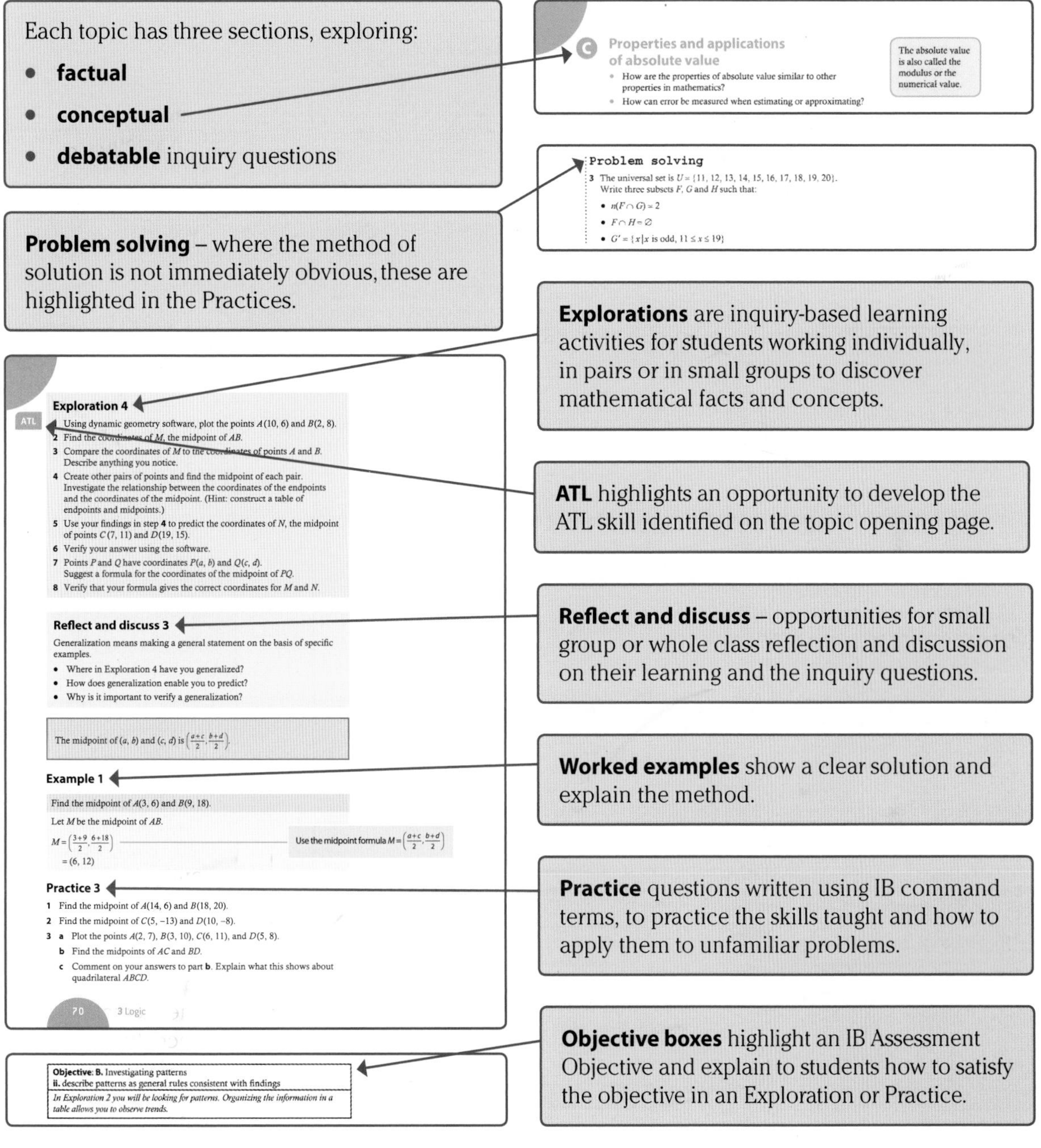

Technology icon

Using technology allows students to discover new ideas through examining a wider range of examples, or to access complex ideas without having to do lots of painstaking work by hand.

This icon shows where students could use Graphical Display Calculators (GDC), Dynamic Geometry Software (DGS) or Computer Algebra Systems (CAS).
Places where students should not use technology are indicated with a crossed-out icon.

The notation used throughout this book is largely that required in the DP IB programs.

Each topic ends with:

Summary

- A **chord** is a line segment with its endpoints on a circle.
- A **diameter** is a chord that passes through the center of a circle.
- A **secant** is a line that intersects a circle at two points.
- A **tangent** is a line that touches a circle at only one point.

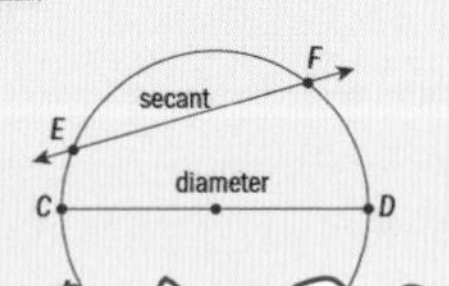

- The larger arc is the **major arc**. The smaller arc is the **minor arc**.

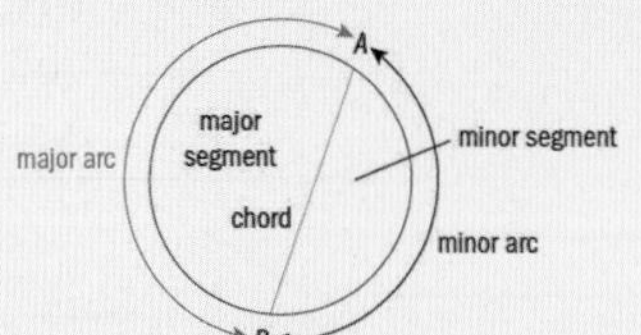

- A **sector** is a 'slice' of a circle between two radii. The center of a circle is usually labelled *O*, so the sector is sector *BOC*. 'Sector' usually means the minor sector.

Summary of the key points

Mixed practice

1 1 EUR = 1.10 USD.

a Convert 950 EUR into USD.

b Convert 2750 USD into EUR.

2 1 Mexican Peso (MXN) = 0.055 US Dollars (USD).

a Convert 4400 MXN into USD.

b A commission of 1% is charged. **Find** the cost of the commission in USD.

c **Calculate** the final number of USD received.

3 1 CHF = 0.72 GBP. A commission of 2% is charged. Convert 1600 CHF into GBP.

4

	Sell RUB	Buy RUB
1 GBP	107.30	105.80

a Jasmine is going on holi...

6 The exchange rate from US Dollars (USD) to Thai Baht (THB) is 1 USD = 33.45 THB. Give the answers to the following correct to two decimal places.

a **Find** the value of 115 US Dollars in Thai Baht.

b **Calculate** the value of 1 THB in USD.

c Alexis receives 600 New Zealand Dollars (NZD) for 14 670 THB. **Calculate** the value of 1 THB in NZD.

d **Calculate** the value of the US Dollar in New Zealand Dollars.

7 A bank sells 1 Australian Dollar (AUD) for 0.72 Euros (EUR) and buys 1 AUD for 0.70 EUR. Frida wants to exchange 800 AUD for EUR.

a **Find** how many Euros Frida will receive.

Mixed practice – summative assessment of the facts and skills learned, including problem-solving questions, and questions in a range of contexts.

Review in context

Globalization and sustainability

1 Twenty-five women from Mexico were asked how many children they had. The results are shown in the frequency histogram.

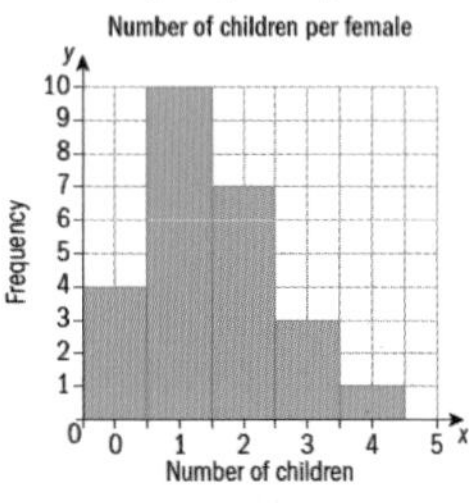

a **Show that** the mean number of children per ...

Number of children	0	1	2	3	4	5
Frequency	4	8	5	4	5	2

Use the results from parts **a** and **b** to **describe** and **compare** the distribution of the numbers of children per woman in Australia and Mexico.

2 The quantity of rainforests in the Amazon region has been rapidly declining over the past decades. This frequency table show the percentage of deforestation for 20 South American countries.

0.4%	0.3%	1.6%	2.1%	0.3%
2.7%	2.2%	3.5%	2.0%	9.2%
3.2%	0%	3.0%	3.3%	4.0%
0.6%	0.5%	4.0%	1.7%	2.6%

Review in context – summative assessment questions within the global context for the topic.

Reflect and discuss

How have you explored the statement of inquiry? Give specific examples.

Reflection on the statement of inquiry

Table of contents

Access your support website:
www.oxfordsecondary.com/myp-mathematics

Problem-solving

Global context: Scientific and technical innovation

LOGIC

Objectives

- Applying Pólya's problem-solving steps to solve any type of problem
- Selecting and applying appropriate mathematical strategies to solve problems
- Checking if a solution makes sense in the context of the problem

Inquiry questions

F
- What are Polya's steps in solving a problem?

C
- Which problem-solving strategies are the most useful?

D
- How can you find the information you need to solve a problem in real-life?
- How useful is it to estimate?

ATL Communication

Use appropriate strategies for organizing complex information

You should already know how to:

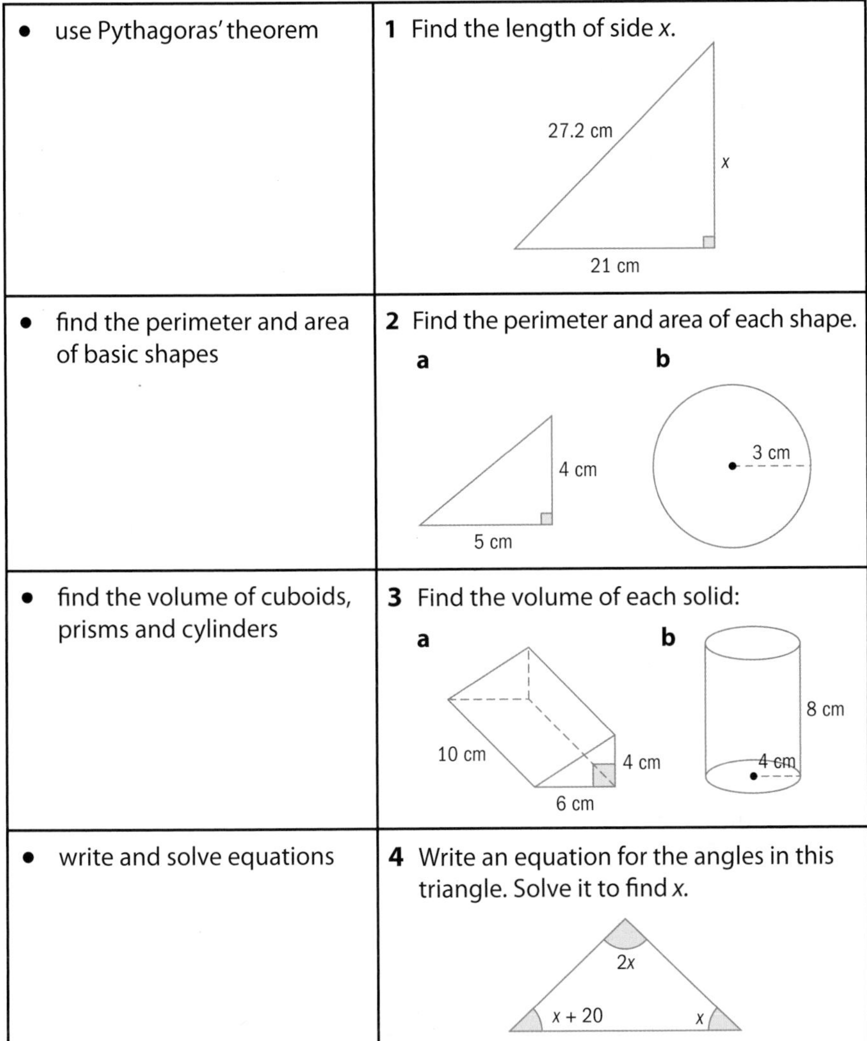

• use Pythagoras' theorem	**1** Find the length of side x.
• find the perimeter and area of basic shapes	**2** Find the perimeter and area of each shape. **a** **b**
• find the volume of cuboids, prisms and cylinders	**3** Find the volume of each solid: **a** **b**
• write and solve equations	**4** Write an equation for the angles in this triangle. Solve it to find x.

Pólya's problem-solving steps

- What are Polya's steps in solving a problem?

Sometimes, the method to solve a problem isn't immediately obvious. Some problems may involve a lot of complex mathematics, but others may be very simple. The same question could be considered easy for some people but challenging for others.

Reflect and discuss 1

Are the following questions easy or challenging? Does everyone agree?

- For $x^2 + 4x = 100$, what is the value of x?
- One liter per minute of water flows into a cylindrical tank of height 1.5 m and radius 30 cm. How long will it take to fill the tank?
- What is the highest common factor of 1752 and 356?
- How many triangles are there in the diagram at the right?

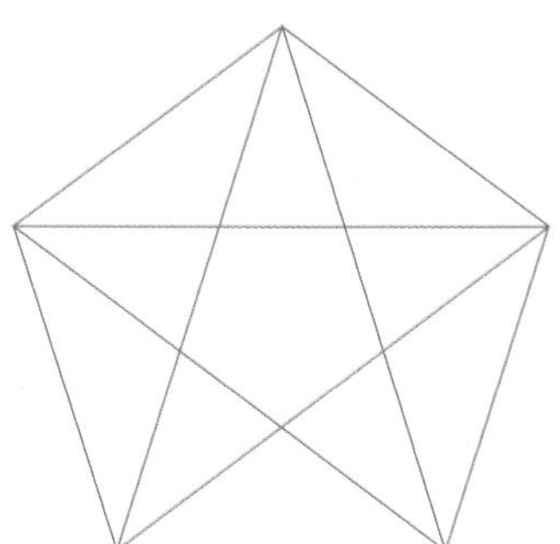

George Pólya (1887–1985) was a Hungarian professor of mathematics. He described the following problem-solving steps to use for all problems in mathematics and in other subjects.

Pólya's problem-solving steps[1]

1 Understand the problem

- **a** Make sure you understand all the words
- **b** State what you need to find
- **c** State what you already know
- **d** Identify the unknowns
- **e** Identify any conditions on the unknowns
- **f** Draw a picture or diagram to help you understand the problem
- **g** Restate the problem in your own words to make sure that you understand what is being asked

2 Devise a plan

- **a** Guess and check
- **b** Look for a pattern
- **c** Eliminate possibilities
- **d** Use a formula
- **e** Solve an equation
- **f** Think of other strategies you could use

3 Carry out your plan

- **a** Show each step so that someone reading your work can follow it
- **b** Check that your intermediate answers make sense
- **c** If you are stuck, try another way
- **d** If you get stuck, devise another plan

4 Look back

- **a** Check your final answer, if possible
- **b** Does your answer make sense in the context of the problem?
- **c** Make sure your final answer answers the question
- **d** What worked? What didn't work?
- **e** Could you have found the answer in another (perhaps better) way?

You already have a range of strategies for solving problems in mathematics. Some you will have been taught, others you may have discovered for yourself.

[1]Pólya, George (1945). *How to Solve It* Princeton University Press.

Exploration 1

Solve these problems using your usual method. For each one, write down the steps you use from Pólya's list.

1 175 − 123 =

2 Find the value of x:

a $5x = 35$ **b** $4x - 2 = 30$

c $\frac{3x-2}{7} = 7$ **d** $\frac{4}{x-4} = \frac{1}{2}$

3 In the diagram to the right, $\angle KLM = 120°$ and $\angle ULM = 30°$.
Find angle KLU.

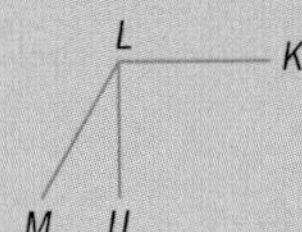

4 Find the unknown angle:

a

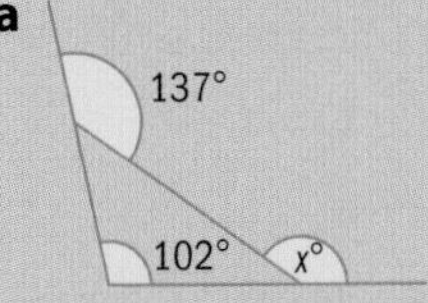

b

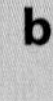
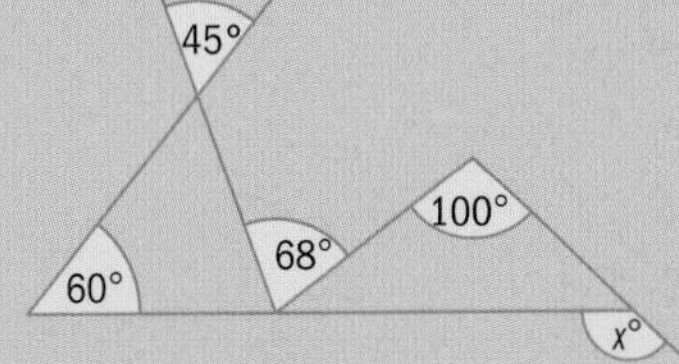

5 Find x:

a

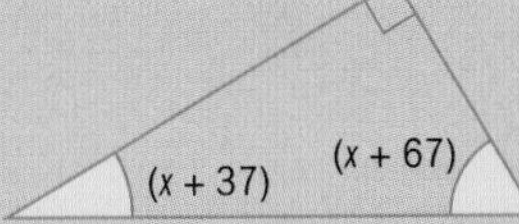

b

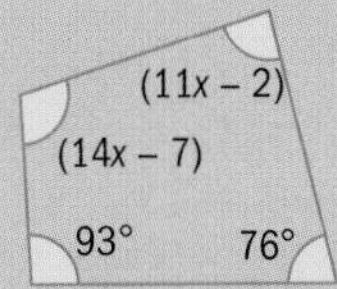

6 Find the area of each triangle. Where necessary, round your final answer to the nearest tenth.

a

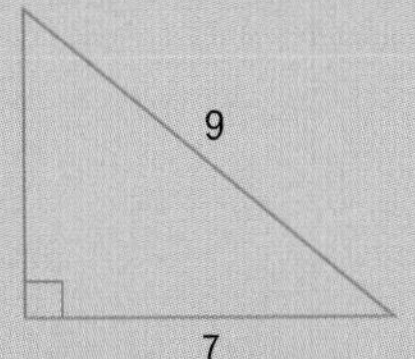

b

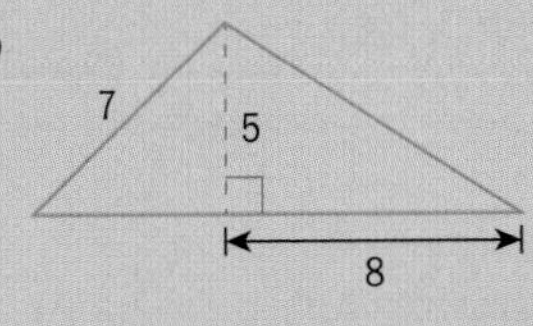

7 Find the volume of each solid. Give your answers to 1 decimal place.

a

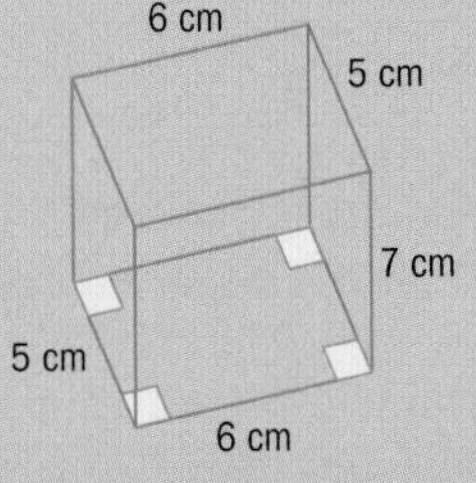

b

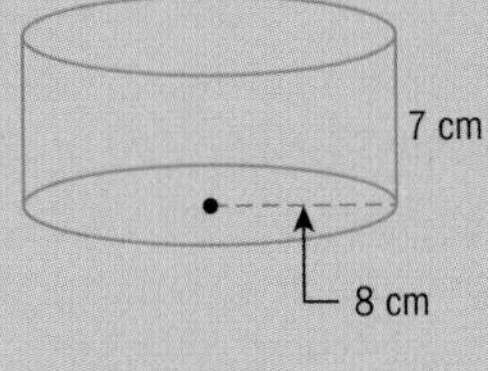

8 Find the volume of a cube that has surface area 384 m^2.

9 In Canada, how many ways can you make a dollar with coins?
The Canadian coins are nickels (5 cents), dimes (10 cents), quarters (25 cents) and dollars (or loonies). One dollar is 100 cents.

> 'The main point in mathematics teaching is to develop tactics of problem solving.'
>
> *George Pólya*

Reflect and discuss 2

- When you were solving the problems in Exploration 1, which of Pólya's steps did you use most often? Were there any you didn't use? Did everyone use the same steps?
- Look back at the four problem-solving steps. Are they equally important? Explain. Would it matter if you missed one?

'If you can't solve a problem, then there is an easier problem you can solve. Find it.'

George Pólya

ATL

Example 1

It takes Anna 20 minutes to wash a car, and Bianka 30 minutes to wash a similar car. How long will they take to wash a car together?

t = time to wash one car together

$t_A = 20$ min

$t_B = 30$ min

1 Understand the problem
b State what you need to find
c State what you already know
d Identify the unknowns

r = rate of washing a car together, 1 car every t minutes.

r_A = Anna's rate

r_B = Bianka's rate

$r = r_A + r_B$

Use $r = \frac{1 \text{ car}}{t}$ to find t.

2 Devise a plan
The time to wash one car depends on the rate of washing – how many cars they wash in a given time.
Their rate of washing together is the sum of their individual rates.

$r_A = \frac{1 \text{ car}}{20 \text{ min}}$

$r_B = \frac{1 \text{ car}}{30 \text{ min}}$

3 Carry out the plan
a Show each step

$r = r_A + r_B = \frac{1 \text{ car}}{20 \text{ min}} + \frac{1 \text{ car}}{30 \text{ min}}$

$= \frac{3 \text{ cars}}{1 \text{ hour}} + \frac{2 \text{ cars}}{1 \text{ hour}}$

Change to fractions with common denominator 1 hour.
b Check that your intermediate answers make sense.

$= \frac{5 \text{ cars}}{1 \text{ hour}}$

Together they wash 5 cars per hour,

which is $\frac{1 \text{ hour}}{5} = 12$ minutes per car.

They wash more cars together than they do on their own.

$t = 12$ minutes

If it takes each girl 20 minutes to wash a car, then together they could wash a car in 10 minutes. If they each took 30 minutes, then together they could wash a car in 15 minutes. Since they take 20 and 30 minutes respectively to wash a car, the answer should be somewhere in between 10 and 15 minutes.

4 Look back
a Check your answer

Anna is the quickest and takes 20 min. Together they should be quicker than that.

b Does your answer make sense in the context of the problem?

They will take 12 minutes to wash one car together.

c Make sure your final answer answers the question

In Example 1, most of the work is in step 2: Devise a plan. The main causes of incorrect solutions are:

- rushing to step 3 and starting to do calculations without a clear plan
- leaving out step 4, and not checking answers in the context of the problem.

If your answer is: a pencil weighs 50 kg or a plane travels at 20 km/h, common sense should tell you something must be wrong in your solution.

Practice 1

Solve these problems using Pólya's steps. Show your plan and your working clearly. Remember to check that your answers make sense in the context of the problem.

1 Light travels six trillion miles in one year. If the radius of the Earth is approximately 4000 miles, find how many times around the Earth light could travel in one year.

2 Under ideal conditions, algae can double in area every day and cover an entire pond in 1 month and 2 days. Determine how many days it will take for the algae to cover just half of the pond.

3 A team can row 40 km in 2 hours when rowing with the current, but only 16 km in 2 hours when rowing against the current. Determine the team's rowing speed when there is no current.

4 Sidney has two different acid solutions: A is a 50% acid solution, and B is an 80% acid solution. Determine how much of each solution he needs to create 200 ml of a 68% solution.

5 Corinne and Justine sold jewelry at a bazaar. They sold some for \$9.50 a piece, and then dropped the price to \$7.50 a piece. In total they sold 90 pieces for \$721. Find how many they sold at \$9.50.

6 Ilhan began her triathlon at 7:22 am. She crossed the finish line at 8:07 pm, after swimming 2.4 miles, cycling 112 miles and running 26.2 miles. Find her average speed for the race.

C Problem-solving strategies

- Which problem-solving strategies are the most useful?

The next examples show some strategies you can use in problem-solving. They may help you get started when you are stuck.

One definition of problem-solving is 'knowing what to do when you don't know what to do'.

Strategy: Take 1

Sometimes you can simplify a problem by replacing one of the numbers with 1. For example, how much does 1 item cost? Or how many hours does it take 1 person to complete a task?

Example 2

Three painters can paint a house in 8 hours. How many painters are needed to paint the house in 6 hours?

n = number of painters to paint the house in 6 hours

3 painters take 8 hours

$n > 3$

1 Understand the problem
- **b** State what you need to find
- **c** State what you already know
- **e** Identify any conditions on the unknowns. Assume all painters paint at the same rate. To paint in less time, you need more painters.

1 painter takes t hours

n painters take 6 hours

2 Devise a plan
Strategy: Take 1
Find out how long it takes one painter to paint the house.
Work out how many painters you need to paint it in 6 hours.

3 painters $\rightarrow$ 8 hours

1 painter $\rightarrow$ 24 hours

4 painters $\rightarrow$ 6 hours

$n = 4$

3 Carry out the plan
One painter takes 3 times as long as 3 painters.
4 painters take $\frac{1}{4}$ the time of 1 painter.

4 painters are needed to paint the house in 6 hours.

4 Look back
The final answer satisfies $n > 3$, which fits with expectations.

Strategy 2: Use easy numbers

Replacing numbers in the problem with 'easy numbers' can help you see the method to use in the original problem.

Example 3

Potatoes cost \$1.56 per kilogram. Shona spent \$1.95 on potatoes. How many kilograms of potatoes did she buy?

x = weight of potatoes

$x > 1$

1 Understand the problem
Shona paid more than \$1.56, so she bought more than 1 kg.

Assume potatoes cost \$2 per kg, and a customer bought \$10 worth.

That means the customer bought 5 kg of potatoes.

$\frac{10}{2} = 5$

2 Devise a plan
Strategy: Use easy numbers
Rewrite with numbers that are easier to work with, to see what operations to use to solve the problem.

▶ Continued on next page

$$\frac{\text{Amount paid}}{\text{price per kg}} = \text{weight in kg}$$

$$\frac{\text{Amount paid}}{\text{price per kg}} = \frac{1.95}{1.56} = 1.25 \text{ kg}$$

3 Carry out the plan

Shona bought 1.25 kg of potatoes.

4 Look back

a Check your answer

b Does it make sense?

1.25 kg satisfies $x > 1$.

\$1.95 is less than \$3.12 for 2 kg, so 1.25 kg is a reasonable answer.

Reflect and discuss 3

- In Example 3, how did substituting easier numbers help solve the problem?
- How are the 'Take one' and 'Use easy numbers' strategies related?

Practice 2

Solve these problems using Pólya's steps. Use one of the two strategies seen so far. Remember to check that your answers make sense in the context of the problem.

1 It takes 6 students 2 hours to stuff envelopes for Back to School Night. Determine how many students would be needed to finish the job in a quarter of an hour.

2 Jordan has an average of 82% on her first four tests.

a Find the score she needs on the next test for her average to be 84%.

b Find the average score she needs on the next two tests for her overall average to be 85%.

c Determine the minimum number of tests that could be left to take if Jordan wanted to have an average of 90%, and assuming that she could score as high as possible on all remaining tests.

3 Two painters work for a painting company. One of them can paint a 200 m^2 area in 5 minutes while the other one takes 8 minutes to paint the same area. Find how long it would take them to paint an area of 1300 m^2 if they painted it together.

4 Light travels 3 000 000 km in ten seconds. If Earth is 150 million kilometers from the sun, find how many minutes it takes for light from the sun to reach our planet.

5 After installing a new pool, Kedrick's family is going to use two pumps to fill it. Pump 1 can fill a pool in 8 hours while pump 2 can fill it in 12 hours.

a Determine how fast the pool can be filled using both pumps.

b If both pumps start at the same time, but pump 1 breaks down after an hour, find how long it will take to fill the pool (in total).

6 Thien lives on the 11th floor of a building. He has to climb 48 steps just to get to the third floor. Determine how many steps he needs to take to get to his floor.

7 There were enough provisions at a space station to feed 20 people for 80 days. After 20 days into the mission, 10 astronauts returned to Earth. Assuming the rate of food consumption doesn't change, determine how many more days the food will last.

8 Diagonals in polygons are line segments that connect the vertices of the polygon, excluding its sides.

a Find how many diagonals there are in a 20-sided polygon.

b Find how many diagonals there are in an n-sided polygon.

Strategy 3: Draw it!

It often helps to draw a diagram, especially if the problem involves geometry.

Example 4

A water sprinkler is placed in the middle of a rectangular lawn 6 m long and 5 m wide. The sprinkler sprays water to a distance of 2 m. What proportion of the lawn is watered by the sprinkler?

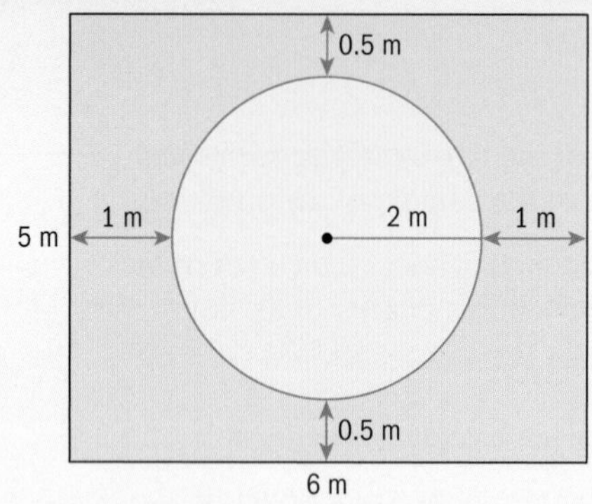

1 Understand the problem

f Draw a diagram to help you understand the problem. A fairly accurate drawing can help you see whether your answer makes sense.

p = proportion of lawn watered by sprinkler.

p = close to $\frac{1}{2}$

b State what you need to find

c Identify any conditions on the unknowns. Based on this fairly accurate diagram, p = close to $\frac{1}{2}$

$$p = \frac{\text{area of the watered circle}}{\text{area of the whole lawn}} = \frac{A_c}{A_l}$$

2 Devise a plan

$A_c = \pi r^2$ where $r = 2$ m

$A_l = l \times w$ where $l = 6$ m and $w = 5$ m

To find the proportion, you first need to find the area of the circle and the area of the rectangle.

$A_c = \pi r^2 = \pi \times 2^2 = 4\pi$

$A_l = l \times w = 6 \times 5 = 30$

$$p = \frac{A_c}{A_l} = \frac{4\pi}{30} = \frac{2\pi}{15}$$

3 Carry out your plan

Always use exact values until the final answer. Do not use 3.14 for π.

$p = 0.4188790205$

$p = 0.42 = 42\%$

4 Look back

In this situation, it is reasonable to round the final answer to the nearest percent. Final answer satisfies p = close to $\frac{1}{2}$

42% of the lawn is watered by the sprinkler.

c Answer the question.

Practice 3

1 A 6 cm long prism has a base that is a right-angled triangle with two sides of length 5 cm. Find its volume.

2 The square base of the Great pyramid at Giza measures 230.4 m on each side. Its slant height is 186.4 m.

a Determine the ratio of the height of the pyramid to the perimeter of its base.

b Explain how this ratio is related to the number π.

c Discuss whether you think this is a coincidence.

Strategy 4: Work backwards

Some problems give you an answer and ask you to find a part of the question. For example, you are given the total price of oranges and are asked how much one orange costs, or you are given the area of a square and asked to find its side length.

Example 5

The diagonal of one side of a cube is 11.3 cm to 3 s.f. Find the cube's volume.

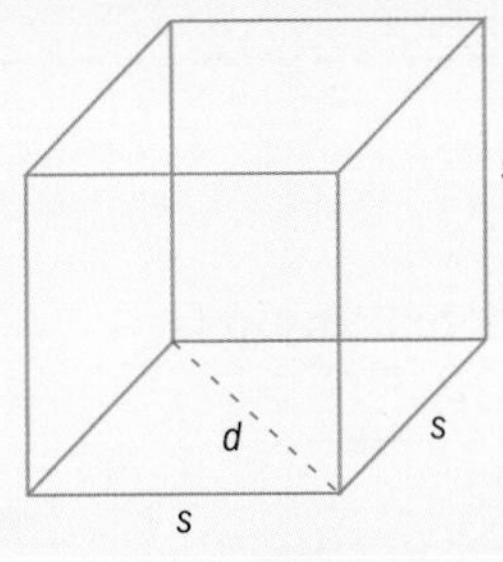

1 Understand the problem
f Draw a diagram

$d = 11.3$ cm

Volume of cube $= s^3$

Need to find s.

c State what you know
b State what you need to find

$s^2 + s^2 = d^2$

$2s^2 = d^2$

2 Devise a plan
Use Pythagoras' theorem with d and s to find s.
Use s to find the volume of the cube.

$2s^2 = 11.3^2$

$2s^2 = 127.69$

$s^2 = 63.845$

$s = 7.990$ cm

Volume $= s^3 = 7.990^2 = 510.1$ cm^3

3 Carry out the plan
The final answer should be rounded to 3 s.f. only at the end. All previous answers must be rounded to *at least* one more place as needed.
b Check your answers make sense.
The side s is shorter than the diagonal.

The volume of the cube is 510 cm^3.

4 Look back
c Answer the question.

Practice 4

1 The height of a cylinder is twice as long as its diameter. Its volume is $500\pi\,\text{cm}^3$. Find the area of its base.

2 The diagonal of a cube (that passes through the middle of the cube, as shown in the diagram at right) measures 17.32 cm. Find the volume of the cube.

3 A box with no top has a height that is six times its length and a width that is $\frac{2}{3}$ of its height. If the surface area is $576\,\text{cm}^2$, find the dimensions of the box.

4 When milk freezes, its volume increases by one-fifteenth. Find the fraction by which its volume will decrease when it melts back into a liquid state.

5 It takes Nathan twice as long to mow the lawn with the pushmower than with the electric mower. One day, he used the electric mower for 30 minutes. Then it broke down. He took 20 minutes to finish mowing the lawn with the pushmower.

Find how long it takes Nathan to mow the whole lawn using only the electric mower.

6 The surface area of a cube-shaped box without a top is $252\,\text{cm}^2$. Find the volume of the box.

7 Suppose you have a 4L container and a 9L container. If you had access to a source of running water, explain how you could measure out exactly 6L of water.

L = liter

Reflect and discuss 4

- Compare the methods you and others used to solve problems so far. Which ones were used most often?
- How can different methods lead to the same solution?
- Is the best strategy always the simplest?

D When is an estimate a solution that is 'good enough'?

- How can you find the information you need to solve a problem in real-life?
- How useful is it to estimate?

In the practice problems you have solved so far, all the information you need to solve the problem was given in the question. In real-life situations, sometimes there is a lot of information and you have to choose what you need. Sometimes there is not enough information and you have to find extra information, or make estimates. Then your answer will also be an estimate.

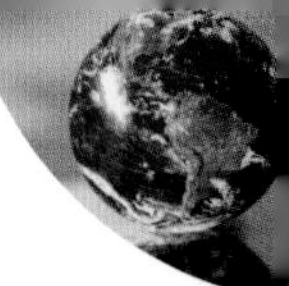

Exploration 2

A square room has four walls, each 3 m long and 2.2 m tall. Paint comes in 10-liter or 5-liter pots. One liter of paint covers an area of 1.5 m^2. The room's one door measures 1.8 m by 0.8 m and its one window is 1.5 m by 1 m.

Pedro calculates the area of the walls and ignores the door and window.

Mia calculates the area of four walls, and subtracts the areas of the door and window.

Work out the number and sizes of tins of paint needed using Mia's and Pedro's areas. Justify whether it is better to calculate accurately or to estimate the amount of paint used in this situation.

Reflect and discuss

In which situations is it better to estimate than to make exact calculations? When is it important to be as exact as possible?

When is it better to over-estimate than to under-estimate?

Practice 5

Some of these problems have too much information and some do not have enough. For each problem, state which information is irrelevant, and write down which information is missing in order for you to solve it.

1 Find the area of this parallelogram.

2 Amy's five friends come to visit her by bus. Each pays \$4 for a ticket. For supper, Amy buys four pizzas at \$5.75 each, 500 g of strawberries at \$2.75 per kilogram, and a tub of icecream for \$4.75. Find how much change she gets.

3 Elizabeth would like to buy a television for her apartment. She knows that a 42-inch television has a diagonal that measures 42 inches. The cabinet where she wants to put the television measures 30 inches wide, 30 inches tall and 24 inches deep.

Determine whether or not the television could fit.

4 Jack gets up at 7:00 am and needs to catch the bus to school at 7:53 am. The bus stop is 400 m from his house. Determine at what time he needs to leave the house.

5 Josip just bought a lightweight cylindrical water bottle where the radius of the bottle is half of its height. When full, it weighs 450 grams and holds a volume of 770 cm^3. Find the dimensions of the water bottle.

6 On her way home, Natalee's plane took off at 8:10 am and flew at an angle of 30° with the ground. If the plane's speed was 285 km/h and it rose to an altitude of 8000 meters, what horizontal distance did it cover during that time?

Summary

Pólya's problem-solving steps

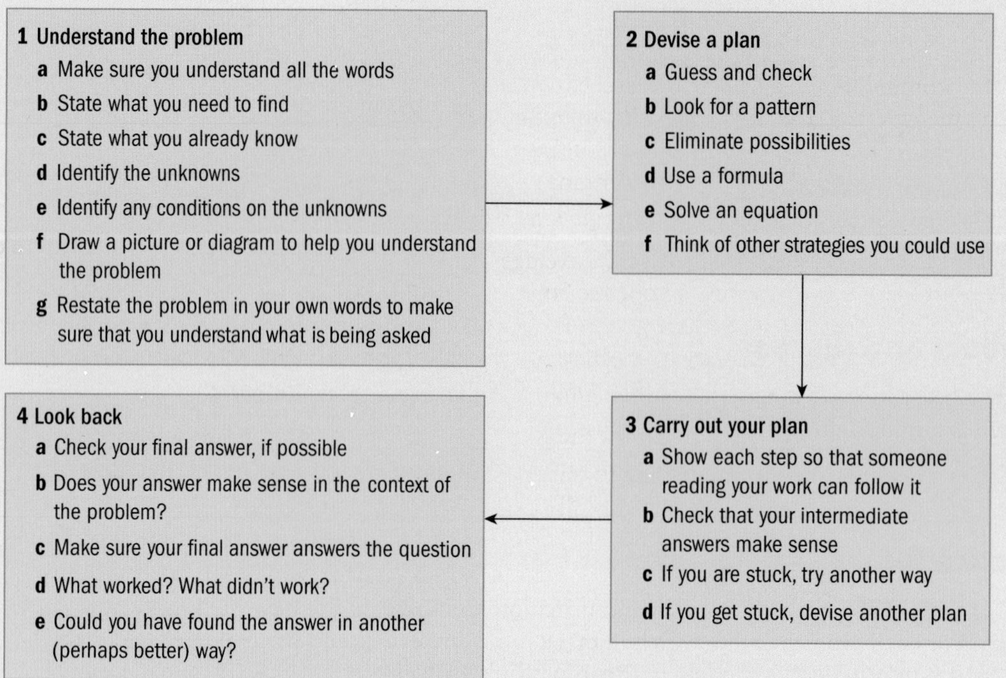

Strategy 1: Take 1

Sometimes you can simplify a problem by replacing one of the numbers with 1. For example, how much does 1 item cost? Or how many hours does it take 1 person to complete a task?

Strategy 2: Use easy numbers

Replacing numbers in the problem with 'easy numbers' can help you see the method to use in the original problem.

Strategy 3: Draw it!

It often helps to draw a diagram, especially if the problem involves geometry.

Strategy 4: Work backwards

Some problems give you an answer and ask you to find a part of the question. For example, you are given the total price of oranges and are asked how much one orange costs, or you are given the area of a square and asked to find its side length.

Mixed practice

1 Three times the reciprocal of a number equals nine times the reciprocal of 6. **Find** the number.

2 Angles x and $\frac{3}{4}x$ are complementary. **Find** x.

3 The denominator of a fraction is 1 less than twice the numerator. When 7 is added to both the numerator and denominator, the resulting fraction is equivalent to $\frac{7}{10}$. **Find** the original fraction.

4 Sophie leaves school and travels east at an average speed of 40 km/h. Jenny leaves school one hour later, and travels west at an average speed of 50 km/h. **Determine** how many hours after Jenny leaves school they will be 400 km apart.

5 Kim and Tom both cycle to work from their homes, at the same speed. Kim lives 20 km away from work, and Tom lives only 15 km away from work. It takes Kim 1 hour more to cycle to work than Tom. **Determine** how long Tom takes to get to work.

6 **Determine** how many milligrams of a metal containing 45% nickel must be combined with 6 milligrams of pure nickel to form an alloy containing 78% nickel.

7 Two different fruit juices are mixed together. 7 liters of the first fruit juice contains 60% apple juice, and 5 liters of the second fruit juice contains 30% apple juice. **Find** the percentage of the new mix that is apple juice.

8 Chloe can type a story in 6 hours. If Molly helps her, they finish typing the story in 4 hours. **Find** how long it would take Molly to type the story on her own.

9 Mary takes 8 days to do a full check of the library's online catalog. Her assistant takes 12 days. If they work together, **determine** the fraction of the catalog that is checked after 4 days.

10 The number of hamsters in a pet store doubles each month. Today, there are 20 hamsters.

- **a** Supposing that no hamster dies and no hamster gets sold, **determine** in how many months the store will have 640 hamsters.
- **b** If 10 hamsters are sold at the end of each month, after the number of hamsters double, **find** how many hamsters are left after 5 months.

11 Sacha picked some apples from a tree in her garden. She then gave some to her friends. To her first friend, she gave half the apples she had, plus 2 more. To her second friend, she gave half the apples she had left, plus 2 more. And she did the same thing for her third friend. After giving away all these apples, she had 1 apple left. **Find** how many apples she originally pick from the tree.

12 A house and a piece of land are sold for €850 000. The house is sold for one and a half times as much as the land. **Find** how much the house sold for.

Review in context

Scientific and technical innovation

Hungarian Paul Erdös (1913–96) was a highly productive, yet highly eccentric mathematician who enjoyed solving and posing mathematical problems, even offering cash prizes for solutions. There are thought to be around 1000 problems still unsolved, with prizes from $25 upward.

Erdös collaborated on academic papers with over 480 co-authors, more than any other mathematician in history.

Many people enjoy mathematical puzzles, not just mathematicians. Some puzzles have led to interesting discoveries in mathematical theory, or other fields entirely. Use problem-solving strategies to solve these famous historical puzzles.

1 Rice on a chessboard

Sissa ben Dahir, Grand Vizier to the Indian King Shirham, invented the game of chess. Shirham liked the game so much, that he asked Sissa what he wanted as a reward. Sissa replied: 'I would like to cover the chessboard with 1 grain of rice on the first square, 2 grains on the second, 4 grains on the third, 8 grains on the fourth, and so on.' The King accepted immediately, thinking he had himself a bargain.

How much rice would there be on the chessboard?

Answer these questions to help work it out.

a **Find** the number of grains of rice on:

i the 8th square

ii the 16th square

iii the 64th square.

How does it help to know that each square has double the grains of rice as the previous square?

b About 50 grains of rice fit in 1 cm^2.
Leaving your answer correct to 2 d.p., **find** how many cm^2 of rice are on:

i the 8th square

ii the 16th square

iii the 24th square.

c 10 000 cm^2 = 1 m^2.

Find how many m^2 of rice are on the 24th square.

d 1 000 000 m^2 = 1 km^2.

Find how many km^2 of rice are on the 64th square. Leave your answer correct to 2 d.p.

e Challenge: **Find** the total number of grains of rice in:

i the first row

ii the first two rows

iii the first three rows

iv the entire chessboard.

2 The Tower of Hanoi

French mathematician Edouard Lucas invented this puzzle in 1883.

This stack of disks in order from the largest to the smallest is called a tower. The aim of the puzzle is to move a tower from one rod to another, obeying the following rules:

Rule 1 Only one disk can be moved at a time.

Rule 2 One move consists of taking the top disk from one stack and placing it on top of a stack on another rod.

Rule 3 A disk can be placed only on top of a larger disk. No disk can be placed on top of a smaller disk.

The minimum number of disks in a tower is 3.
The original problem had 64 disks.

a **Find** the minimum number of moves needed to move a tower of:

i 3 disks

ii 4 disks

iii 5 disks.

You could use different sized coins or counters to model the tower.

b Suppose each move takes 1 second. **Find** the minimum time it would take to move a tower of:

i 3 disks

ii 5 disks

iii 6 disks.

c The minimum number of moves for a tower of n disks is $2^n - 1$. **Find** the minimum time it would take to move a tower of:

i 10 disks

ii 20 disks

3 Rope around the Earth

The English philosopher and mathematician William Whiston first posed this in 1702.

Part 1: Wrap a rope around the equator of a basketball. What length of rope would need to be added to make the rope hover 1 inch away from the basketball's equator at all points?

Part 2: Wrap a very, very long rope around the equator of the earth. What length of rope would need to be added to make the rope hover 1 inch above the earth's equator at all points?

You may find this information useful.

- the diameter of a basketball is about 24 cm.
- the radius of the earth at the equator is 6378 km.

a Before calculating the amount of rope to be added in each part of the problem, **suggest** an amount that seems reasonable to you.

b Write down an expression to represent the circumference of a sphere with radius r.

c Write down an expression to represent the circumference of a sphere with radius $(r + 1\text{in})$. Simplify this expression.

d Hence, **find** how much string needs to be added to the circumference when adding 1in to the radius of a sphere.

e **Determine** whether the size of the sphere matters in part **d**. Justify your answer. Now you can answer the riddle. Are you surprised?

4 24 Game

In 1988, inventor Robert Sun created the 24 Game to make mathematics appealing and accessible to children. The rules of the game are:

Rule 1 Choose four numbers between 1 and 13 at random.

Rule 2 Use all four numbers with mathematical operations and brackets, to make a total of 24.

Use problem-solving strategies to make 24 from each set:

a 7, 2, 1, 1

b 2, 3, 2, 4

c 2, 3, 4, 6

d 2, 2, 7, 12

The Columbus Problem Puzzle

In 1882, chess enthusiast and recreational mathematician Sam Loyd (1841–1911) offered a $1000 prize for the answer that best showed how to arrange the seven figures and the eight "dots" below which would add up to 82.

. 4 . 5 . 6 . 7 . 8 . 9 . 0 .

Of the several million answers submitted, only two were found to be correct. One is shown here; can you find another solution?

$$\begin{array}{r} 80 \\ .\dot{5} \\ .\dot{9}\dot{7} \\ +\ .\dot{4}\dot{6} \\ \hline 82 \end{array} \qquad .\dot{9}\dot{7} \equiv 80 + \frac{55}{99} + \frac{97}{99} + \frac{46}{99} = 82$$

1 Form

Form is the shape and underlying structure of an entity or piece of work, including its organization, essential nature and external appearance.

The language of mathematics

Mathematics is written using:

- numbers
- words
- symbols
- diagrams
- letters
- shapes

What forms help you recognize mathematics when you see them?

Hilbert's Hotel Infinity

Are there different *forms* of infinity denoting different *sizes* of infinity? David Hilbert's thought experiment imagined a hotel with an infinite number of rooms – with the motto 'always room for one more!'

- What if a guest arrives, but all the infinite number of rooms in the hotel are already occupied? No problem – the guest in room 1 moves to room 2, the guest in room 2 moves to room 3, and so on. Room 1 is now free, and everyone is happy.
- And what if an *infinite* number of guests arrive at the same time but all rooms are occupied? Still no problem – the guest in room 1 moves to room 2, the guest in room 2 moves to room 4, and so on, so that the one in room n moves to room $2n$. All the even-numbered rooms are now occupied, but all the odd numbered rooms are free.

The symbol for infinity is called a lemniscate. The word lemniscate is of Latin origin, and means 'pendant ribbon'.

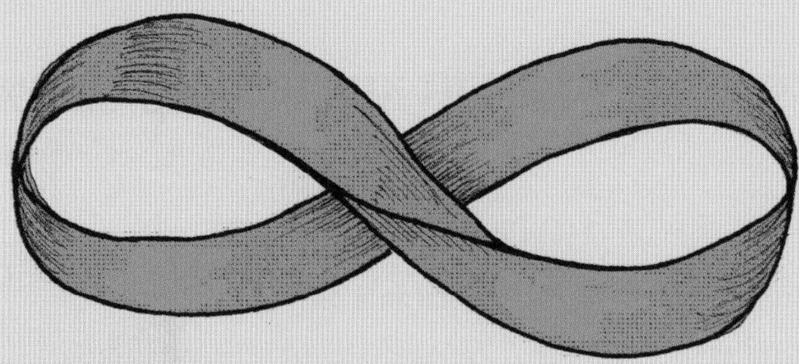

- Why do you think this form was chosen to symbolize infinity?
- Can you design an alternative symbol that would effectively signify infinity?

This curve is known as the Lemniscate of Bernoulli; on the Cartesian plane its equation is:

$$(x^2 + y^2)^2 = 2a^2(x^2 - y^2)$$

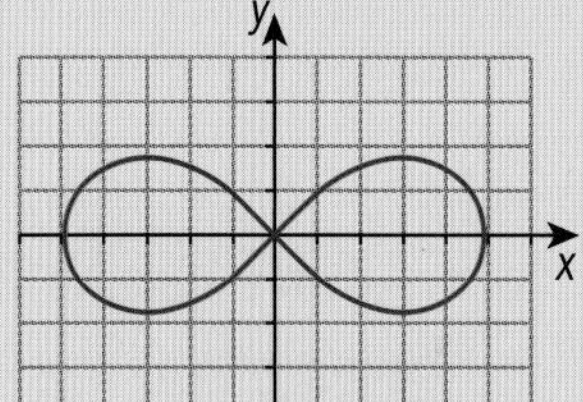

- Use a GDC to graph this equation and create a slider for the variable a from -5 to 5. What effect does the variable a have on the graph?

Form and function

In design, form is the shape and appearance of an object. Function is the purpose of the object; does it do the job it was designed for? One definition of a good design is one that balances form and function – it does the job it was designed to do and is also attractive.

- Which do you think is the best chair design?

- Which do you think is the best graphic design?

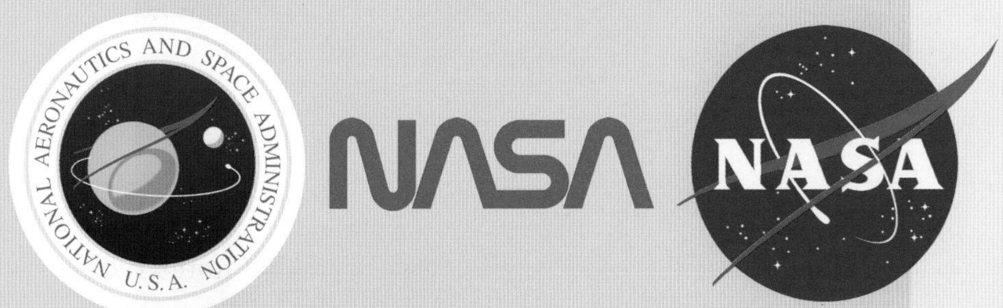

1.1 Mathematically speaking

Global context: Scientific and technical innovation

FORM

Objectives

- Classifying the different kinds of real numbers
- Representing the different kinds of real numbers
- Drawing and interpreting Venn diagrams to solve problems in real-life contexts
- Applying the language of sets to different areas of mathematics
- Using the language of sets to model real-life problems

Inquiry questions

F
- What is a set?
- What kinds of sets are there?

C
- What are set operations?
- How do you represent sets and their operations?

D
- How useful are sets and Venn diagrams in solving real-life problems?
- How does form influence function?

ATL **Communication**

Organize and depict information logically

You should already know how to:

• describe the different types of real numbers	**1** For each number, state which number set(s) it is a member of (e.g. natural numbers, irrational numbers, etc.). **a** 7 **b** −7 **c** 0 **d** 1.5 **e** π **f** −0.1 **g** $\sqrt{14}$ **h** $\sqrt{121}$

Sets

- What is a set?
- What kinds of sets are there?

Some examples of sets are: the set of students in your class, the set of all your subject teachers this year, the set of your favorite books, the set of parking spaces at your local supermarket.

> A **set** is a collection of objects. Each member of a set is called an **element** of the set.

You use capital letters to denote a set, and put the members of the set in curly brackets. You can describe a set using words, or by writing a list of its elements.

Sets described in words:	**Sets described by listing their elements:**
A = {my three favorite colors}	A = {blue, orange, green}
B = {vowels in the English alphabet}	B = {a, e, i, o, u}
C = {seven largest moons in the solar system}	C = {Ganymede, Titan, Callisto, Io, Earth's Moon, Europa, Triton}
D = {integers between 1 and 21}	D = {2, 3, 4, …, 20}

> The symbol $\in$ means 'is an element of'.
>
> The symbol $\notin$ means 'is not an element of'.

For example, if the set $A = \{16, 25, 36, 49, 64\}$, then $49 \in A$ but $81 \notin A$.

The order in which you write the elements of a set is not important.

B = {a, e, i, o, u} is the same as B = {e, o, u, i, a}.

In a class there are ten students aged 16, and four students aged 15.

The set of *students* in the class has 14 elements (the students' names). The set of their *ages* has only two elements: {15, 16}, not 16 written ten times and 15 written four times.

> The number of elements in a set A is written $n(A)$. It is called the **cardinality** of the set.

> …
>
> The three ellipsis dots mean 'and so on' or 'continue the pattern'. It was first used by the ancient Greeks. With sets, the ellipsis is used when writing out all the elements of a set is impractical.

For example if the set $A = \{\text{knife, fork, spoon}\}$ then $n(A) = 3$. If $n(A)$ is a real number, then the set A is a finite set.

You can write common number sets using set notation.

Natural numbers $\mathbb{N} = \{0, 1, 2, 3, \ldots\}$

Integers $\mathbb{Z} = \{\ldots, -3, -2, -1, 0, 1, 2, 3, \ldots\}$

Positive integers $\mathbb{Z}^+ = \{1, 2, 3, \ldots\}$

These are infinite sets, as they contain an infinite number of elements.

The symbols for numbers sets are often referred to as 'double struck' because they could be produced by double-striking a character on a typewriter, or 'blackboard bold' as they were written with the edge of the chalk on a blackboard instead of the point, to differentiate them from other bold characters.

Practice 1

1 List the elements of these sets, and find the number of elements in each set.

- **a** A is the set of the days of the week.
- **b** B is the set of months of the year not containing the letter r.
- **c** C is the set of factors of 12.
- **d** D is the set of positive integers less than 30 that are multiples of 4.

2 Describe each set in words.

- **a** $J = \{1, 3, 5, 7, 9\}$
- **b** $K = \{\text{isosceles, equilateral, right-angled, scalene}\}$
- **c** $L = \{\text{right angle, obtuse, acute, reflex}\}$
- **d** $M = \{4, 8, 12, 16, \ldots, 40\}$

ATL

3 Given the sets $A = \{4, 6, 8, 10\}$, $B = \{1, 8, 27, 64\}$, $C = \{1, 3, 4, 7, 11\}$ and $D = \{0, \pm1, \pm2, \pm3, \pm4\}$, state whether each statement is true or false. If the statement is false, write the correct statement.

- **a** $4 \in A$
- **b** $7 \notin C$
- **c** $1 \in A$
- **d** $27 \in A$
- **e** $8 \in D$
- **f** $n(C) = n(D)$

Set builder notation

In addition to describing a set in words or by listing its elements it can be described using set builder notation. This uses curly brackets enclosing a variable, a vertical line, and any restrictions on the variable. It can be used for finite or infinite sets. For example, the set $A = \{1, 2, 3, \ldots 9\}$ can be written in set builder notation, which would look and read like this:

A	$=$	$\{x$	$\mid$	$x \in \mathbb{N}$	,	$1 \le x \le 9\}$
A	is the set of	all values of x	such that	x is a natural number	and	x is between 1 and 9 inclusive.

Tip

For natural numbers, $1 \le x \le 9$ is equivalent to $0 < x < 10$.

Using set builder notation you can define another number set that you need to know: the rational numbers

Rational numbers: $\mathbb{Q} = \left\{\frac{p}{q} \mid p, q \in \mathbb{Z}, q \neq 0\right\}$

In words, $\mathbb{Q}$ is the set of rational numbers (numbers of the form $\frac{p}{q}$) such that both p and q are integers, and q does not equal 0.

All of the integers, rational numbers and irrational numbers together form the set of real numbers, $\mathbb{R}$. The real numbers can be represented on the real number line.

$-\frac{7}{4}$ $\sqrt{8}$ π $\sqrt{22}$

−5 −4 −3 −2 −1 0 1 2 3 4 5 x

We don't have a symbol for irrational numbers. They are numbers that cannot be expressed as a fraction, for example: π or $\sqrt{2}$.

Example 1

Write these sets in set builder notation.

a S is the set of real numbers between 0 and 1

b $P = \{2, 3, 5, 7, \ldots, 37\}$

c $M = \{2, 4, 6, 8, \ldots\}$

a $S = \{x \mid x \in \mathbb{R}, 0 < x < 1\}$ — Use the 'less than' symbol because 0 and 1 are not included.

b $P = \{x \mid x \in \text{primes}, 2 \leq x \leq 37\}$ — There is no special symbol for prime numbers.

c $M = \{x \mid x = 2n, n \in \mathbb{Z}^+\}$ — Even numbers are multiples of 2, so they can be written as $2n$.

Example 2

Write out the elements of each set in list form.

a $E = \{x \mid x \in \mathbb{Z}, -3 < x < 2\}$

b $F = \left\{\frac{1}{n} \mid n \in \mathbb{Z}^+\right\}$

Describe set G in words.

c $G = \{x \mid x \in \mathbb{R}, 0 < x < 1\}$

a $E = \{-2, -1, 0, 1\}$ — −3 and 2 are not included.

b $F = \left\{1, \frac{1}{2}, \frac{1}{3}, \frac{1}{4} \ldots\right\}$

c G is the set of real numbers greater than 0 and less than 1. — There is no first real number greater than 0, or last real number less than 1, so you cannot list the elements of this set.

ATL

Practice 2

1 Write out the elements of each set in list form. State whether each set is finite or infinite. If the set is finite, state its cardinality.

a $\{x | x \in \mathbb{Z}, -2 < x < 5\}$

b $\{y | y \in \mathbb{Z}, y > 0\}$

c $\{a | a \in \mathbb{N}, a \text{ is a multiple of } 5\}$

d $\{b | b \in \mathbb{R}, b \text{ is a factor of } 28\}$

e $\{c | c \in \mathbb{N}, c + 3 < 8\}$

f $\{p | p \in \text{primary colors}\}$

2 Write each set using set builder notation.

a $S = \{1, 9, 25, 49, 81, \ldots\}$

b $T = \{\ldots, -10, 0, 10, 20, \ldots\}$

c U is the set of real numbers between 1 and 2, including 2.

d V is the set of rational numbers between 0 and 1.

Problem solving

3 Write in set builder notation:

a The set of odd numbers.

b The set of multiples of 3.

c $W = \{1, 2, 4, 8, 16, 32\}$

The **universal set**, U, is the set that contains all sets being considered.

The **empty set**, { } or $\varnothing$, is the set with no elements, so $n(\varnothing) = 0$.

For example, for the set W of winter months and set S of summer months, the universal set U is the set of months of the year.

From the universal set U = {yellow, red, blue} you can make these sets:

$J = \varnothing$ | P = {yellow, red}

K = {yellow} | S = {yellow, blue}

L = {red} | T = {red, blue}

M = {blue} | U = {yellow, red, blue}

In examinations, the universal set will be given to you in the question, when necessary.

All these sets, which can be made from the elements of U, including the empty set and U itself, are called subsets of U.

The set A is a **subset** of a set B if every element in A is also in B. The symbol for subset is $\subseteq$. Written in mathematical form:
if for all $x \in A \Rightarrow x \in B$, then $A \subseteq B$.

The **empty set** is a subset of any set. So, for any set A, $\varnothing \in A$.

Every set is a subset of itself.

'$x \in A \Rightarrow x \in B$, then $A \subseteq B$' can be read as 'If for all elements in A the elements are also in B, then A is a subset of B.'

Two sets are equal if they contain exactly the same elements.

Some further examples of subsets are:

- If $A = \{1, 5\}$ and $B = \{1, 2, 3, 4, 5\}$, then $A \subseteq B$.
- If $C = \{3^n \mid n \in \mathbb{N}\}$ and $D = \{1, 3, 9, 27\}$, then $D \subseteq C$, because $D = \{3^0, 3^1, 3^2, 3^3\}$
- If $E = \{x \mid x^2 = 1\}$ and $F = \{-1, 1\}$, then $E \subseteq F$ and $F \subseteq E$ are both true. In other words, $E = F$.

You can always use the symbol $\subseteq$ to denote subset. If two sets are not equal, (when there is at least one element of A that is not in B) then you can use the symbol $\subset$ without a line underneath it. $A \subset B$ means A is a **proper subset** of B.

If $A = B$, then A is an improper subset of B, or B is an improper subset of A, and in this case you use the symbol $\subseteq$.

The symbol $\subseteq$ and $\subset$ are similar in the way they work to $\leq$ and $<$. If in doubt use $\subseteq$ as it can be used for an improper or proper subset.

Exploration 1

1 Consider a set containing two elements, for example, $A = \{1, 2\}$. Write down all of its subsets.

2 Now consider a set containing three elements, for example $B = \{1, 2, 3\}$. Write down all of its subsets.

3 Do the same again for a set containing four elements.

4 Based on your findings, suggest a rule for determining the number of subsets of a set with n elements.

5 Test your rule on a set containing five elements.

The set itself and the empty set are subsets of any given set.

Reflect and discuss 1

- For the set $A = \{1, 2\}$, is every element of $\varnothing$ a member of A?

  Is there any element of $\varnothing$ that is not in A?

  Use your answers to explain why $\varnothing$ is a subset of A, and of any set.

A generalization or general rule is 'a general statement made on the basis of specific examples'.

- What specific examples did you use in Exploration 1?
- Compare the rule you found in Exploration 1 with others in your class. Did you all get the same result? Is that enough to say that this rule is always true for any number of elements in the original set?
- Why should you be cautious when generalizing?

Later in your mathematics course you will learn about the key concept of Logic and the related concept of Justification. Until you have justified your general statement, you cannot be certain it will always be true.

ATL

Practice 3

1 Determine whether or not these pairs of sets are equal.

a {1, 2, 3} ; {2, 3, 1}

b {1, 2, 3, 5} ; {prime numbers less than 6}

c {16, 17, 18, 19, ...} ; {rational numbers greater than 15}

2 Determine whether these statements are true or false. If the statement is false, give a reason why.

a $2 \in \{2, 3, 4\}$

b $\{2\} \subseteq \{2, 3, 4\}$

c $\{3\} \in \{2, 3, 4\}$

d $4 \subseteq \{2, 3, 4\}$

e Given $A = \{x \mid x \in \mathbb{Z}, 1 < x < 5\}$ and $B = \{2, 3, 4\}$, then $A = B$.

Problem solving

3 Determine if the statements **a** to **d** below are true or false. For those that are false, give a counter-example (find an example that makes the statement false). You may find drawing a diagram helpful.

a If $A \subseteq B$ and $B \subseteq C$, then $A \subseteq C$.

b If $R \subseteq S$ then $S \subseteq R$.

c If $p \in P$ and $P \subseteq Q$, then $p \in Q$.

d If $A = B$, then $B \subseteq A$.

4 Write a set that has 32 subsets.

5 Determine whether or not a set can have exactly 24 subsets.

C Set operations

- What are set operations?
- How do you represent sets and their operations?

Just as you use arithmetic operations to manipulate numbers, sets also have their own operations.

Union and intersection of sets

- Union: $A \cup B = \{x \mid x \in A \text{ or } x \in B\}$
- Intersection: $A \cap B = \{x \mid x \in A \text{ and } x \in B\}$

If $A = \{1, 2, 3\}$ and $B = \{2, 3, 4, 5\}$, the set that contains all the elements that are in both A and B without repeating any of them is $C = \{1, 2, 3, 4, 5\}$. Set C is the **union** of sets A and B, and is written $A \cup B$.

The set that contains the elements that both sets have in common is $D = \{2, 3\}$. D is the **intersection** of sets A and B, and is written $A \cap B$.

The **complement** of a set A is denoted by A', and described in set notation as $A' = \{x \mid x \in U, x \notin A\}$.

For the universal set $U = \{1, 2, 3, 4, 5, 6\}$, and set $E = \{4, 6\}$, the complement of the set E is made up of the elements in the universal set that are *not* in set E. So $E' = \{1, 2, 3, 5\}$.

Example 3

If $U = \{1, 2, 3, \ldots, 10\}$, $A = \{2, 4, 6, 8\}$, $B = \{2, 3, 5, 7\}$ and $C = \{1, 5, 9\}$, find:

a $A \cap C$ **b** $A \cup B'$

c $(A' \cap B)'$ **d** $(A \cup B \cup C)'$

a $A \cap C = \varnothing$ — A and C have no elements in common.

b $B' = \{1, 4, 6, 8, 9, 10\}$ — First find B'.

$A \cup B' = \{1, 2, 4, 6, 8, 9, 10\}$

c $A' = \{1, 3, 5, 7, 9, 10\}$

$A' \cap B = \{3, 5, 7\}$ — First find $A' \cap B$. $(A' \cap B)'$ is made up of the elements in U that are not in $A' \cap B$.

$(A' \cap B)' = \{1, 2, 4, 6, 8, 9, 10\}$

d $A \cup B \cup C = \{1, 2, 3, 4, 5, 6, 7, 8, 9\}$

$(A \cup B \cup C)' = \{10\}$

ATL

Practice 4

1 If $U = \{x \mid x \in \mathbb{Z}^+, 1 \le x \le 20\}$, $A = \{2, 4, 8, \ldots, 20\}$, $B = \{1, 3, 5, \ldots, 19\}$ and $C = \{x \mid x \in \text{primes}, 1 \le x \le 20\}$, find:

a $A \cap B$ **b** $A \cup B$ **c** $A' \cap C$ **d** $(A \cap C)'$ **e** $(A' \cup B)'$

2 If $U = \mathbb{Z}$, $R = \{x \mid x \in \mathbb{Z}^+, x < 10\}$, $S = \{x \mid -5 < x < 5\}$ and $T = \{x \mid x \in \mathbb{N}; x \le 15\}$, find:

a $R \cap S$ **b** $R \cup T$ **c** $R' \cap T$ **d** $S' \cap T$ **e** $(R \cap S)' \cap T$

Problem solving

3 The universal set is $U = \{11, 12, 13, 14, 15, 16, 17, 18, 19, 20\}$. Write three subsets F, G and H such that:

- $n(F \cap G) = 2$
- $F \cap H = \varnothing$
- $G' = \{x \mid x \text{ is odd}, 11 \le x \le 19\}$

Reflect and discuss 2

For any two sets, A and B, explain why the intersection $A \cap B$ is a subset of their union $A \cup B$.

Tip

Try with any two sets A and B. Can you generalize your result?

Venn diagrams

In a Venn diagram, a rectangle represents the universal set.

Circles represent subsets of the universal set.

It is important that you use the forms of representation correctly – remembering to include the universal set in your Venn diagrams, for example, or using curly brackets when listing the elements of a set.

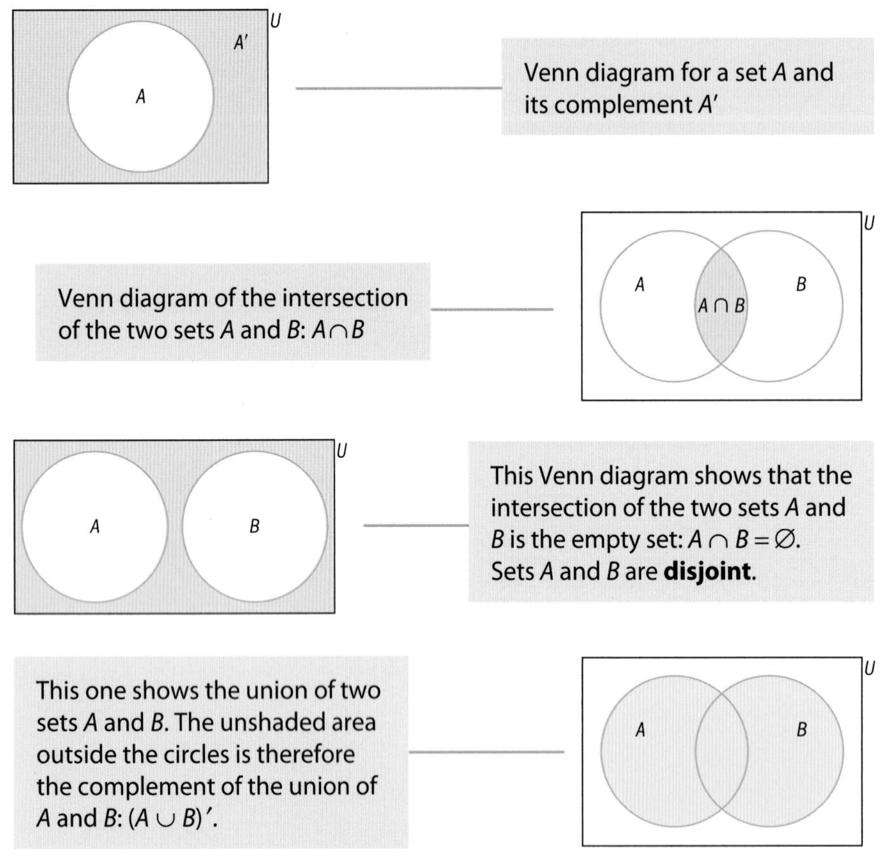

Exploration 2

1 Draw two Venn diagrams: in the first one shade the region that represents the set $A' \cap B'$; in the second diagram shade the region that represents the set $(A \cup B)'$. Describe the relationship between these two sets mathematically.

2 Draw two Venn diagrams: in the first one shade the region that represents the set $(A \cap B)'$; in the second diagram shade the region that represents the set $A' \cup B'$. Describe the relationship between these two sets mathematically.

3 Draw a Venn diagram to show that for any two sets, A and B, the intersection $A \cap B$ is a subset of their union $A \cup B$. Which do you think was easier – proving this using reasoning in Reflect and discuss 2, or by drawing a diagram?

Reflect and discuss 3

- How many different ways have you represented the set $A' \cup B'$ in Exploration 2?
- What can you say about $A' \cup B'$ and $(A \cap B)'$?
- How has representing sets using a Venn diagram helped you discover new information about sets?

The relationships you have discovered so far by using Venn diagrams are called De Morgan's Laws, named after the British mathematician Augustus De Morgan. They are used extensively in the fields of circuitry and electronics, as well as in the field of logic.

Example 4

Draw Venn diagrams to represent these sets or relationship between sets.

a $B \subseteq A$ **b** $A' \cup B$ **c** $A \cap B'$ **d** $A \cap B \cap C$ **e** $(A \cap B) \cup C$

a $B \subseteq A$

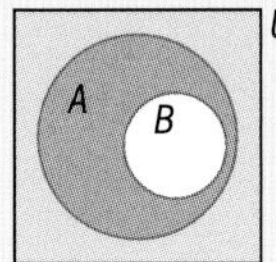

Draw a rectangle representing the universal set. Set B is completely contained in A.

b $A' \cup B$

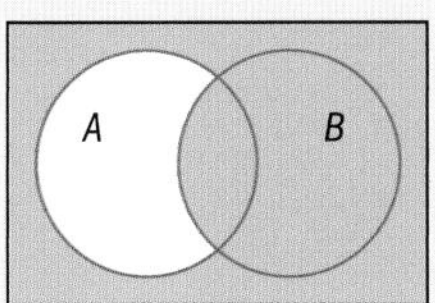

Shade the area that contains everything that is not in A, together with all of B.

c $A \cap B'$

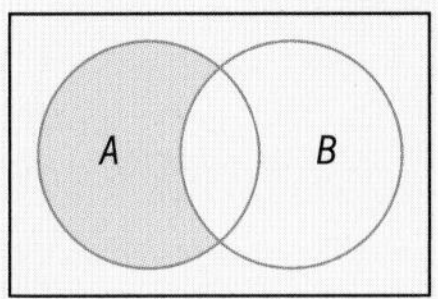

The shaded area represents what A and the complement of B have in common.

d $A \cap B \cap C$

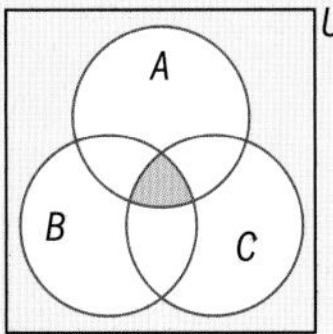

The shaded area shows the intersection of all three sets.

e $(A \cap B) \cup C$

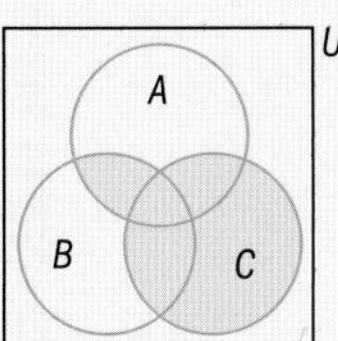

Shade the intersection of A and B first, then shade all of C.

Tip

When drawing Venn diagrams it can be useful to shade one set using vertical lines and the other using horizontal lines. The union will be the total area shaded and the intersection of the sets is where the lines cross.

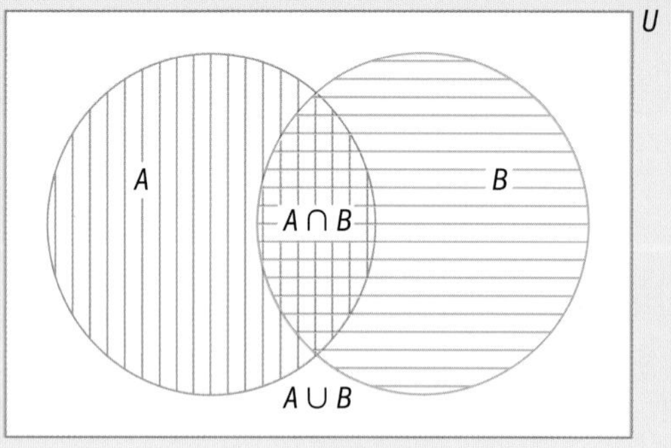

ATL

Practice 5

1 Draw Venn diagrams to represent each set.

a $(A \cap B)'$

b $A \cup (A \cap B)$

c $(A' \cap B)'$

d $(A' \cup B')'$

e $A \cup (B \cap C)$

f $(A \cap B)' \cup C$

g $A \cap (B \cup C)'$

h $(A \cap B) \cup (A \cap C) \cup (B \cap C)$

Problem solving

2 Write the set that the shaded part of each Venn diagram represents.

a

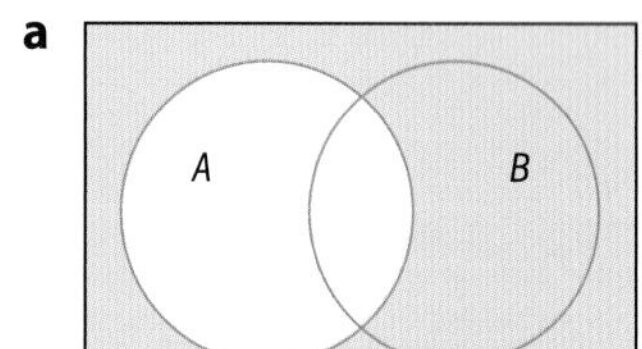

b

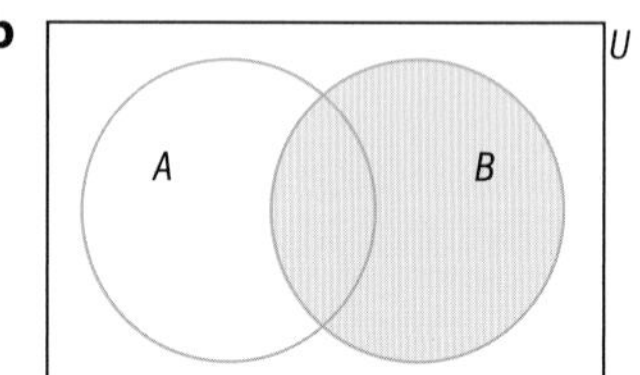

c

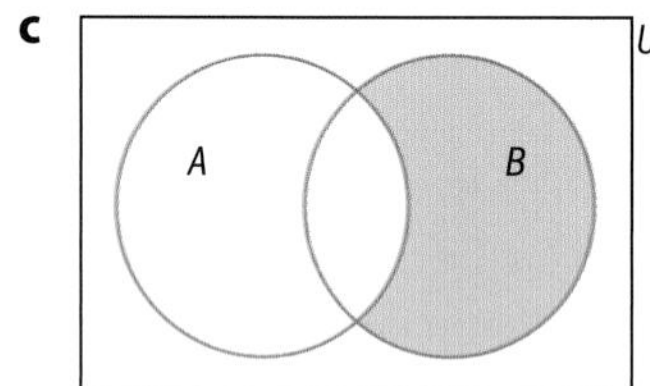

d

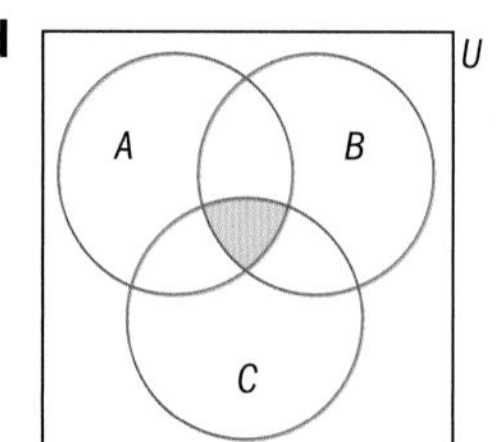

Venn diagrams are named after an English mathematician, John Venn. This stained glass window in Cambridge University commemorates his achievements.

e

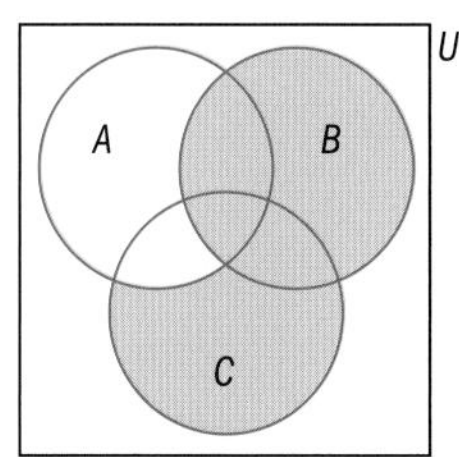

f

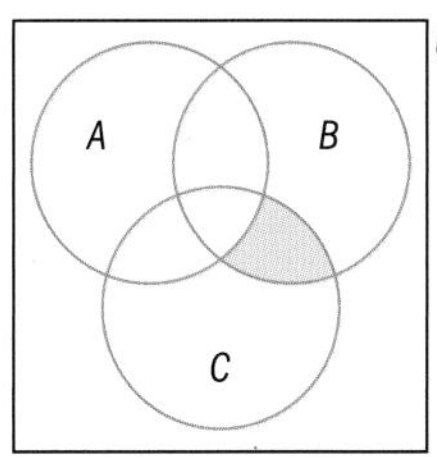

g

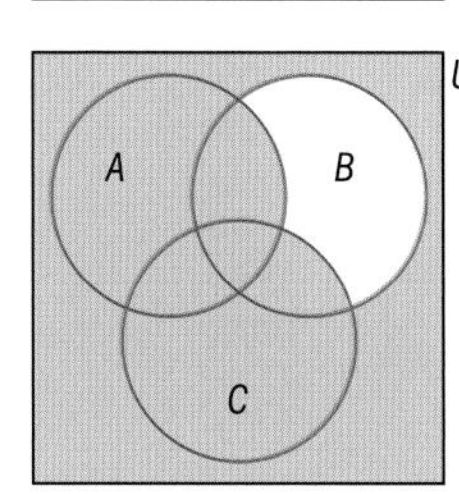

h

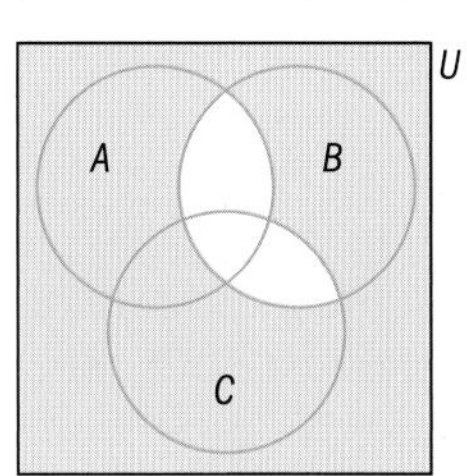

You are already familiar with the properties of real numbers under the binary operations of addition and multiplication. In the following exploration you will determine which of these properties are also valid under the set operations of intersection and union.

Exploration 3

1 By drawing a Venn diagram for each side of the equals sign in the statements below, show that the following properties hold for set operations:

a $A \cap B = B \cap A$ **b** $A \cup B = B \cup A$ — Commutative laws

c $A \cap (B \cap C) = (A \cap B) \cap C$ **d** $A \cup (B \cup C) = (A \cup B) \cup C$ — Associative laws

e $A \cap (B \cup C) = (A \cap B) \cup (A \cap C)$ **f** $A \cup (B \cap C) = (A \cup B) \cap (A \cup C)$ — Distributive laws

2 The additive identity of the real numbers is 0 since $n + 0 = 0 + n = n$ and the inverse of n is $-n$ since $n + (-n) = (-n) + n = 0$.

The multiplicative identity of m is 1 since $m \times 1 = 1 \times m = m$ and the inverse of m is $\frac{1}{m}$ since $m \times \frac{1}{m} = \frac{1}{m} \times m = 1$ $(m \neq 0)$.

Determine whether or not there is an identity and inverse for the operations $\cup$ and $\cap$.

Find sets B and C such that $A \cup B = B \cup A = A$ and $A \cap C = C \cap A = A$. B and C are the two identities.

Find if sets E and F exist such that $A \cup E = E \cup A = B$ and $A \cap F = F \cap A = C$. E and F, if they exist, are the inverses.

3 Summarize the properties of the set operations union and intersection.

D Modelling real-life problems using Venn diagrams

- How useful are sets and Venn diagrams in solving real-life problems?
- How does form influence function?

Venn diagrams are used to model and solve problems in fields such as market research, biology, and social science, where 'overlapping' information needs to be sorted.

Example 5

A market research company surveys 100 students and learns that 75 of them own a television (T) and 45 own a bicycle (B). 35 students own both a television and a bicycle. Draw a Venn diagram to show this information.

a Find how many students own either a television or a bicycle, or both.

b Find how many own neither a television nor a bicycle.

c Explain how your diagram shows that 50 students own either a television or a bicycle, but not both.

Set theory is used by online shopping search engines. Data on products for sale is split into sets so that your search query will get a response more quickly.

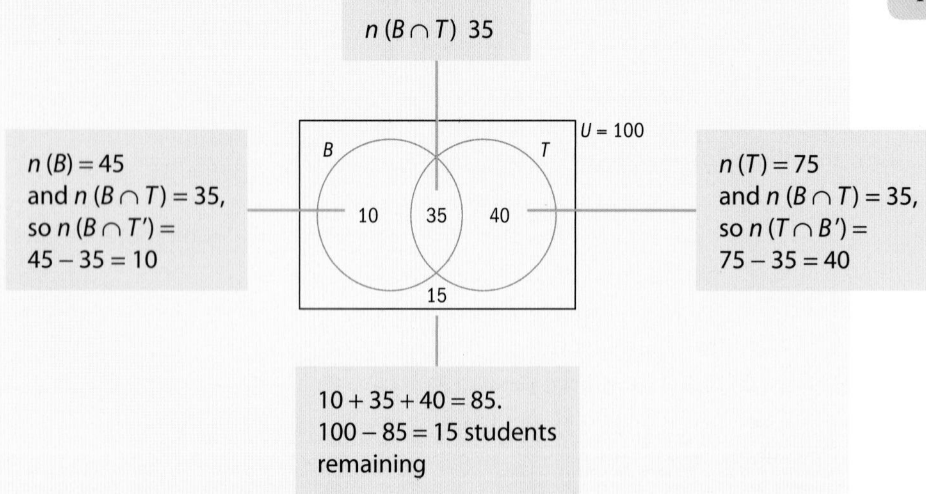

a The number of students who own either a television or a bicycle, or both is $n(B \cup T) = 85$.

b The number of students who own neither a television nor a bicycle is $n(B \cup T)' = 15$.

c The number of students who own either a television or a bicycle but not both can be seen as the union of the two sets less the intersection, or $10 + 40 = 50$.

Practice 6

Objective: **C.** Communicating **v.** organize information using a logical structure
These questions encourage students to use Venn diagrams (and a table in Question 1) to organize information logically. The students should be able to draw and interpret Venn diagrams.

1 a i For the Venn diagram, describe the characteristics of only whales, of only fish, and those of both whales and fish.

ii Explain how the diagram has helped you with your descriptions and whether a different form could be more useful.

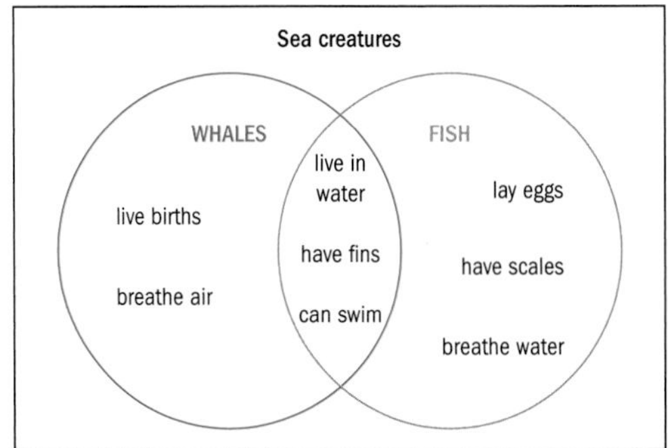

b i Below is a table describing some characteristics of whales, fish, and shrimp. Create a Venn diagram that illustrates these characteristics. Make sure to choose an appropriate universal set.

	live births	fins	have scales	lay eggs	live in water	breathe air	breathe water	have legs
whales	✓	✓			✓	✓		
fish		✓	✓	✓	✓		✓	
shrimp				✓	✓		✓	✓

ii Explain whether the table form or the Venn diagram form is best to illustrate the given information.

2 In a year group of 50 students, 18 are enrolled in Music, 26 in Art, and 2 in both Music and Art. By drawing a Venn diagram, determine how many students are not enrolled in either Music or Art.

3 In a school of 350 students, 75 are involved in community service projects, 205 in athletics, and 62 are involved in both. By drawing a Venn diagram, determine how many students are involved in either one of these two activities.

4 In a school cafeteria at lunchtime, 93 students chose a soft drink and 47 students chose bottled water. 25 students chose both drinks. If each student chose one of these drinks, determine the total number of students in the cafeteria by drawing a Venn diagram.

5 Of 150 new university students, 85 signed up for Mathematics and 70 for Physics, while 50 signed up for both subjects. By drawing a Venn diagram, determine how many students signed up for:

a only Mathematics

b only Physics

c neither Mathematics nor Physics.

6 There are 30 students enrolled in three different school clubs: chess, archery and cookery. Of these, 5 students are in all three clubs, 6 of them are only in the cookery club, 2 are in chess and cookery but not archery, 15 belong to cookery in total, 2 are only in chess, and 3 are only in archery. By drawing a Venn diagram, determine how many students are in:

a the archery and cookery clubs only

b the chess club.

7 In a class of 32 students, 5 live in the school town and travel to school by bus, and they have school lunches. 3 live in the school town and travel to school by bus, but do not have school lunches. 9 students do not live in the school town, do not travel to school by bus, and do not have school lunches. 11 students live in the school town and have school lunches. A total of 16 students live in the school town. 9 students travel to school by bus and have school lunches. 13 students travel to school by bus. By drawing a Venn diagram, determine how many students in total have school lunches.

Problem solving

8 The Venn diagrams below represent participants in an after school sports program. The students can choose to select table tennis (T), basketball (B), or squash (S).

a 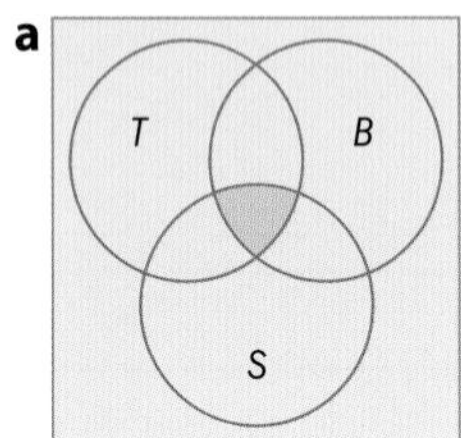

b 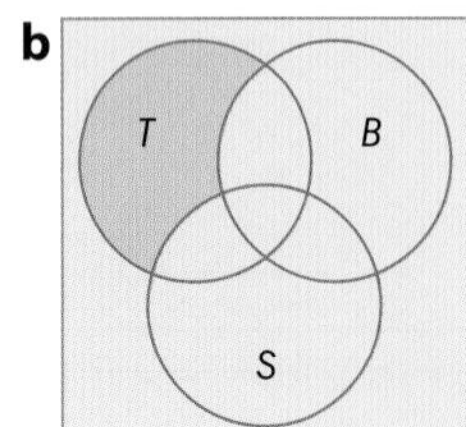

c 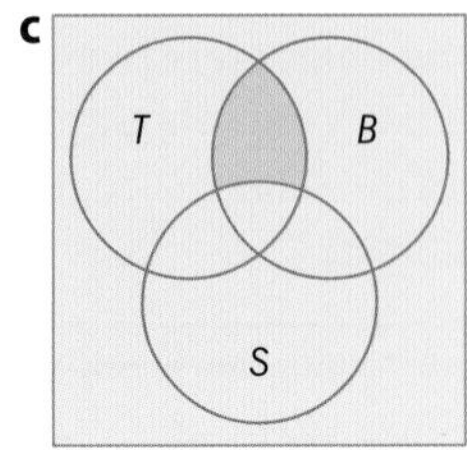

Write in words what each diagram represents.

Exploration 4

1 $n(U) = 105$, $n(A) = 37$, and $n(B) = 84$. All of the elements of U are either in A or B or both.

a Draw a Venn diagram showing A, B and U.

b Find a rule connecting the number of elements in $A \cup B$ with the number of elements in sets A, B, and the intersection of A and B.

c Justify your rule.

2 Apply your rule to the following situation. Your student council held a vote on which charity to support. A mobile library received 47 votes, and a neighborhood watch committee received 36 votes. 24 students voted for both charities. Determine how many students actually voted.

Reflect and discuss 4

In Exploration 4:

- How have you used sets to model the situation in step **2**?
- Why was it easier to consider sets of students rather than considering the students individually?
- What set operation models the fact that some students voted for both charities?

Reflect and discuss 5

Working with one or more of your classmates, discuss how Venn diagrams can be used to organize and categorize some of the following situations: your school timetable, your household chores, your homework for different subjects, your test schedules. Create two Venn diagrams with at least three circles to illustrate the different situations you have chosen.

Summary

A **set** is a collection of objects. Each member of a set is called an **element** of the set.

The symbol $\in$ means 'is an element of'.

The symbol $\notin$ means 'is not an element of'.

The number of elements in a set A is written $n(A)$. It is called the **cardinality** of the set.

The **universal set**, U, is the set that contains all sets being considered.

The **empty set**, { } or $\varnothing$, is the set with no elements, so $n(\varnothing) = 0$.

The set A is a **subset** of a set B if every element in A is also in B. The symbol for subset is $\subseteq$.

Written in mathematical form, if for all $x \in A \Rightarrow x \in B$, then $A \subseteq B$.

The **empty set** is a subset of any set. So, for any set A, $\varnothing \in A$.

Every set is a subset of itself.

Two sets are equal if they contain exactly the same elements.

Union and intersection of sets

- Union: $A \cup B = \{x \mid x \in A \text{ or } x \in B\}$
- Intersection: $A \cap B = \{x \mid x \in A \text{ and } x \in B\}$

The complement of a set A is denoted by A', and described in set notation as $A' = \{x \mid x \in U, x \notin A\}$.

Mixed practice

1 **Write down** the elements of each set in list form.

a A is the set of names of all continents.

b B is the set of names of the three tallest mountains in the world.

c C is the set of names of the three tallest buildings in the world.

d D is the set of positive integers between 20 and 50, including 20 and 50, that are multiples of 5.

2 **State** whether each set is finite or infinite. If the set is finite, **state** its cardinality.

a $\{x \mid x \in \mathbb{N}, x \text{ is a multiple of } 4\}$

b $\{x \mid x \in \mathbb{R}, -2 \le x \le 5\}$

c $\{y \mid y \in \mathbb{Z}, -3 < y \le 2\}$

d $\{b \mid b \in \mathbb{N}, b < 7\}$

3 **Write down** these sets using set builder notation:

a F is the set of integers greater than zero

b $G = \{3, 6, 9, \ldots, 21\}$

c $H = \{1, 4, 9, 16, \ldots\}$

d $J = \{1, 8, 27, \ldots, 1000\}$

4 $U = \{x \mid x \in \mathbb{N}\}$,
$A = \{x \mid x \in \mathbb{N}, x \text{ is a multiple of } 7\}$,
$B = \{1, 3, 5, 7, 9, \ldots, 19\}$ and
$C = \{x \mid x \in \text{primes } 5 < x \le 20\}$.

Find:

a $A \cap B$

b $A' \cap C$

c $A \cap B'$

d $(A \cap C)'$

5 **Create** Venn diagrams to represent the sets:

a $A \cap B'$ **b** $A' \cup B$

6 The table shows the facilities at three hotels in a resort.

	Hotel A	Hotel B	Hotel C
Pool	✓		✓
Bar	✓	✓	✓
Jacuzzi		✓	
Sauna	✓		
Tennis		✓	✓
Gym			✓

Draw a Venn diagram to represent this information.

7 Create a Venn diagram to illustrate the relationships of the following number sets: $\mathbb{R} = \{\text{real numbers}\}$, $\mathbb{Z} = \{\text{integers}\}$, $\mathbb{Q} = \{\text{rational numbers}\}$, $I = \{\text{irrational numbers}\}$, $\mathbb{N} = \{\text{natural numbers}\}$, and $P = \{\text{prime numbers}\}$.

8 Create a Venn diagram illustrating the relationships of the following geometric figures: $U = \{\text{all quadrilaterals}\}$, $P = \{\text{parallelograms}\}$, $R = \{\text{rectangles}\}$, $S = \{\text{squares}\}$, $K = \{\text{kites}\}$, $T = \{\text{trapezoids}\}$.

9 If $U = \{-10, -9, \ldots, 9, 10\}$, $A = \{0, 1, 2, \ldots, 9\}$, $B = \{-9, -8, \ldots, 0\}$ and $C = \{-5, -4, \ldots, 4, 5\}$, **list** the elements of the following sets:

a $A \cap B$

b $(B \cup C)'$

c $(A \cup B) \cap C$

d $A' \cap (B \cup C)$

e $(A \cap B) \cup (A \cap C)$

10 Determine whether the following statements are true or false. If false, explain why.

a $0 \in \mathbb{Q}$

b $\{\text{primes}\} \subseteq \{\text{odd integers}\}$

c If $U = \mathbb{N}$, then $(\mathbb{Z}^+ \cap \mathbb{N})' = \{0\}$

d $2 \subseteq \{\text{primes}\}$

Problem solving

11 In a school sports day, medals were awarded as follows: 36 in running, 12 in high jump, and 18 in discus. The medals were awarded to a total of 45 students, and only 4 students received medals in all three events. **Determine** how many students received medals in exactly two out of three events.

12 A survey of 39 university students found:

- 10 worked part-time while studying
- 18 received financial help from home
- 19 withdrew money from their savings as needed
- 2 financed themselves from all three sources
- 12 received financial help from home only
- 5 received financial help from home and withdrew money from savings
- 6 financed themselves only by working part-time and withdrawing money from savings.

Determine how many students:

a did not finance themselves using any of the three resources

b worked part-time and received money from home

c received financial help from home and withdrew money from their savings

d financed themselves using only one of the three ways surveyed.

Review in context

Scientific and technical innovation

1 Below is a Venn diagram used by medical researchers showing the genes associated with different brain diseases.

The sets represent the number of genes associated with:

$A = \{\text{Alzheimer's disease}\}$
$M = \{\text{multiple sclerosis}\}$
$S = \{\text{stroke}\}$
$G = \{\text{brain diseases}\}$

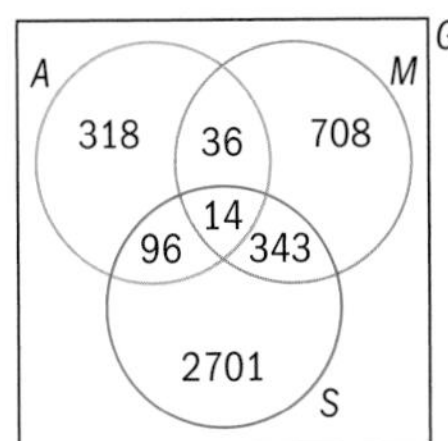

Using this diagram:

a **Calculate** how many genes in total are associated with each of the three diseases: Alzheimer's, multiple sclerosis, and stroke.

b **State** how many genes all three diseases share in common.

c **Explain** how this diagram might be useful to medical researchers.

Problem solving

2 Venn diagrams are widely used to show 'overlapping' concepts. Here are two fields from Sociology and Environmental Science where a Venn diagram can be used to illustrate the interrelation of key concepts.

In each example, **identify** sets and **draw** a Venn diagram to illustrate the information given.

a According to Plato there are many propositions. Some of these are true and some are beliefs (some may be neither). Only true beliefs can be justified and those he defines as knowledge.

b There are three main types of development: environmental, social and economic. Development that is both social and economic is equitable, that which is economic and environmental is viable, and that which is environmental and social is bearable. Development that is all three of these is sustainable.

3 Scientists have studied the genome of a number of organisms. The genome contains all the information used to build and maintain that organism. For medical researchers, knowing that different species share particular genes will enable them to unlock the secrets of countless diseases.

The Venn diagram shows the genes of four species: human, mouse, chicken, and zebrafish.

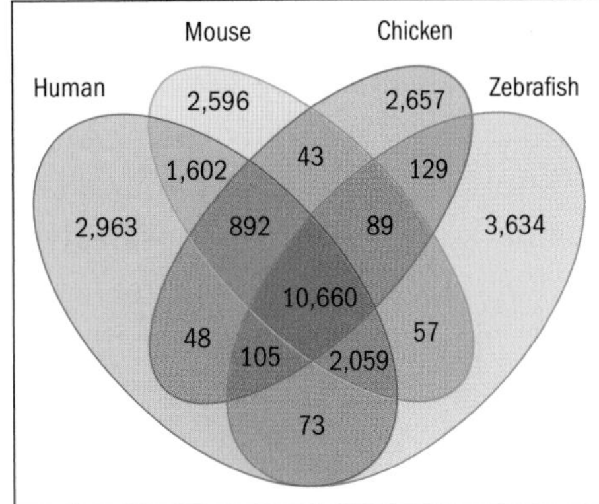

Using the diagram:

a **Determine** how many genes are shared by all four species.

b **Determine** how many genes does a human have in total.

c **Determine** how many genes are shared by human and mouse, by human and chicken, and by human and zebrafish.

d **Determine** the percentage of genes shared by the human and each of these three species.

e If you were a medical scientist and you wanted to conduct research into a species that was genetically closest to the human, which of the three – mouse, chicken or zebrafish – would you choose?

2 Relationships

Relationships are the connections and associations between properties, objects, people and ideas – including the human community's connections with the world in which we live.

Subitizing

How many penguins do you see here? Subitizing is immediately knowing how many items there are without actually counting them. Animals subitize too – knowing that one or two predators is less than a pack of predators is a primitive survival tool for estimating danger. Subitizing involves creating a simple mathematical relationship between amounts.

- How many items can you subitize? Work with a partner, trying different numbers of small objects like coins or buttons.

Expressing relationships

Some mathematical relationships are given as rules using numbers; others use algebra, like Pythagoras' theorem.

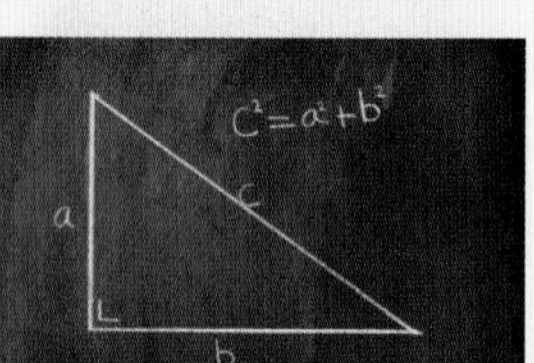

In mathematics, relationships express connections between quantities, properties or concepts and these connections may be expressed as models, rules or statements.

Any algebraic equation or inequality expresses a relationship between numbers and variables.

- $4x = 16.2$ states that there is a relationship of equality between $4x$ and 16.2, which helps you deduce that x must be 4.05.
- $x \geq 7$ states that the relationship between x and 7 is that x is equal to 7 or greater than 7, but cannot be less than 7.

Cartesian graphs show a relationship between the x- and y-coordinates. What relationships do these graphs show?

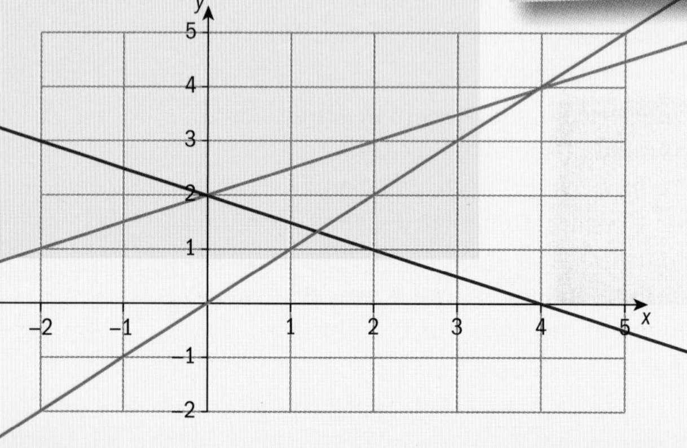

Biological relationships

To reproduce and survive, living organisms must coexist and interact with each other. Ecological relationships between living organisms are either oppositional or symbiotic.

In oppositional relationships, one organism eats others (predation) or organisms compete for the same resources (competition).

In symbiotic relationships, one organism lives in or on another. Sometimes this relationship is beneficial to both (mutualism), such as the clownfish and anemone (below, left). The anemone protects the clownfish (which is immune to its toxin) and the clownfish's waste provides food for the anemone.

The suckerfish (below, right) attaches itself to a host like this turtle and eats its parasites.

A symbiotic relationship may benefit one organism but have no effect on the other (commensalism). Corals and anemones provide camouflage for commensal shrimps and pygmy seahorses but do not benefit from them.

2.1 Are we related?

Global context: Fairness and development

RELATIONSHIPS

Objectives

- Understanding the difference between a relation and a function
- Understanding mapping diagrams
- Knowing how to find ordered pairs in a relation
- Understanding domain and range
- Manipulating functions using the correct notation

Inquiry questions

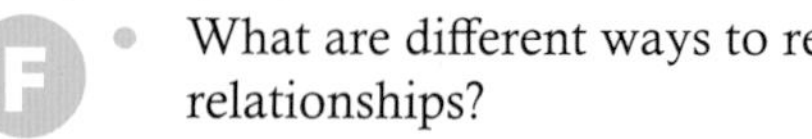

F
- What are different ways to represent relationships?
- What is a function?

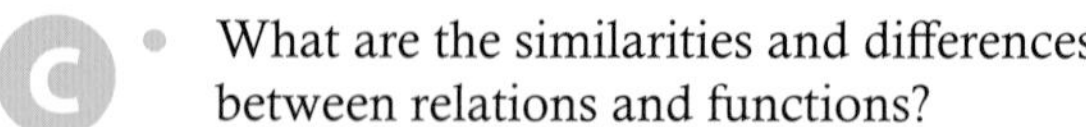

C
- What are the similarities and differences between relations and functions?
- How are the different ways of determining if a relation is a function related to one another?

D
- What do relations that are not functions look like?
- Can inequality be justified?

ATL **Communication**

Organize and depict information logically

You should already know how to:

• substitute values into an equation	**1** Find the value of $y = x^3$ when **a** $x = 9$ **b** $x = 0.25$ **2** Find the value of $y = \frac{2}{x^2}$ when **a** $x = 6$ **b** $x = \frac{1}{3}$
• interpret and use the basic language of sets, including set builder notation	**3** Use set builder notation to write: **a** the set of *x*'s such that *x* is any real number **b** the set of *x*'s such that *x* is any integer greater than zero.

Relations, mappings, and functions

- What are different ways to represent relationships?
- What is a function?

There are many situations where you pair objects or things. For example, a cause can be paired with an effect, a piece of art can be paired with the artist, a color can be paired with its HTML color code.

In mathematics, if the first object is x and the second object is y, you can write this pair as (x, y).

> In an **ordered pair** (x, y), the first term represents an object from a first set and the second term represents an object from a second set.

The HTML color code for this shade of blue-gray is 608FB0. The number of possible color codes represented by the HTML system is 16 777 216.

To write an ordered pair, first give the sets that the pairs come from. Examples:

$A = \{\text{house pets}\}$ and $B = \{\text{number of legs}\}$, then (dog, 4) is an ordered pair.

$A = \{\text{person}\}$ and $B = \{\text{age}\}$, then (Jack, 14) and (Juanita, 15) are ordered pairs.

Exploration 1

1 Set A contains all the students in your class. Set B contains the sports available at your school.

a List the members of set A.

b List the members of set B.

c Write down 5 ordered pairs relating set A to set B, describing which sports different students play.

Continued on next page

'Set' has many meanings outside the sphere of mathematics, as do many other terms used in mathematics. In French the word for 'set' is 'ensemble' and in German it is 'Menge', which both have different meanings than 'set' outside the scope of mathematics.

2 Choose other pairs of sets A and B that describe students in your class.

a List the members of each set and list the ordered pairs.

b Compare them to other ordered pairs created by your classmates.

3 Consider these ordered pairs, where A = {student} and B = {class}: (George, Science), (Ann, History), (Enrico, Science), (Matt, Music), (Donald, History), (Cindy, Science).

a Suggest what these ordered pairs could represent.

b This diagram shows the members of sets A and B. Copy the diagram. For each ordered pair in part **a**, draw a line mapping the element of set A to the element in set B. This is called a mapping diagram.

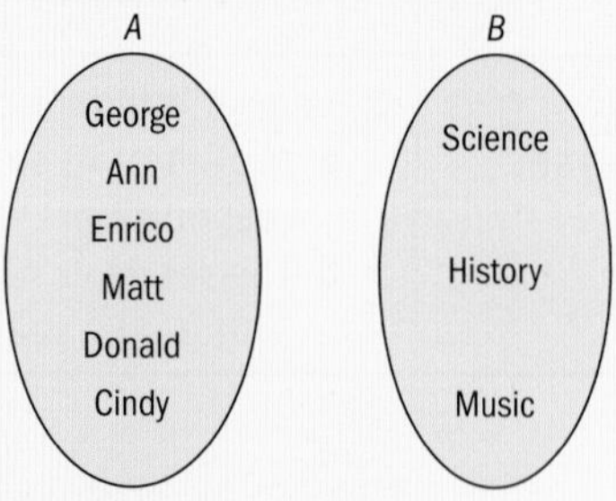

4 Construct a mapping diagram for your ordered pairs in step **1c**.

5 Decide on two sets that you encounter in real life. Draw a mapping diagram for ordered pairs from these sets. Write down 4 ordered pairs.

A **mapping diagram** shows how the elements in a relation are paired. Each set is represented by an oval, and lines or arrows are drawn from elements in the first set to elements in the second set for each ordered pair in the relation.

These mapping diagrams representing four different types of relations.

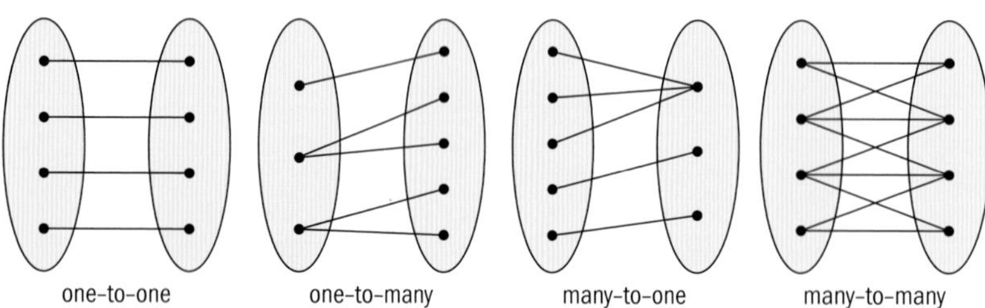

How do you think each mapping diagram got its name?

So far, you have looked at ordered pairs that describe relationships in real life. However, mathematics is also full of relationships that can be described in the same way.

Exploration 2

1 Consider the equation $y = x + 2$, where x is an element of the set $A = \{0, 1, 2, 3, 4, 5\}$ and y is an element of set B.

a List the elements of set B.

b Draw a mapping diagram to represent the relation between set A and set B.

▶ Continued on next page

c List all the ordered pairs (x, y) that represent this relation.

d Using the ordered pairs as coordinates, plot these points on a coordinate plane.

2 Consider the same equation $y = x + 2$, where x is an element of the set A, where $A = \mathbb{R}$, and y is an element of set B.

a Define set B.

b Determine if you can draw a mapping diagram to represent all the ordered pairs in this equation.

i If possible, draw the mapping diagram.

ii If not possible, explain why not.

c Draw the graph of this equation on a coordinate plane, where x is any real number.

3 Consider the equation $x^2 = y^2$, where x is an element of the set $A = \{-2, -1, 0, 1, 2\}$ and y is an element of set B.

a List all the ordered pairs (x, y) that represent this relation.

b Hence, list the elements of set B.

c Draw a mapping diagram to represent the relation between set A and set B.

d Using these ordered pairs as coordinates, plot these points on a coordinate plane.

4 Consider the same equation $x^2 = y^2$, where x is an element of the set A, where $A = \mathbb{R}$, and y is an element of set B.

a Use your knowledge of the graph in step **3** to draw the graph of this equation on a coordinate plane, where x is any real number.

b Define set B.

A relation can be defined either by listing the set of ordered pairs that make a relation (as in Exploration 1) or by giving a specific rule (as in Exploration 2).

Reflect and discuss 1

- What are the advantages and limitations of using a mapping diagram to represent a relation?
- What are the advantages and limitations of using a graph to represent a relation?
- In which situations is a mapping diagram better than a graph?
- In which situations is a graph better than a mapping diagram?

A **relation** is a set of ordered pairs $\{(x, y) | x \in A, y \in B\}$.

It has three components:

- a relation or rule that maps x onto y for each ordered pair in the relation
- a set A that contains all the x elements of each ordered pair
- a set B that contains all the y elements.

A relation maps set A onto set B.

ATL

Practice 1

1 Determine which type of mapping diagram (one-to-one, one-to-many, many-to-one, or many-to-many) best matches each of these relations:

a A = {students in your class}, B = {number of siblings}, and the relation maps each student in A to their number of siblings in B.

b A = {students in your class}, B = {mothers}, and the relation maps each student in A to their mother in B.

c A = {couples}, B = {grandchildren}, and the relation maps each couple in A to each of their grandchildren in B.

2 Find a real-life situation for each type of mapping diagram.

a Define the sets A and B as in question **1**.

b Write down the relation that maps the elements in set A to the elements in set B.

3 For the two equations $y = x + 2$ and $x^2 = y^2$, determine the type of mapping diagram that best describes the relation between set $A = \mathbb{R}$ and set $B = \mathbb{R}$. Hint: Look back at your answers to Exploration 2.

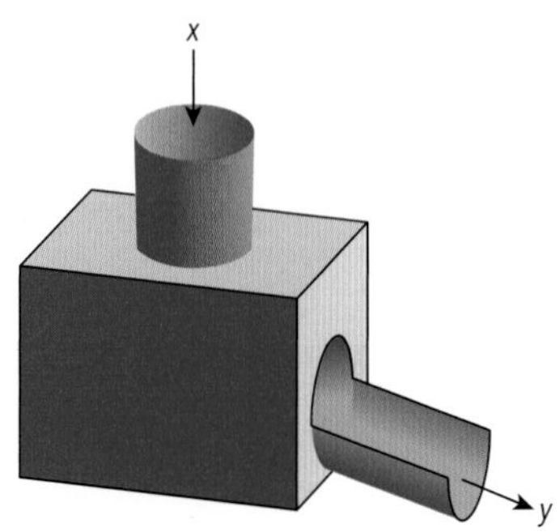

A **function** is a relation where each element in set A maps to one and only one element in set B.

A function can be thought of as a number machine, where for every input value there is only one possible output value. Such a number machine is called a **function machine**.

A function can be written as $f(x) = y$, where:

- x is the input value, $x \in A$
- y is the output value, $x \in B$
- f is the name of the function that maps x to y.

$f(x) = y$ is read 'the function f of x is y' or just 'f of x is y'.

Another way of writing $f(x) = y$ is $f : x \mapsto y$.

The letter f is most often used to represent a function, but other letters such as g and h are often used in problems involving more than one function.

Linear equations, such as $y = x + 2$, are functions. You can write $f(x) = x + 2$ or $f : x \mapsto x + 2$. Using this notation, you can say that $f(3) = 5$, meaning that when the input x to the function f is 3, the output is 5. You could also write $f(4) = 6$ and $f(-5) = -3$, for example. Function notation uses both elements of an ordered pair in one equation.

If a function maps elements of set A onto elements of set B, then set A is called the domain and set B is called the range.

The **domain** of a function is the set of input values that the function can take.

The **range** of a function is the set of all output values that the function generates. The range is also called the set of images of the elements in the domain.

In a mapping diagram, all the elements of the domain are in one set, and all the elements of the range are in the other set.

Usually, the domain is specified in the question. If not, it is assumed to be the set of real numbers, $\mathbb{R}$. If a function is defined as a set of ordered pairs $(x, f(x))$, then the domain is simply the set of input values (all the values of x) and the range is the set of output values (all the values of $f(x)$).

Example 1

Consider this mapping diagram.

a Write in words what the relation could represent.

b Justify why the relation is a function.

c State the domain and range of the function.

d Write the function for one of the ordered pairs in this diagram.

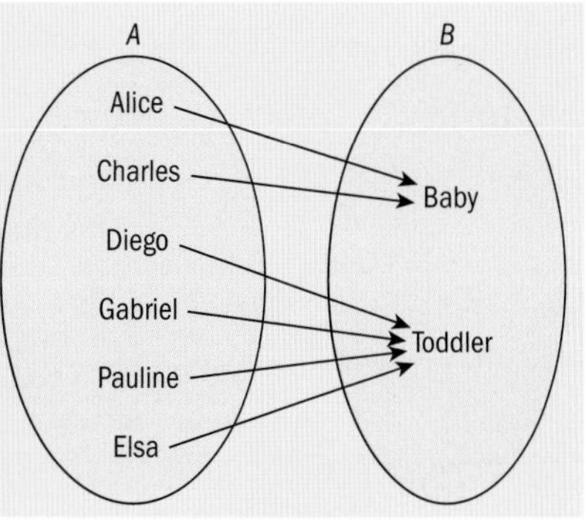

In the 18th century, Leonhard Euler, one of the most prolific mathematicians in history, introduced the notation $f(x)$ to represent a function of x.

a This mapping diagram could represent which age group children belong to at a day care center.

b This relation is a function because each child is in only one age group.

In order to be a function, each element from the domain must map to one and only one element in the range.

c Domain: {Alice, Charles, Diego, Gabriel, Pauline, Elsa}
Range: {Baby, Toddler}

d $g(\text{Elsa}) = \text{Toddler}$
Elsa is a toddler.

g was chosen for the function, since it represents the *group* a child belongs to.

Practice 2

1 Consider this mapping diagram.

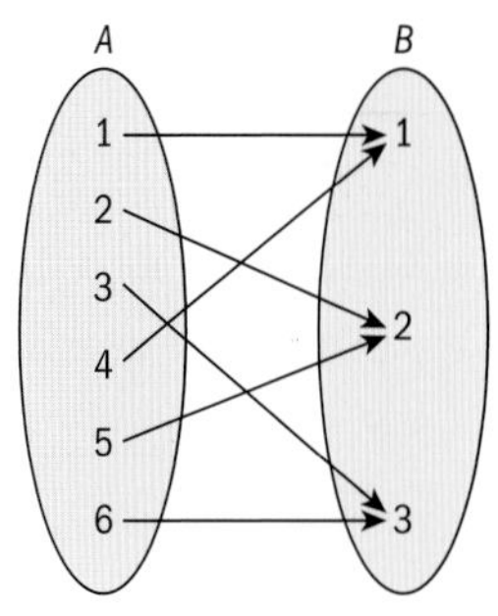

a Determine if this relation is a function. Justify your answer.

b Write the domain and range using set notation.

c Find:

i $f(1)$ **ii** $f(2)$ **iii** $f(5)$

2 For each of the following mapping diagrams:

i Write in words what the relation could represent.

ii Justify why the relation is a function.

iii State the domain and range.

iv Write down one specific ordered pair in the function.

a

Chocolate brand

Country

Neuhaus

Côte d'Or

Leonidas

Toblerone

Lindt

Hershey's

Baci

Belgium

Switzerland

USA

Italy

b

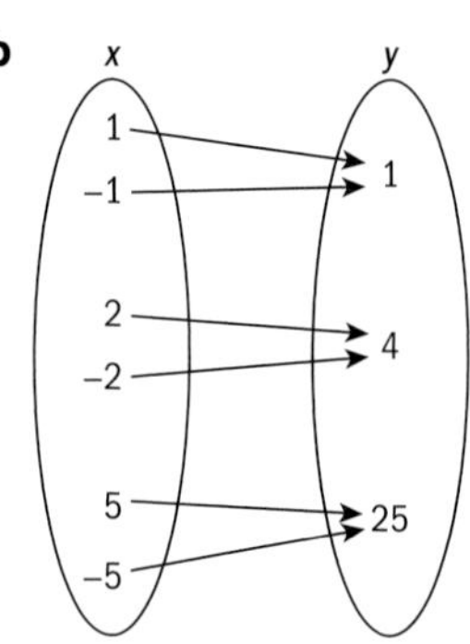

c

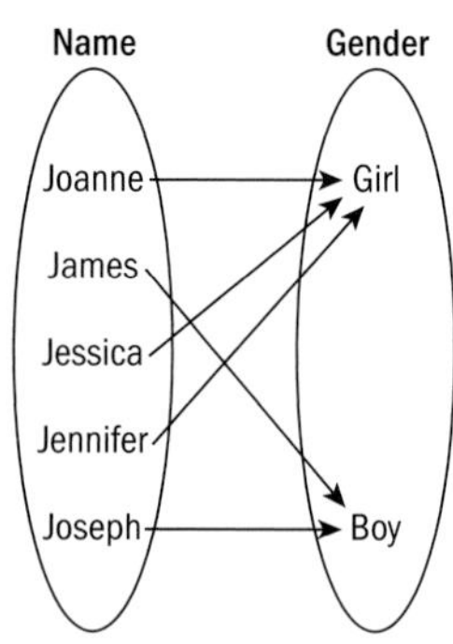

Example 2

State the domain and range of the relation:

$\{(1, 2), (2, 4), (3, 6), (4, 8), (5, 10)\}$

Domain = $\{1, 2, 3, 4, 5\}$

In a list of ordered pairs, the domain is the list of *x*-values. The domain is written in set notation.

Range = $\{2, 4, 6, 8, 10\}$

In a list of ordered pairs, the range is the list of *y*-values. The range is written in set notation.

Practice 3

1 State the domain and range for each relation.

a $\{(7, 4), (2, 9), (4, 6), (8, 1), (5, 2)\}$

b $\{(6, 2), (8, 8), (3, 7), (4, 4), (4, -4)\}$

c $\{(-3, -1), (5, 1), (-2, -1), (3, 1), (-7, -1), (8, 1)\}$

d $\{(a, g), (t, c), (s, g), (k, w), (j, k), (p, p)\}$

e $\{(a, x), (a, y), (a, z), (b, x), (b, z)\}$

2 Consider the set $A = \{-2, -1, 0, 1, 2\}$. For each relation **a** to **c**:

i List all the ordered pairs (x, y) that represent this relation.

ii Hence, list the elements of the range (set B).

iii Draw a mapping diagram to represent the relation between set A and set B.

iv Use your mapping diagram to determine which type of relation it is.

v Determine whether or not the relation is a function.

a $y = x$ **b** $y = 3x + 3$ **c** $y = x^2$

Activity

With some functions, it is possible to determine the domain and range by simply thinking about the kinds of values that can be used as input and the kinds of values that are produced as output. Using this method, copy and complete this table.

Function	Domain	Range
$f(x) = 2x$	x can be any value, since you can double any value, so the domain of f is $\mathbb{R}$.	
$g(x) = \sqrt{x}$		By convention, the square root function gives only the positive square root. So the range of g is $\{y \mid y \geq 0, \in \mathbb{R}\}$.
$h(x) = x^2$	You can square any real number, so the domain of h is $\mathbb{R}$.	
$f(x) = \frac{1}{x}$		

The four functions in the Activity show the different possible restrictions you can have on the domain or range for different functions.

A **natural** domain is the largest possible set of values that a function can take.

A **restricted** domain is a subset of the natural domain of the function.

Practice 4

State the largest possible domain and the corresponding range for each function.

1 $y = 5x - 3$

2 $y = -4x + 9$

3 $y = \frac{2}{5}x + 12$

4 $y = 2x^2$

5 $y = x^2 + 5$

6 $y = 3x^2 - 3$

7 $y = \frac{32}{x}$

8 $y = \frac{15}{2x}$

9 $y = \frac{1}{x+1}$

10 $y = \frac{1}{x} + 1$

Reflect and discuss 2

Recall the definitions of a relation and a function.

- Are all functions relations?
- Are all relations functions?
- If you answered 'no' to either question above, explain why.

C Using functions to express relationships

- What are the similarities and differences between relations and functions?
- How are the different ways of determining if a relation is a function related to one another?

So far, you have seen that:

- a function is a special type of relation, where for each element of the domain there exists one and only one element in the range. (Each input value generates exactly one output value.)
- a function is a one-to-one or a many-to-one relation.

Exploration 3 investigates further how you can determine whether or not a relation is a function.

Exploration 3

1 For each mapping diagram, draw a set of points on a coordinate plane. Label each graph with its relationship.

Use a new coordinate plane for each mapping diagram.

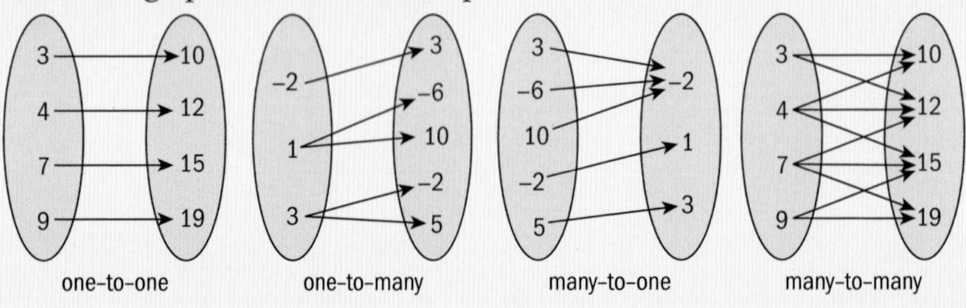

If you are stuck, try the following steps to guide you:

a Determine which axis represents the domain, x.

b Determine which axis represents the range, $y = f(x)$.

2 Determine of your graphs show a function. Suggest how you could determine from a graph whether or not a relation is a function.

3 **a** Draw the graph of $y = 4x - 9$.

b Looking at your graph, determine the type of relation this is.

Continued on next page

c Draw a mapping diagram to represent $y = 4x - 9$. Select only a few ordered pairs for your mapping diagram, until you are satisfied with the type of relationship it represents. You do not need to list every possible ordered pair for this mapping diagram.

d Decide whether or not the mapping diagram confirms the type of relationship that your graph represents.

e Hence, decide whether $y = 4x - 9$ is a function.

4 Here is the graph of $y^2 = x$.

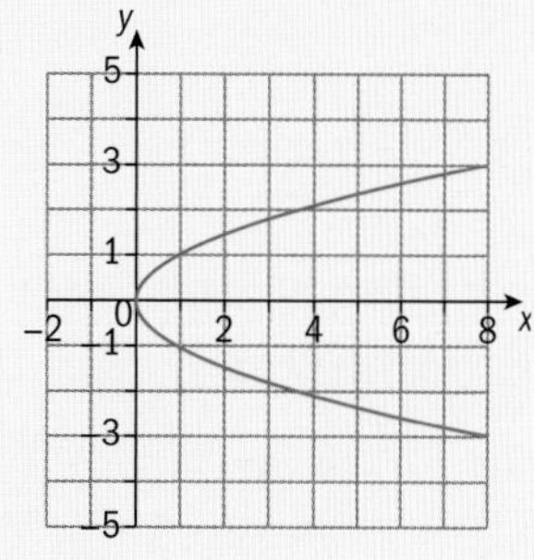

a Determine which of the four types of relationship this graph represents.

b Hence, decide whether or not $y^2 = x$ is a function.

You have used two different methods for deciding whether or not a relation is a function (sometimes called 'a function from x to y'):

- with a mapping diagram: if a mapping diagram is one-to-one or many-to-one, then the relation is a function
- with a graph: if each x-value has only one corresponding y-value, then the relation is a function.

> The **vertical line test**: if no vertical line intersects a graph at more than one point, the relation is a function. If a vertical line intersects the graph at more than one point, then the relation is not a function.

$y = 0.5x - 1.5$ is a function, because no vertical line intersects the graph more than once.

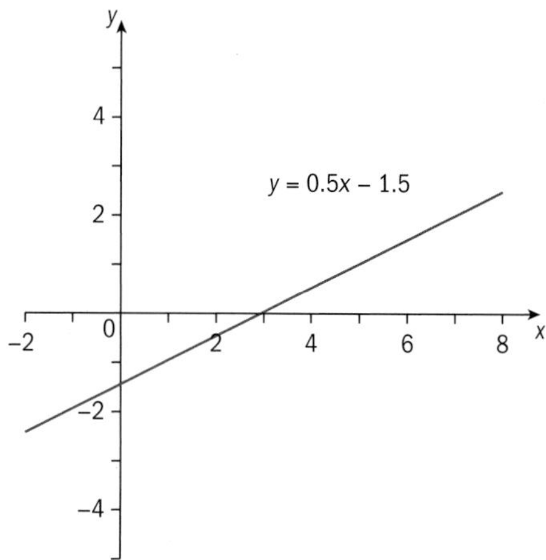

$y^2 = x$ is not a function, because at least one vertical line intersects the graph twice.

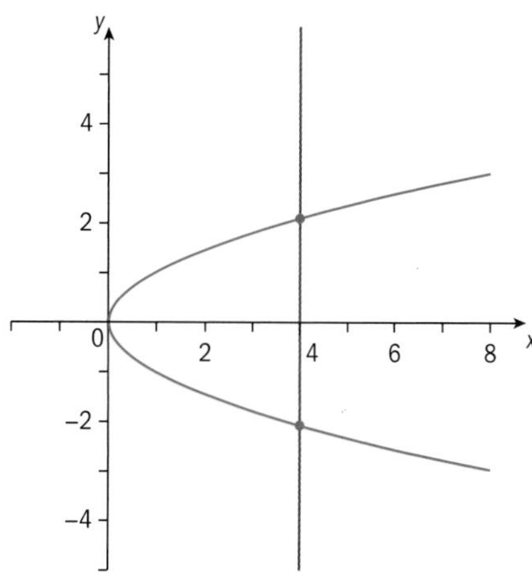

Practice 5

1 Use the vertical line test to determine which of these graphs show functions.

a
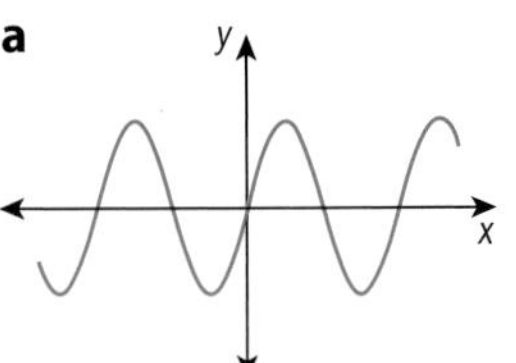

b
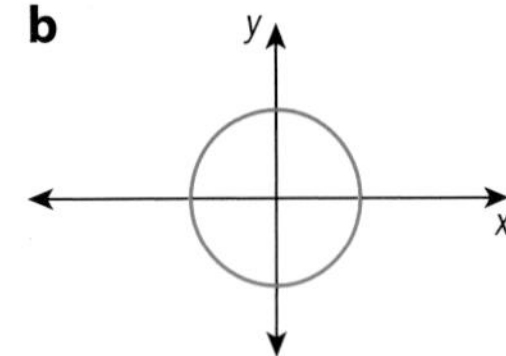

c
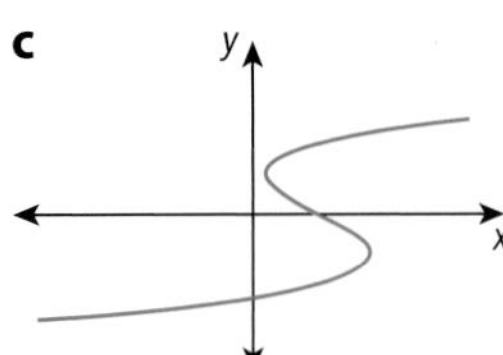

d
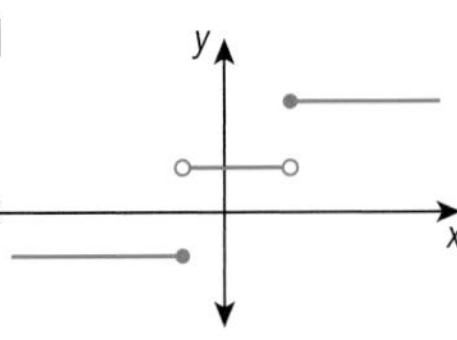

Example 3

Determine whether or not these relations represent a function from x to y.

a $\{(2, 5), (5, 6), (2, -5), (4, -6)\}$ **b** $\{(1, -3), (2, -4), (3, -3), (4, -4)\}$

c $y = x - 3$ **d** $x^2 + y^2 = 4^2$

a

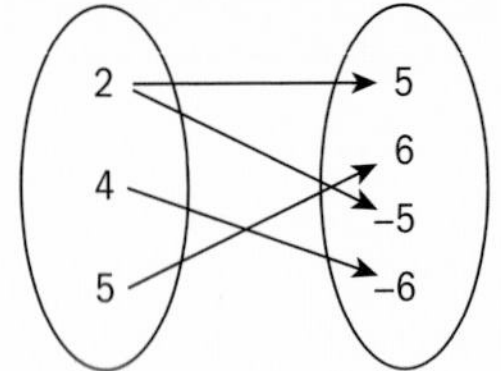

Draw a mapping diagram for the ordered pairs.

This relation is not a function.

2 maps to both 5 and –5.
This is a one-to-many relation.

b

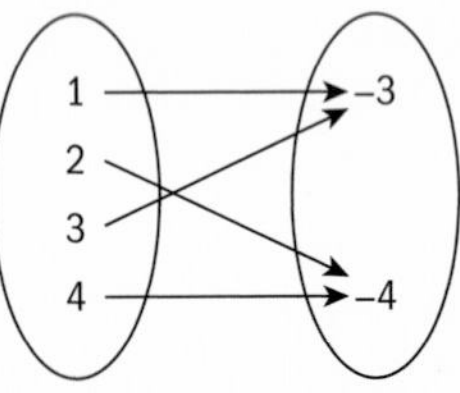

This relation is a function.

This is a many-to-one function.

c

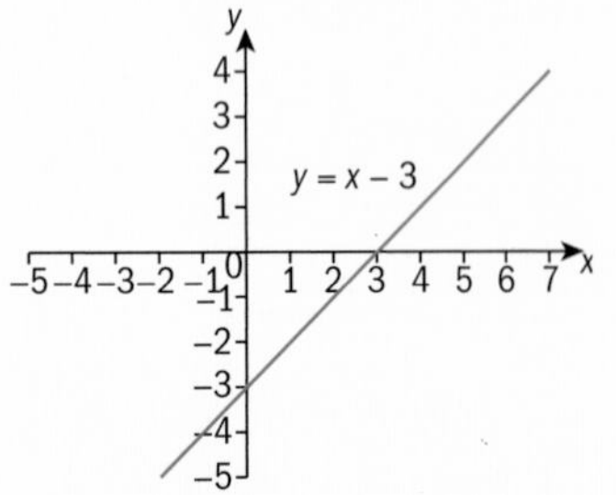

Draw a graph and use the vertical line test.

This relation is a function. Therefore, you use the notation $y = x - 3$ or $f(x) = x - 3$.

No vertical line would cross the graph at more than one point.

Continued on next page

d

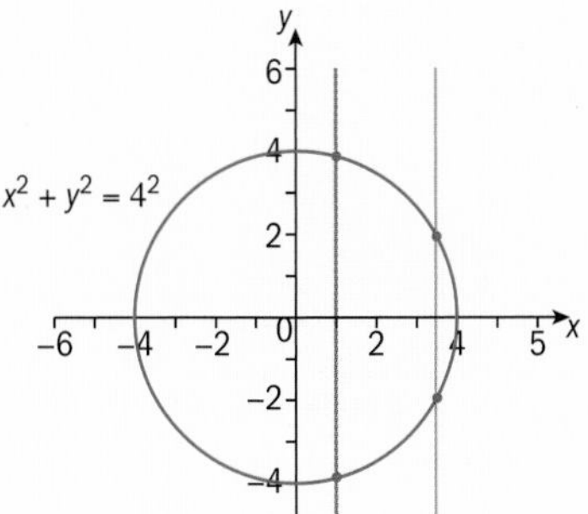

Draw a graph and use the vertical line test.

This relation is not a function.

You can draw vertical lines that cross the graph of this circle at more than one point.

Reflect and discuss 3

Representation is an MYP related concept. It is defined as 'the manner in which something is presented'.

- How did you represent relations and functions in Example 3?
- How does representation enable you to determine whether or not a relation is a function?
- Why is it important to distinguish between relations and functions?

Practice 6

1 Decide which of the following relations are functions. Give a reason for your answer.

a

A	1	2	3	4
B	4	2	0	2

b

A	−1	0	−1	1
B	5	5	7	5

c {(3, 12), (−16, 10), (3, 12), (15, 3), (3, 12), (−15, −3)}

d {(9, 9), (8, 7), (7, 5), (5, 9), (9, 5), (6, 12)}

e

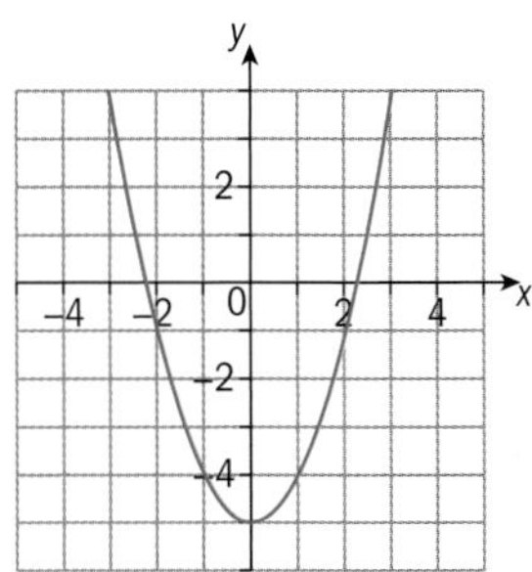

f

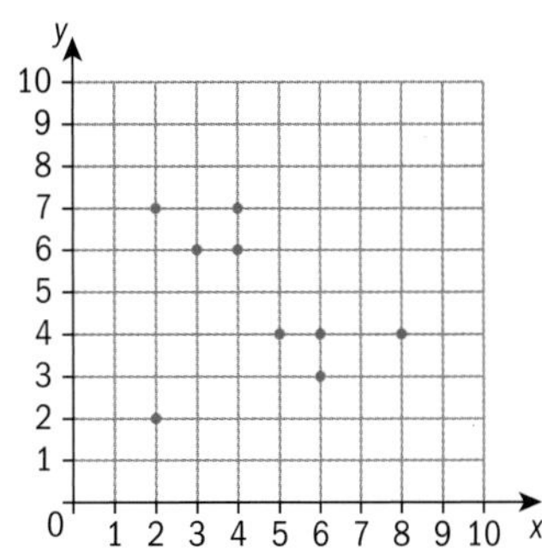

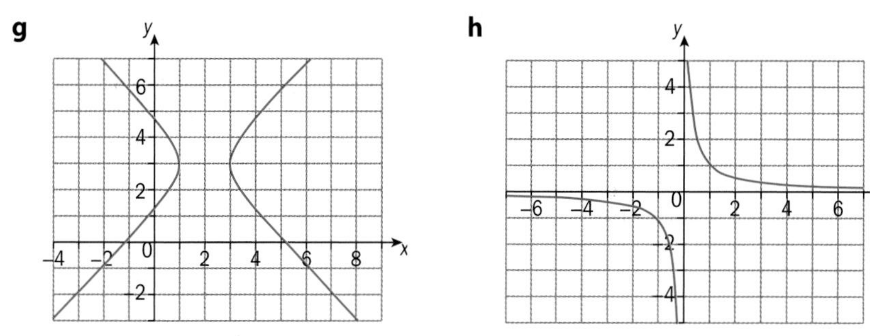

Reflect and discuss 4

Justification is an MYP related concept. It is defined as 'valid reasons or evidence used to support a statement'.

- Where in the previous example and practice problems have you used justification?
- How does justification enable you to support whether or not a relation is a function?
- Is it important to justify whether a relation is a function?
- Can you think of some equations you have worked with that are not functions?

Exploration 4

'I think of a number, double it, add 3 and then square the result.'

1 Explain why this is a function.

2 If x is an element of the domain of this function, write this function mathematically (i.e. $f(x)$ = something).

3 Make up your own 'I think of a number' puzzle. Write the function for your puzzle in mathematical notation. Then share it with others and see if they write the same function definition as you have.

Writing a function mathematically is called defining a specific function.

> A function is **defined** by a mathematical expression that specifies the relationship between a domain (input values) and range (output values).

For example, if a function f is defined as: $f(x) = 3x + 2$, then:

- f is the name of the function
- x is an element of the domain
- $f(x)$ is read as 'f of x'
- '$3x + 2$' tells you what the function does to each element of the domain to get the corresponding element from the range.

This is the definition of this specific function f. When defining a function, the domain is usually specified. If it isn't, it is assumed to be the set of real numbers.

Example 4

A recipe states that chicken needs to be cooked for 15 minutes per kg plus an extra 40 minutes. Express the relationship between weight and cooking time using function notation, and define the domain and range.

x: the weight of the chicken (in kg)
$f(x)$: the cooking time (in minutes)

Set the input and output variables in the context of the problem.

Domain: $x \geq 0$
Range: $f(x) \geq 40$

Neither the weight nor the cooking time can be negative. The chicken will take *at least* 40 minutes to cook.

$f(x) = 15x + 40$

15 minutes per kg (x) plus 40 more minutes.

Check:

Check with easy values. Do the results seem sensible?

$f(1) = 15 \times 1 + 40 = 55$

A 1 kg chicken needs 15 min + 40 min = 55 min.

$f(2) = 15 \times 2 + 40 = 70$

A 2 kg chicken needs 30 min + 40 min = 70 min.

Reflect and discuss 5

Think what input values are sensible for the function in Example 4. If you had no chicken at all, meaning that $x = 0$, does it make sense that the cooking time is 40 minutes? Yet, $f(0) = 15 \times 0 + 40 = 40$. For what values of x is this function definition intended?

Sometimes input values are mathematically possible, but they do not make sense in a real-life context (as with cooking 0 kg of chicken).

A real-life situation may restrict the elements of the domain in order to make sense of the problem. These restrictions are called *constraints* on the domain.

When setting variables to represent a worded (or real-life) situation, it is important that each variable represents a *number* or an *amount* within the context of the problem. Good variable definition examples are x = Tina's height and y = Tina's age. Why do you think writing x = Tina would be a bad variable definition?

Practice 7

Objective: C. Communicating
i. use appropriate mathematical language (notation, symbols and terminology) in both oral and written explanations.

In these questions you need to use function notation to express the relation, set builder notation for the domain and range, and words such as 'domain', 'range', 'one-to-one' to explain how the relation is a function.

For each question **1** to **3**:

a Express the relation using function notation.

b Define the domain and range. Decide what the constraints are on the domain and range based on the real-life context.

c Explain how the relation is a function.

1 A carpenter charges his clients a rate of €30 an hour, plus a single €40 fee for each job. Express the relationship between time spent on a job and the fee charged to the client.

2 A bathtub is filled with 120 liters of water. The drain plug is pulled and the water empties out at a rate of 25 liters per minute. Express the relationship between time elapsed since the plug was pulled, and the amount of water in the bathtub.

3 On a tropical island, the cost of a parcel of land is \$200 per square meter. Taxes and fees account for an additional 15%. Express the relationship between the size of a parcel of land, and the final price that the client pays for it.

The placement of the € symbol is often based on what countries did with their old currency. So it might be €37 in one country, but 37€ in another.

Evaluating a function means finding the element of the range that corresponds to a given element in the domain.

Example 5

If $f(x) = 3x + 2$, find the value of $f(4)$.

$f(x) = 3x + 2$ — Start with the original function definition.

$f(4) = 3 \times 4 + 2$ — Substitute 4 for *x*.

$f(4) = 12 + 2 = 14$

Practice 8

1 $f(x) = 4x - 2$

a Describe in words what this function does to the input value.

b Draw a table of values that shows at least 5 different input values and the corresponding output values.

2 $g(x) = x^2 + 2$

a Describe in words what this function does to the input value.

b Draw a table of values that shows at least 5 different input values and the corresponding output values.

3 $h(x) = \frac{2}{x}$

a Describe in words what this function does to the input value.

b Draw a table of values that shows at least 5 different input values and the corresponding output values.

4 If $f(x) = 7x - 3$, find:

a $f(3)$ **b** $f(-1)$ **c** $f(0)$ **d** $f(20)$

5 If $p(3)$ = triangle and $p(5)$ = pentagon, find:

a $p(4)$ **b** $p(6)$ **c** $p(8)$ **d** $p(10)$

In a function, an input value generates one specific output value. And in certain cases, two or more input values generate the same output value. In Exploration 5, you will see when it is possible to find the input value if you are given only the output value. This is called solving the function.

Solving an equation that is a function means finding the element of the domain that corresponds to a given element in the range.

Exploration 5

Consider the two functions $f(x) = 7x - 12$ and $f(x) = x^2$.

1 For each function, find $f(3)$. This is called *evaluating* the function at $x = 3$.

2 For each function, find the value of x such that $f(x) = 16$. This is called *solving* the equation when $f(x) = 16$.

3 For each function, determine whether there is only one possible input value for each output value.

4 Use your findings to decide if a function is always a one-to-one relationship between an input value and an output value. Justify your answer.

Evaluating a function (finding the value of $f(x)$) only gives one answer.

Solving a function (finding the value of x, given $f(x)$) can give more than one answer.

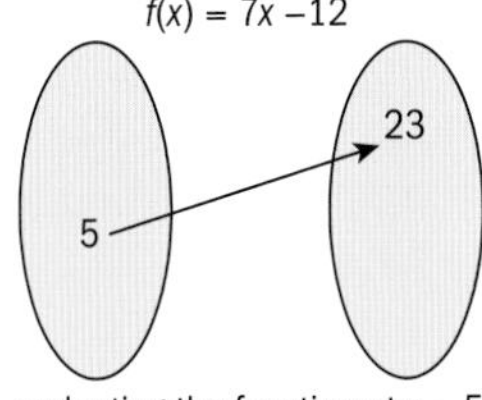

evaluating the function at $x = 5$

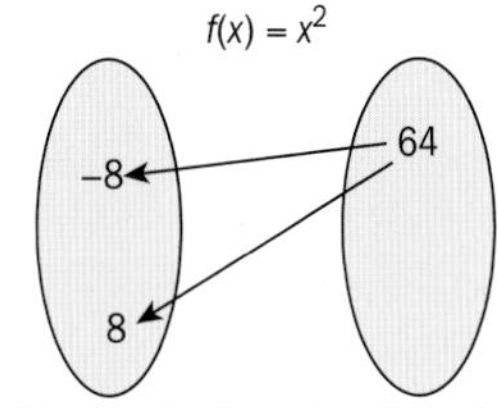

solving the function when $f(x) = 64$

Practice 9

In questions **1** to **7**, evaluate the function at the given values.

> In **1e**, substitute $-a$ for x in the function.

1 $h(x)=2-4x$

a $h(0)$ **b** $h(1)$ **c** $h(-1)$

d $h(20)$ **e** $h(-a)$ **f** $h(2x)$

2 $f(x)=3x-5$

a $f(0)$ **b** $f(4)$ **c** $f(-4)$

d $f(12)$ **e** $f(3x)$ **f** $f(x+1)$

3 $f(x)=13$

a $f(0)$ **b** $f(19)$ **c** $f(20{,}000)$

d $f(-25)$ **e** $f(-x)$ **f** $f(12x)$

4 $f(x)=x^2+1$

a $f(0)$ **b** $f(5)$ **c** $f(-5)$

d $f(9)$ **e** $f(x+1)$ **f** $f(4x)$

5 $f(x)=x^2-2x+1$

a $f(1)$ **b** $f(2)$ **c** $f(0)$

d $f(7)$ **e** $f(-10)$ **f** $f(3x)$

6 $f(x)=4x^3-2x+1$

a $f(2)$ **b** $f(-1)$ **c** $f(0.5)$

d $f(3)$ **e** $f(-2)$ **f** $f(-x)$

7 $f(x)=3x+5$

a $f(5)$ **b** $f(10)$ **c** $f(2x)$

d $f(3x)$ **e** $f(x+5)$ **f** $f(3x-2)$

8 If $f(x)=8x-4$, find the value of a such that $f(a)=12$.

9 If $f(x)=x+12$, find the value of b such that $f(b)=12$.

10 If $h(x)=15-2x$, find the value of c such that $h(c)=3$.

11 If $r(x)=x^2$, find the value of d such that $r(d)=36$.

12 If $f(x)=x^2+5$, find the value of e such that $f(e)=54$.

13 If $k(x)=\frac{1}{x+2}$, find the value of p such that $k(p)=\frac{1}{4}$.

14 If $f(x)=\frac{1}{x-5}$, find the value of q such that $f(q)=\frac{1}{5}$.

> In **11** and **12** there are two possible values.

Problem solving

15 If $p(135)$ = south-east and $p(180)$ = south, find the value of s such that $p(s)$ = north-west.

16 If child(Olimpia) = Viola and child(Arienne) = William, state the condition(s) that must be satisfied in order for child(x) to be a function. Suggest a value of m such that child(m) = you, and determine if in your family child(m) = you is a function. Explain your reasoning. Share your ideas with others.

> In **16**, how many children can a parent have if child(x) is a function?

D Recognizing relations that are not functions

- What do relations that are not functions look like?
- Can inequality be justified?

So far, you know that:

- a function is a specific relation between the domain and range
- if an input to a relation generates more than one output, the relation is not a function.

You have already looked at how to decide if a mapping diagram or a graph represents a function. However, is it possible to tell if a relation is a function just by looking at its equation?

Exploration 6

Look at the lists of equations below. For each relation, $x \in \mathbb{R}$ and $y \in \mathbb{R}$.

FUNCTION	**NOT A FUNCTION**
$y = 3x - 4$	$x^2 + y^2 = 16$
$2x + 5y = 9$	$x = 3(y + 1)^2 - 7$
$y = 2$	$2x + 3y^2 - 4y = 5$
$x^2 + 3x - 4y - 5 = 0$	$x - y^2 = 1$
$y = 2(x + 1)^3 - 8$	$\frac{(x-1)^2}{4} + \frac{(y-2)^2}{25} = 1$
$y = \sqrt{x-4},\ x \geq 4, y \geq 0$	$x = \pm\sqrt{y-4},\ y \geq 4$

1 Based on the lists above, make a conjecture about the types of equations that are functions and those that are not.

2 If possible, rewrite each equation to look like $y = f(x)$.

3 Based on your results, suggest which type of equation represents a relation, but is not the equation of a function.

Reflect and discuss 6

- What sort of relations generate more than one output value for a single input value?
- Why do those relations generate more than one output value?
- Are there other relations that have similar properties (and are therefore not functions)?
- How does this relate to the 'vertical line test'?

Summary

In an **ordered pair** (x, y), the first term represents an object from a first set and the second term represents an object from a second set.

A **relation** is a set of ordered pairs: $\{(x, y) | x \in A, y \in B\}$.

It has three components:

- a relation or rule that maps x onto y for each ordered pair in the relation
- a set A that contains all the x elements of each ordered pair
- a set B that contains all the y elements.

A relation maps set A onto set B.

A **mapping diagram** shows how the elements in a relation are paired. Each set is represented by an oval, and lines or arrows are drawn from elements in the first set to elements in the second set for each ordered pair in the relation.

A **function** is a relation where each element in set A maps to one and only one element in set B.

A function can be written as $f(x) = y$, where:

- x is the input value, $x \in A$
- y is the output value, $y \in B$
- f is the name of the function that maps x to y.

$f(x) = y$ is read 'the function f of x is y' or just 'f of x is y'.

Another way of writing $f(x) = y$ is $f : x \mapsto y$.

The **domain** of a function is the set of input values that the function can take.

The **range** of a function is the set of all the output values that the function generates. The range is also called the set of images of the elements in the domain.

A **natural** domain is the largest possible set of values that a function can take.

A **restricted** domain is a subset of the natural domain of the function.

The **vertical line test**: if no vertical line intersects a graph at more than one point, the relation is a function. If a vertical line intersects the graph at more than one point, then the relation is not a function.

A function is **defined** by a mathematical expression that specifies the relationship between a domain (input values) and range (output values).

Evaluating a function means finding the element of the range that corresponds to a given element in the domain.

Solving an equation that is a function means finding the element of the domain that corresponds to a given element in the range.

Mixed practice

1 **Determine** whether each diagram represents a relation or a function. **Justify** your answer.

a

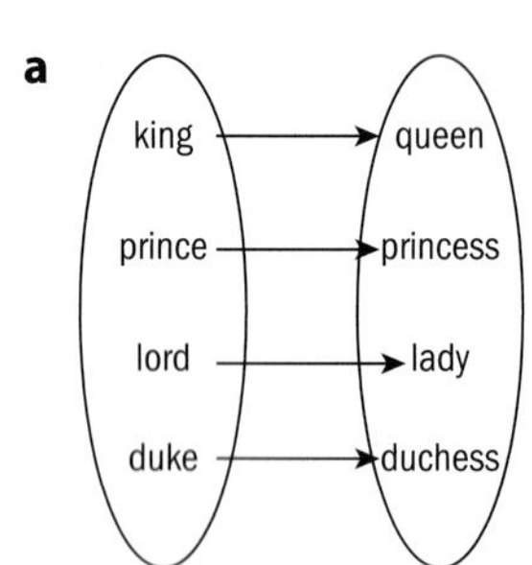

b

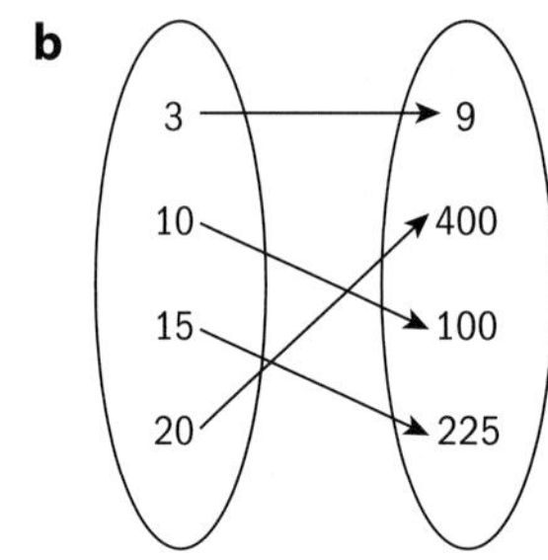

c

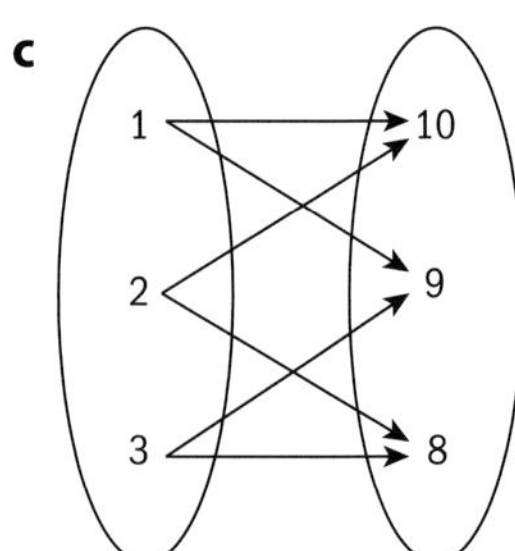

d

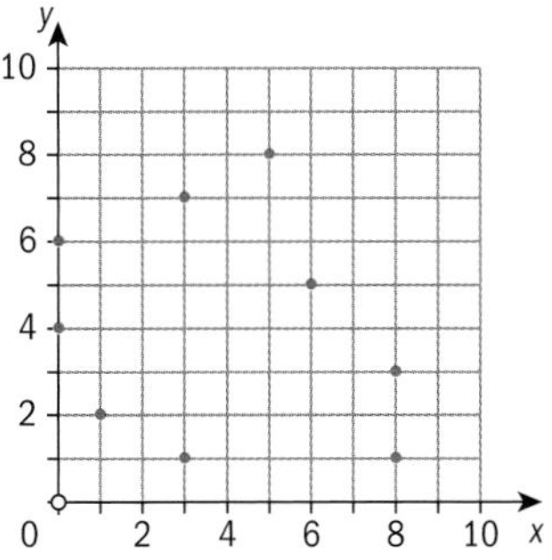

e

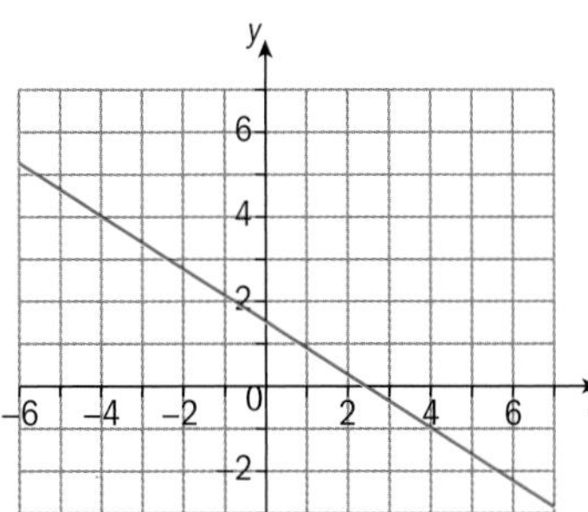

f

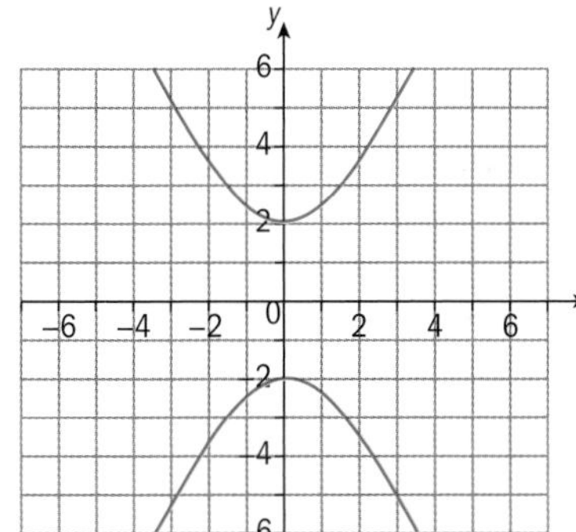

g

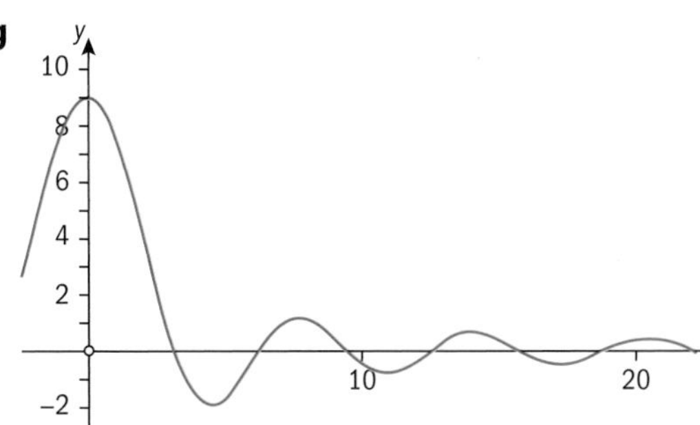

h

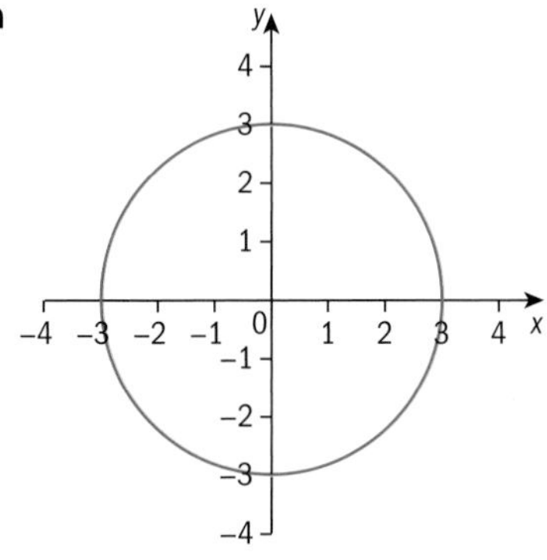

i

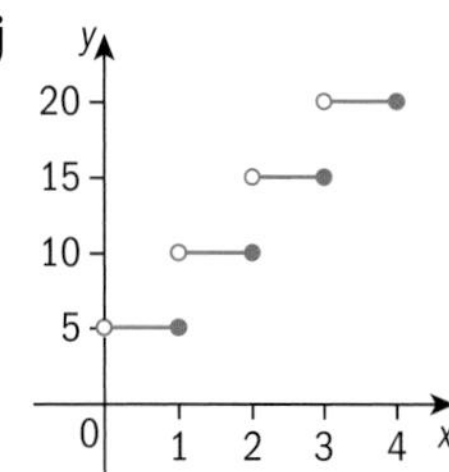

j

2 State the domain and range:

a $\{(1, 1), (2, 4), (-2, -4)\}$

b $\{(3, 5), (4, 5), (5, 6)\}$

c $\{(-2, -3), (-3, -2), (-2, 5)\}$

d $\{(1, 1), (2, 3), (5, 8)\}$

3 Find the largest possible domain and range:

a $f(x) = 3x - 8$ **b** $f(x) = 3\sqrt{x}$

c $f(x) = \frac{x}{4}$ **d** $f(x) = \frac{4}{x}$

e $f(x) = \frac{4}{x-4}$ **f** $f(x) = x^2 - 1$

4 Evaluate each function at the given values:

a $a(x) = 2 - x^2$

i $a(2)$ **ii** $a(0)$ **iii** $a(-2)$

b $b(x) = 8x - 4$

i $b(4)$ **ii** $b(3)$ **iii** $b(1)$

c $c(x) = \sqrt{x-1}$

i $c(1)$ **ii** $c(5)$ **iii** $c(0)$

d $d(x) = 2x + 3$

i $d(4)$ **ii** $d(4x)$ **iii** $d(2 - x)$

5 **Solve** each function at the given values:

a $f(x) = 4x - 2$

i $f(x) = 18$ **ii** $f(x) = 0$ **iii** $f(x) = 2$

b $f(x) = \frac{1}{x}$

i $f(x) = \frac{1}{5}$ **ii** $f(x) = -\frac{1}{4}$ **iii** $f(x) = 2$

c $f(x) = \sqrt{x}$

i $f(x) = 5$ **ii** $f(x) = 1$ **iii** $f(x) = 9$

6 A computer program asks you for any word and returns the number of letters in that word.

a **Justify** why this computer program is a function.

b **Write down** what this computer program does in function notation, where the function is called 'p' for program.

c **Evaluate**:

i p(horse) **ii** p(Mississippi) **iii** p(hi)

d **Find** x such that

i $p(x) = 3$ **ii** $p(x) = 8$

e What is p(three)?

f For what numbers $0 \le x \le 10$ does $p(x) = x$?

7 A hospital has 9 floors above ground (including ground floor) and 3 floors below ground. $T(n)$ gives the average number of times the elevator stops at the nth floor each day.

a **Justify** why $T(n)$ is a function.

b **Find** the largest possible domain of $T(n)$.

c For each floor in the domain, **suggest** the number of times the elevator stops at that floor during 24 hours. Write your answer as a list of ordered pairs.

d Hence, **suggest** a possible range of $T(n)$. **Justify** your choice.

8 The table shows postage rates for letters and parcels.

Letters	
Weight less than	Price
50 g	£0.65
100 g	£0.92
200 g	£1.20
Small parcels	
Weight less than	Price
100 g	£1.90
200 g	£2.35
500 g	£3.40
750 g	£4.50
1 kg	£5.00

a **Draw** a mapping diagram to represent the weights of items and the delivery charge.

b Decide whether the weight determines the price or the price determines the weight. Add arrows to your mapping diagram to show this.

c **Determine** whether this relation is a function.

Review in context

Fairness and development

1 Many universities use a test in their admissions decisions. The results of one such test were paired with family income as reported by test-takers, shown in the table here:

Family income ($)	Total test score (max. 2400)
10000	1326
30000	1402
50000	1461
70000	1497
90000	1535
110000	1569
130000	1581
150000	1604
180000	1625
More than 180000	1714

a **Draw** a mapping diagram to represent the family income and the total test score.

b **Determine** whether or not this relation is a function.

c **Suggest** reasons why this relationship may exist.

d Should universities use this information in their admission decisions? If so, how?

2 The table shows the percentage of world population in different countries and their use of natural resources.

Country	Percentage of world population	Percentage of world's resources used
Brazil	3	2
China	19	20
India	17	5
Indonesia	4	1
Japan	3	4
Pakistan	3	0.5
Russian Federation	3	6
United States	5	18

a **State** the input variable and the output variable and the domain and range of this relation.

b Is this relation a function? **Explain**.

c **Draw** a mapping diagram of this relation.

d Does this demonstrate an equitable or fair relationship? **Explain** your reasoning.

e What action, if any, do you think should be taken?

3 Car insurance rates vary greatly depending on your age, the kind of car you drive and other factors. For example, male drivers under the age of 22 in one city can calculate their average annual rate by multiplying their age by $1000 and subtracting that from $24000.

a **Write down** the equation for the average annual cost of insurance (C) for a young male based on his age (a).

b **State** what $C(18)$ represents. Then find $C(18)$.

c For males under the age of 22, is the annual average cost of car insurance a function of age?

d Is charging different rates depending on the age of the driver a fair practice? **Explain**.

e **State** what $C(21)$ represents. Then find $C(21)$.

f **Suggest** reasons why this function is only valid for drivers under the age of 24.

3 Logic

A method of reasoning and a system of principles used to build arguments and reach conclusions

Transport puzzle

A farmer is travelling to market with a bag of grain, a chicken, and a wolf. While he is with them, he can prevent the chicken from eating the grain, and the wolf from eating the chicken.

He needs to cross a river in a small boat that has room for him and only one other passenger at a time.

How can the farmer get himself, the grain, the chicken and the wolf safely to the other side of the river? Can you work it out using the fewest number of trips?

Of course, sometimes even logicians overlook sensible questions: Why would the farmer be travelling with a wolf in the first place?

Lateral thinking

Three friends have lunch in a bistro and the bill is $25. They each give the waiter $10. The waiter doesn't know how to divide up the $5 in change evenly between three people, so he gives each person $1 back, and keeps $2 for himself as a tip.
Now, each customer contributed $10 and got $1 back, so effectively they each paid $9, which means they paid $27 altogether. So, $27 plus the $2 that the waiter kept adds up to a total of $29.
Where has the remaining $1 gone from the original $30 that they gave the waiter?

The mathematics here seems logical, but the outcome does not. Which do you think is in error?

Logic in philosophy

The study of logic is a core part of philosophy. There are different types of logical reasoning:

Inductive reasoning involves finding evidence that supports a statement. The more supporting evidence you find, the more you can be sure that the statement is true.

Statement: Every swan I have seen is white, so that means **all** swans are white.

Test: Would you be happy to wager a lot of money that all swans are white? If not, how could you test your hypothesis?

Deductive reasoning involves generating new facts by manipulating existing ones. The transport puzzle on the facing page uses deductive reasoning.

Abductive reasoning involves looking at known facts and finding the most likely explanation for them. If you saw a tanned man with a ring of white skin on his finger, you might abduce that he had once been married but later got divorced. If you saw him holding hands with woman, would that change your reasoning, or reinforce it?

3.1 But can you prove it?

Global context: Orientation in space and time

LOGIC

Objectives

- Measuring the distance between two points, and between a line and a point
- Finding the distance between two points
- Finding the shortest distance from a point to a line
- Finding the midpoint between two points
- Proving the midpoint formula
- Proving the distance formula

Inquiry questions

F
- How do you find the distance between two points?
- How do you find the shortest distance from a point to a line?
- How do you find the midpoint between two given points?

C
- How do you form logical arguments?
- How can you prove the distance formula and/or the midpoint formula?

D
- Does it matter how proof is presented?
- How much 'proof' is enough?

ATL Communication

Make inferences and draw conclusions

You should already know how to:

• plot points on the Cartesian plane	**1** Plot the points $A(2, 4)$, $B(4, -2)$, $C(-5, -2)$, $D(-5, 4)$, and join them up in order. Describe the shape you have drawn.
• apply Pythagoras' theorem	**2** Find the missing length, h, in this triangle. Give your answer to the nearest millimeter. h; 5 cm; 9 cm
• construct a perpendicular from a point to a line with a ruler and compasses	**3** On plain paper, draw a point P, and a line L which does not pass through P. Use a ruler and compasses to construct a line from point P that is perpendicular to line L.

Using logic to make geometric deductions

- How do you find the distance between two points?
- How do you find the shortest distance from a point to a line?
- How do you find the midpoint between two given points?

Logic in mathematics is the process of creating a rigorous argument.

There are three parts to any logical argument: *premises*, *process*, and *conclusion*.

- The premises are facts that you know or are given at the start.
- The process is a way of manipulating those facts, according to rules.
- The conclusion is a new fact, which you were not certain was true before.

The principles of logic state that if you agree with the premises and the process, you must also agree with the conclusion.

Once something has been proved logically, you can accept it as true and use it to derive other results in the future. This is known as deductive reasoning.

Logic has traditionally been a concept studied in mathematics and philosophy but is now also used in other fields such as computer science, linguistics and psychology.

Deductive reasoning is the process of taking previously known or proven facts and putting them together to make new ones.

Exploration 1

1. On 10 mm square graph paper, draw axes for $0 \le x \le 15$ and $0 \le y \le 10$.
2. Plot the points $A(2, 1)$, $B(8, 9)$, and $C(14, 6)$.
3. Using a ruler, measure the distances AB, AC, and BC.
4. Now plot the point $D(14, 1)$.
5. Describe, as thoroughly as you can, the shape ACD.
6. Using triangle ACD, calculate the length of AC and compare it to the value you measured in step **3**.

▶ Continued on next page

7 Add a point to your diagram to calculate the length of BC using a similar process.

8 Calculate the length of BC and compare it your measurement in step **3**.

9 Calculate the length of AB and compare it to your measurement in step **3**. Explain which you think are more accurate: your measurements or your calculations.

Reflect and discuss 1

Measurement is an MYP Related Concept, defined as 'a method of determining quantity, capacity or dimension using a defined unit'.

- How does Exploration 1 relate to the concept of Measurement?
- To what degree of accuracy did you measure the lengths?
- What is meant by 'a defined unit' in the context of Exploration 1?
- Have you obtained accurate measurements?

In Exploration 1 you used Pythagoras' theorem to find the distance between two points on a coordinate grid. First you constructed a right-angled triangle. A sketch can help you find the lengths of the shorter sides of the triangle.

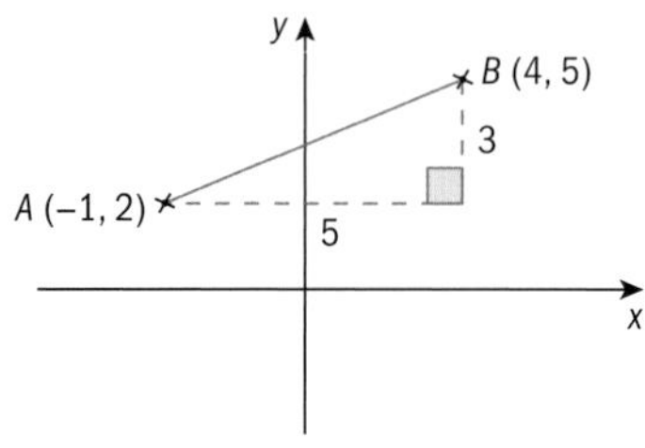

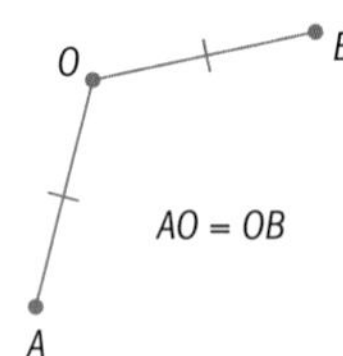

A and *B* are **equidistant** from *O* if the distances *OA* and *OB* are equal.

Practice 1

1 Find the distance from (4, 2) to (7, 6).

2 Find the distance from (–3, –1) to (2, 11).

Make a sketch.

3 For the points $P(10, 13)$, $Q(17, 37)$, $R(24, 13)$, and $S(17, -11)$:

  a Find the distances PQ, QR, RS and SP.

  b Robin says: 'the lengths PQ, QR, RS and SP are all equal, so the shape $PQRS$ must be a square.' Determine whether Robin is correct.

  c Robin's logical process was:

Premises:	The four lengths are equal.
	If a quadrilateral has four equal sides, then it is a square.
Reasoning process:	Since PQRS has four equal sides, it is a square.
Conclusion:	PQRS is a square.

  Explain the fault in Robin's logic.

4 a On 10 mm square graph paper, plot the points $O(0, 0)$, $U(22, 8)$, and $V(4, 23)$.

b Zach claims that the triangle OUV is equilateral.

Without measuring, determine whether or not his claim is true.

To 'show' means to demonstrate using examples or diagrams; to 'prove' is a much more rigorous procedure.

5 a On suitable axes, plot the points $A(41, 99)$, $B(99, 41)$, $C(99, -41)$, $D(41, -99)$, $E(-41, -99)$, $F(-99, -41)$, $G(-99, 41)$, and $H(-41, 99)$.

b Show that all eight points are equidistant from the origin, $(0, 0)$.

c Mark any lines of symmetry of the octagon $ABCDEFGH$ on your diagram.

d Describe the rotational symmetry of the octagon $ABCDEFGH$.

e Verify that $ABCDEFGH$ is not a regular octagon.

f To what extent would it be appropriate to use the coordinates of the points $ABCDEFGH$ to represent a regular octagon in print or on a computer screen?

Problem solving

6 In the constellation of Pegasus, the horse's body is represented by four stars called 'The Great Square'. The stars at the corners of The Great Square can be drawn on a Cartesian plane at $(-1, 5)$, $(3, 2)$, $(-4, 1)$, and $(0, -2)$.

Show that The Great Square of Pegasus has these properties of a square:

a All sides are the same length

b All angles are 90°

c Opposite sides are parallel

The constellation of Pegasus is named after the winged horse in Greek mythology. According to the myth, when Perseus killed the Gorgon Medusa (whose hair was made of snakes), Pegasus was born from her blood.

Exploration 2

The diagram shows the point $A(4, 7)$.

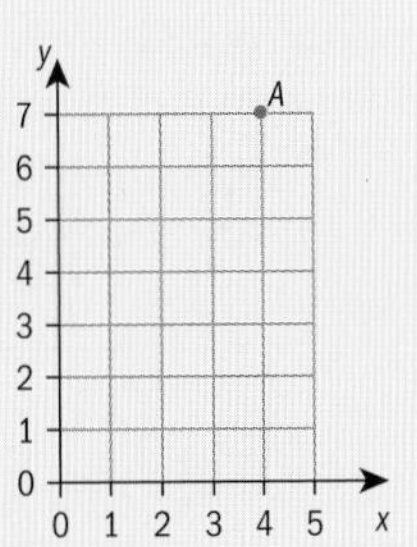

1 Write down the distance from A to the x-axis.

2 Libby says, 'The point $(2, 0)$ lies on the x-axis. Pythagoras' theorem tells me that the distance from A to $(2, 0)$ is $\sqrt{53} \approx 7.3$ units.

Therefore the distance from A to the x-axis is 7.3 units.'

Explain whether or not this is a logical argument.

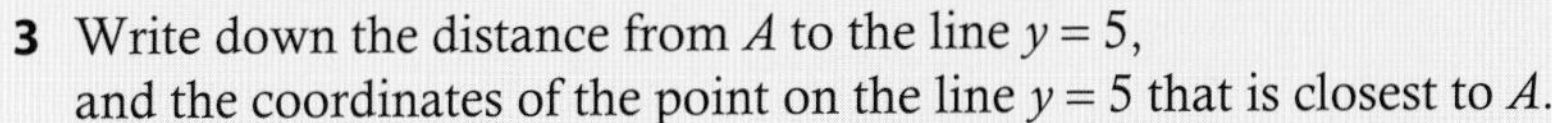

3 Write down the distance from A to the line $y = 5$, and the coordinates of the point on the line $y = 5$ that is closest to A.

4 Write down the distance from A to the line $x = 1$, and the coordinates of the point on the line $x = 1$ that is closest to A.

Continued on next page

5 Find the coordinates of the point on the line $y = x + 1$ that is closest to A.

Find the shortest distance from A to the line $y = x + 1$.

6 Find the coordinates of the point on the line with equation $y = 4 - \frac{1}{2}x$ that is closest to A.

Find the shortest distance from A to the line $y = 4 - \frac{1}{2}x$.

7 Find the coordinates of the point on the line with equation $y = 7 - 2x$ that is closest to A.

Find the shortest distance from A to the line $y = 7 - 2x$.

Reflect and discuss 2

For Exploration 2:

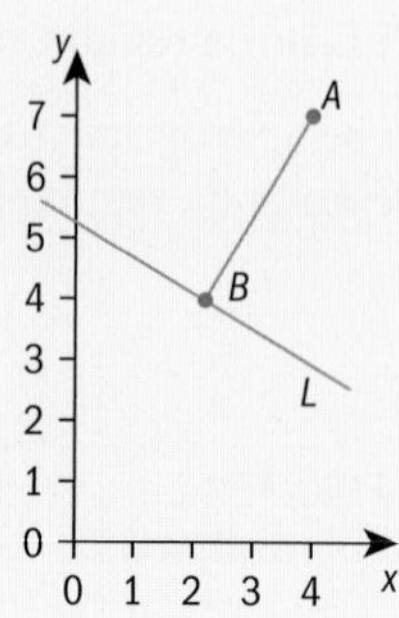

- What difficulties did you encounter in step **7** compared to steps **3** to **6**?
- Suppose you are trying to find the shortest distance from A to a general line L whose equation you don't know.

  If B is the point on L closest to A, what can you say about the line L and the line AB?

The shortest distance between a given line and a point not on that line is measured along a line perpendicular to the line.

You can use logic to establish that this is true. Once you have determined that it is true, you can use this fact in the future.

Suppose you make an assumption and then some valid deductions based on that assumption. If you discover that some of your deductions are untrue, this tells you that your first assumption was not true. This is called proof by contradiction.

'Proof' means very different things in Science and Mathematics. In Mathematics, a proof is a robust logical argument, which guarantees that a result will always be true. It is permanent, and if a proof is valid, it cannot subsequently be disproved. In Science, 'proof' is a body of evidence that supports a hypothesis, but which could subsequently be overturned if contradictory evidence is found.

ATL

Exploration 3

1 On plain paper, draw a line L and a point A that is not on the line.

2 Using compasses and a ruler, construct a second line L' – that is perpendicular to L and passes through A.

▶ Continued on next page

3 Label the point where the two lines meet P.

To show that P is the closest point to A on L, start by supposing that P is *not* the closest point.

4 If P is not the closest point, then the closest point must lie somewhere else. Draw another point on line L. Label it B.

5 Copy and complete the following sentences:

Triangle ABP is a ____________ triangle, and angle $\angle APB$ is ____________ because ____________.

Since $\angle APB$ is ____________, side AB is the ____________ of triangle ABP, which means it is the longest side.

Therefore $AP < AB$, which contradicts the assumption that B is the point on L closest to P.

6 Explain in your own words how this shows that the point on L which is closest to A lies on a line perpendicular to L through A.

A proof by contradiction involves applying a logical process to a hypothesis and producing a result that is obviously false, or contradicts your assumption. If this occurs, then the original assumption must have been incorrect.

Practice 2

1 Point A has coordinates (5, 8). Construct the line through A that is perpendicular to the line $y = 4x + 2$.

Find the shortest distance from A to the line $y = 4x + 2$.

2 Line L_1 has equation $y = 3x - 3$.

Line L_2 has equation $y = 8 - 2x$.

Point B has coordinates (2, 9).

Determine which line passes closer to point B.

Justify your answer.

3 Draw the line L with equation $y = 4 + \frac{1}{2}x$.

Add the points $A(2, 9)$ and $B(7, 3)$ to your diagram.

Determine which point is closer to L.

Justify your answer.

You can use dynamic geometry software or a GDC for questions **2** and **3**.

Problem solving

4 Air traffic controllers are in charge of planes taking off and landing at their airport and any others that fly overhead on the way to another destination. When planes come into their airspace, they track them using a coordinate system.

Plane A is at (12, 5), flying towards (21, 12). Plane B is at (9, 8).

a Find the shortest distance from Plane B to Plane A's flight path.

b Suggest what other information (not included in the coordinate system) the air traffic controller needs to prevent the planes getting too close to each other.

Make sure you use logic to justify your answers. Your justifications should leave no doubt as to whether your answer is correct.

ATL

Exploration 4

1 Using dynamic geometry software, plot the points $A(10, 6)$ and $B(2, 8)$.

2 Find the coordinates of M, the midpoint of AB.

3 Compare the coordinates of M to the coordinates of points A and B. Describe anything you notice.

4 Create other pairs of points and find the midpoint of each pair. Investigate the relationship between the coordinates of the endpoints and the coordinates of the midpoint. (Hint: construct a table of endpoints and midpoints.)

5 Use your findings in step **4** to predict the coordinates of N, the midpoint of points $C(7, 11)$ and $D(19, 15)$.

6 Verify your answer using the software.

7 Points P and Q have coordinates $P(a, b)$ and $Q(c, d)$. Suggest a formula for the coordinates of the midpoint of PQ.

8 Verify that your formula gives the correct coordinates for M and N.

Reflect and discuss 3

Generalization means making a general statement on the basis of specific examples.

- Where in Exploration 4 have you generalized?
- How does generalization enable you to predict?
- Why is it important to verify a generalization?

The midpoint of (a, b) and (c, d) is $\left(\frac{a+c}{2}, \frac{b+d}{2}\right)$.

Example 1

Find the midpoint of $A(3, 6)$ and $B(9, 18)$.

Let M be the midpoint of AB.

$M = \left(\frac{3+9}{2}, \frac{6+18}{2}\right)$ — Use the midpoint formula $M = \left(\frac{a+c}{2}, \frac{b+d}{2}\right)$

$= (6, 12)$

Practice 3

1 Find the midpoint of $A(14, 6)$ and $B(18, 20)$.

2 Find the midpoint of $C(5, -13)$ and $D(10, -8)$.

3 a Plot the points $A(2, 7)$, $B(3, 10)$, $C(6, 11)$, and $D(5, 8)$.

b Find the midpoints of AC and BD.

c Comment on your answers to part **b**. Explain what this shows about quadrilateral $ABCD$.

4 Triangle *PQR* has vertices $P(-2, -3)$, $Q(2, 3)$, and $R(6, -3)$.

a Show that triangle *PQR* is isosceles.

L, *M*, and *N* are the midpoints of *PQ*, *QR* and *RP* respectively.

b Write down the coordinates of *L*, *M*, and *N*.

c Show that triangle *LMN* is isosceles.

Problem solving

5 Show that the midpoint of points $U(7, 11)$ and $V(-2, 1)$ lies on the line with equation $y = 2x + 1$.

6 As you saw in Practice 1 question **6**, the corners of The Great Square of Pegasus are at $(-1, 5)$, $(3, 2)$, $(-4, 1)$, and $(0, -2)$.

Show that The Great Square of Pegasus has these properties of a square:

a Diagonals are perpendicular

b Diagonals bisect each another

In question **5**, show that the midpoint's coordinates satisfy the equation of the line.

Using logic to prove claims and general rules

- How do you form logical arguments?
- How can you prove the distance formula and/or the midpoint formula?

You will encounter logical arguments in many different situations; not just in mathematics, but also in everyday life. Any thought process with the structure '*If… then … therefore…*' is a form of logical argument. Thinking logically will help you to make sensible decisions. The ability to explain your logic will help you to persuade others that your ideas are valuable.

Each step of a logical argument should contain a simple justification for why the step is valid. If a step cannot be justified, you won't know if it is necessarily true.

The process of justifying a claim by providing step-by-step reasons is known as **deduction**.

A **deductive argument** is any proof or collection of reasoning that uses deduction to draw a conclusion.

Reflect and discuss 4

Justification is defined as 'Valid reasons or evidence used to support a statement'.

- What different types of justification have you seen so far in **3.1**?
- Why is justification important in mathematics?

Exploration 5

ATL

You have already shown that Pythagoras' theorem can be used to find the distance between two points. In this exploration, you will find a formula for the distance between two points $P_1(x_1, y_1)$ and $P_2(x_2, y_2)$.

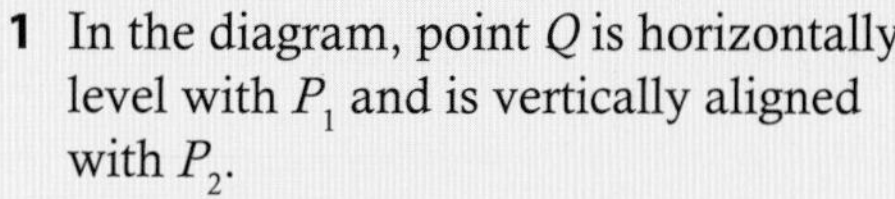

1 In the diagram, point Q is horizontally level with P_1 and is vertically aligned with P_2.

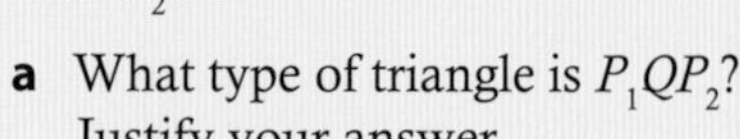

a What type of triangle is P_1QP_2? Justify your answer.

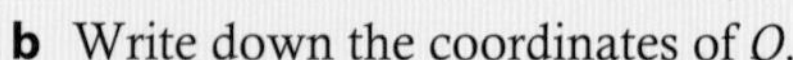

b Write down the coordinates of Q.

c Hence write down the length of P_1Q and P_2Q in terms of x_1, x_2, y_1, and y_2.

d Use these lengths to find the length d of P_1P_2.

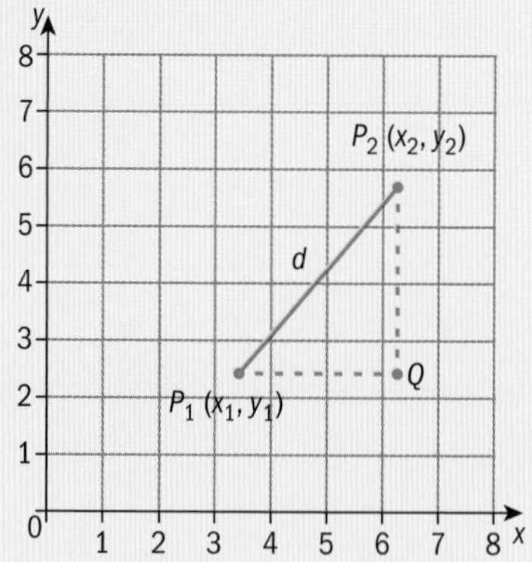

2 In the first diagram, you assumed that P_2 was above and to the right of P_1. Suppose instead that P_2 was below and to the left of P_1. Show that the distance formula would not change.

3 Suggest other possible positions that P_1 and P_2 could take relative to each other. Explain why, for any positions of P_1 and P_2, the distance formula remains the same.

You could sketch the two other possible arrangements of P_1 and P_2.

The distance formula states that the distance, d, between two points (x_1, y_1) and (x_2, y_2) is given by $d = \sqrt{(x_2 - x_1)^2 + (y_2 - y_1)^2}$.

Reflect and discuss 5

When P_1 and P_2 are level horizontally:

- Does the distance formula still work?
- What happens to point Q, and the triangle P_1P_2Q?

Practice 4

1 Find the distance between each pair of points.

a (3, 11) and (6, 15)

b (–3, –8) and (2, –16)

c (103, 9) and (–17, 208)

Problem solving

2 Show that $P(7, 14)$, $Q(11, 12)$, and $R(13, 8)$ are all the same distance from point $C(6, 7)$.

Hence find the area of the circle that passes through P, Q, and R.

3 Show that the points $F(-6, 5)$, $G(-4, 5)$, and $H(-5, 9)$ form an isosceles triangle.

Activity: Proving the midpoint formula

Objective: C. Communicating
iv. communicate complete, coherent and concise mathematical lines of reasoning

The proof outline will help you give a complete and coherent proof of the midpoint formula, if you give clear and concise explanations and details in the sections you complete.

Complete the gaps in this proof that claims the midpoint of points $P(a, b)$ and $Q(c, d)$ is $M\left(\frac{a+c}{2}, \frac{b+d}{2}\right)$.

The proof has two parts: first, to prove that M lies on the line PQ; second, to prove that M is equidistant from P and Q.

Copy and complete the proof outline below.

Theorem:

The midpoint of points $P(a, b)$ and $Q(c, d)$ is $M\left(\frac{a+c}{2}, \frac{b+d}{2}\right)$.

Draw a diagram to show P and Q on a grid.

The position of P and Q on the graph is not important – just don't put them in the same place!

Proof:

PMQ is a straight line if the gradient of PM and the gradient of MQ are the same.

Show clearly that PM and MQ have the same gradient.

Since the gradients are equal, PMQ is a straight line.

Use the distance formula to show that $MP = MQ$.

Therefore M is equidistant from P and Q.

Since ______________________

and ______________________,

M is the midpoint of PQ. □

A square marks the end of a proof.

Practice 5

At the end of a proof you might also see 'Q.E.D.' short for *quod erat demonstrandum*, meaning 'that which was to be demonstrated'.

1 **a** Find the midpoint of $A(3, 9)$ and $B(7, -3)$.
  b Find the midpoint of $A(14, -2)$ and $B(6, 11)$.

2 **a** Show that the points $A(2, 5)$, $B(4, 19)$, $C(14, 29)$, and $D(12, 15)$ form a rhombus.
  b Find the coordinates of E, F, G, and H, the midpoints of AB, BC, CD, and DA respectively.
  c Show that $EFGH$ forms a parallelogram.
  d Find the lengths of EG and FH.
  Hence describe the shape $EFGH$.

3 **a** Plot the points $A(0, 8)$, $B(-4, 20)$, $C(16, 28)$, and $D(28, 28)$.

i Show that none of the sides of $ABCD$ are the same length, and that none of the sides are parallel to each other.

ii Find the coordinates of E, F, G, and H, the midpoints of AB, BC, CD, and DA respectively.

iii Show that $EFGH$ forms a parallelogram.

b Consider four points $P(x_1, y_1)$, $Q(x_2, y_2)$, $R(x_3, y_3)$, and $S(x_4, y_4)$.

i Find the coordinates of T, U, V, and W, the midpoints of PQ, QR, RS, and SP respectively.

ii Prove that $TUVW$ forms a parallelogram.

Reflect and discuss 6

- How is finding the midpoint related to finding the average or arithmetic mean?
- What is the relationship between **3a** and **3b**?

4 When archaeologists excavate a site, they mark it off with a coordinate grid so that they can easily identify where artefacts are found and keep a record of the site forever. At the West Kennet Long Barrow in England, a grave site dating back to 3600 BCE, there are three chambers at the coordinates $(7, 9)$, $(17, 33)$, and $(-36, 41)$. It has been said that they form an isosceles triangle with height twice the length of its base.

Determine whether the statement is true. Justify your answer.

Problem solving

5 Given the points $A(a, b)$ and $B(c, d)$, prove that the point $\left(\frac{2a+c}{3}, \frac{2b+d}{3}\right)$ divides the line segment AB in the ratio $1 : 2$.

6 Copy and complete the following skeleton proof.

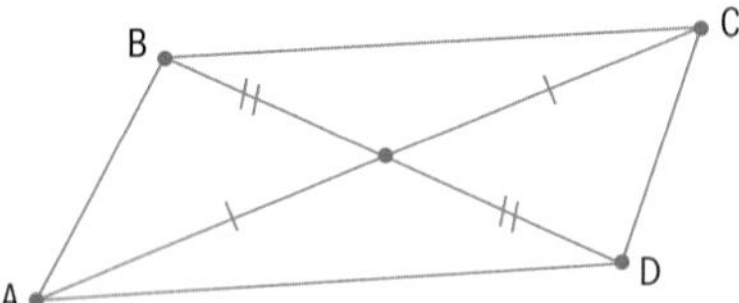

Given: line segments AC and BD; the midpoint of AC is coincident with the midpoint of BD.
Prove: $ABCD$ is a parallelogram.

Tip

Two points are said to be *coincident* if they are in the same place, this is, they have the same coordinates.

Theorem:

If the midpoint of AC is coincident with the midpoint of BD, then $ABCD$ forms a parallelogram.

Proof:

If lines AB and CD are parallel and the same length, then $ABCD$ is a parallelogram.

Let the midpoint of AC have coordinates (X, Y).

Therefore the coordinates of the midpoint of BD are .... — Think about the information you are given.

Let the coordinates of A be (x_A, y_A), B be (x_B, y_B) — Provide sensible notation for C and D.

By the midpoint formula, ...
and .... — Use the midpoint formula to express the x-coordinate of C in terms of x_A and X.

Similarly, ... — Now find three more results which describe the x and y values of C and D.

The gradient of AB is given by ....

The gradient of CD is given by ....

Therefore ....

You will probably need to write a complete expression for the gradient of CD and simplify it carefully to show that it matches the gradient of AB.

.... is given by $\sqrt{(x_A - x_B)^2 + (y_A - y_B)^2}$.

.... — Again, you may need to simplify a complicated expression.

Therefore AB and CD ....

Since .... and ...., we know that .... □

Problem solving

7 Prove that the points $O(0, 0)$, $A(a, b)$, $B(a + c, b + d)$, and $C(c, d)$ form a parallelogram.

8 Assuming $a, b, c, d \neq 0$, and $\frac{b}{a} \neq \frac{c}{d}$, prove that the points $A(a, b)$, $B(2a, 2b)$, $C(2c, 2d)$ and $D(c, d)$ form a trapezoid.

D Presentation of proof

- Does it matter how proof is presented?
- How much 'proof' is enough?

Here are two different proofs of *the Isosceles triangle theorem*, which you will meet later. They are presented as you might find them in a textbook. You are not yet expected to understand the proof – for now, just look at how the proofs are presented. Don't worry if some words are unfamiliar – you may not have encountered the term *congruent* before, for example. What similarities and differences are there?

Theorem: If two sides of a triangle are equal then the angles opposite to those sides are equal.

Proof 1

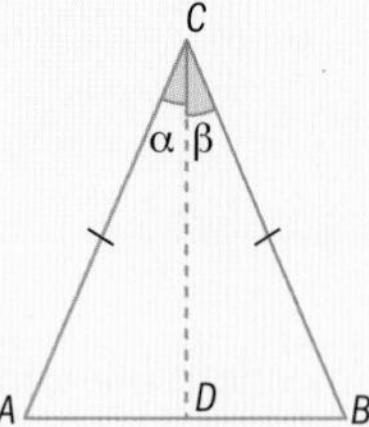

Given ΔABC with $AC = BC$, prove that $\hat{A} = \hat{B}$.

Let the bisector of $A\hat{C}B$ meet AB at D.

In ΔCAD and ΔCDB, $AC = BC$, (Given)

$CD = CD$, (Common)

$\alpha = \beta$;

$\therefore$ the triangles are congruent

$\therefore \hat{A} = \hat{B}$ Q.E.D.

Proof 2

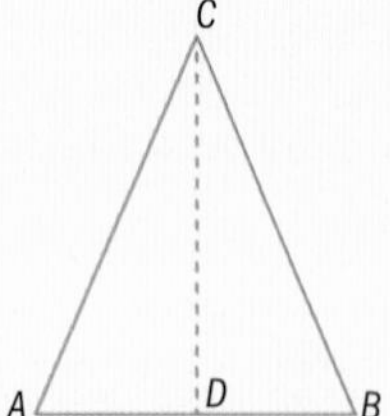

The triangle represents any isosceles triangle, with vertex C and sides $CA = CB$. Prove that $\angle A = \angle B$.

Draw CD so as to bisect $\angle C$, forming two triangles, CDA and CDB.

Then in the two triangles, $\angle ACD = \angle BCD$, by definition of bisection; $AB = AC$ is given; and $CD = CD$, as this side is common to both triangles.

That is, the two sides and the included angle of ΔCDB are equal to two sides and the included angle of ΔCDA, and hence the two triangles are congruent.

Therefore, as corresponding parts of congruent triangles, $\angle A = \angle B$, as was to be proved. □

Reflect and discuss 7

Discuss your observations with others.

- What structural elements do both proofs have?
- Which proof do you think would be easier to understand?
- How do you know when the proof is finished?

Both proofs include a diagram, but the presentation of the two proofs is very different. Proof 2 is very wordy; Proof 1 has far fewer words.

A good proof describe the logic used to move from one step to the next, will include justification (for example, previously proven results) and will be laid out so that it is easy to follow.

Summary

- **Deductive reasoning** is the process of taking previously known or proven facts and putting them together to make new ones.
- *A* and *B* are **equidistant** from *O* if the distances *OA* and *OB* are equal.

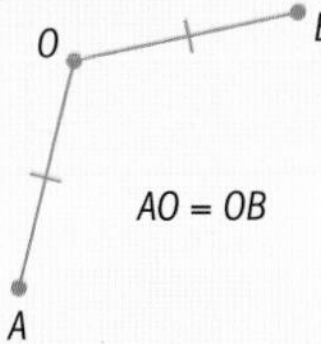

- The shortest distance between a point and a line is measured along a line perpendicular to the line.
- A proof by contradiction involves applying a logical process to a hypothesis and producing a result that is obviously false, or contradicts your assumption. If this occurs, then the original assumption must have been incorrect.
- The midpoint of (a, b) and (c, d) is $\left(\frac{a+c}{2}, \frac{b+d}{2}\right)$.
- The process of justifying a claim by providing step-by-step reasons is known as **deduction**.
- A **deductive argument** is any proof or collection of reasoning that uses deduction to draw a conclusion.
- The distance formula states that the distance, d, between two points (x_1, y_1) and (x_2, y_2) is given by $d = \sqrt{(x_2 - x_1)^2 + (y_2 - y_1)^2}$.

Mixed practice

1 Find the distance between each pair of points.

- **a** (0, 0) and (5, 12)
- **b** (1, 4) and (4, 8)
- **c** (5, 7) and (11, 7)
- **d** (−3, −12) and (4, 12)

2 Find the distance between each pair of points, giving your answer correct to 3 significant figures.

- **a** (4, 8) and (10, 13)
- **b** (14, 8) and (17, −3)
- **c** (1.5, 4.6) and (2.3, −1.8)

3 Find the distance between each pair of points, giving your answer in the form $\sqrt{n}$, where n is a whole number.

- **a** (2, 7) and (6, 15)
- **b** (−3, 9) and (5, −2)
- **c** (−3, −5) and (−9, −14)

4 Find the midpoint of each pair of points.

- **a** (4, 2) and (6, 12)
- **b** (−3, 9) and (5, −2)
- **c** (−3, −5) and (−8, −14)
- **d** (2.3, 5.9) and (−3.6, 11.2)
- **e** $\left(\frac{1}{4}, \frac{1}{3}\right)$ and $\left(\frac{1}{3}, \frac{1}{2}\right)$

5 **Prove** that the points (0, 0), (6, 0), and (3, 4) form an isosceles triangle.

6 Three points have coordinates $A(-4, -1)$, $B(1, 4)$, and $C(-5, 6)$.

Show that the points form an isosceles triangle.

7 Point A has coordinates (3, 5). **Construct** the line through A that is perpendicular to the line $y = -2x + 3$.

Find the shortest distance from A to the line $y = -2x + 3$. Give your answer to 1 dp.

8 A has coordinates (−1, 3) and B has coordinates (4, 1).

Determine which point is closer to the line $y = x$.

9 Given the points $A(a, b)$ and $B(c, d)$, **prove** that the point $\left(\frac{3a+c}{4}, \frac{3b+d}{4}\right)$ divides the line segment AB in the ratio 1 : 3.

10 Point A has coordinates (a, b) and point B has coordinates (c, d).

M is the midpoint of AB.

- a **Find** the coordinates of M.
- b **Find** the distance AB.
- c **Find** the distance AM.
- d Hence, **prove** that $AM = \frac{1}{2}AB$

11 ABC is a triangle with vertices $A(1, 0)$, $B(5, 9)$ and $C(9, 0)$.

- a **Show** that triangle ABC is isosceles.
- b **Find** the coordinates of the midpoints of the three sides of ABC.
- c **Show** that the midpoints of the three sides of ABC are the vertices of an isosceles triangle.
- d **Show** that the isosceles triangle formed from the midpoints of the sides of ABC has area 8 square units.

12 In a Sudoku puzzle, players work with a set of digits. Most puzzles use the set {1, 2, 3, 4, 5, 6, 7, 8, 9}. Every row, every column, and every medium-sized square must contain all of the digits in the set.

The Sudoku puzzles in this question use the set {1, 2, 3, 4}.

a By considering the positions of the 1s and 2s, **show** that it is not possible to complete this puzzle:

1	4	2	
	2	1	
3			
2	1		

b Copy this grid:

1	4	3	
			4
2	1		

- i **Explain** how you can be certain that the top right-hand corner must contain a 2.
- ii Hence **determine** the possible positions of the other two 2s in the grid.

  Justify your answer.
- iii **Deduce** the digits that should appear in the grids remaining cells.

Problem solving

13 Point X has coordinates (4, 7) and is the midpoint of AB. Given that A has coordinates (−3, 11), **find** the coordinates of B.

Review in context

Orientation in space and time

You are lucky enough to have found the only surviving copy of a treasure map left by the notorious pirate Short James Platinum - scourge of the seven seas. He has buried many pieces of treasure on the island - each in a different location. Can you find them all? (On the map, one unit equals one league.)

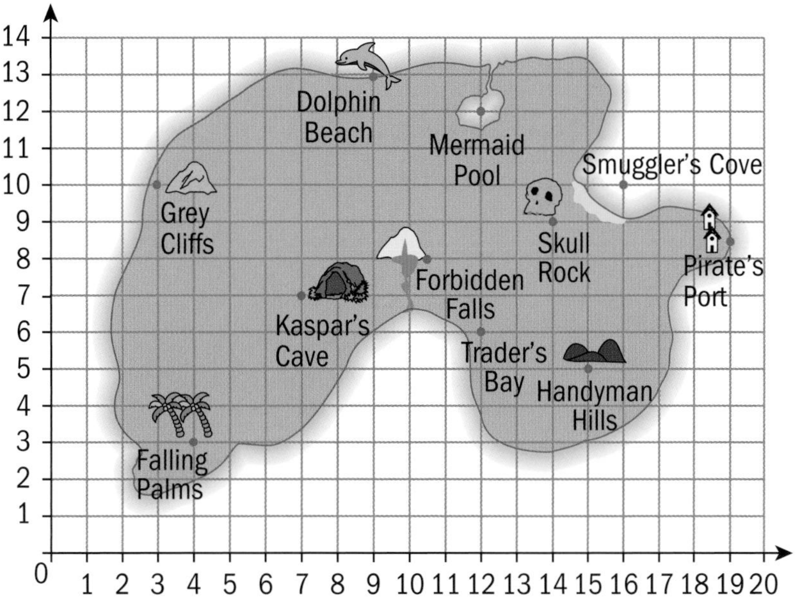

1 The Tsarina's Tiara is buried due west of Skull Rock. It is as far from Skull Rock to the Tiara as it is from Falling Palms to Kaspar's Cave.

Find the coordinates of the place where the Tsarina's Tiara is buried.

2 The wreck of the Purple Porpoise, a ship that sank carrying 200 gold doubloons, is exactly halfway between Falling Palms and Handyman Hills.

Find the coordinates of the Purple Porpoise.

3 If Skull Rock is closer to Trader's Bay than Forbidden Falls, then my prized Silver Cutlass is buried two leagues east of Mermaid Pool. If not, it is buried two leagues west of Mermaid Pool.

Determine the position of the Silver Cutlass.

4 The Emerald Crown, which I looted from an ancient tomb, has been hidden in a ruined building at one of the marked sites on the map. If you can find the correct site, you will find the treasure.

The true location of the crown is as far away from Kaspar's Cave as the Falling Palms are from the Grey Cliffs. Determine the location of the Emerald Crown.

Explain how you know this is the only possible correct location.

5 To find the Cursed Medal of Caracas, send your best navigator to the point halfway from Kaspar's Cave to the Grey Cliffs. From there, travel halfway to Pirate's Port, and there the medal is to be found.

Find the coordinates of the Cursed Medal of Caracas. Justify your answer.

6 The Bounty from Belize is buried deep - so be sure to pick the right place.

On the route straight from Falling Palms to the Grey Cliffs, stop at the point that is closest to Kaspar's Cave. Head one league west and dig deep!

a **Find** the distance from Falling Palms to Kaspar's Cave.

b **Find** the distance from the Grey Cliffs to Kaspar's Cave.

c Hence **show** that Grey Cliffs, Kaspar's Cave and Falling Palms form an isosceles triangle.

d **Explain** how this tells you that the point closest to Kaspar's Cave that lies on the line from Falling Palms to Grey Cliffs is the midpoint of the line.

e Hence **determine** the location of the Bounty from Belize.

4 Representation

The manner in which something is presented

Which representation is best?

The best representation may depend on what you want to show or to find out.

Cartesian coordinate plane

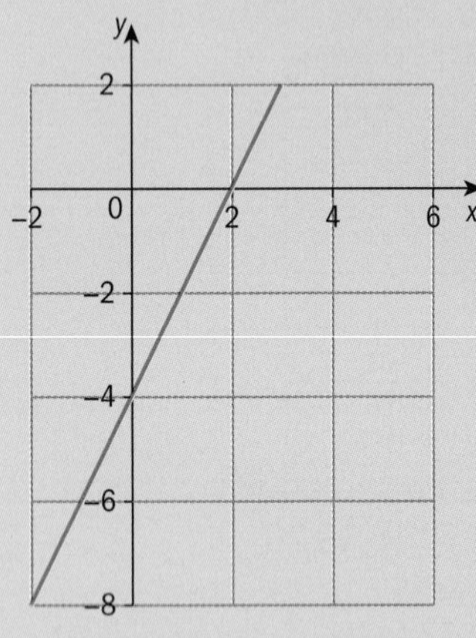

Mapping diagram

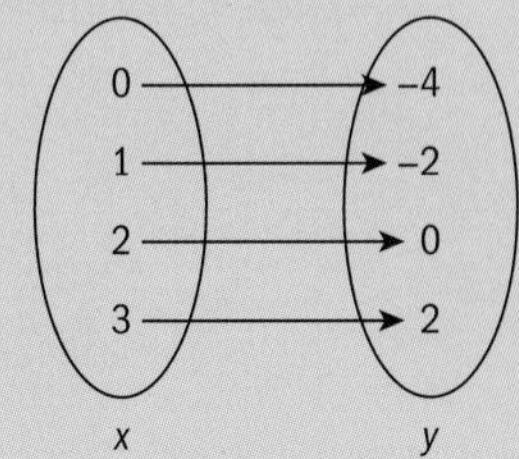

Linear equation

$y = 2x - 4$

Table of values

x	0	1	2	3	4	5
y	−4	−2	0	2	4	6

Which representation best shows:

- that the function is linear
- that $f(1) = -2$
- that the y-coordinates increase by 2 for every increase in x?

Is a picture worth a thousand words?

ABCD is a quadrilateral. *AB* is parallel to *CD* and perpendicular to *BC*.

Angle $ADC = 70°$. Find angle DAB.

- How does drawing a diagram help you solve this problem?
- Could you solve it without a diagram?

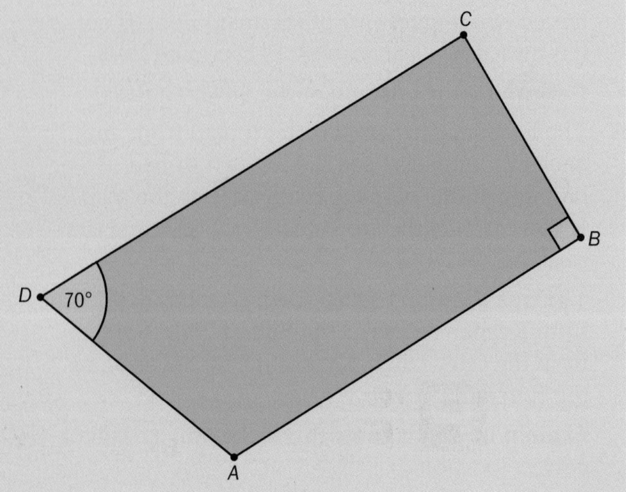

Representation in symbols

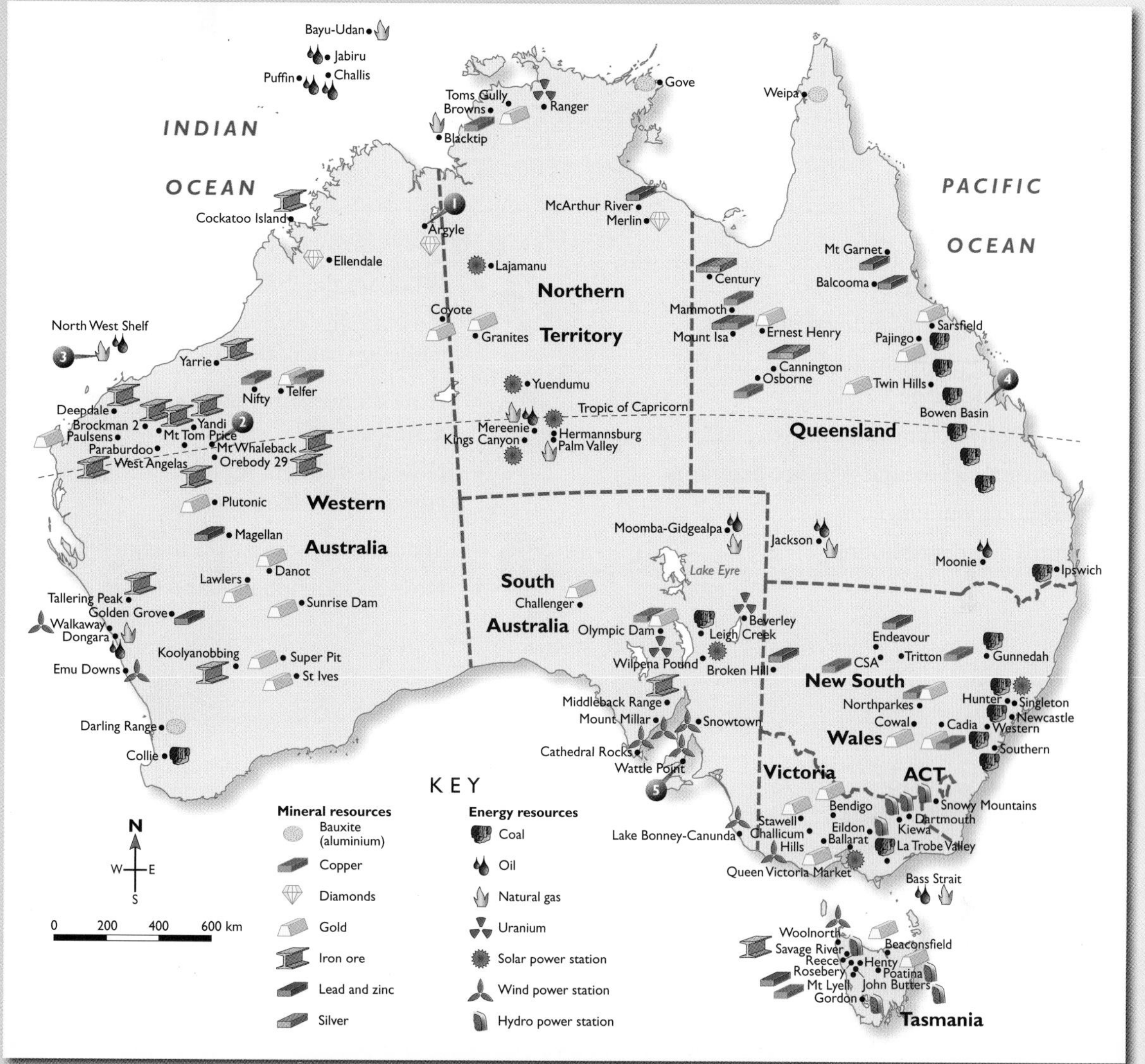

INDIAN OCEAN

PACIFIC OCEAN

Bayu-Udan, Jabiru, Challis, Puffin, Toms Gully, Browns, Ranger, Gove, Weipa, Blacktip, Cockatoo Island, McArthur River, Merlin, Argyle, Ellendale, Lajamanu, Mt Garnet, Balcooma, Century, Mammoth, Mount Isa, Ernest Henry, Pajingo, Sarsfield, Cannington, Osborne, Twin Hills, Bowen Basin, North West Shelf, Yarrie, Coyote, Granites, Yuendumu, Nifty, Telfer, Deepdale, Brockman 2, Yandi, Paulsens, Mt Tom Price, Paraburdoo, Mt Whaleback, West Angelas, Orebody 29, Mereenie, Kings Canyon, Hermannsburg, Palm Valley, Tropic of Capricorn, Plutonic, Magellan, Moomba-Gidgealpa, Jackson, Moonie, Ipswich, Lake Eyre, Lawlers, Danot, Sunrise Dam, Challenger, Tallering Peak, Golden Grove, Walkaway, Dongara, Olympic Dam, Beverley, Leigh Creek, Endeavour, CSA, Tritton, Gunnedah, Koolyanobbing, Super Pit, St Ives, Emu Downs, Wilpena Pound, Broken Hill, Middleback Range, Mount Millar, Snowtown, Northparkes, Hunter, Singleton, Newcastle, Cowal, Cadia, Western, Southern, Darling Range, Collie, Cathedral Rocks, Wattle Point, Bendigo, Snowy Mountains, Stawell, Dartmouth, Challicum Hills, Eildon, Kiewa, Ballarat, La Trobe Valley, Lake Bonney-Canunda, Queen Victoria Market, Bass Strait, Woolnorth, Beaconsfield, Savage River, Reece, Henty, Rosebery, Poatina, Mt Lyell, John Butters, Gordon

Northern Territory

Western Australia

South Australia

Queensland

New South Wales

ACT

Victoria

Tasmania

N W E S

0 200 400 600 km

KEY

Mineral resources
- Bauxite (aluminium)
- Copper
- Diamonds
- Gold
- Iron ore
- Lead and zinc
- Silver

Energy resources
- Coal
- Oil
- Natural gas
- Uranium
- Solar power station
- Wind power station
- Hydro power station

Maps contain symbols that represent features of the landscape or terrain. Which of the symbols below do you recognize? Are they ones that are used in places other than maps?

4.1 A whole range of things

Global context: Globalization and sustainability

RELATIONSHIPS

Objectives

- Categorizing data
- Constructing stem-and-leaf diagrams
- Calculating quartiles, the range and the interquartile range
- Giving a five-point summary of a set of data
- Constructing box-and-whisker diagrams
- Identifying outliers
- Comparing distributions

Inquiry questions

F
- What are the different types of data?
- How are the different measures of central tendency calculated?

C
- How do measures of dispersion help you describe data?
- How do different representations help you compare data sets?

D
- Should we ignore results that aren't typical?
- How do individuals stand out in a crowd?

ATL Communication

Organize and depict information logically

Statement of Inquiry:

How quantities are represented can help to establish underlying relationships and trends in a population.

4.1

4.2

6.2

You should already know how to:

• find the mode, median, mean and range of a set of data	**1** Find the mode, median, mean and range of these data sets. **a** 1, 3, 5, 3, 6, 9, 8, 3, 5, 6, 4 **b** 22, 21, 19, 20, 18, 19, 21, 19, 20, 19, 19, 22, 19, 21, 17, 16, 14, 22

Representing sets of data

- What are the different types of data?
- How are the different measures of central tendency calculated?

Statistics involves *collecting* raw data, *organizing* the data into visual representations, and *analyzing* the data.

Data is the plural of the Latin word *datum* meaning 'a piece of information'.

Qualitative data describes a certain characteristic (for example: color or type) using words. **Quantitative data** has a numerical value. There are two types:

- **Discrete data** can be counted (for example: number of goals scored) or can only take certain values (such as shoe size).
- **Continuous data** can be measured (for example: weight and height) and can take any numerical value.

Tip

In order to perform statistical analysis, continuous data must be rounded to a certain degree of accuracy.

Practice 1

Categorize each set of data as qualitative or quantitative. For quantitative data, state whether it is discrete or continuous.

1 Number of red cars passing through an intersection

2 Time taken for a car to cross an intersection

3 Color of each student's eyes in your class

4 Numbers of children in each student's family in your class

5 Maximum daily temperatures in Qingdao, China

6 Heights of students in your class

7 Dress sizes of girls in your class

8 Daily intake of protein, in grams, for members of a sporting team

9 Number of milkshakes sold in a cafeteria

10 Flavors of milkshakes sold in a cafeteria

A stem-and-leaf diagram is a visual representation of ordered raw data, which can then be analyzed.

Stem-and-leaf diagrams are used in Japanese train timetables.

Example 1

Here are the numbers of climate change surveys carried out by 23 students.

0, 1, 14, 11, 0, 6, 10, 1, 0, 39, 1, 13, 10, 26, 7, 4, 17, 12, 58, 22, 15, 20, 17

Construct an ordered stem-and-leaf diagram to represent this data.

0 0 0 1 1 1 4 6 7 10 10 11 12
13 14 15 17 17 20 22 26 39 58

Sort the data in ascending order. The data ranges from 0 to 58, so use the tens digit to form the stem.

Stem	Leaf
0	0 0 0 1 1 1 4 6 7
1	0 0 1 2 3 4 5 7 7
2	0 2 6
3	9
4	
5	8

Key: 3 | 9 represents 39 surveys

Construct a stem-and-leaf diagram from your ordered data. Remember to include a key.

Reflect and discuss 1

- Is a stem-and-leaf diagram similar to a bar chart? Explain.
- What are the advantages of keeping raw data?

In a stem-and-leaf diagram:

- the **stem** represents the **category** figure
- the **leaves** represent the **final** digit(s) of each data point
- the **key** tells you how to read the values.

ATL

Practice 2

1 Here are the numbers of croissants a baker sold each day during a three-week period:

35, 47, 34, 46, 62, 41, 35, 47, 51, 59, 56, 73, 38, 41, 44, 51, 45, 60, 25, 35, 46

Construct an ordered stem-and-leaf diagram to represent the data.

2 The masses (in grams) of 11 Chinese striped hamsters are:

21.6, 22.4, 27.2, 30.5, 25.2, 23.1, 25.3, 21.3, 20.9, 24.5, 25.2

Construct a stem-and-leaf diagram to represent the data.

Use the whole number part as the stem and the decimal part as the leaf.

Problem solving

3 The stem-and-leaf diagram shows the distances in kilometers students in one class travel to school.

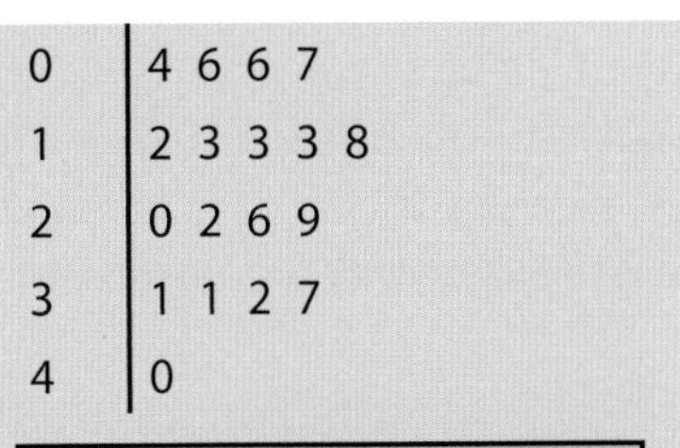

Stem	Leaf
0	4 6 6 7
1	2 3 3 3 8
2	0 2 6 9
3	1 1 2 7
4	0

Key: 2 | 0 represents 20 km

a Write down the:

i least distance travelled

ii greatest distance travelled.

b Find the number of students in the class.

c Write down how many students travel 13 km.

d Students who travel more than 18 km get cheaper bus fares. Find the percentage of students who get cheaper bus fares.

4 Here are the heights of one-year-old apple trees in centimeters.

180 184 195 177 175 173 169 167 197

173 166 183 161 195 177 192 161 166

Organize and represent the data in a stem-and-leaf diagram.

Use the first two digits as the stem.

Reflect and discuss 2

- What problems might arise when using stem-and-leaf diagrams for large data sets?
- What about data sets with a small range?

For any set of data there are two categories of **summary statistics**:

- **Measures of central tendency (location)** summarize a set of data with a single value that is most typical of the data set. The mean, median and mode are all measures of central tendency.
- **Measures of dispersion (spread)** measure how spread out a set of data is. Range is a measure of dispersion.

Example 2

The stem-and-leaf diagram shows the number of emails received by Kirsty every day for 17 days. Find the mode, median and range of this data.

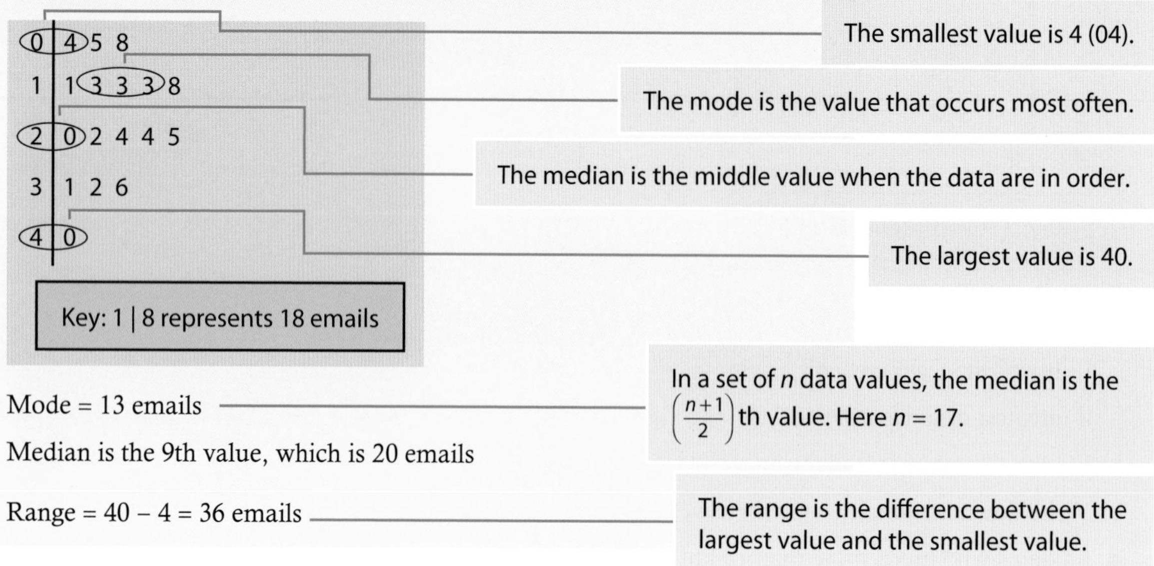

Mode = 13 emails

Median is the 9th value, which is 20 emails

Range = 40 − 4 = 36 emails

Practice 3

1 The stem-and-leaf diagram shows the number of goals scored by the Welsh Wizards during a netball tournament.

Find the mode, median and range of this data.

Stem	Leaf
2	7
3	0 3 5 7 9
4	1 2 2 3 3 5 7 8 9 9 9
5	2 5

Key: 4 | 1 represents 41 goals

2 The stem-and-leaf diagram shows the number of cars parked in a car park every day over a period of 20 days.

Find the mode, median and range of this data.

Stem	Leaf
1	5 7 8
2	0 2 2 5 7 7 8
3	0 0 0 3 6 8
4	0 1 1 5

Key: 1 | 5 represents 15 cars

Problem solving

3 The back-to-back stem-and-leaf diagram shows the marks for a group of students following their English and Mathematics exams.

English		Mathematics
2 2	4	0 3
0	5	1 1 6 7 9
7 1	6	3 3 3
5 4 2 0	7	5
8 3	8	

Key:
2 | 4 represents 42 marks on the English exam

4 | 0 represents 40 marks on the Mathematics exam

a Find the number of students who took the English and Mathematics exams.

b Write down the mode of the Mathematics marks.

c Write down the mode of the English marks.

d Find the median of the English marks.

e Find the range of the English and Mathematics marks.

f Justify whether the English or Mathematics marks are the more consistent.

ATL

4 The number of minutes a sample of 13 people had to wait to see a doctor in her office are shown.

16, 28, 47, 21, 32, 19, 25, 39, 20, 12, 36, 16, 23

a Construct a stem-and-leaf diagram for this data.

b Determine how many people waited more than 30 minutes.

c Find the mode, median and range of the data.

Reflect and discuss 3

- What does the shape of a back-to-back stem-and-leaf diagram show you?
- Which statistics are easier to find from the original data and which statistics are easier to extract from the stem-and-leaf diagram?

Quartiles divide an ordered set of data values into quarters:

- The **median** is another name for the second quartile, Q_2.
- The **lower/first quartile** (Q_1) is the median of the observations to the left of the median in an arranged set of observations.
- The **upper/third quartile** (Q_3) is the median of the observations to the right of the median in an arranged set of observations.
- The **interquartile range** (IQR) is the difference between the lower quartile and the upper quartile ($Q_3 - Q_1$).

Example 3

Find the median, lower and upper quartiles and interquartile range of this data set: 19, 5, 10, 15, 11, 2, 4, 20, 6, 13, 9, 8

2, 4, 5, 6, 8, 9, 10, 11, 13, 15, 19, 20 — Write the data in order. There are 12 data values, so $n = 12$.

The median, $Q_2 = 9.5$ — The median is $\frac{n+1}{2} = \frac{12+1}{2} = 6.5$th value, which is the mean of 9 and 10.

Q_2

9.5

2 4 5 | 6 8 9 | 10 11 13 15 19 20

The lower quartile, $Q_1 = 5.5$ — The lower quartile is the median of the values less than Q_2.

Q_1 Q_2

5.5 9.5

2 4 5 | 6 8 9 | 10 11 13 | 15 19 20

The upper quartile, $Q_3 = 14$ — The upper quartile is the median of the values greater than Q_2.

Q_1 Q_2 Q_3

5.5 9.5 14

2 4 5 | 6 8 9 | 10 11 13 | 15 19 20

Interquartile range, $\text{IQR} = Q_3 - Q_1$

$= 14 - 5.5 = 8.5$

In Example 3, the number of data values *n* was a multiple of 4. In this case none of the quartiles were values in the data set.

Exploration 1

1 Find Q_1, Q_2 and Q_3 for these data sets:

a 2, 4, 5, 6, 8, 9, 10, 11, 13, 15, 19, 20, 22 $n = 13$, (multiple of 4) + 1

b 2, 4, 5, 6, 8, 9, 10, 11, 13, 15, 19, 20, 22, 23 $n = 14$, (multiple of 4) + 2

c 2, 4, 5, 6, 8, 9, 10, 11, 13, 15, 19, 20, 22, 23, 27 $n = 15$, (multiple of 4) + 3

2 Summarize your findings in a table:

n	multiple of 4	(multiple of 4) +1	(multiple of 4) +2	(multiple of 4) +3
median Q_2	not a value in original data set			
Q_1 and Q_3	not a value in original data set			

3 Determine when Q_1, Q_2 and Q_3 are values in the original data set.

Reflect and discuss 4

- A data set has n values, where n is (multiple of 4) – 1.
  Will Q_1, Q_2 and Q_3 be values in the original data set?
- In any data set, what fraction of the data is less than or equal to:
  - the median
  - the lower quartile
  - the upper quartile?

  Why are they are called quartiles?
- What fraction of the data lies between the lower and upper quartiles? What is this as a percentage?
- Is the IQR a measure of central tendency or a measure of dispersion? What does the IQR tell you?

Practice 4

1 Cara compared her SAT Mathematics score with ten of her friends. Their scores were:

650, 750, 700, 670, 420, 720, 750, 730, 780, 750, 780

a Calculate the median, and the upper and lower quartiles.

b Find the range and the interquartile range.

2 A baker records the number of doughnuts she sells each day during a three-week period:

35, 47, 34, 46, 62, 41, 35, 47, 51, 59, 56, 73, 38, 41, 44, 51, 45, 43

a Calculate the median, and the upper and lower quartiles.

b Find the range and the interquartile range.

3 The masses (in grams) of 11 newly hatched chicks are:

31.6, 28.4, 37.2, 31.5, 45.2, 43.1, 33.1, 35.3, 41.3, 49.9, 44.5

Find the median, range and IQR.

4 The stem-and-leaf diagram shows the time taken (in minutes) by a class of students to get to school.

Find the median, range and interquartile range of the times.

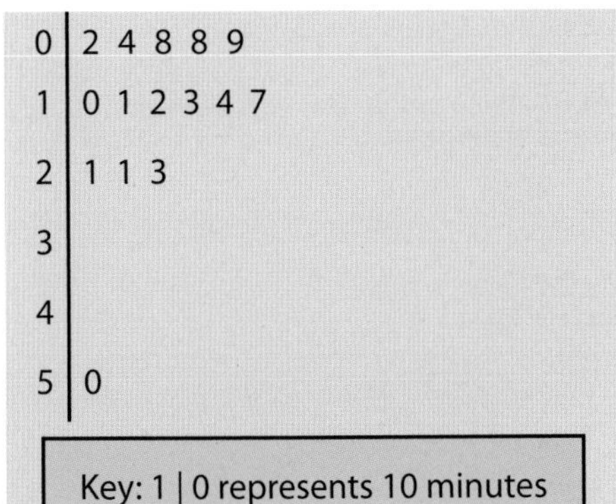

Stem	Leaf
0	2 4 8 8 9
1	0 1 2 3 4 7
2	1 1 3
3	
4	
5	0

Key: 1 | 0 represents 10 minutes

Problem solving

5 The annual salaries of a sample of 9 senior and 9 junior employees at a law firm are shown in the back-to-back stem-and-leaf diagram.

Senior		Junior
	3	5 8
	4	0 2 8 9
9 2 0	5	0 1 3
5	6	
8 0 0	7	
2 0	8	

Key: 9 | 5 represents a salary of £59 000 for a senior employee

5 | 0 represents a salary of £50 000 for a junior employee

a Find the median salary of a senior employee.

b Find the median salary of a junior employee.

c Find the range of both the senior and junior salaries.

d Find the interquartile range of both the senior and junior salaries.

e Compare the salaries of the senior and junior employees.

Compare the medians and the measures of spreads.

6 Here are the ages of 12 people attending an evening class, in order:

22, 24, 29, 30, 30, x, 36, 45, y, 47, 53, z

The median age is 35.

The mean age is 37.5.

The upper quartile is 46.5.

Find the values of x, y and z.

7 Write two different data sets with median 5 and IQR 7.

The five-point summary of a data set is:

- the minimum value
- the lower quartile (Q_1)
- the median (Q_2)
- the upper quartile (Q_3)
- the maximum value.

Exploration 2

A box-and-whisker diagram is a visual representation of the five-point summary. Each vertical line represents one of the five-point summary values.

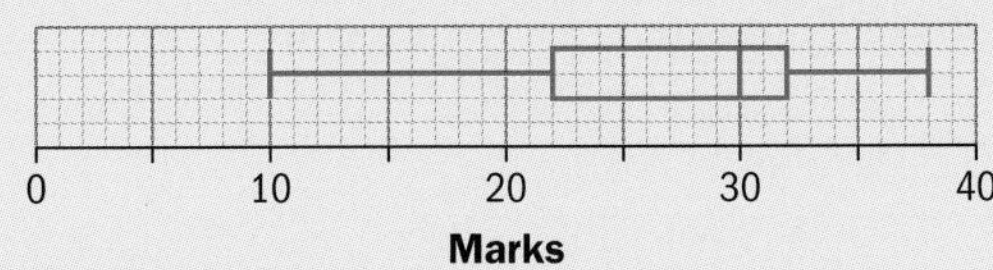

1 Write down the 5 values of the five-point summary.

2 From the information write down:

  a the median

  b the upper quartile.

3 Explain how you can calculate the range from this diagram.

4 Explain how you calculate the IQR from this diagram.

5 Determine what the box represents. What do the whiskers represent?

6 Describe a situation this diagram could represent.

American mathematician John Wilder Tukey contributed much to the field of statistics. He came up with both box-and-whisker and stem-and-leaf diagrams.

ATL

Practice 5

1 The box-and-whisker diagram represents the numbers of strawberries harvested from 30 strawberry plants.

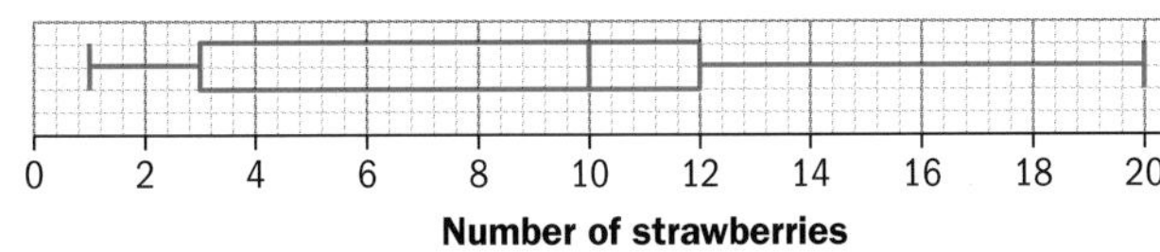

  a Find the median number of strawberries.

  b Find the maximum number of strawberries.

  c Find the IQR (interquartile range).

2 The masses of 100 male Amazon dolphins were represented in a box-and-whisker diagram.

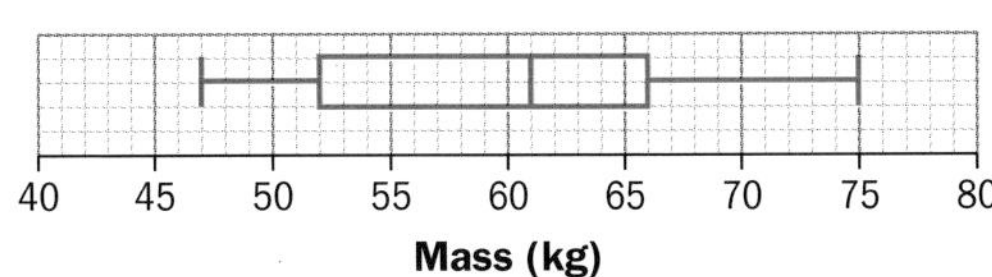

  a Write down the five-point summary of the data.

  b Calculate the range and IQR.

  c Find the number of dolphins who weigh 52 kg or less.

3 Write a five-point summary for each data set:

  a 1, 1, 2, 3, 3, 3, 4, 4, 4, 4

  b 10, 18, 15, 24, 15, 16, 21, 17, 15, 13, 12

  c 10 000, 64 000, 12 000, 11 500, 11 750, 12 250, 10 300

4 Construct a box-and-whisker diagram for this data on the number of lost property items found on a train.

Minimum	Q_1	Median	Q_3	Maximum
15	22	33	37	41

5 The number of minutes that 10 people wait to be connected to a helpline are recorded.

5, 8, 13, 19, 20, 22, 22, 27, 30, 34

a Find the median time.

b Find the lower and upper quartiles.

c Construct a box-and-whisker diagram for the data.

6 The heights, in meters, of 12 trees are:

29.2, 30.1, 36.5, 32.4, 20.0, 28.5, 32.0, 34.6, 39.1, 28.9, 34.4, 24.7

Construct a box-and-whisker diagram for the data.

7 The stem-and-leaf diagram shows the number of diners at a restaurant during a two-week period.

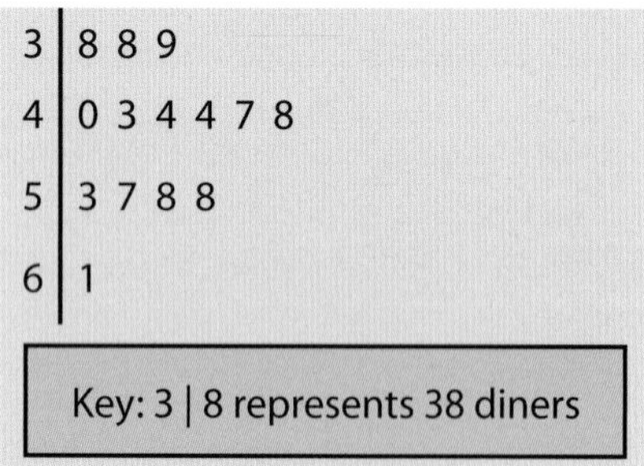

a Write down the minimum number of diners.

b Find the median number of diners.

c Find the interquartile range.

d Construct a box-and-whisker diagram for the data.

Problem solving

8 Construct a box-and-whisker diagram for a data set with:

median 32, minimum 18, Q_1 23, IQR 12, range 21, and find the values of A, B, C, D and E in the stem-and-leaf diagram here.

Stem	Leaf
1	A 9
2	B 6 8
3	C 3 4 D 8 E

C Describing and comparing sets of data

- How do measures of dispersion help you describe data?
- How do different representations help you compare data sets?

Exploration 3

A doctor is investigating how lack of sleep affects a person's ability to do simple tasks. In one task, participants were instructed to press a button as soon as they saw a red light come on. The response times were measured in seconds (to 2 d.p.). Here are two sets of data from the task:

Set 1: 0.58, 0.53, 0.58, 0.43, 0.58

Set 2: 0.66, 0.10, 0.58, 0.58, 0.78

1 Find the mean, mode, median, and interquartile range for each set.

2 The people were fully rested for one set of readings. For the other set, they had been awake for 24 hours. Decide which set is for 'fully rested' and which is for 'awake for 24 hours'. Explain the reasons for your decision.

Reflect and discuss 5

In Exploration 3:

- You could say that a typical reaction time for Set 1 was 0.58 seconds. Why would we choose this value? What does this single value tell you about the times for Set 1?
- The two data sets have the same mean. How well does the mean describe the real situation in Set 1 and in Set 2? What other measure could you use to compare the two data sets?

Distributions with one clear peak are called unimodal. Distributions with two clear peaks are called bimodal.

Example 4

The back-to-back stem-and-leaf diagram shows how much time a group of students spent studying Mathematics in one week. All times are in minutes.

a Describe the study times of the girls and boys.

b Compare the study times of the girls and boys.

Girls		Boys
	0	0 0
5 0 0	1	0 0 0
5 5 0 0	2	0 0 0 0 0 0 0 0 0 5 5
5 5 0 0 0 0 0	3	0 0 0
0 0 0 0	4	0 0 5
0 0	5	0
0 0 0	6	0
5 0	7	
	8	5

Key: 5 | 2 | 0 represents a girl studying for 25 minutes and a boy studying for 20 minutes

▶ Continued on next page

a The median is the $\frac{25+1}{2} = 13$th value.

n = 25 for the girls and for the boys.

The median time for the girls is 35 minutes.

The median time for the boys is 20 minutes.

For the girls:

$Q_1 = 25$ minutes, $Q_3 = 50$ minutes

The interquartile range = 50 − 25 = 25 minutes

For the boys:

$Q_1 = 20$ minutes, $Q_3 = 35$ minutes

The interquartile range = 35 − 20 = 15 minutes

Use the median and interquartile range to describe the center and spread of each set of data.

Both the distributions have only one mode (unimodal). About two-thirds of the boys spent 25 minutes or less studying Mathematics; more than two-thirds of the girls spent 30 minutes or more studying.

Use the stem-and-leaf diagram to describe the shape of each distribution.

b The median time for the girls is longer than the median time for the boys. On average, the girls spent more time studying Mathematics. The girls' IQR is greater than the boys' IQR, so the middle 50% of the girls' times has a larger spread.

Compare the measure of central tendency and the measure of dispersion for the girls and boys.

Objective: C. Communicating
iii. move between different forms of mathematical representation

In this practice you will be changing the representation of the data from tables to stem-and-leaf diagrams, and to box-and-whisker diagrams. These are examples of moving between different forms of representation.

ATL

Practice 6

1 The back-to-back stem-and-leaf diagram shows the times taken (in seconds) by some boys and girls to complete a simple jigsaw.

Girls		Boys
9 9	1	
7 5 5 4 2	2	3 5 7 7 7 9
8 8 8 5 1 0	3	1 2 3 4 4 7 8
9 6 4 3	4	5 5 6 9
7 2	5	1 2
2 1	6	9 9

Key: 3 | 4 | 5 represents a girl who took 43 seconds and a boy who took 45 seconds

a Calculate the median and interquartile range of times for the girls and boys.

b Describe the shape of the distributions.

c Compare the times taken by the girls and the boys.

Problem solving

2 The average monthly temperatures recorded over a year in the US cities of Sante Fe, New Mexico, and Saint Paul, Minnesota, are given below. All temperatures are in Fahrenheit.

a Represent the data in a back-to-back stem-and-leaf diagram.

b Describe the temperatures in the two cities.

c Compare the temperatures.

On the Fahrenheit scale, zero represents the temperature produced by mixing equal weights of snow and common salt.

	J	F	M	A	M	J	J	A	S	O	N	D
Sante Fe	44	48	56	65	74	83	86	83	78	67	53	43
Saint Paul	26	31	43	58	71	80	85	82	73	59	42	29

3 The heights (in centimeters) of one-year-old apple and pear trees are shown below.

Apple trees: 180, 184, 195, 177, 175, 173, 169, 167, 197, 173, 166, 183, 161, 195, 177, 192, 161, 166

Pear trees: 171, 160, 182, 168, 194, 177, 192, 160, 165, 178, 183, 190, 172, 174, 170, 165, 166, 193

a Represent the data in a back-to-back stem-and-leaf diagram.

b Describe the heights of the apple and pear trees.

c Compare the heights.

4 The box-and-whisker diagrams show the delayed departure of two trains, A and B.

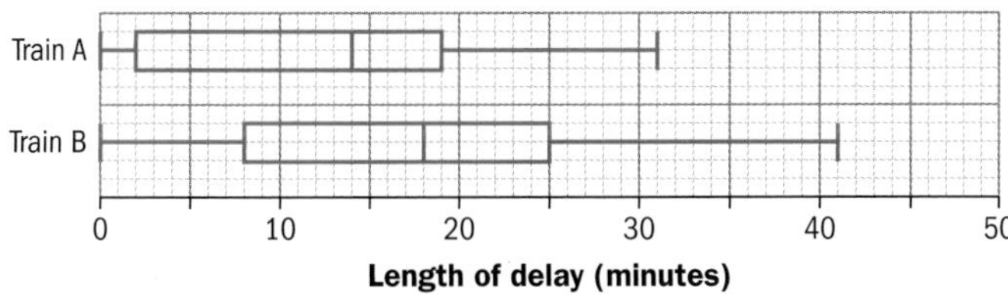

Compare the lengths of the delays.

5 The ages of a sample of subscribers to two German newspapers are shown below.

Süddeutsche Zeitung: 65, 36, 44, 25, 37, 29, 27, 19, 60, 46, 24, 35, 20, 55, 64, 30, 31, 22, 48, 53, 67

Der Tagesspiegel: 46, 18, 35, 20, 27, 25, 40, 24, 31, 29, 20, 63, 18, 30, 19, 28, 21, 34, 54, 22, 27

a Construct two box-and-whisker diagrams for the two data sets.

b Compare the ages.

Use the same scale, and draw one box-and-whisker diagram above the other.

6 Patricia and Ray work in a mobile phone store. The stem-and-leaf diagram shows Patricia's monthly sales for the previous year.

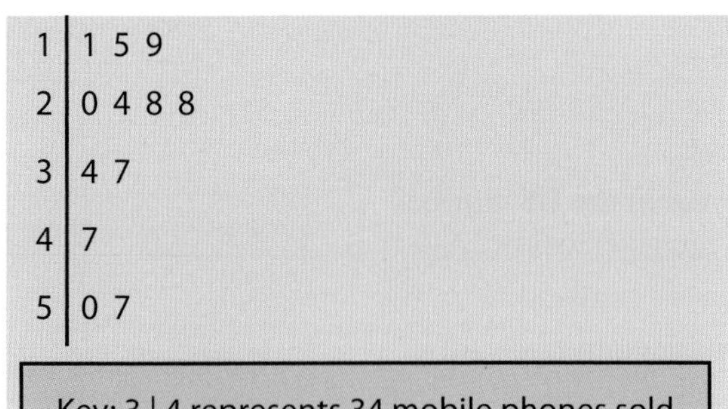

The box-and-whisker diagram shows Ray's monthly sales for the same year.

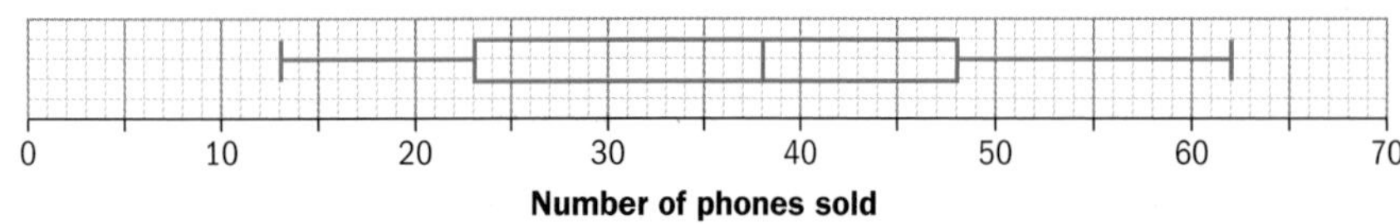

Compare Patricia's and Ray's sales.

D The effect of outliers

- Should we ignore results that aren't typical?
- How do individuals stand out in a crowd?

An **outlier** is a member of a data set which does not fit with the general pattern of the rest of the data. A data point is an outlier if it is less than $1.5 \times$ IQR below Q_1, or greater than $1.5 \times$ IQR above Q_3.

Exploration 4

The foot lengths (measured to the nearest centimeter) of 19 students are given below.

20, 23, 22, 17, 24, 27, 27, 24, 23, 46, 18, 26, 25, 21, 24, 26, 30, 24, 20

1 a Find the lower quartile, upper quartile and interquartile range.

b Find $1.5 \times$ interquartile range (IQR).

c Use the definition of an outlier to determine if any of the data points can be considered outliers.

2 Find the mean, median, mode, range and interquartile range of the foot length data:

a including any outliers

b excluding an outlier.

Record your results in a table.

Which of the statistics are affected by the outlier? Justify this result.

Reflect and discuss 6

In Exploration 4:

- What might account for any outliers?
- When the outlier is included, which measure of central tendency and which measure of dispersion would you use to describe the distribution?
- What are some potential results of ignoring outliers?

When a set of data has outliers, you need to decide whether or not to use them when calculating the measures of central tendency and spread. If you think a result occurred by human error (typing error, misreading a measuring device, broken measuring device) then you can ignore the outlier. If it is a genuine data point, then decide whether or not to keep it.

Generally speaking, the range is a good indication of the dispersion of a set of data, but it can be greatly affected by just one extreme value. You may need to exclude any outliers before calculating the range.

To describe a distribution:

(1) identify any outliers

(2) use a measure of central tendency (center) and a measure of dispersion (spread) and

(3) describe the shape.

The median and interquartile range are not affected by outliers.

Practice 7

1 A group of 17 young children were tested to see how long it took each of them to assemble a set of interconnecting bricks. Their times (in seconds) are shown here.

8, 12, 16, 23, 24, 25, 25, 27, 28, 30, 31, 31, 32, 32, 40, 48, 51

a Find the mean, median and mode of the times.

b Determine whether it is better to use the mean, median or mode for this data set. Give reasons for your answer.

2 The number of text messages sent by 17 adults in one week are shown below.

36, 40, 22, 8, 16, 19, 48, 62, 27, 22, 34, 31, 36, 28, 30, 12, 20

a Find the mean and median number of texts sent.

b Find the interquartile range.

c Determine if any of the data values can be considered outliers.

d Find the range. Give a reason why any outliers have been included or excluded in your calculation.

e Which measure of central tendency is more appropriate in this case? Explain.

3 Look back at the data on Mathematics study times in Example 4.

a Determine if any of the data values can be considered outliers.

b Find the range of study times for the girls and boys. Give a reason why any outliers have been included or excluded in your calculations.

Problem solving

4 The back-to-back stem-and-leaf diagram shows the masses (to the nearest kg) of 30 male and 30 female northern hairy-nosed wombats.

Male wombat		Female wombat
	0	8
9 8 5 2 0	1	0 5 7 8 9
8 8 7 7 1 0 0 0	2	0 0 0 0 5 7 8 8
8 8 7 5 2 2 2 1 1 0	3	0 0 1 1 1 2 5 5 6 8 8
4 3 2 2 0	4	2 2 3 4
0	5	0
	6	
	7	
	8	
	9	
0	10	

Key: 0 | 4 | 2 represents a male wombat with mass 40 kg and a female wombat with mass 42 kg

a Decide whether there are any outliers.

b Describe the masses of the male and female wombats.

c Compare the masses of the male and female wombats.

5 The table shows the percentage of households that speak English in the 26 cantons of Switzerland.

a Construct a stem-and-leaf diagram and determine whether there are any outliers.

b Suggest how any outliers could have occurred.

c Describe the distribution.

ZH	BE	LU	UR	SZ	OW	NW	GL	ZG
6.4	3.0	2.9	1.6	3.8	2.8	2.6	1.2	8.4
SO	BS	BL	SH	AR	AI	SG	GR	AG
2.5	8.6	4.2	3.6	2.1	2.4	2.5	2.5	3.5
TI	VD	VS	NE	GE	JU	FR	TG	
3.1	7.4	2.6	3.3	10.7	1.5	2.5	2.4	

Switzerland has four official languages. From 2010 the Swiss Federal Statistical Office has allowed citizens to indicate more than one language as their main language so the total for all languages exceeds 100%.

Summary

- In a stem-and-leaf diagram:
  - the **stem** represents the **category** figure
  - the **leaves** represent the **final** digit(s) of each data point
  - the **key** tells you how to read the values.

stem	leaf
0	0 0 0 1 1 1 4 6 7
1	0 0 1 2 3 4 5 7 7
2	0 2 6
3	9

Key: 1|0 represents 10

- For any set of data there are two categories of **summary statistics**:
  - **Measures of central tendency (location)** describe where most data lies. They answer the question 'What is an average data value?'
  - **Measures of dispersion (spread)** describe how spread out the data is. They answer the question 'How much variation is there between the values?'
- The **lower/first quartile (Q_1)** is the median of the observations to the left of the median in an arranged set of observations.
- The **upper/third quartile (Q_3)** is the median of the observations to the right of the median in an arranged set of observations.
- The **interquartile range (IQR)** is the difference between the lower quartile and the upper quartile ($Q_3 - Q_1$).
- The median is another name for the second quartile, Q_2.
- A five-point summary of a data set is:
  - the minimum data point (min)
  - the lower quartile (Q_1)
  - the median (Q_2)
  - the upper quartile (Q_3)
  - the maximum data point (max).
- A box-and-whisker diagram represents the five-point summary:

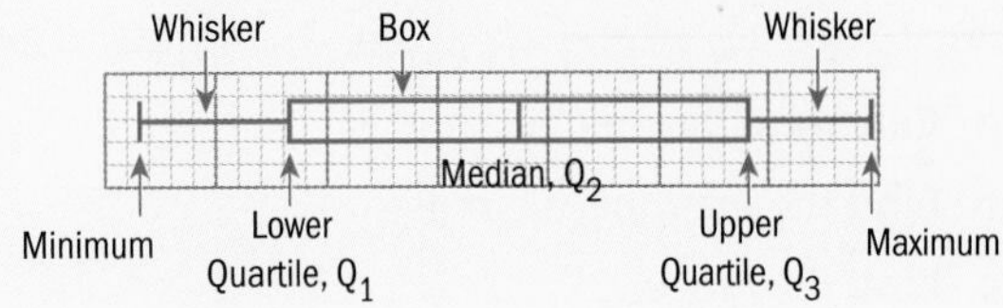

- **Qualitative data** – describes a certain characteristic using words (color, animal).
- **Quantitative data** – has a numerical value.

Quantitative data has a numerical value. There are two types:

- **Discrete data** – can be counted (number of goals scored) or can only take certain values (shoe size)
- **Continuous data** – can be measured (weight, height) and can take any value.

- The **distribution** of a data set describes the behavior of all the data points in the set.
- An **outlier** is a member of the data set which does not fit with the general pattern of the rest of the data. It could be an anomaly in the data or an inaccurate reading.
- A data point is an outlier if it is less than $1.5 \times \text{IQR}$ below Q_1, or greater than $1.5 \times \text{IQR}$ above Q_3.

Mixed practice

1 Write down whether these are qualitative or quantitative data.

a The color of flowers

b The prices of graphic calculators

c The areas of public parks

d The numbers of people at a store

2 Write down whether these are discrete or continuous data.

a The time taken to run a marathon

b The number of carriages on a train

c The heights of Humboldt penguins

d The temperature at midday in Iceland

3 The masses (in kilograms) of a group of people are given below.

ATL

46, 52, 64, 60, 82, 48, 72, 61, 70, 75, 59

a **Construct** a stem-and-leaf diagram to represent the data.

b **Write down** the number of people who weigh less than 60 kg.

c **Find** the median mass.

d **Find** the range.

4 During a one-hour study break, the amount of time students spent on social media sites was recorded. Their times, in minutes, were:

20, 45, 37, 29, 48, 52, 41, 32, 26, 50, 32, 44

a **Calculate** the median time.

b **Find** the interquartile range.

c **Find** the range.

Problem solving

5 Sebastian collected data on the number of pairs of shoes owned by every student in his class. The values are in order.

5, 6, 7, 7, 9, 9, r, 10, s, 13, 13, t

He calculated that the median of the data set was 9.5 and the upper quartile Q_3 was 13.

a **Write down** the value of r and s.

b The mean of the data set is 10.

Find the value of t.

ATL

6 All the dogs attending a veterinary surgery were weighed. Their masses, in kilograms, were:

26, 18, 54, 32, 30, 25, 6, 32, 43,
90, 16, 5, 27, 18, 3, 23, 27

a **Find** a five-point summary for the data.

b **Construct** a box-and-whisker diagram for the data.

7 This box-and-whisker diagram represents the results from a survey that asked: 'How old were you when you got your first smartphone?'

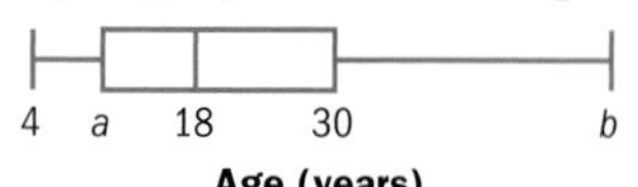

The interquartile range is 20 and the range is 40.

a **Write down** the median value.

b **Find** the values of a and b.

c 160 people were surveyed. **Find** the number of people who were 30 or older.

Problem solving

8 The box-and-whisker diagrams show the ages of people watching two different films at a cinema.

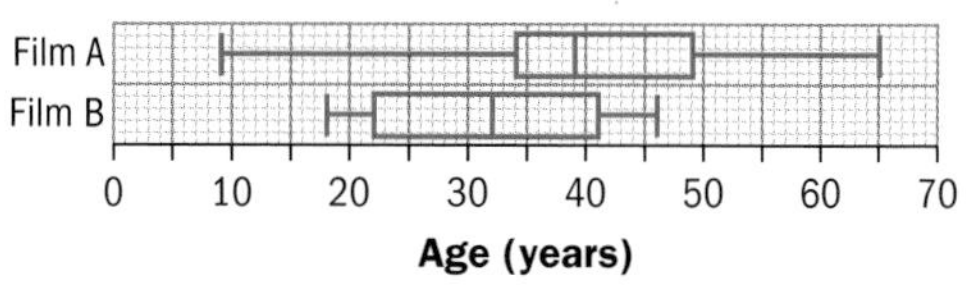

a **Compare** the ages.

b **Deduce** which film is categorized as a family film. Give reasons for your answer.

9 The profits made by a small business during the last 11 years are as follows:

\$45 000, \$560 000, \$1000, \$85 000, \$160 000, \$170 000, \$62 000, \$250 000, \$3100, \$120 000, \$38 000

a **Find** the interquartile range.

b **Determine** if any of the data values can be considered outliers.

c **Find** the range. Give a reason why any outliers have been included or excluded in your calculation.

d **Find** the measure of central tendency which is most appropriate in this case. **Justify** your choice.

Problem solving

10 The table shows the masses (in kilograms) of Blackface sheep on two farms.

ATL

Farm A	63	48	60	55	45	49	19	55
Farm B	54	71	68	57	62	70	54	49
Farm A	65	49	57	56	64	43	64	48
Farm B	66	68	72	50	56	49	70	64

a **Construct** a back-to-back stem-and-leaf diagram to represent the data.

b **Compare** the masses of the Blackface sheep on the two farms.

c The table shows the average adult bodyweight of Blackface sheep for different grazing conditions.

Poor hill	45–50 kg
Average/good hill	50–65 kg
Upland	70 kg

Deduce the grazing conditions available to the sheep on the two farms. Give reasons for your answer.

Review in context

Globalization and sustainability

1 The table shows the percentage of land covered by forest in 19 European Union states in 2010.

Austria	46.7%
Belgium	22.0%
Czech Republic	34.3%
Denmark	11.8%
Estonia	53.9%
Finland	73.9%
France	28.3%
Germany	31.7%
Greece	29.1%
Hungary	21.5%
Ireland	9.7%
Italy	33.9%
Lithuania	33.5%
Netherlands	10.8%
Poland	30.0%
Portugal	41.3%
Slovakia	40.1%
Sweden	66.9%
United Kingdom	11.8%

a **Construct** a box-and-whisker diagram to represent the data.

b **Determine** if there are any outliers in this data.

2 The box-and-whisker diagram shows the forest coverage for the same EU states in 2012.

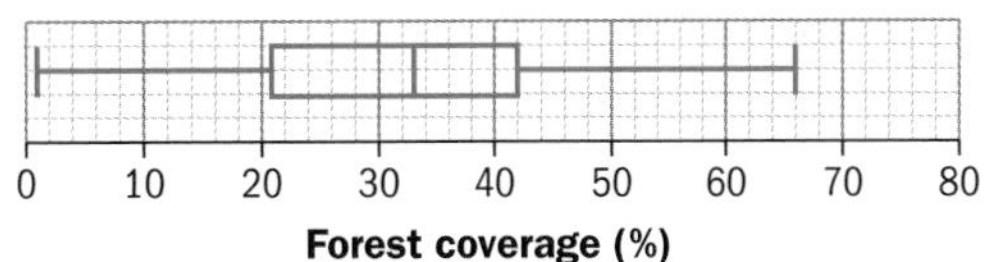

Compare the data for 2010 and 2012 and **identify** any changes between the two years.

3 **Use** the UNECE (United Nations Economic Commission for Europe) website to find more forestry data. **Use** an appropriate form of representation and prepare a short report to present to your classmates, and show how you can move between different forms of representation. **Explain** the conclusions you can draw from your diagram.

Reflect and discuss

How have you explored the statement of inquiry? Give specific examples.

Statement of Inquiry:

How quantities are represented can help to establish underlying relationships and trends in a population.

4.2 Getting your ducks in a row

Global context: Globalization and sustainability

RELATIONSHIPS

Objectives

- Finding the mean, median, mode and range from a grouped frequency table
- Representing grouped data in a cumulative frequency curve
- Finding the five-point summary from a cumulative frequency curve
- Constructing a box-and-whisker diagram from a cumulative frequency curve

Inquiry questions

F
- How can you find the measures of central tendency from a grouped frequency table?
- How can you represent grouped data in a cumulative frequency curve?

C
- How does the type of data affect the way it can be represented and analyzed?

D
- How can real data ever be misleading?
- How do individuals stand out in a crowd?

ATL **Communication**

Organize and depict information logically

Statement of Inquiry:

How quantities are represented can help to establish underlying relationships and trends in a population.

4.1

4.2

6.2

You should already know how to:

<table>
<tr><td>• understand inequality notation</td><td>1 Which of these values satisfy the inequality $3 < x \leq 6$?
a 4.2 b 3.0 c 6.4 d 6.0</td></tr>
<tr><td>• recognize discrete and continuous data</td><td>2 Is each data set discrete or continuous?
a Numbers of eggs laid by 6 hens
b Heights of sunflower plants
c Shoe sizes
d Lap times in a Formula 1 car race</td></tr>
<tr><td>• calculate the mean from a frequency table</td><td>3 The frequency table shows the ages of children in a playgroup. Calculate the mean age.<table><tr><td>Age</td><td>1</td><td>2</td><td>3</td><td>4</td><td>5</td></tr><tr><td>Freq</td><td>0</td><td>2</td><td>6</td><td>4</td><td>3</td></tr></table></td></tr>
<tr><td>• write a five-point summary and construct a box-and-whisker diagram</td><td>4 a Construct a box-and-whisker diagram for this data on the ages of a Scout group.
11, 12, 13, 13, 13, 14, 15,
15, 17, 18, 18, 20, 21
b Find the interquartile range.</td></tr>
<tr><td>• identify how quartiles divide a data set</td><td>5 In any data set, what fraction of the data is less than or equal to:
a the lower quartile
b the median
c the upper quartile?
Write each fraction as a percentage.</td></tr>
</table>

F Working with grouped data

- How can you find the measures of central tendency from a grouped frequency table?
- How can you represent grouped data in a cumulative frequency curve?

When you have a large set of data it is often convenient to organize the data into groups before doing any statistical analysis.

ATL **Example 1**

The lengths of 50 Eelgrass leaves were measured. The results are given below, to the nearest tenth of a centimeter.

46.3	35.6	75.2	43.7	49.0	42.6	47.1	50.3	50.7	55.6
56.0	52.7	57.3	57.3	34.2	58.0	37.0	59.3	68.1	59.9
53.6	34.0	74.4	70.9	44.2	41.3	64.9	39.3	44.2	68.3
54.2	40.7	76.8	38.2	70.3	33.5	64.3	64.6	68.7	69.9
48.9	51.6	51.7	46.0	52.0	45.6	64.1	63.4	45.1	62.0

Measuring a leaf gives continuous data; this has to be rounded to a certain degree of accuracy to be used in statistical analysis.

a Construct a grouped frequency table to represent the data.

b Find the class interval that contains the median.

c Write down the modal class.

a minimum = 33.5 cm, maximum = 76.8 cm

range = 76.8 – 33.5 = 43.3 cm

$\frac{43.3}{10} = 4.33$ so use a class width of 5 cm

Choose a class width to give between 5 and 15 equal width classes. (In this case, 10.)

Write the class intervals using inequality notation.

Length, x (cm)	Freq	Cumulative frequency
$30 < x \le 35$	3	3
$35 < x \le 40$	4	7
$40 < x \le 45$	6	13
$45 < x \le 50$	7	20
$50 < x \le 55$	8	(28)
$55 < x \le 60$	7	35
$60 < x \le 65$	6	41
$65 < x \le 70$	4	45
$70 < x \le 75$	3	48
$75 < x \le 80$	2	50

Add a Cumulative frequency column to help find the class interval that contains the median.

The cumulative frequency is the sum of the frequencies for every class, up to and including the current one. Here, the cumulative frequency is $3 + 4 + 6 = 13$.

The 25th and 26th values are in this class.

b $n = 50$, so the median is the $\frac{n+1}{2} = \frac{50+1}{2} = 25.5$th data value.

The class interval that contains the median is $50 < x \le 55$.

c The modal class is $50 < x \le 55$.

The modal class is the class interval with the highest frequency.

Reflect and discuss 1

- What does cumulative frequency represent?
- Why can't you find the exact value of the median from a grouped frequency table?
- How would you estimate the range from a grouped frequency table? Estimate the range of lengths of Eelgrass leaves.

The **class width** is the difference between the maximum and the minimum possible values in a class interval.

The class $18 < x \leq 20$ has class width 2, because $20 - 18 = 2$.

An estimate for the range from a grouped frequency table is (upper bound of highest class interval) – (lower bound of lowest class interval).

When the class widths are equal, the **modal class** is the class containing the most data. It has the highest frequency.

For grouped data, you can find the class interval that contains the median.

Add a cumulative frequency column to the frequency table to find which class interval contains the $\left(\frac{n+1}{2}\right)$th value.

Grouped data has a modal class instead of one mode value.

Practice 1

1 The table shows the heights of a group of students on their 14th birthdays.

Height, x (cm)	Frequency
$1.20 < x \leq 1.30$	4
$1.30 < x \leq 1.40$	6
$1.40 < x \leq 1.50$	8
$1.50 < x \leq 1.60$	6
$1.60 < x \leq 1.70$	6

a Write down the modal class.

b Find the class interval that contains the median.

ATL

2 Here are the times taken for some students to complete a 1500 m race (measured to the nearest whole second).

325	580	534	500	532	328	600	625	450	435
450	340	357	370	401	456	388	626	532	399

a Construct a grouped frequency table for this data. Use class widths of equal size and first class interval $300 < x \leq 350$.

b Find the modal class.

c Determine which class interval contains the median value.

ATL

3 The heights of one-year-old apple trees are shown below, measured to the nearest cm.

180 184 195 177 175 173 169 167 197

173 166 183 161 195 177 192 161 166

a Construct a grouped frequency table for this data.

b Write down the modal class.

c Determine which class interval contains the median value.

Problem solving

4 The table shows the weights of apples picked from one tree.

Weight, w (g)	Frequency	Cumulative frequency
$30 < w \leq 40$	a	5
$40 < w \leq 50$	7	12
$50 < w \leq 60$	5	b
$60 < w \leq 70$	c	19
$70 < w \leq 80$	8	d
$80 < w \leq 90$	3	30

a Calculate the values of a, b, c and d.

b Verify that the median lies in the class $50 < w \leq 60$.

Exploration 1

1 Look back at the grouped frequency table for Eelgrass leaves in Example 1. Imagine you had not received the raw data.

a Assume that all the data in each class interval has the value of the upper class boundary. For example, in the class $30 < x \leq 35$ assume all the values are 35. Using this assumption, calculate an estimate for the mean length of the leaves.

b Now assume that all the data in each class interval has the value of the lower class boundary. For example, in the class $30 < x \leq 35$ assume all the values are 30. Using this assumption, calculate an estimate for the mean length of the leaves.

The lower boundary of the interval is 30 (not 31 or 30.1, etc.).

c Compare your two estimates. How would choosing a different value of the class interval affect the estimate of the mean?

d What value would be better than the upper or lower class boundaries to represent all the data in a class interval?

To calculate an estimate of the mean from a grouped frequency table:

- Find the midpoint or mid-interval value of each class interval by adding the upper and lower class boundaries and dividing by 2.
- Use the midpoint of each class interval to calculate the mean.

Continued on next page

2 To calculate an estimate of the mean for the Eelgrass leaves, add a column for mid-interval value and a column for mid-interval value × frequency.

Length, x (cm)	Frequency	Mid-interval value	Mid-interval value × frequency
$30 < x \leq 35$	3	32.5	$32.5 \times 3 = 97.5$
$35 < x \leq 40$	4	37.5	$37.5 \times 4 = 150$
$40 < x \leq 45$	6	42.5	
$45 < x \leq 50$	7	47.5	
$50 < x \leq 55$	8	52.5	
$55 < x \leq 60$	7	57.5	
$60 < x \leq 65$	6	62.5	
$65 < x \leq 70$	4	67.5	
$70 < x \leq 75$	3	72.5	
$75 < x \leq 80$	2	77.5	
Total frequency (number of observations)		Sum of mid-interval value × frequency	

For grouped data the mean is defined as:

$$\bar{x} = \frac{\text{sum of (mid-interval value} \times \text{frequency)}}{\text{number of observations}}$$

Copy and complete the table, and calculate an estimate for the mean.

For individual points in a set of data the mean is defined as:

$$\bar{x} = \frac{\text{sum of observations}}{\text{number of observations}}$$

You can also calculate an estimate of the mean using a GDC.

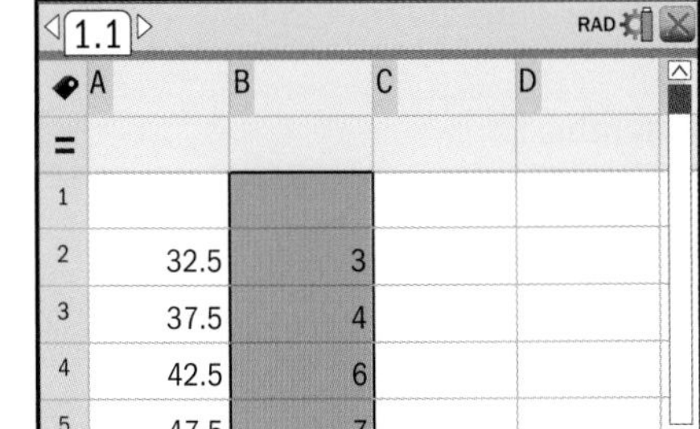

Enter the mid-interval values and frequencies into your GDC.

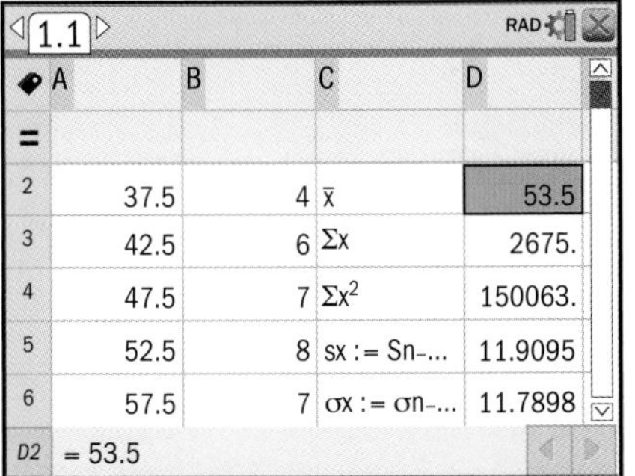

$\bar{x}$ is the mean.

Reflect and discuss 2

- In Exploration 1, what does the mid-interval value represent for the Eelgrass leaves? What does the mid-interval × frequency value represent?
- Why can you calculate only an *estimate* for the mean from a grouped frequency table? Is the estimate for the mean likely to be one of the original data values?
- Use the raw data in Example 1 to calculate the mean length of the Eelgrass leaves. Compare the calculated mean with your estimate for the mean from Exploration 1.

Practice 2

1 The table shows the number of hours a group of 50 students spent watching online videos in one week.

Time, t (hours)	Frequency	Mid-interval value	Mid-interval value × frequency
$0 < t \leq 10$	8	5	40
$10 < t \leq 20$	12		
$20 < t \leq 30$	16		
$30 < t \leq 40$	11		
$40 < t \leq 50$	3		
Total			

To find the mid-interval value, add the upper and lower class boundaries and divide by 2.

a Copy and complete the table.

b Calculate an estimate for the mean time each student spent watching online videos.

2 The weights of the Irish Wolfhounds at a dog show are shown in the table.

Weight, w (kg)	Frequency
$45 < w \leq 48$	2
$48 < w \leq 51$	5
$51 < w \leq 54$	12
$54 < w \leq 57$	13
$57 < w \leq 60$	8

a Calculate an estimate for the mean weight of the dogs.

b Write down the modal class.

c Find the number of dogs who weighed 54 kg or less.

d Determine the class interval that contains the median.

Problem solving

3 The table below shows the weights (w) of fish caught one morning.

Weight, w (kg)	Frequency (f)
$0.6 < w \le 0.8$	16
$0.8 < w \le 1.0$	35
$1.0 < w \le 1.2$	44
$1.2 < w \le 1.4$	23
$1.4 < w \le 1.6$	10

a Estimate, correct to the nearest 0.1 kg, the mean weight of the fish.

b Find the modal class.

c Any fish caught that weighs no more than 1.0 kg must be returned to the sea. Find the number of fish returned.

d Find the class interval that contains the median.

e Estimate the range of the weights.

f A fish weighing 1.2 kg was incorrectly recorded in the class $1.0 \le w < 1.2$. Without doing any calculations, explain whether or not the mean and median will stay the same.

Many types of scale can be used to weigh fish but only some are recognized by the International Game Fish Association for potential records. If the scale reading is between two marks the angler must round down to the heaviest known weight.

Cumulative frequency curves

A cumulative frequency curve is a graph with the upper class boundaries plotted on the x-axis, and the cumulative frequencies on the y-axis.

You can find the five-point summary from a cumulative frequency curve.

For a set of n data values on a cumulative frequency curve, to find the estimate for:

- the lower quartile Q_1, find the value on the x-axis which corresponds to 25% of n
- the median Q_2, find the value on the x-axis which corresponds to 50% of n
- the upper quartile Q_3, find the value on the x-axis which corresponds to 75% of n.

The cumulative frequency curve here represents the grouped frequency table of the Eelgrass data in Example 1.

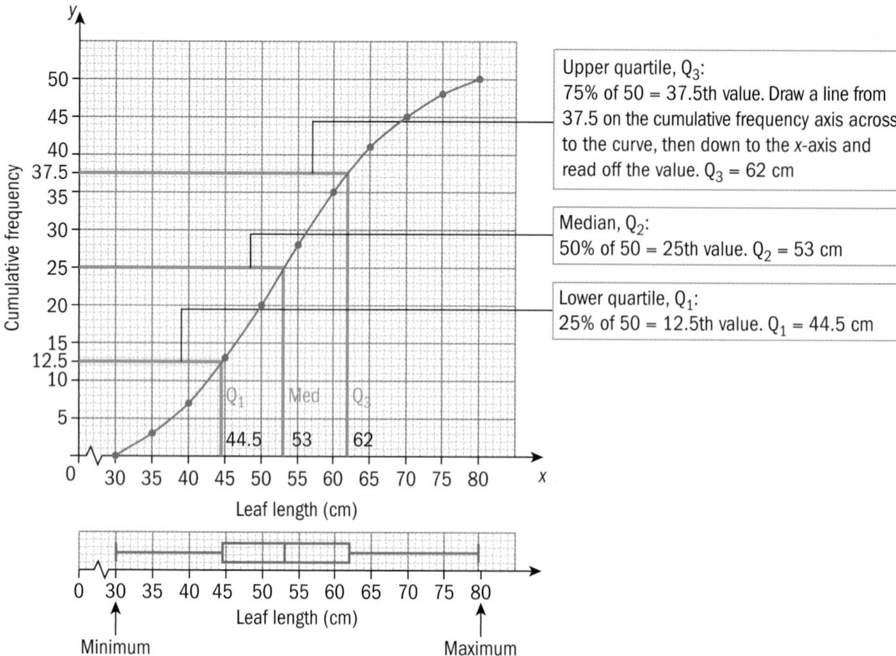

You can take the five-point summary from the cumulative frequency curve and use it to draw the box-and-whisker diagram at the same scale.

Reflect and discuss 3

- Why are the percentiles 25%, 50% and 75% used to find estimates for Q_1, Q_2 and Q_3 respectively?
- Find the five-point summary from the raw Eelgrass data in Example 1.
- Do the values from the cumulative frequency curve agree with the values from the raw data in Example 1? Which values are the most accurate?
- How could you use the cumulative frequency curve to estimate the number of leaves with lengths less than or equal to 62 cm? Or greater than 62 cm? Or between 50 and 62 cm?

Practice 3

1 The cumulative frequency curve shows the heights of the basketball players in a university league.

  a Find an estimate for the median height of the players.

  b Find an estimate for the upper and lower quartiles.

  c Find an estimate for the interquartile range.

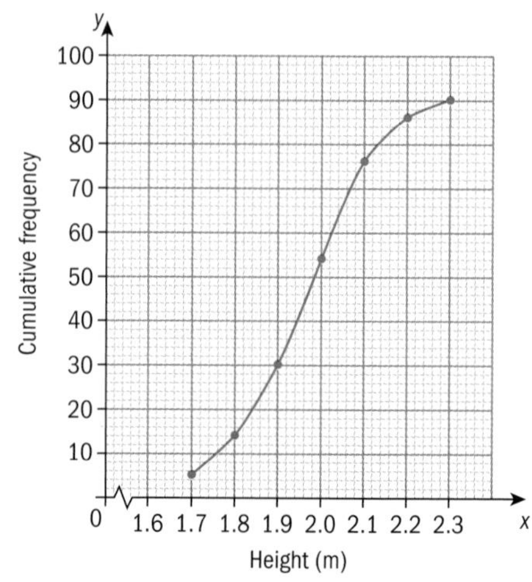

The shortest player in National Basketball Association history measured 1.6 meters while the tallest was 2.31 meters.

ATL

2 The table shows the distances walked (in kilometers) by members of a walking group in one week.

Distance, d (km)	Frequency	Cumulative frequency
$0 < d \leq 3$	4	4
$3 < d \leq 6$	14	18
$6 < d \leq 9$	6	
$9 < d \leq 12$	8	
$12 < d \leq 15$	4	

a Copy and complete the table.

b Construct a cumulative frequency curve to represent this data.

c Find estimates for the median and quartiles.

d Find estimates for the range and interquartile range.

Plot the points (3, 4), (6, 18) etc. and join with a smooth curve.

ATL

3 The table shows the masses of the Great Danes at a rehoming shelter.

Mass, m (kg)	$65 < m \leq 70$	$70 < m \leq 75$	$75 < m \leq 80$	$80 < m \leq 85$	$85 < m \leq 90$
Frequency	7	13	9	12	9

a Construct a cumulative frequency curve to represent this data.

b Write a five-point summary for the data.

c Construct a box-and-whisker diagram.

ATL

Problem solving

4 For the fish data in **Practice 2, Q3:**

a Construct a cumulative frequency curve.

b Construct a box-and-whisker diagram.

c Use your cumulative frequency curve to decide whether these statements are true or false.

i 90 fish weigh 1.1 kg or less.

ii 60 fish weigh between 1.1 kg and 1.5 kg.

d 25% of the fish are classed as overweight. By using the calculated upper quartile, find the minimum weight at which a fish is classed as overweight.

ATL

5 The table shows the masses (in grams) of apples in a box.

Mass, m (kg)	$30 < m \leq 40$	$40 < m \leq 50$	$50 < m \leq 60$	$60 < m \leq 70$	$70 < m \leq 80$	$80 < m \leq 90$
Frequency	5	7	5	2	8	3

a Construct a cumulative frequency curve for this data.

b Apples weighing less than 45 g cannot be sold.

Find an estimate for the number of apples that cannot be sold.

c 10% of the apples weigh more than x grams. Find x.

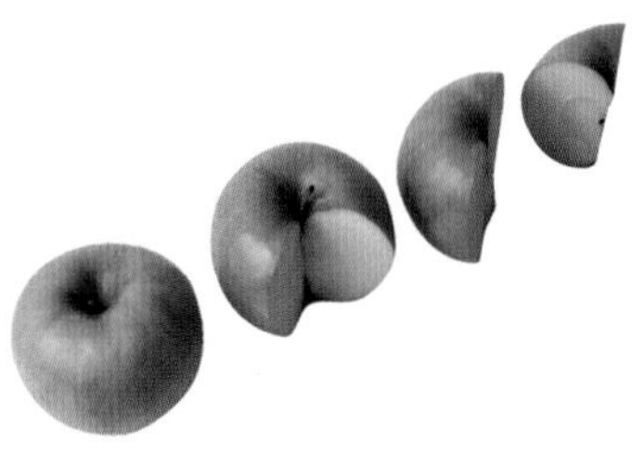

ATL

6 This box-and-whisker diagram shows the times taken by 100 people to complete a simple word puzzle. Construct a cumulative frequency curve for the data.

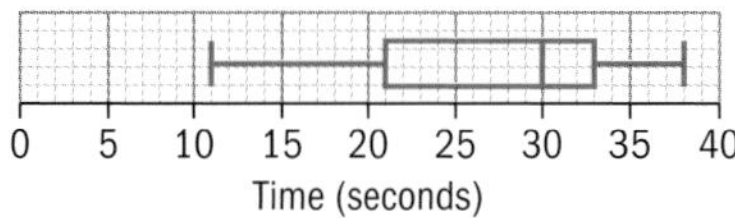

ATL

Problem solving

7 The cumulative frequency curve here shows the speeds (in kilometers per hour) of some cars on motorway A.

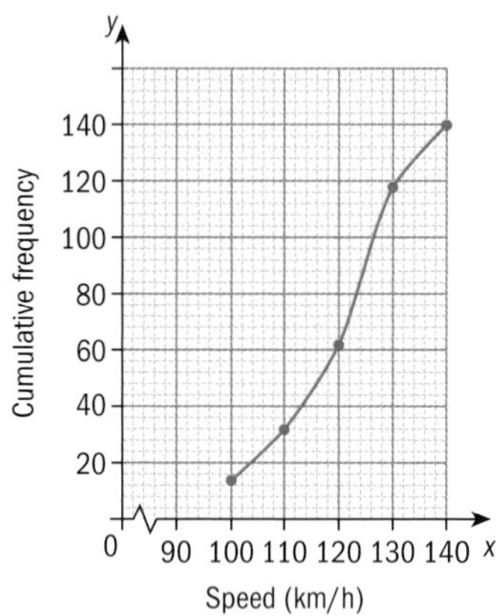

Only three countries on the world give the motorway speed limit in miles per hour rather than kilometers per hour: the Great Britain, the United States and Burma.

a Find an estimate for the median speed.

b The speed limit on the motorway is 130 km/h.

Find an estimate for the percentage of cars exceeding the speed limit.

c Construct a box-and-whisker diagram for the speed data for motorway A.

The box-and-whisker diagram below shows the speeds of some cars on motorway B.

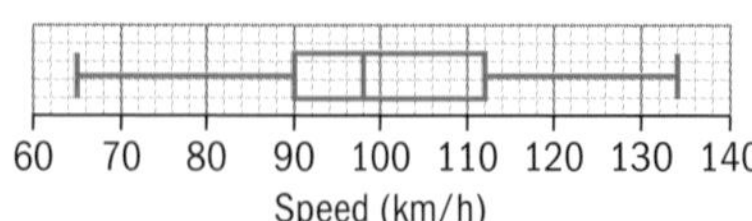

d Compare the speeds of the cars on motorway A and motorway B.

Grouping discrete data

- How does the type of data affect the way it can be represented and analyzed?

When you represent data in a frequency table, you first need to decide if the data is discrete or continuous, as this affects how you write the class intervals.

Continuous data has overlapping boundary values and is written with continuous intervals, with one < sign and one ≤, for example:	Discrete data has no overlapping boundary values and there are gaps between the class intervals. Both inequalities are ≤, for example:
$35 < x \leq 40$	$35 \leq x \leq 40$
$40 < x \leq 45$	$41 \leq x \leq 45$

When you have decided on the class intervals, you can calculate an estimate of the mean in the same way as for grouped continuous data.

For continuous data, for example $35 < x \leq 40$, $40 < x \leq 45$, plot the cumulative frequencies against the upper class boundaries 40 and 45.

For discrete data, for example $35 \leq x \leq 40$, $41 \leq x \leq 45$, plot the cumulative frequencies against the point halfway between the upper bound of one interval (40) and next lower bound (41), at 40.5.

All values < 40.5 round down to 40, and all values ≥ 40.5 round up to 41.

ATL

Example 2

Dieter recorded the number of trucks driving past his house in 5-minute intervals over a period of 3 hours. His results are shown below.

3	14	6	16	21	6	20	14	4	11
12	19	1	12	19	14	11	16	24	20
6	18	8	27	7	23	2	7	11	12
11	7	27	24	12	22	15	29	9	25
14	10	10	16	15	19	1	17	8	12

Construct a cumulative frequency curve for this data.

Number of trucks	Freq	New upper boundary	Cumulative frequency
$1 \leq x \leq 5$	5	$\frac{5+6}{2} = 5.5$	5
$6 \leq x \leq 10$	11	$\frac{10+11}{2} = 10.5$	16
$11 \leq x \leq 15$	15	$\frac{15+16}{2} = 15.5$	31
$16 \leq x \leq 20$	10	$\frac{20+21}{2} = 20.5$	41
$21 \leq x \leq 25$	6	$\frac{25+26}{2} = 25.5$	47
$26 \leq x \leq 30$	3	$\frac{30+31}{2} = 30.5$	50

Find the new upper boundary for each class interval – the value halfway between the upper class boundary of one interval and the lower class boundary of the next.

Add a Cumulative frequency column to the table.

▶ Continued on next page

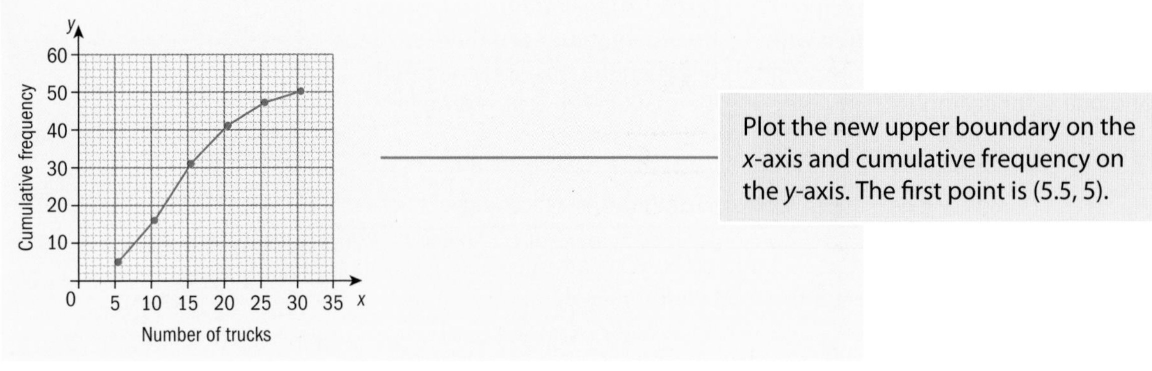

You can obtain a five-point summary and draw a box-and-whisker diagram for discrete data in exactly the same way as for continuous data.

Practice 4

1 **a** State whether these class intervals represent continuous or discrete data, and give reasons for your answers.

b Write down the class width of each interval.

i	$10 < x \le 20$	**ii**	$14 \le x \le 18$	**iii**	$6 < x \le 9$	**iv**	20 – 39
	$20 < x \le 30$		$19 \le x \le 23$		$9 < x \le 12$		40 – 64

2 The table here shows the number of words in each of 100 sentences in a book.

Number of words	Frequency
$0 \le x \le 4$	17
$5 \le x \le 9$	25
$10 \le x \le 14$	30
$15 \le x \le 19$	12
$20 \le x \le 24$	10
$25 \le x \le 29$	3
$30 \le x \le 34$	2
$35 \le x \le 39$	0
$40 \le x \le 44$	1

a Calculate an estimate for the mean number of words in each sentence.

Does your answer make sense in the context of this question?

b Find the class interval that contains the median.

c Write down the modal class.

3 Beatrice records the number of goals scored by her unihockey team for each of the 25 matches in the season.

Number of goals scored	$0 \le x \le 3$	$4 \le x \le 7$	$8 \le x \le 11$	$12 \le x \le 15$	$16 \le x \le 19$
Frequency	3	5	6	6	5

a Calculate an estimate for the mean number of goals scored per match.

b Explain why this data is bimodal.

c Find the class interval that contains the median.

d Estimate the range.

ATL

4 The table shows the Extended Essay scores of 50 DP students.

Score	Frequency
$1 \le x \le 5$	1
$6 \le x \le 10$	9
$11 \le x \le 15$	15
$16 \le x \le 20$	15
$21 \le x \le 25$	5
$26 \le x \le 30$	5

a By drawing a new table with the correct upper boundary for discrete data and the cumulative frequency, construct a cumulative frequency curve.

b Using your cumulative frequency curve, write down the median score.

c Find an estimate for the upper and lower quartiles.

d Draw a box-and-whisker diagram to represent the information.

ATL

Problem solving

5 Brooke recorded the number of text messages she sent every day for a month in this table:

Number of messages sent	$0 \le x \le 4$	$5 \le x \le 9$	$10 \le x \le 14$	$15 \le x \le 19$	$20 \le x \le 24$	$25 \le x \le 29$	$30 \le x \le 34$
Frequency	10	7	3	4	3	2	1

a Calculate the values of a and b in this cumulative frequency table for the text message data.

≤ 9.5 means 'all the values up to and including 9.5'.

Upper boundary	Cumulative frequency
≤ 4.5	10
≤ 9.5	17
≤ 14.5	a
≤ 19.5	24
≤ 24.5	27
≤ 29.5	b
≤ 34.5	30

b Draw a cumulative frequency curve for this data.

c Use your curve to calculate each value in the five-point summary. Hence represent the data in a box-and-whisker plot.

ATL

6 The ages (in years) of the first 30 people visiting the Tuileries Gardens in Paris on a certain day are shown.

25	65	34	48	4	55
32	45	23	43	23	37
45	36	26	39	43	45
29	15	15	20	64	37
61	25	17	23	45	36

a Construct a grouped frequency table for this data.

b Construct a cumulative frequency curve.

c Use your cumulative frequency curve to estimate the percentage of visitors who are 25 or under.

d Estimate the percentage of visitors who are older than 40.

7 The cumulative frequency curves give information about the ages of pedestrians and pedal cyclists killed in Great Britain in 2014.

a Draw a box-and-whisker diagram for the pedestrians.

b Draw a box-and-whisker diagram for the pedal cyclists.

c Compare the ages of the pedestrians and pedal cyclists.

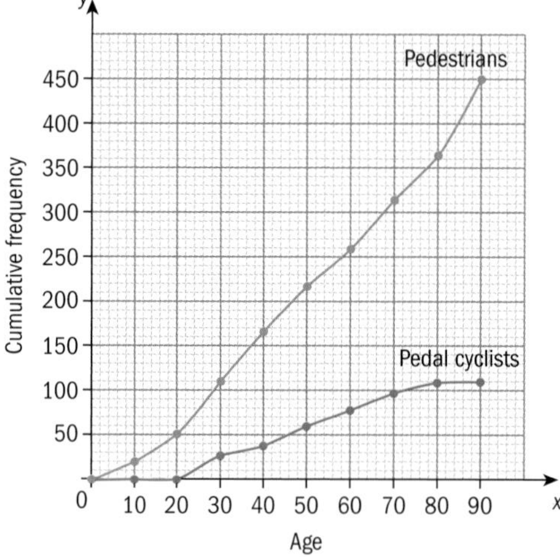

Use the same scale and draw one box-and-whisker diagram above the other.

Exploration 2

1 In your class, record 500 random numbers between 1 and 30 generated in the following way:

Use technology to produce random 3-digit numbers and record the last two digits of each, ignoring any over 30.

For example:

Random number	Action
307	record 7
951	51 > 30 so ignore
823	record 23
430	record 30

Construct a grouped frequency table with carefully chosen intervals. As a class, try different sized class intervals. Use technology to calculate the measures of central tendency.

2 Compare your results with other students who chose different class intervals. What are the differences?

Use the random number button on your calculator, a spreadsheet formula, or random number tables. You could find 15 random numbers each and combine them into one class set.

▶ Continued on next page

3 Discuss how choosing different class intervals affects the mean, median class and modal class.
4 Why would it not be sensible to represent this data in a stem-and-leaf diagram?
5 Represent the data in a cumulative frequency curve and box-and-whisker diagram.
6 Compare your results with others.

Reflect and discuss 4

If a set of data is grouped in equal class width intervals:

- Does the choice of class interval affect the estimate of the mean? What about the median class and modal class? Explain.
- Does the choice of class interval affect the cumulative frequency curve and box-and-whisker diagram? Explain.
- Discuss the advantages and disadvantages of having too few classes. Do likewise for too many classes.

D Misleading statistics

- How can real data ever be misleading?
- How do individuals stand out in a crowd?

Exploration 3

The table gives the ages of fathers at the birth of their first child, in Switzerland in 2013.

Age	Freq	Age	Freq	Age	Freq	Age	Freq	Age	Freq
≤ 12	0	22	388	32	5522	42	2203	52	189
13	0	23	598	33	5766	43	1773	53	146
14	0	24	958	34	5612	44	1487	54	128
15	0	25	1414	35	5516	45	1187	55	95
16	2	26	1815	36	5225	46	856	56	86
17	13	27	2365	37	4703	47	717	57	61
18	26	28	3073	38	4126	48	579	58	45
19	69	29	3792	39	3705	49	442	59	40
20	132	30	4382	40	3210	50	365	≥ 60	174
21	251	31	5108	41	2703	51	290		

1 Draw a grouped frequency table to represent this data. Use equal class width intervals $1 \le x \le 10$, $11 \le x \le 20$, etc. Assume that all the fathers were under 70. What will the highest class interval be?
2 Calculate an estimate for the mean and the five-point summary of the age of fathers from your table.

▶ Continued on next page

3 Peter uses the class intervals below so that the frequencies in all the classes are roughly equal.

$0 \le x \le 25$, $26 \le x \le 29$, $30 \le x \le 34$, $35 \le x \le 38$, $39 \le x \le 43$, $44 \le x \le 50$, $51 \le x \le 100$

Construct a grouped frequency table using his class intervals. Calculate an estimate for the mean and the five-point summary from this new table.

Reflect and discuss 5

- In Exploration 3, which grouping of the data gives a five-point summary that best represents the data?
- How should you group data so that it accurately represents a data set? Consider:
  - the number of classes
  - class width
  - the upper and lower boundaries of the class intervals
- Do you need to consider the size of the data set?

Summary

The **class width** is the difference between the maximum and the minimum possible values in a class interval.

When the class widths are equal, the **modal class** is the class interval containing the most data. It has the highest frequency.

For grouped data, you can find the class value that contains the median. Add a cumulative frequency column to the frequency table to find which class interval contains the $\left(\frac{n+1}{2}\right)$th value.

To calculate an estimate of the mean from a grouped frequency table:

- Find the midpoint or mid-interval value of each class interval by adding the upper and lower class boundaries and dividing by 2.
- Use the midpoint of each class interval to calculate the mean.

For grouped data the mean is defined as:

$$\bar{x} = \frac{\text{sum of (mid-interval value} \times \text{frequency)}}{\text{number of observations}}$$

An estimate for the range from a grouped frequency table is (upper bound of highest class interval) – (lower bound of lowest calls interval).

A cumulative frequency curve is a graph with the upper class boundaries plotted on the horizontal axis, and the cumulative frequencies on the vertical axis.

- For continuous data, for example: $35 < x \le 40$, $40 < x \le 45$, plot the cumulative frequencies against the upper class boundaries 40 and 45.
- For discrete data, for example: $35 \le x \le 40$, $41 \le x \le 45$, plot the cumulative frequencies against the point halfway between the upper bound of one interval (40) and next lower bound (41), at 40.5.

Objective: C. Communicating
v. organize information using a logical structure.

In the mixed practice you will be specifically looking at organizing information using a logical structure to represent the data given.

Mixed practice

1 The masses of 35 desert hedgehogs are listed here. All masses are to the nearest gram.

290	455	342	465	480	400	500
325	460	328	284	436	280	370
450	368	295	310	390	435	450
315	505	510	495	310	400	375
347	450	474	298	380	463	360

a **Construct** a grouped frequency table to represent the data.

b **State** the modal class.

c **Determine** the class interval that contains the median.

d **Calculate** an estimate for the mean mass.

2 The table shows learner drivers' marks in a hazard perception test.

Mark, m	Frequency
$16 \le m \le 27$	9
$28 \le m \le 39$	21
$40 \le m \le 51$	18
$52 \le m \le 63$	23
$64 \le m \le 75$	19

a **Estimate** the range of the marks.

b **Construct** a cumulative frequency table.

c **Construct** a cumulative frequency curve.

3 The tails of a random sample of 200 adult foxes were measured in cm. The results are represented in the cumulative frequency curve.

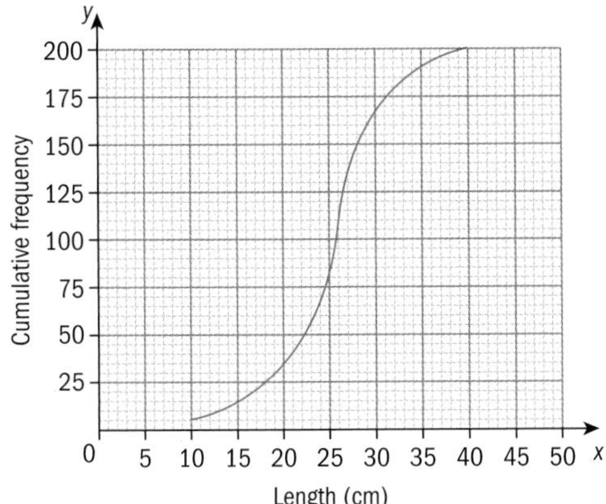

a **Estimate** the median length of fox tail in the sample.

b **Estimate** the interquartile range for the length of fox tails in the sample.

c Given that the shortest length was 11 cm and the longest 37 cm, **draw** and label a box-and-whisker plot for the data.

4 The box-and-whisker diagram shows the masses of 80 frogs at a zoo. **Construct** a cumulative frequency curve for the data.

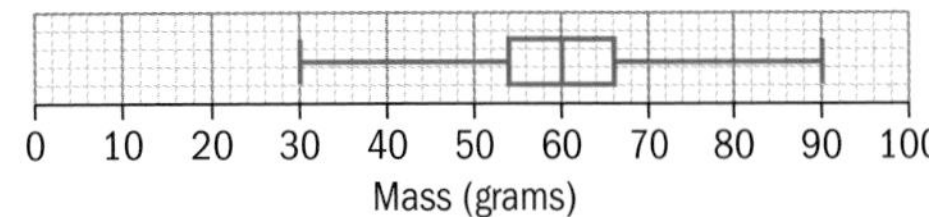

Problem solving

5 The speeds of cars passing a speed camera on a highway are recorded in this table.

Speed, v (km/h)	Number of cars
$v \le 60$	0
$60 < v \le 70$	10
$70 < v \le 80$	22
$80 < v \le 90$	61
$90 < v \le 100$	74
$100 < v \le 110$	71
$110 < v \le 120$	39
$120 < v \le 130$	17
$130 < v \le 140$	6

a **Calculate** an estimate for the mean speed of the cars.

b Here is a cumulative frequency table for the same data.

Speed, v (km/h)	Number of cars	Cumulative frequency
$v \le 60$	0	0
$v \le 70$	10	10
$v \le 80$	22	32
$v \le 90$	61	93
$v \le 100$	74	a
$v \le 110$	71	238
$v \le 120$	39	b
$v \le 130$	17	294
$v \le 140$	6	300

Write down the values of a and b.

c On graph paper, **construct** a cumulative frequency curve to represent this information.

d 25% of cars exceed the speed limit.
By calculating the upper quartile, **estimate** the speed limit on this stretch of highway.

6 The cumulative frequency curve shows the weights, in grams, of a selection of oranges.

a Use the graph to **estimate**:

i the median

ii the upper quartile.

Give your answers to the nearest gram.

b 10% of the oranges weigh more than x grams. **Find** x.

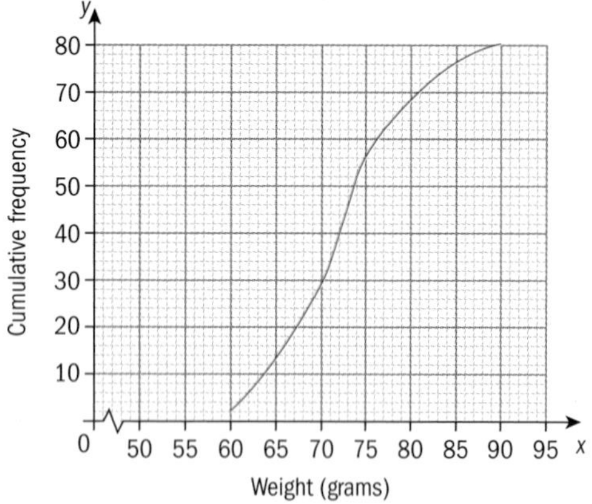

7 A fisherman catches 200 halibut. The table shows the lengths of these halibut to the nearest cm.

Length, x (cm)	Frequency
$10 < x \le 15$	24
$15 < x \le 20$	38
$20 < x \le 25$	53
$25 < x \le 30$	39
$30 < x \le 35$	29
$35 < x \le 40$	10
$40 < x \le 45$	7

a **Calculate** an estimate for the mean length of the halibut.

b The cumulative frequency diagram shows the lengths of the halibut.

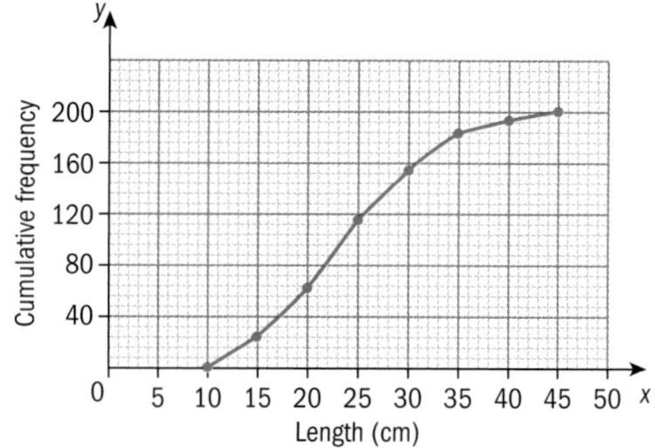

Estimate the interquartile range.

c The fisherman returns any fish smaller than 20 cm to the river. **Calculate** an estimate for the number of fish he returns.

d Fish greater than or equal to 20 cm but less than 28 cm are classified as small fish. Fish that are 28 cm or longer are classified as large fish.

Estimate the number of fish in each category.

e The fisherman sells small fish for \$6 and large fish for \$10. **Estimate** how much money he will earn if he sells all the fish.

8 The cumulative frequency curves show the ages of foreign male and foreign female residents in Switzerland in 2013.

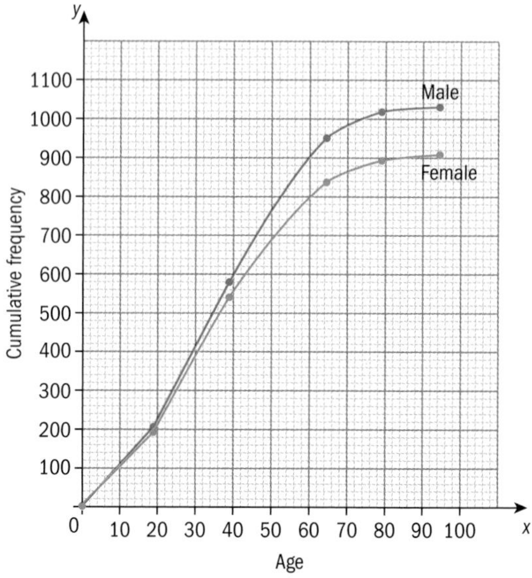

By constructing a double box-and-whisker plot, **compare** the ages of the foreign male and foreign female residents.

Review in context

Globalization and sustainability

1 A marine biologist is studying rainbow trout in a local river. To the nearest centimeter, she records the lengths of a sample of 100 rainbow trout from the river.

Rainbow trout	
Length, x (cm)	Frequency
$25 < x \leq 27$	1
$27 < x \leq 29$	5
$29 < x \leq 31$	9
$31 < x \leq 33$	21
$33 < x \leq 35$	29
$35 < x \leq 37$	19
$37 < x \leq 39$	16

a **Construct** a cumulative frequency table for this data.

b **Draw** a cumulative frequency curve.

c **Use** the cumulative frequency curve to write a five-point summary for the data.

d Copy and complete this statement:

75% of the rainbow trout are over ____ cm in length.

e The rainbow trout feed on smaller fish of other species, so the biologist records the lengths of 100 female fish and 100 male fish of other species, from the same river.

These are shown in the box-and-whisker diagrams.

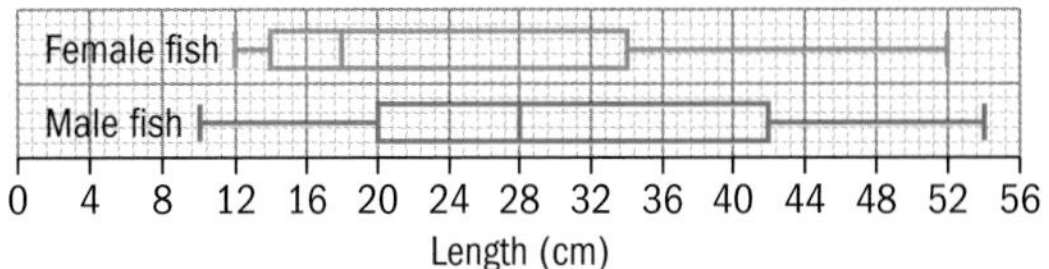

i **Estimate** the number of female fish that are smaller than 75% of the rainbow trout in the river.

ii **Estimate** the number of male fish that are smaller than 75% of the rainbow trout.

iii Hence **estimate** the percentage of the fish of other species in the river that 75% of the rainbow trout can feed on.

Reflect and discuss

How have you explored the statement of inquiry? Give specific examples.

Statement of Inquiry:

How quantities are represented can help to establish underlying relationships and trends in a population.

How did *that* happen?

Global context: Identities and relationships

RELATIONSHIPS

Objectives

- Drawing a scatter diagram for bivariate data
- Drawing a line of best fit (regression line) by eye
- Understanding and interpreting the correlation between two sets of data
- Using technology to obtain the equation of a line of best fit

Inquiry questions

F
- How can you describe a relationship between two sets of data?
- How do you draw and use a line of best fit?

C
- How does the way data is represented affect our ability to make predictions?

D
- Does correlation indicate causation?
- Do I want to be like everybody else?

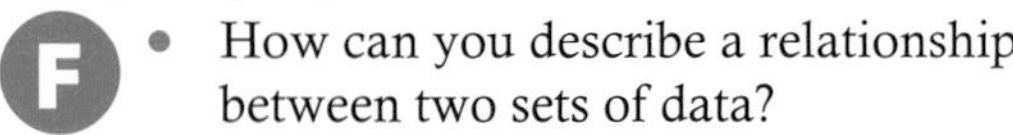

ATL **Information literacy**

Understand and use technology systems

Statement of Inquiry:

Generalizing and representing relationships can help to clarify trends among individuals.

4.3

12.3

E4.1

You should already know how to:

• represent data by plotting points on the Cartesian plane	**1** The table shows the cost of different weights of potatoes. Plot points on a Cartesian plane to represent this data. Weight (kg) | Price (£) 0.5 | 0.40 3.4 | 2.70 5 | 4 2.1 | 1.70 4.5 | 3.60 3.8 | 2.90 2 | 1.60
• calculate the mean of a set of data	**2** Calculate the mean of this set of numbers: 2, 4, 6, 2, 7, 1, 9, 4, 6, 2
• identify an outlier in a set of data	**3** Determine whether any of these data values can be classified as outliers. Explain your answer. 1, 6, 6, 6, 8, 8, 9, 9, 11, 15, 20

Introduction to correlation

- How can you describe a relationship between two sets of data?
- How do you draw and use a line of best fit?

Investigating the relationship between variables is a key topic in Statistics. If you can identify a relationship between two variables, you can often use the value of one variable to predict the value of the other.

Exploration 1

The table shows 12 students' marks for Mathematics and Physics tests. The maximum possible score on each test was 8.

Student	A	B	C	D	E	F	G	H	I	J	K	L
Mathematics	8	6	7	5	4	8	5	6	7	7	3	2
Physics	8	7	6	4	4	7	5	5	7	6	3	3

1 Draw a graph with Mathematics scores on the x-axis and Physics scores on the y-axis. Use a sensible scale, and label the axes. Plot points (Mathematics score, Physics score) to represent each student's scores. For example, for student A, plot the point (8, 8). For student B, plot (6, 7), and so on. Do not join the points.

2 Explain whether your graph suggests that there might be a relationship between the students' Mathematics and Physics marks?

▶ Continued on next page

3 Describe any patterns you see in the data. Determine whether these patterns apply to every student in the class.

4 The teacher claimed that students who got a good mark in one subject also got a good mark in the other. Is this claim justified?

5 Draw a straight line on your graph that best shows the relationship between the marks for Mathematics and Physics. Which way does the line slope? Discuss how close the points are to the line.

A scatter diagram shows the relationship between two quantitative variables. The independent variable is represented on the x-axis and the dependent variable on the y-axis. Data from two variables is called **bivariate data**.

Scatter diagrams are also called scatter plots or scatter graphs.

Correlation is a measure of the association between two variables.

When one variable increases as the other increases, there is positive correlation.

When one variable decreases as the other increases, there is negative correlation.

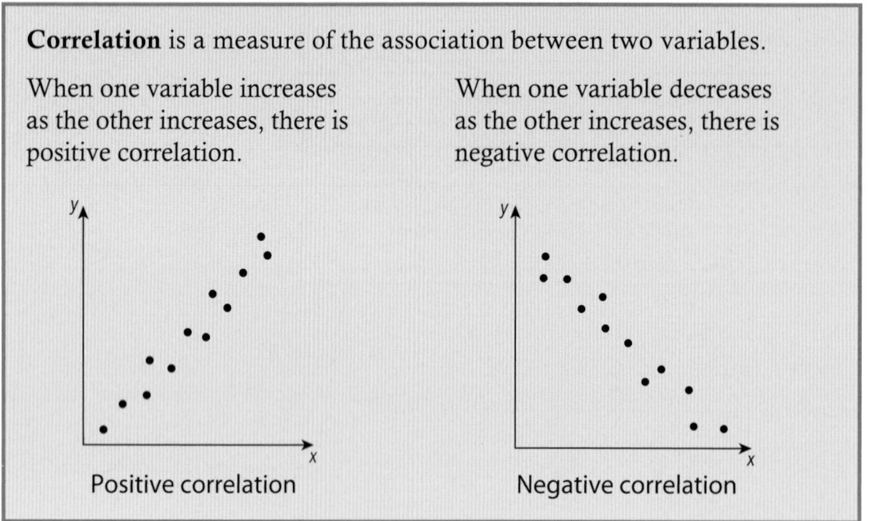

When the plotted points are close to a straight line drawn through the middle of the points, you say there is a linear correlation between the variables.

The closer the points are to a straight line, the stronger the correlation between the variables.

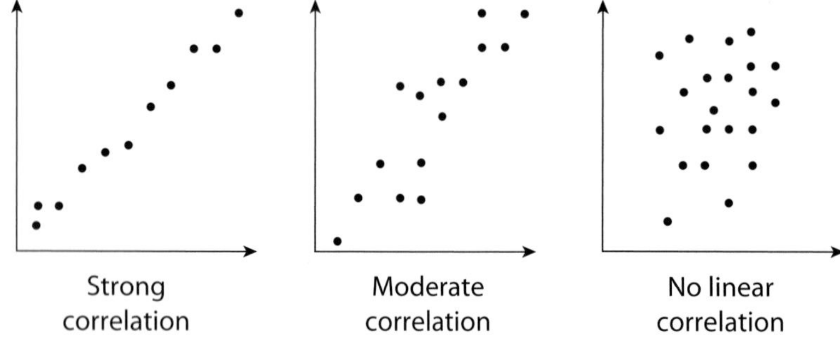

The scatter graph you drew in Exploration 1 shows strong linear correlation between Mathematics and Physics scores. The higher the Mathematics score, the higher the Physics score.

Practice 1

For each scatter plot, describe the form, direction and strength of the association between the variables. Interpret your description in the context of the data.

'Form' is linear or non-linear. 'Direction' is positive or negative.

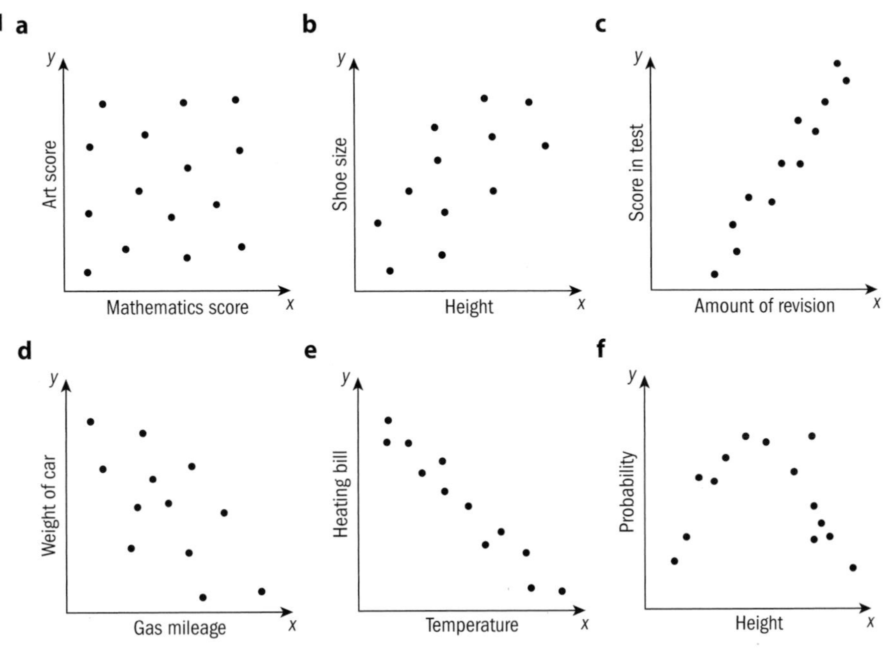

> A line of best fit is a straight line drawn through the middle of a set of data points so that they are equally distributed on either side of the line. The line of best fit passes through the point $(\overline{x}, \overline{y})$ where $\overline{x}$ is the mean of the x-values and $\overline{y}$ is the mean of the y-values.

The line of best fit can also be called a *trend line*.

You can draw a line of best fit by eye. A GDC or graph plotting software draws a similar line, called the regression line, by calculating the points mathematically.

You can use of the line of best fit to predict a y-value when you know an x-value, and vice versa.

Example 1

The table shows the average daily temperature in °C for a week and the daily sales of ice cream in dollars.

	Mon	Tue	Wed	Thu	Fri	Sat	Sun
Temperature, x (°C)	17	22	18	23	25	19	22
Sales, y ($)	408	445	421	544	614	412	522

a Draw a scatter diagram of this data and draw a line of best fit through the points.

b Use the line of best fit to predict the sales of ice cream on a day when the temperature is 20°C.

a $\bar{x} = \frac{17+22+18+23+25+19+22}{7} = 20.9$ — Calculate the mean temperature ($\bar{x}$).

$\bar{y} = \frac{408+445+421+544+614+412+522}{7} = 480.9$ — Calculate the mean sales ($\bar{y}$).

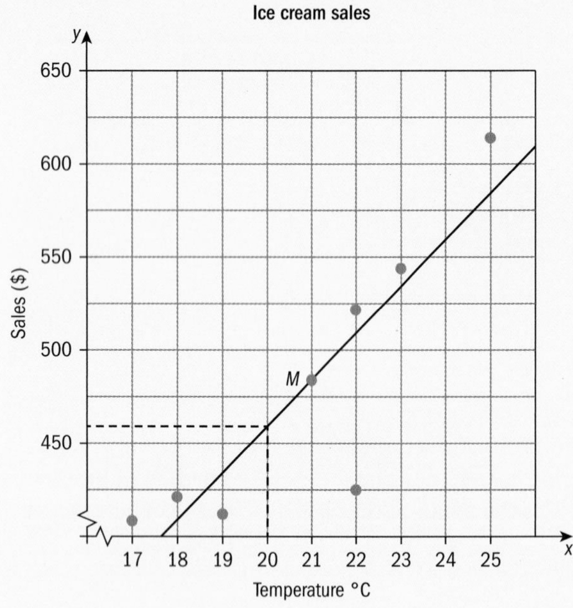

Plot the points from the table and the point $M = (\bar{x}, \bar{y})$.

Draw a line through M and through the middle of the other points.

Tip

Use a transparent ruler so you can see all the points above and below it. Pivot the ruler through the point $(\bar{x}, \bar{y})$ to find the best position.

b When the temperature is 20°C, predicted sales are $460. — Draw a line from $x = 20$ and read off the y value.

To draw a line of best fit by eye:

- Calculate the mean of x ($\bar{x}$) and the mean of y ($\bar{y}$).
- Plot the point $M = (\bar{x}, \bar{y})$ on the scatter diagram.
- Draw a line through $M(\bar{x}, \bar{y})$ and through the middle of the other points. The points should be evenly distributed above and below the line.

Practice 2

1 The table shows data on 10 students:

- Reaction distance, in a test to measure visual responses
- Height, in meters
- Weight, in kilograms
- Time, in seconds, taken to fasten their shoelaces.

	A	B	C	D	E	F	G	H	I	J
Reaction distance (cm)	5.3	4.6	7.7	3.7	4.6	3.6	4.2	4.8	5.3	4.6
Height (m)	1.64	1.50	1.67	1.75	1.45	1.35	1.87	1.67	1.56	1.70
Weight (kg)	60	60	64	76	48	45	75	75	52	60
Fastening time (s)	9.1	10.5	13.4	9.6	10.5	9.2	8.8	8.5	10.4	6.3

a Draw four separate scatter diagrams:

i Plot height on the x-axis and weight on the y-axis.

ii Plot weight on the x-axis and reaction distance on the y-axis.

iii Plot reaction distance on the x-axis and fastening time on the y-axis.

iv Plot height on the x-axis and fastening time on the y-axis.

b Draw a line of best fit on each scatter diagram.

c Comment on the form, direction and strength of the correlation in each scatter diagram.

d Use the line of best fit to estimate the weight of a person who is 1.68 m tall.

e Use the line of best fit to estimate the reaction distance of a student who took 12 seconds to fasten their shoelaces.

f Explain which of your graphs would give the most accurate predictions.

2 The table shows information on road deaths and vehicle ownership in ten countries.

Country	Vehicles per 100 population	Road deaths per 100 000 population
UK	31	14
Belgium	32	29
Denmark	30	22
France	47	32
Germany	30	25
Ireland	19	20
Italy	36	21
Netherlands	40	22
Canada	47	30
USA	58	35

a Draw a scatter diagram with vehicles per 100 population on the x-axis and road deaths per 100 000 population on the y-axis. Does the graph show any correlation?

b Draw a line of best fit to represent the pattern in the data.

c The road death data for Switzerland is missing. Switzerland has 25 vehicles per 100 population. Use your line of best fit to estimate the number of road deaths per 100 000 population in Switzerland.

C Interpreting correlation

- How does the way data is represented affect our ability to make predictions?

ATL

You can use a GDC, graphing software or a spreadsheet to plot scatter diagrams, add a line of best fit and produce an equation for the line that models the relationship between the variables.

Technology uses linear regression to find the line of best fit. This method finds the vertical displacements of the points from the line and then uses a process called 'least squares' to find a line that will predict y for a given value of the independent variable x.

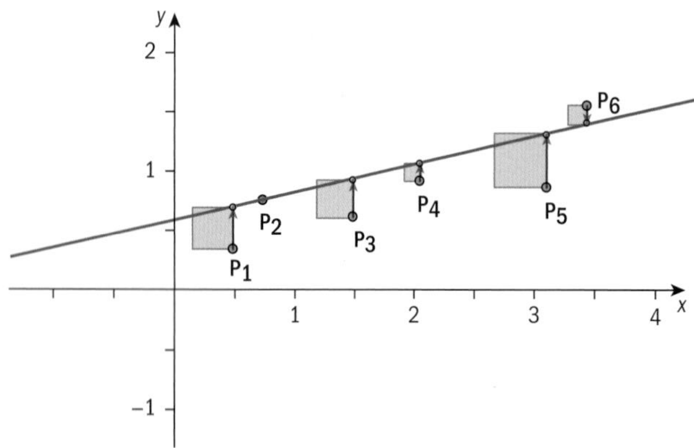

Each square in the diagram is the square of the vertical distance from a point to the line. The least squares method finds the line that gives the minimum value for the sum of these squares. With some GDCs or graphing software you can try to find the line of best fit yourself by plotting a line through $(\bar{x}, \bar{y})$, showing the squares and minimizing the sum of their areas.

Example 2

The table gives the weight and height of 10 students.

a Use technology to draw a scatter diagram and find the line of best fit.

b Comment on the form, direction and strength of the correlation.

c Use the graph to estimate the weight of a 162 cm tall student.

Name	Height (cm)	Weight (kg)	Name	Height (cm)	Weight (kg)
Alex	168	77	Frida	159	87
Bea	149	45	George	195	94
Chip	173	65	Harry	144	53
Dave	166	59	Inge	176	65
Erin	185	79	Jepe	180	87

a

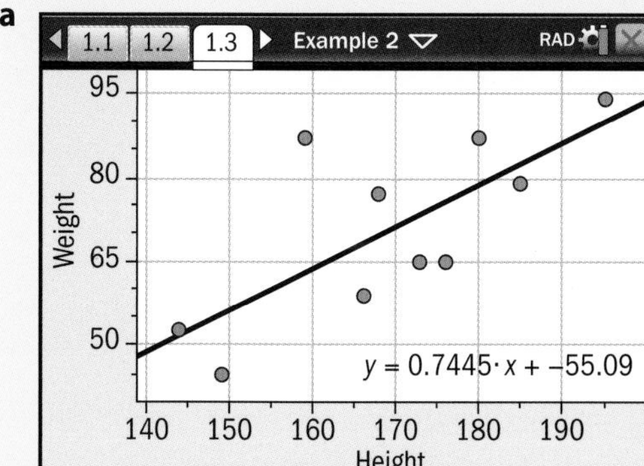

User your GDC to create a line of best fit from the data.

b There is moderate positive linear correlation between the heights and weights.

c Using the graph, a height of 162 cm would correspond to a weight of approximately 65 kg.

Practice 3

ATL

1 The table gives the height and weight of 14 adult mountain zebras.

Height x (m)	1.37	1.17	1.2	1.34	1.42	1.42	1.37	1.48	1.51	1.23	1.57	1.29	1.30	1.44
Weight y (kg)	257	171	185	214	315	271	242	329	314	185	356	228	230	285

a Use technology to draw a scatter diagram.

b Find the equation of the line of best fit.

c Comment on the correlation between the height and weight of the zebras.

d Use the line of best fit to predict the weight of an adult mountain zebra with height 1.38 m.

2 There are twelve 15-year-old girls in an athletics club. The table shows their leg length and time to run 1000 m.

Leg length x (cm)	82	91	85	93	86	95	70	87	90	75	91	80
Time to run y (s)	245	210	243	220	240	225	250	243	230	248	220	240

a Using technology, draw a scatter diagram.
b Find the equation of the line of best fit.
c Comment on the correlation between leg length and the time taken to run 1000 m.
d Use the line of best fit to predict the time to run 1000 m for a student with leg measurement 84 cm.

3 The table shows the diameter of the cylinder in a washing machine and the cost of the machine when purchased new at the same shop.

Size, x (cm)	33	36	41	48	52	55	62	66
Cost, y (\$)	230	350	400	550	550	600	620	700

a Plot the points on a scatter diagram.
b Find the equation of the line of best fit.
c Use your line of best fit to estimate the cost of a washing machine with a cylinder diameter of 50 cm.

D Making predictions

- Does correlation indicate causation?
- Do I want to be like everybody else?

An outlier can arise in a variety of ways.

It could be due to human error, categorized by:

- Measurement error (wrong units)
- Clerical error (writing the results incorrectly or misunderstanding the question)
- Instrumental error (faulty instruments)

However, it could be a legitimate point which caused these results.

In 4.1 you used the IQR to determine the outliers; now it is your decision whether or not to include the outliers in the data set.

If you think the outlier is due to human error, remove the data point; if you think it is a legitimate point, it must be kept in.

An **outlier** is an anomaly in the data. To identify outliers in bivariate data, look at the distance of the data point from the line of best fit.

Exploration 2

The Olympic Games are held every four years. The table shows the gold medal distance for the men's long jump event over a period of 28 years.

Year	1956	1960	1964	1968	1972	1976	1980	1984
Long jump (m)	7.83	8.12	8.07	8.90	8.25	8.35	8.54	8.54

Use 'number of years after 1956' on the x-axis. (0, 4, 8, ...)

1 Use a GDC or graphing software to draw a scatter diagram and find the line of best fit for the data.

2 Describe the relationship that the graph shows.

3 One jump stands out from the others. Is it much further from the line of best fit than the other points? Is it an outlier?

4 How does this outlier affect the line of best fit and any predictions that are based on it?

5 Find the line of best fit without the anomalous jump. Use this line to estimate the gold medal long jump distance for that year.

Discover more about this jump by researching the Olympic long jump gold medal for this year, in particular search for a video called Bob Beaman's Long Olympic Shadow by *The New York Times*.

A scatter graph may show a relationship between two variables (correlation), but it does not show that a change in one variable causes a change in the other (causation). For example, the number of pirates and global average temperature measured over a period of years have a negative correlation. But it is unlikely that the fall in the number of pirates has caused higher temperatures. Demonstrating cause and effect is much harder than showing a strong linear correlation.

Reflect and discuss 1

- In Exploration 2, you used the line of best fit to predict a value within the range of data you were given. Was it appropriate to do this?
- The modern Olympics began in 1896. Could you use your line of best fit to predict the winning jump in this year? What about the distance jumped in the original Greek Olympic Games 1500 years earlier, or Olympic Games 100 years in the future?

You cannot know that the relationship you can see between two data sets will still hold for higher or lower data values than those given.

Predicting values inside the range of the given data is called **interpolation**. Predicting values outside the range of the given data is called **extrapolation**.

In general, interpolation is more likely to give an accurate prediction than extrapolation.

Exploration 3

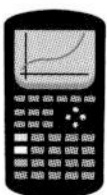

The table shows the number of visitors to the Bahamas for a seven-year period and the revenue from registered nail salons in the US during the same time period.

	2008	2009	2010	2011	2012	2013	2014
Visitors to the Bahamas (millions)	4.6	4.6	5.3	5.6	5.9	6.2	6.3
Revenue from nail salons (billions of US dollars)	6.3	6.0	6.2	7.3	7.5	8.2	8.5

1 Draw a scatter diagram to illustrate these results, with visitors to the Bahamas on the x-axis and revenue from nail salons on the y-axis.

2 Describe the correlation between the two variables.

3 Draw a line of best fit and use it to predict the revenue from nail salons when there are 5 million visitors to the Bahamas.

4 Does it make sense to make predictions from the line of best fit?

Exploration 3 shows that relationships can occur between the most obscure variables. Obviously there is an association between the visitors to the Bahamas and the revenue of nail salons, but an increase in one does not *cause* an increase in the other. They are both dependent on another factor, time.

You cannot use correlation to show that two variables are related to each other.

ATL

Practice 4

1 The average prices of a liter of milk in January in the UK tabulated for six consecutive years are given here.

a Use technology to draw a scatter diagram and line of best fit. Put the dependent variable on the x-axis, with January 2009 at zero. Then January 2010 is 12 months, January 2011 is 24 months, etc.

b Use the line of best fit to estimate the average price of milk in:

i July 2009

ii July 2012

iii January 2015

c In January 2015, the average price of milk was 24.46 pence per liter. Compare this to your prediction. Explain any differences.

Month	Pence per liter
Jan 2009	23.73
Jan 2010	24.67
Jan 2011	27.36
Jan 2012	28.08
Jan 2013	31.64
Jan 2014	31.52

2 Discuss what the correlation analyses below show. Can you draw any logical conclusions?

a Over a period of 10 years, the correlation between the annual number of human births and the population of Labradoodles in Denmark was strong and positive.

b The correlation between the amount of time a person spent looking at a computer screen each day and the quality of sleep that they reported that night was medium and negative.

c Between the years 2005–2015 there was a strong, positive linear correlation between the number of TV licenses bought in the UK and the number of young offenders sent to prison.

3 Do you think that tall people have longer arms than short people? The table here lists height and arm span for 10 students.

Arm span (cm)	155	156	160	165	170	177	177	188	196	200
Height (cm)	182	162	180	166	167	173	176	182	184	186

a Use technology to draw a scatter diagram with arm span on the x-axis and height on the y-axis. Find the line of best fit.

b Use the line of best fit to predict (interpolate) the height of a student whose arm span is 171 cm.

c Are there any outliers? If so, remove them from the list and find the new line of best fit. Use this new line to predict the height of a student whose arm span is 171 cm.

d Discuss the differences between **b** and **c**.

Problem solving

4 Students weigh sets of drawing pins and record the number of pins and their total weight in grams. Here is their data:

Number of pins	7	13	17	20	24	24	28	29	31	33
Weight (g)	5	7	11	10	14	15	16	9	11	20

The students work in two groups to calculate the mean weight of a drawing pin.

a Group 1 finds the total weight of the drawing pins and divides by the number of drawing pins. Use this method to calculate the mean weight.

b Group 2 draws a scatter diagram. Use technology to draw a scatter diagram with number of pins on the x-axis and weight on the y-axis. Find the line of best fit.

c Group 2 decide that two of the weights are unreliable. Remove the unreliable data and find a new line of best fit. Use it to find the mean weight of a drawing pin.

d Determine which group's method gives the most reliable estimate of the mean weight of a drawing pin.

Summary

A scatter diagram shows the relationship between two quantitative variables. The independent variable is represented on the x-axis and the dependent variable on the y-axis. Data from two variables is called **bivariate data**.

Correlation is a measure of the association between two variables. When one variable increases as the other increases, there is positive correlation. When one variable decreases as the other increases, there is negative correlation.

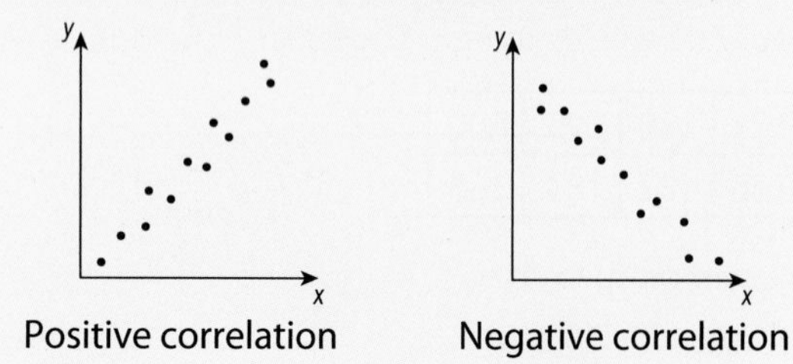

Positive correlation Negative correlation

The closer the points are to a straight line, the stronger the correlation between the variables.

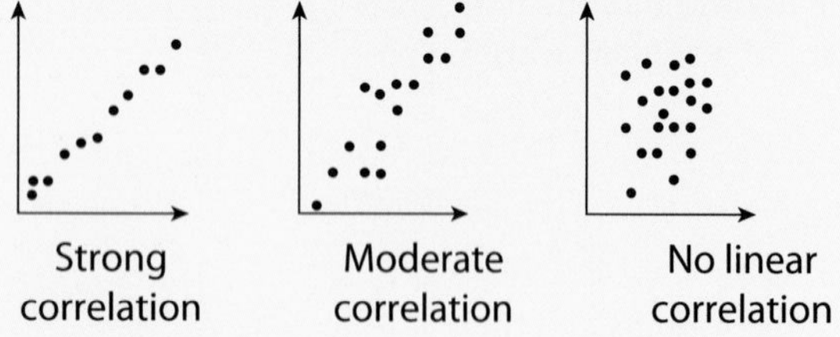

Strong correlation Moderate correlation No linear correlation

A **line of best fit** is a straight line drawn through the middle of a set of data points so that they are equally distributed on either side of the line. The line of best fit passes through the point $(\bar{x}, \bar{y})$ where $\bar{x}$ is the mean of the x-values and $\bar{y}$ is the mean of the y-values.

To draw a line of best fit by eye:

- Calculate the mean of x $(\bar{x})$ and the mean of y $(\bar{y})$.
- Plot the point $M(\bar{x}, \bar{y})$ on the scatter diagram.
- Draw a line through M and through the middle of the other points. The points should be evenly distributed above and below the line.

An **outlier** is an anomaly in the data. To identify outliers in bivariate data, look at the distance of the data point from the line of best fit.

Predicting values inside the range of the given data is called **interpolation**. Predicting values outside the range of the given data is called **extrapolation**. In general, interpolation is more likely to give an accurate prediction than extrapolation.

Mixed practice

1 The table gives the length and width of 10 oak leaves that fell to the ground.

Length (mm)	Width (mm)
103	38
146	44
119	38
149	53
89	36
135	38
151	51
147	43
123	33
128	42

a Represent the data on a scatter diagram, with length on the x-axis and width on the y-axis.

b **Comment** on the correlation between length and width of the leaves.

c By finding the mean length and mean width, **draw** a line of best fit.

Problem solving

2 Two of these statements about correlation contain a mistake. **Explain** what the mistakes are and how the third statement could be correct.

a 'There is a high correlation between the income of workers and their gender.'

b 'There is a strong correlation between the age of Miss America and the number of murders using hot objects between 1999 and 2009.'

c 'There is a positive correlation between the average speed and the time taken by a train between two stations.'

Objective: **D.** Applying mathematics in real-life contexts
ii. select appropriate mathematical strategies when solving authentic real-life situations

In this review you will have to use the mathematical strategies of drawing scatter diagrams, using technology to extract information and determining the reasons for outliers, in real-life situations.

Review in context

Identities and relationships

1 Florence's parents were concerned that she seemed short for her age and they recorded her height over a 36-month period:

Age (months)	36	48	51	58	66	72
Height (cm)	86	90	91	93	94	95

a **Draw** a scatter diagram of this data.

b **Use** this scatter diagram to **estimate** her height at age 42 months.

c Could you use this data and the scatter diagram to **predict** her height at age 18?

2 There are 12 students training for a charity 10 km run. The table shows the average number of hours of training per week and the time taken to complete the run.

Training time (hours)	Time to complete run (minutes)
10	56
9	54
13	53
4	62
26	42
7	66
11	54
6	66
7	68
22	39
5	70
10	67

a **Comment** on the correlation between training time and the time to complete the run.

b **Estimate** how long the run would take a student who trains 18 hours per week.

c **Determine** whether you could use this data to **predict** how long the run would take a student who trained 50 hours per week.

Reflect and discuss

How have you explored the statement of inquiry? Give specific examples.

Statement of Inquiry:

Generalizing and representing relationships can help to clarify trends among individuals.

What are the chances?

Global context: Identities and relationships

LOGIC

Objectives

- Representing sample spaces in tables, lists and diagrams
- Drawing tree diagrams, Venn diagrams and two-way tables
- Calculating probabilities from Venn diagrams and two-way tables
- Using tree diagrams to calculate probabilities with and without replacement
- Understanding informal ideas of randomness

Inquiry questions

F
- What are the different ways of representing a sample space?
- How do you calculate the probability of an event?

C
- What are the advantages and disadvantages of the different probability representations?

D
- Does randomness affect the decisions we make?

ATL **Communication**

Understand and use mathematical notation

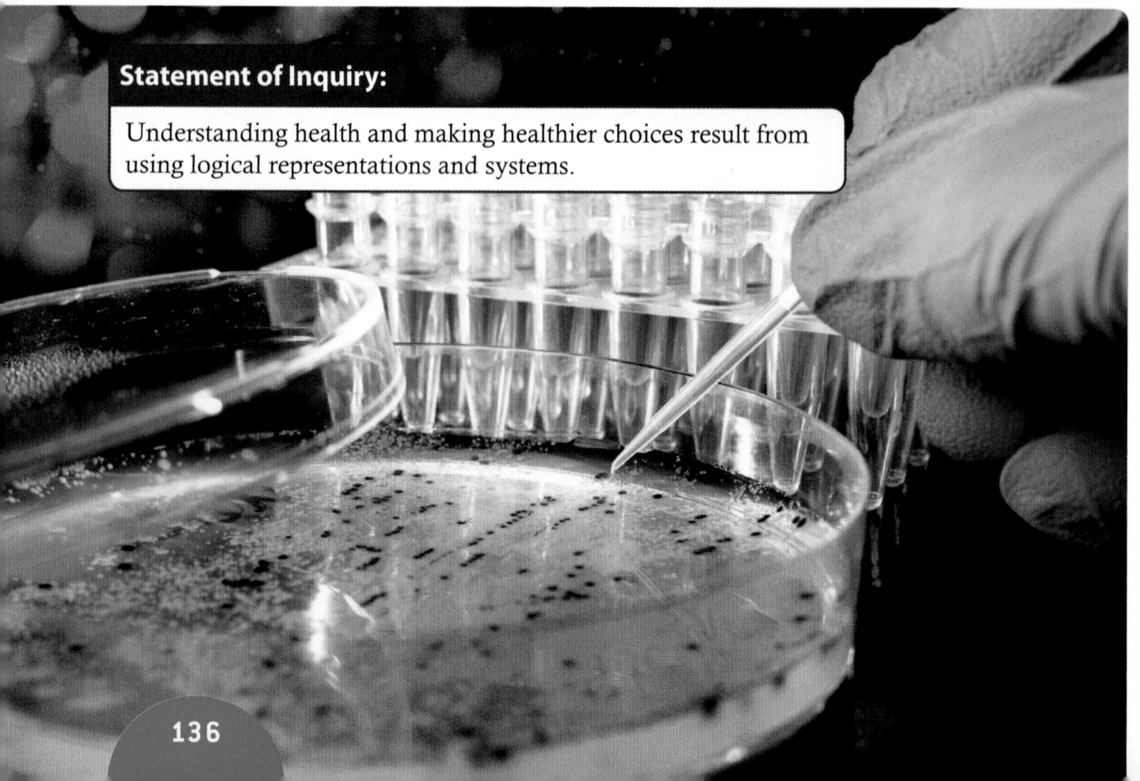

Statement of Inquiry:

Understanding health and making healthier choices result from using logical representations and systems.

4.4

15.1

E15.1

You should already know how to:

• calculate simple probabilities	**1** This spinner is spun once. What is the probability that it lands on **a** blue **b** yellow **c** blue or white **d** blue or yellow **e** not yellow?
• use set notation, specifically the complement of a set	**2 a** Using set notation, write the set P of prime numbers less than 10. **b** Let the Universal set $\mathbb{U} = \{1, 2, 3, 4, 5, 6, 7, 8, 9, 10\}$. Write down the elements of P'. **c** Write down $n(P)$ and $n(P')$.
• draw a Venn diagram	**3** In a group of 24 children, 10 of them like bananas, 16 of them like apples and 7 children like both. Draw a Venn diagram to represent this information.

Representing outcomes

- What are the different ways of representing a sample space?
- How do you calculate the probability of an event?

The mathematics behind events involving chance is called Probability. By looking for patterns as events are repeated you will begin to see how to predict and calculate probabilities.

Exploration 1

A new restaurant in town runs a promotion to generate business. Before a person pays their bill, they are given a sealed envelope with a message inside that either says 'Your meal today is free' or 'Sorry, better luck next time'.

The restaurant owners claim that the chance of a winning message is 1 in 6.

During a one-hour period on the first day, eight customers pay their bill, and three of them get a winning message.

To test that the restaurant's claim is correct, you can set up a simulation. Assume that they are correct and a winning message indeed appears in 1 out of every 6 sealed envelopes. The probability of winning is therefore the same as rolling a six on a fair die.

1 Roll a die 8 times to simulate 8 people paying their bills.

2 Record the number of 6s you roll.

> The probability of 1 in 6 is true for any number on a standard die; you could record 3s or 4s, as long as you were consistent.

▶ Continued on next page

3 Repeat this process 30 times. Combine the results for others in a spreadsheet.

Based on the combined results of the simulation:

4 Calculate the percentage of outcomes where 8 bills paid resulted in 3 winners.

5 Calculate the percentage of outcomes with no winners.

6 Deduce the most common number of winners in 8 bills paid.

7 Discuss whether the 8 people who had 3 winners were just lucky. Do you think the restaurant printed too many winning messages?

Your GDC or other software can run simulations that perform the experiment thousands or even millions of times. Try this and compare the results with your findings in Exploration 1.

Reflect and discuss 1

In Exploration 1:

- Aside from rolling a die, what other way(s) could you have used to simulate the probability of 1 winning message in 6?
- What number of winners from 8 people is most likely?

Simulation is a powerful tool to investigate the probabilities of events and predict what might happen in the future. You can also find probabilities by constructing a sample space for the experiment or situation.

The sample space S, is a representation of the complete set of all possible outcomes from an experiment. It can be a list, a table or a diagram.

When you toss a coin, there are two possible outcomes: Heads (H) and Tails (T).

ATL

Describing the sample space using set notation gives: $S = \{H, T\}$.

H and T are the only possible outcomes so they cover the whole sample space.

For any sample space S, $P(S) = 1$, because for each experiment one of the outcomes in the sample space must occur.

In Example 2, A and A' together cover the whole sample space, because either A occurs or it does not.

So $P(A) + P(A') = P(S) = 1$

Using the probabilities in Example 2: $P(A) + P(A') = \frac{1}{6} + \frac{5}{6} = 1$

So $P(A') = 1 - P(A) = 1 - \frac{1}{6} = \frac{5}{6}$

$P(A)$ represents the probability of event A occurring. $P(A')$ is the probability of A not occurring.

$$P(A) + P(A') = 1$$

Example 1

Two fair 6-sided dice are rolled. One die is green and the other is orange.

Represent all the possible outcomes as:

a a list

b a table

c a diagram.

a $S = \{(1,1),(2,1),(3,1),(4,1),(5,1),(6,1),(1,2),(2,2),(3,2),$
$(4,2),(5,2),(6,2),(1,3),(2,3),(3,3),(4,3),(5,3),(6,3),$
$(1,4),(2,4),(3,4),(4,4),(5,4),(6,4),(1,5),(2,5),(3,5),$
$(4,5),(5,5),(6,5),(1,6),(2,6),(3,6),(4,6),(5,6),(6,6)\}$

As a set

b

	1	2	3	4	5	6
1	(1,1)	(1,2)	(1,3)	(1,4)	(1,5)	(1,6)
2	(2,1)	(2,2)	(2,3)	(2,4)	(2,5)	(2,6)
3	(3,1)	(3,2)	(3,3)	(3,4)	(3,5)	(3,6)
4	(4,1)	(4,2)	(4,3)	(4,4)	(4,5)	(4,6)
5	(5,1)	(5,2)	(5,3)	(5,4)	(5,5)	(5,6)
6	(6,1)	(6,2)	(6,3)	(6,4)	(6,5)	(6,6)

As a table

The table is the most common representation of a sample space.

c

As a diagram

Practice 1

Objective C: Communicating
iii Move between different forms of mathematical representation

Make sure you list outcomes systematically and include all the possible outcomes in all your representations of sample spaces.

1 The Scandinavian board game Daldøs uses 4-sided (tetrahedral) dice. The roll of one such die results in a score of I, II, III or IV.

Represent the sample space for rolling two of these dice as a list and as a table.

2 A fair coin is tossed three times.

Define the sample space for this experiment.

For Q2, make a list of the possible outcomes.

3 Copy and complete this sample space table for the sum of the scores from rolling two 4-sided dice, each with faces numbered 1, 2, 3 and 4.

		2nd die			
1st die		1	2	3	4
	1	2			
	2			5	
	3				
	4			7	

In 2009 a woman threw a pair of dice 154 times without throwing a 7, setting a new world record. The chances of this are 1 in 1.56 trillion.

4 A street vendor sells two types of sandwich – a South African Gatsby or a Greek Gyro – and three types of fruit – apple, orange or banana. List the sample space for the possible choices of a sandwich and a piece of fruit.

5 A tetrahedral die (4-sided) and a normal die (6-sided) are rolled simultaneously. Construct a sample space to represent the possible outcomes.

6 Each of two boxes contains five cards numbered 1 to 5. A card is drawn from each box. Draw a sample space to represent the possible outcomes.

7 There are two sets of five cards. One contains the numbers 0 to 4 and the other contains the numbers 2 to 6. You draw a card from each set and multiply the numbers together. Draw a sample space to represent this information.

Problem solving

8 The sample space for an experiment is:

$S = \{(1, \text{H}), (2, \text{H}), (3, \text{H}), (4, \text{H}), (1, \text{T}), (2, \text{T}), (3, \text{T}), (4, \text{T})\}$

Describe the experiment.

Representing sample spaces allows you to count specific outcomes.

An **event** is a subset of the possible outcomes listed in the sample space.

From the sample space of scores on the two dice in Example 1 you can see how many ways the event 'getting the same number on both dice' can occur.

Games of chance have been played since ancient times; archaeologists have discovered dice in Turkey dating from 3000 BCE and Mahjong tiles dating from 500 BCE in China.

ATL

Use capital letters for events, like A, B, C and define the event. For example, A = getting the same number on both dice.

$$\text{Probability of event } A = \frac{\text{number of ways event } A \text{ can occur}}{\text{total number of possible outcomes}}$$

$$\text{P}(A) = \frac{n(A)}{n(S)}$$

Example 2

You roll two 6-sided dice. Calculate the probability of getting:

a the same number on both dice

b two different numbers.

a A = getting the same number on both dice. — Define the event.

	1	2	3	4	5	6
1	(1,1)	(1,2)	(1,3)	(1,4)	(1,5)	(1,6)
2	(2,1)	(2,2)	(2,3)	(2,4)	(2,5)	(2,6)
3	(3,1)	(3,2)	(3,3)	(3,4)	(3,5)	(3,6)
4	(4,1)	(4,2)	(4,3)	(4,4)	(4,5)	(4,6)
5	(5,1)	(5,2)	(5,3)	(5,4)	(5,5)	(5,6)
6	(6,1)	(6,2)	(6,3)	(6,4)	(6,5)	(6,6)

Identify the outcomes with the same number on both dice.

$\text{P}(A) = \frac{6}{36} = \frac{1}{6}$

b A' = getting different numbers on the two dice. — The outcomes with two different numbers are in the set A', the complement of A.

$\text{P}(A') = \frac{30}{36} = \frac{5}{6}$ — Count all the other outcomes …

$\text{P}(A') = 1 - \frac{1}{6} = \frac{5}{6}$ — … or use $\text{P}(A) + \text{P}(A') = 1$

Practice 2

Use your sample spaces from Practice 1 to answer these questions.

1 a Calculate the probability of getting two 4s when you roll two 4-sided dice.

b Calculate the probability of not getting two 4s when you roll two 4-sided dice.

2 Two tetrahedral dice are rolled and their scores are added.

a Calculate the probability of getting a sum of 5.

b Calculate the probability of getting a sum not equal to 5.

3 A fair coin is tossed three times. Calculate the probability of getting:

a three Heads

b two Heads in any order

c no Heads.

Non-cubical dice are commonly used in role-playing games and trading card games. The numerals 6 and 9 often have a dot or line below them to indicate what value they represent.

4 Calculate the probability that the sum of the scores will be a prime number when you roll two tetrahedral dice.

5 Calculate the probability that the sum of the scores will be a prime number when you roll a tetrahedral die and a normal (6-sided) die.

6 Here is the sample space for choosing a card from a standard deck of 52 playing cards.

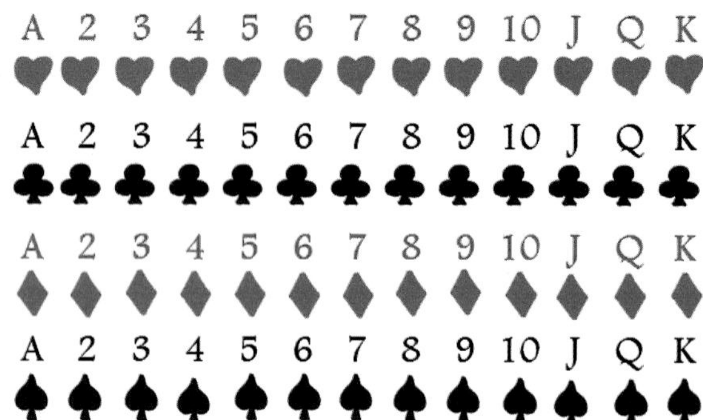

A, K, Q and J are called the picture cards.

Let A be the event 'getting a picture card'.

Let B be the event 'getting a heart'.

Let C be the event 'getting an even number'.

a Calculate the probability of 'getting a picture card', or P(A).

b Calculate the probability of 'getting a heart', or P(B).

c Calculate P(C).

d Calculate P(A').

e Calculate P(B').

f Calculate P(C').

Problem solving

7 A bag contains 20 clip-on bicycle bells. Some are red and some are blue. A bell is chosen at random from the bag. R is the event 'picking a red bell'.

If $P(R') = \frac{1}{4}$, find the number of red bicycle bells in the bag.

C Probability diagrams

- What are the advantages and disadvantages of the different probability representations?

Exploration 2

1 a Write down the sample space for rolling a 4-sided die and tossing a fair coin.

b Based on your sample space, calculate the probability of rolling a 1 and tossing Tails.

▶ Continued on next page

c How does your answer to **1b** relate to the probability of rolling a 1 and the probability of tossing Tails?

2 a Write down the sample space for rolling two 4-sided dice.

b Based on your sample space, calculate the probability of rolling an even number on the first die and a 3 on the second one.

c State how your answer to **2b** relates to the probability of rolling an even number and the probability of rolling a 3.

3 Determine whether the probability of either event is affected by the outcome of the other event.

The probability of two independent events A and B occurring is:
$P(A \text{ and } B) = P(A) \times P(B)$.

A tree diagram is a representation of a sample space, so you can see all the possible outcomes and calculate the probabilities of more than one event.

Example 3

By drawing a tree diagram, calculate the probability of rolling an odd number on both of two 6-sided dice.

1st event: rolling the 1st die.

2nd event: rolling the 2nd die.

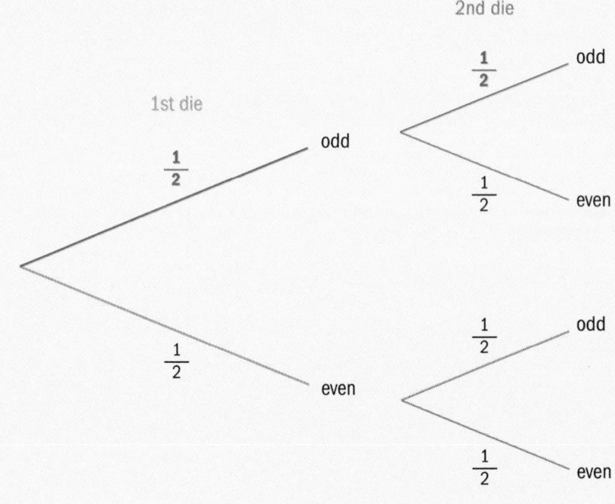

Draw a pair of branches to represent the two outcomes (odd or even) for rolling the 1st die.

For each outcome on the 1st die, draw the two possible outcomes for the 2nd die.

There are three odd numbers (1, 3 and 5) and six possible outcomes.

$P(\text{odd}) = \frac{3}{6} = \frac{1}{2}$

$P(\text{even}) = 1 - \frac{1}{2} = \frac{1}{2}$

$P(\text{odd, odd}) = \frac{1}{2} \times \frac{1}{2} = \frac{1}{4}$

Identify the branches that represent the event P(odd, odd), shown in red in the tree diagram. Multiply the probabilities along those branches.

Reflect and discuss 2

- Would it change the answer in Example 3 if you put the branches for the 2nd die first in the tree diagram?
- What do you notice about the sum of the probabilities on any pair of branches? Explain why this happens.
- Why do we multiply the probabilities on each branch?

Half of all possible outcomes have an odd number on the 1st die, and half of these have an odd number on the 2nd die as well.

Practice 3

1 A box contains three brand-A batteries and seven brand-B batteries.

The probability that a brand-A battery is faulty is 0.3, and the probability that a brand-B battery is faulty is 0.4.

a Copy and complete the tree diagram to show all the probabilities.

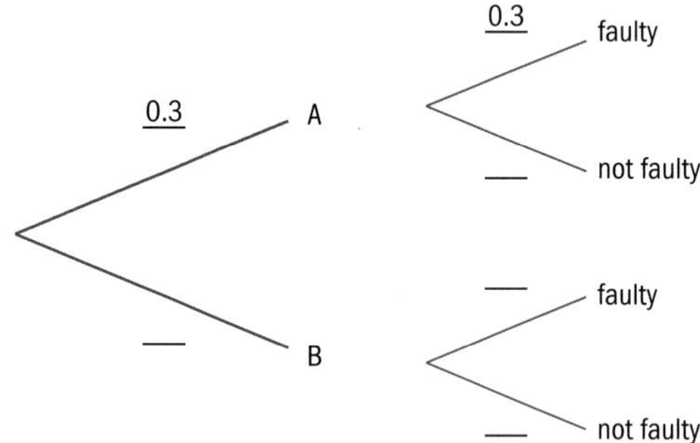

b A battery is selected at random from the box. Calculate the probability that the battery is:

i a faulty brand-A

ii not a faulty brand-A.

2 Two fair coins are tossed.

a Draw a tree diagram to represent the possible outcomes.

b Calculate the probability that both coins show Heads.

3 James wants to find the probability of rolling two sixes on two fair dice.

He starts to draw this tree diagram.

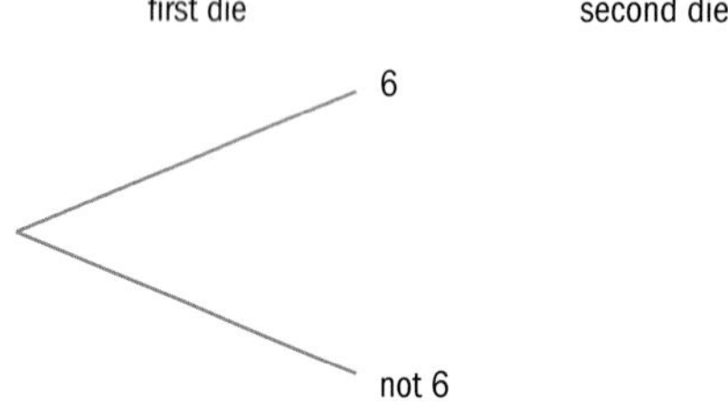

Copy and complete the tree diagram.

Calculate the probability of rolling two sixes.

4 Nienke and Merel both play netball. The probability that Nienke will score a goal on her first attempt is 0.75. The probability that Merel will score a goal on her first attempt is 0.82.

By drawing a tree diagram, calculate the probability that:

a Nienke and Merel will both score a goal on their first attempt

b neither Nienke nor Merel will score a goal on their first attempt.

5 A drawer contains 3 green and 2 yellow candles. G is the event picking a green candle, and Y is the event picking a yellow candle.

a Calculate P(G) and P(Y).

b A candle is picked from the drawer. Its color is noted and it is replaced in the drawer. A second candle is then picked.

Draw a tree diagram to represent the sample space.

Because the candle is replaced, there are still 5 candles to pick from the second time.

Problem solving

c Verify that picking two green candles is more likely than picking two yellow candles.

Exploration 3

1 Calculate the probability of picking a red card from this set.

2 a If you pick a red card and then replace it, calculate the probability of picking a red card the next time.

b If you pick a blue card and then replace it, calculate the probability of picking a red card the next time.

3 a If you pick a red card but do not replace it, calculate the probability of picking a red card from the remaining cards.

b If you pick a blue card but do not replace it, calculate the probability of picking a red card from the remaining cards.

4 Determine what happens to the probabilities for the second event when you do not replace the card.

Tree diagrams show each outcome and its associated probability clearly. Here are two types of tree diagram:

- With replacement – these represent situations where the probability does not change for the second event.
- Without replacement – these represent situations where the probability changes for the second event.

Tip

Situations 'with replacement' produce independent events; 'without replacement' events are dependent.

Example 4

A bag contains 3 lemons and 2 limes. A piece of fruit is picked at random from the bag and not put back. Then a second piece of fruit is picked at random.

a Draw a tree diagram to represent the outcomes and their probabilities.

b Find P(Lemon, Lemon) and P(Lime, Lime).

This is a 'without replacement' situation.

a

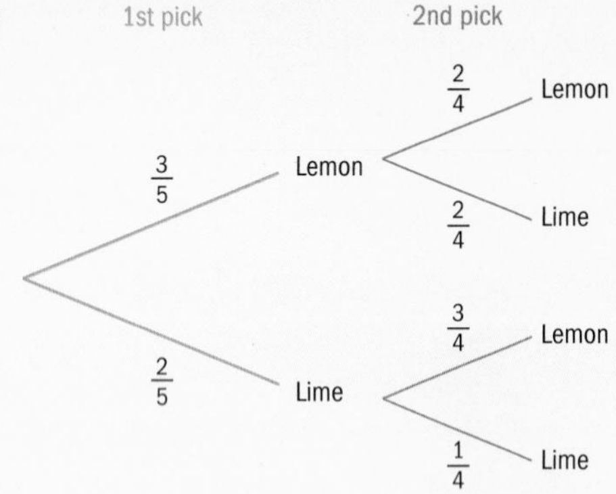

The upper arm of the first branch shows the probability of selecting a lemon; the lower arm represents selecting a lime. Since there is no replacement, the first pick affects the probabilities for the second pick , as you can see in the second set of arms.

b P(Lemon, Lemon) $= \frac{3}{5} \times \frac{2}{4} = \frac{6}{20} = \frac{3}{10}$

P(Lime, Lime) $= \frac{2}{5} \times \frac{1}{4} = \frac{2}{20} = \frac{1}{10}$

ATL

Example 5

In a box are 3 red pens, 4 black pens and 2 green pens. You pick a pen at random and then put it back. Your friend then picks one at random. Calculate the probability that you both select the same color pen.

This is a 'with replacement' situation.

R = choosing a red pen

B = choosing a black pen

G = choosing a green pen

Define the events.

▶ Continued on next page

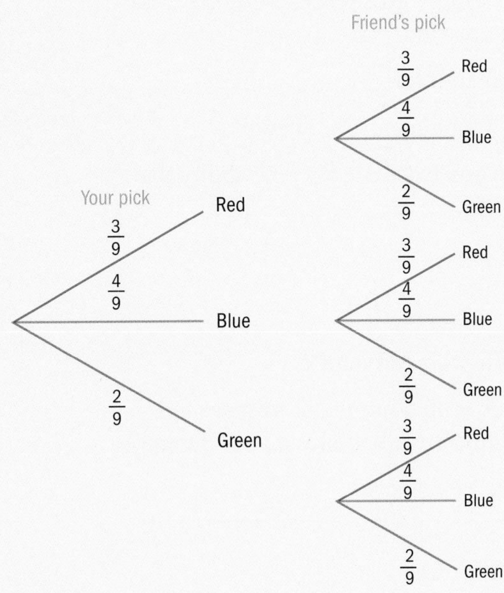

$$P(R, R) = \frac{3}{9} \times \frac{3}{9} = \frac{9}{81}$$

$$P(B, B) = \frac{4}{9} \times \frac{4}{9} = \frac{16}{81}$$

$$P(G, G) = \frac{2}{9} \times \frac{2}{9} = \frac{4}{81}$$

Find the probability for getting the same color pen, for each of the three colors.

$$P(R, R) + P(B, B) + P(G, G) = \frac{9}{81} + \frac{16}{81} + \frac{4}{81} = \frac{29}{81}$$

Probability of (R, R) **or** (B, B) **or** (G, G) is P(R, R) + P(B, B) + P(G, G)

The probability that you both select the same color is $\frac{29}{81}$.

Practice 4

1 A bag contains three red peppers and four green peppers. You take a pepper from the bag and eat it. You then take another pepper.

a Copy and label the tree diagram below to show all the probabilities.

1st pick | 2nd pick

$\frac{3}{7}$ Red — Red, Green

— Green — $\frac{3}{6}$ Red, — Green

b Calculate the probability you eat two red peppers.

c Calculate the probability you do not eat a red pepper.

d Calculate the probability you eat at least one red pepper.

2 A bag contains 1 white, 4 red and 2 blue counters. Rose picks a counter and does not replace it. She then picks another counter.

a Draw a tree diagram to show all the probabilities.

b Calculate the probability that she picks 1 white and 1 red counter.

In Q2b, there are two possible outcomes: (white, red) and (red, white).

3 Three coins are rolled. Draw a tree diagram and calculate the probability that:

a only one Head is rolled

b at least one Head is rolled

c no Heads are rolled.

d Explain the relationship between your answers to **b** and **c**.

4 There are 9 numbered cards on a table, each with a number 1 to 9 printed on one side. The other side is blank. The cards are all blank side up. Serena picks two cards from the set.

a Copy and complete this tree diagram.

Find the probability that:

b both cards are even

c at least one card is odd.

5 Of a group of five students, two will be selected for a school trip by picking names out of a hat. The five students are Jack, Mary, Rafa, Harry and Pietronella.

'Mary and Jack' is the same as 'Jack and Mary'.

a With the aid of a tree diagram or a table of outcomes, find the number of **different** possible combinations of students that could be selected.

b Find the probability that Mary and Pietronella will go on the trip.

Problem solving

6 Alix has a box of red, yellow and green marbles. She picks a marble from the box, without looking.

The table shows the probabilities of picking the different colors.

Color	Probability
red	0.4
yellow	0.25
green	

a What is the probability that Alix picks a green marble?

b There are 5 yellow marbles in the box. How many red marbles are there?

c Alix picks a marble, then replaces it in the box and picks another.

Draw a tree diagram to show the probabilities.

d What is the probability that at least one of the marbles is red?

Reflect and discuss 3

Compare your tree diagrams for Practice 4, questions 2, 3, 5 and 6 with others.

- Did you all get the same answers?
- Did you all draw the same tree diagrams?
- For question 6, do you need to draw a branch for each color?

Venn diagrams

Example 6

In a group of 30 children, 9 children like only vanilla ice cream, 13 like only strawberry ice cream and 5 children like both. The remaining children do not like either flavor.

a Draw a Venn diagram to represent this information.

b What is the probability that a child picked at random:

i likes strawberry ice cream but not vanilla

ii doesn't like strawberry or vanilla?

You can use Venn diagrams to represent information and then calculate probabilities from the diagram.

a

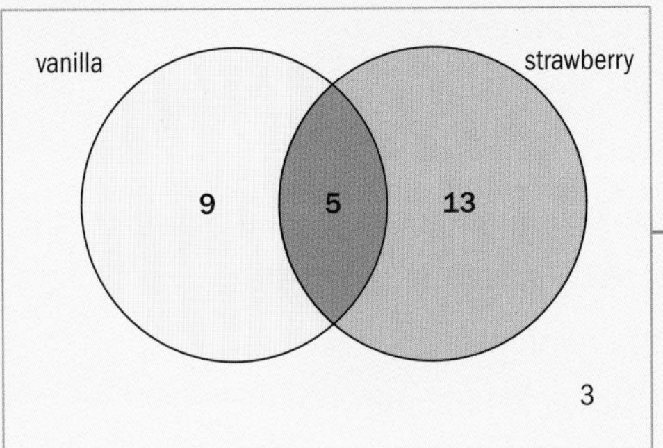

Draw the Venn diagram. Remember to include the 3 children who like neither flavor.

b i P(likes strawberry but not vanilla) $= \frac{13}{30}$

$\frac{n(\text{strawberry but not vanilla})}{n(S)}$

ii P(doesn't like strawberry or vanilla) $= \frac{3}{30} = \frac{1}{10}$

Practice 5

1 In a group of 40 people, 35 choose a main course, 10 choose a starter and 7 choose both.

a Draw a Venn diagram to represent this information.

b What is the probability that a person picked at random chooses a main course but no starter?

2 A group of 30 children are asked if they play lacrosse (L), basketball (B), volleyball (V) or none of these sports. The results are:

- 3 children do not play any of these sports
- 2 children play all three sports
- 6 play volleyball and basketball
- 3 play lacrosse and basketball
- 6 play lacrosse and volleyball
- 16 play basketball
- 12 play volleyball

a Draw a Venn diagram to display this information.

b Calculate the probability that a child selected at random:

i plays volleyball and basketball but not lacrosse

ii plays only lacrosse

iii plays only volleyball.

3 A group of 90 students filled in a questionnaire about their free-time activities. The three most popular choices were: go to the cinema (C), read (R) and watch television (T). The results were:

- 26 students watch television, read and go to the cinema
- 20 students watch television and go to the cinema only
- 18 students read and go to the cinema only
- 10 students read and watch television only
- 60 students watch television, 60 students read, 70 students go to the cinema.

a Draw a Venn diagram to display this information.

Calculate the probability that a student picked at random:

b only watches television

c only goes to the cinema.

Problem solving

4 Of 150 new university students, 65 of them study Arabic, 80 study Chinese, and 50 students study both languages. Calculate the probability that a student selected at random studies:

a only Arabic

b only Chinese

c neither Arabic or Chinese.

Two-way tables

A two-way table represents information in rows and columns.

This two-way table shows the genders and degree subjects of 150 first-year university students.

	Science	Arts	Linguistics
Male	40	18	33
Female	15	20	24

From the **rows**:

- There are 91 male students (40 + 18 + 33)
- There are 59 female students (15 + 20 + 24)

A total of 150 students (91 + 59)

From the **columns**:

- There are 55 Scientists (40 + 15)
- There are 38 Artists (18 + 20)
- There are 57 Linguists (33 + 24)

A total of 150 students (55 + 38 + 57)

There are two ways of calculating the total number of students – adding row totals or adding column totals.

Example 7

From the two-way table of 150 students and the subjects they study:

	Science	Arts	Linguistics
Male	40	18	33
Female	15	20	24

a Find the probability that a student chosen at random:

i is male

ii is either male or studies Science.

b A female student is picked at random. Find the probability that she studies Arts.

a i P(male) $= \frac{91}{150}$ — 91 male students. Total 150 students.

ii P(male or studies Science) $= \frac{106}{150} = \frac{53}{75}$ — 91 males, plus 15 females studying Science makes 106.

b P(female, studies Arts) $= \frac{20}{59}$ — 59 female students in total, of which 20 study Arts. Note that since it is given that a female student is being picked at random, the sample space is now 59, not 150.

Practice 6

1 The table shows customers' menu choices in a restaurant.

	Miso-glazed salmon	Chicken stir-fry	Lamb kibbeh
Male	9	12	8
Female	1	14	6

From this two-way table calculate:

a the number of male diners in the restaurant

b the number of diners who ordered chicken stir-fry

c the probability that a randomly chosen dish was chicken stir-fry for a female diner.

2 a Complete a two-way table to represent this information on snowboarding groups. The students were categorized according to gender (M/F) and ability (advanced/intermediate/beginner).

- There were 60 students in total.
- Half of them were male.
- There were 16 male beginners.
- There were 28 beginners in total.
- There were 12 advanced females.
- There were 10 intermediate students in total.

b A student is selected at random. Calculate the probability that this student is an advanced male snowboarder.

3 The table shows staff preferences for mid-morning drink, grouped by age.

	Tea	Coffee	Water
Under 40	4	13	10
40 or over	12	12	6

Calculate:

a the percentage of staff under 40

b the fraction of staff who drink water

c the probability that a randomly chosen staff member was under 40 and drank tea.

Problem solving

4 A choir has 51 members in three age groups: under 15, 15 to 20, and over 20.

- There are 4 male under-15s.
- $\frac{2}{3}$ of the under-15s are female.
- 26 females are over 15.
- 7 males and 11 females are over 20.

a Calculate the probability that a choir member picked at random is:

i female

ii male aged 15 to 20

iii over 15

b The conductor chooses a male soloist at random. Calculate the probability that the soloist is under 15.

D Randomness

- Does randomness affect the decisions we make?

What do you understand by the word random in the following sentence: 'Choose one piece of fruit at random'? In probability, it means that each piece of fruit has an equal chance of being chosen.

Exploration 4

1 **a** Pretend that you are tossing a coin. Without actually tossing the coin, imagine the first outcome (H or T) and write it down.

b Now, imagine tossing the coin again and record the result.

c Repeat this until you have recorded 50 outcomes.

2 Next, take a real coin and toss it 50 times recording the results.

3 Simulate the tossing of a coin using technology; let the number of tosses, $n = 100$, 1000, 10 000 etc.

4 The simulation of tossing a real coin give random results. Determine whether your imaginary results are random.

5 Find the longest run of Heads and longest run of Tails in each set of data. State whether there is much difference in the lengths of run in random and non-random data.

6 Count the number of Heads in each set of data.
For a fair coin P(Head) $= \frac{1}{2}$, so you would expect roughly half the results in each set of data to be a Head.
In which set of data is the number of Heads closest to half the results?

> The probability of an event happening is the proportion of times the event would occur in a large number of trials.

Probabilities describe what will happen in the long run. Sometimes events may not look random because they do not show the regularity that occurs after many repetitions.

These ideas lead to the differences between experimental and theoretical probability. If you take a fair die and roll it 100 times or 1000 times and record the number of 6s, you can calculate the experimental probability of 'rolling a 6'. As the number of trials increases, this probability gets closer to the theoretical probability value of $\frac{1}{6}$.

Practice 7

1 Amelia wants to pick 5 students at random from her class for a probability experiment. Which of these methods will ensure a truly random selection?

   a Picking the first 5 students to walk into the room.

   b Picking the last 5 students on the register.

   c Putting the names of all the students into a hat, and picking 5 out.

2 Jo, Amy and Sam play a board game where you need to throw a 6 to start. After five rounds, Jo rolled a 6 twice, Amy rolled a 6 three times, and Sam did not roll a 6 at all. Sam thinks Jo and Amy must be cheating. Use the ideas of randomness and probability to explain how these results are due to chance.

3 Max flips a coin 10 times and gets 7 Tails.

   a Explain whether or not this shows his coin is biased.

   b He keeps flipping the coin and counting the Tails. The table shows his results.

Number of flips	Number of Tails
50	24
100	52
500	265
1000	580

   Do you think the coin is fair? Justify your answer.

Biased means unfair.

4 a Hadley and Morgan each roll a 6-sided die, then sum their scores. Hadley wins if the sum is even, and Morgan wins if the sum is odd.

   Draw a sample space to represent the outcome. Is the game fair?

   b Hadley and Morgan change the rules. If the sum of the two dice is a multiple of 3, Hadley wins. If it is a multiple of 4 then Morgan wins. If neither multiple, they roll again.

      i Determine if this game is fair.

      ii Suggest what would happen if they both rolled a 6.

5 Peter and Eliott are playing a coin game.

   If Peter throws a Head he wins; if he throws a Tail, it is Eliott's turn. If Eliott throws a Head he wins; if he throws a Tail it is Peter's turn.

   a Draw a tree diagram to represent this game.

   b Calculate the probability that:

      i Peter wins on his first turn

      ii Eliott wins on his first turn

      iii Peter wins on this third turn.

   c Discuss whether or not you think this is a fair game.

Reflect and discuss 4

Design a game for two players using dice, coins or spinners, so that:

- each player has an equal chance of winning
- one player has a greater chance of winning.

Summary

The sample space S, is a representation of the complete set of all possible outcomes from an experiment. It can be a list, a table or a diagram.

A single **event** is a subset of the possible outcomes listed in the sample space.

P(A) represents the probability of event A occurring. P(A') is the probability of A not occuring.

$$P(A) + P(A') = 1$$

The probability of an event happening is the proportion of times the event would occur in a large number of trials.

$$\text{Probability of event } A = P(A) = \frac{\text{number of ways event } A \text{ can occur}}{\text{total number of possible outcomes}} = \frac{n(A)}{n(S)}$$

Mixed practice

1 Miloš is taking two summer classes at the local college. One course is 'pass/fail' (those are the only two grades) while the other has a grading system of A, B, C, F (with F being the only failing grade). Assume that the probability of each course is equal.

a **Write** the sample space as a list and as a table.

b **Use** your diagrams to **find** the probability that:

i Miloš passes both classes

ii Miloš passes exactly one class

iii Miloš fails both classes.

2 There are four main blood types: A, B, AB and O. These are paired with something called a Rhesus factor, which is either '+' or '–'. For example, your blood type could be B+.

a **Write** the sample space for the different blood types that are possible.

If all blood types are equally likely, what is the probability that you have:

b type AB– blood

c type O blood

d a blood type other than A or B

e a 'positive' blood type?

3 At the school picnic, one of the coolers contains 12 cans of juice and 10 cans of soda. Rhona reaches into the cooler to grab a drink for herself and one for her friend Marco.

Draw a tree diagram and **calculate** the probability that:

a she grabs two juices

b she grabs two drinks that are the same

c she grabs two drinks that are different

d neither person gets a juice.

4 Olivia rolls two 6-sided dice at the same time. One die has three red sides and three black sides. The other die has the sides numbered from 1 to 6. By means of a tree diagram, table of outcomes or otherwise:

a **Find** how many different possible combinations she can roll.

b **Calculate** the probability that she will roll a red and an even number.

c **Calculate** the probability that she will roll a red or black and a 5.

d **Calculate** the probability that she will roll a number less than 2.

5 Ann has a bag containing 3 blue whistles, 4 red whistles and 1 green whistle.

Simon has a bag containing 2 blue whistles and 3 red whistles.

The whistles are identical except for the color.

Ann chooses a whistle at random from her bag and Simon chooses a whistle at random from his bag.

a **Draw** a tree diagram to represent this information and **write down** the probability of each of the events on the branches of the tree diagram.

b **Calculate** the probability that both Ann and Simon will choose a blue whistle.

c **Calculate** the probability that the whistle chosen by Ann will be a different color to the one chosen by Simon.

6 A recent study of 24 sodas revealed that 8 have high amounts of caffeine, 12 have high amounts of sugar and 6 have both.

a **Draw** a Venn diagram to represent this information.

What is the probability that a soda picked at random from the group in the study:

b is high in sugar but not in caffeine

c is high in caffeine only

d is not high in caffeine or sugar?

7 A bag contains four calculators (C) and six protractors (P). One item is taken from the bag at random and *not replaced*. A second item is then taken at random.

a Complete the tree diagram by writing probabilities in the spaces provided.

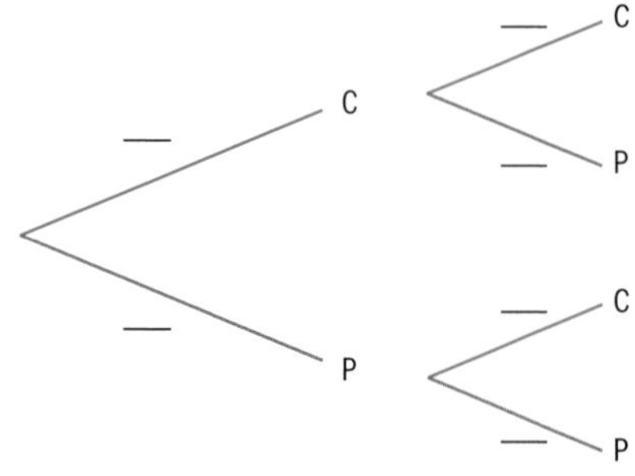

b **Calculate** the probability that one protractor and one calculator are taken from the bag.

8 Repeat **Q7**, but this time the item *is* replaced.

Calculate how the probabilities change.

9 A typical teenager should consume approximately 2000 calories per day. A survey of 120 students, three-fifths of which were male, revealed the following:

- 80 students ate more than the recommended amount.
- Half of the girls ate less than the recommended amount.
- Five-sixths of the boys ate more than the recommended amount.
- The same number of boys as girls ate the recommended amount of calories.

a Complete a two-way table to represent this information.

Calculate the probability that a student selected at random:

b ate the recommended amount of calories

c is male and ate less than the recommended amount

d ate more than the recommended amount given that they are female.

10 In a group of 50 people, 10 are healthy and the rest have either high blood pressure, high cholesterol or both; 23 people have high blood pressure and 28 have high cholesterol. **Find** the probability that a person selected at random:

a has high blood pressure

b has high blood pressure and high cholesterol

c has high blood pressure or high cholesterol

d has high cholesterol only.

Review in context

Identities and relationships

Problem solving

1 The chart shows the risk of developing Type 2 diabetes for different body mass index (BMI) values. BMI is a way of measuring the amount of fat in the body. For adults, a healthy BMI is between 18.5 and 25.

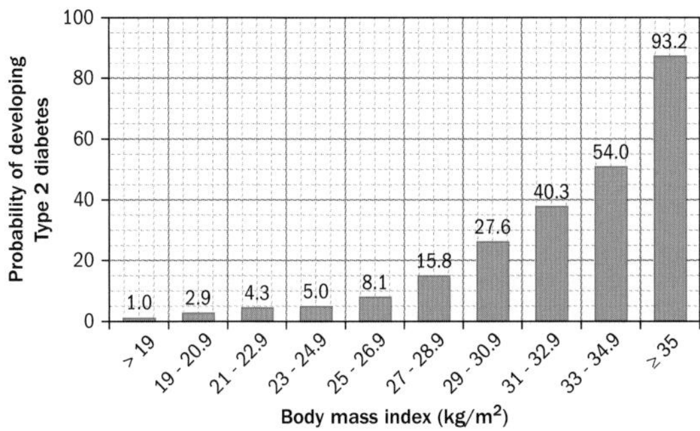

a An adult has BMI 28. What is their risk of developing Type 2 diabetes?

b **Describe** how the risk of developing diabetes changes as BMI increases.

c An adult has BMI 31. He loses weight and has BMI 26. How many times smaller is his risk of Type 2 diabetes now?

d **Write** some advice for reducing the risk of developing Type 2 diabetes. **Use** probabilities to justify your comments.

2 Vincent knows that there is a 15% chance of his children inheriting a disease that runs in his family. Suppose he has three children. **Find** the probability of:

a all three of his children having the disease

b at least two of his children having the disease

c none of his children having the disease.

3 Probability can be used in genetics to predict the likelihood of children inheriting a condition from their parents. Suppose the gene for normal sight is represented by 'M' while that of short sightedness (myopia) is represented by 'm'. A child inherits one allele (gene) from each parent. For instance, if the parents are MM and Mm, then the child would receive an 'M' from the first parent and an 'M' or 'm' from the second.

a Copy and complete the sample space for the different outcomes.

			Mother					
			MM		Mm		mm	
			M	M	M	m	m	m
Father	MM	M						
		M						
	Mm	M						
		m						
	mm	m						
		m						

b If a child inherits myopia only when they inherit 'mm', calculate the probability that a child born will have myopia.

c **Calculate** the probability that the child will carry the myopia gene (m).

Suppose two parents, both of whom are Mm, have two children.

d **Calculate** the probability that the first child will have myopia (mm).

e **Calculate** the probability that only one child will have myopia.

f **Calculate** the probability that neither child will have myopia.

g **Determine** which genes the parents would have to have in order for the probability that their first child has myopia to be $\frac{1}{2}$.

Reflect and discuss

How have you explored the statement of inquiry? Give specific examples.

Statement of Inquiry:

Understanding health and making healthier choices result from using logical representations and systems.

5 Simplification

The process of reducing to a less complicated form

Programming code

In computer programming, a block is a section of code that is grouped together. Experienced programmers write code in blocks for at least three reasons: it makes it simpler for others to understand, each block can perform just one simple task, and debugging is made simpler. (The term 'debugging' here means the act of working through a computer program to identify and remove errors that cause the program to fail or give incorrect results. It is much easier to do this when the code has been written in blocks that can be tested independently.)

Get a sheet of graph paper and see if you figure out what the code here written using Scratch is designed to do. The result is part of a graphic 'proof' of the infinite series sum below, which you will learn about in IB DP Mathematics.

$$\frac{1}{2} + \frac{1}{4} + \frac{1}{8} + \frac{1}{16} + \ldots = 1$$

Scratch is developed by the Lifelong Kindergarten Group at the MIT Media Lab. See http://scratch.mit.edu

```
go to x: -160 y: -160
pen down
point in direction 90
repeat 4
    move 320 steps
    turn ↺ 90 degrees
pen up
set n to 160
set q to 0
point in direction 90
repeat 8
    go to x: q y: q
    repeat 4
        pen down
        move n steps
        pen up
        go to x: q y: q
        turn ↺ 90 degrees
    set n to n / 2
    set q to q + n
```

Smartphones

Smartphones (and the applications that run on them) have replaced the need for other devices and equipment, such as: a camera, a laptop, diary, maps, flashlight, calendar, clock, alarm, etc.

- Suggest some ways that smartphones have simplified our lives, but also suggest some ways that they've made life more complicated.

Simplified Chinese

In the writing of the Chinese language, many traditional characters are very complex, making the written language hard to learn. Beginning in 1949, in an effort to increase literacy, simplified Chinese was introduced. Simplified Chinese is probably easier to learn, because it has fewer strokes per character. Pinyin is the conversion of Chinese characters to a Roman script, based on pronunciation.

The table here shows the Pinyin, Simplified and Traditional characters for three nouns and three verbs.

English	Pinyin	Simplified character(s)	Traditional character(s)
Mathematics	shù xué	数学	數學
To count	shù	数	數
To learn	xué	学	學
Book	shū	书	書
Horse	mǎ	马	馬
To thank	xiè	谢	謝

The first character of Cuandixia, the name of a village near Beijing, is one of the most complicated traditional Chinese characters, made up of 30 strokes.

Are you saying I'm irrational?

FORM

Objectives

- Simplifying irrational numerical expressions
- Approximating radicals
- Applying rules of radicals to simplify them
- Performing operations on radicals to simplify expressions that contain radicals

Inquiry questions

F
- What is the difference between a rational number and an irrational number?
- What is a radical (surd)?
- How do you approximate a radical?

C
- How are the rules of radicals related to the rules for combining terms in algebra?
- How is simplifying radicals similar to simplifying fractions?

D
- Can irrational numbers be combined to form rational numbers?

ATL **Communication**

Draw reasonable conclusions and generalizations

5.1

10.3

E5.1

Conceptual understanding:

Forms can be changed through simplification.

You should already know how to:

• identify radicals and understand what they represent	**1** Simplify each expression. **a** $\sqrt{9+16}$ **b** $\sqrt{9}+\sqrt{16}$ **c** $\sqrt{3}\times\sqrt{3}$ **d** $11\div\sqrt{11}$ **e** $\sqrt{3}\times\sqrt{6}\times\sqrt{8}$ **f** $\sqrt{\frac{1}{9}}$

Muhammad ibn Musa al-Khwarizmi (c. 780–c. 850 AD) referred to rational numbers as *audible*, and irrational numbers as *inaudible*. This later led to the Arabic word 'أصم' (asamm, meaning 'deaf' or 'dumb') for irrational number, which was then translated into Latin as 'surdus'.

Reviewing radicals (surds)

- What is the difference between a rational number and an irrational number?
- What is a radical (surd)?
- How do you approximate a radical?

A **square root** (also known as a **radical** or **surd**) of a positive number x is a number which, when multiplied by itself, gives the original number x. Any positive number has two square roots: one positive and one negative.

The **principal square root** of a positive number x is the *positive* square root of x, and is written as $\sqrt{x}$.

In general, 'the square root' means the *positive* square root.

For example, $\sqrt{25}=5$, but the square root of 25 can also be –5. This is normally written down as $-\sqrt{25}=-5$.

Exploration 1

Here are some examples of what we call 'rational numbers':

$\frac{1}{2}$ 1.45   7.262626...   –3 $\frac{5}{6}$ 18.2 $\sqrt{9}$

And here are some examples of what we call 'irrational numbers':

π 2.39841...   17.41002368...   0.83126674...

1 State the differences between a 'rational number' and an 'irrational number'.

2 Define each of them in your own words.

3 How would you classify a number like $\sqrt{5}$? Explain your reasoning.

A **rational number** is a number that can be written as a fraction $\frac{p}{q}$ where $p, q \in \mathbb{Z}$ and $q \neq 0$. For example: $\frac{1}{2}, \frac{5}{6}, \frac{9}{5}, 4, -\frac{12}{51}$ 1.2, $-0.\bar{3}$, and 0 are all rational numbers.

An **irrational number** is a number that cannot be written as a fraction. Irrational numbers *cannot* be represented as terminating or repeating decimals. $\sqrt{2}$ and π are examples of irrational numbers.

We have given names to only a few special irrationals, like π, φ, and Euler's number, e.

Here are some important facts about square roots:

- The square roots of positive square numbers are rational numbers because their values are whole numbers. $\sqrt{4}=2$, $\sqrt{25}=5$, $\sqrt{100}=10$, and $\sqrt{2025}=45$ are rational numbers.
- The square roots of positive non-square numbers are irrational. The numbers $\sqrt{2}$, $\sqrt{24}$, $\sqrt{247}$ and $\sqrt{1000}$ are irrational because their values cannot be written as a fraction. A radical (or a surd) is another word for the square root of a number that has an irrational value.
- Some irrational numbers (such as π, for example) can't be written as a square root, yet they are still irrational numbers.

Tip

The terms 'radical' and 'surd' are both used to refer to an irrational square root. You may see either word (or both) used in mathematics books.

Reflect and discuss 1

- Try to find the value of $\sqrt{-25}$.
- Is it possible to take the square root of a negative number? If not, why not?

ATL

Exploration 2

In this exploration, you will estimate a square root without using a calculator.

For the purpose of this exploration, you can ignore negative square roots.

1 In a copy of this table, complete the values of *just the rational* square roots.

$\sqrt{1}=$	$\sqrt{2}=$	$\sqrt{3}=$	$\sqrt{4}=$
$\sqrt{5}=$	$\sqrt{6}=$	$\sqrt{7}=$	$\sqrt{8}=$
$\sqrt{9}=$	$\sqrt{10}=$	$\sqrt{11}=$	$\sqrt{12}=$
$\sqrt{13}=$	$\sqrt{14}=$	$\sqrt{15}=$	$\sqrt{16}=$

Tip

1, 4, 9, and 16 are the first few square numbers.

2 Look at the square roots that are still blank. $\sqrt{2}$ and $\sqrt{3}$ are in between $\sqrt{1}$ and $\sqrt{4}$, so they are both between 1 and 2. Assuming that they are evenly spaced between 1 and 2, estimate the values of $\sqrt{2}$ and $\sqrt{3}$.

3 Square your estimated values from step **2** using a calculator. Comment on the accuracy of your estimates. Can you find a better estimate to 1 decimal place?

4 Repeat for the square roots that are between 2 and 3, and between 3 and 4. Assume that all square roots in between pairs of consecutive whole numbers are evenly spaced, and estimate the value of each one.

5 Square your estimated values using a calculator. Comment on the accuracy of your estimates. Does assuming that the square roots are evenly spaced give you more accurate estimates for small square roots or large square roots?

6 Using similar reasoning, suggest a method for estimating any square root.

Practice 1

1 Without using a calculator, write down or estimate the square roots of each of these numbers.

a 121 **b** 64 **c** 15

d 12 **e** 27 **f** 50

2 Without using a calculator, estimate the value of the principal square roots.

a $\sqrt{24}$ **b** $\sqrt{55}$ **c** $\sqrt{82}$

d $\sqrt{90}$ **e** $\sqrt{56}$ **f** $\sqrt{99}$

Problem solving

3 By estimating the values of any radicals or otherwise, determine which inequality, less than (<) or greater than (>), should replace the □ to complete these statements.

a $\sqrt{7}\ \square\ \sqrt{11}$ **b** $3+\sqrt{5}\ \square\ \sqrt{38}$ **c** $\sqrt{23}\ \square\ 1+\sqrt{10}$

d $7-\sqrt{15}\ \square\ \sqrt{8}$ **e** $5\times\sqrt{3}\ \square\ 4\times\sqrt{6}$ **f** $4\times\sqrt{50}\ \square\ 25+\sqrt{15}$

$\sqrt{2}$ is irrational, so it cannot be represented by a terminating decimal or a recurring decimal. A calculator can approximate $\sqrt{2}$ to a decimal, such as 1.41421356. Even if you wrote $\sqrt{2}$ correct to a large number of decimal places, it could only ever be an approximation. The only way to write the *exact* value of the square root of 2 is to write $\sqrt{2}$. Unlike the decimal form of, for example, $\frac{4}{11} = 0.363636...$ which is recurring, the decimal form of $\sqrt{2}$ does not recur.

Numbers like $\frac{3}{8}$ can be represented as a terminating decimal, in this case 0.375. Other numbers such as $\frac{4}{15}$ can be represented as recurring infinite decimals, in this case 0.26666666…

Hippasus of Metapontum, a follower of Pythagoras, is said to have discovered irrational numbers. He showed that $\sqrt{2}$ cannot be written as the ratio of two integers (i.e. as a rational number), and that therefore it must be irrational.

This discovery of the irrational numbers upset and terrified Pythagoras so much that he sentenced Hippasus to death by drowning at sea (according to legend).

Exploration 3

Consider the following list of numbers:

$$3 \quad \frac{4}{3} \quad \sqrt{5} \quad \frac{3}{4} \quad 1 \quad \sqrt{4} \quad 2 \quad \frac{5}{2} \quad \sqrt{3}$$

1 Without using a calculator, order the numbers from smallest to largest.

2 Use a calculator to check your answer.

3 Suggest a strategy that doesn't involve using a calculator to compare:

a a fraction to a whole number

b a radical to a whole number

c a radical to a fraction.

Tip

Your strategy could involve common denominators and squaring numbers. Does a number always become larger when you square it?

▶ Continued on next page

4 Without using a calculator, use your strategies to compare the following pairs of numbers. Replace the □ by <, > or =.

a $\sqrt{5}\ \square\ 5$ **b** $\sqrt{5}\ \square\ \frac{5}{2}$ **c** $6\ \square\ \frac{18}{3}$

d $\sqrt{6}\ \square\ \frac{17}{3}$ **e** $\sqrt{2}\ \square\ \frac{7}{5}$ **f** $\sqrt{8}\ \square\ \frac{141}{50}$

g $\sqrt{3}\ \square\ \frac{433}{250}$ **h** $\sqrt{\frac{1}{2}}\ \square\ \frac{1}{2}$ **i** $\sqrt{\frac{1}{2}}\ \square\ \frac{7}{10}$

Reflect and discuss 2

The only way to write the *exact* value of a radical number is by using the $\sqrt{\ }$ sign.

- What are the advantages of using such notation for radicals? Are there any possible disadvantages?
- How does this notation affect the way you perform operations on an expression containing radical numbers? Do radicals behave differently when they are written using the $\sqrt{\ }$ sign than when they are approximated to a decimal?

The square root sign, $\sqrt{\ }$, is called the radical sign. Any number or term underneath it is called the radicand.

C How radicals behave

- How are the rules of radicals related to the rules for combining terms in algebra?
- How is simplifying radicals similar to simplifying fractions?

ATL

Exploration 4

You are going to explore the three basic rules of radicals.

Rule 1

1 Evaluate these amounts.

a $\sqrt{16}\times\sqrt{16}$ **b** $\sqrt{49}\times\sqrt{49}$ **c** $\sqrt{12}\times\sqrt{12}$

d $(\sqrt{36})^2$ **e** $(\sqrt{20})^2$ **f** $(\sqrt{1001})^2$

2 Deduce a general rule from what you found in step **1**.

Rule 2

3 Use a calculator to evaluate these amounts.

a $\sqrt{4\times 9}$ **b** $\sqrt{4}\times\sqrt{9}$ **c** $\sqrt{5\times 6}$

d $\sqrt{5}\times\sqrt{6}$ **e** $\sqrt{22\times 38}$ **f** $\sqrt{22}\times\sqrt{38}$

4 Deduce a general rule from what you found in step **3**.

▶ Continued on next page

According to the National Oceanic and Atmospheric Administration, the speed s (in meters per second) at which a **tsunami** moves is determined by the depth d (in meters) of the ocean, using the formula $s = \sqrt{gd}$, where g (the acceleration due to gravity) is 9.8 m/s².

Rule 3

5 Use a calculator to evaluate these radical expressions.

a $\sqrt{\frac{4}{9}}$ **b** $\frac{\sqrt{4}}{\sqrt{9}}$ **c** $\sqrt{\frac{2}{13}}$

d $\frac{\sqrt{2}}{\sqrt{13}}$ **e** $\sqrt{\frac{28}{9}}$ **f** $\frac{\sqrt{28}}{\sqrt{9}}$

6 Deduce a general rule from what you found in step **5**.

The rules of radicals that you discovered in Exploration 4 relate to multiplying and dividing radicals.

Example 1

a Simplify $3\sqrt{2}\times5\sqrt{5}$ **b** Simplify $3\sqrt{12}\div4\sqrt{3}$

a $3\sqrt{2}\times5\sqrt{5}=3\times\sqrt{2}\times5\times\sqrt{5}$ — $3\sqrt{2}=3\times\sqrt{2}$

$=3\times5\times\sqrt{2}\times\sqrt{5}$ — When multiplying, the order doesn't matter (associative property).

$=15\times\sqrt{10}$ — $\sqrt{2}\times\sqrt{5}=\sqrt{10}$ (Rule 2)

$=15\sqrt{10}$ — Leave the final answer without the $\times$ sign.

b $3\sqrt{12}\div4\sqrt{3}=\frac{3\sqrt{12}}{4\sqrt{3}}$ — Rewrite a division as a fraction.

$=\frac{3}{4}\times\frac{\sqrt{12}}{\sqrt{3}}$ — Recall the rules of fractions: $\frac{a}{b}\times\frac{c}{d}=\frac{ac}{bd}$

$=\frac{3}{4}\times\sqrt{\frac{12}{3}}$ — $\frac{\sqrt{12}}{\sqrt{3}}=\sqrt{\frac{12}{3}}$ (Rule 3)

$=\frac{3}{4}\times\sqrt{4}$ — $\frac{12}{3}=4$

$=\frac{3}{4}\times2$ — $\sqrt{4}=2$

$=\frac{3}{2}$

Rules of radicals

For any real numbers $a,\ b,\ c\geq0$:

Rule 1 $\sqrt{a^2}=a$ **Rule 2** $\sqrt{a\times b}=\sqrt{a}\times\sqrt{b}$ **Rule 3** $\sqrt{\frac{a}{b}}=\frac{\sqrt{a}}{\sqrt{b}}\ b\neq0$

Practice 2

1 Simplify each expression.

a $\sqrt{26}\times\sqrt{5}$ **b** $3\sqrt{21}\times 21\sqrt{5}$ **c** $\sqrt{7}\times\sqrt{11}$ **d** $\frac{\sqrt{18}}{\sqrt{2}}$

e $\frac{3\sqrt{5}}{\sqrt{75}}$ **f** $\frac{\sqrt{42}}{\sqrt{6}}$ **g** $\frac{\sqrt{12}\times\sqrt{35}}{\sqrt{20}}$ **h** $\frac{\sqrt{2}}{\sqrt{4}\times\sqrt{8}}$

Problem solving

2 Fill in the blank to make the expressions equal.

$$\frac{\sqrt{18}}{\sqrt{9}\times\sqrt{\square}}=\frac{\sqrt{6}}{3}$$

ATL

Exploration 5

1 Simplify the fraction $\frac{\sqrt{18}}{3\sqrt{2}}$.

2 When a fraction is equal to 1, determine what this tells you about the numerator and the denominator.

3 Hence, write down (as an equality) what you can deduce from $\frac{\sqrt{18}}{3\sqrt{2}}=1$.

4 Using the Pythagorean theorem, find the hypotenuse of this triangle.

5 Suppose you aligned three such triangles along the hypotenuse to make a new, larger right-angled triangle like this:

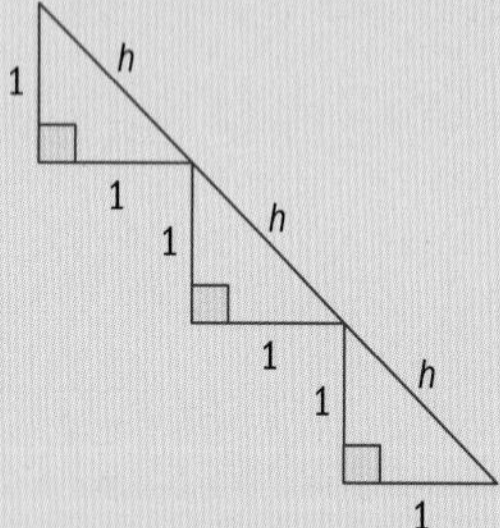

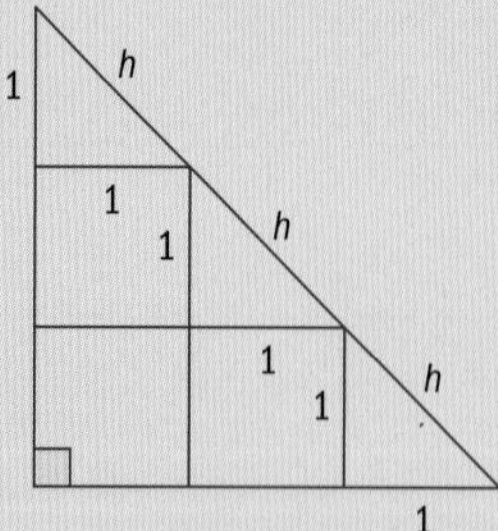

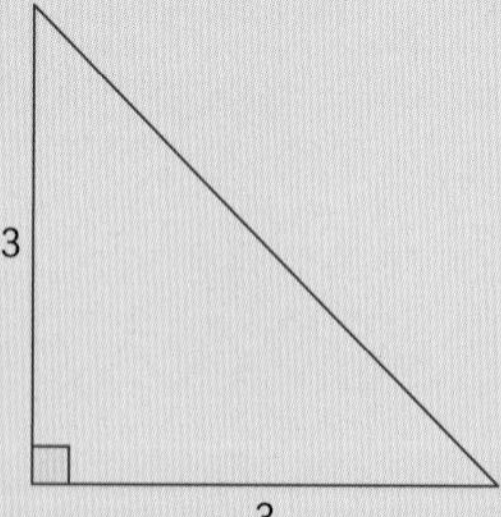

Write down an expression for the length of the hypotenuse of this larger triangle.

6 Now, using the same triangle, use the Pythagorean theorem to find the length of the hypotenuse.

7 What can you deduce from the previous two steps? Draw a reasonable conclusion or generalization.

Reflect and discuss 3

- Compare your findings in step **3** to your findings in step **7** of Exploration 5. Are they equivalent?
- How does this diagram illustrate your findings?

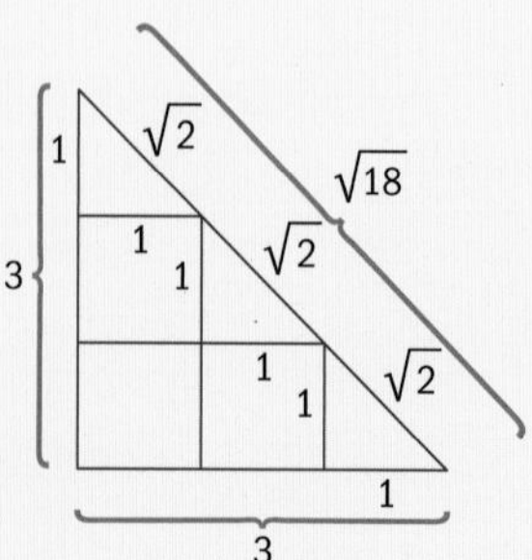

In Exploration 5, you found that $3\sqrt{2} = \sqrt{18}$. A way of saying this is that $\sqrt{18}$ *simplifies* to $3\sqrt{2}$.

Exploration 6

Below are some radicals and their simplified form.

Unsimplified form	Simplified form
$\sqrt{90}$	$3\sqrt{10}$
$\sqrt{12}$	$2\sqrt{3}$
$\sqrt{75}$	$5\sqrt{3}$
$\sqrt{200}$	$10\sqrt{2}$
$\sqrt{28}$	$2\sqrt{7}$
$\sqrt{32}$	$4\sqrt{2}$

1 Explain the process for simplifying a radical.

2 Explain how you would simplify $\sqrt{48}$.

3 Determine in how many ways you can answer question **2**. State which form is the most simplified.

4 Explain how you know when a radical *can* be simplified.

Exploration 6 demonstrated one method of simplifying radicals. Example 2 will illustrate it and a different method.

Example 2

Simplify $\sqrt{18}$.

Method 1:

$\sqrt{18} = \sqrt{9 \times 2}$ — 9 is a square number factor of 18.

$= \sqrt{9} \times \sqrt{2}$ — $\sqrt{9 \times 2} = \sqrt{9} \times \sqrt{2}$ (Rule 2)

$= 3 \times \sqrt{2}$

$= 3\sqrt{2}$ — Write the final answer without the × sign.

Method 2:

$\sqrt{18} = \sqrt{2 \times 3 \times 3}$ — The prime factorization of 18 is $2 \times 3 \times 3$.

$= \sqrt{2 \times 3^2}$ — $\sqrt{2 \times 3^2} = \sqrt{2} \times \sqrt{3^2}$ (Rule 2)

$= \sqrt{2} \times \sqrt{3^2}$

$= \sqrt{2} \times 3$ — $\sqrt{3^2} = 3$ (Rule 1)

$= 3\sqrt{2}$

Reflect and discuss 4

- When simplifying a square root using Method 1 from Example 2, why is it important to find factors that are square numbers? Can you simplify a square root by using factors that are *not* square numbers?
- Compare the two methods used in Example 2. Would you say that they are equivalent methods?
- How is simplifying radicals like simplifying fractions?

Sometimes the square root cannot be simplified.

Example 3

Determine whether or not these radicals can be simplified.

a $\sqrt{15}$ **b** $\sqrt{45}$ **c** $\sqrt{30}$

a $\sqrt{15}=\sqrt{1\times 15}$ — 15 has factors 1, 3, 5, and 15. Only 1 is a square factor.

$=\sqrt{3\times 5}$

$=\sqrt{3}\times\sqrt{5}$ — You could rewrite 15 as a product of 3 and 5. However, 3 and 5 are not square numbers.

No, $\sqrt{15}$ cannot be simplified.

b $\sqrt{45}=\sqrt{9\times 5}$ — 45 has factors 1, 3, 5, 9, 15 and 45. 9 is the largest square factor.

$=\sqrt{3\times 3\times 5}$

$=3\times\sqrt{5}$

$=3\sqrt{5}$

Yes, $\sqrt{45}$ can be simplified.

c $\sqrt{30}=\sqrt{1\times 15}$ — 30 has factors 1, 2, 3, 5, 6, 10, 15, 30. Only 1 is a square factor.

No, $\sqrt{30}$ cannot be simplified

There are two situations when a square root cannot be simplified:

- the square root of a prime number cannot be simplified
- the square root of a composite number that has no square factors other than 1 cannot be simplified.

The condition to simplify a square root is that the number has at least one square factor (other than 1).

Practice 3

Express each radical in its simplest form.

1 $\sqrt{24}$ **2** $\sqrt{32}$ **3** $\sqrt{72}$ **4** $\sqrt{125}$ **5** $\sqrt{135}$

6 $\sqrt{960}$ **7** $\sqrt{675}$ **8** $\sqrt{864}$ **9** $\sqrt{991}$ **10** $\sqrt{992}$

Problem solving

11 $\sqrt{99a^6}$ **12** $\sqrt{200b^4c^2}$ **13** $\sqrt{288x^8y^3}$

ATL

Exploration 7

When adding and subtracting radicals, the same rules apply as when you are adding or subtracting whole numbers, fractions or variables.

1 Write down each of the following statements in mathematical notation, and then find the answer.

a Three twos plus four twos makes □ twos.

b Two sixths plus three sixths makes □ sixths.

c Five x's plus six x's makes □ x's.

d Three square roots of two plus five square roots of two makes □ square roots of two.

2 Write down each of the following expressions using mathematical notation and determine if they can be further simplified.

a Three twos plus four fives.

b One third plus three quarters.

c Six x's plus five y's.

d Two square roots of two plus two square roots of three.

3 Based on steps **1** and **2**, deduce a rule for adding and subtracting square roots.

4 Compare and contrast these four expressions. State in what ways are they similar and in what ways are they different.

$3x+8x$ $\quad$ $3\times5+8\times5$

$\frac{3}{10}+\frac{8}{10}$ $\quad$ $3\sqrt{2}+8\sqrt{2}$

5 Simplify each of these four expressions.

$\sqrt{5}+\sqrt{5}$ $\quad$ $2\sqrt{5}$

$\sqrt{10}$ $\quad$ $\sqrt{20}$

a Without using a calculator, determine which of the expressions are equal.

b Use your calculator to verify your answer.

c Of the expressions that are equal, suggest which one is written in its simplest form.

6 Each of these expressions has been simplified.

$2\sqrt{5}+7\sqrt{5}=9\sqrt{5}$ $\quad$ $6\sqrt{3}-2\sqrt{3}=4\sqrt{3}$

$2\sqrt{7}-10\sqrt{7}=-8\sqrt{7}$ $\quad$ $7\sqrt{10}+3\sqrt{10}-4\sqrt{10}=6\sqrt{10}$

Deduce a rule for adding/subtracting radical expressions.

A *vinculum* is a horizontal line used in mathematical notation for grouping things together.

In 1637, René Descartes was the first to combine the German radical sign $\surd$ with the vinculum to create the radical symbol we use today.

See how it is used to distinguish $\sqrt{2}88x^8y^3$ from $\sqrt{288x^8y^3}$.

You can add or subtract two numbers only when you are adding multiples of the same amount. For example:

- $2a + 3a$ can be simplified to $5a$, since a can be factored out to give $(2 + 3)a$ or $5a$.
- $2a + 3b$ cannot be simplified since a and b are not like terms, that is, a and b have no common factors. Thus, $2a + 3b$ is written in its *simplest form.*

The same is true for square roots:

- $2\sqrt{5}+3\sqrt{5}=5\sqrt{5}$
- $2\sqrt{5}+3\sqrt{7}$ cannot be simplified.

Example 4

Where possible, simplify these expressions that involve square roots.

a $4\sqrt{6}+2\sqrt{6}$ **b** $\sqrt{8}+\sqrt{2}$ **c** $\sqrt{5}+\sqrt{7}$ **d** $\sqrt{48}+\sqrt{32}$

a $4\sqrt{6}+2\sqrt{6}=6\sqrt{6}$ — Just like $4x + 2x = 6x$.

b $\sqrt{8}+\sqrt{2}$ — Convert to equivalent square roots.

$\sqrt{8}=\sqrt{4\times 2}=\sqrt{4}\times\sqrt{2}=2\sqrt{2}$ — Simplify $\sqrt{8}$.

$\sqrt{8}+\sqrt{2}=2\sqrt{2}+\sqrt{2}$ — Substitute $2\sqrt{2}$ for $\sqrt{8}$ in the original expression.

$=3\sqrt{2}$ — Adding terms involving square roots is possible only when they have the same square root.

c $\sqrt{5}+\sqrt{7}$ — There are no like terms, so this is the simplest form.

d $\sqrt{48}+\sqrt{32}=\sqrt{16\times 3}+\sqrt{16\times 2}$ — The largest square factor of 48 is 16. The largest square factor of 32 is 16.

$=4\sqrt{3}+4\sqrt{2}$ — This cannot be simplified any further.

Reflect and discuss 5

- How does simplification help you add or subtract unlike radicals?
- In part **d** of Example 4, why is it better to simplify $\sqrt{48}+\sqrt{32}$ to $4\sqrt{3}+4\sqrt{2}$?
- How is adding and subtracting square roots similar to adding and subtracting polynomials?
- How is simplifying square roots similar to simplifying fractions?

Rule 4 of radicals

For $c \geq 0$:

$a\sqrt{c} + b\sqrt{c} = (a+b)\sqrt{c}$

Similarly, you can also subtract radicals in the same way:

$a\sqrt{c} - b\sqrt{c} = (a-b)\sqrt{c}$

Practice 4

Simplify these expressions.

1 $7\sqrt{3} - 9\sqrt{3}$

2 $\sqrt{17} + 16\sqrt{17}$

3 $12\sqrt{a+3} - 9\sqrt{a+3}$

4 $2\sqrt{pq} + 4\sqrt{pq}$

5 $\sqrt{27} - \sqrt{12}$

6 $2\sqrt{12} + \sqrt{48}$

7 $2\sqrt{75} - 5\sqrt{27}$

8 $\sqrt{8} + \sqrt{128} + \sqrt{48} + \sqrt{18}$

9 $\sqrt{54} - \sqrt{20} + \sqrt{45} - \sqrt{24}$

10 $\sqrt{12} + \sqrt{32} + \sqrt{75} + \sqrt{162}$

Problem solving

11 $4\sqrt{12x} + 3\sqrt{27x}$

12 $\sqrt{3x^2} - \sqrt{48}$

13 $\sqrt{27x} - \sqrt{3x^3}$

14 $2\sqrt{4x+8y} - 5\sqrt{9x+18y}$

The period T in seconds of a simple pendulum of length L in feet is given by $T = 2\pi\sqrt{\frac{L}{32}}$.

Another way to simplify radicals is to rationalize the denominator.

Reflect and discuss 6

Until now, you have probably dealt with fractions only where the denominator is a whole number, such as $\frac{1}{4}$ or $\frac{10}{3}$.

You can think of $\frac{1}{4}$ as $1 \div 4$ and $\frac{10}{3}$ as $10 \div 3$.

There are two possible types of decimal numbers: those with a finite number of decimal places or an infinite number of decimal places that repeat (rational numbers), and those with an infinite number of decimal places that don't repeat (irrational numbers).

- How difficult is it to divide by rational numbers?
- How difficult would it be to divide by an irrational number?

Exploration 8

Consider the fraction $\frac{4}{\sqrt{2}}$.

1. Describe the amount that this fraction represents.
2. Simplify $\frac{\sqrt{2}}{\sqrt{2}}$.
3. Determine whether $\frac{4}{\sqrt{2}} \times \frac{\sqrt{2}}{\sqrt{2}}$ is equal to $\frac{4}{\sqrt{2}}$.
4. Simplify $\frac{4}{\sqrt{2}} \times \frac{\sqrt{2}}{\sqrt{2}}$.
5. Compare your answer to $\frac{4}{\sqrt{2}}$. Is it now easier to describe the amount that this fraction represents?

Tip

For step **4**, use the rule $\frac{a}{b} \times \frac{c}{d} = \frac{ac}{bd}$

ATL

Reflect and discuss 7

- Performing an operation so that a fraction with a radical in the denominator has a rational denominator instead is known as rationalizing the denominator. Is this similar to simplifying an irrational number?
- Does rationalizing the denominator make it easier to compare numbers? If so, explain why.

Rule 5 of radicals

For $b > 0$:

$$\frac{a}{\sqrt{b}} = \frac{a}{\sqrt{b}} \times \frac{\sqrt{b}}{\sqrt{b}} = \frac{a\sqrt{b}}{b}$$

Example 5

Simplify:

a $\frac{2}{\sqrt{5}}$ **b** $\frac{12}{\sqrt{3}}$

a $\frac{2}{\sqrt{5}} = \frac{2}{\sqrt{5}} \times \frac{\sqrt{5}}{\sqrt{5}}$

$= \frac{2\sqrt{5}}{\sqrt{5}}$

To simplify $\frac{a}{\sqrt{b}}$ multiply by $\frac{\sqrt{a}}{\sqrt{b}}$

b $\frac{12}{\sqrt{3}} = \frac{12}{\sqrt{3}} \times \frac{\sqrt{3}}{\sqrt{3}}$

$= \frac{12\sqrt{3}}{3}$

$= 4\sqrt{3}$

Sometimes, you can use different methods to simplify radical expressions:

- Simplify the numerator and the denominator separately
- Use Rule 3: $\sqrt{\frac{a}{b}} = \frac{\sqrt{a}}{\sqrt{b}}$ to simplify the fraction
- Rationalize the denominator

Example 6

Simplify:

a $\frac{8}{\sqrt{40}}$ **b** $\frac{\sqrt{15}}{\sqrt{6}}$

a $\frac{8}{\sqrt{40}} = \frac{8}{2\sqrt{10}} = \frac{4}{\sqrt{10}}$ — Simplify the denominator and cancel.

$= \frac{4}{\sqrt{10}} \times \frac{\sqrt{10}}{\sqrt{10}} = \frac{4\sqrt{10}}{10}$ — Rationalize the denominator.

$= \frac{2\sqrt{10}}{5}$

b $\frac{\sqrt{15}}{\sqrt{6}} = \frac{\sqrt{3}\sqrt{5}}{\sqrt{3}\sqrt{2}} = \frac{\sqrt{5}}{\sqrt{2}}$ — Simplify the numerator and denominator separately, then cancel.

$= \frac{\sqrt{5}}{\sqrt{2}} \times \frac{\sqrt{2}}{\sqrt{2}}$

$= \frac{\sqrt{10}}{2}$

Practice 5

Simplify the following expressions. There should be no radical in the denominator in your final answer.

1 $\frac{\sqrt{40}}{\sqrt{10}}$ **2** $\frac{\sqrt{45}}{\sqrt{5}}$ **3** $\frac{1}{\sqrt{3}}$ **4** $\frac{1}{\sqrt{8}}$

5 $\frac{15}{\sqrt{3}}$ **6** $\frac{25}{\sqrt{5}}$ **7** $\frac{10}{\sqrt{2}}$ **8** $\frac{6\sqrt{2}}{\sqrt{5}}$

9 $\frac{9}{\sqrt{27}}$ **10** $\frac{14}{\sqrt{28}}$ **11** $\frac{\sqrt{12}}{\sqrt{18}}$ **12** $\frac{\sqrt{45}}{\sqrt{150}}$

Problem solving

13 $\frac{\sqrt{108x^3}}{\sqrt{3x}}$ **14** $\frac{\sqrt{3xy}}{\sqrt{15x^3y}}$ **15** $\frac{5}{\sqrt{10xy}}$ **16** $\frac{8}{\sqrt{32x}}$

D How radicals behave

- Can irrational numbers be combined to form rational numbers?

Irrational numbers were discovered around 520 BCE, during the time of Pythagoras. However, it wasn't until around 300 BCE that the Greek mathematician Euclid proved the existence of irrational numbers in his book *The Elements*.

Objective B: Investigating patterns
ii. Describe patterns as general rules consistent with findings

To find a general rule, make sure that the rule applies to all the specific cases which are similar. If the rule applies only to some specific cases but not others, then it is not consistent with your findings.

Exploration 9

Some (but not all) square roots are irrational numbers:

$\sqrt{3}$ is irrational

$\sqrt{4} = 2$ is rational

1 Here is a list of irrational numbers:

$$\sqrt{5} \quad 1+\sqrt{2} \quad \sqrt{27} \quad 3+\sqrt{3} \quad \sqrt{20} \quad 2-\sqrt{2} \quad \sqrt{45}$$

Find pairs such that:

a their sum is rational

b their sum is irrational

c their product is rational

d their product is irrational.

2 Suggest values for the length and width of a rectangle such that:

a its perimeter is rational and its area is irrational

b its perimeter is irrational and its area is rational

c both its perimeter and area are irrational.

Determine if all these combinations are possible. Justify your answer.

3 The simplified form of $\sqrt{20}$ is $2\sqrt{5}$. Explain whether or not you think there are times when $\sqrt{20}$ is better to use than $2\sqrt{5}$.

Some irrational numbers found in nature such as π (= 3.14159...) and e (= 2.718...) cannot be expressed in terms of radicals. They are part of a set of numbers known as transcendental numbers.

The Greek letter phi (φ) represents the 'Golden ratio', a geometric relationship. The value of φ is given by: $\varphi = \frac{1+\sqrt{5}}{2}$.
The Golden ratio appears not just in mathematics, but in architecture, music, painting, nature, design and many other areas.

Summary

- A **square root** (also known as a **radical** or **surd**) of a positive number x is a number which, when multiplied by itself, gives the original number x. Any positive number has two square roots: one positive and one negative.
- The **principal square root** of a positive number x is the *positive* square root of x, and is written as $\sqrt{x}$.
- A **rational number** is a number that can be written as a fraction $\frac{p}{q}$ where $p,\ q \in \mathbb{Z}$ and $q \neq 0$.

  For example, $\frac{1}{2}, \frac{5}{6}, \frac{9}{5}, 4, -\frac{12}{51}, 1.2, -0.\overline{3}$, and 0 are all rational numbers.
- An **irrational number** is a number that cannot be written as a fraction. Irrational numbers cannot be represented as terminating or repeating decimals. $\sqrt{2}$ and π (pi) are examples of irrational numbers.

Rules of radicals

For any real numbers a, b, $c \geq 0$:

Rule 1 $\sqrt{a^2} = a$

Rule 2 $\sqrt{a \times b} = \sqrt{a} \times \sqrt{b}$

Rule 3 $\sqrt{\frac{a}{b}} = \frac{\sqrt{a}}{\sqrt{b}}$ $b \neq 0$

Rule 4 For $c \geq 0$:

$a\sqrt{c} + b\sqrt{c} = (a+b)\sqrt{c}$

You can also subtract radicals in the same way:

$a\sqrt{c} - b\sqrt{c} = (a-b)\sqrt{c}$

Rule 5 For $b > 0$:

$\frac{a}{\sqrt{b}} = \frac{a}{\sqrt{b}} \times \frac{\sqrt{b}}{\sqrt{b}} = \frac{a\sqrt{b}}{b}$

Mixed practice

Compare the following pairs of values and replace the □ by using $<$, $>$ or $=$.

1 $\sqrt{84}$ □ $2\sqrt{20}$

2 $\sqrt{20}$ □ $\frac{9}{6}$

3 $\frac{\sqrt{6}}{7}$ □ $\frac{\sqrt{7}}{6}$

4 $\sqrt{15}$ □ 3.75

Without using a calculator, **estimate** the value of the principal square roots correct to 1d.p.

5 $\sqrt{45}$

6 $\sqrt{17}$

7 $\sqrt{54}$

8 $\sqrt{12}$

Rationalize the denominator for each radical expression:

9 $\frac{1}{\sqrt{2}}$

10 $\frac{6}{\sqrt{3}}$

11 $\frac{3}{4\sqrt{8}}$

12 $\frac{15}{\sqrt{75}}$

Simplify these expressions completely. Remember to rationalize the denominator when necessary.

13 $\sqrt{12}$

14 $\sqrt{18}$

15 $\sqrt{125}$

16 $\sqrt{121x^3}$

17 $\sqrt{10} \times \sqrt{360}$

18 $\sqrt{3} \times \sqrt{12}$

19 $\sqrt{xy^3} \times \sqrt{x^5 y}$

20 $\frac{\sqrt{98}}{\sqrt{2}}$

21 $\frac{\sqrt{15}}{\sqrt{3x}}$

22 $\frac{6}{\sqrt{24}}$

23 $\frac{10}{\sqrt{12}}$

24 $12\sqrt{13} - 2\sqrt{13}$

25 $5\sqrt{8} - 3\sqrt{50}$

26 $6\sqrt{4x} + 3\sqrt{9x}$

27 $2\sqrt{20t} + 4\sqrt{45t}$

28 $\sqrt{200a} + \sqrt{128a} - \sqrt{8a}$

Reflect and discuss

How have you explored the statement of conceptual understanding? Give specific examples.

Conceptual understanding:

Forms can be changed through simplification.

6 Quantity

An amount or number

How many fish?

Suppose you want to know the quantity of fish in a lake. How would you go about this? You could drain the lake and count all the dead fish; that would give you an accurate count, but it's not very practical, and not too considerate of the fish either!

One way to estimate the quantity is called the capture-recapture method:

Capture 100 fish, tag them, and then release them back into the lake. Assume that all these fish distribute themselves evenly and randomly throughout the lake.

Later, you capture 100 fish again and find that, of these, 5 are ones that you tagged previously. You know that there are 100 tagged fish in the lake. So how many batches of 100 fish would you need to capture all 100 tagged fish? If you assume you'd catch 5 tagged fish for every 100, then you would have to catch $100/5 = 20$ batches. At 100 fish per batch that would mean there are an estimated $20 \times 100 = 2000$ fish in the lake.

- How many fish would you estimate are in the lake if you recaptured 11 tagged fish?
- What are all the assumptions that are being made in this method?
- What conclusion(s) could you draw if none of the fish you caught later on were tagged?

Skating by numbers

In the Olympics, how do the complex figure skating routines boil down to just a single number from each judge? It's more complicated than you might think.

Skaters accumulate points in a system designed to prevent judges from fixing a competition and also to make scoring less subjective.

First, skaters get a basic score for every move they execute, regardless of how its executed.

The judges then add to or subtract from this basic score, based on how well it was performed.

Skaters are also judged on choreography, transition, execution, interpretation and skating skills.

The final score is tallied using a computer which randomly selects the scores awarded by 7 of the 9 judges. Of those 7 scores, the highest and lowest scores are disregarded and the remaining 5 are added up for the final score.

How hot is hot?

Is it possible to create a scale of spiciness? In 1912, American pharmacist Wilbur Scoville devised a method to do just that. In Scoville's method, dried pepper is dissolved and then diluted in a solution of sugar water. Varying concentrations are given to a panel of trained tasters who rate the heat level in Scoville Heat Units.

The method is naturally imprecise, given that it is based on human subjectivity, but it does give a good estimation of the amount of heat (and pain!) experienced when eating various types of chili peppers.

Bell peppers barely register on the scale, with a rating of 0. The current champion of heat is the Carolina Reaper, weighing in at 2.2 million Scoville units.

Bell pepper	Bird's eye	Red Savina habanero	Ghost pepper
0	100 000 – 350 000	350 000 – 580 000	over 1 000 000

Can I exchange this please?

Global context: Globalization and sustainability

RELATIONSHIPS

Objectives

- Converting between different currencies
- Recognizing conversions in real-life contexts
- Solving word problems involving currencies
- Understanding the meaning of different rates and commission charges
- Calculating commission charges

Inquiry questions

F
- Why do we need to convert currencies?
- What is an exchange rate?

C
- When are quantities in different currencies equivalent?

D
- How do systems influence communities?

ATL **Information literacy**

Access information to be informed and inform others

Statement of Inquiry:

Quantities and measurements illustrate the relationships between human-made systems and communities.

6.1 7.1 7.2

You should already know how to:

• calculate the percentage of an amount	**1** Find: **a** 25% of 386 kg **b** 1% of 83 mm **c** 20% of 78 Ω **d** 8% of 75 liters
• calculate percentage change	**2** The cost of a package of noodles increases from \$1.75 to \$2.15. What is the percentage increase in cost?

Introduction to currency conversion

- Why do we need to convert currencies?
- What is an exchange rate?

The official monetary standard of a country is called the *currency*.

The number of people travelling around the world is constantly increasing. As there is no single global currency, you need to be able to convert between currencies to understand the purchasing power of a particular currency.

Example 1

Convert 1500 Swiss Francs (CHF) into Euros (EUR). The exchange rate is 1 CHF = 1.20 EUR.

1 CHF = 1.20 EUR — Write out the exchange rate.

Method 1:

$1500 \times 1\text{ CHF} = 1500 \times 1.20\text{ EUR}$

$= 1800\text{ EUR}$

Think: 'What mathematical operation do I need to perform to get from 1 to 1500?' Multiply both sides by 1500.

Method 2:

$1500\text{ CHF} \times \frac{1.20\text{ EUR}}{1\text{ CHF}} = 1800\text{ EUR}$ — Multiply by the exchange rate as a fraction, to cancel the Swiss Francs.

Example 2

Convert 800 Euros into Swiss Francs. The exchange rate is 1 CHF = 1.20 EUR.

Method 1:

1 CHF = 1.20 EUR — Write out the exchange rate.

$\frac{1}{1.20}\text{ CHF} = 1\text{ EUR}$ — Express the exchange rate in terms of Euros.

$\frac{1}{1.20} \times 800\text{ CHF} = 800\text{ EUR}$ — Multiply both sides by 800.

$\Rightarrow 800\text{ EUR} = 666.67\text{ CHF}$

Method 2:

$800\text{ EUR} \times \frac{1\text{ CHF}}{1.20\text{ EUR}} = 666.67\text{ CHF}$ — Multiply by an equivalent fraction.

Reflect and discuss 1

- Divide both sides of 1 CHF = 1.20 EUR by 1.20. What have you found?
- What do your results mean in the context of currency exchange?

In Switzerland, the smallest coin is 0.05 CHF, so for a conversion of 666.67 (as in Example 2) you may be given 666.65 CHF or 666.70 CHF.

A **rate of exchange**, or **exchange rate**, gives the value of one currency in terms of another currency.

Objective: **A.** Knowing and understanding
ii. apply the selected mathematics successfully when solving problems

In this practice set you will use the skill of converting currencies to solve the following problems.

Practice 1

1 1 US Dollar (USD) = 0.90 Euros (EUR). Convert 750 USD into EUR.

2 1 CHF = 1.02 USD.

a Convert 1200 CHF into USD.

b Convert 1734 USD into CHF.

3 The exchange rate from Canadian Dollars (CAD) into Great British Pounds (GBP) is 1 CAD = 0.52 GBP. Convert 870 GBP into CAD.

4 a You want to convert 600 US Dollars (USD) into Australian Dollars (AUD). The exchange rate is 1 USD = 1.26 AUD. Calculate how much your 600 USD will be worth in AUD.

b Using the same exchange rate, convert 500 AUD into USD.

5 The table below shows some exchange rates for the Japanese Yen (JPY).

Currency	1 JPY
Mexican Peso	0.12
Chinese Yuan	0.053
Euro	0.0076
Norwegian Krone	0.066

a You have 1250 Japanese Yen to exchange for Chinese Yuan. Calculate how many Yuan you will receive. Give your answer to the nearest Yuan.

b Your friend has 855 Mexican Pesos to exchange for Japanese Yen. Calculate how many Yen he will receive.

c Find how many Mexican Pesos there are to the Euro. Give your answer correct to the nearest Peso.

Problem solving

6 **a** The exchange rate from Turkish Lira (TRY) to Euros (EUR) is 1 TRY = 0.25 EUR. What is the value of 1 Euro in Turkish Lira?

b Victor receives 500 Swiss Francs (CHF) for 400 EUR. Use this information to find the value of 1 Euro in Swiss Francs.

c Find the value of 1 Turkish Lira in Swiss Francs.

7 Use the exchange rate table here for this question. Give answers correct to 4 s.f.

	USD	BRL	CRC
1 USD	1	2.580	536.1
1 BRL	0.3976	1	207.8
1 CRC	0.001865	0.004812	1

a Andrew started a trip in the US. He converted 750 US Dollars (USD) into Brazilian Real (BRL). Calculate how many BRL he received.

b In Brazil, he spent 350 BRL. Then he went to Costa Rica. How many Costa Rican Colón (CRC) did he receive for his remaining BRL?

c In Costa Rica he spent 200 000 CRC. Then he returned to the US. He converted his remaining money back into USD. Calculate the amount that Andrew received.

8 Copy the table below and find the exchange rates for the three currencies. Give answers correct to 3 s.f.

	AOA (Angolan Kwanzy)	BGN (Bulgarian Lev)	CLP (Chilean Peso)
1 AOA	1	0.0170	6.03
1 BGN			
1 CLP			

On January 1 1999, the concept of the Euro was introduced and coins and notes were recognized as legal tender from January 1 2001.

However, not all members of the European Union (EU) have adopted the Euro as their common currency.

ATL

Exploration 1

1. You start with €100. Use a currency converter site or app to find the value of €100 in US Dollars (USD).
2. You spend 75% of your converted €100, so you have only 25% of your original amount left. Calculate how much money you have left in USD.
3. Convert the money you have left back into Euros. Decide whether the amount you have left is 25% of your initial €100.
4. Repeat the steps above for: £ (Great British Pounds/GBP); $ (Australian Dollar/AUD); ¥ (Japanese Yen/JPY); $ (Mexican Peso/MXN) and Bs (Venezuelan Bolivar/VEF).

C Financial mathematics

- When are quantities in different currencies equivalent?

Exploration 2

Claudia lives in Germany. She is planning a holiday to Canada. She exchanges 2000 Euros into Canadian Dollars at a rate of 1 EUR = 1.39 CAD.

1 Calculate how many Canadian Dollars she receives.

Before she even leaves Germany, Claudia becomes ill and has to cancel her holiday. She sells the Dollars back to the bank at a rate of 1 EUR = 1.48 CAD.

2 Calculate how many Euros she receives.

3 Determine how much money the bank makes from the two transactions.

4 Calculate Claudia's loss as a percentage of her original 2000 EUR.

Selling (Going on holiday)	Buying (On return from holiday)
If you have currency from your home country (HC) and you want to change it to a foreign currency (FC) then the bank is **selling** you the foreign currency.	If you have foreign currency (FC) and you want to change it to your home currency (HC) then the bank is **buying** your foreign currency.

The buying and selling rates are from the point of view of the bank.

The **commission** is a fee that foreign-exchange providers charge for exchanging one currency into another. The commission is charged in the currency that they buy and sell.

Example 3

A bank's exchange rate from Canadian Dollars (CAD) to US Dollars (USD) is 1 CAD = 0.73 USD. The bank charges 1% commission.

Convert 5000 CAD to USD, taking into account the bank's commission.

1 CAD = 0.73 USD — Write out the exchange rate.

5000×1 CAD = 5000×0.73 USD — First multiply both sides by 5000 to work out the conversion.

5000 CAD = 3650 USD

3650 USD $\times$ 0.01 = 36.5 USD — Find 1% of the conversion to work out the cost of the commission.

3650 USD − 36.5 USD = 3613.50 USD — Subtract the cost of the commission from the conversion to find out how much you would receive.

Practice 2

1 1 US Dollar (USD) = 0.90 Euros (EUR).

a Convert 840 USD into EUR.

b A commission of 1% is charged for the conversion. Find the cost of the commission in EUR.

c Calculate the final number of Euros received.

2 1 CHF = 1.02 USD. A commission of 1% is charged.

Convert 3500 CHF into USD.

3 The exchange rate from Great British Pounds (GBP) into South African Rand (RND) is 1 GBP = 21.33 RND. The bank charges 2% commission.

Convert 800 GBP into RND.

More than $2500 worth in coins are thrown into the Trevi fountain in Rome every day.

Problem solving

4 You want to change British Pounds (GBP) into Swiss Francs (CHF) at your bank. The exchange rate is 1 GBP = 1.30 CHF. There is also a fixed bank charge of 3 GBP for each transaction.

a Calculate how many Swiss Francs you could buy with 133 GBP.

b Your friend exchanged some GBP and received 430 CHF. Calculate how many British Pounds she exchanged.

c Let S be the number of Swiss Francs received in exchange for B British Pounds. Express S in terms of B.

5 The exchange rate from GBP into EUR is 1 GBP = 1.27 EUR.

A commission of 1% is charged per transaction.

Nathan wants at least 320 Euros of spending money for his holiday.

Calculate the minimum amount in GBP he needs to convert. Give your answer to the nearest Pound.

Exploration 3

You are going to compare the rates offered by two banks: Bank 1 and Bank 2.

For each stage of the journey below, work out the conversion using the rates offered by the two banks. Decide on the best conversion and then use this for the next stage of the journey.

1 Seren started her trip in Australia. She converted 1600 Australian Dollars (AUD) into Great British Pounds (GBP). Calculate how many GBP she received.

2 In the UK she spent 520 GBP. Then she went to France. Find how many Euros she received for her remaining GBP.

▶ Continued on next page

3 In France, Seren spent 190 EUR. Then she returned to Australia. She converted her remaining money back into AUD. Calculate the amount that Seren received.

Bank 1

	AUD sells	AUD buys	GBP sells	GBP buys	EUR sells	EUR buys
1 AUD			0.485	0.515	0.62	0.66
1 GBP	1.94	2.06			1.271	1.299
1 EUR	1.465	1.565	0.753	0.787		

Bank 2 (charges 1% commission)

	AUD buys/sells	GBP buys/sells	EUR buys/sells
1 AUD		0.52	0.668
1 GBP	2.0		1.323
1 EUR	1.553	0.793	

Reflect and discuss 2

- Why do banks have different buying and selling rates?
- Many currency conversion sites only offer a single conversion rate. Why is this?
- What is the difference between sites that provide information for people that work in financial institutions and those aimed at people needing a foreign currency to go on holiday?
- How might you decide which bank or foreign exchange center to use?

Practice 3

1 a Dafydd is going on holiday. He wants to exchange 700 USD for AUD. Use the table to calculate how many AUD he will receive.

	Buy AUD	Sell AUD
1 USD	1.395	1.379

b Bernice has returned from holiday. She wants to exchange 980 AUD back into USD. Use the table to calculate how many USD she will receive.

2 The table shows a bank's exchange rate between 1 Danish Krone (DKK) and five other currencies.

	Sell DKK	Buy DKK
Albania (ALL)	18.75	19.25
Hong Kong (HKD)	1.17	1.28
France (EUR)	0.13	0.15
Canada (CAD)	0.19	0.22
Mexico (MXN)	2.21	2.24

a Your friends go to Mexico. Before leaving, they exchange 600 DKK into Mexican Pesos (MXN). Calculate the number of Mexican Pesos they receive for 600 DKK.

b Your friends spend 300 MXN in Mexico, and on returning to Denmark they exchange their remaining Pesos into Krone. Calculate the number of DKK they receive, correct to two decimal places.

c A Canadian family exchange 1500 CAD into DKK before their holiday. They are unable to travel so they exchange their money back into CAD. Calculate how many CAD they have lost on the transaction.

Problem solving

3 The table below shows part of a currency conversion chart at a bank. The bank charges a 1% commission after the conversion.

	GBP	USD	CAD
1 GBP	1	p	1.90
1 USD	0.65	1	q
1 CAD	0.53	0.81	1

For all calculations in this question give your answers correct to 2 d.p.

a Calculate the value of:

i p **ii** q

b Stefan is taking a trip from the USA to the UK and then home again via Canada. He wants to exchange 1500 USD. How many GBP will he receive?

c In the UK Stefan spends 890 GBP. He then exchanges his remaining money into CAD. In Canada he plans to stay in a hotel for three nights. Determine whether he has enough CAD for a hotel room charged at 45 CAD per night.

4 Christine lives in Tanzania. She wants to exchange 350 000 Tanzanian Shillings (TZS) into US Dollars (USD).

Bank 1 buys 1 TZS for 0.0005 USD, and sells 1 TZS for 0.0004 USD.

Bank 2 has an exchange rate of 1 TZS = 0.00051 USD, and charges a 1% commission.

At which bank would you advise Christine to exchange her money?

D Currency as a quantity

- How do systems influence communities?

ATL

Exploration 4

1 Compare the currencies for each group of countries in the tables. Decide why the countries have been grouped in this way.

2 Research the bitcoin and answer these questions.

   a What is the bitcoin and how does it work?

   b Who created the bitcoin? Why was the bitcoin developed?

   c Is the bitcoin secure? Is it legal?

3 Exchange rates are constantly changing.

   a Research five currencies and analyze the percentage change in the last year.

   b Determine which have changed the most. Think about why this might be.

Group A
France Germany Spain Greece Italy

Group B
Denmark United Kingdom

Group C
Switzerland Sweden Serbia

DETECT

1 **D**efine the task
2 **E**xamine information
3 **T**rack down sources and find information in those sources
4 **E**xtract meaningful information
5 **C**reate a product to inform others
6 **T**ask reflection

Reflect and discuss 3

Single currency:

- Why have some European countries decided not to join the EU?
- Why do some EU countries opt out of the Euro?
- What are the advantages and disadvantages of a single European currency?

New currencies:

- Does the bitcoin have a future?
- Does every person have the right to create their own currency? To what extent does everyone have equal access to these opportunities?

Exchange rates:

- What factors influence the exchange rate?

Of the 20 most traded world currencies, 17 of them use a strikethrough or double-strikethrough in their symbol.

Why do you think the struck-through letters are used rather than just plain capital letters?

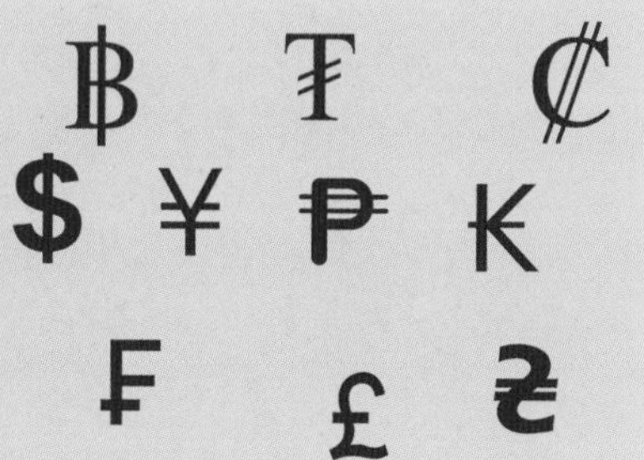

Summary

A **rate of exchange**, or **exchange rate**, gives the value of one currency in terms of another currency.

The **commission** is a fee that foreign-exchange providers charge for exchanging one currency into another. The commission is charged in the currency that they buy and sell.

The **buying** and **selling rates** are from the point of view of the bank, who is buying or selling currencies.

Mixed practice

1 1 EUR = 1.10 USD.

a Convert 950 EUR into USD.

b Convert 2750 USD into EUR.

2 1 Mexican Peso (MXN) = 0.055 US Dollars (USD).

a Convert 4400 MXN into USD.

b A commission of 1% is charged.

Find the cost of the commission in USD.

c **Calculate** the final number of USD received.

3 1 CHF = 0.72 GBP. A commission of 2% is charged. Convert 1600 CHF into GBP.

4

	Buy RUB	Sell RUB
1 GBP	107.30	105.80

a Jasmine is going on holiday.

She wants to exchange 1400 GBP for Russian Rubles (RUB).

Calculate how many RUB she will receive.

b Kali has returned from holiday.

She wants to change 1420 RUB back to GBP.

Calculate how many GBP she will receive.

5 The table below show part of a currency conversion chart. **Find** the value of p and q.

	EUR	GBP	AUD
EUR	1	p	1.53
GBP	1.27	1	1.92
AUD	q	0.52	1

6 The exchange rate from US Dollars (USD) to Thai Baht (THB) is 1 USD = 33.45 THB. Give the answers to the following correct to 4 s.f.

a **Find** the value of 115 US Dollars in THB.

b **Calculate** the value of 1 THB in USD.

c Alexis receives 600 New Zealand Dollars (NZD) for 14 670 THB. **Calculate** the value of 1 THB in NZD.

d **Calculate** the value of the US Dollar in New Zealand Dollars.

7 A bank buys 1 Australian Dollar (AUD) for 0.72 Euros (EUR) and sells 1 AUD for 0.70 EUR. Frida wants to exchange 800 AUD for EUR.

a **Find** how many Euros Frida will receive.

b Frida has to cancel her trip and changes her money back later when the rates are 'buys 1 AUD = 0.73 EUR, sells 1 AUD = 0.69 EUR'. **Find** how many Australian Dollars she receives.

c **Determine** how many Australian Dollars she has lost on the transaction.

8 Ed had to change British Pounds (GBP) into Swiss Francs (CHF) at a bank. The exchange rate was 1 GBP = 1.4 CHF. The bank charged 2% commission.

a **Determine** how many Swiss Francs Ed bought with 200 GBP.

b Ed has 100 CHF of his initial amount left. **Find** how many British Pounds he could buy.

9 Santiago lives in Mexico. He wants to change 67000 Mexican Pesos (MXN) into Euros (EUR).

A bank buys 1 MXN for 0.051 EUR and sells 1 MXN for 0.0531 EUR.

An exchange center has an exchange rate of 1 MXN = 0.054 EUR and charges a 2% commission.

Find the difference between the number of EUR he would receive from the bank and the exchange center.

10 Pavla travels from the Czech Republic to Thailand. She changes 29000 Czech Koruna (CZK) to Thai Baht. The exchange rate is 1 CZK = 1.34 THB.

a **Calculate** the number of THB Pavla buys.

Pavla leaves Thailand and travels to New Zealand. She has 20 000 THB and uses these to buy New Zealand Dollars (NZD). The exchange rate is 24 450 THB = 1000 NZD.

b **Calculate** the total number of New Zealand Dollars Pavla receives.

c **Find** an approximate exchange rate between New Zealand Dollars and Czech Koruna.

Give your answer in the form 1 NZD = x CZK, correct to 2 d.p.

Review in context

Globalization and sustainability

1 The headquarters of Jim's company are in the UK. He is expanding and starting a company in Singapore. He will be exchanging British Pounds (GBP) into Singaporean Dollars (SGD). The exchange rate is 1 GBP = 2.07 SGD, and there is a bank charge of 10 GBP for each transaction.

a **Determine** how many SGD Jim could buy with 1330 GBP.

b Let s be the number of SGD received in exchange for b GBP. Express s in terms of b.

c At the end of the second year, Jim needs to transfer 15000 SGD from Singapore to his UK bank account. He has two ways of converting SGD to GBP. He can use the British bank (the exchange rate is 1 GBP = 2.06 SGD, and there is a bank charge of 10 GBP for each transaction), or he can use the Singaporean bank (the exchange rate is 1 SGD = 0.49 GBP, and there is a bank charge of 2% for each transaction).

Decide which is the better option.

2 Andy, Neil and Jelena each had 3000 USD to invest in 2005.

Andy used his 3000 USD to buy Swiss Francs.

Neil used his 3000 USD to buy Brazil Reals.

Jelena used his 3000 USD to buy Swedish Krone.

The table shows the exchange rates per 1 USD at the time of conversion and at five-year intervals.

2005	2010	2015
1.32 CHF	0.97 CHF	0.85 CHF
2.50 BRL	1.52 BRL	3.29 BRL
7.25 SEK	6.40 SEK	9.30 SEK

Ignoring any bank buying and selling costs:

a **Calculate** the amount each person received in their chosen currency in 2005.

b **Calculate** the amount in USD each person would receive if they chose to sell in 2010.

c **Calculate** the amount each person would receive if they chose to convert it back to USD in 2015.

Reflect and discuss

How have you explored the statement of inquiry? Give specific examples.

Statement of Inquiry:

Quantities and measurements illustrate the relationships between human-made systems and communities.

6.2 City skylines

Global context: Globalization and sustainability

Objectives

- Constructing and interpreting frequency and relative frequency histograms with equal class widths
- Constructing and interpreting frequency density histograms with unequal class widths
- Describing distributions

Inquiry questions

F
- What are the differences between a bar chart and a histogram?

C
- How do you accurately analyze a data distribution from a histogram?

D
- How can real data ever be misleading?
- How do individuals stand out in a crowd?

RELATIONSHIPS

ATL **Critical thinking**

Revise understanding based on new information and evidence

Statement of Inquiry:

How quantities are represented can help to establish underlying relationships and trends in a population.

4.1

4.2

6.2

You should already know how to:

• represent continuous and discrete data in a grouped frequency table	**1** The heights in cm of 24 seedlings (rounded to 1 d.p.) are: 2.4, 3.1, 5.2, 2.6, 5.8, 2.2, 4.9, 3.0, 4.7, 5.3, 2.6, 4.5, 3.7, 2.3, 5.4, 5.7, 3.5, 2.1, 4.0, 4.2, 3.6, 2.5, 2.8, 4.1 Construct a grouped frequency table with equal class widths for this data.
• find measures of central tendency and dispersion from a grouped frequency table	**2** From your grouped frequency table, find the modal class and the class that contains the median. Calculate estimates for the mean and the range.
• recognize qualitative and quantitative data	**3** Classify as qualitative or quantitative data: **a** shoe size **b** shoe color **c** shoe price

F Bar charts and frequency histograms

- What are the differences between a bar chart and a histogram?

Before now, you may have drawn bar charts for both qualitative and quantitative discrete data. However, strictly speaking, a chart with bars representing quantitative data is a histogram.

A **bar chart** represents qualitative data.

- All the bars have equal width.
- There are spaces between the bars.
- The *height* of the bar represents the frequency.
- Frequency is on the vertical axis.

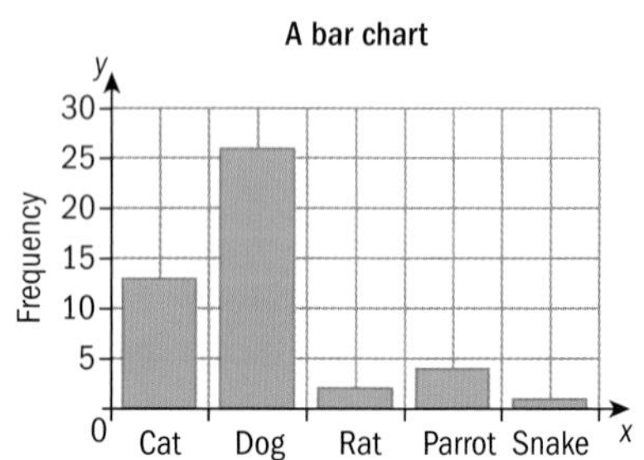

A **histogram** represents quantitative discrete data and all continuous data.

- The bars may or may not be of equal width.
- There are no spaces between the bars.
- The *area* of the bar represents the frequency.
- Frequency, relative frequency or frequency density are on the vertical axis.

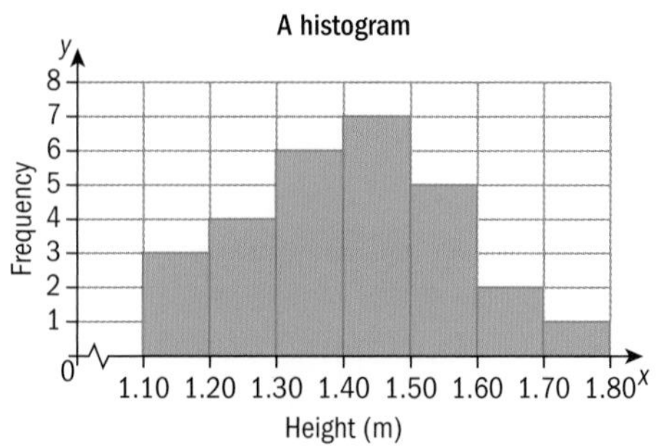

ATL

Reflect and discuss 1

Decide which of these graphs are bar charts, which are histograms and which are neither. Justify your answers.

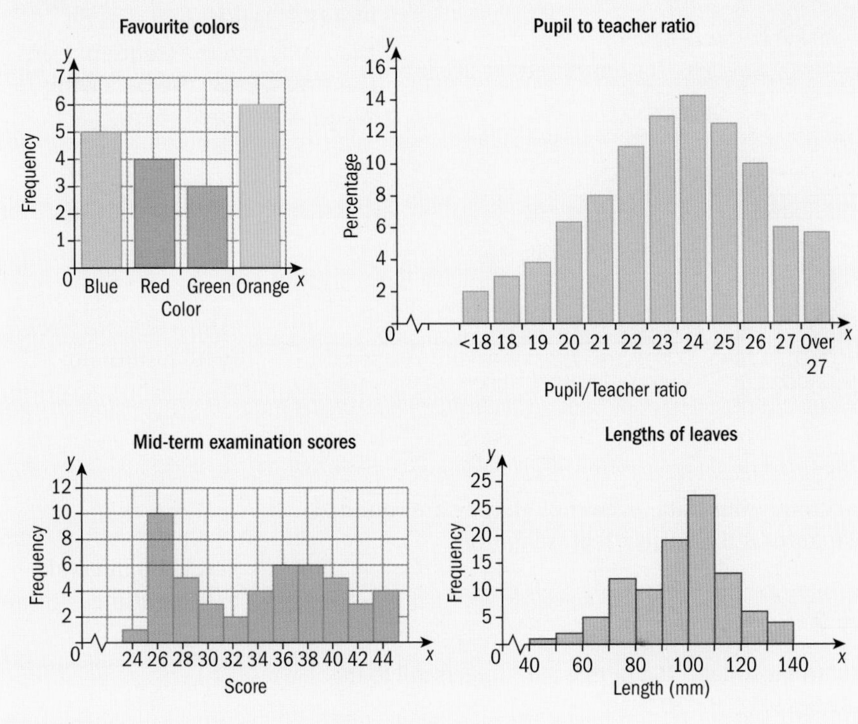

The word *histogram* originates from Greek words: *histos* meaning 'stands upright' and *gram* meaning 'a drawing'.

As the scores do not start at zero, the zigzag symbol indicates that some values are not included on the scores axis.

Frequency histograms for continuous data

Example 1

The table shows the weights in grams of 50 baby chicks hatched at a farm in one week. Draw a histogram to represent the data.

Weight, x (grams)	Freq	Weight, x (grams)	Freq
$19.5 < x \leq 20$	3	$22.5 < x \leq 23$	4
$20 < x \leq 20.5$	5	$23 < x \leq 23.5$	6
$20.5 < x \leq 21$	1	$23.5 < x \leq 24$	5
$21 < x \leq 21.5$	6	$24 < x \leq 24.5$	5
$21.5 < x \leq 22$	2	$24.5 < x \leq 25$	6
$22 < x \leq 22.5$	6	$25 < x \leq 25.5$	1

▶ Continued on next page

Use your frequency table to draw the histogram on graph paper.

Plot the class boundaries on the x-axis and the frequency on the y-axis. Draw the bars with no gaps between them.

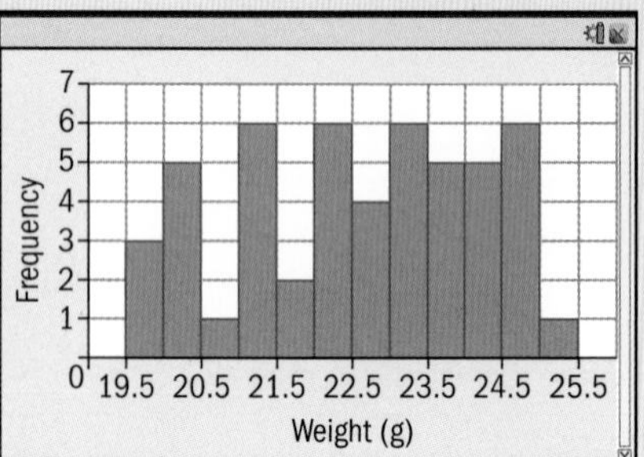

Use a GDC to check your histogram.

In Example 1, the weights are continuous data, even though they are rounded to one decimal place, so a histogram is the correct representation.

The class intervals are sometimes called *bins*.

ATL

Reflect and discuss 2

Ages are rounded differently to most data. If a fir tree is 10 meters tall to the nearest meter, its height h satisfies $9.5 \leq h < 10.5$.

If a woman is 40 years old, her age a satisfies $40 \leq a < 41$, where a is a continuous quantity.

The table gives the ages of musicians in an orchestra.

Age	18–25	26–33	34–41	42–50
Frequency	5	8	10	7

- How would you draw bars on a histogram to show these ages?
- What are the class boundaries for each class?
- How would you write the class interval, using ≤ and <, for the class 18–25?

Practice 1

1 The heights of the chicks from Example 1, measured to the nearest centimeter, are recorded in this grouped frequency table:

Height (cm)	Frequency
$1.5 \leq x < 2.5$	1
$2.5 \leq x < 3.5$	3
$3.5 \leq x < 4.5$	7
$4.5 \leq x < 5.5$	14
$5.5 \leq x < 6.5$	18
$6.5 \leq x < 7.5$	5
$7.5 \leq x < 8.5$	2

Draw a frequency histogram to represent the data.

2 The ages of 100 shoppers randomly chosen for a survey are given in the table.

Age	$14 \le x < 24$	$24 \le x < 34$	$34 \le x < 44$	$44 \le x < 54$	$54 \le x < 64$
Number of users	41	30	15	8	6

Draw a frequency histogram to represent the data.

Problem solving

3 The incomplete table and frequency histogram give some information about the masses (to the nearest kilogram) of some female polar bears.

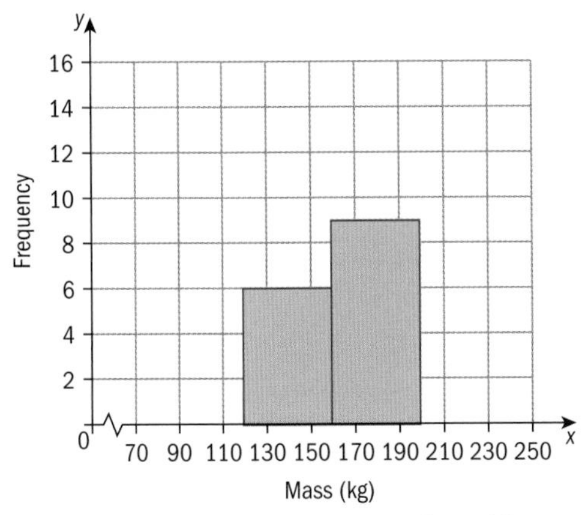

Class interval	Class boundaries	Frequency
80–119	79.5–119.5	15
120–159		
160–199		
200–239		8

a Use the histogram to complete the table.

b Complete the histogram.

Frequency histograms for discrete data

You can represent ungrouped discrete data with a vertical line graph.

This type of graph is not suitable for grouped discrete data. A single vertical line to represent the category 1–3 eggs could be misleading, but separate lines at 0, 1, 2 and 3 could be confusing.

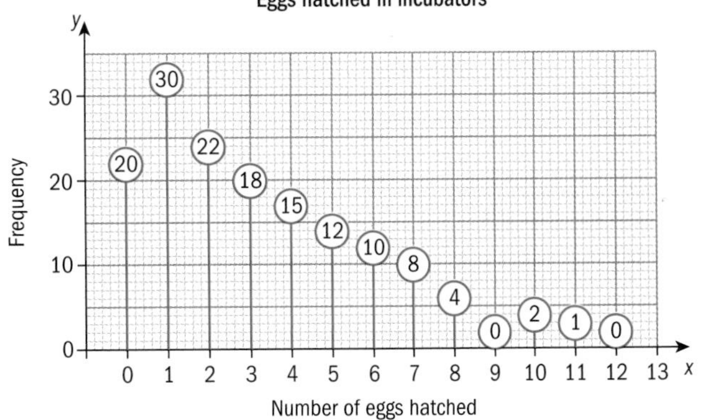

A histogram for grouped discrete data uses bars to represent each group.

For discrete data, 1–3, 4–6, 7–9, etc., plot the bars from the point halfway between the upper boundary of one class interval and the lower boundary of the next.

For 4–6, plot the bar between 3.5 and 6.5. Start the first bar at 0.5, to make all the bars the same width.

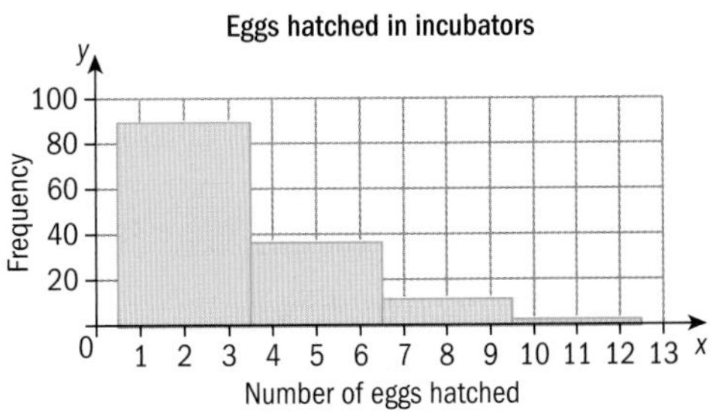

A histogram for ungrouped discrete data 1, 2, 3, 4 … uses bars from 0.5 to 1.5, 1.5 to 2.5, and so on.

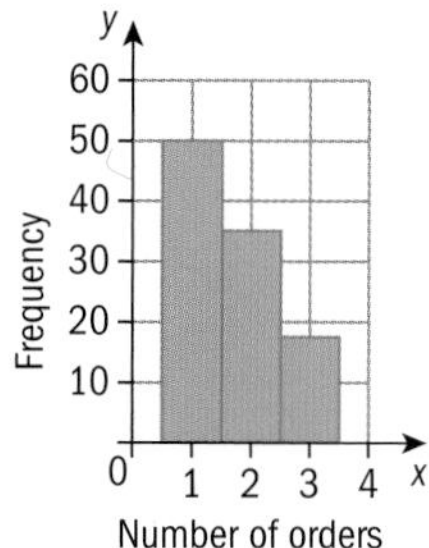

You plot the class intervals of cumulative frequency curves in a similar way.

Practice 2

1 The table shows the numbers of people travelling in 33 cars.

Number of people	1	2	3	4	5	6
Frequency	8	11	6	4	2	2

a Classify the data as discrete or continuous, grouped or ungrouped.

b Represent the data in a frequency histogram.

2 There were 25 trays of eggs in an incubator. Each tray started with 24 unhatched eggs. The numbers of eggs that hatched from each tray were:

12 18 17 13 9

21 2 5 12 1

5 14 11 13 15

18 21 14 15 9

12 14 15 16 24

a Construct a grouped frequency table for the data.

b Construct a frequency histogram for this data.

3 The lengths of 30 Swiss cheese plant leaves were measured to the nearest cm. The table shows the results.

Length of leaf (to the nearest cm)	$10 \le x < 15$	$15 \le x < 20$	$20 \le x < 25$	$25 \le x < 30$
Frequency	3	8	12	7

a Determine whether this is discrete or continuous data.

b Draw a frequency histogram to represent the data.

4 A taxi driver records the distances of his journeys, to the nearest km, over the course of a long weekend shift.

Distance (km)	1–5	6–10	11–15	16–20	21–25
Frequency	10	6	8	5	2

a Write down the modal class.

b Determine the class boundaries for each class.

c Draw a frequency histogram for this data.

For grouped continuous data, 9–12, 13–16 etc. are the class intervals. The class boundaries are 8.5–12.5 or $8.5 < x \le 12.5$, and so on.

Problem solving

ATL

5 This histogram has no title or axis labels.

a Determine whether it represents discrete or continuous data.

b Suggest what the data could be.

c Do you think that the data represented here contains any outliers? If not, why not?

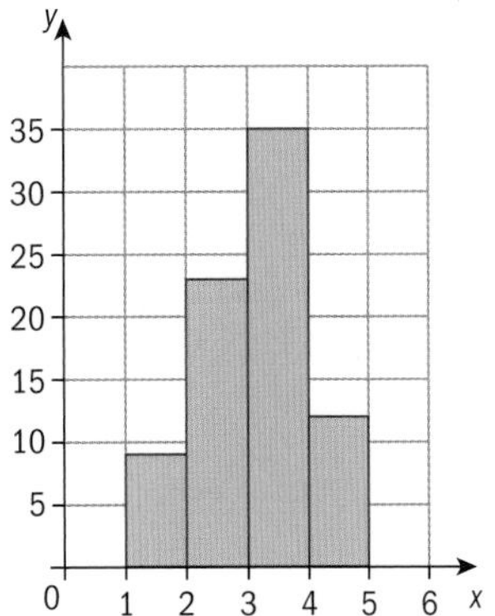

6 The histogram shows the number of hours that Macey spent doing homework each day in June.

a Use the information in the frequency histogram to construct a grouped frequency table.

b Use your frequency table to calculate an estimate for the mean time that Macey spent doing homework.

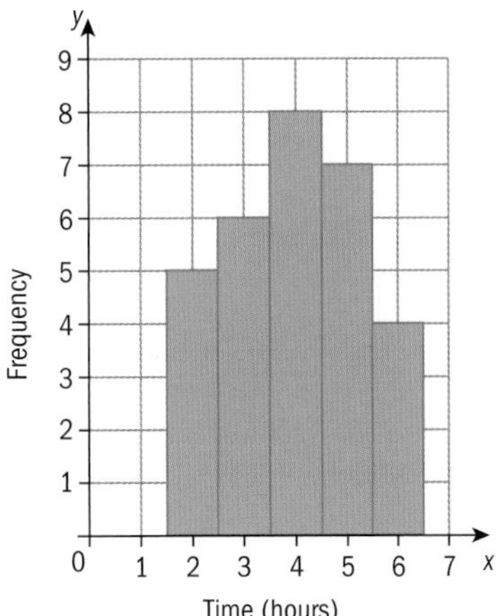

C Distributions in histograms

- How do you accurately analyze a data distribution from a histogram?

Examining a histogram can help you visualize certain characteristics of the data set it represents.

To comment on the distribution of data, describe:

Center	measure of central tendency
Spread	measure of dispersion
Outliers	extreme data values that don't fit the pattern
Shape	unimodal, bimodal or multimodal

Example 2

This frequency histogram shows the lengths in centimeters of 57 fish caught in the Kispiox River, British Colombia, Canada.

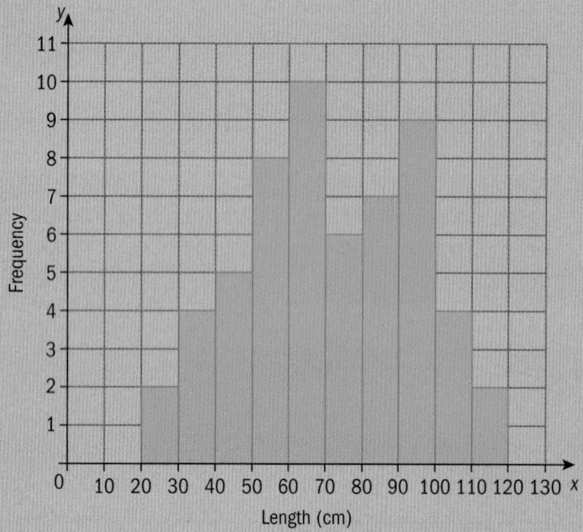

a Find:

i the modal class

ii the class interval that contains the median

iii an estimate for the range.

b Describe the distribution.

a i The modal class is $60 \leq x < 70$.

ii 57 data items

The median $= \frac{57+1}{2} = 29$th item of data.

The class $60 \leq x < 70$ contains the median.

Count the number of data items in each bar, until you reach the 29th.

iii Estimate for the range $= 120 - 20 = 100$ cm

b The median lies in the class $60 \leq x < 70$ and this is also the modal class. The lengths of the fish caught are widely spread from 20 cm to 120 cm. There are no outliers. The distribution is bimodal. Most fish caught have length $60 \leq x < 70$ cm or $90 \leq x < 100$ cm.

To describe a distribution, use a measure of central tendency (center) and a measure of dispersion (spread). State if there is an outlier, and describe the shape.

These three histograms all show the weights, to the nearest gram, of the same 50 eggs measured one week before hatching. The first histogram, for ungrouped data, appears to be a multimodal distribution.

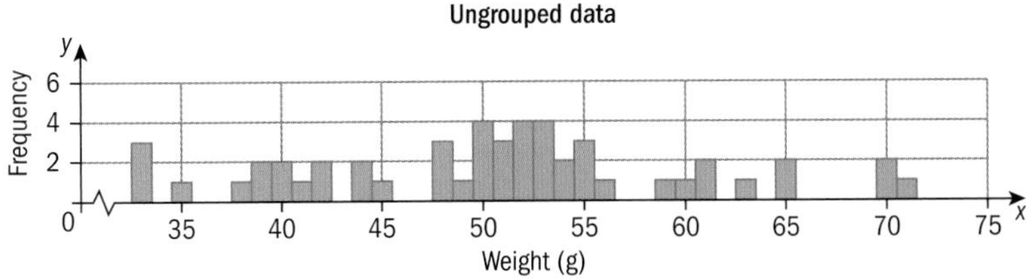

This appears to be a unimodal distribution (at $43 \le w < 53$) that is not symmetrical.

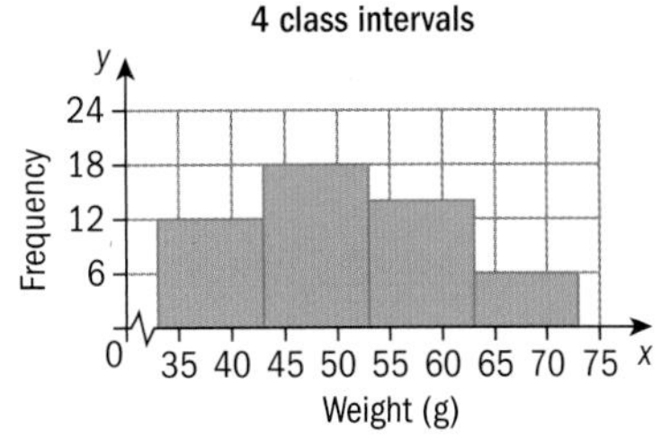

This appears to be a bimodal distribution that is fairly symmetrical.

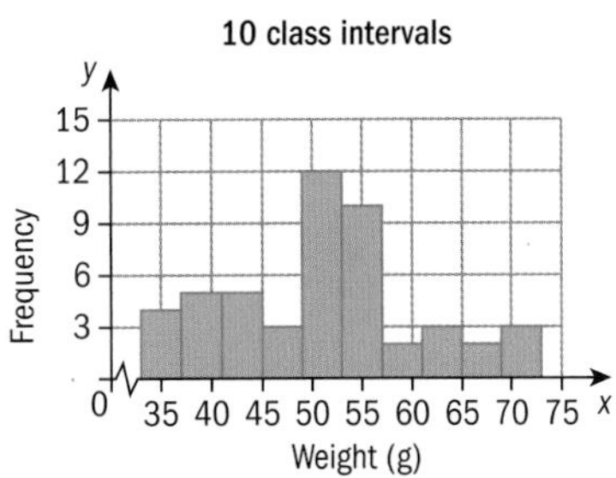

Reflect and discuss 3

- How can changing the class intervals change the shape of a distribution?
- What are the effects of having too few classes? What about too many classes?
- How many class intervals should you use? Do you need to consider the size of the data set when deciding on the number of classes?

Practice 3

1 Determine whether each histogram:

i is symmetrical **ii** is unimodal, bimodal or multimodal **iii** contains outliers

a

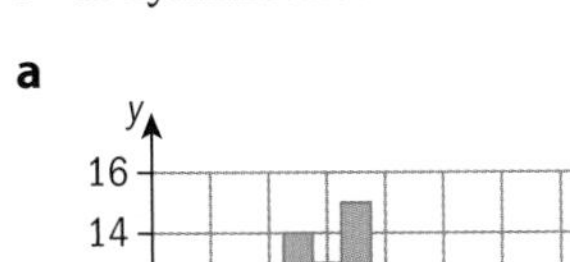
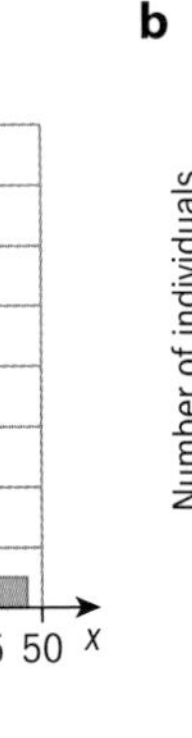

b

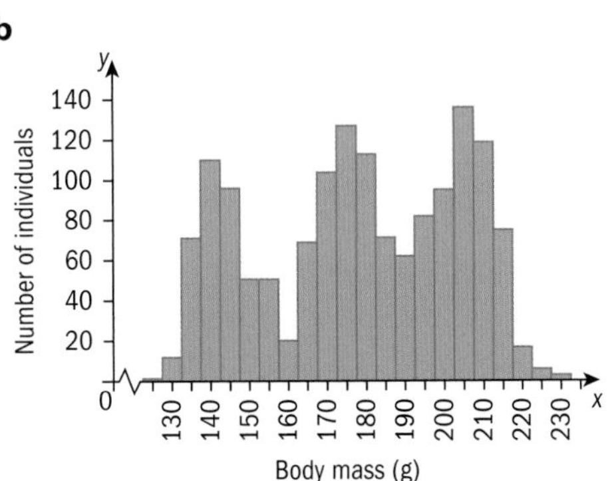

c

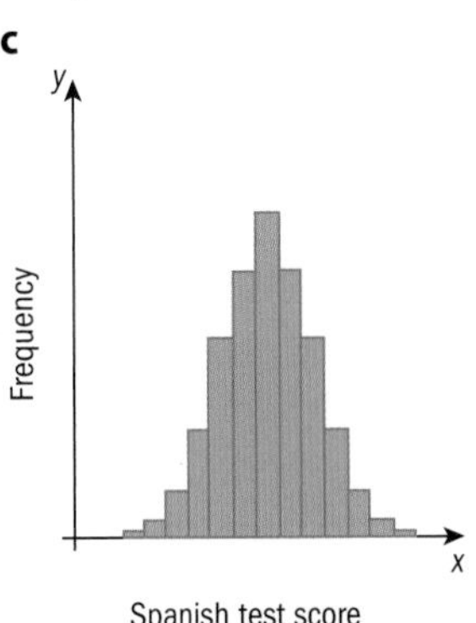

2 The histogram shows the lengths of 26 black rats.

a Write down the modal class.

b Find the class interval that contains the median.

c Calculate an estimate for the range.

d By first constructing a grouped frequency table, calculate an estimate for the mean length.

e Describe the distribution.

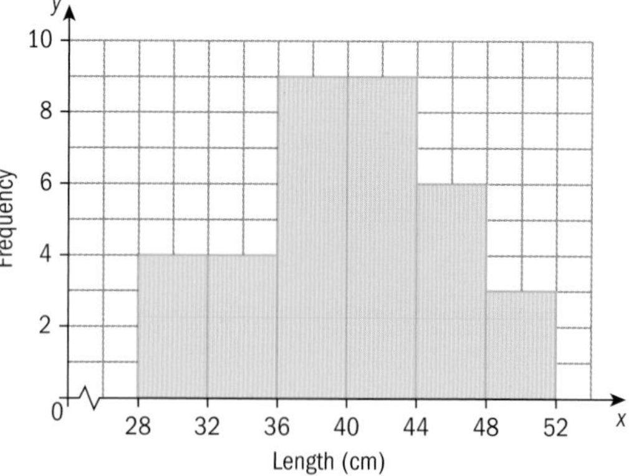

Problem solving

3 The histogram shows the amounts of money, to the nearest \$10, some families spend on food each week.

a Write down the total number of families.

b Calculate an estimate for the mean amount spent on food.

c Comment on the distribution of the data.

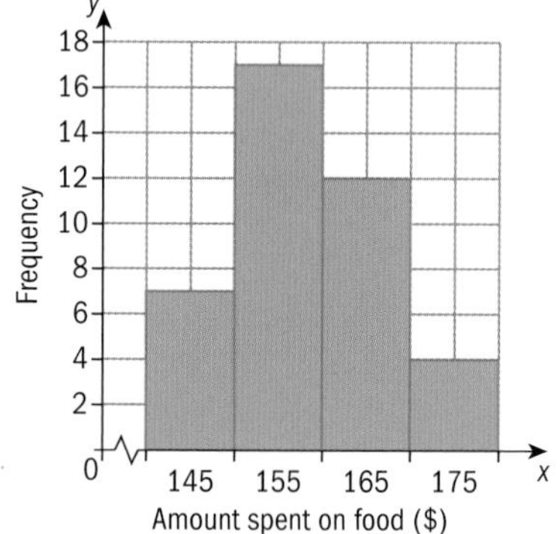

4 Imogen asked everyone in her class to tell her their birthday month.

a Explain why she should use a bar chart to show the results.

b Sketch a bar chart to predict what the distribution would look like.

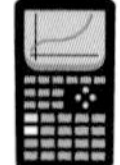

5 Here are the weights (to the nearest whole kg) of 50 emus in an Australian nature reserve in the Snowy Mountains.

33	19	24	35	36	24	29	29	29	34
38	35	35	35	36	60	35	50	34	48
41	41	51	42	35	36	32	61	30	40
41	19	33	34	17	35	35	38	35	42
20	29	50	33	37	28	49	58	45	40

a Construct a frequency histogram for this data.

b Describe the distribution.

Relative frequency histograms

In order to make fair comparisons between data sets with different numbers of data points, you can use a *relative frequency histogram*. Relative frequency shows the proportion of the total frequency in each class interval. It is plotted on the vertical axis.

Objective: **D.** Applying mathematics in real-life contexts **v.** justify whether a solution makes sense in the context of the authentic real-life situation
This exploration encourages students to see mathematics as a tool for solving problems in an authentic real-life context. The students should be able to justify whether or not the comparison of data makes sense in this specific case.

ATL

Exploration 1

The frequency tables show the lengths of fish caught in the Au Sable and Kalamazoo rivers.

Au Sable River	
Length, x (cm)	Frequency
$10 < x \leq 20$	2
$20 < x \leq 30$	52
$30 < x \leq 40$	93
$40 < x \leq 50$	30
$50 < x \leq 60$	25
$60 < x \leq 70$	33
$70 < x \leq 80$	30
$80 < x \leq 90$	30
$90 < x \leq 100$	40
$100 < x \leq 110$	25
$110 < x \leq 120$	30
$120 < x \leq 130$	5
$130 < x \leq 140$	7
Total frequency	

Kalamazoo River	
Length, x (cm)	Frequency
$10 < x \leq 20$	0
$20 < x \leq 30$	8
$30 < x \leq 40$	15
$40 < x \leq 50$	30
$50 < x \leq 60$	10
$60 < x \leq 70$	11
$70 < x \leq 80$	10
$80 < x \leq 90$	7
$90 < x \leq 100$	7
$100 < x \leq 110$	8
$110 < x \leq 120$	1
$120 < x \leq 130$	2
$130 < x \leq 140$	4
Total frequency	

1 Draw a frequency histogram for each data set. Use the same scale for both, with the y-axis from 0 to 100.

2 **a** Copy each table, adding a third column for 'Relative frequency'.

b Find the total frequency of each data set.

c For each class interval, calculate the relative frequency: $\left(\frac{\text{frequency}}{\text{total frequency}}\right)$ Round the values to 2 decimal places and add to your tables.

d Draw a relative frequency histogram for each data set. Plot the class boundaries on the x-axis and relative frequency on the y-axis.

3 Look at your histograms.

a Which pair of histograms should you use to compare the distribution of fish lengths in the two rivers? Explain fully.

b Which pair of histograms should you use to find the measures of central tendency? What about the measures of dispersion?

Relative frequency is the proportion (or percentage) of the data set belonging in the class interval.

For a data set with n members, a class interval with frequency f has relative frequency $\frac{f}{n}$.

Practice 4

ATL

1 The tables show the length of time some men and women spent on their mobile phones one day.

Time spent in minutes (men)	Frequency
$0 \le x < 15$	5
$15 \le x < 30$	8
$30 \le x < 45$	10
$45 \le x < 60$	5
$60 \le x < 75$	2

Time spent in minutes (women)	Frequency
$0 \le x < 15$	4
$15 \le x < 30$	5
$30 \le x < 45$	7
$45 \le x < 60$	14
$60 \le x < 75$	20

a Explain why you need to use a relative frequency histogram to compare these data distributions.

b Calculate the relative frequencies for each class interval.

c Draw relative frequency histograms for the two sets of data.

d Describe each distribution.

e Compare the length of time spent on the phone by the men and women.

Tip

When comparing two distributions, use descriptive words such as: wider, narrower, more varied, less varied.

2 The tables show the masses of some male and female Siberian huskies.

Male Siberian husky	
Mass, m (kg)	Frequency
$17 \le m < 19$	3
$19 \le m < 21$	6
$21 \le m < 23$	6
$23 \le m < 25$	11
$25 \le m < 27$	9

Female Siberian husky	
Mass, m (kg)	Frequency
$17 \le m < 19$	5
$19 \le m < 21$	8
$21 \le m < 23$	4
$23 \le m < 25$	2
$25 \le m < 27$	0

a Calculate the relative frequency of each class interval as a percentage.

b Draw relative frequency histograms for each data set.

c Comment on the distribution of the masses of the male and female Siberian huskies.

Problem solving

3 The histogram shows the relative frequency of items sold at different prices at a community fundraising event. There were 32 items in total.

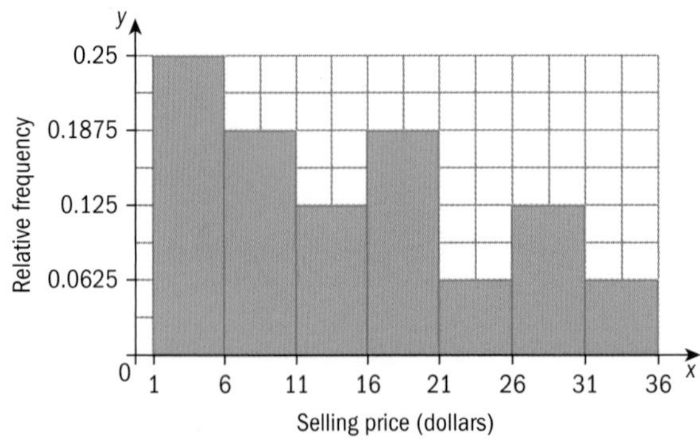

Calculate how many items sold for less than 16 dollars.

Histograms with unequal class widths

- How can real data ever be misleading?
- How do individuals stand out in a crowd?

This stem-and-leaf diagram represents the weights, measured to the nearest gram, of fish caught in a river.

stem	leaf
0	8 8 9
1	
2	1 2 2 2 3 4 8
3	2 3 5 5 5 6 7 7
4	1 2 2 3 3 3 3 3 3 3 3 3 3 3 3 3 3 4 4
	4 5 6 6 6 6 7 7 7 7 8 8 8 9 9 9 9 9
5	1 1 1 2 2 3 4 4 4 5 5 5 6 6 6 6 6 7 7
	7 7 7 7 8 8 8 8 8 8 8 8 9 9 9 9 9 9 9
6	0 0 0 0 0 0 1 1 1 2 2 2 3 3 3 4 4
	4 4 4 4 5 5 6 6 7 7 8 8 8 8 9 9 9
7	
8	
9	0 0 1 2 4 5 5 5 7 8

Key: 2 | 5 means 25 g

Both these histograms show the weights, to the nearest gram, of the same 136 fish.

This histogram represents the data grouped into small class intervals of equal width. This grouping has resulted in three class intervals containing no data.

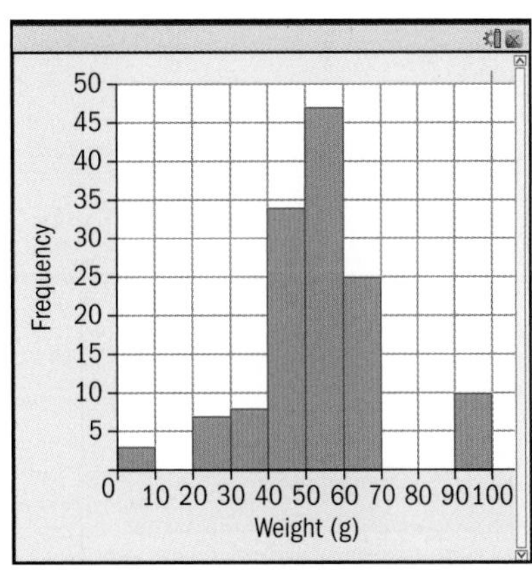

This histogram represents the data grouped into fewer, wider classes of equal width.

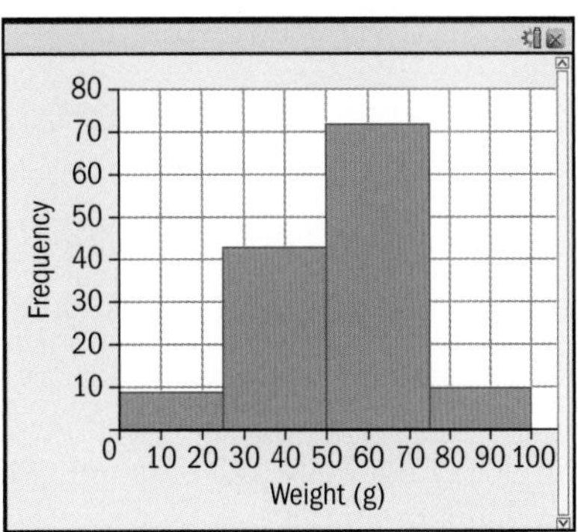

Reflect and discuss 4

Why does using a small number of wide classes make it difficult to comment on the shape of a distribution?

Both of the histograms are accurate representations of the data. Wide classes are appropriate when data is very spread out, and narrow classes help when the data is concentrated together. You can combine these ideas to create a better histogram with wide classes where the data is spread out, and narrow classes where the data is concentrated together.

Using unequal class widths, the grouped frequency table for the fish caught in the river could look like this:

Weight (g)	Frequency
$0 < w \leq 30$	10
$30 < w \leq 40$	8
$40 < w \leq 45$	20
$45 < w \leq 50$	16
$50 < w \leq 55$	12
$55 < w \leq 57$	11
$57 < w \leq 59$	15
$59 < w \leq 61$	9
$61 < w \leq 65$	14
$65 < w \leq 70$	11
$70 < w \leq 100$	10

Class width 30 g

Class width 10 g

Class width 5 g

Class width 2 g

A frequency histogram to represent this frequency table looks like this:

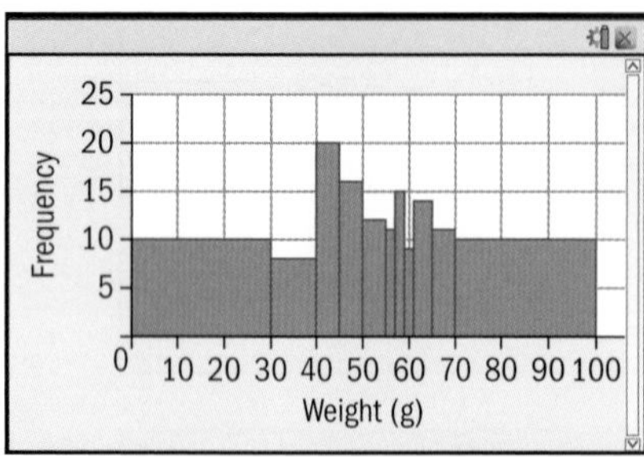

Reflect and discuss 5

- Do you think this histogram represents the data well?
- Does it obey the principle that in a histogram, the area of each bar is proportional to the frequency?

Because the height of each bar is its frequency, the bars representing wider classes have ended up looking much too large.

Frequency density histograms

When data is grouped in class intervals of unequal widths, a frequency density histogram represents the data so that the area of each bar is proportional to the frequency.

> In a **frequency density histogram**, the area of each bar is equal to the frequency for that class interval.
>
> The **frequency density** is plotted on the vertical axis.

For the fish weight data, the first class interval $0 < w \leq 30$ has frequency 10. The bar to represent this class interval has width 30 and area 10.

Area = frequency = 10

height (frequency density)

class width = 30

What is the height of this rectangle?

For any rectangle, area = height × width, so:

$$\text{frequency} = \text{frequency density} \times \text{class width}.$$

Rearranging this formula gives the definition of frequency density:

> $$\text{Frequency density} = \frac{\text{frequency}}{\text{class width}}$$

Example 3

The weights, to the nearest whole gram, of 50 Blue Heron chicks are given in this frequency table.

Weight, x (g)	Freq
$14 \leq x < 19$	14
$19 \leq x < 21$	10
$21 \leq x < 22.5$	9
$22.5 \leq x < 25$	10
$25 \leq x < 28$	7

Construct a frequency density histogram for this data.

Weight, x (g)	Freq	Class width	Frequency density
$14 \leq x < 19$	14	5	2.8
$19 \leq x < 21$	10	2	5
$21 \leq x < 22.5$	9	1.5	6
$22.5 \leq x < 25$	10	2.5	4
$25 \leq x < 28$	7	3	2.3

Add columns for class width and frequency density to the table.

$$\text{Frequency density} = \frac{\text{frequency}}{\text{class width}} = \frac{14}{5} = 2.8$$

Round decimals to a suitable degree of accuracy.

▶ Continued on next page

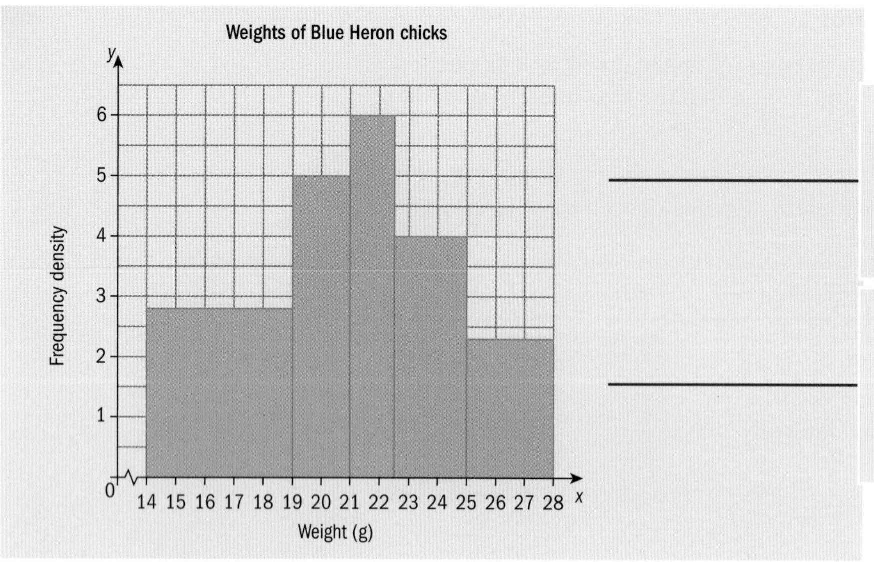

Draw the histogram with weight on the horizontal axis and frequency density on the vertical axis.

The first bar goes from 14 to 19 and has a height of 2.8. Draw the bars with no gaps between them.

Reflect and discuss 6

In the histogram in Example 3:

- What is the modal class? Why is this?
- How would you work out an estimate for the number of chicks who weighed less than 20 g? What about an estimate for the number who weighed between 22 g and 24 g? Why are they only estimates?

The mode of a data set is the most commonly occurring data point. With continuous data, such as the Blue Heron chick weight data, even if you had the weight of each chick recorded correct to the nearest 0.1 g, it's very likely that no two chicks would have exactly the same weight. You would probably have 50 chicks, all with different weights, so there would be no mode.

With continuous data, what you are really interested in is where the data is most dense – in other words, where the weights are most closely clustered together. In this case, that is the class interval with the greatest frequency density: $21 \le x < 22.5$.

The **modal class** is the class with the greatest frequency density.

Practice 5

1 The lengths, in centimeters, of some horned vipers are recorded in the table.

Length, x (cm)	Frequency	Class width	Frequency density
$45 \le x < 55$	13	10	1.3
$55 \le x < 60$	20		
$60 \le x < 65$	26		
$65 \le x < 72$	19		

a Copy and complete the table.

b Construct a frequency density histogram for this data.

Problem solving

2 The ages of people visiting the Eiffel Tower are recorded in the table.

Ages, x (years)	Frequency
$0 \le x < 15$	15
$15 \le x < 25$	28
$25 \le x < 40$	30
$40 \le x < 60$	42
$60 \le x < 100$	20

In 2015 there were 6.91 million visitors to the Eiffel Tower. The upper deck is 276 meters above ground, the highest in the European Union.

a Construct a frequency density histogram for this data.

b From the graph, Voleta estimates that there were ten visitors between 60 and 80 years old, and ten visitors between 80 and 100 years old.

i Explain the assumption Voleta has made.

ii State whether you think this is a reasonable assumption. Justify your answer.

c State the modal class for this data.

3 The doctors at a medical center see 80 patients in total during morning surgery. The table gives information about the lengths of these appointments in minutes.

Length, L (min)	Frequency
$0 < L \le 5$	7
$5 < L \le 7$	15
$7 < L \le 9$	24
$9 < L \le 13$	19
$13 < L \le 16$	9
$16 < L \le 20$	6

a Construct a frequency density histogram for the information.

b Calculate an estimate for the number of patients who saw their doctor for between 10 and 15 minutes.

4 The table shows the times taken by members of a cycling club to finish a 100 km charity ride.

Time, t (min)	$110 < t \le 130$	$130 < t \le 145$	$145 < t \le 160$	$160 < t \le 185$	$185 < t \le 195$	$195 < t \le 250$
Frequency	6	30	45	55	28	16

a Construct a frequency density histogram for the information.

b State the modal class.

c Calculate an estimate for the number of cyclists who took between $2\frac{1}{2}$ and $3\frac{1}{2}$ hours to complete the ride.

5 The histogram represents the distances students threw the javelin on school sports day.

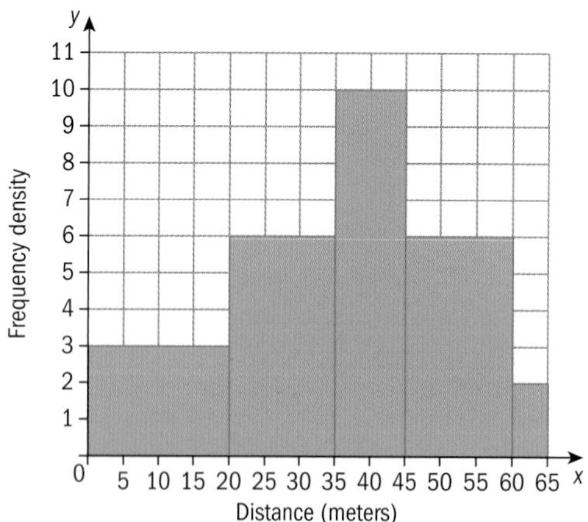

In javelin the distance of the throw is rounded down to the nearest centimeter.

a Construct a frequency table for the distribution.

b Find the number of throws.

c Calculate an estimate for the mean distance.

d Find the class interval that contains the median.

6 This histogram represents children's heights. Fifteen children measured between 110 cm and 125 cm.

Find the number of children in the sample.

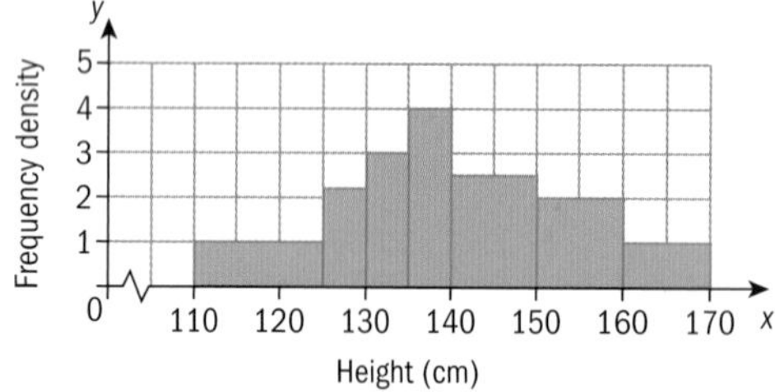

7 In a histogram, a bar of width 2 cm and height 5 cm represents a frequency of 50. A second bar is 4 cm wide and represents a frequency of 50. Calculate the height of this bar.

Summary

A **bar chart** represents qualitative data.

- All the bars have equal width.
- There are spaces between the bars.
- The **height** of the bar represents the frequency.
- Frequency is on the vertical axis.

A **histogram** represents quantitative discrete data and all continuous data.

- The bars may or may not be of equal width.
- There are no spaces between the bars.
- The **area** of the bar represents the frequency.
- Frequency, relative frequency or frequency density are on the vertical axis.

$$\text{Frequency density} = \frac{\text{frequency}}{\text{class width}}$$

The modal class is the class with the greatest frequency density.

To comment on the distribution of data, describe:

Center measure of central tendency

Spread measure of dispersion

Outliers extreme data values that don't fit the pattern

Shape unimodal, bimodal or multimodal.

Relative frequency is the proportion (or percentage) of the data set belonging in the class interval.

For a data set with n members, a class interval with frequency f has relative frequency $\frac{f}{n}$.

In a **frequency density histogram**, the area of each bar is equal to the frequency for that class interval.

The **frequency density** is plotted on the vertical axis.

$$\text{Frequency density} = \frac{\text{frequency}}{\text{class width}}$$

The modal class is the class with the greatest frequency density.

Mixed practice

1 The table shows the heights of 24 students.

Height, x (cm)	Frequency
$140 \le x < 150$	6
$150 \le x < 160$	7
$160 \le x < 170$	6
$170 \le x < 180$	4
$180 \le x < 190$	1

a **Explain** why a frequency histogram can be used to represent this set of data.

b **Construct** the frequency histogram.

2 A restaurant records the number of pizzas served from 8pm to 9pm on consecutive nights during one week. The table shows the results.

Number of pizzas served	5	6	8	9	10
Frequency	3	8	12	4	5

Draw a frequency histogram for this data.

3 The weights of 50 hoglets are recorded to the nearest gram.

93	88	105	90	92	89	90	88	108	86
103	96	94	100	104	98	100	120	115	94
84	130	110	125	115	112	105	129	118	105
95	124	112	96	114	128	132	95	129	122
120	85	93	108	105	88	95	105	123	97

a **Construct** a grouped frequency table for this data.

b **Construct** a histogram to represent this data.

4 Fifty students recorded how much they spent on eating out in one month (to the nearest dollar). The results are shown in the partially complete frequency table:

Amount spent (dollars)	Class boundaries	Frequency
1–10	0.50–10.50	14
11–20	10.50–20.50	13
21–30		15
31–40		0
41–50		4
51–60		3
61–70		1

a **Draw** and clearly **label** a frequency histogram to represent this information.

b **State** the modal class.

c **State** the class that contains the median.

d **Describe** the distribution.

5 The table shows the age distribution of teachers in a school.

Age, x (years)	Number of teachers
$20 < x \le 30$	6
$30 < x \le 40$	4
$40 < x \le 50$	3
$50 < x \le 60$	2
$60 < x \le 70$	2

a **Calculate** an estimate for the mean age.

b **Construct** a histogram to represent this data.

c **Describe** the distribution.

6 The heights of 14-year-old students from Sri Lanka and Peru are given in the frequency tables below.

Heights of students in Sri Lanka (measured to the nearest cm)	Frequency
120–129	3
130–139	11
140–149	14
150–159	8
160–169	6

Heights of students in Peru (measured to the nearest cm)	Frequency
120–129	22
130–139	41
140–149	32
150–159	6
160–169	6

a Explain why it is necessary to construct a relative frequency histogram to compare these data sets.

b Comment on the distributions of the heights of 14-year-old students in Sri Lanka and Peru.

7 The frequency distribution for the masses (to the nearest kg) of 35 parcels is given in the table.

Mass (kg)	6–8	9–11	12–17	18–20	21–29
Frequency	4	6	10	3	12

a Determine the class boundaries for each class.

b Explain why you cannot draw a frequency histogram for this data.

c Copy and complete the frequency density table.

Mass (kg)	Frequency	Class width	Frequency density
6–8	4	3	
9–11	6	3	
12–17	10	6	
18–20	3	3	
21–29	12	9	

d Construct a frequency density histogram.

8 In New Zealand, 90 mothers were asked how old they were when they had their first child. The frequency table gives the results.

Age, x (years)	Frequency
$15 < x \le 20$	6
$20 < x \le 22$	14
$22 < x \le 25$	19
$25 < x \le 30$	21
$30 < x \le 40$	30

a Determine which type of histogram would give the best representation.

b Construct the histogram.

c Calculate an estimate for the mean and **determine** the modal class.

Problem solving

9 The histogram shows the lengths of a sample of European otters.

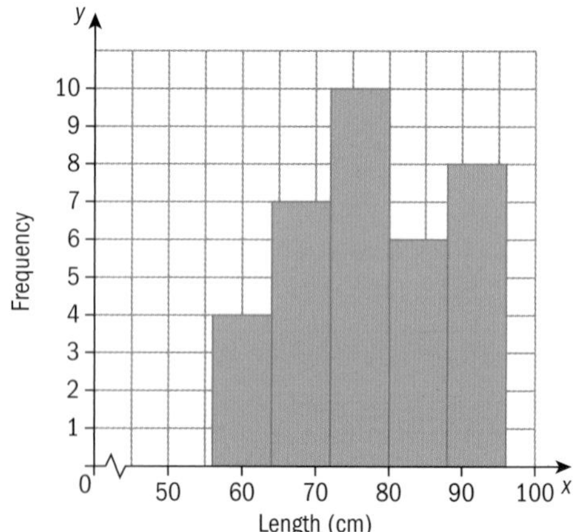

a Find the class interval that contains the median.

b Calculate an estimate for the mean length.

10 The histogram shows the waiting times for some customers at a supermarket.

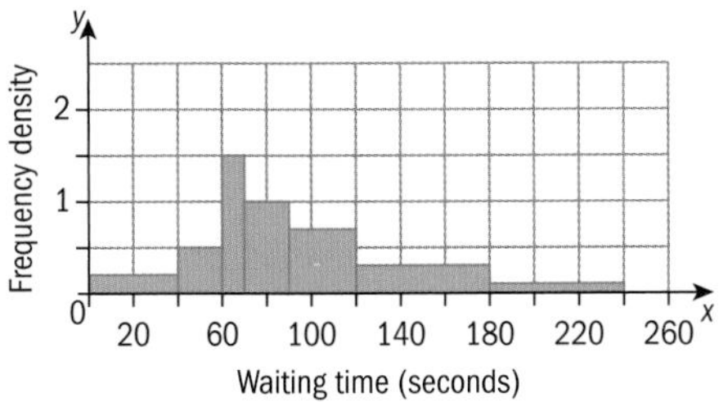

a **State** the modal class.

b **Calculate** an estimate for the mean waiting time.

c **Find** the total number of customers.

d **Find** an estimate for the number of customers who had to wait:

i between 60 and 90 seconds

ii longer than 2 minutes.

Review in context

Globalization and sustainability

1 Twenty-five women from Mexico were asked how many children they had. The results are shown in the frequency histogram.

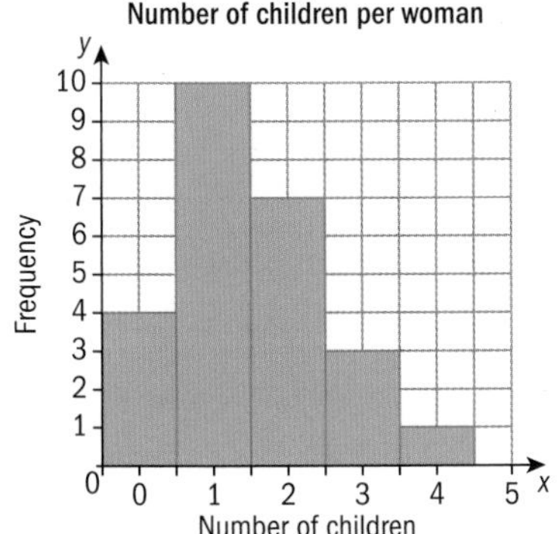

a **Show that** the mean number of children per woman is 1.48.

b A group of 25 women from Australia were asked the same question. The results are given in this table.

Number of children	0	1	2	3	4	5
Frequency	4	8	5	4	5	2

Use the results from parts **a** and **b** to **describe** and **compare** the distribution of the numbers of children per woman in Australia and Mexico.

2 The amount of land covered by rainforest in the Amazon region has been rapidly declining over the past decades. This frequency table show the percentage of deforestation for 20 South American countries.

0.4%	0.3%	1.6%	2.1%	0.3%
2.7%	2.2%	3.5%	2.0%	9.2%
3.2%	0%	3.0%	3.3%	4.0%
0.6%	0.5%	4.0%	1.7%	2.6%

a **Sketch** a frequency histogram for this data.

b **Comment** on the distribution.

c **Determine** any data that could be outliers.

Reflect and discuss

How have you explored the statement of inquiry? Give specific examples.

Statement of Inquiry:

How quantities are represented can help to establish underlying relationships and trends in a population.

7 Measurement

A method of determining quantity, capacity or dimension using a defined unit

Is 10 always less than 100?

Would you rather

- wait 10 minutes or 100 seconds
- pay 100 cents or $10
- walk 10 km or 100 meters
- have 100 g or 10 kg of gold
- carry 10 liters or 100 ml of water?

Quantity + unit = measurement

Why do we need standard units?

Humans devised systems of measurement to describe and compare quantities. Early measurements were based on body parts (thumbs, hands, forearms) or other objects that were readily available, for example: 3 barleycorns make 1 inch.

US and UK shoe sizes are still based on barleycorns. The difference between size 6 and size 7 is 1 barleycorn (one-third of an inch).

Standard units such as feet, pounds and miles defined measurements more precisely, but varied from country to country. The idea of the metric system was to be a system 'for all people, for all time'. Metric units are understood worldwide, although some countries still use their own systems.

- What are the disadvantages of using non-standard units such as hands, paces or barleycorns?
- What are the advantages of using standard units? Are there any disadvantages?

How big is it?

Although we have sophisticated measuring tools and standard units, most people are not very good at visualizing or interpreting measurements.

- Would a sofa measuring 160 cm × 80 cm × 70 cm fit in a family car?

As big as a whale

Media reports use comparisons such as 'an area the size of two football pitches' or 'an area twice the size of Belgium.'

A blue whale can be up to 30m long and weigh as much as 200 tonnes.

- Which of these two comparisons gives you the best idea of the size of a blue whale?

A blue whale can be as long as 9 family cars and weigh as much as 33 elephants.

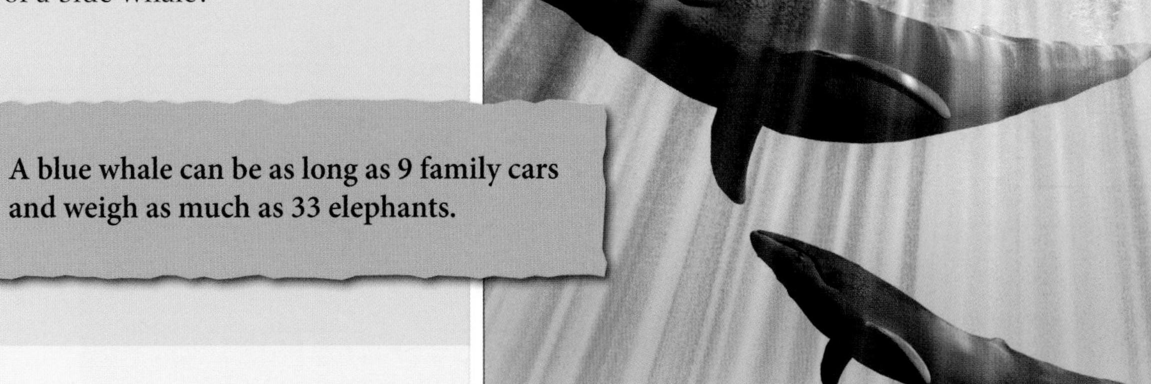

Yes, I'm absolutely positive

Global context: Globalization and sustainability

RELATIONSHIPS

Objectives

- Knowing different definitions for the absolute value of a number
- Understanding the properties of the absolute value of a number

Inquiry questions

F
- What is an absolute value?

C
- How are the properties of absolute value similar to other properties in mathematics?
- How can error be measured when estimating or approximating?

D
- How can error best be measured?
- Can you ever know the exact value of a measurement?

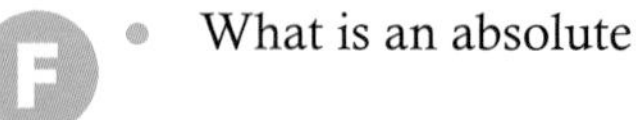

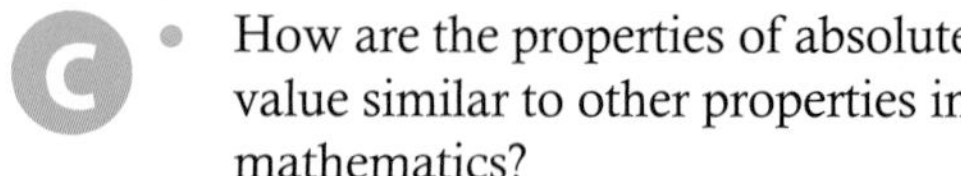

ATL **Critical thinking**

Draw reasonable conclusions and generalizations

Statement of Inquiry:

Quantities and measurements illustrate the relationships between human-made systems and communities.

6.1

7.1

7.2

You should already know how to:

• calculate with negative integers	**1** Work out these calculations. **a** $4-6$ **b** $-3+7$ **c** -5×3 **d** $18\div -2$
• find the square root of a number	**2** Work out these expressions. **a** $\sqrt{121}$ **b** $\sqrt{7^2}$ **c** $\sqrt{\frac{1}{4}}$ **d** $\sqrt{0.01}$
• substitute positive and negative numbers into expressions	**3** When $a=4$ and $b=-2$, find: **a** $a+b$ **b** $a-b$ **c** a^2 **d** b^2 **e** $\sqrt{a}$ **f** $-ab$

F Introduction to absolute value

- What is an absolute value?

Activity

State the temperature difference between:

1 sunrise and noon

2 sunrise and sunset

3 sunrise and night

4 noon and sunset

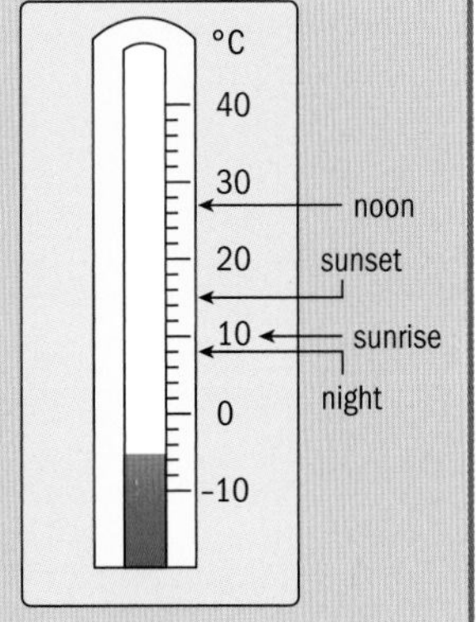

To calculate a difference between two quantities a and b, you subtract one from the other. In this Activity, did you always calculate $a-b$ or $b-a$? Or did you choose the one that gave a positive answer?

Here is one way of calculating the temperature differences:

1 Between sunrise and noon, the temperature *rose* by 17°, so the difference is 27° – 10° = 17°.

2 Between sunrise and sunset, the temperature *rose* by 5°, so the difference is 15° – 10° = 5°.

3 Between sunrise and night, the temperature *dropped by* 2°, so the difference is 8° – 10° = –2°

4 Between noon and sunset, the temperature *dropped* by 12°, so the difference is 15° – 27° = –12°.

Reflect and discuss 1

When we talk about differences in real-life situations, why do we usually use the positive differences?

After playing a game, asking the question 'Who won?' tells you who had the highest score. Asking 'By how much?' gives you the difference between the two scores. The winners may say 'We won by 7 points'. The losers may say 'We lost by 7 points'. It would be highly unusual for someone to say 'We had −7 points'.

The negative difference is implied in the context. When a team loses by 7 points, it has obviously scored 7 points *less* than the other team, and not 7 points *more*. If the temperature dropped by 20°, it obviously went *down* and not *up*. The numbers used are always positive, because we use the absolute value.

The **absolute value** of a number is the distance on the number line between that number and 0. We write $|x|$ to mean 'the absolute value of x'.

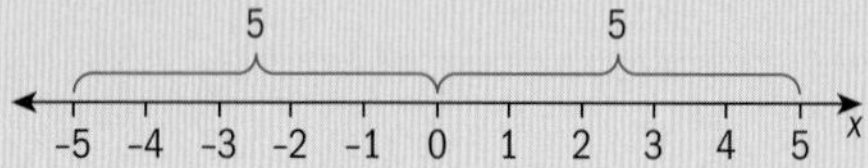

For example, $|5| = 5$, and $|-5| = 5$.

Tip

On some calculators the absolute value function button is abs(x).

Using absolute value notation, the temperature differences you calculated in the Activity are:

1 Between sunrise and noon:
$|27° - 10°| = |17°| = 17°$

2 Between sunrise and sunset:
$|15° - 10°| = |5°| = 5°$

3 Between sunrise and night:
$|8° - 10°| = |-2°| = 2°$

4 Between noon and sunset:
$|15° - 27°| = |-12°| = 12°$

Exploration 1

1 Let $x = a$ if $a \geq 0$, and $x = -a$ if $a < 0$. Find x for each value of a.

a $a = 12$ **b** $a = 0$ **c** $a = -5$ **d** $a = \frac{3}{5}$

e $a = -\frac{3}{5}$ **f** $a = 3.14159$ **g** $a = -3.14159$ **h** $a = -0.003$

Describe what you notice about your answers.

State your findings as a general rule.

2 Give the positive square root of:

a $\sqrt{3^2}$ **b** $\sqrt{5^2}$ **c** $\sqrt{(-2)^2}$ **d** $\sqrt{12^2}$

e $\sqrt{(-12)^2}$ **f** $\sqrt{\left(\frac{1}{2}\right)^2}$ **g** $\sqrt{\left(-\frac{1}{3}\right)^2}$ **h** $\sqrt{(-a)^2}$

Describe what you notice about your answers.

State your findings as a general rule.

Definition of absolute value

1 $|x| = \begin{cases} -x, \text{ if } x < 0 \\ x, \text{ if } x \geq 0 \end{cases}$

2 $|x| = \sqrt{x^2}$

The first definition is an example of a *piecewise function*. The function has a different definition for the different values of x.

Reflect and discuss 2

- Why can't you take the absolute value of the square root of a negative number, for example: $\sqrt{-4}$ or $\sqrt{-9}$?
- Can you take the square root of the absolute value of any real number ($\sqrt{|a|}$, if $a \in \mathbb{R}$)?
- Can you take the absolute value of any real number?

Practice 1

1 Write down the value of the following. Leave you answer as a simplified surd when possible.

a $|35|$ **b** $-|35|$ **c** $|-234|$ **d** $-|-234|$

e $|-5.6|$ **f** $|2.8|$ **g** $|0|$ **h** $\left|\frac{4}{8}\right|$

i $\left|-\frac{2}{5}\right|$ **j** $|10^{-2}|$ **k** $|-4^3|$ **l** $|\sqrt{36}|$

m $-|\sqrt{25}|$ **n** $|-\sqrt{32}|$ **o** $-|-\sqrt{24}|$ **p** $-|(-10)^3|$

2 Find the value of:

a $|4+9|$ **b** $|-2+5|$ **c** $|12-9|$ **d** $|4-9|$

e $|5.1-7.2|$ **f** $|3\times 6|$ **g** $|-7.5\times 2|$ **h** $\left|-\frac{3}{5}\right|\times 10$

i $|-3|+|8|$ **j** $|-3|-|8|$ **k** $|-4|\times|-2|$ **l** $|-4\times -2|$

Problem solving

3 a Write two numbers that have an absolute value of 4.

b Write two calculations whose answers have an absolute value of 6.

4 On a highway, driving too fast or too slow can cause accidents. For safe highway driving, there should be no more than a 20 km/h difference between cars' speeds.

a Gauthier is driving at 97 km/h. Nathan is driving at 113 km/h on the same section of highway. Determine whether or not they are driving safely relative to each other.

b The average speed on a stretch of highway is 110 km/h. If Louis is travelling at the average speed, determine the minimum and maximum speeds for others to be driving safely relatively to Louis on this stretch of highway.

C Properties and applications of absolute value

- How are the properties of absolute value similar to other properties in mathematics?
- How can error be measured when estimating or approximating?

The absolute value is also called the modulus or the numerical value.

Objective: **B.** Investigating patterns
ii. describe patterns as general rules consistent with findings

In Exploration 2 you will be looking for patterns. Organizing the information in a table allows you to observe trends.

ATL

Exploration 2

1 Determine which of these real numbers could be an absolute value.

a 5 **b** -392 **c** -2.625 **d** 1.001

e -0.005 **f** $\sqrt{25}$ **g** $\sqrt{17}$ **h** $-\sqrt{2}$

Write down your findings as a general property.

2 Find $|a \times b|$ and $|a| \times |b|$ for each pair of values a and b.

a $a = 11, b = 7$ **b** $a = -6, b = 4$ **c** $a = 8, b = -\frac{1}{4}$

d $a = -0.2, b = -15$ **e** $a = -\sqrt{3}, b = -3$ **f** $a = -\frac{1}{6}, b = 6$

Describe what you notice.

Write down your findings as a general property.

3 Find $\left|\frac{a}{b}\right|$ and $\frac{|a|}{|b|}$ for each pair of values of a and b.

a $a = 3, b = 10$ **b** $a = -2, b = 4$ **c** $a = 1, b = -5$

d $a = -3, b = -30$ **e** $a = -\sqrt{2}, b = -2$ **f** $a = -0.2, b = -0.8$

Describe what you notice.

Write down your findings as a general property.

4 Find $|a^2|$, $|a|^2$ and a^2 for the following values of a:

a $a = 5$ **b** $a = -3$ **c** $a = \frac{1}{2}$

d $a = \sqrt{6}$ **e** $a = -\sqrt{11}$

Describe what you notice.

Write down a general rule consistent with your findings.

5 Copy and complete this table:

a	$\lvert a^3\rvert$	$\lvert a\rvert^3$	a^3	$\lvert a^4\rvert$	$\lvert a\rvert^4$	a^4	$\lvert a^5\rvert$	$\lvert a\rvert^5$	a^5	$\lvert a^n\rvert$	$\lvert a\rvert^n$	a^n
2												
−2												
−3												

Describe any generalizations that you notice.

Continued on next page

Write down your findings as a general rule using the last three columns to help you.

6 Copy and complete this table:

a	b	$\|a+b\|$	$\|a\|+\|b\|$	$\|a-b\|$	$\|b-a\|$	$\|a\|-\|b\|$
8	3					
6	−4					
−1	9					
−5	−7					
$\frac{1}{4}$	$-\frac{1}{2}$					

Draw conclusions from the table.

Use the symbols ≥, ≤ and = to complete these general properties:

a $|a-b|$ □ $|b-a|$ **b** $|a+b|$ □ $|a|+|b|$

c $|a+b|$ □ $|a|-|b|$ **d** $|a-b|$ □ $|a|-|b|$

Properties of absolute value

For any numbers $a, b \in \mathbb{R}$:

1 $|a| \ge 0$ **2** $|a \times b| = |a| \times |b|$ **3** $\left|\frac{a}{b}\right| = \frac{|a|}{|b|}$ **4** $|a|^2 = a^2$

5 $|a^n| = |a|^n$ **6** $|a-b| = |b-a|$ **7** $|a+b| \le |a|+|b|$ **8** $|a-b| \ge ||a|-|b||$

Practice 2

Substitute the values for a and b to find the value of each expression.

1 $a = 7, b = -4$:

a $|a+2b|$ **b** $|a-2b|$ **c** $|-2a+b|$

d $|-2a-b|$ **e** $|3b-3a|$ **f** $|b^3|$

2 $a = -5, b = 3$:

a $a+|a+b|$ **b** $-a+|a-b|-b$ **c** $a+|a|+b+|b|$

d $3a-|a|$ **e** $3b-|b|$ **f** $|a^4|$

3 $a = 1, b = -1$:

a $a + |a| + b + |b|$ **b** $-b - |b| - a - |a|$ **c** $\frac{a + |b|}{b - |a|}$

d $\left|\frac{a + |b|}{b - |a|}\right|$ **e** $\frac{a + |b|}{a - |b|}$ **f** $\left|a^3 - |b^{11}|\right|$

Problem solving

4 Determine values of b for which $|8 - b| - |b - 3|$ is a positive number.

Any measurement, such as a length, is made to a certain degree of accuracy. For example, you might measure the distance between two cities in kilometers, and the height of a person in meters and centimeters. You choose a degree of accuracy that is 'good enough' to give an idea of the actual distance or height. Would it make sense to calculate the distance between cities to the nearest centimeter? Or a person's height to the nearest kilometer?

Every measurement you make is accurate only to a certain degree, so it is not an exact value. You can weigh an object to the nearest milligram, and this is more accurate than weighing it to the nearest gram or to the nearest centigram, but not as accurate as weighing it to the nearest microgram.

length = 17.4 cm

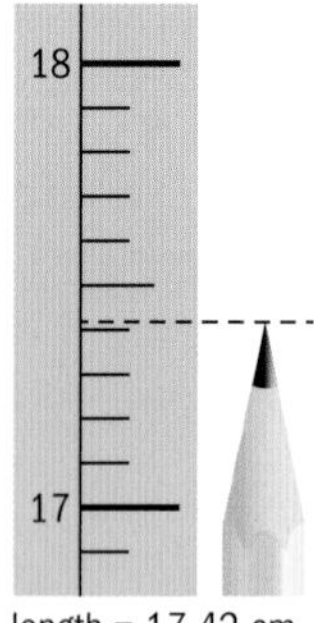

length = 17.42 cm

The **true value** of a measurement is the value that would be obtained by a perfect measurement. This is, by its nature, not really possible to obtain. **Accuracy** describes how close a measurement is to the true value.

In the following exploration, example, and practice we use the most accurate possible measurement as the true value.

ATL

Exploration 3

1 Measure your height, the length of a piece of paper, and the width of an eraser to the nearest centimeter. Copy the table and write your measurements in the first column.

2 Now measure your height, the length of a piece of paper, and the width of an eraser as accurately as you can, to the nearest millimeter. Consider these to be the true values for these measurements. Write them in the second column of your table.

	Measurement to the nearest cm	True value – most accurate measurement (to the nearest mm)	Absolute error	Relative error	Percentage error
Your height					
The length of a piece of paper					
The width of an eraser					

3 Calculate the absolute error for each measurement using this definition:

> The **absolute error**, Δx, of a measurement is the actual amount of error in a measurement.
>
> $\Delta x = |\text{true value} - \text{measured value}|$

Write the absolute errors in your table.

4 Compare the absolute errors. Write down what you notice.

5 Calculate the relative error for each measurement using this definition:

> The **relative error** is the error in a measurement compared to the size of the thing being measured.
>
> Relative error $= \frac{\Delta x}{x} = \frac{\text{absolute error}}{\text{true value}}$

6 Write the relative errors in your table.

7 Calculate the percentage error for each measurement using this definition:

> The **percentage error** is the relative error of a measurement written as a percentage.

8 Write the percentage errors in the table.

9 Compare the percentage errors. Write down what you notice.

Example 1

Frederico measured his height to be 132 cm. His doctor then measures him and says that Frederico has made an error of 5% in his measurement.

a Find Frederico's true height.

b Explain why there are two possible values for his true height.

a An error of 5% means a relative error of 0.05. — Write the percentage error as a relative error.

Let x = true value

$|132 - x|$ = absolute error

$\text{Relative error} = \frac{\text{absolute error}}{\text{true value}}$ — Substitute into the definition.

$0.05 = \frac{|132 - x|}{x}$ — Solve for x.

$0.05x = |132 - x|$

$0.05x = 132 - x$ or $0.05x = x - 132$

$1.05x = 132$ $\quad -0.95x = -132$

$x = 125.714$ $\quad x = 138.947$

Frederico's true height is either 126 cm or 139 cm. — Use the same level of accuracy as in the question.

b Absolute error does not tell us whether Frederico's measurement is above or below his true height – we would need more information to know which.

Practice 3

1 Elias estimated that 20 people would attend an evening lecture. In fact, 23 people attended. Find the percentage error in his estimate.

2 Breanna measured the capacity of a mug five times. Her results were: 240.5 ml, 240.9 ml, 240.2 ml, 239.5 ml, 241.2 ml.

- **a** Calculate the mean value for the capacity.
- **b** The true capacity of the mug is 240.1 ml. Calculate the percentage error of the mean result.
- **c** Do you think this is an acceptable percentage error? Explain.

3 Markéta bought a 250 g pack of butter. She weighed it to find its actual mass was 240 g. Find the percentage difference in mass. Take 250 g as the true value.

4 Amelia made an 18% error in a calculation. Her incorrect answer is $27.50. Find two possible correct answers.

5 A factory produces cereal bars with mass 58 g, and percentage error up to 2%. Find the range of possible masses of cereal bars from this factory.

'To err is human; to forgive, divine.'
Alexander Pope, 1711

'To err is human; but to really foul things up you need a computer.'
Paul Ehrlich, 1969

Problem solving

6 Huýnh tests all the scales in the science lab before an experiment. A scale fails the test if it has a percentage error of 12% or more. Huýnh uses a 500 g mass to test each scale. Complete this statement: Scales that measure a 500 g mass as less than ___g or more than ___g will fail the test.

7 Find the measurement that has the largest absolute error.

- 500 kg measured with a percentage error of 1%
- 30 kg measured with a percentage error of 10%

D When are absolute values useful?

- How can error best be measured?
- How do systems influence communities?

Absolute value measures the error in a measurement. These errors either come from the accuracy of the measuring tool or from your accuracy in reading it.

Reflect and discuss 3

- Which type of error (absolute, relative or percentage) is the best to use? Justify your arguments.
- Absolute error is the absolute value of the difference between the true value and the measured value. Could you measure the absolute error without taking the absolute value?
- Think of situations when it would be better *not* to use the absolute value to describe an error. When is it important to know whether the error has a positive or negative value?

Some websites claim to give the true world population, plus the number of births and deaths so far that year, so naturally this population number is changing constantly. If the population figure varied between 7 379 930 858 and 7 379 930 871 over a one-minute period, what would be a good approximation for the world population? Could anyone ever know the exact value of the world population?

The world's population density is 47 people per square kilometer. The most densely populated country is Monaco at 16 205 people per square kilometer; the least populated is Mongolia with 2 per square kilometer.

Reflect and discuss 4

- Why do we often use approximations instead of exact amounts? Is it always appropriate to use approximations?
- If you measure a length to the nearest cm and then to the nearest mm, you can still make a more exact measurement to the exact 0.1 mm or 0.01 mm. Can you ever know the *exact* length of something?
- Is it important to know *exact* values, or can you always use approximations? How can knowing the size of the error in a measurement help you decide whether an approximation is acceptable?

Summary

The **absolute value** of a number is the distance on the number line between that number and 0.

$|x|$ means 'the absolute value of x'.

For example, $|5| = 5$, and $|-5| = 5$

Properties of absolute value

For any numbers $a, b \in \mathbb{R}$:

1 $|a| \geq 0$

2 $|a \times b| = |a| \times |b|$

3 $\left|\frac{a}{b}\right| = \frac{|a|}{|b|}$

4 $|a|^2 = a^2$

5 $|a^n| = |a|^n$

6 $|a-b| = |b-a|$

7 $|a+b| \leq |a| + |b|$

8 $|a-b| \geq ||a|-|b||$

The **true value** of a measurement is the value that would be obtained by a perfect measurement.

Accuracy describes how close a measurement is to the true value.

The **absolute error**, Δx, of a measurement is the actual amount of error in a measurement.

$\Delta x = |\text{true value} - \text{measured value}|$

The **relative error** is the error in a measurement compared to the size of the thing being measured.

Relative error $= \frac{\Delta x}{x} = \frac{\text{absolute error}}{\text{true value}}$

The **percentage error** is the relative error of a measurement written as a percentage.

Mixed practice

1 **Find** the value of these expressions.

a $|-33|$

b $-|4-6.5|$

c $-|-2.3+8.05|$

d $|12.3-7.8|$

e $\left|-\frac{2}{9}\right|$

f $|(-2)^5|$

g $|\sqrt{27}|$

h $-|-\sqrt{81}|$

2 **Find** the value of these expressions.

a $|4| \times |-5|$

b $-|-(-6)^2|$

c $\frac{|6|}{-|12|}$

d $\left|\frac{-27}{-6}\right|$

e $\left|-\frac{2}{5}\right| \times \left|\frac{9}{-8}\right|$

f $\left|\frac{1}{12}\right| \times \left|-\frac{144}{5}\right|$

g $\frac{|-56|}{|7.2-8|}$

h $\frac{|15| \times |-3|}{|4-6.25| \times |7-3|}$

3 The exact length of a bent stick is 22.2 cm. Daniel measured it with a straight ruler and estimated its length to be 21.5 cm. **Find** the percentage error of his estimate.

4 Soraya bought 3 meters of ribbon. When she measured it, she found there were 3.15 meters of ribbon. **Find** the percentage error in the 3 meter measurement.

5 Moa charges €10 to iron 8 shirts. On average, she irons 8 shirts per hour.

a **Determine** how much she charges to iron 20 shirts.

b **Determine** her average hourly rate.

c On Monday, Moa ironed 20 shirts in 2 hours and 15 minutes. She charged the customer €25.

i **Determine** Moa's hourly rate for these shirts.

ii **Find** the percentage difference between her average hourly rate (in part **b**) and her hourly rate for these shirts.

d On Friday, she took 3 hours and 12 minutes to iron 20 shirts. **Find** the percentage difference between her average hourly rate (in part **b**) and her hourly rate for these shirts.

Which rate is the exact rate? Which one is an estimate?

Review in context

Globalization and sustainability

1 The most recent census in China was in July 2015 and the population was reported in official government statistics to be 1 376 048 943.

In two separate articles, this figure was quoted to be 1 376 049 000 and 1.38 billion.

a **Calculate** the absolute error in each of these figures compared to the official figure.

b **Calculate** the percentage errors.

c **Discuss** whether it is better to use absolute or percentage errors.

d Government sources suggest that, due to the reliability of the way it is measured, the official figure may be inaccurate by plus or minus 1.8%.

Calculate what this is as an absolute error.

e **Calculate** the possible range for the population.

f **Discuss** what might be a reliable figure to use to report the population of China.

g Here are the populations of some countries, also given in 2015:

United States	321 442 019
Germany	80 688 545
Malaysia	30 331 007
Australia	23 968 973
Monaco	37 731

If the figures could also be inaccurate by 1.8%, **find** the absolute errors. Use these to give reliable values to quote for these populations, and **justify** why the values are reliable.

2 **UK population grows by half a million in a year**

Between 2013 and 2014 the UK population rose from 64 106 779 to 64 510 376.

a **Find** the exact change in population.

b **Find** the percentage error in the headline figure of half a million.

c In 2014, 17.7% of the population were aged 65 and over. In 2013 the percentage was 17.4%.

Find the actual change in the number of people aged 65 and over between 2013 and 2014.

d Is this figure an increase or a decrease?

e **Suggest** how these figures could be useful for planning healthcare in the UK.

3 Tobias went grocery shopping for himself and his roommate Felix. They usually split the cost of groceries in half, and since they never have exact change, they usually round the amount paid before splitting it in half. Today, Tobias paid exactly €95.74, and he rounded the amount up to an even €100 before telling Felix that he owed Tobias €50.

a **Calculate** the percentage error between the true amount Tobias paid and the rounded amount.

b **Discuss** whether or not it is reasonable for Felix to pay €50.

c **Calculate** the exact amount Felix should have reimbursed Tobias. Then **calculate** the percentage error between what Felix did pay and what he should have paid.

d **Discuss** what you notice about the two percentage errors. Would it have made sense to just calculate absolute errors in this case?

Reflect and discuss

How have you explored the statement of inquiry? Give specific examples.

Statement of Inquiry:

Quantities and measurements illustrate the relationships between human-made systems and communities.

7.2 How do they measure up?

Global context: Globalization and sustainability

RELATIONSHIPS

Objectives

- Converting between metric units, including metric units of area and volume
- Converting between metric and imperial units
- Using units correctly in problem solving
- Solving problems involving compound measures
- Deciding if the answer to a problem is reasonable

Inquiry questions

- F: What are the different systems of measurement?
- C: How do we convert measures of area and volume into different units?
- D: Should we adopt a single global system of measurement?
- How do systems influence communities?

ATL **Communication**

Use intercultural understanding to interpret communication

Statement of Inquiry:

Quantities and measurements illustrate the relationships between human-made systems and communities.

You should already know how to:

• recognize units for length, mass and volume	**1** Group these measurements into lengths, masses and volumes. Which is the odd one out? 2 kg 34 mm 15 g 24 s 5 ml 9 km 13 miles 6 liters 8.4 cm^3
• write powers of ten in full	**2** Write down these powers of ten in full. **a** 10^3 **b** 10^6 **c** 10^{-2}
• multiply and divide by powers of ten	**3** Calculate. **a** 280×0.001 **b** $34\,500 \div 10^2$
• find the area of a square	**4** Calculate the area of a square with side length 12 cm.
• find the surface area and volume of a cuboid	**5** A cuboid measures 3 m × 4 m × 5 m. Calculate its surface area and volume.

Units for measurement

- What are the different systems of measurement?

Exploration 1

1 Add to this list by writing down all the units of measurement you can think of. Group similar units together.
- kilograms (kg)
- minutes (min)
- kilometers per hour (km/h)
- degrees (°)

Compare your list with others. Did anyone group units differently?

2 Add to this list all the types of measurement you can think of.
- length
- volume
- mass

Are length and distance the same measurement? What about volume and capacity?

3 Compare your two lists. Match each unit to the type of measurement it is used for. Are there any types of measurement that have only one unit?

Think of the units you use in mathematics and science, in cooking, when travelling, etc.

Think about how the words volume and capacity are used. To what space (1D, 2D, 3D) does each refer?

Reflect and discuss 1

- Does every measurement have units? Explain why this is the case.
- Why are there often different units for a single type of measurement?
- Can units measure more than one type of measurement? Explain fully.

Practice 1

1 Suggest sensible units for these specific measurements.

  a A matchbox measures $40 \times 25 \times 12$

  b The length of an A4 sheet of paper is 297

  c The length of an A4 sheet of paper is 29.7

  d A shoebox has a capacity of 6307

  e A shipping container measures $6.06 \times 2.44 \times 2.59$

  f The floor area of a house is 250

  g The length of a swimming pool is 50

2 Select appropriate units for these.

  a The weight of a farm tractor

  b The dimensions of a tennis court

  c The area of a wall in a house

  d The speed of an ambulance

  e The height of a giraffe

  f The dimensions of a piece of land

  g The speed of a javelin, when thrown by an athlete

The 'A' series of paper sizes is used worldwide. Only the US and Canada do not currently conform to the ISO standard although in practice Mexico, Venezuela, Colombia, Chile and the Philippines favour the US letter format.

Systems of measurement

The two most widely used systems of measurement are the metric system and the imperial system.

> A **system of measurement** is a system of measures based on a set of base units. All other units in the system are derived from the base units.

The metric system

The metric system is used widely in Europe and most of the rest of the world.

The International System of Units (or SI) defines seven base units:

- meter (m) for measuring length
- kilogram (kg) for measuring mass
- second (s) for measuring time
- ampere (A) for measuring electric current
- Kelvin (K) for measuring thermodynamic temperature
- mole (mol) for measuring amounts of a substance
- candela (cd) for measuring the intensity of light.

The International System of Units (abbreviated SI from the French 'Le Système International d'Unités') officially came into being at the 11th General Conference on Weights and Measures in Paris in October 1960.

The French brought in the metric system after the French Revolution. In 1793 they even tried to decimalize time with 10 days in a week, 10 hours in a day, 100 minutes in an hour and 100 seconds in a minute. They returned to traditional timekeeping in 1806.

All other units are derived from these and are called *derived units*. For example, the unit of speed, in terms of the base units, is m/s.

In some books, m/s is written ms^{-1}.

The SI (or metric system) allows other units to be created by using prefixes which act as multipliers. The unit prefixes specify a power of ten:

- a *kilo*meter is 10^3 times larger than a meter (standard unit),
  $1 \text{ km} = 1 \times 10^3 \text{ m} = 1000 \text{ m}$
- a *nano*second is 10^{-9} times smaller than a second (standard unit),
  $5 \text{ nanoseconds} = 5 \times 10^{-9} \text{ seconds} = 0.000\,000\,005 \text{ seconds}$.

Dividing by the factor allows you to convert a standard unit into another unit:

- a gram (standard unit) is smaller than a *kilo*gram,
  $500 \text{ g} = 500 \text{ g} \div 10^3 = 0.5 \text{ kg}$
- a meter (standard unit) is larger than a *micro*meter,
  $1 \text{ m} = 1 \text{ m} \div 10^{-6} = 1\,000\,000 \text{ m}$

Prefix	Letter	Power
tera-	T	10^{12}
giga-	G	10^9
mega-	M	10^6
kilo-	k	10^3
deci-	d	10^{-1}
centi-	c	10^{-2}
milli-	m	10^{-3}
micro-	μ	10^{-6}
nano-	n	10^{-9}

Some derived units are named after a person closely associated with them. For example, the Newton (N) is a measure of force. $1 \text{ N} = 1 \text{ kg m/s}^2$.

Practice 2

1 Find these conversions.

a 3.5 megatonnes to tonnes

b 8 gigabytes to bytes

c 257 mm to meters

d 6.5 dl to liters

2 How many:

a nanoseconds are in a second

b micrometers are in a meter

c watts are in a kilowatt

d centiliters are in a liter?

'How many millimeters are in a meter' is the same as 'Convert meters to millimeters'.

3 Find these conversions.

a 2 m to μm

b 3.2 g to mg

c 2500 watts to microwatts

d 0.5 seconds to nanoseconds

4 Find these conversions.

a 285 cm = _____ m

b 923.5 mg = _____ g

c 4358 m = _____ km

d 4358 m = _____ mm

e 7.263 kg = _____ cg

f 12.45 l = _____ cl

g 18 655 ml = _____ l

h 560 mm = _____ cm

i 0.05 km = _____ cm

j 380 506 mg = _____ kg

Tip

Check that your answer makes sense. Should the converted unit be smaller or larger than the original unit?

Problem solving

5 A human ovum has diameter 0.1 mm. The head of a human sperm is roughly 5 μm across. Calculate how many times larger the ovum is compared to the sperm.

6 The cells of the bacterium Staphylococcus aureus are spheres with diameter 1 μm. How many of these cells could fit on a rectangular 75 mm by 25 mm microscope slide?

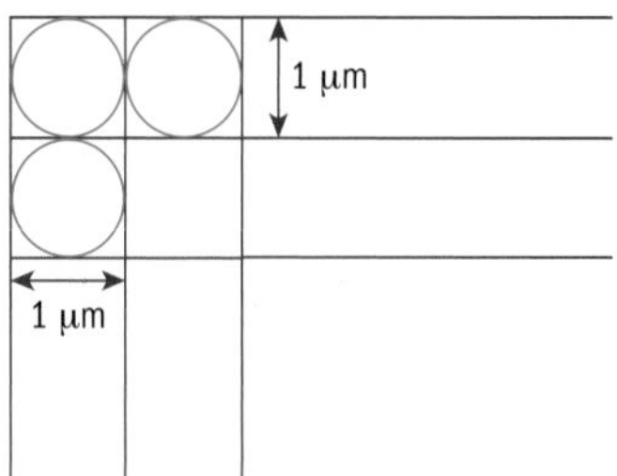

On the slide, each cell takes up the same area as a square with side length 1 μm.

The imperial system

The imperial system is now chiefly used in the US (US customary units).

Originally, many systems of measurement had their own standard on which the other measures were based. Small distances were counted in number of feet and paces (yards) and larger distances in miles (1000 paces). Capacity was measured using kitchen items, such as cups, because it was mainly used in cooking. Eventually a standard was set so that all measurements were the same for everyone.

There are differences between imperial and US customary units, but they both are derived from the same system.

Reflect and discuss 2

- What difficulties might have arisen in measuring with units such as paces and cups?
- What types of problems might have occurred from the use of non-standardized units?

The SI (or metric) standards are now accepted worldwide and all other measures are defined in terms of that standard.

For example, 1 yard ≡ 0.9144 meters, where ≡ means 'is equivalent to'.

Some conversions for metric and imperial units are shown in the tables. Those marked * are exact; all others are defined to be exact by the International Yard and Pound Agreement.

Metric	Imperial
1 km	0.625 mile
1 m	39.4 inches
1 kg	2.205 pounds (lbs)
1 liter	1.76 pints (UK)
1 liter	2.113 pints (US)

Imperial	Metric
1 mile	1.609 km
1 foot	30.48 cm (*)
1 inch	2.54 cm (*)
1 ounce (oz)	28.35 g
1 gallon (UK)	4.546 liters
1 gallon (US)	3.785 liters

In 1999, the Mars Climate Orbiter was lost because the NASA team used metric units while a contractor used imperial units. Failing to convert the measurements for such crucial calculations caused the probe to come too close to Mars, where it is thought to have been destroyed by the planet's atmosphere.

Example 1

Find these conversions.

a 8 UK gallons to liters

b 13 liters to UK gallons

a 1 UK gallon $\simeq$ 4.5 liters

$8 \times 4.5 = 36$ liters

×4.5

1 UK gallon $\simeq$ 4.5 liters

To convert UK gallons to liters, multiply by 4.5.

b $13 \div 4.5 = 2.888\ldots$

≈ 2.9 UK gallons (1 d.p.)

1 UK gallon $\simeq$ 4.5 liters

÷4.5

Use the inverse operation.

To convert liters to UK gallons, divide by 4.5.

Round to a sensible degree of accuracy.

Objective: D. Applying mathematics in real-life contexts
iv. justify the degree of accuracy of a solution

In Practice 3, justify the degree of accuracy of each of your solutions.

Practice 3

1 Find these conversions.

a 1 inch$\simeq$ ___ m

b 1 g $\simeq$ ____ ounces

c 1 pound $\simeq$ ____ kg

d 1 liter $\simeq$ ____ gallons (UK)

e 1 UK pint $\simeq$ ____ liters

f 1 cm = ___ feet

2 Find how many:

a kilometers is 50 miles
b miles is 50 kilometers
c pounds is 70 kilograms
d centimeters is 6 inches
e centimeters is 3 feet
f liters is 5 UK pints
g UK pints is 2 gallons
h centimeters is 1 mile

For **2h**, first convert miles to kilometers.

Problem solving

3 Determine which measurement is largest for each pair.

a 58 miles or 72 km
b 25 kg or 60 lbs
c 50 cl or 1 UK pint
d 4 oz or 100 g
e 6 in or 20 cm
f 6 ft or 180 cm

4 In *This is Spinal Tap*, members of a fictional rock group asked for an 18-inch model of a Stonehenge megalith (large standing stone). The model was intended for their stage show.

a By converting the height to an appropriate metric measure, determine whether the model would have been effective.

b Decide what imperial measure of length the band members should have asked for, given that they intended to use the model on a live stage.

Problem solving

5 In 1983, an Air Canada flight ran out of fuel halfway through its journey due to a mix-up between pounds and kilograms. The plane needed 22 300 kg of fuel for the journey. There were 7682 liters of fuel already in the tanks.

a One liter of jet fuel has a mass of 0.803 kg. Calculate how much fuel should have been loaded. Give your answer in liters.
b Instead of 0.803 kg per liter, a conversion of 1.77 was used. (One liter of jet fuel has a mass of 1.77 pounds). Calculate how much fuel was actually loaded. Give your answer in liters.
c Find how many liters short of fuel the plane was.

C What units tell us

- How do we convert measures of area and volume into different units?

ATL

Exploration 2

1 The diagram shows a square with side length 1 m and a square with side length 100 cm.

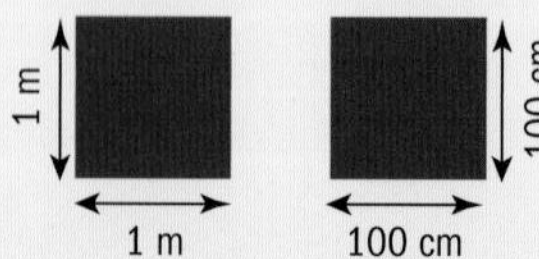

a Explain why these two squares have the same area.

b Calculate the area of each square.

Continued on next page

c Copy and complete:

$1 \text{ m}^2 = \_\_\_\_ \text{ cm}^2$

d Sketch a square with side length 1 cm and a square with side length 10 mm. How do you convert from cm^2 to mm^2? How do you convert from mm^2 to cm^2?

2 Sketch a cube with side length 1 m and a cube with side length 100 cm.

a Calculate the volume of each cube.

b Copy and complete:

$1 \text{ m}^3 = \_\_\_\_ \text{ cm}^3$

c Work out the volume of a cube with side length 1 cm and a cube with side length 10 mm.

d How do you convert from cm^3 to mm^3? How do you convert from mm^3 to cm^3?

Capacity refers to the amount of space available to hold something. **Volume** refers to the amount of space actually occupied.

Volume is generally measured in cm^3, and capacity in liters.

$1 \text{ cm}^3 = 1 \text{ ml}$

$1000 \text{ cm}^3 = 1 \text{ liters}$

ATL

Practice 4

1 Find these conversions.

a 20 cm^2 to mm^2 **b** 600 mm^2 to cm^2

c 5000 cm^2 to m^2 **d** 4.5 m^2 to cm^2

e 2.9 cm^2 to mm^2 **f** 0.7 m^2 to mm^2

For **Q1f,** convert m^2 to cm^2 then to mm^2.

2 Find these conversions.

a 40 m^3 to cm^3 **b** 5000 mm^3 to cm^3

c $2\,400\,000 \text{ cm}^3$ to m^3 **d** 3 cm^3 to ml

e 2.5 liters to cm^3 **f** 10 m^3 to liters

For **Q2f,** first convert m^3 to cm^3.

3 A piece of foam has volume 500 cm^3. Find its volume in m^3.

4 **a** 1 decimeter (dm) = 10 cm

Copy and complete:

$1 \text{ dm}^3 = \_\_\_\_ \text{ cm}^3$

b Find the number of dm^3 in 1 m^3.

c A removal truck has a capacity of 6 m^3. Find its capacity in dm^3.

5 Find the capacity in liters of a swimming pool that measures $33 \text{ m} \times 16 \text{ m}$, with an average depth of 2.15 m.

Problem solving

6 Suggest suitable dimensions for a cuboid-shaped carton designed to hold 1 liter of juice.

Exploration 3

A cuboid is 30 cm tall, 50 cm long and 1 m wide.

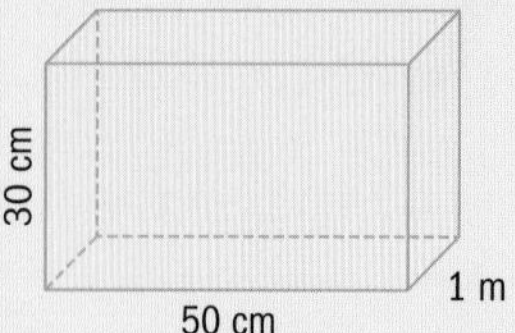

1 **a** Use the lengths in cm to calculate the volume of the cuboid in cm^3.

b Use the lengths in m to calculate the volume of the cuboid in m^3.

c Show that your two results are equal.

2 **a** Calculate the surface area of the cuboid in cm^2.

b Use step **a** to find the surface area of the cuboid in m^2.

3 A box in the shape of a cuboid is 250 cm tall, 472 cm long and 545 cm wide. Calculate the volume of this cuboid in m^3.

Reflect and discuss 3

In Exploration 3:

- Which units would you use to calculate the surface area and volume of the first cuboid? Which units would you use to give the final answers?
- Option 1: Find the volume in cm^3, and convert to m^3. Option 2: Convert the dimensions to meters, then find the volume. Which is the easiest way to find the volume of the box in m^3?

ATL

Practice 5

1 The dimensions of a tennis court are 2377 cm × 8.23 m. Find its area in m^2.

2 A soccer field measures 0.110 km × 73 m. Find its area in m^2 and in km^2. Which measure is easier to visualize?

3 A container has length 9750 mm, width 6400 mm and height 5640 mm.

a Find its floor area in m^2.

b Find its volume in m^3.

Problem solving

4 1 hectare (ha) is the area of a square 10 000 m^2.

A square field has an area of 0.36 km^2.

Find the area of the field in hectares.

You can use the imperial-to-metric conversion factors for lengths to convert units of area and volume.

Example 2

Find the number of cubic meters (m^3) in 10 cubic yards.

Use the conversions: 1 inch = 2.54 cm, 1 yard = 36 inches

1 inch = 2.54 cm

1 yard = 36 inches — Write down the conversions.

10 yards = 36 inches × 10 = 360 inches

360 inches = 2.54 cm × 360 = 914.4 cm — Convert 10 yards to inches, then centimeters.

10 yards = 9.144 m — Convert 10 yards to meters.

10 cubic yards = $(10 \text{ yards})^3 = (9.144 \text{ m})^3$

$= 764.554\ldots$

10 cubic yards = 764.55 m^3 (to 2 d.p.) — Check that the answer makes sense: 10 yards is *shorter* than 10 m (10 yards = 9.144 m) so 10 cubic yards should be *smaller* than $(10 \text{ m})^3$.

Practice 6

1 A rectangular piece of land is 2 km long and 3 km wide.

a Find the dimensions of the piece of land in miles.

b Find the area of the land in square miles.

2 a A box measures 4 in × 5 in × 6 in. Find **i** its volume in cm^3, and **ii** its surface area in cm^2.

b Another box measures 3 in × 5 in × 8 in. Find **i** its volume in cm^3, and **ii** its surface area in cm^2.

Write down what you notice about your answers to parts **a** and **b**.

3 Find the number of cubic centimeters (cm^3) in 5 cubic inches.

4 Find the number of square meters (m^2) in 10 square yards.

1 yard = 0.9144 m

5 A truck has a capacity of 25 m^3. Find the capacity of the truck in cubic feet.

6 A cube has volume 64 in^3.

a Find the side length of the cube in inches.

b Find the side length in cm.

Problem solving

7 A dressage arena for horse riding is a rectangle twice as long as it is wide. Its area is 8611.13 ft^2. Find its length in meters.

Converting compound measures

Compound measures involve more than one unit. For example, speed can be measured in meters per second and density in grams per cubic centimeter. When compound measures are involved, you may need to convert both units.

Density is the mass of substance contained in a certain volume. The density of solids and liquids is measured in g/cm^3, and the density of gases is measured in g/l.

Example 3

The density of iron is 7.87 g/cm^3. Find the density of iron in kg/m^3.

1 kg = 1000 g — Write down the conversions you will need.

1 m^3 = 1 000 000 cm^3

Method 1:

7.87 g/cm^3 × 1 000 000 = 7 870 000 g/m^3 — First convert g/cm^3 to g/m^3.

7 870 000 g/m^3 ÷ 1000 = 7870 kg/m^3 — Now convert g/m^3 to kg/m^3.

Method 2:

$$7.87\frac{\text{g}}{\cancel{\text{cm}^3}}\times\frac{1000000\cancel{\text{cm}^3}}{1\text{m}^3}=7870000\frac{\text{g}}{\text{m}^3}$$

Write the conversion factor for volume as a fraction and multiply.
Multiplying by $\frac{1000000\text{ cm}^3}{1\text{ m}^3}$ is equivalent to multiplying by 1, since 1 000 000 cm^3 = 1 m^3.
The cm^3 units cancel.

$$7870000\frac{\cancel{\text{g}}}{\text{m}^3}\times\frac{1\text{kg}}{1000\cancel{\text{g}}}=7870\frac{\text{kg}}{\text{m}^3}$$

Write the conversion factor for mass as a fraction and multiply.
Multiplying by $\frac{1\text{ kg}}{1000\text{ g}}$ is equivalent to multiplying by 1, since 1 kg = 1000 g.
This time, the grams cancel.

Example 4

The speed limit on motorways in the UK is 70 miles per hour (mph).
In some parts of Europe, the speed limit is 110 kilometers per hour (km/h).
Which is faster?

1 mile = 1.609 km — Write down the conversion you will need.

70 mph × 1.609 = 112.63 km/h — Convert 70 miles per hour to kilometers per hour.

70 mph is faster than 110 km/h.

Practice 7

1 Convert these speeds from kilometers per hour (km/h) to meters per hour (m/h).

a 10 km/h b 54 km/h c 4.8 km/h d 280 km/h

2 Convert these speeds from meters per hour (m/h) to meters per second (m/s).

a 3600 m/h b 43 200 m/h c 9000 m/h d 2160 m/h

3 Oliver is driving at 65 km/h. Find his speed in m/s.

4 A sea turtle can swim at a speed of 2 m/s. Find its speed in km/h.

5 Aluminium has density 2.70 g/cm^3. Find its density in kg/m^3.

6 Oxygen has a density of 1.43 g/l. Find its density in mg/cl.

For **Q2**, first convert meters per *hour* to meters per *minute*, then to meters per *second*.

Problem solving

7 Lead has density 11.34 g/cm^3. Find the mass of 1 m^3 of lead.

8 Gold has density 19.32 g/cm^3. Find the mass of 0.5 m^3 of gold.

9 The speed limit in urban areas is 30 mph in the US, and 50 km/h in most of Europe. Are these speeds the same? Explain your answer.

Problem solving

10 A domestic cat can reach a top speed of 48 km/h. The sprinter Usain Bolt has reached a top speed of 12.27 m/s. Determine which is faster.

11 Oxygen has a density of 1.43 g/l. Enzo says that 120 cl of oxygen has a mass of 17.16 grams. Justify whether or not he is correct.

12 Timeo is travelling at a speed of 25 m/s. He claims that it will take him 1 hour to travel 110 km. Justify whether or not he is correct.

Tip

Units of a compound measure can tell you what to do: $\frac{\text{meters}}{\text{seconds}}$ will mean distance divided by time.

Problem solving

13 In 1997, a world land-speed record of 763.035 miles per hour was set by the car *Thrust SSC*. The sound barrier is reached when an object moves at a speed of about 340 meters per second. Determine if *Thrust SSC* broke the sound barrier. Explain your answer.

D More conversions and units

- Should we adopt a single global system of measurement?
- How do systems influence communities?

Conversions

Activity

1 Euro (EUR) = 1.10 US Dollar (USD)

1 USD = 0.996 Swiss Franc (CHF)

1 Convert 650 EUR into USD.

2 Convert 1800 CHF into USD.

3 Calculate an approximate value of 900 CHF in EUR.

Reflect and discuss 4

- How is the process of converting from one unit to another similar to converting from one currency to another? Are both the unit conversions and currency conversions standardized? Explain fully.
- Quantity is defined as an amount or number. Measurement is defined as a method of determining quantity, capacity or dimension using a defined unit. Is currency conversion more to do with quantity, and unit conversion more to do with measurement? Explain fully.

Exploration 4

1 Research natural units and answer these questions.

a Determine what natural units are.

b Which systems of natural units are used?

c What are the advantages of the natural system of units? Are there any disadvantages?

2 Systems of man-made units need a prototype, that is, an object that defines a unit. Other measurements are compared to the prototype.

a Determine what the prototype is for the kilogram.

b Decide whether a natural unit needs a prototype.

Reflect and discuss 5

- Is the day a man-made unit or natural unit?
- Why is there no single global system of measurement?
- What would be the advantages and disadvantages of having a single system of measurement?
- If a single system of measurement was to be adopted, which should be chosen? Explain fully.

Some unusual units of measurement include:

- The Hand (= 4 inches) for measuring the height of a horse
- The Board foot (= 1 inch × 1 foot × 1 foot) for measuring lumber
- The Shake (= 10 nanoseconds) for referring to very short periods of time
- The Foe (= 10^{44} joules) for measuring the enormous amount of energy produced when a star goes supernova

Summary

A **system of measurement** is a system of measures based on a set of base units. All other units in the system are derived from the base units.

The two most common systems of measurement are the metric system and the imperial system.

The metric system allows other units to be created by using prefixes, which act as multipliers. Each unit prefix specifies a power of ten; for example, the prefix *kilo-* specifies 10^3.

Some common conversions:

To convert from m^2 to cm^2, multiply by 10 000.
To convert from mm^2 to cm^2, divide by 100.
To convert from m^3 to cm^3, multiply by 1 000 000.
To convert from mm^3 to cm^3, divide by 1000.

Capacity refers to the amount of space available to hold something. **Volume** refers to the amount of space actually occupied. Both capacity and volume can be measured in cm^3 and liters: $1\ cm^3 = 1$ ml, $1000\ cm^3 = 1$ liter.

Compound measures such as speed and density combine measures of two quantities.

Density is the mass of substance contained in a certain volume.

$$\text{Density} = \frac{\text{mass}}{\text{volume}}$$

The density of solids and liquids is measured in g/cm^3; the density of gases is measured in g/l.

Mixed practice

1 Determine which unit of measurement you would use to measure:

a the volume of water in a swimming pool

b a person's walking speed

c the capacity of a storage box

d the weight of a person

e the amount of flour in a cake

f the distance from Chicago to Los Angeles

2 Find these conversions.

a 3 m to cm

b 4.8 kg to g

c 760 cm to m

d 9845 m to km

e 0.01 km to cm

f 400 600 mg to kg

3 Find these conversions.

a 4 ounces ≃ ___ g

b 15 inches ≃ ___ cm

c 8 m ≃ ___ feet

d 6 liters ≃ ___ UK gallons

e 10 UK pints ≃ ___ gallons

f 800 g ≃ ___ pounds

4 Find these conversions.

a 50 cm^2 to mm^2

b 9500 cm^2 to m^2

c 0.5 m^2 to mm^2

d 6400 mm^3 to cm^3

e 10 cm^3 to ml

f 100 m^3 to liters

5 Find these conversions.

a 950 meters per hour (m/h) to km/h

b 36 km/h to m/h

c 10 meters per second (m/s) to m/h

d 7200 m/h to m/s

e 54 km/h to m/s

Problem solving

6 A bottle contains 240 cl of medicine.

Calculate the number of 25 ml doses that can be poured from it.

7 The table gives the heights and weights of two people.

	Height	Weight
Person A	6 ft 2 in	190 lbs
Person B	1.82 m	90 kg

a Who is taller? **b** Who is heavier?

8 An airline's hold baggage allowances are:
SIZE: length + width + height must be less than 275 cm.
WEIGHT: free up to 20 kg. Each extra kg (or part of a kg) is charged at 13 Euros.

a Jo's suitcase measures 18 in × 22 in × 8 in. Decide whether he can take it as hold luggage.

b Suzanne's bag weighs 48 lbs. **Calculate** how much she would have to pay to take it as hold baggage.

9 A water tank has length 2250 mm, width 5800 mm and height 3860 mm. **Calculate** the capacity of the tank in liters.

10 **Find** the number of cubic inches in 9 cm^3.

11 **Find** the number of m^2 in 6 square yards.

1 in = 2.54 cm
1 yd = 36 inches

12 A square field has area 2323.24 ft^2.

Find the side length of the field in meters.

13 The density of nickel is 8900 kg/m^3.

Find the density of nickel in g/cm^3.

Problem solving

14 Platinum has a density of 21 450 kg/m^3. Gold has a density of 19.32 g/cm^3.

Determine which metal is denser.

15 A car travels at 20 m/s in a village where the speed limit is 50 km/h. **Determine** if the car is exceeding the speed limit.

16 A top speed of 231.523 mph was set by a Formula 1 racing car in 2005. A Chinook helicopter has a maximum speed of 87.5 m/s.

Determine which is faster.

Review in context

Globalization and sustainability

1 The Golden Gate Bridge in San Francisco opened for vehicular traffic in 1937.

a Below are some facts about the bridge. **Find** the equivalent measurements in metric units.

i The towers are 746 feet tall.

ii Each of the two main cables is 7650 feet long.

iii The total length of wire used in both main cables is 80 000 miles.

b Carbon steel is used for the main cables. 1 cm^3 of carbon steel has a mass of about 7.85 g. **Find** the density of carbon steel in kg/m^3.

c When the bridge was built, 389 000 cubic yards of concrete was used.

1 yard = 0.9144 m

After the original concrete roadway deck was replaced, there was 6.4% less concrete. **Calculate** the total quantity of concrete in the refurbished bridge in m^3.

d A bolt of diameter 2.125 inches was used in the construction.

i **Find** the diameter of the bolt in mm.

ii Imagine that a bolt of diameter 2.125 cm was mistakenly used. **Calculate** the error in mm.

2 Railway tracks in different countries have different gauges (spacing between the rails).

a Approximately 60% of the world's railways today use the standard gauge, which is 4 ft $8\frac{1}{2}$ in. **Find** this measurement in metric units, correct to the nearest mm.

b Originally, English trains used a 4 ft 8 in gauge for coal transport trains. Half an inch was added to the gauge for passenger trains, as this made the trains run more smoothly on the curves.

i **Find** the length of the coal-mining gauge in metric units, to the nearest mm.

ii **Find** the difference between the two gauge sizes, in mm.

iii **Determine** whether your answer in part **ii** corresponds to $\frac{1}{2}$ in. If not, find the percentage error.

c Some countries use the meter gauge, which is exactly 1000 mm. **Find** this measurement in imperial units, in feet and inches.

d Until 2011, Spain used a 1668 mm gauge, and France used the standard gauge. **Suggest** why Spain modified its railway tracks and engines.

Reflect and discuss

How have you explored the statement of inquiry? Give specific examples.

Statement of Inquiry:

Quantities and measurements illustrate the relationships between human-made systems and communities.

7.3 Going around and around

Global context: Personal and cultural expression

RELATIONSHIPS

Objectives

- Knowing the terms chord, arc, segment and sector
- Finding the length of an arc of a circle
- Finding the angle in a sector of a circle
- Finding the perimeter and area of a sector of a circle
- Finding the length of a chord

Inquiry questions

- F: How can you find the measurements for the different parts of a circle?
- C: How are the parts of a circle related?
- D: Is it advantageous to generate your own formulae?

ATL Communication

Use and interpret a range of discipline-specific terms and symbols

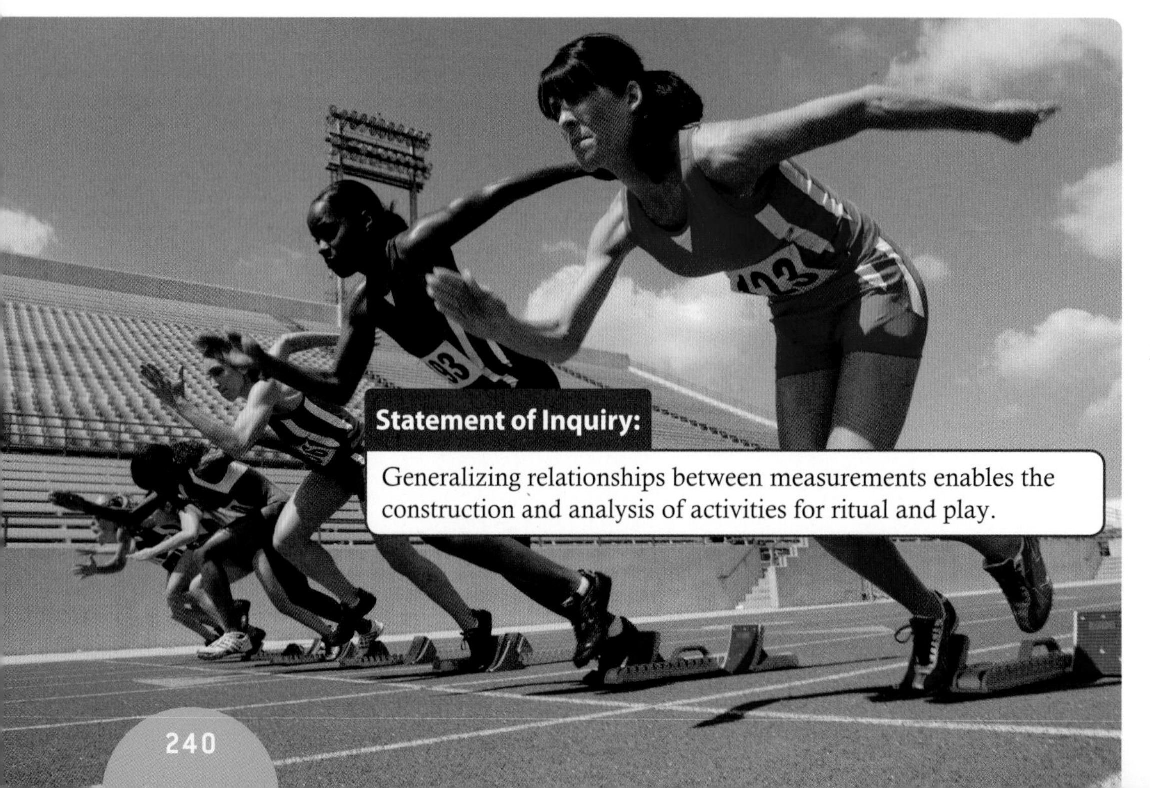

7.3

9.1

E9.1

Statement of Inquiry:

Generalizing relationships between measurements enables the construction and analysis of activities for ritual and play.

You should already know how to:

• find the circumference of a circle using $C = 2\pi r$ or πd	**1** Find the circumference of a circle: **a** of radius 2.4 cm **b** of diameter 5.75 m. Give your answers to 1 d.p.
• find the area of a circle using $A = \pi r^2$	**2** Find the area of the two circles above. Give your answers to 1 d.p.
• find missing lengths in a right-angled triangle using trigonometric ratios	**3** Find the missing length x in this triangle. 5 cm x 30°

Parts of a circle

- How can you find the measurements for the different parts of a circle?

ATL

Activity

1 Match each definition with a part of the circle.

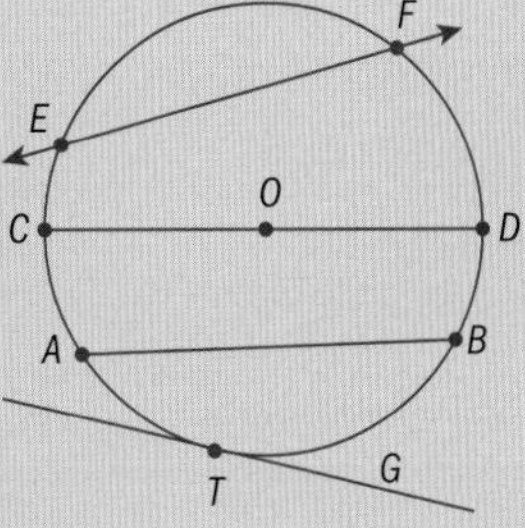

- A **diameter** is a chord that passes through the center of a circle.
- A **chord** is a line segment with its endpoints on a circle.
- A **secant** is a line that intersects a circle at two points.
- A **tangent** is a line that touches a circle at only one point.

2 Match the following definitions with the parts of the circle in the diagram.

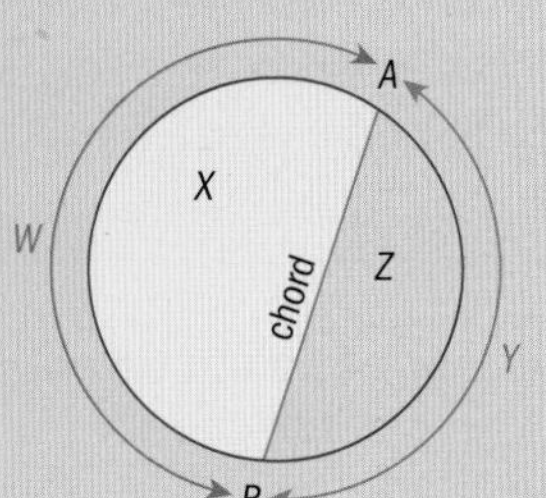

The **major segment** is the larger segment.

The **minor segment** is the smaller segment. ('Segment' usually refers to the minor segment.)

The larger arc is the **major arc**.

The smaller arc is the **minor arc**.

A **sector** is a 'slice' of a circle between two radii. The center of a circle is usually labelled O, so the sector here is BOC. 'Sector' usually means the minor sector.

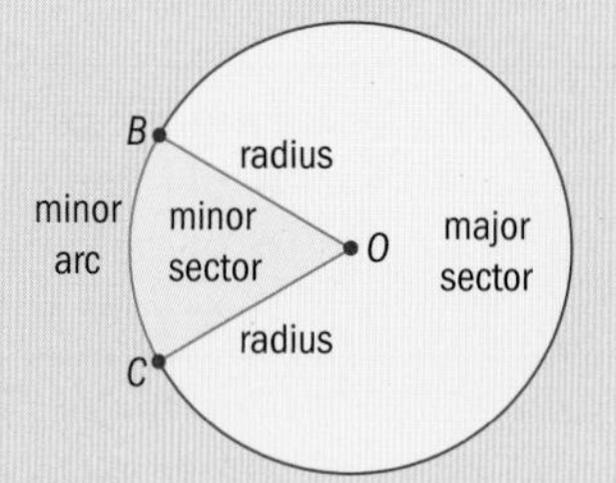

Lorenzo Mascheroni proved in 1797 that any point you can construct with a compass and ruler can be constructed with just a compass. It was then discovered that George Mohr, a Danish mathematician, had already proven this in 1672 so it is known as the Mohr–Mascheroni theorem.

ATL

Reflect and discuss 1

How are secants and chords similar? How are they different?

Exploration 1

1 Use a set of compasses to draw a circle. Clearly mark its center. Draw a chord that is **not** the diameter of the circle.

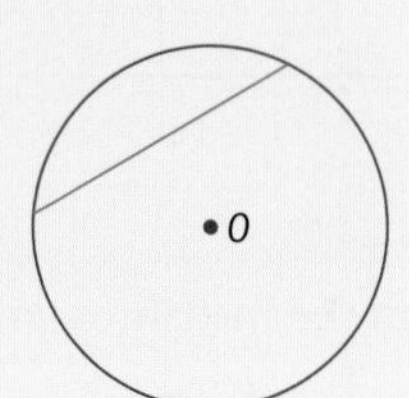

2 Draw radii from the center of the circle to each endpoint of the chord. Determine which type of triangle you have drawn.

3 Use a protractor to measure the angle of the sector you have drawn.

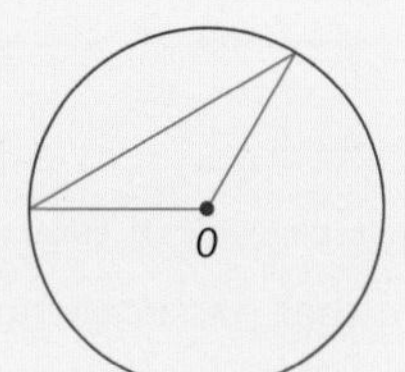

4 Draw a line segment from the center of the circle, O, to the midpoint of the chord, M.

a Explain why this line segment:

- divides the triangle in half
- divides the angle of the sector in half (measure to check that it does).

b Calculate the angle that the new line segment makes with the chord. Measure to check.

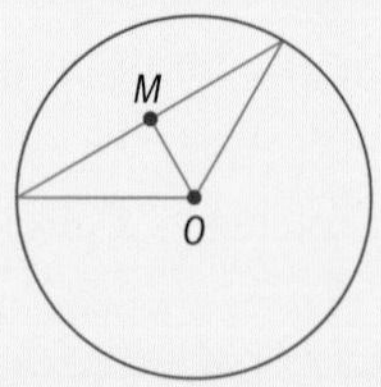

Tip

Use symmetry.

In a circle, two radii and a chord form an isosceles triangle. A line from the center of the circle to the midpoint of the chord forms two right-angled triangles.

When you know the angle at the center, you can use trigonometry in one of the right-angled triangles to find the length of the chord.

Example 1

Find the length of chord XY.

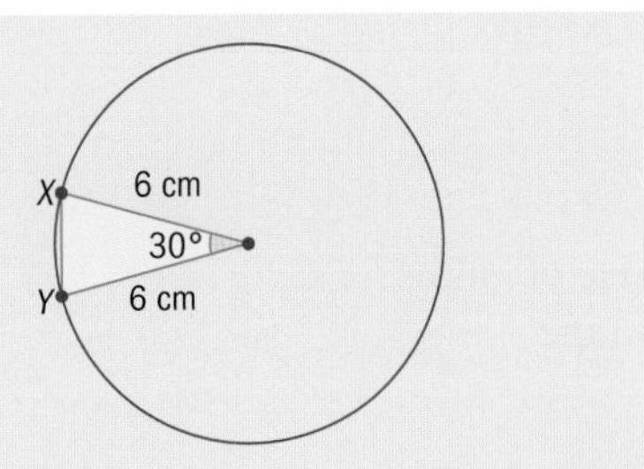

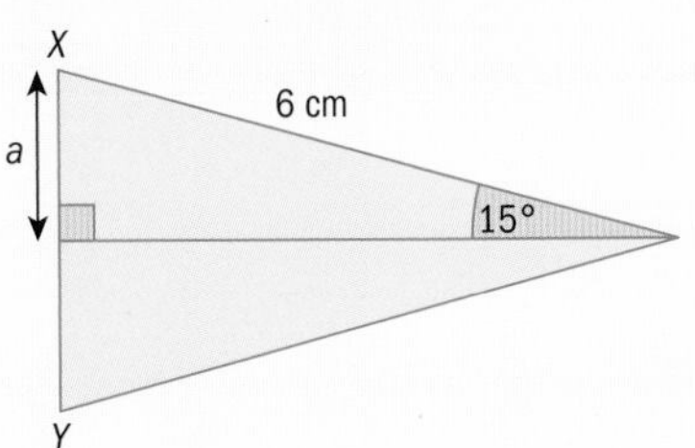

Divide the isosceles triangle into two right-angled triangles.

$\sin 15° = \frac{a}{6}$ — $\sin x = \frac{\text{opposite}}{\text{hypotenuse}}$

$a = 6 \times \sin 15°$

$= 1.5529\ldots$ cm

XY is $1.5529\ldots \times 2 = 3.11$ m (3 s.f.) — The length of the chord is twice the length of a.

Practice 1

Give your answers to 1 decimal place.

1 Find the length of chord JK.

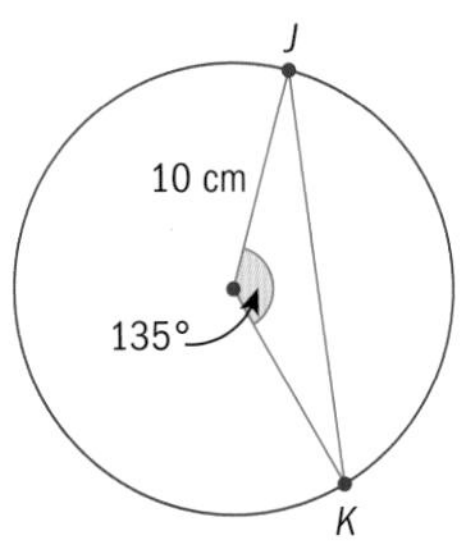

2 Find the length of chord LM.

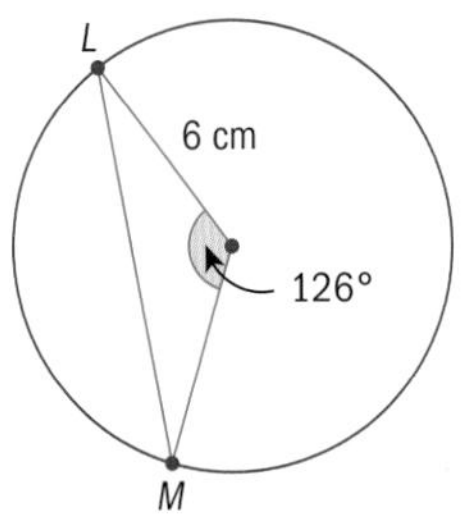

Problem solving

3 The sector AOB is one third of the circle. Find the length of chord AB.

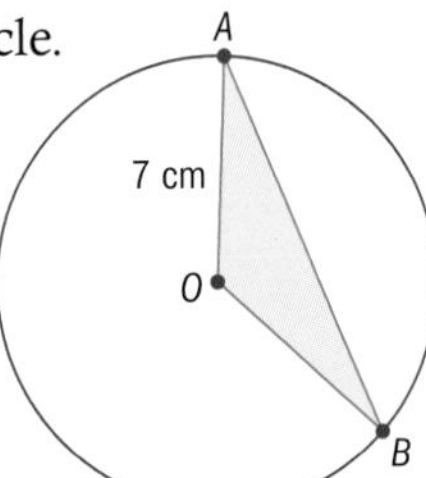

Tip

First find the angle of the sector.

4 *POQ* is a sector of a circle. The angle of the sector is 75°. Chord *PQ* has length 8 cm. Find the radius of the circle.

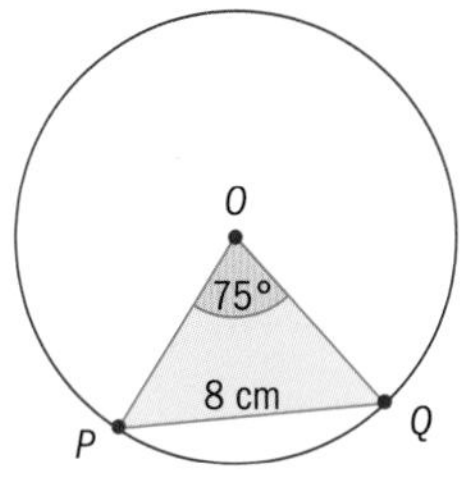

The Mesopotamians used a sexagesimal system, that is, base 60. We use this today when measuring time – 60 seconds in a minute, 60 minutes in an hour – and angles – 360° = 60 × 6.

5 *ROS* is a sector of a circle of radius 5 cm. Chord *RS* is 8 cm. Find the angle of sector *ORS*.

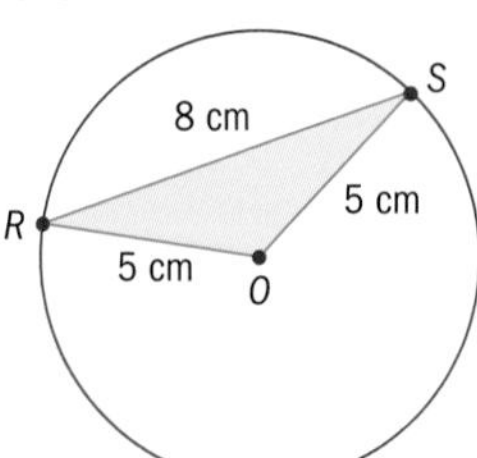

C Relationships between measurements in a circle

- How are the parts of a circle related?

Exploration 2

1. Draw a circle with a set of compasses and then draw a chord that is not the diameter of the circle. Label the chord *c*.
2. Draw radii from the center of the circle to each endpoint of the chord. Indicate the angle, θ, in the center of the sector.
3. By drawing a line segment from the center of the circle to the midpoint of the chord *c*, use trigonometry to show that the relationship between θ, r and c can be expressed as $\theta = 2\arcsin\left(\frac{c}{2r}\right)$.

   Verify that your formula works for questions **1**, **2** and **3** in Practice 1.

   Explain what changes you would have to make to your formula for questions **4** and **5** in Practice 1.

When sharing a pizza, the more equal slices you cut, the smaller each slice will be. To describe the size of a pizza slice you could measure:

1. the fraction of the pizza
2. the **central angle** θ (angle at the center of the sector), sometimes called the **measure of the sector**
3. the length of the crust, or **arc length**, l
4. the perimeter of the sector, P
5. the area of the sector, A.

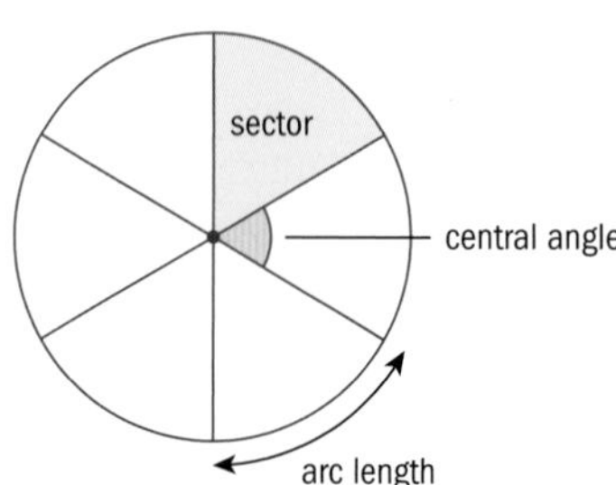

'You better cut the pizza in four pieces because I'm not hungry enough to eat six.'

Lawrence Peter "Yogi" Berra

Objective: **B.** Investigating patterns
ii. describe patterns as general rules consistent with findings

In this Exploration you will search for patterns and generalize the relationships between the different measurements in a circle.

ATL

Exploration 3

In this Exploration you will generalize the relationship between the measurements.

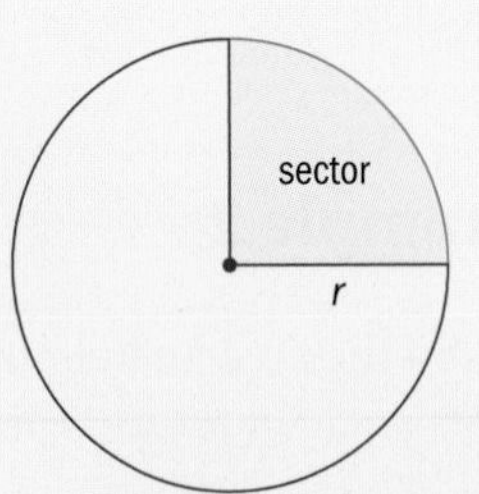

For a pizza of radius r cut into four equal slices or sectors:

1 State the fraction of the pizza in each sector.

2 State the central angle for each sector.

3 What fraction of the circumference is the arc length of each sector? Write down an expression for the arc length of each sector.

4 Describe how you can use the arc length of a sector to find the perimeter of a sector. Write down an expression for the perimeter of each sector.

5 What fraction of the area of the circle is the area of each sector? Write an expression for the area of each sector.

6 Copy and complete the table.

When answering questions about the length of arcs and sectors, your final answer should always include the appropriate units.

Number of equal sectors	Fraction	Central angle	Arc length	Sector perimeter	Sector area
1 (the whole circle)	$\frac{1}{1}=1$	360°			
2	$\frac{1}{2}$	$\frac{360°}{2}=180°$	$\frac{2\pi r}{2}$	$\pi r+2r$	$\frac{\pi r^2}{2}$
3	$\frac{1}{3}$	$\frac{360°}{3}=120°$	$\frac{2\pi r}{3}$		$\frac{\pi r^2}{3}$
4					
	$\frac{1}{5}$				
		60°			
			$\frac{2\pi r}{8}$		
					$\frac{\pi r^2}{9}$
			$\frac{2\pi r}{10}$		
		12°			
		10°			
		9°			
n					
$360/\theta$					

Arc length and sector area formulae

- **arc length** $= \frac{\theta}{360°} \times 2\pi r$
- **sector area** $= \frac{\theta}{360°} \times \pi r^2$

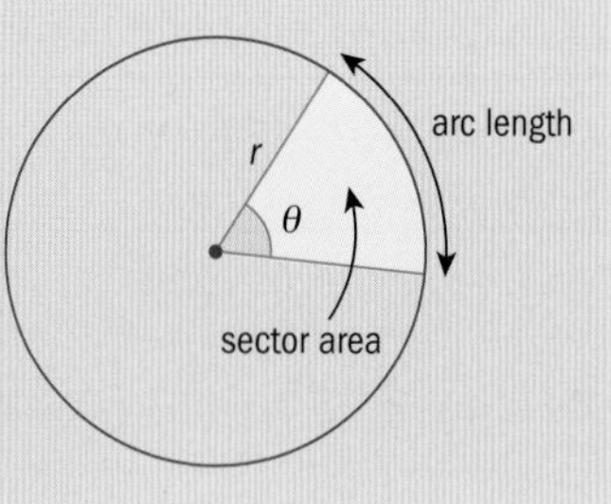

Example 2

Find the area and circumference of this circle in terms of π.

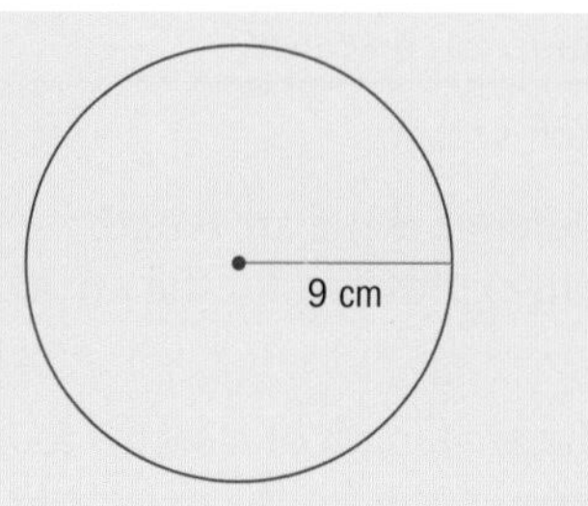

'In terms of π' means leave π in your answer.

$\text{Area} = \pi r^2 = \pi \times 9^2 = 81\pi \text{ cm}^2$

$\text{Circumference} = 2\pi r = 2 \times \pi \times 9 = 18\pi \text{ cm}$

Remember to include the correct units.

Example 3

Find the length of arc XY in terms of π, and as a number to 2 decimal places.

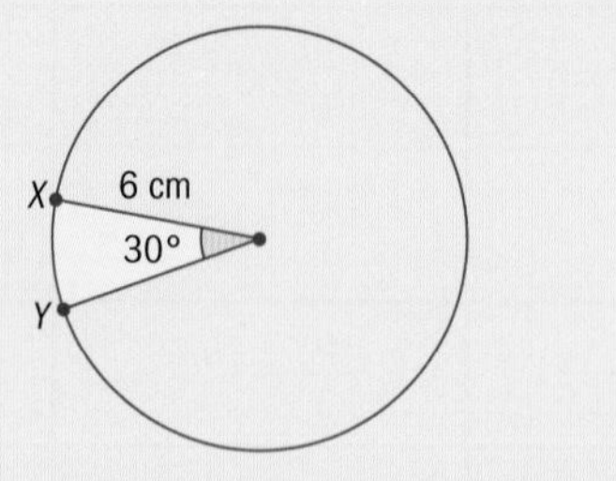

$\theta = 30°$

$r = 6 \text{ cm}$

$\text{Arc length } XY = \frac{\theta}{360°} \times 2\pi r$

Use the arc length formula.

$= \frac{30}{360} \times 2 \times \pi \times 6$

$= \pi \text{ cm}$

$= 3.14 \text{ cm}$

Reflect and discuss 2

Circle and sector problems may ask for lengths and areas in terms of π, or to a given number of decimal places.

- Which gives the most accurate value?
- What degree of accuracy do you think is necessary?

Example 4

Find the area of sector PQ.
Give your answer to 3 s.f.

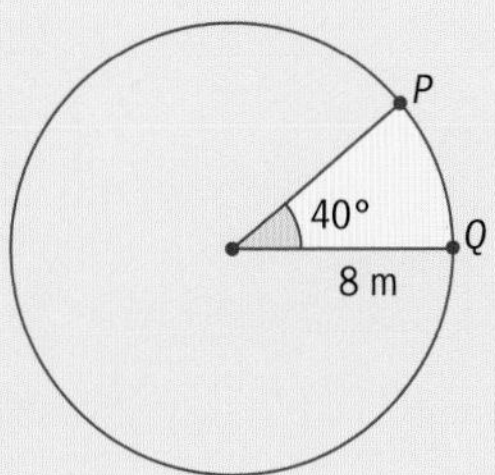

$\theta = 40°$

$r = 8$ m

$$\text{Area of sector } PQ = \frac{\theta}{360°} \times \pi r^2$$ Use the formula for the area of a sector.

$$= \frac{40}{360} \times \pi \times 8^2$$

$$= \frac{1}{9} \times \pi \times 64$$

$$= 22.3402\ldots$$

Area of sector $PQ = 22.3$ m^2 (3 s.f.) Remember to include the units.

Practice 2

1 Find the area and circumference of each circle. Give your answers in terms of π.

a

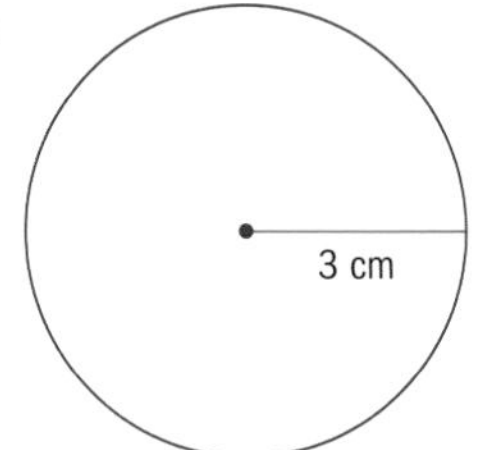

b

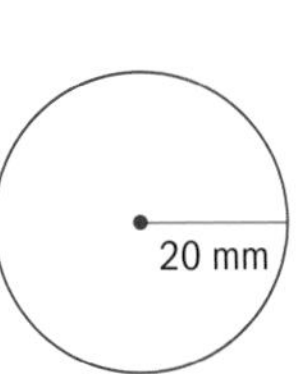

c

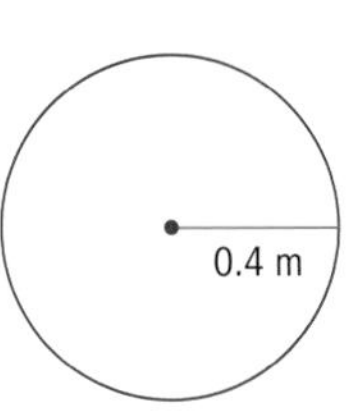

2 Find the area and perimeter of the half circle and quarter circle. Give your answers to 2 d.p.

a

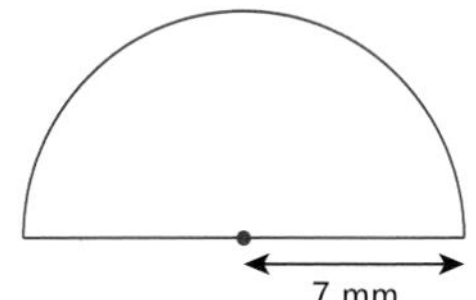

b

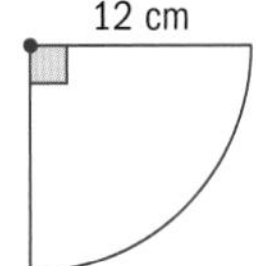

3 The radius of this circle is 8 cm.

a Find the arc length *HI*.

b Hence find the perimeter of sector *HI*.

c Find the perimeter of sector *KJ*.

d Find the area of sector *HK*.

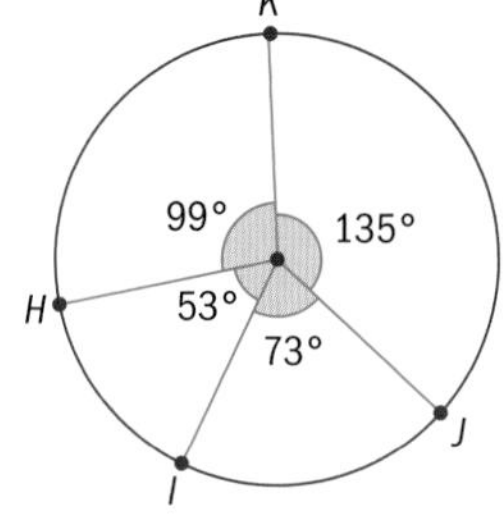

Tip

Perimeter = arc length + 2*r*
($P = l + 2r$)

4 In this question, give all your answers in terms of π.

a Find the length of arc *AB*.

b Find the perimeter of the shaded sector.

c Find the area of the shaded sector.

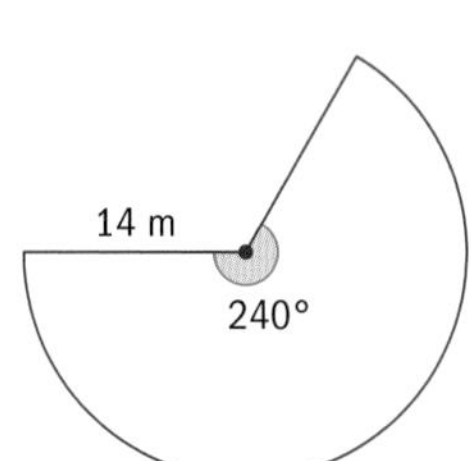

5 This is a sector of a circle.

a Find the perimeter of the sector.

b Find the area of the sector.

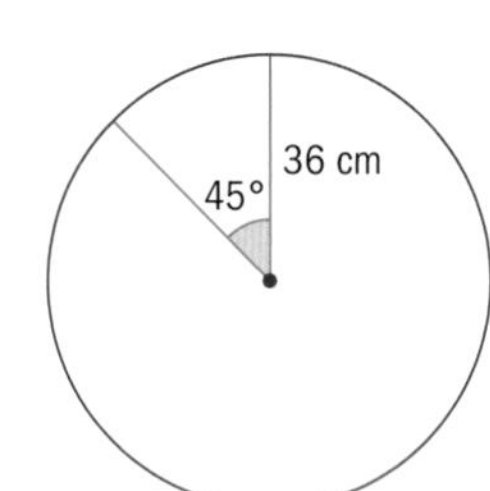

6 **a** Find the length of the major arc.

b Find the perimeter of the major sector.

c Find the area of the major sector.

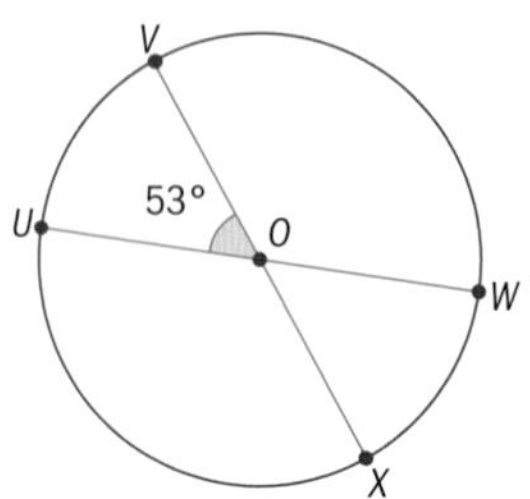

Problem solving

7 *VX* and *UW* are both diameters of the circle; $VX = UW = 12$ cm.
Find the perimeter of sector *XOW*.

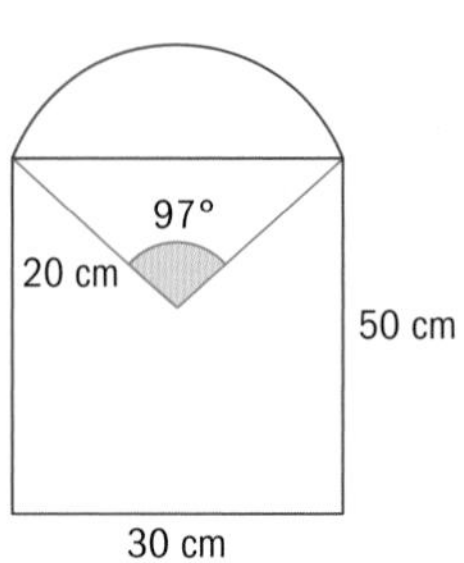

8 A frame for a mirror is in the shape of a rectangle with a curved top as shown in the diagram. The vertical sides of the frame are 50 cm long, and the top is an arc with central angle 97° and radius 20 cm. Calculate the perimeter of this frame.

9 Calculate the volume of this solid. The end face is a sector of a circle, radius 4 cm.

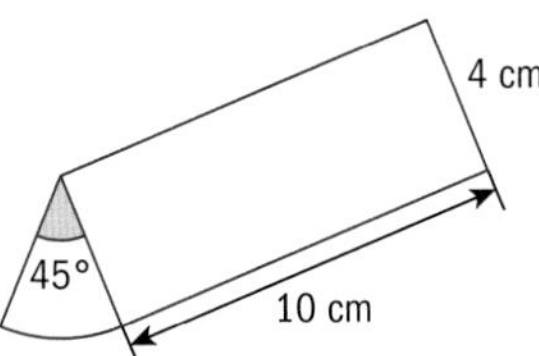

10 A light hardwood dowel has a semi-circular cross-section of radius 3 cm, and is 2 m in length. Calculate the volume of wood in the dowel.

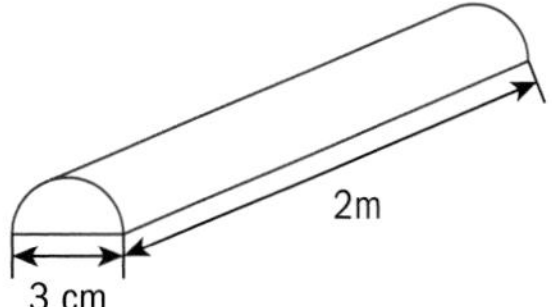

D Finding unknown values in circles

- Is it advantageous to generate your own formulae?

In this section you will generate more of your own formulae.

ATL

Exploration 4

1 Calculate the circumference of each whole circle in terms of π.

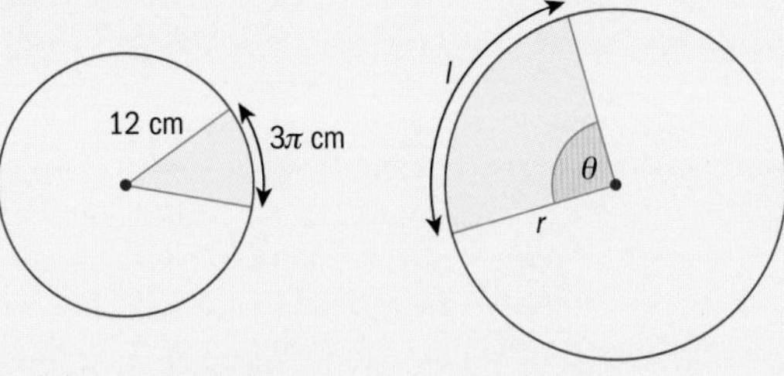

Use the arc length to work out the fraction of each circle that is shaded. Write down a general formula for finding the angle of sector θ given the arc length l and the radius r.

2 Calculate the area of each whole circle in terms of π.

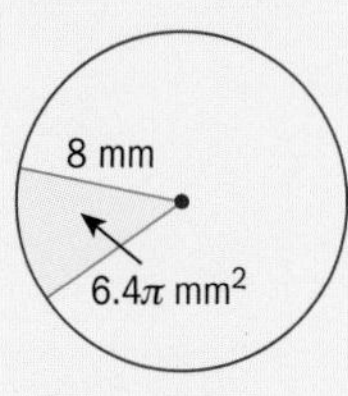

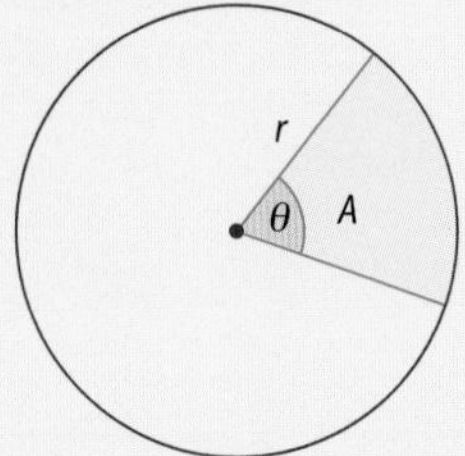

Use the sector area to work out the fraction of the circle that is shaded. Write down a general formula for finding the angle of the sector θ given the area A and the radius r.

3 Rearrange the formula you found in step **1** to make arc length l the subject.

4 Make r the subject of the formula.

5 Rearrange the formula you found in step **2** to make area A the subject.

6 Solve the formula for r.

π has fascinated people for centuries. The Great Pyramid at Giza has dimensions that relate very closely to it. The base is a square, the perimeter of which is equal to the circumference of a circle with a radius equal to the pyramid's height.

Example 5

Find the measure of arc θ, to the nearest degree.

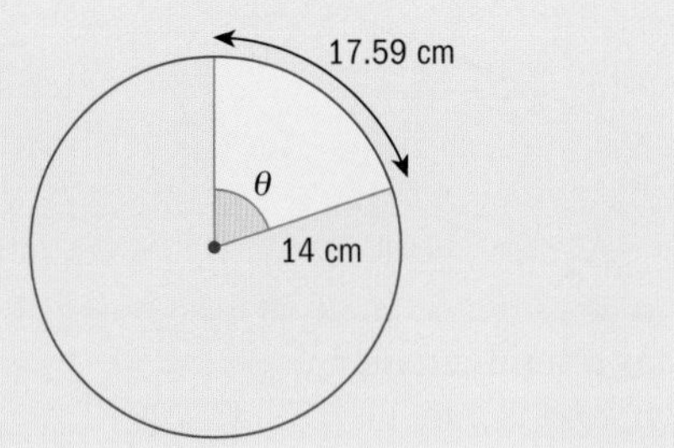

$$\text{arc length} = 17.59 = \frac{\theta}{360} \times 2\pi r$$

Use the arc length formula.

$$17.59 = \frac{\theta}{360} \times 2 \times \pi \times 14$$

$$\frac{17.59 \times 360}{28\pi} = \theta$$

Rearrange and solve to find θ.

$$\Rightarrow \quad \theta = 71.988\ldots$$

The measure of θ to the nearest degree is 72°.

Example 6

Find the radius of the circle to 3 s.f.

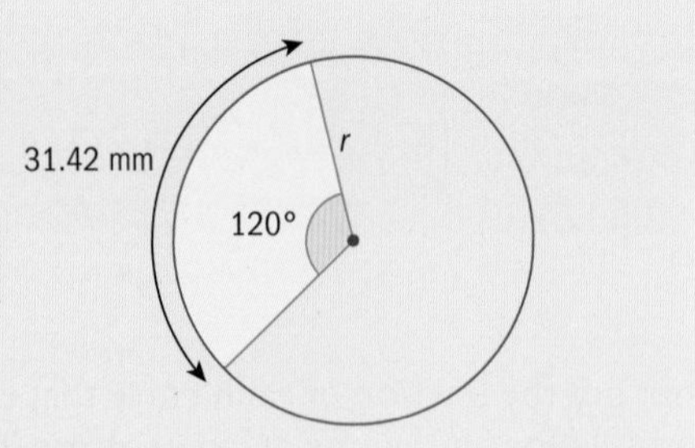

$$\text{arc length} = 31.42 \text{ mm}$$

$$= \frac{\theta}{360°} \times 2\pi r$$

Use the formula for the arc length.

$$31.42 = \frac{120}{360} \times 2\pi r$$

$$31.42 = \frac{1}{3} \times 2\pi r$$

$$r = \frac{31.42 \times 3}{2\pi}$$

Rearrange and solve for r.

$$r = 15.0019\ldots$$

The radius of the circle is 15.0 mm (3 s.f.)

Practice 3

1 Find the measure of the central angle of the sector in each circle.

a

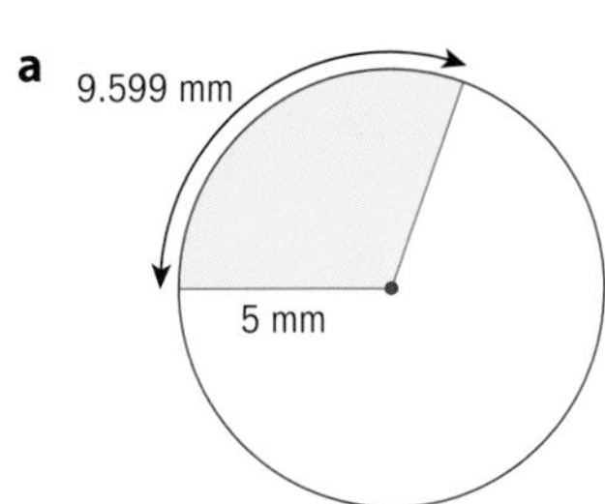

b

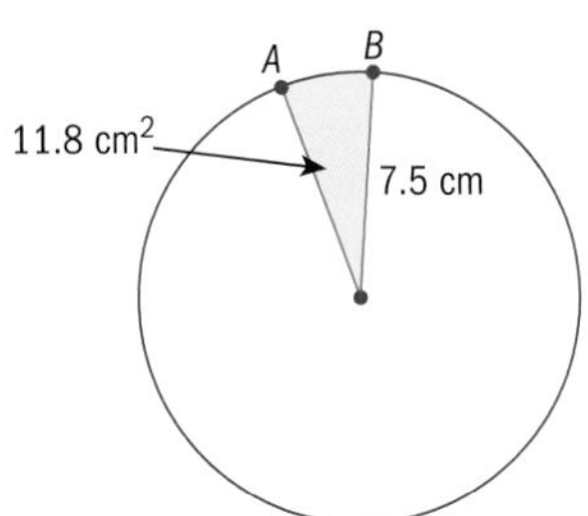

2 **a** Find the angle in the sector.

b Find the length of the arc AB.

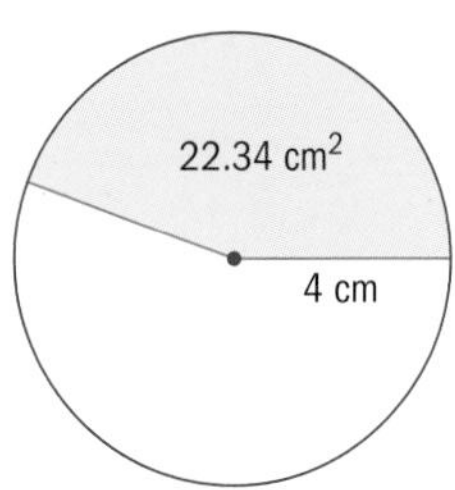

3 **a** Find the measure of the central angle of the major sector.

b Find the length of the major arc.

c Find the perimeter of the major sector.

4 Find the radius of each circle. Give your answers to 3 s.f.

a

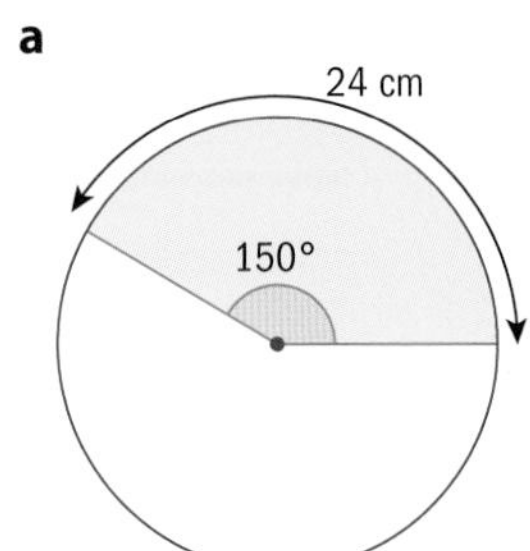

b

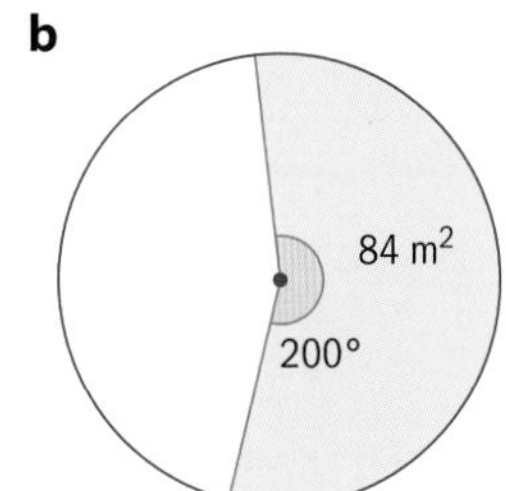

5 Find the perimeter of the major sector.

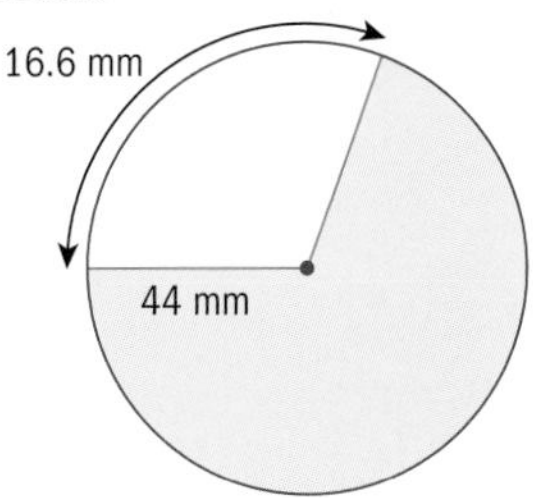

6 Find the area of the minor sector.

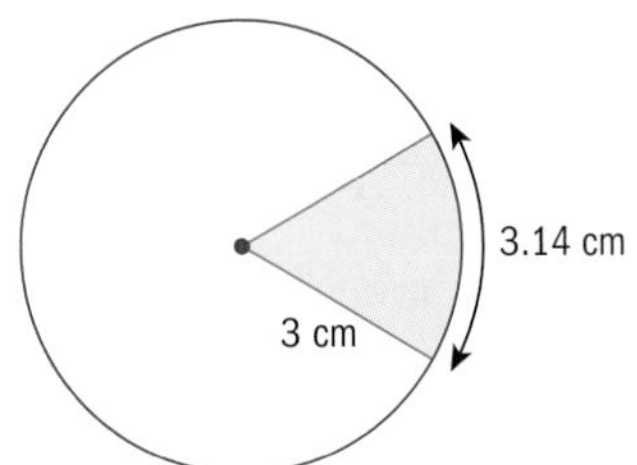

In question 6, first find the angle of the sector.

Problem solving

7 Ellie plans to make a paper cone from a sector of a circle with angle 220°. What size circle does she need to start with to make a cone with curved surface area 120 cm²? Give your answer in terms of the circle's radius, accurate to 1 d.p.

8 Find the length of chord *KM*.

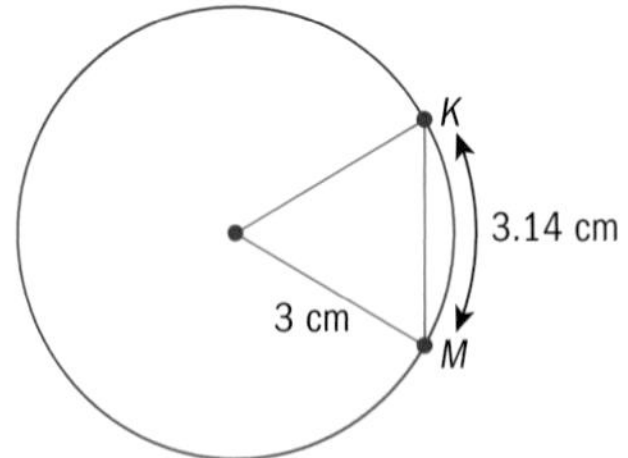

9 Find the length of chord *PQ*.

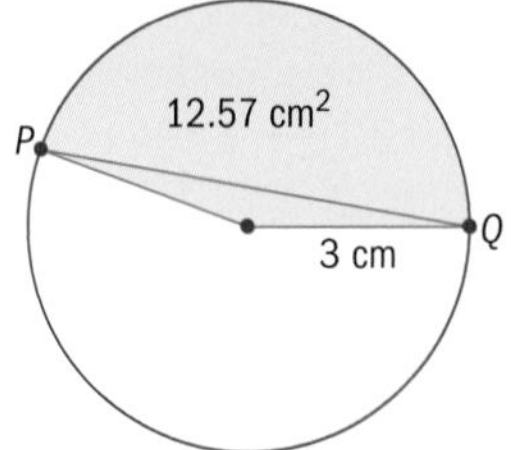

10 The landing zone for a javelin competition is the sector of a circle with radius 120 m. The central angle of the sector is 40°.

a Draw a diagram to represent this landing zone.

b Find the perimeter of the landing zone.

c Find the area of the landing zone.

d Find the shortest distance between the two outer corners of the landing zone.

Summary

- A **chord** is a line segment with its endpoints on a circle.
- A **diameter** is a chord that passes through the center of a circle.
- A **secant** is a line that intersects a circle at two points.
- A **tangent** is a line that touches a circle at only one point.

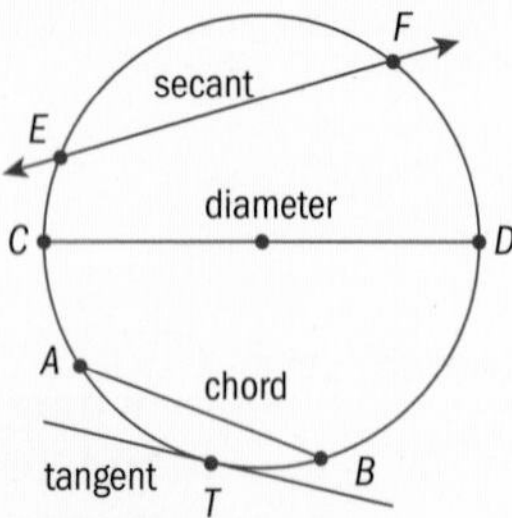

- A chord divides a circle into two **segments**.
- The larger segment is the **major segment**. The smaller segment is the **minor segment**. 'Segment' usually means the minor segment.
- An **arc** is a part of the circumference of a circle.
- The larger arc is the **major arc**. The smaller arc is the **minor arc**.

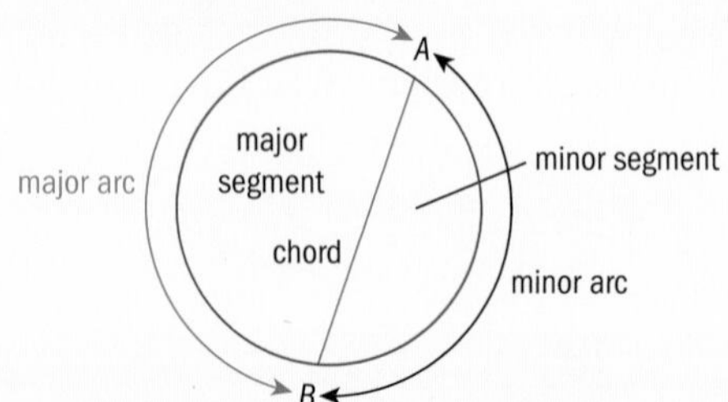

- A **sector** is a 'slice' of a circle between two radii. The center of a circle is usually labelled *O*, so the sector is sector *BOC*. 'Sector' usually means the minor sector.

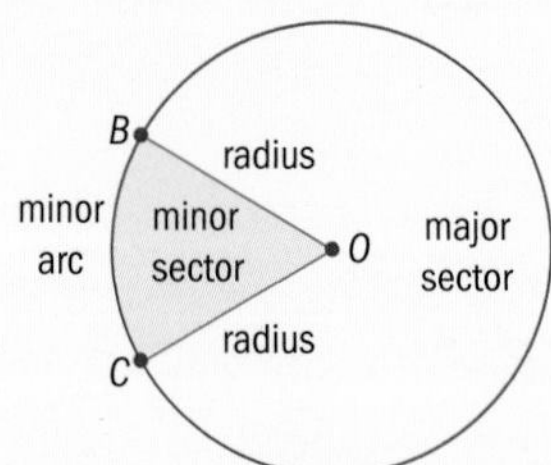

- Two radii and a chord form an isosceles triangle. A line from the center of the circle to the midpoint of the chord forms two right-angled triangles.

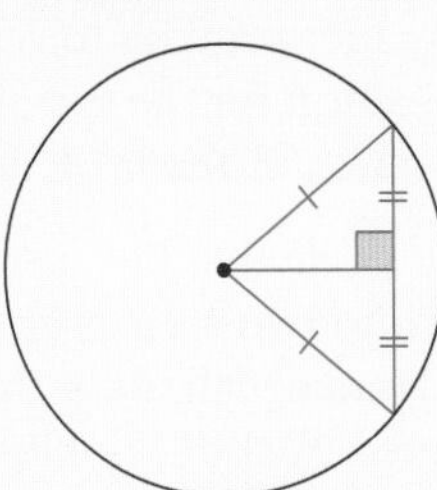

- The formula for the length of a chord, c, in a sector radius r with angle θ is $c = 2r\sin\frac{\theta}{2}$ or $\theta = 2\arcsin(\frac{c}{2r})$.

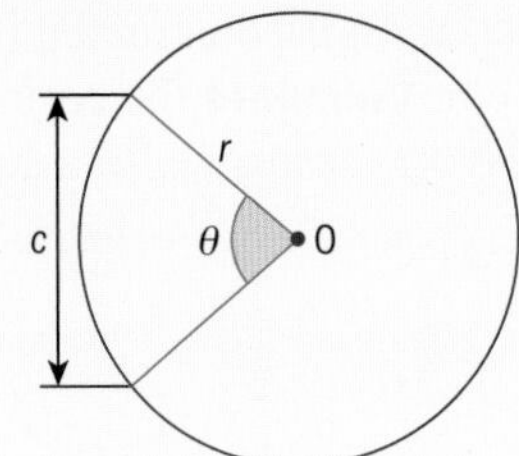

- **Arc length** $= \frac{\theta}{360°} \times 2\pi r$

- **Sector area** $= \frac{\theta}{360°} \times \pi r^2$

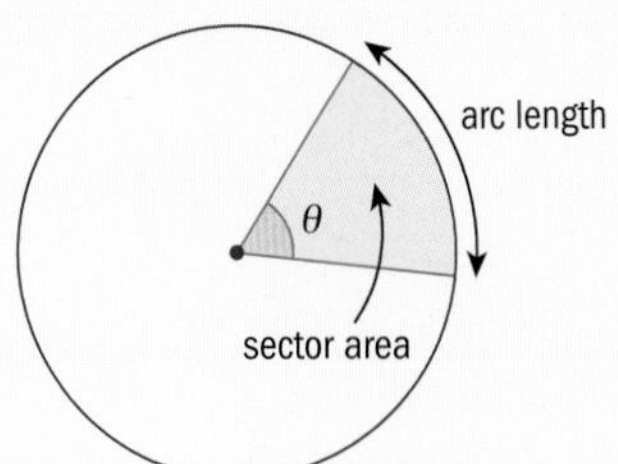

$\theta = \frac{180}{\pi r}$ $\qquad \theta = \frac{360A}{\pi r^2}$

$r = \frac{180}{\theta r}$ $\qquad r = \sqrt{\frac{360A}{\theta\pi}}$

$= 2\pi r\frac{\theta}{360}$ $\qquad A = \pi r^2\frac{\theta}{360}$

Mixed practice

1 a Find the arc length of sector *DC*.

b Find the perimeter of sector *BD*.

c Find the area of sector *EB*.

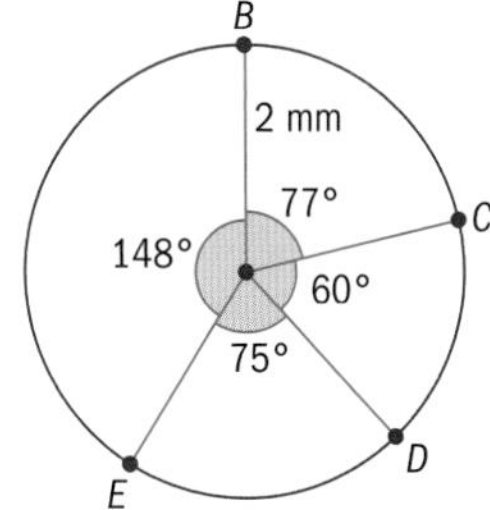

2 a Find the length of the major arc *QR*.

b Find the area of the major sector.

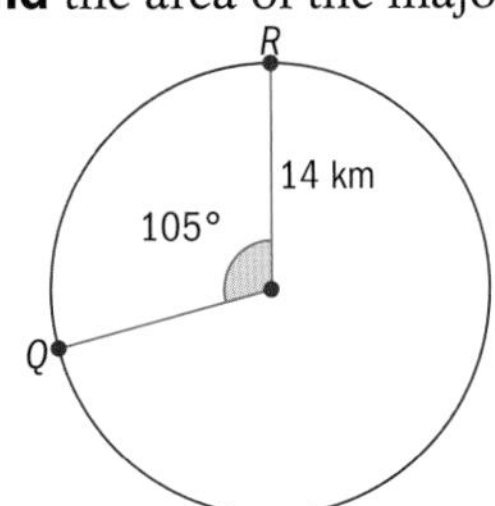

3 Find the perimeter and area of:

a the minor sector

b the major sector.

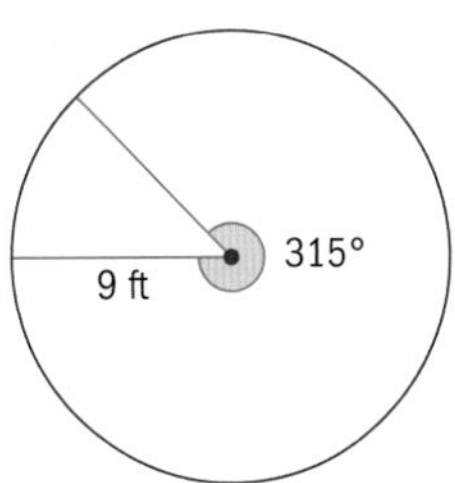

4 a Find the length of the minor arc.

b Find the perimeter of the minor sector.

c Find the area of the major sector.

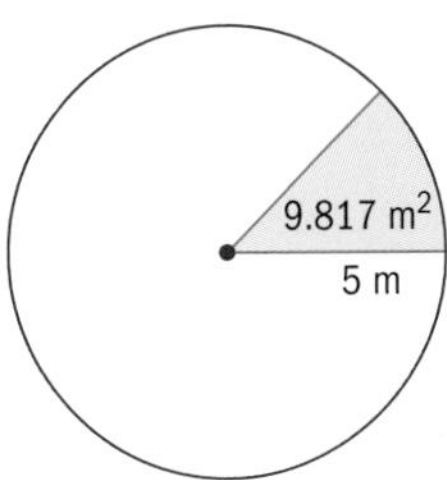

5 a **Find** the measure of the central angle.

b **Find** the length of the minor arc.

c **Find** the perimeter of the shaded sector.

d **Find** the length of the chord MN.

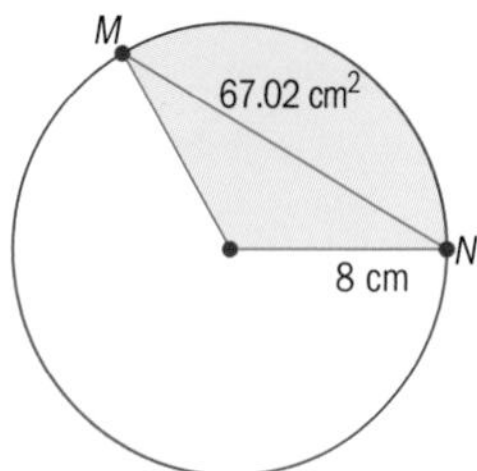

6 a **Find** the measure of the central angle.

b **Find** the perimeter of the shaded sector.

c **Find** the length of the chord UV.

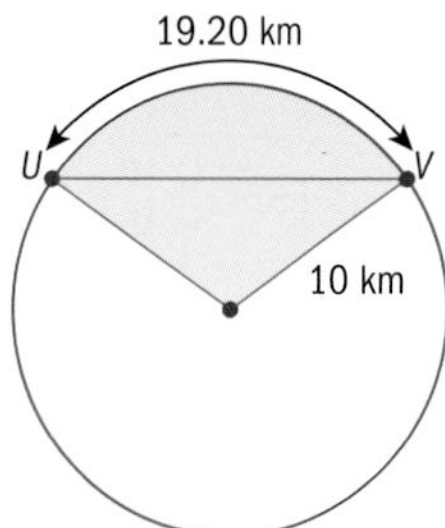

7 Toni has a rectangular lawn 8 m by 10 m. Her lawn sprinkler sprays water up to 5 m.

a **Draw** and label a scale diagram of the lawn.

b By using compasses to draw the region watered by the sprinkler to scale, **determine** the best position for the sprinkler so that it waters as much of the lawn as possible.

c Divide the circle that lies inside the garden boundary into two sectors and two isosceles triangles. **Hence, calculate** the area of the lawn watered by the sprinkler.

d **Calculate** the percentage of the lawn that is not watered by the sprinkler.

Problem solving

8 A regular hexagon is constructed in a circle of radius 5 cm. **Calculate** the perimeter of the hexagon. Give your answer to a suitable degree of accuracy.

Review in context

Personal and cultural expression

1 'Pendulum dowsing' as a means of finding water, gold or even answers to questions has been used throughout history and has been recorded as far back as the time of the pharaohs in Egypt. In one version, a pendulum is held above a cloth with 'yes' or 'no' written on it and the degree to which the pendulum sways to one side or the other is an indication of how likely or unlikely the event in question is going to happen.

a If the pendulum has a total length of 30 cm, **calculate** the distance travelled if it moved a total of 45 degrees.

b If the pendulum starts facing straight down and it travels an arc of 10 cm, **calculate** the turned angle.

2 The diagram shows the landing area for the shot put. The throwing circle has a diameter of 2.135 m, and the landing area is a sector with angle 34.92° and sector lines 25 m long, starting from the center of the circle. The distance thrown is measured from the circumference of the throwing circle to the imprint made in the soil by the shot in the landing area.

a **Find** the area enclosed by the entire sector.

b **Find** the area enclosed by the sector inside the throwing circle.

c Hence, **find** the area of the landing area.

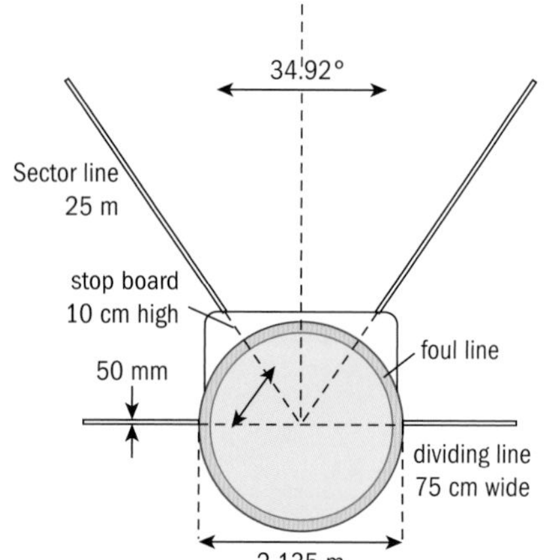

3 'Medicine wheels' were used by native Americans for a variety of rituals. The Big Horn Medicine Wheel is a circle of stones with a diameter of about 24 meters. It is divided into 28 approximately equal sections, likely to represent the 28 days of the lunar calendar.

a **Calculate** the approximate central angle of each sector.

b **Calculate** the area of each sector.

c Different spots along the circle have been found to correspond to specific astronomical events (winter solstice, rising of stars). Two spots related to the rising of the star Sirius are 20 meters apart. **Calculate** their approximate distance apart along the circle.

d The rising of the star Aldebaran happens between two points whose central angle measures 80 degrees. **Calculate** the shortest distance between them.

Reflect and discuss

How have you explored the statement of inquiry? Give specific examples.

Statement of Inquiry:

Generalizing relationships between measurements enables the construction and analysis of activities for ritual and play.

7.4 Which triangle is just right for you?

Global context: Scientific and technical innovation

RELATIONSHIPS

Objectives

- Solving problems in right-angled triangles using trigonometric ratios
- Knowing the properties of trigonometric ratios
- Solving problems that include angles of elevation and angles of depression

Inquiry questions

F
- How do you find measurements of immeasurable objects?
- What are the relationships in the special triangles?

C
- How do relationships between sides and angles in right-angled triangles help you find real-life measurements?

D
- How does understanding the trigonometric ratios help you create and understand mathematical models?

ATL **Critical-thinking skills**

Propose and evaluate a variety of solutions

7.4

14.2

Statement of Inquiry:

Generalizing relationships between measurements can lead to better models and methods.

You should already know how to:

• use the Pythagorean theorem to find the missing side in a right-angled triangle	**1** Find the missing side in each triangle, accurate to 3 s.f. All measurements are in cm. 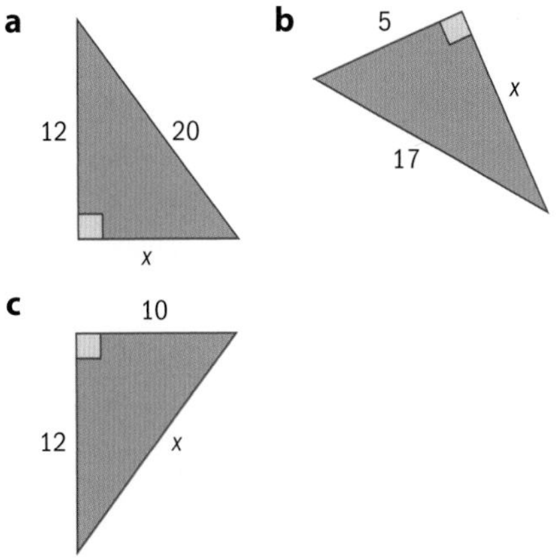 **a** 12, 20, x **b** 5, x, 17 **c** 10, 12, x
• relate angles and sides of a right-angled triangle using the trigonometric ratios (sine, cosine and tangent)	**2** Find $\sin A$, $\cos B$, $\tan A$ and $\tan B$. 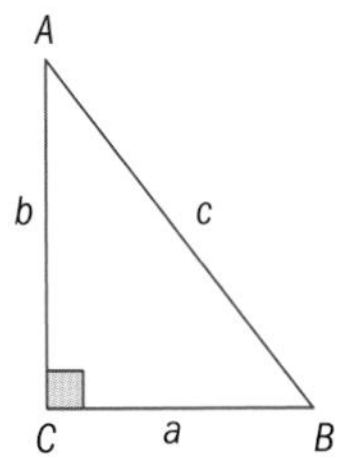 A, b, c, C, a, B
• solve simple problems using the trigonometric ratios	**3** Find the value of x in each triangle. 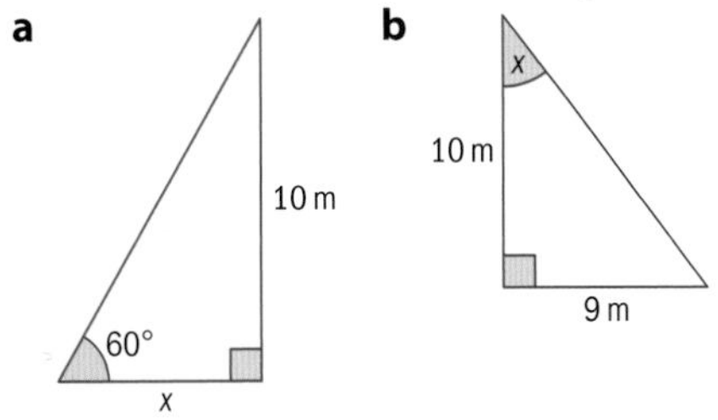 **a** 10 m, 60°, x **b** x, 10 m, 9 m

F Right-angled triangles and trigonometric ratios

- How do you find measurements of immeasurable objects?
- What are the relationships in the special triangles?

The right angle is the most common angle in our world – after all, we stand at right angles to our earth, and the homes we live in are built at right angles to their foundations. This fundamental angle is at the heart of the study of geometry and trigonometry.

ATL

Activity

1 Make a list of different methods you could use to find the height of a very tall building. For each method, state where you are in relation to the building.'

2 Write down the advantages and disadvantages of using each method.

3 Discuss these methods with others and rank them starting with the best method first. Explain your ranking order.

4 If you were standing on the ground in front of a building, what measurements would you need to take in order to calculate its height? Draw a diagram and write an equation relating these measurements.

Land surveyors determine property boundaries and prepare maps and survey plots accordingly. In their job they always are measuring immeasurable distances and angles.

5 Research what kind(s) of tools land surveyors use to measure far distances and angles between the line of vision and an object.

6 How would land surveyors use their instruments, together with the trigonometric ratios of sine, cosine, and tangent, to determine immeasurable heights and angles?

You have already learned about the trigonometric ratios in a right-angled triangle, but did you know that they can help you determine immeasurable distances? In order to do so, you will need to review how to find the trigonometric ratios with a calculator.

Practice 1

1 Use your calculator to evaluate these, either exact or to three significant figures.

a $\sin 47°$ **b** $\cos 32°$ **c** $\tan 17°$ **d** $\sin 2°$

e $\tan 89°$ **f** $\cos 60°$ **g** $\tan 45.8°$ **h** $\sin 0.789°$

i $\cos 0°$ **j** $\tan 30°$

2 Find the angle in each expression.

a $\sin A = 0.467$ **b** $\cos B = \frac{\sqrt{2}}{2}$ **c** $\tan C = 1$

d $\cos \alpha = 0.898$ **e** $\sin \beta = 0.5$ **f** $\tan \theta = \sqrt{3}$

g $\cos \varphi = 0.123$ **h** $\tan A = 27.3$

Tip

Make sure that your calculator is in degree mode (not in radian mode) when you are using degrees to measure angles.

Reflect and discuss 1

Take a closer look at the values of the trigonometric ratios in the questions of Practice 1.

- What do you notice about the range of values for each?
- Can you explain the reason for your observation?

Some people use the memory aid **SOH-CAH-TOA** for remembering the trigonometric ratios in a right-angled triangle. Given a right-angled triangle with acute angle α:

SOH — $\textbf{Sin}\,\alpha = \dfrac{\text{side } \textbf{O}\text{pposite to } \alpha}{\textbf{H}\text{ypotenuse}}$

CAH — $\textbf{Cos}\,\alpha = \dfrac{\text{side } \textbf{A}\text{djacent to } \alpha}{\textbf{H}\text{ypotentuse}}$

TOA — $\textbf{Tan}\,\alpha = \dfrac{\text{side } \textbf{O}\text{pposite to } \alpha}{\text{side } \textbf{A}\text{djacent to } \alpha}$

Practice 2

1 Use your knowledge of the trigonometric ratios to find the side marked x in each triangle. Round your answers to the nearest hundredth where necessary.

a

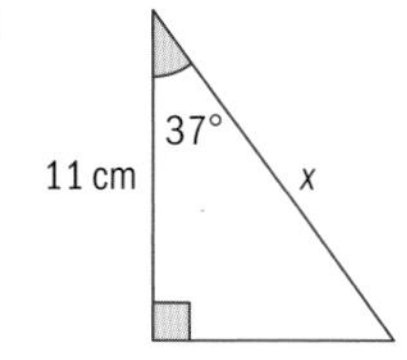

b

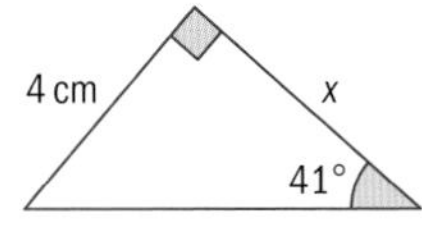

c

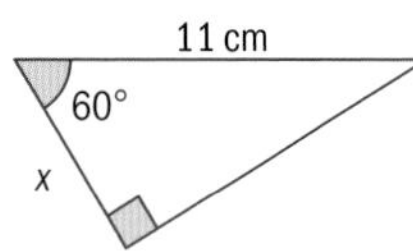

d

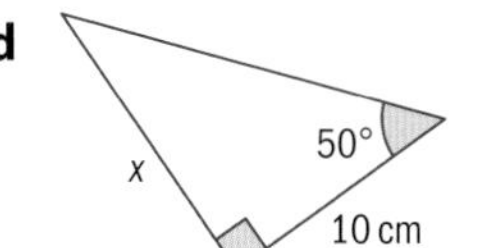

e

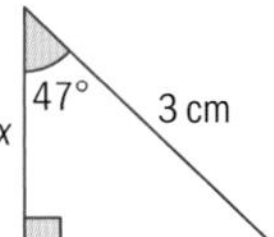

f

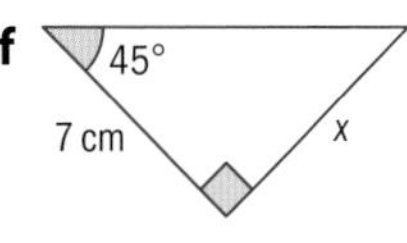

2 Find the angle θ in each triangle. Round your answers to the nearest tenth. All lengths are in meters.

a

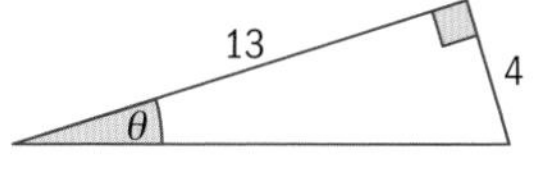

b

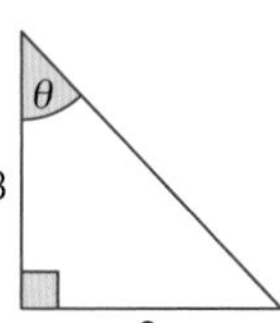

c

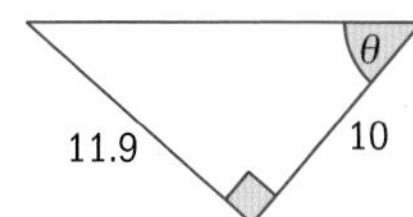

d

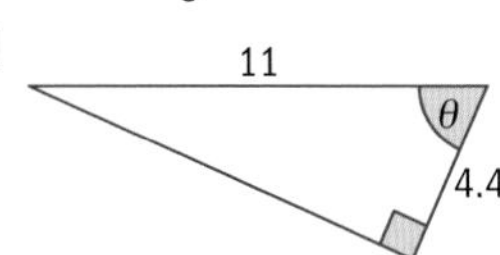

3 Find all missing sides and angles in these right-angled triangles.

a

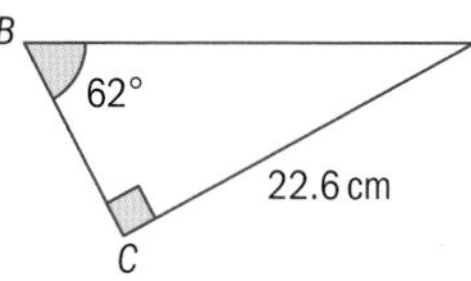

b

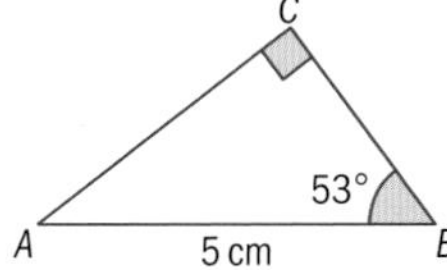

c

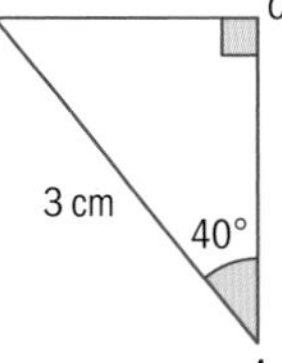

d

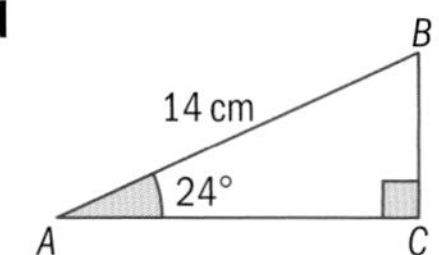

e

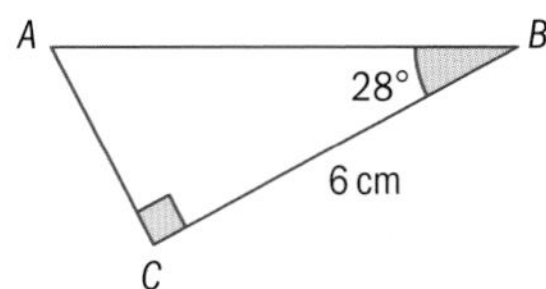

Objective: **C.** Communicating
i. use appropriate mathematical language (notation, symbols and terminology) in both oral and written explanations

In this Exploration, you will need to use appropriate and accurate notation in your sketches to be able to convince others that your answers are justified.

Exploration 1

One method you may have discussed in the Activity is to use the shadow of a building to measure its height. You are now going to explore two methods to find out how tall your school is without actually measuring its height.

Method 1: Shadow

1 Sketch the building and its shadow in your notebook, and draw the right-angled triangle that is represented by this situation. Label the sides of the triangle.

2 Determine what measurements you would need to make in order to use trigonometric ratios to calculate the height of the building. If you can, make these measurements.

3 Use your sketch to determine which trigonometric ratio will help you find the height of the building. Write down the trigonometric equation that represents the situation. (Solve it if you have actual measurements.)

Method 2: Mirror

1 Put a small mirror on the ground and stand far enough away that you can see the top of the building in the middle of the mirror.

2 Measure your distance to the mirror, your height and the distance of the building to the mirror.

3 Sketch the situation and indicate the measurements in your diagram.

4 Justify how the two triangles in your diagram are related.

5 Explain how you could calculate the height of the building **a** using trigonometric ratios and **b** without using trigonometric ratios.

Reflect and discuss 2

- Which method do you think is more accurate, the shadow or the mirror method? Justify your answer.

Exploration 2

1 **a** Using the Pythagorean theorem, find the length of the hypotenuse of this triangle. Leave your answer as a radical.

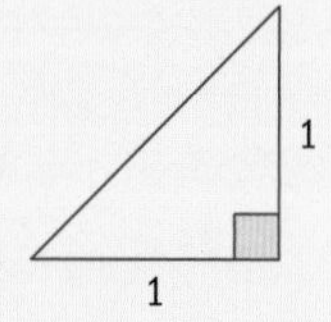

b Identify what type of triangle this is, and thus find the missing angles.

▶ Continued on next page

c Hence, use your knowledge of trigonometric ratios in a right-angled triangle to find the following values (leaving your answers as radicals):

- sin 45°
- cos 45°
- tan 45°

Now consider the equilateral triangle shown here:

2 a Using your knowledge of equilateral triangles, find angle α.

b Assume that the dotted line is the perpendicular bisector of the base. Using your knowledge of the angle sum in a triangle, find angle β.

c Using the Pythagorean theorem, find the length of the dotted line.

d Hence, use your knowledge of trigonometric ratios in a right-angled triangle to find the following values (leave your answers in surd form):

- sin 30°
- cos 30°
- tan 30°
- sin 60°
- cos 60°
- tan 60°

3 Check your results in **1 c** and **2 d** by comparing the surd form with the values from your calculator

4 In step **2**, which was the longest side of the right-angled triangle and which was the greatest angle? Where were these in relation to each other? What about the shortest side and the smallest angle?

The results of this exploration are very useful to memorize, as are the methods used to find these angles.

These triangles are known as 'special triangles'. Recognizing these ratios can help you solve problems.

Practice 3

By using the special triangles, find the unknown measurements. Leave your answers in radical form where appropriate.

1

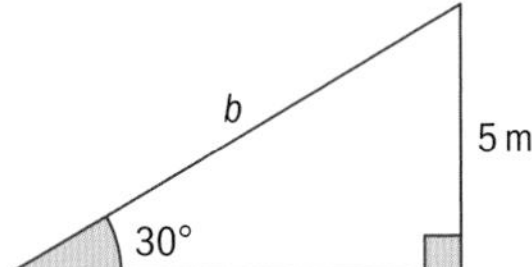

2

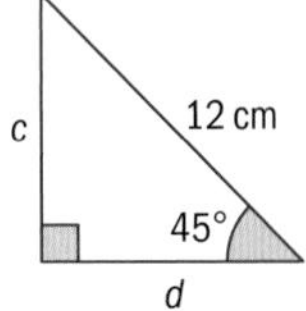

3

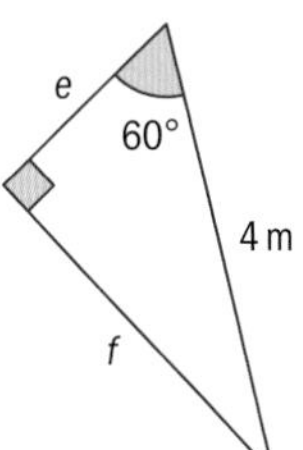

You can check your answers using Pythagoras' theorem.

4

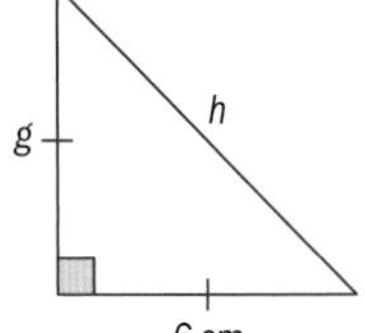

5

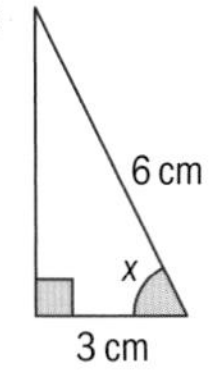

C Real-life applications

- How do relationships between sides and angles in right-angled triangles help you find real-life measurements?

Example 1

You are standing 9 m away from a building, and your eyes are 160 cm from the ground. Using a surveyor, you read the angle of elevation of the top of the building is 50°. Find the height of the building to the nearest meter.

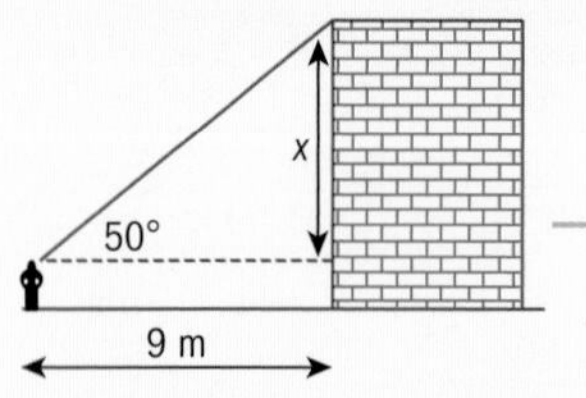

Draw a sketch and label the given information.

$\tan 50° = \frac{x}{9} \Rightarrow x = 9 \times \tan 50° = 10.726$ — In relation to the angle, you are given the side adjacent and you are looking for the opposite side, hence use tangent.

$1.6 + 10.726 = 12.326$ — Add your height to the answer and round to the nearest meter.

The building is about 12 m tall.

An **angle of depression** is the angle between the horizon and an object below the horizon.

An **angle of elevation** is the angle between the horizon and an object above the horizon.

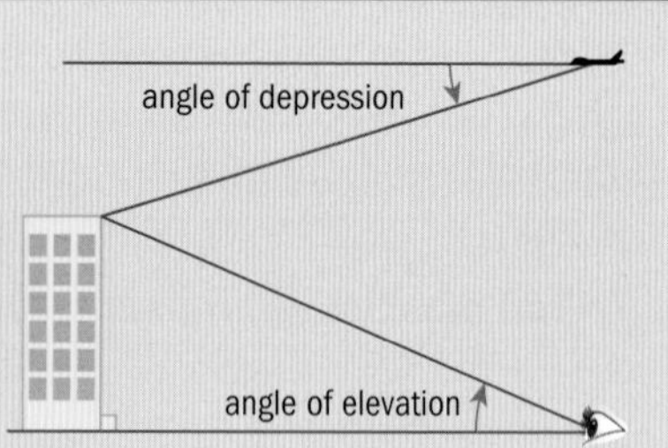

Reflect and discuss 3

- Why do we refer to sine, cosine and tangent as trigonometric *ratios*?
- The angles and sides of right-angled triangles are measured in specific units. What are the units of the trigonometric ratios? Justify your answer.
- Why are the angles of elevation and depression equal?

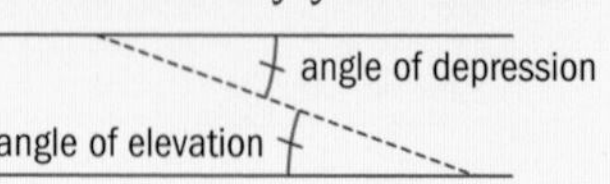

Practice 4

Problem solving

1 The famous Leaning Tower of Pisa 'leans' at an angle of 4°, and its top is 55.86 m above the ground. Determine the height of its top above the ground when it was originally built (standing at right angles to the ground).

2 To a person standing 130 m away from the base of a building, determine the angle of elevation of the top of a building that is 58 m high. Explain the limitations on the accuracy of your answer.

3 Gina is flying a kite that makes a 50° angle with the ground. Determine how high the kite is above the ground when she has let out 150 m of string.

4 A vertical stick 1m long casts a shadow which is 1.3 m long. Determine the altitude of the sun (its angle of elevation).

5 A ladder that is 6 m long is leaning against the side of a building making an angle of 60° with the ground. Determine how far the ladder's base is from the building, and how far up it is on the building.

6 A ship is on the surface of the water, and its radar detects a submarine at a distance of 238 m away underwater, at an angle of depression of 23°. Find the depth of the submarine.

Reflect and discuss 4

- For Practice 3, question 2, what happens to the angle of elevation as you get closer to the building, and as you get farther from the building?
- For Practice 3, question 4, describe what happens to the length of the shadow of the stick as the sun's altitude changes.

It is important to understand, at this point, that a trigonometric ratio is a relationship between specific sides of a right-angled triangle. That ratio is normally represented as either a fraction or decimal.

Example 2

Find the length of the side labelled x to 3 s.f.

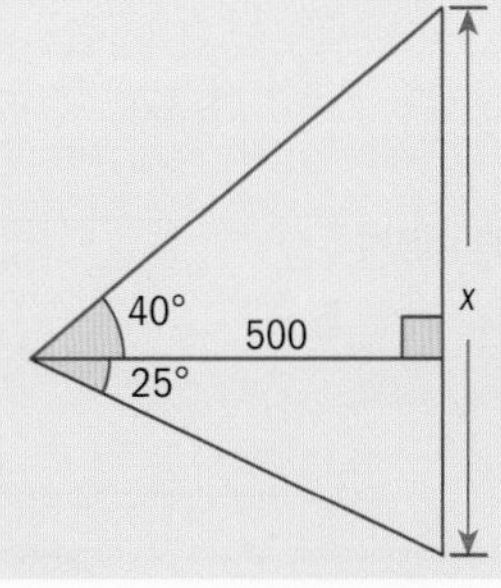

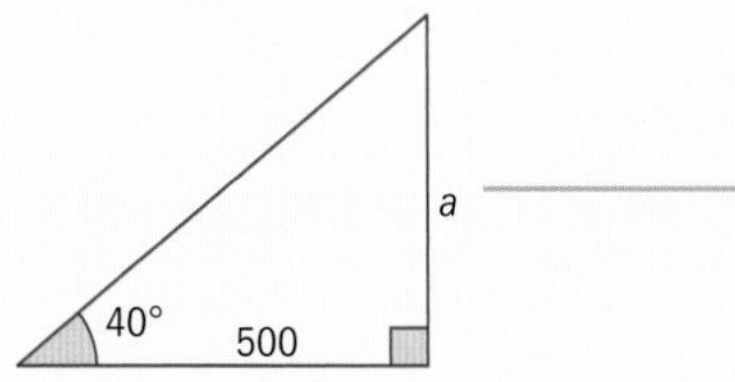

Separate the triangle into two right-angled triangles. Label the unknown side in the upper triangle and use the trig ratios to find it. Since the final answer is required to 3 s.f., all calculations before the final answer must be carried out to at least one more s.f.

$\tan 40° = \frac{a}{500} \Rightarrow a = 500 \times \tan 40° = 419.5$

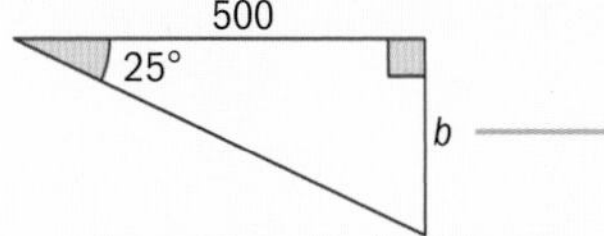

Label the unknown side in the lower triangle and use trig ratios to find it.

$\tan 25° = \frac{b}{500} \Rightarrow b = 500 \times \tan 25° = 233.2$

$x = a + b$

$= 419.5 + 233.2 = 652.7 = 653$ (3 s.f.)

Add the two lengths that make up the unknown side, and round to the required degree of accuracy.

Example 3

In the diagram, $AC = 52$ m. The angle of depression from C to B is 28°, and the angle of depression from C to D is 42°. Find the distance between B and D.

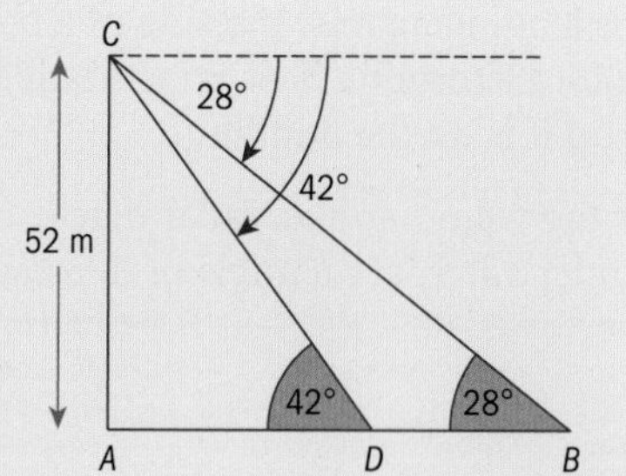

In ΔABC, $\tan 28° = \frac{52}{AB} \Rightarrow AB = \frac{52}{\tan 28°} = 97.80$

Use the trig ratios to find the required sides.

In ΔABD, $\tan 42° = \frac{52}{AB} \Rightarrow AB = \frac{52}{\tan 42°} = 57.75$

$\Rightarrow BD = 97.80 - 57.75 = 40.0$ m (3 s.f.)

Practice 5

1 Find the missing length x in each diagram.

a

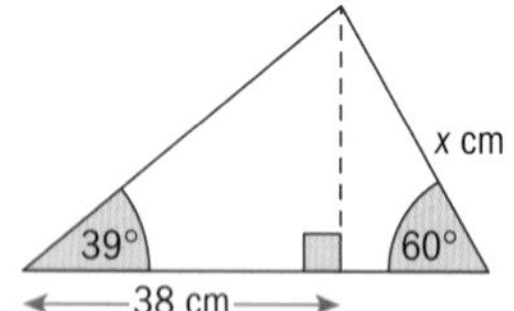

b

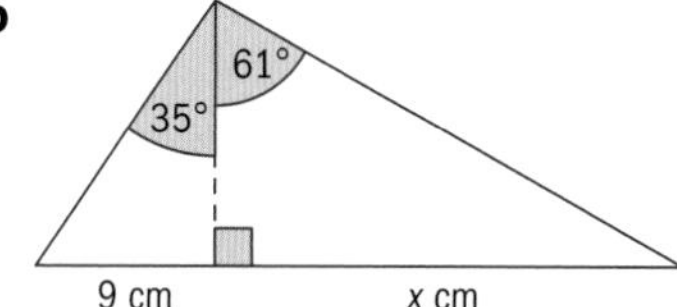

2 Find the side length x in each triangle.

a

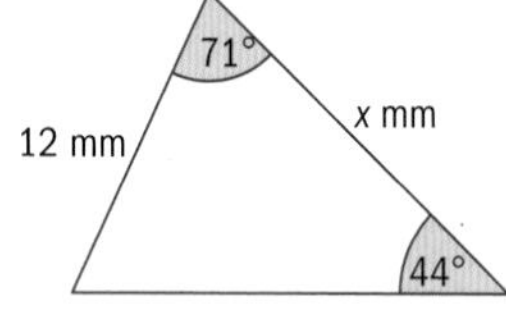

b

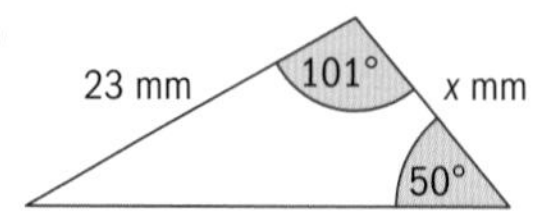

In **2**, you will need to decide where to draw a dotted line in order to solve the problem.

3 Find the angle θ.

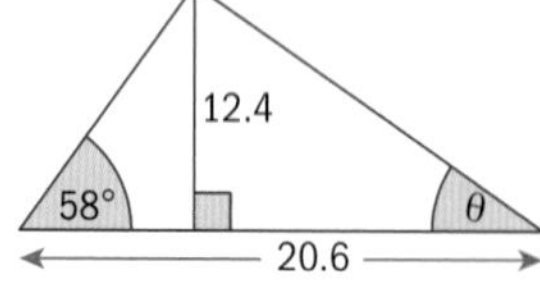

4 Find the distance m in the diagram.

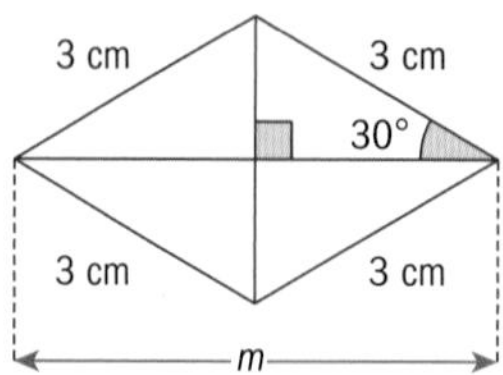

5 Find the heights from the ground of A, B and C.

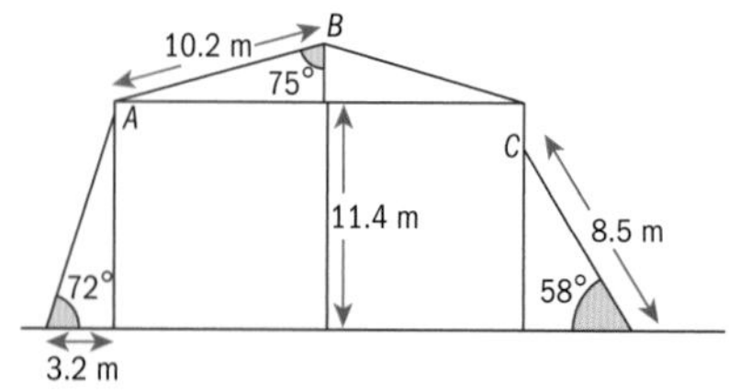

6 Find the side length x in this diagram.

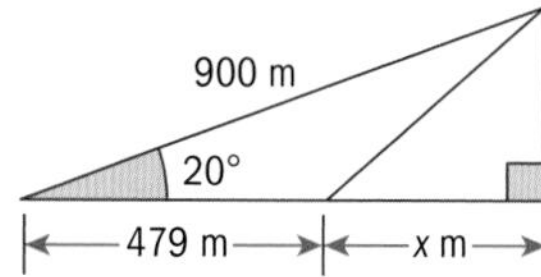

Problem solving

7 A single engine private plane flies due North for an hour and 18 minutes at an average speed of 110 km/h. It then turns and flies at the same average speed for an hour and a half on a bearing of 130°. At the end of this time, determine **a** how far the plane has travelled, and **b** its bearing from its initial position.

8 Two ships leave port at 10:00 and each one travels on a straight line course. The first one travels on a bearing of 60° at 23 km/h and the second one travels on a bearing of 115° at 28 km/h. Determine how far the two ships are from one another at 12:00 noon.

9 A lookout post is on top of a hill that is 100 m above sea level. From the base of the lookout post, the angle of depression of a boat out at sea is 30°, and from the top of the lookout post, the angle of depression of the boat is 60°. Find the height of the lookout post.

Reflect and discuss 5

- How does trigonometry help determine immeasurable distances?
- How do you think we solve problems when the triangle isn't a right-angled triangle?

D More on the trigonometric ratios

- How does understanding the trigonometric ratios help you create and understand mathematical models?

Exploration 3

1 Using your calculator, find the sines, cosines and tangents of several angles between 0° and 90°.

2 What is the range of values of each of the three ratios?

3 Summarize your findings in a table.

Reflect and discuss 6

- Considering right-angled triangles, give reasons why the sines and cosines of angles cannot be greater than 1.
- What was the greatest value you found for a tangent ratio? By choosing a bigger angle, could you have found an even greater value?
- What does your calculator say when you enter tan 90°? Why is this? Why is there no maximum value for the tangent ratio?

So far, you should have been able to make the following generalizations related to right-angled triangles:

- The lengths of the sides of a right-angled triangle are proportional to the measures of their opposite angles. The longest side will therefore be opposite the largest angle, and the shortest side will be opposite the smallest angle. Since the largest angle in a right-angled triangle is 90°, the longest side of the triangle has to be the hypotenuse.
- The sine and cosine ratios are the length of one of the legs divided by the length of the hypotenuse. Since the hypotenuse is always the longest side, the denominator of the fraction is always larger than the numerator; hence the ratio is always less than 1.
- The tangent of an angle is the ratio of the leg opposite the angle to the leg adjacent to the angle. The legs that make up the right angle are either equal, or one is larger than the other. The tangent can therefore be less than 1 (but greater than zero), greater than 1, or equal to 1.

Exploration 4

1 Draw a right-angled triangle where one of the legs is very short (very close to length 0) and the hypotenuse is 5 cm. Label it like this:

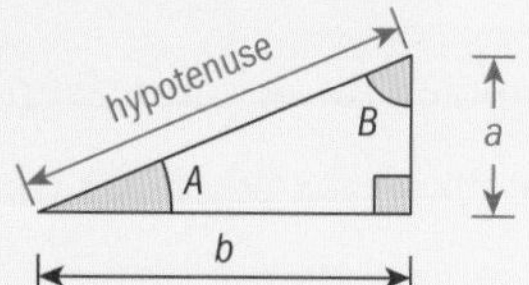

2 Measure lengths a and b.

3 Using the definition of sine and cosine, find $\sin A$, $\cos A$, $\sin B$ and $\cos B$.

4 Explain what happens to sin A as side a gets smaller.
Generalize the effect on $\sin A$ and $\cos A$ as angle A gets closer to zero.
Generalize the effect on $\sin B$ and $\cos B$ as angle B gets closer to 90°.

You might want to look at the effects on $\tan A$ and $\tan B$ as angle A gets closer to zero.

In Exploration 5 you will see how the tangent ratio is related to another mathematical concept you have studied.

Exploration 5

1 Plot the points (5, 0) and (10, 4) on graph paper. Calculate the gradient of the line segment joining these two points.

2 Measure the angle between the line and the x-axis and use your calculator to find the tangent of the angle.

3 Write down the relationship you notice between the gradient of the line and the tangent of the angle.

4 Plot the points (0, –2) and (3, 4), and find the gradient of the line segment joining them. Find the tangent of the angle between the line segment and the x-axis. Then, find the tangent of the angle between the line segment and the line $y = -2$. Write down what you notice between the two tangent values, and the relationship between the gradient of the line and the tangent of the angle formed by the line segment and a line parallel to the x-axis.

5 Justify the relationships you noticed in steps **3** and **4**.

The gradient of a straight line between (x_1, y_1) and (x_2, y_2) is $m = \frac{y_2 - y_1}{x_2 - x_1}$.

Continued on next page

6 Consider again the line segment in step **1**. Write down what happens to the gradient of the line segment as you move the point (10, 4) closer to the y-axis. Reflect on the change of the tangent of the angle formed by the line and the x-axis as you move the point closer to the y-axis. Determine the gradient of the line segment when you move this point on to the y-axis? What is the tangent of the angle when the point is on the y-axis?

7 Plot the points (5, 0) and (0, 4) and find the gradient of the line segment between the two points. Find the tangent of the angle formed between the line segment and the x-axis. Explain why should the tangent of the angle be a negative value.

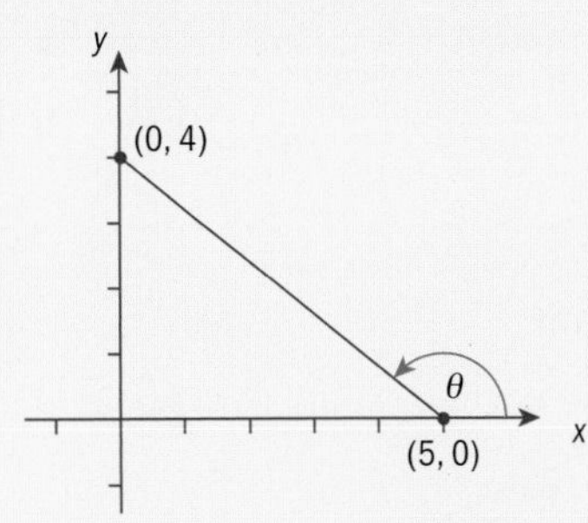

Note that you find the angle between the line and the *positive* x-axis. So θ is obtuse (>90°).

The idea of trigonometric ratios for angles that cannot be represented in a right-angled triangle will be explored further in the MYP 4/5 Extended book.

Summary

- The trigonometric ratios in a right-angled triangle help you make measurements that you couldn't easily make otherwise.
- The largest angle in a right-angled triangle is the right angle. The two other angles are always acute angles.
- The longest side in a right-angled triangle is the hypotenuse (opposite the right angle) and the other sides, called legs, are shorter.
- In a right-angled triangle, the values of sin and cos are always between 0 and 1. However, the value of tan can be any value greater than 0.
- The following results hold true for sin, cos and tan:

$\sin 0° = 0$	$\cos 0° = 0$	$\tan 0° = 0$
$\sin 90° = 1$	$\cos 90° = 0$	$\tan 90° = $ *undefined*

- Some results worth knowing:

1, $\sqrt{2}$, 45°, 1	$\sqrt{3}$, 30°, 2, 60°, 1	$\sqrt{3}$, 60°, 2, 30°, 1
$\sin 45° = \frac{\sqrt{2}}{2}$	$\sin 30° = \frac{1}{2}$	$\sin 60° = \frac{\sqrt{3}}{2}$
$\cos 45° = \frac{\sqrt{2}}{2}$	$\cos 30° = \frac{\sqrt{3}}{2}$	$\cos 60° = \frac{1}{2}$
$\tan 45° = 1$	$\tan 30° = \frac{\sqrt{3}}{2}$	$\tan 60° = \frac{\sqrt{3}}{1}$

- The gradient of a line is equal to the tangent of the angle that the line makes with the positive x-axis.
- Angles of depression and elevation:

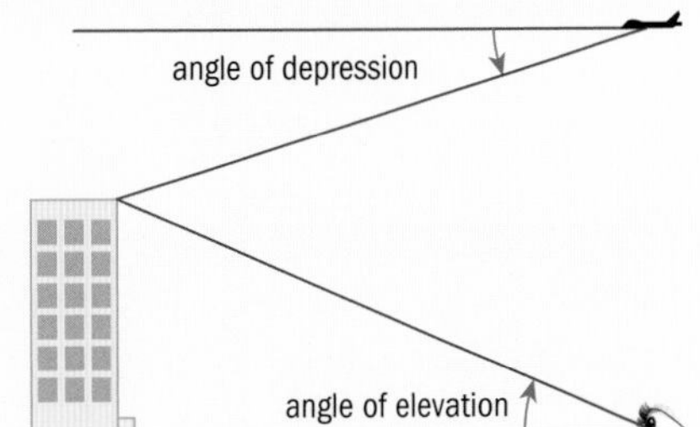

Mixed practice

1 A kite string is 48 m long. As the wind blew, the angle between the kite and the ground went from 27 degrees to 54 degrees. **Determine** the increase in the vertical height of the kite above the ground.

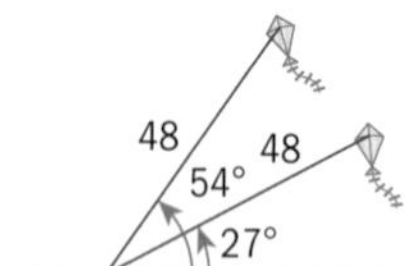

2 **Find** length x in each diagram.

a

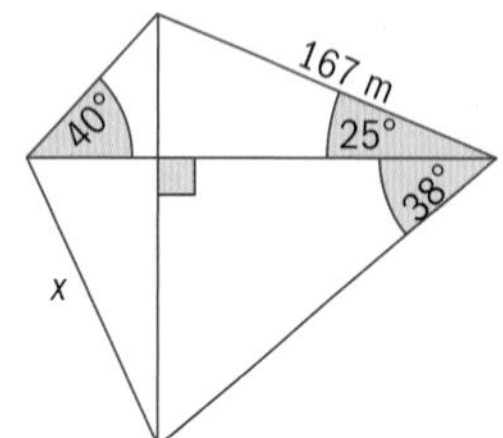

b

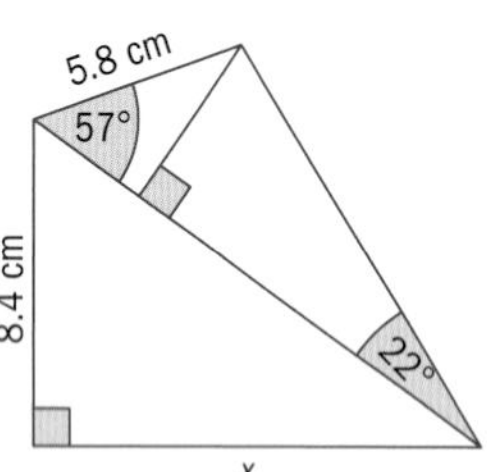

3 $ABCD$ is a square, where $AP = 5$ cm, $QC = 7$ cm and $PB = 12$ cm.

Calculate:

a the size of the angle marked x

b the length of AB

c the length of DQ

d the length of PD

e the size of $\angle BQD$.

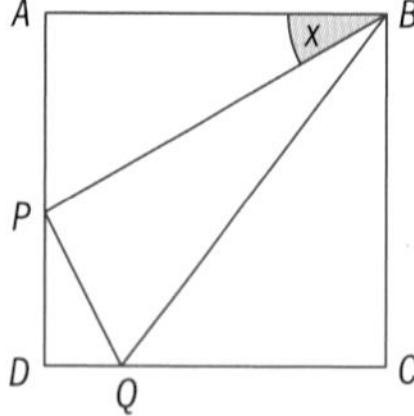

4 In a house that is being renovated, two corridors meet at right angles, as shown in the diagram. Workmen attempting to take a ladder around the corner get the ladder stuck when it is at an angle of 45° to each wall.

a **Determine** the length of the ladder.

b **Determine** the length of a 2nd ladder that got stuck when it was at an angle of 60° with the wall of the 1 m corridor.

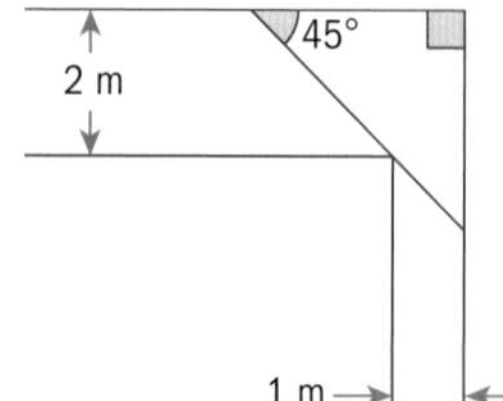

Problem solving

5 A 5 m ladder is resting against a vertical wall. The base of the ladder is 2 m from the wall. Keeping its base fixed, the ladder is rotated so that it now rests against the opposite wall which is 1 m away. **Determine** the angle through which the ladder has been rotated.

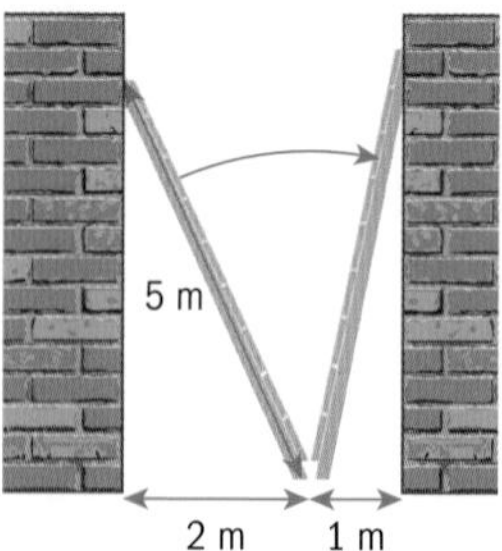

6 A coast guard attendant atop a cliff 120 m high observes two boats in line with him at angles of depression of 45 degrees and 69 degrees.

Determine how far apart the boats are.

Review in context

Scientific and technical innovation

1 In 1852, an Indian mathematician, Radhanath Sikdar, used measurements and trigonometry to calculate the height of Peak XV in the Himalayas (later to be named Mount Everest). This required the use of a device that measured angles from the ground to the top of an object. By measuring the angle to the top of Peak XV from two different spots, and knowing the distance between these spots, Sikdar may have come up with a drawing like the following:

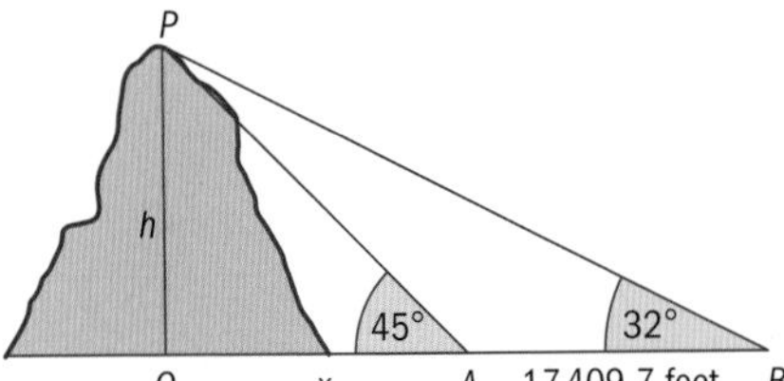

The angle of elevation to the top of the peak (from point B) was measured to be 32°. From a point 17 409.7 feet closer (point A), the angle of elevation was 45°.

Find the height Sikdar calculated for Peak XV.

Legend has it that the value Sikdar calculated was so perfect that he thought nobody would believe him. Supposedly, he added 2 feet to make it more believable! History books record his calculated height to be 29 002 feet.

2 You are the pilot of a light aircraft, returning to the airfield after a parachuting trip. Your altitude is 3050 m and you are heading towards the landing strip which is currently 14 km away. What angle of depression should you use so that you touch down on the landing strip at the closest point?

Reflect and discuss

How have you explored the statement of inquiry? Give specific examples.

Statement of Inquiry:

Generalizing relationships between measurements can lead to better models and methods.

8 Patterns

Sets of numbers or objects that follow a specific order or rule

Patterns in language

Patterns play an important role in the learning and understanding of language. Simple sentences follow patterns – and those patterns often vary from language to language. In English, and in other Romance languages, the SVO structure is common: simple sentences are built using a Subject, a Verb and an Object, in that order.

Many languages use the SOV structure. In Turkish, for example, the sentence 'İskender elmayı yedi' would translate in word order as 'İskender the apple ate', but the English translation of the sentence would be 'İskender ate the apple'. The study of patterns in language structure is known as linguistics.

Patterns in music

Steve Reich is an American composer who experiments with pattern and structure in his music. One of his famous works is titled 'Clapping Music for two people', and consists of a single rhythm repeated over 150 times, but with variation created by one musician starting half a beat later every 8 bars.

Search online for 'Steve Reich clapping music' to listen to a live performance and see if you can follow any patterns.

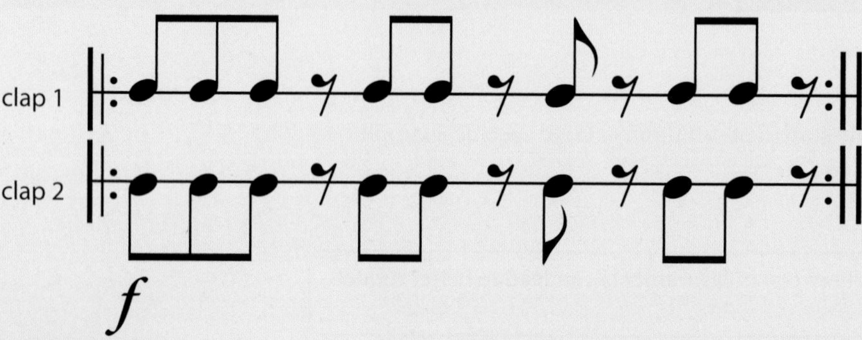

Patterns in physics

Johannes Kepler (1571–1630) discovered physical laws that explain the motion of the planets around the sun. His third law tells us that if T is the time it takes a planet to orbit the sun, and r is the radius of its orbit, then $T \propto r^{\frac{3}{2}}$.

What does it mean to have an exponent of $\frac{3}{2}$? By looking for patterns when working with exponents, we can give fractional exponents a sensible definition.

Patterns in finance

When you invest money at a bank, the bank usually awards interest on your savings; many banks compound the interest annually.

What pattern is generated by the amount in the bank account at the end of each year?

Understanding the pattern at the end of each year can help you to work out what would be a fair value for an investment if you needed to withdraw money partway through the year.

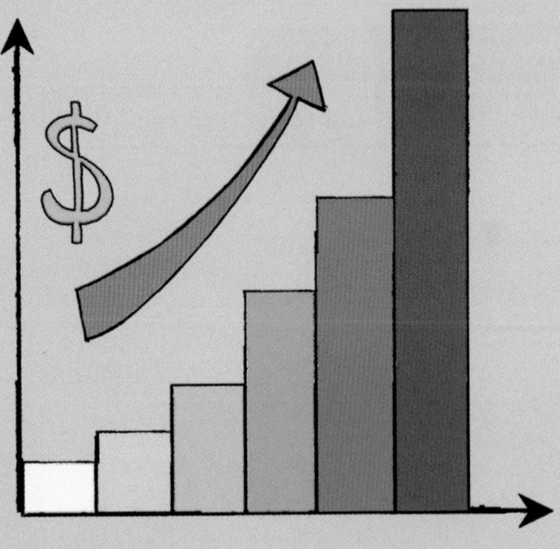

8.1 What comes next?

Global context: Scientific and technical innovation

FORM

Objectives

- Understanding and using recursive and explicit formulae for sequences
- Recognizing linear and quadratic sequences
- Finding a general formula for a linear or quadratic sequence
- Recognizing patterns in real-life contexts
- Solving problems involving sequences in real-life contexts

Inquiry questions

F
- What is a sequence?
- How can you describe the terms of a sequence?
- What is a general formula for a sequence?

C
- What types of sequence are there?
- How can you identify sequences?
- How do you find a general formula for a sequence?

D
- Can you always predict the next terms of a sequence?

ATL **Critical-thinking**

Identify trends and forecast possibilities

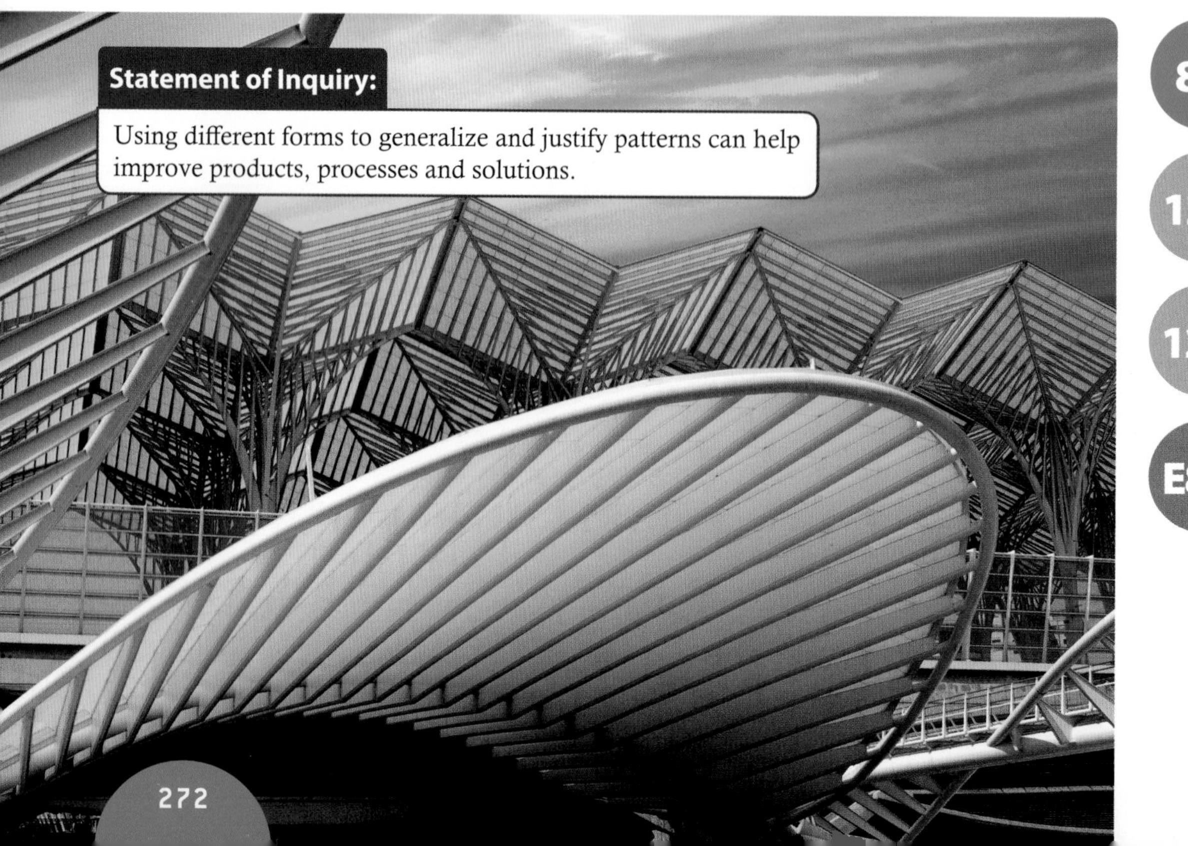

Statement of Inquiry:

Using different forms to generalize and justify patterns can help improve products, processes and solutions.

8.1

12.1

12.2

E8.1

You should already know how to:

• identify the next term in a sequence	**1** Write down the next two terms in each sequence. **a** 2, 4, 6, 8, … **b** 4, 9, 14, 19, … **c** 0, 2, 6, 12, …
• substitute into formulae	**2** Find the value of: **a** $6n + 3$, when $n = 4$ **b** $-2n + 7$, when $n = 10$ **c** $n^2 - 1$, when $n = 2$ **d** $4n^2 - 2n + 11$, when $n = 1$
• solve simple linear and quadratic equations	**3** Solve for n. **a** $3n + 7 = 19$ **b** $5n - 2 = 14$ **c** $5n^2 + 1 = 181$
• write a general formula for simple number patterns	**4** Find a formula for the nth term of each sequence. **a** 4, 8, 12, 16, 20, … **b** 12, 24, 36, 48, … **c** −3, −6, −9, −12, −15, …

F Introduction to sequences

- What is a sequence?
- How can you describe the terms of a sequence?
- What is a general formula for a sequence?

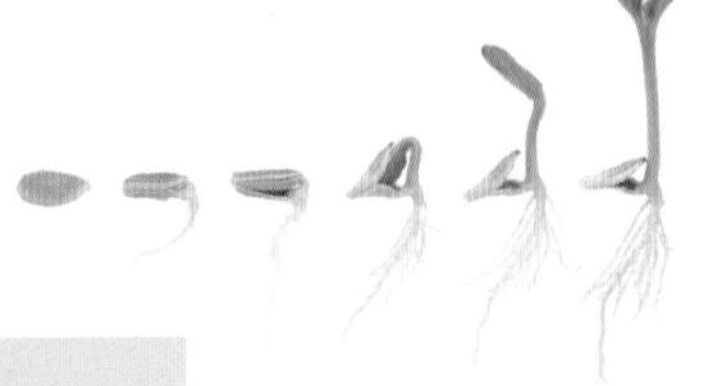

Exploration 1

1 Describe each list of numbers. Compare your descriptions with others.

a 1, 3, 5, 7, 9 **b** 1, 4, 9, 16, 25 **c** 7, 10, 13, 16, 19

d 1, 2, 1, 2, 1 **e** 6, 12, 24, 48, 96 **f** 0, 1, 8, 6, 5

Predict the next three numbers in each list. Compare your results with others.

2 Here are two different ways of continuing the list 3, 1, 4, 1, 5, …

3, 1, 4, 1, 5, 1, 6, 1, 7, 1, 8

3, 1, 4, 1, 5, 9, 2, 6, 5, 3, 5

Explain the pattern in each one.

Tip

Search online for '3 1 4 1 5 9 2 6 5 3 5' if you need help identifying the pattern.

Reflect and discuss 1

Here is a way of continuing the list from **1f** in Exploration 1:

0, 1, 8, 6, 5, 5, 5, 6, 7, 6, 7

Is there a pattern?

Try searching online for the numbers grouped like this: 01865 556 767.

- Does this list of numbers count as a pattern?
- Could you have identified the sequence without an internet search?
- If one digit was missing, would you be able to find out what it was?

A **sequence** is an ordered list of numbers. Each number in the list is called a **term**.

In some sequences the terms follow a pattern, specific rule or order. All the sequences in Exploration 1 were generated by a pattern. The list in Reflect and discuss 1, however, is not a mathematical pattern.

Describing terms in a sequence

You can use u_1 to represent the first term of the sequence, u_2 to represent the second term, and so on. The subscripts 1, 2, 3, match the term number. For the sequence 1, 3, 5, 7, 9, ...

1st term	2nd term	3rd term	4th term	5th term
$u_1 = 1$	$u_2 = 3$	$u_3 = 5$	$u_4 = 7$	$u_5 = 9$

You could also use u_0 for the first term, so that:

1st term	2nd term	3rd term	4th term	5th term
$u_0 = 1$	$u_1 = 3$	$u_2 = 5$	$u_3 = 7$	$u_4 = 9$

Sequences starting from u_0 are often used in computer programs. When u_0 starts the sequence, the subscripts do not match the term number.

The subscript is sometimes called the **index** of the term. As the word 'index' is also used to mean 'exponent', you need to work out its meaning from the context.

Tip

Be careful when you talk about 'the first term' – do you mean u_0 or u_1?

Reflect and discuss 2

- Why is the notation u_1, u_2, u_3, u_4, u_5, ... more useful than labelling the terms *a*, *b*, *c*, *d*, *e*, ...?
- What does u_n mean?

Describing a sequence using an explicit formula (position-to-term rule)

An **explicit formula** uses the term's position number, *n*, to calculate its value.

The formula

$u_n = 2n - 1$ for $n \geq 1$

tells us that the value of the nth term (u_n) is given by $2n - 1$ for any value of n greater than or equal to 1. When you are working with sequences, n is always an integer.

Using the above formula, letting $n = 1$ gives the following:

$$u_1 = 2 \times 1 - 1$$
$$= 1$$

Similarly, letting $n = 2$ gives:

$$u_2 = 2 \times 2 - 1$$
$$= 3$$

and so on.

Example 1

A sequence is given by the explicit formula $u_n = 3n^2 - 2$ for $n \geq 1$.

a Find:

i the first, second and tenth terms of the sequence

ii the term of the sequence with value 673.

b Determine whether or not 524 is a term of the sequence.

a i Since $n \geq 1$, the terms required are u_1, u_2 and u_{10}.

$u_1 = 3 \times 1^2 - 2 = 1$ — Substitute $n = 1$ for the 1st term.

$u_2 = 3 \times 2^2 - 2 = 10$ — Substitute $n = 2$ for the 2nd term.

$u_{10} = 3 \times 10^2 - 2 = 298$ — Substitute $n = 10$ for the 10th term.

ii $3n^2 - 2 = 673$ — Solve the explicit formula for the specific value of 673.

$3n^2 = 675$

$n^2 = 225$

$n = \pm 15$

As $n \geq 1$, $n \neq -15$, so $n = 15$.

When $n = 15$, $u_n = 3 \times 15^2 - 2 = 3 \times 225 - 2 = 673$ ✓ — Check your solution.

b If 524 is a term in the sequence, then

$3n^2 - 2 = 524$

$3n^2 = 526$

$n^2 = 175.333\ldots$

n is not an integer, so 524 is not a term in the sequence.

Practice 1

1 Find the first five terms of each sequence.

a $u_n = 4n - 1$ for $n \geq 1$

b $u_n = 6n^2 + 2$ for $n \geq 1$

c $u_n = 10 - \frac{1}{4}n^2$ for $n \geq 1$

d $u_n = 2^n - 2$ for $n \geq 1$

e $u_n = (n + 4)(n - 3)$ for $n \geq 1$

f $u_n = 4n - 5$ for $n \geq 0$

g $u_n = \frac{25}{24}n^4 - \frac{155}{12}n^3 + \frac{1307}{24}n^2 - \frac{1051}{12}n + 45$ for $n \geq 1$

2 Find the tenth term of the sequence given by $u_n = 5n - 3$ for $n \geq 1$.

3 Determine if 54 is a term in the sequence $u_n = 4n - 1$ for $n \geq 1$.

4 Determine which term of the sequence $u_n = 3n - 5$ has value 61.

5 Find the fifteenth term of the sequence given by $u_n = 10 - \frac{10}{n}$ for $n \geq 1$.

6 Find the value of the fourth term of the sequence given by $u_n = 2n + 12$ for $n \geq 0$.

7 **a** Find the term of the sequence $u_n = 2n^2 + 4$ that has value 246.

b Show that 396 is a term in this sequence.

8 Find the eleventh term of the sequence given by $u_n = 100 - \frac{n^2}{5}$ for $n \geq 0$.

Problem solving

9 Identify which explicit formula, **a** to **f**, corresponds to each sequence, **i** to **vi**.

a $u_n = 4n + 1$ for $n \geq 0$

b $u_n = 2n + 3$ for $n \geq 1$

c $u_n = 5n - 4$ for $n \geq 1$

d $u_n = 3n$ for $n \geq 1$

e $u_n = 3n - 2$ for $n \geq 0$

f $u_n = 2n + 1$ for $n \geq 0$

i 3, 6, 9, 12, 15, …

ii 1, 3, 5, 7, 9, …

iii −2, 1, 4, 7, 10, …

iv 1, 5, 9, 13, 17, …

v 5, 7, 9, 11, 13, …

vi 1, 6, 11, 16, 21, …

> **Tip**
>
> The command term **identify** requires you to state briefly how you have made your decisions.

Describing a sequence using a formula (term-to-term rule)

> A **recursive formula** gives the relationship between consecutive terms. When you know one term, you can work out the next.

With this recursive formula you are given the value of u_1:

$$u_{n+1} = u_n + 2,\ u_1 = 1 \text{ for } n \geq 1$$

$$u_1 = 1$$

Substituting $n = 1$ gives the 2nd term:

$$u_{1+1} = u_1 + 2$$

$$u_2 = 1 + 2 = 3$$

Substituting $n = 2$ gives the 3rd term:

$$u_{2+1} = u_2 + 2$$

$$u_3 = 3 + 2 = 5$$

and so on.

> **Tip**
>
> u_n is the nth term of a sequence, so u_{n+1} is the next term or $(n + 1)$th term.
> u_{n-1} is the term before the nth term.

Example 2

A sequence has term-to-term rule $u_{n+1} = \frac{1}{2}u_n + 8$ for $n \geq 1$, and $u_1 = 12$.
Find the 2nd and 3rd terms of the sequence.

$u_1 = 12$ — The first term is given with the formula.

$u_2 = \frac{1}{2}u_1 + 8$ — The 2nd term is found when $n = 1$.

$= \frac{1}{2} \times 12 + 8$

$= 6 + 8$

$= 14$

$u_3 = \frac{1}{2}u_2 + 8$ — Use the value you just got for u_2 to find the 3rd term.

$= \frac{1}{2} \times 14 + 8$

$= 7 + 8$

$= 15$

You can use a GDC to plot explicit and recursive formulae. Here are the graphs for the formulae in Examples 1 and 2.

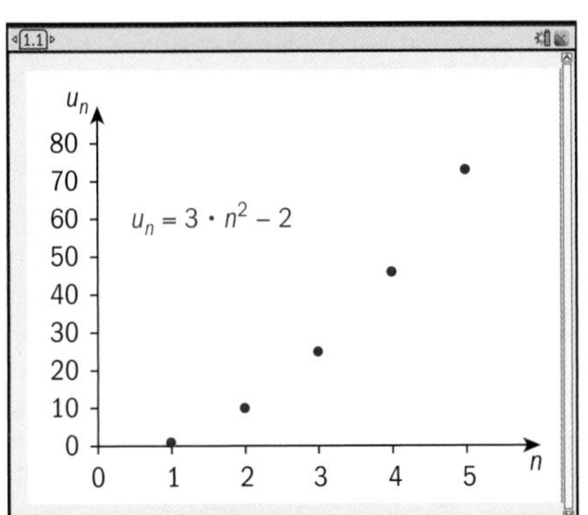

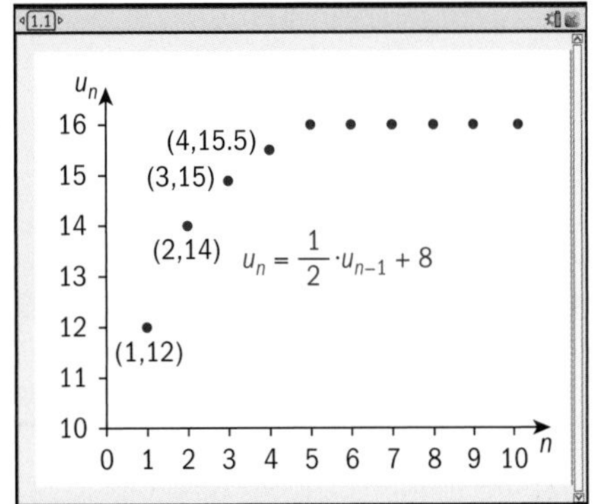

Tip

Some GDCs use u_n and u_{n-1} rather than u_{n+1} and u_n. In this example, the formula is input as $u_n = \frac{1}{2}u_{n-1} + 8$.

Practice 2

1 Find the first five terms of each sequence.

a $u_{n+1} = 4u_n - 1$, $u_1 = 1$ for $n \geq 1$

b $u_{n+1} = 2u_n + 1$, $u_1 = 2$ for $n \geq 1$

c $u_{n+1} = u_n + 7$, $u_1 = -4$ for $n \geq 1$

d $u_{n+1} = \frac{1}{2}u_n + 1$, $u_1 = 3$ for $n \geq 1$

e $u_{n+1} = 2u_n(1 - u_n)$, $u_1 = 0.8$ for $n \geq 1$

f $u_{n+1} = 4u_n(1 - u_n)$, $u_1 = 0.8$ for $n \geq 1$

g $u_n = u_{n-1} + 3$, $u_1 = 4$ for $n \geq 2$

Tip

Look for patterns when you work with sequences. Do the terms of the sequence seem to increase, decrease, or bounce around chaotically?

2 A sequence is given by $u_{n+1} = 3 - u_n$ and $u_0 = -3$.
Find the first four terms of the sequence.

3 A sequence is given by $u_{n+1} = \frac{u_n - 1}{u_n}$, $u_1 = 2$ for $n \geq 1$.

a Find the first six terms of the sequence.

b Describe any patterns you notice.

c Predict the next few terms of the sequence.

Tip

It can be useful to list a few terms to explore an unfamiliar sequence.

Problem solving

4 A sequence is given by $u_{n+1} = 2u_n - 1$, $u_1 = 2$ for $n \geq 1$.

Find the term of the sequence that has value 257.

5 A sequence is given by $u_{n+1} = 3u_n - 2$, $u_1 = 2$ for $n \geq 1$.

Find the value of the largest term in the sequence that is less than 10 000.

The structure of sequences

- What types of sequence are there?
- How can you identify sequences?
- How do you find a general formula for a sequence?

ATL

Exploration 2

1 For each sequence, write down the first five terms.

$a_n = 3n + 7$ $\quad c_n = 5n - 2$

$b_n = 3 - 2n$ $\quad d_n = 4n + 5$

2 **a** Copy and complete the diagram for the sequence given by $a_n = 3n + 7$, filling in missing terms in the sequence and the first difference row.

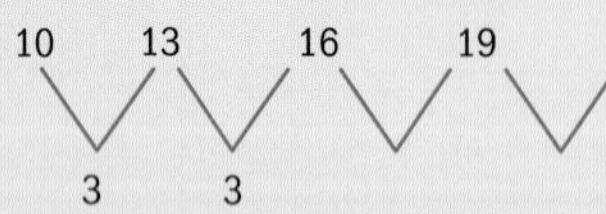

The first difference is the difference between consecutive terms.

b Make a similar diagram for the other sequences in step **1**.

3 Look at your diagrams.

a Describe anything you notice.

b Compare the first difference for each sequence with its explicit formula. Describe anything you notice. Suggest a relationship between the explicit formula and the value of the first difference.

c What do the explicit formulae of these sequences have in common?

Sequences like the ones in Exploration 2, where the terms increase or decrease by a constant number, are called linear sequences.

For a **linear sequence**, the difference between consecutive terms is constant and the explicit formula is of the form $u_n = a + bn$.

Reflect and discuss 3

- If you plotted the terms of a sequence from Exploration 2, what type of graph would you get?
- Plot the graph of $a_n = 3n + 7$ to check your answer. Plot coordinate pairs as (n, a_n).

ATL

Exploration 3

1 a Copy and complete this diagram for the sequence given by $e_n = n^2 + n + 5$, filling in missing terms in the sequence and the difference rows.

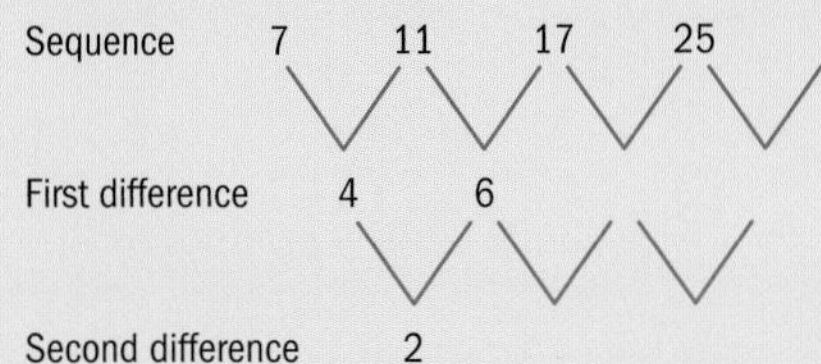

The second difference is the difference between consecutive first differences.

b Make a similar diagram for each of these sequences:

$f_n = n^2 + 5n + 6$

$g_n = 4 + 5n + n^2$

$h_n = n(n + 4) + 1$

2 Look at the first and second differences in your diagrams.

a Describe the differences between these sequences and the sequences in Exploration 2.

b Describe anything you notice about the first and second differences of each sequence.

c Describe anything that the explicit formulae for these sequences have in common.

For a **quadratic sequence**, the second difference is constant and the explicit formula is of the form $u_n = a + bn + cn^2$.

Reflect and discuss 4

- The sequences in Exploration 3 are quadratic sequences. What is the highest power of n in a quadratic sequence?
- Plot the graph of the sequence $e_n = n^2 + n + 5$. What type of graph do quadratic sequences produce?
- Other than linear and quadratic sequences, what other types of sequence can you describe?

Practice 3

1 By constructing a diagram, show that the sequence 6, 13, 23, 36, 52, 71, ... is a quadratic sequence.

2 Complete these diagrams by filling in any missing terms in the sequence and in the difference rows.

a

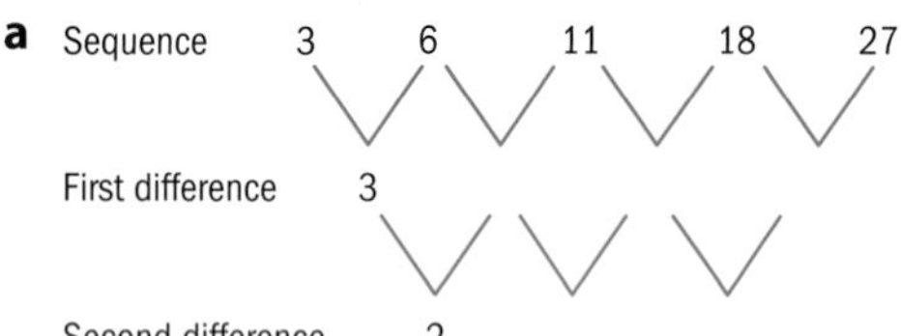

Problem solving

b

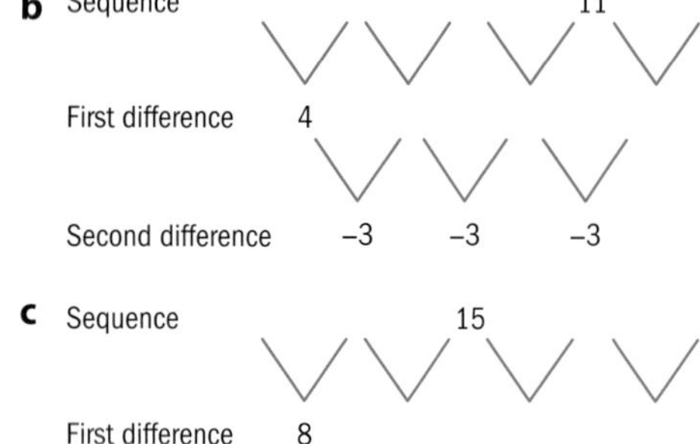

c Sequence 15

First difference 8

Second difference 4 5 6

3 Determine whether each sequence is linear, quadratic or neither.

a 4, 11, 18, 25, 32, …

b 104, 97, 90, 83, 76, …

c 91, 84, 75, 64, 51, …

d 2, 4, 7, 11, 16, …

e 5, 11, 17, 23, 29, …

f 4, 12, 36, 108, 324, …

A **general formula** for a sequence is a rule that can be used to generate each term. Usually the general formula is an explicit formula.

Example 3

Find a general formula for u_n, the nth term of the sequence 11, 17, 23, 29,...

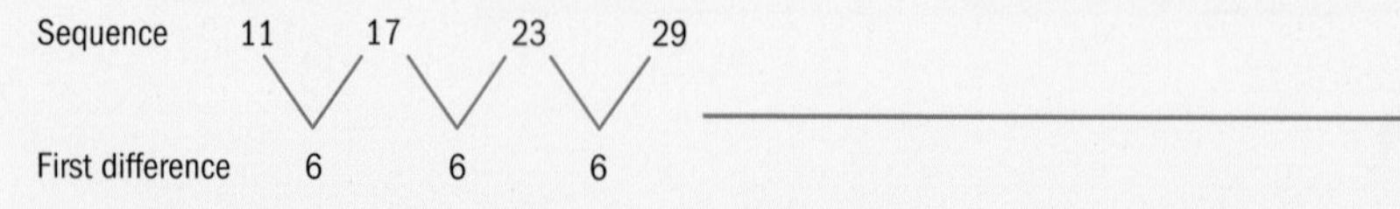

Find the first differences.

The first difference is 6, so compare to the sequence $u_n = 6n$.

The first difference is constant so this is a linear sequence.

n	1	2	3	4
u_n	11	17	23	29
$6n$	6	12	18	24

+5

Look for a pattern connecting u_n and $6n$. Adding 5 to $6n$ gives u_n.

The general formula is $u_n = 6n + 5$.

Notice that the general formula you find describes the terms you were given. It doesn't necessarily mean that the pattern will continue. To know that, you need more information about the sequence, for example, if it is linear.

Practice 4

ATL

1 Find a general formula to describe the terms in each sequence.

a $a_1 = 8, a_2 = 16, a_3 = 24, a_4 = 32$

b $b_1 = 14, b_2 = 17, b_3 = 20, b_4 = 23$

c $c_1 = -14, c_2 = -28, c_3 = -42, c_4 = -56$

d $d_1 = 49, d_2 = 38, d_3 = 27, d_4 = 16$

e $e_1 = 11, e_2 = 29, e_3 = 47, e_4 = 65$

f $f_1 = 17, f_2 = 17.5, f_3 = 18, f_4 = 18.5$

g $g_1 = 8\frac{1}{3}, g_2 = 7\frac{2}{3}, g_3 = 7, g_4 = 6\frac{1}{3}$

h $h_1 = -1.4, h_2 = 3.2, h_3 = 7.8, h_4 = 12.4$

i $i_1 = 15, i_2 = -3, i_3 = -21, i_4 = -39$

Tip

You can check your general formula is correct by working out the first few terms.

2 Each set of terms below is part of a linear sequence.

Find a general formula describing the terms of each sequence.

a $a_5 = 9, a_6 = 11, a_7 = 13, a_8 = 15$

b $b_{11} = 17, b_{12} = 14, b_{13} = 11, b_{14} = 8$

c $c_9 = 13.5, c_{10} = 14, c_{11} = 14.5, c_{12} = 15$

d $d_5 = 7, d_7 = 15, d_9 = 23, d_{11} = 31$

Problem solving

3 A string of decorative lights are wired as shown.

a Explain why it takes 500 cm of cable to connect the first bulb, and then another 20 cm to connect the next bulb.

b Let d_n be the total length of cable needed to connect the nth bulb. Find a general formula for d_n.

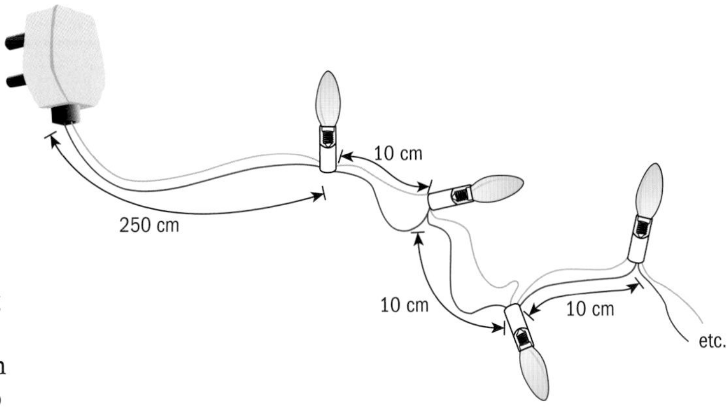

4 A railway train is stored overnight in a depot which is 6 km from its starting station. It then spends all day moving from its starting station, back and forth along a section of track 14 km long, so in each complete journey it travels 28 km.

a Explain why after its first complete journey of the day it has travelled 34 km.

b Find a general formula for the distance it has travelled after n journeys up and down the track.

c The train can travel 400 km before refuelling. Find the number of complete journeys it can make before refuelling.

5 The first five terms of a linear sequence are –3, 4, 11, 18, 25.

a Find the 100th term of the sequence.

b Show that 102 is a member of the sequence and find its term number.

6 A sequence has formula $u_{n+1} = u_n + 5$ and $u_1 = -1$.

Find a formula for the nth term of the sequence.

Problem solving

7 A linear sequence has terms $u_{10} = 25$ and $u_{15} = 70$.

a Find the difference between consecutive terms.

b Write down the value of **i** u_9 and **ii** u_1.

8 A linear sequence begins 191, 173, 155, …

Find the first term that is less than 65.

9 A linear sequence has first term 108 and second term 103.

Find its first negative term.

The first step in finding the general formula for a sequence is to identify the type of sequence. Having done that, you can conjecture a formula and modify it until it describes the sequence exactly.

For the sequence –2, –2, 0, 4, 10, 18, 28, 40 the first and second differences are:

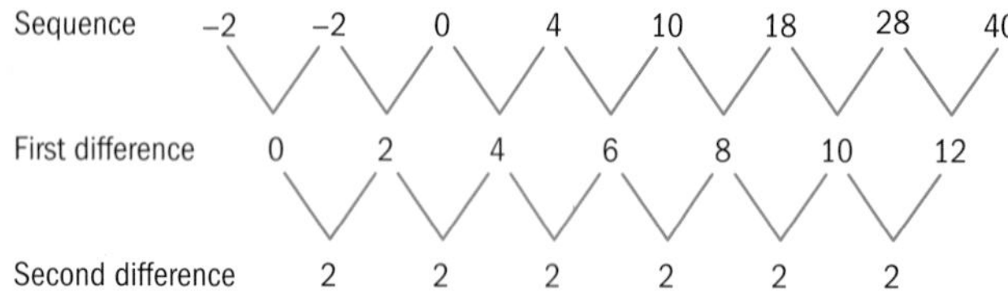

The second difference is constant so this is a quadratic sequence and is related to the sequence for n^2. Add a row for n^2 to the table:

n	1	2	3	4	5	6	7	8
u_n	–2	–2	0	4	10	18	28	40
n^2	1	4	9	16	25	36	49	64

Add another row for the differences between n^2 and u_n.

n	1	2	3	4	5	6	7	8
u_n	–2	–2	0	4	10	18	28	40
n^2	1	4	9	16	25	36	49	64
$n^2 - u_n$	3	6	9	12	15	18	21	24

$n^2 - u_n$ gives multiples of 3. This is a linear sequence with general formula $3n$.

So $n^2 - u_n = 3n$ and rearranging gives $u_n = n^2 - 3n$.

Check: the first five terms of $u_n = n^2 - 3n$ are –2, –2, 0, 4, 10 ✓

Practice 5

1 Find a general formula for each sequence:

a 4, 7, 12, 19, 28, 39, …

b −7, −4, 1, 8, 17, 28, …

c 3.5, 6.5, 11.5, 18.5, 27.5, 38.5, …

d 2, 6, 12, 20, 30, 42, …

e −1, 0, 3, 8, 15, 24, …

f 6, 14, 24, 36, 50, 66, …

g −0.5, 1, 4.5, 10, 17.5, 27, …

h 1, 2, 5, 10, 17, 26, …

i 3, 9, 17, 27, 39, 53, …

j 13, 22, 33, 46, 61, 78, …

k 0, 1.5, 5, 10.5, 18, 27.5, …

l −4, −3, 0, 5, 12, 21, …

2 A quadratic sequence begins −18, −20, −20, −18, −14, −8, 0, …

a Find a general formula for the sequence.

b Use your formula to predict the 15th term of the sequence.

3 A quadratic sequence begins −3, −4, −3, 0, 5, 12, …

a Find a formula for the *n*th term.

b Find the value of the 20th term of the sequence.

c Find the value of u_{50}.

4 A quadratic sequence begins −25, −38, −49, −58, −65, −70, ….

a Show that the 16th term of the sequence is negative.

b Show that the 17th term is positive.

c Explain why all the subsequent terms will be positive.

Problem solving

5 A quadratic sequence begins 98, 100, 100, 98, 94, …
Find its first negative term.

ATL

Exploration 4

1 Construct difference diagrams for these quadratic sequences.

a $u_n = 3n^2, n \geq 1$

b $u_n = 2n^2 + 4, n \geq 1$

c $u_n = 5n^2 - n, n \geq 1$

d $u_n = 2n^2 + 3n, n \geq 1$

e $u_n = n^2 + 7, n \geq 1$

f $u_n = 2 - 4n^2, n \geq 1$

2 Look at the difference diagrams.

a Describe the relationship between the explicit formula and the second difference.

b Suggest a rule linking the coefficient of n^2 to the value of the second difference.

The coefficient of n^2 is the number in front of it. The coefficient of $3n^2$ is 3.

▶ Continued on next page

3 Use your rule to predict the second difference of the sequence with formula:

a $u_n = 7n^2 - 2n + 1$

b $u_n = 3 - 8n^2$

c $u_n = \frac{1}{2}n^2 - 4n + 2$

Check your predictions by constructing a difference diagram for the first six terms of each sequence.

Example 4

Find a general formula for the nth term of the quadratic sequence which begins $u_1 = 4$, $u_2 = 12$, $u_3 = 26$, $u_4 = 46$, $u_5 = 72$, $u_6 = 104$.

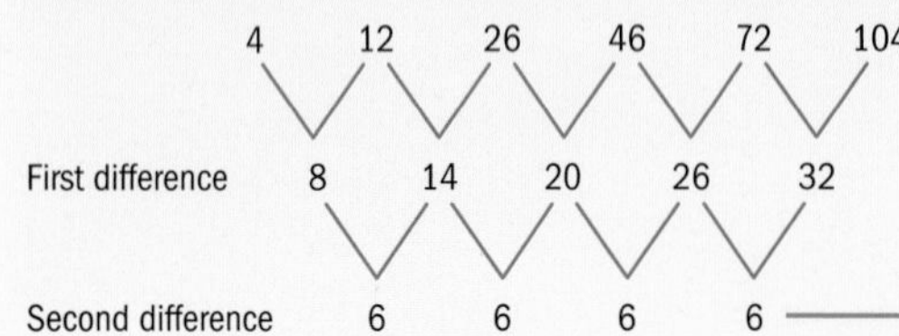

The constant second difference is 6, so the coefficient of n^2 is $6 \div 2 = 3$.

n	1	2	3	4	5	6
u_n	4	12	26	46	72	104
$3n^2$	3	12	27	48	75	108
$u_n - 3n^2$	1	0	−1	−2	−3	−4

Write out the values of $3n^2$ and compare them to the values of u_n.

The sequence 1, 0, −1, −2, −3, −4, ... has nth term $2 - n$.

Find the general formula for the linear sequence 1, 0, −1, −2, −3, −4, ...

Therefore, $u_n - 3n^2 = 2 - n$

Therefore $u_n = 3n^2 - n + 2$

Rearrange.

In a quadratic sequence, the coefficient of n^2 in the general formula is always half the value of the second difference.

Practice 6

1 Find a general formula for the nth term of each sequence.

a 2, 8, 18, 32, 50, …

b 0.5, 2, 4.5, 8, 12.5, …

c 3, 12, 27, 48, 75, …

d 3, 9, 19, 33, 51, …

e −1, 8, 23, 44, 71, …

f −4.5, 0, 7.5, 18, 31.5, …

g 7, 24, 51, 88, 135, …

h 2, 7, 15, 26, 40, …

i 2, 6.5, 13.5, 23, 35, …

j 6, 17, 34, 57, 86, …

k −1, 1, 7, 17, 31, …

l 3, 8, 15.5, 25.5, 38, …

2 A quadratic sequence begins 10, 16, 26, …

a Find the next three terms.

b Find a formula for the nth term.

Problem solving

Draw a difference diagram.

3 The nth triangular number, T_n, is the number of dots in a triangular grid of dots n dots wide.

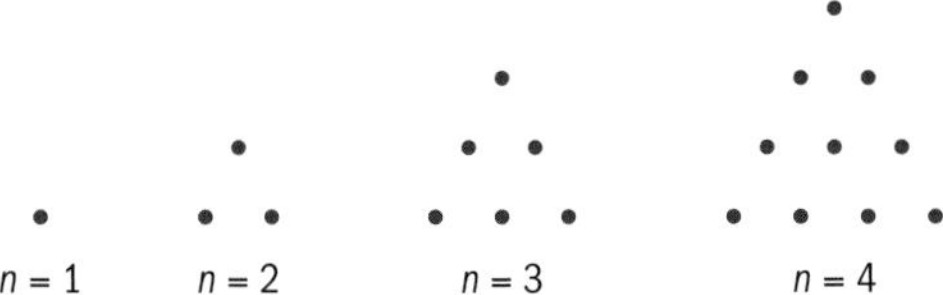

a Write down the value of T_4.

b Find the value of T_5.

c Find a general formula for T_n.

d Use your formula to find the value of T_7 and T_8. Draw diagrams to check your answers.

e Find the value of T_{15}.

f Verify that T_{49} is a square number.

4 A quadratic sequence has first term 4 and second term 10.
The coefficient of n^2 in its general formula is 2.
Find the next three terms.

5 A quadratic sequence begins 3, x, 29, 51, ...
Find the value of x.

Objective: D. Applying mathematics in real-life contexts
ii. select appropriate mathematical strategies when solving authentic real-life situations.

'Select appropriate mathematical strategies' means, for example, that it is up to you how you tackle Q6.

6 In the group stage of a volleyball competition, all the teams in the group have to play each other. When there are four teams in a group, six games are needed for each team to play each of the other teams exactly once.

Find the number of games needed for each team to play each of the other teams exactly once in a group of n teams.

Tip

Think about whether you need to define any variables, draw diagrams, and make any lists or tables.

D Continuing sequences

- Can you always predict the next terms of a sequence?

ATL

Exploration 5

In these circles, the dots on the circumference are connected by line segments, and this divides the circle into regions.

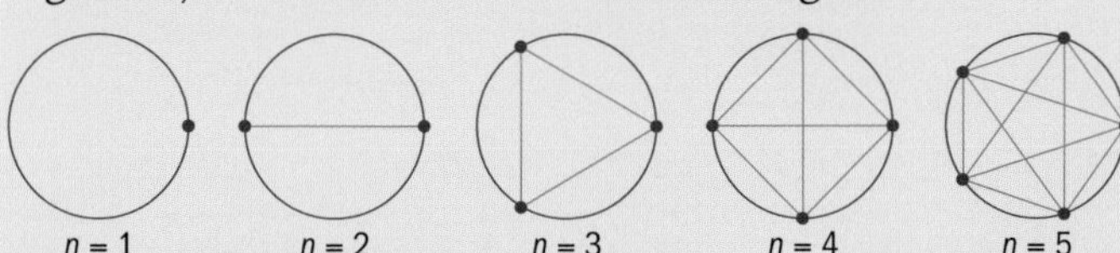

The first diagram has 1 dot, 1 region.
The second diagram has 2 dots, 2 regions.

Continued on next page

The third diagram has 3 dots, 4 regions.

1 Write down the number of regions for $n = 4$ and $n = 5$.

2 Predict the number of regions you would expect when $n = 6$. Give reasons for your answer.

3 Here are two sensible ways you could draw the circle for $n = 6$.

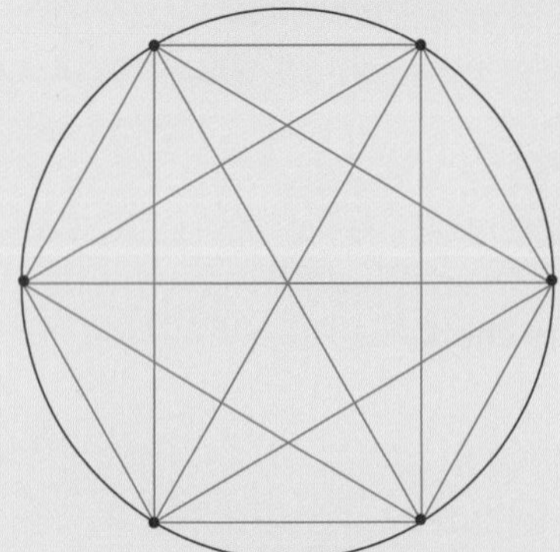

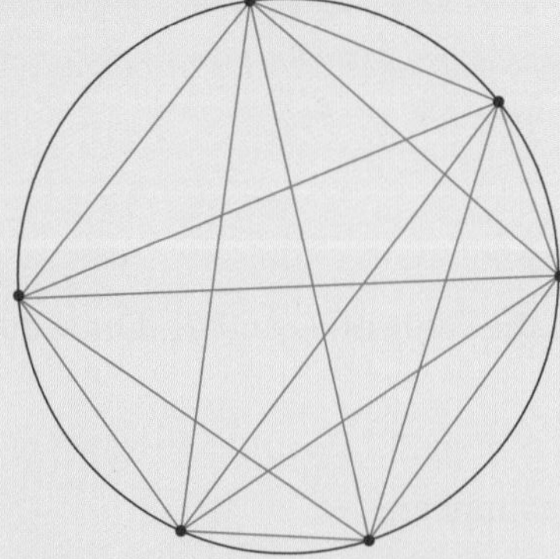

a What is the difference between the number of regions in each diagram?

b Explain which diagram you think is the correct way to continue the sequence. Explain why.

Reflect and discuss 5

- Do either of the diagrams fit your prediction? If not, can you form a prediction that does hold true? Does it help to look at the next term in the sequence?
- What are the difficulties in trying to find a general formula for the sequence in Exploration 5?

Summary

- A **sequence** is an ordered list of numbers. Each number in the list is called a **term**.
- An **explicit formula** uses the term's position number, n, to calculate its value.
- A **recursive formula** gives the relationship between consecutive terms. When you know one term, you can work out the next.
- A **general formula** for a sequence is a rule that can be used to generate each term. Usually the general formula is an explicit formula.
- Sequences where the terms increase or decrease by a constant number, are called **linear sequences**.
- For a **linear sequence**, the difference between consecutive terms is constant and the explicit formula is of the form $u_n = a + bn$.
- For a **quadratic sequence**, the second difference is constant and the explicit formula is of the form $u_n = a + bn + cn^2$.
- In a quadratic sequence, the coefficient of n^2 in the general formula is always half the value of the second difference.
- The triangular numbers describe the number of dots needed to make simple triangular grids. They form a quadratic sequence with general formula $T_n = \frac{n(n+1)}{2}$.

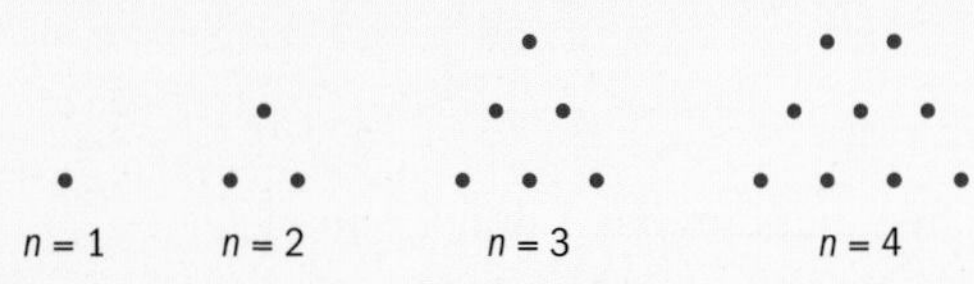

Mixed practice

1 **Find** a formula for the nth term of each sequence.

a 6, 10, 14, 18, 22, 26, …

b 42, 35, 28, 21, …

c 2, 5, 10, 17, 26, 37, …

d 3, 8, 15, 24, 35, …

e 2.5, 7, 11.5, 16, 20.5, …

f −0.5, 4, 13.5, 28, 47.5, …

g 9, 13, 13, 9, 1, …

h 6, 7, 7, 6, 4, …

2 A linear sequence begins 14, 17, 20, …

Find its hundredth term.

3 A sequence has formula $u_n = u_{n-1} - 5$ and $u_1 = 12$.

a **Write down** the first five terms.

b **Find** a formula for the nth term of the sequence.

Problem solving

4 A sequence has formula $u_{n+1} = u_n + 7$ and $u_0 = 15$.

a **Write down** the first five terms of the sequence.

b **Show that** 85 is a member of the sequence.

5 A linear sequence begins 18, 33, 48, ...

Find the first term greater than 1000.

6 A quadratic sequence begins 4, 10, 18, 28, …

a **Find** a formula for the nth term.

b **Write down** the value of the seventh term.

c **Show that** 460 is a member of the sequence and find its term number.

7 A quadratic sequence begins 2, 4, $\frac{20}{3}$, 10, 14, …

a **Find** a formula for u_n, the nth term of the sequence.

b **Find** the value of u_{20}.

Problem solving

8 A linear sequence has terms $u_3 = 6$ and $u_6 = 27$.

a **Find** the difference between successive terms.

b **Write down** the value of u_7.

9 A quadratic sequence begins 3, x, 15, 25.5, …

Find the value of x.

10 Consider the honeycomb patterns below, which have side length 1, 2 and 3 hexagons respectively.

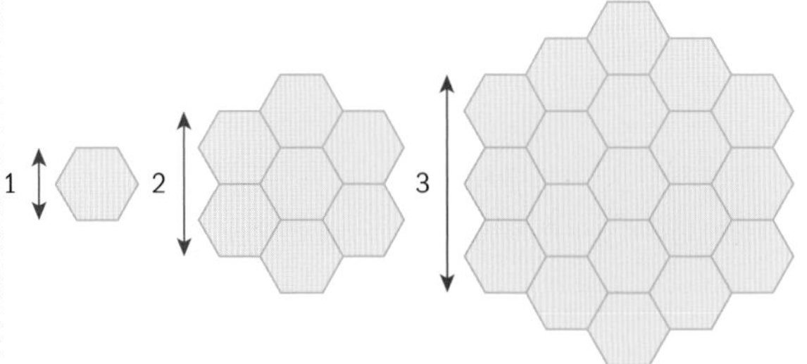

Find the side length of the first pattern that contains over 1000 hexagons.

11 In a football stadium, the first row of seats has 610 seats. The second row has 620 seats, the third has 630 seats, and so on. The stadium sells seats from the middle of a row outwards. Each row must be completely sold before they sell tickets for the next row.

a **Find** the number of seats that can be sold if only the front two rows are used.

b **Find** the number of seats that can be sold if only the front three rows are used.

c **Create** a formula for the total number of seats that can be sold if the front n rows are used.

d Hence **find** the number of seats that can be sold if the front 30 rows are used.

e **Find** the number of rows needed for the first 10 000 tickets.

12 A 12-storey tall building has an elevator that stops at every floor.

The elevator takes 8 seconds to travel one floor, 14 seconds to travel two floors, 20 seconds to travel three floors and 26 seconds to travel four floors.

Let T_n be the time taken to travel n floors.

a **Show that** T_n forms a linear sequence.

b **Find** a formula for T_n.

c **Find** the greatest number of floors that the elevator could travel in under a minute.

Review in context

Scientific and technical innovation

1 Engineers are laying signalling cable alongside a railway track. They place fixed lengths of cable separated by junction boxes.

The total length of cable for 1, 2, 3 or 4 junction boxes is 12 m, 18 m, 24 m or 30 m.

Let u_n be the cable length for n junction boxes.

a **Find** a formula for u_n.

b **Use** your formula to **predict** the length of cable needed for 30 junction boxes.

2 The number of passengers, P_n, who will comfortably fit into an n-carriage train is given by the following table for small values of n:

n	1	2	3	4
P_n	40	110	180	250

a **Show** that P_n forms a linear sequence.

b **Find** a formula for P_n.

c **Predict** the number of passengers that an eight-carriage train could comfortably hold.

d **Suggest** a reason why a two-carriage train might hold more than twice the number of passengers of a one-carriage train.

3 In a subway network of stations, every journey consists of a start point and a different end point.

If you consider just two of the stations, call them A and B, then the only two possible journeys are AB and BA.

If you add in a third station, C, there are six possible journeys.

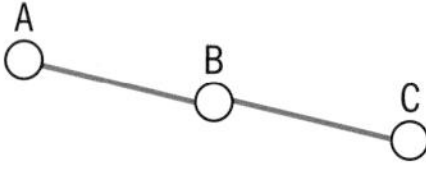

a **List** the six journeys that are possible with three stations, A, B and C.

b **List** the journeys that would be possible with four stations, A, B, C and D.

Hence **write down** the total number of possible journeys with four stations.

c **List** the journeys that would be possible with five stations, A, B, C, D and E.

Hence **write down** the total number of possible journeys with five stations.

d **Create** a formula linking the number of stations on the network to the total number of possible journeys.

The *Metropolitano de Lisboa*, Lisbon's underground railway, has 55 stations.

e **Predict** the total number of possible journeys.

f **Predict** the number of additional possible journeys that there would be if two extra stations were built.

Reflect and discuss

How have you explored the statement of inquiry? Give specific examples.

Statement of Inquiry:

Using different forms to generalize and justify patterns can help improve products, processes and solutions.

8.2 Back to the beginning

Objectives

- Factorizing quadratic expressions, where the coefficient of x^2 is 1, including the difference of two squares.
- Factorizing quadratic expressions where the coefficient of x^2 is not 1.

Inquiry questions

F
- What does 'expanding brackets' mean?
- What does 'factorize a quadratic expression' mean?

C
- How do the patterns in expanding brackets help you factorize quadratic expressions?
- How can patterns help you write quadratic expressions in a form that is easier to factorize?

D
- Can everything be written in a different form?

FORM

ATL **Reflection**

Draw reasonable conclusions and generalizations

Conceptual understanding:

Patterns can be represented in equivalent forms.

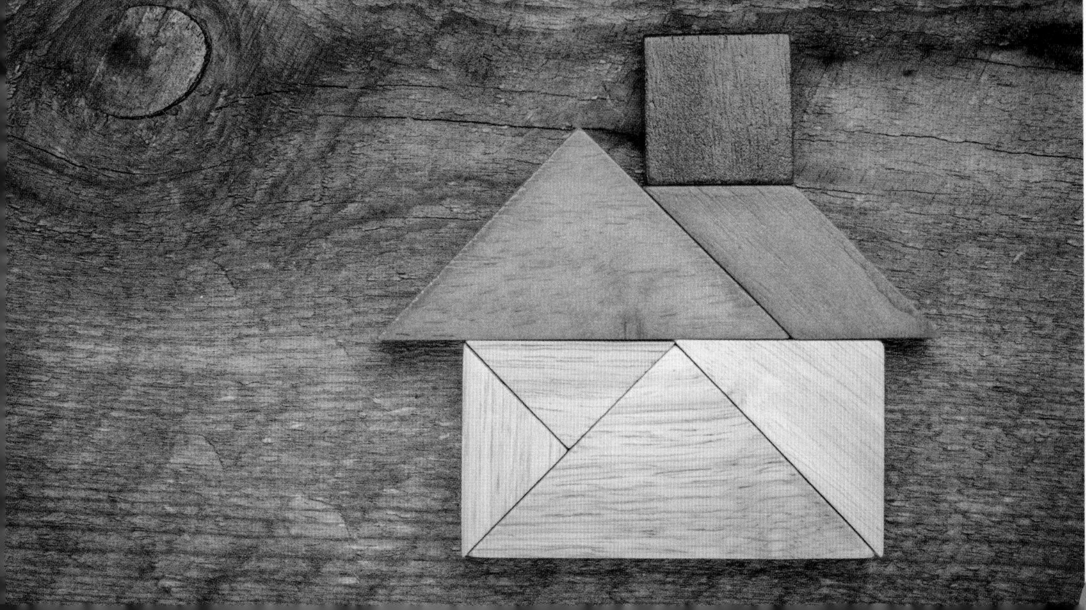

8.2

9.2

11.2

You should already know how to:

• expand brackets	**1** Expand: **a** $(x+3)(x+4)$ **b** $(x-5)(x+1)$ **c** $(x-3)(x-2)$ **d** $(x+2)(x-2)$
• factorize expressions by taking out a common factor	**2** Factorize: **a** $3x+12$ **b** x^2+5x **c** $6x^2+3x+12$ **d** $4x(3x-5)+7(3x-5)$

F Expanding and factorizing

- What does 'expanding brackets' mean?
- What does 'factorize a quadratic expression' mean?

To 'expand' an algebraic expression means to multiply out each of the terms and then rewrite the expression without brackets.

For example:

$$(x-7)(x+2) \equiv x^2 - 7x + 2x - 14 \equiv x^2 - 5x - 14$$

The symbol $\equiv$ means 'is identically equal to'. It shows that two expressions take the same value for any possible value of the unknown variable (in this case, x).

For example, $3(x-2) \equiv 3x-6$, because the statement is always true for any value of x.

$3x+5=17$ is an equation, not an identity, because it is true only when $x=4$.

Exploration 1

Look at the following expressions:

Quadratic expressions	**NOT quadratic expressions**
$x^2-5x-14$	$x^3-5x-14$
$3x^2+2x$	$8t+12$
$-7w^2+21$	$x^2+3y-25$
$\frac{1}{2}m^2+m+3$	z^4+3z^2-2z+1
$2y-5y^2+7$	17^2

1 Based on your observations, suggest what is meant by 'quadratic expression'.

2 Explain why each of the expressions in the right-hand column do not satisfy your description.

3 Explain whether or not you would consider $(x+3)(x-5)$ to be a quadratic expression.

$x^2-5x-14$ is a **quadratic expression** because it contains a single variable (x) whose highest exponent is 2.

- $x^2-5x-14$, x^2-4 and $3x^2$ are all quadratic expressions.
- $x^3-5x-14$ is not quadratic because its highest exponent is greater than 2.
- $5x-14$ is not quadratic because its highest power is less than 2.

An expression with highest exponent 3, like $x^3-5x-14$, is a cubic expression.

Quadratic expressions can be written in a variety of forms. To 'factorize' means to write as a product of factors. 'Factorize a quadratic' means 'write it as a product of two expressions in brackets'.

ATL

Exploration 2

1 Copy the table. For each row, expand the brackets to help you complete the table.

Factorized expression $(x+p)(x+q)$	p	q	Expanded expression ax^2+bx+c	a	b	c
$(x+3)(x+5)$	3	5	$x^2+8x+15$	1	8	15
$(x+2)(x+9)$						
$(x-3)(x-6)$						
$(x+3)(x-4)$						
	−3	4				
$(x+2)(x-5)$						
$(x-8)(x+4)$						
$(x\quad)(x+6)$	−1					

2 Look carefully at your table. Describe any patterns you notice in the values and signs of a, b, c, p and q.

3 Copy the table below. Use the patterns you observed in step **2** to predict the values of a, b and c.

Factorized expression $(x+p)(x+q)$	p	q	Expanded expression ax^2+bx+c	a	b	c
$(x+1)(x+7)$						
$(x-3)(x+8)$						
$(x-1)(x-2)$						
$(x+5)(x-5)$						

In step **3**, you shouldn't need to expand the brackets. If you can't see a pattern yet, go back to step **2** and discuss your results.

4 Verify your predictions by expanding each factorized expression.

5 Copy and complete this table.

Factorized expression $(x+p)(x+q)$	p	q	Expanded expression ax^2+bx+c	a	b	c
$(x+2)(x+2)$						
$(x+4)(x+4)$						
$(x-1)(x-1)$						
$(x-6)(x-6)$						

6 Describe any patterns you notice in the values and signs of a, b, c, p and q.

Practice 1

1 Copy and complete this table based on the patterns you observed assuming that $(x + p)(x + q) = ax^2 + bx + c$.

p	*q*	*a*	*b*	*c*
2	−3	1		
4		1		−8
8		1	0	
	−2	1	−5	
	−7	1		21
9		1	0	
		1	−2	−24

2 Find values such that $a = 1$, $p = q$ and $b = c$.

3 N'nyree has concluded that if $c = 0$, then either p or q is zero. Explain whether you agree or disagree with this statement.

C Factorizing expressions

- How do the patterns in expanding brackets help you factorize quadratic expressions?
- How can patterns help you write quadratic expressions in a form that is easier to factorize?

In Exploration 2 you found that in all the expansions of $(x + p)(x + q) = ax^2 + bx + c$, $a = 1$, $c = pq$ and $b = p + q$.

You can use these facts to factorize expressions such as $x^2 + x - 12$.

Suppose $x^2 + x - 12 = (x + p)(x + q)$, where p and q are whole numbers.

Using the fact that $c = pq = -12$ suggests these possible pairs of values for p and q:

−12 and 1 12 and −1 −6 and 2 6 and −2 −4 and 3 4 and −3

Trying each pair in turn gives:

$(x - 12)(x + 1) \equiv x^2 - 11x - 12$

$(x + 12)(x - 1) \equiv x^2 + 11x - 12$

$(x - 6)(x + 2) \equiv x^2 - 4x - 12$

$(x + 6)(x - 2) \equiv x^2 + 4x - 12$

$(x - 4)(x + 3) \equiv x^2 - x - 12$

$(x + 4)(x - 3) \equiv x^2 + x - 12$ ✓

$(x + 4)(x - 3)$ gives the correct quadratic expression.

Also using the fact that $b = p + q$ makes this process more efficient. Having found pairs of numbers which have product c, you just need to find a pair which has sum b.

Tip

Always check your factorization by expanding the brackets.

In $x^2 + x - 12$, $b = p + q = 1$

$-12 + 1 = -11$

$12 + (-1) = 11$

$-6 + 2 = -4$

$6 + (-2) = 4$

$-4 + 3 = -1$

$4 + (-3) = 1$ ✓

So $x^2 + x - 12 \equiv (x + 4)(x - 3)$.

> To factorize a quadratic $x^2 + bx + c \equiv (x + p)(x + q)$, find two numbers p and q which have product c and sum b.

In Exploration 2 steps **5** and **6** you expanded brackets with $p = q$:

$(x + p)(x + p) = x^2 + 2px + p^2$

You can also write this as:

$(x + p)^2 = x^2 + 2px + p^2$

Recognizing this pattern can help you to factorize special quadratics called perfect squares.

Example 1

Factorize $x^2 + 10x + 25$.

$x^2 + 10x + 25$ — $25 = 5^2$ and $10 = 2 \times 5$

$\equiv (x + 5)(x + 5)$

$\equiv (x + 5)^2$ — This is a *perfect square.*

If all three terms have a common factor, taking this common factor out first makes the factorization simpler.

Example 2

Factorize $2a^2 + 6a + 4$.

$2a^2 + 6a + 4$

$\equiv 2(a^2 + 3a + 2)$ — Take out the common factor 2.

$\equiv 2(a + 1)(a + 2)$

Practice 2

1 Factorize each quadratic.

a $x^2 + 7x + 12$ **b** $x^2 + 8x + 15$ **c** $x^2 + 3x - 18$

d $x^2 + 7x - 18$ **e** $x^2 - 7x - 18$ **f** $x^2 + 5x - 14$

g $x^2 - 13x + 36$ **h** $x^2 - 11x + 24$ **i** $x^2 - 3x$

2 Factorize each quadratic.

a $x^2 + 14x + 49$ **b** $x^2 + 22x + 121$ **c** $x^2 - 12x + 36$

3 Factorize each quadratic.

a $3a^2 + 6a + 3$ **b** $4b^2 + 2b$ **c** $5c^2 - 10c - 75$

d $6d^2 - 3d$ **e** $4e^2 + 20e - 144$ **f** $3f^2 - 24f + 45$

4 Find an expression for the unknown side in these rectangles:

a

b

Problem solving

5 Copy and complete, using integer values:

a $x^2 - \square x + 12 \equiv (x - 3)(x - \square)$

b $x^2 \square\square x + 16 \equiv (x - 8)(x \square\square)$

c $x^2 - 5x \square\square \equiv (x + \square)(x - 7)$

d $x^2 \square 9x + 20 \equiv (x + 5)(x \square\square)$

e $x^2 \square 10x \square 16 \equiv (x + \square)(x + \square)$

f $x^2 - 5x \square 24 \equiv (x \square\square)(x \square\square)$

6 A rectangle has area $x^2 - 10x + 21$.

Its perimeter is $4x - 20$. Find, in terms of x, the lengths of each of its sides.

Exploration 3

1 Copy and complete this table.

Factorized expression $(x+p)(x+q)$	p	q	Expanded expression $ax^2 + bx + c$	a	b	c
$(x+3)(x-3)$						
$(x-9)(x+9)$						
$(x+6)(x-6)$						
$(x+4)(x-4)$						

2 Describe any patterns you notice in the values and signs of a, b, c, p and q.

In Exploration 3 you found that expanding brackets in the form $(x + p)(x - p)$ gave a special pattern. When they are expanded, the coefficient of the x term is 0 so there is just an x^2 term and a constant, which is negative.

In general, expanding $(a + b)(a - b) \equiv a^2 - ab + ab - b^2 \equiv a^2 - b^2$.

Any expression of the form $a^2 - b^2$ is the **difference of two squares**, because it is one squared quantity subtracted from another.

$a^2 - b^2 \equiv (a + b)(a - b)$

Reflect and discuss 1

Explain how this diagram illustrates the factorization of $a^2 - b^2$.

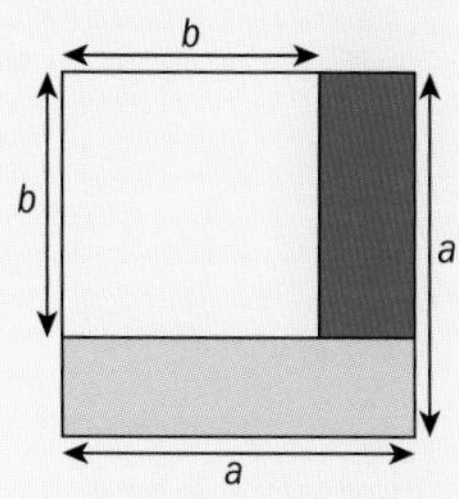

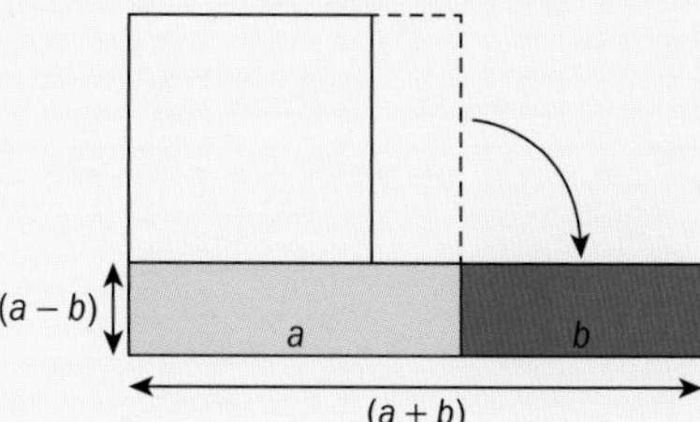

If you recognize an expression as the difference of two squares, you can factorize it easily.

Example 3

Factorize $x^2 - 64$.

$x^2 - 64$ — Difference of two squares, $x^2 - 8^2$

$\equiv (x-8)(x+8)$

'Fully factorize' means write as a product of expressions which cannot be factorized any further.

Example 4

Fully factorize $16x^4 - 81$.

$16x^4 - 81$ — $(4x^2)^2 - 9^2$

$\equiv (4x^2 - 9)(4x^2 + 9)$ — This is not fully factorized, as $4x^2 - 9$ is the difference of two squares: $(2x)^2 - 3^2$.

$\equiv ((2x)^2 - 3^2)(4x^2 + 9)$

$\equiv (2x - 3)(2x + 3)(4x^2 + 9)$

Practice 3

1 Fully factorize each expression.

a $x^2 - 25$ **b** $x^2 - 121$ **c** $4x^2 - 9$

d $x^2 - y^2$ **e** $9x^2 - 1$ **f** $x^4 - 1$

g $16x^2 - 169$ **h** $81x^2 - 9$ **i** $25u^2 - 16v^2$

j $16 - 49x^2$ **k** $16 - x^4$ **l** $1 - 81y^4$

Problem solving

2 Copy and complete each identity, using integer values.

a $\square x^2 - 100 \equiv (3x - \square)(3x + \square)$

b $25y^2 - \square \equiv (\square y - 4)(\square y + 4)$

c $16a^2 - \square b^2 \equiv (\square a - \square b)(\square a + 7b)$

d $\square u^2 - \square \equiv (3u + 2)(3u + \square)$

e $\square t^2 - \square \equiv (3t + 5)(\square t - 20)$

f $\square x^2 - \square \equiv (3x + 4)(6x - \square)$

Reflect and discuss 2

Aishah tried to factorize $3x^2 + 13x + 12$ using the methods you have learned. Here is a sample of her work:

In the quadratic $3x^2 + 13x + 12$, $a = 3$, $b = 13$ and $c = 12$.
In the expansion $(x + p)(x + q)$, the product $pq = c$, $\Rightarrow c = 12$.
The sum $p + q = b$, $\Rightarrow b = 13$, and so $p = 1$ and $q = 12$.

Therefore, $3x^2 + 13x + 12 \equiv (x + 1)(x + 12)$

- Is Aishah's factorization correct? Is it equivalent to the original expression?
- Why doesn't this method work for this expression?
- How is this expression different from the other expressions you have factorized?

Quadratics where the coefficient of x^2 is 1 are called *monic*. If the coefficient is not 1, the quadratic is *non-monic*.

To have $3x^2$ as the first term of a quadratic expression, the factors need to start with terms that multiply to make $3x^2$.

The factors of 3 are 1 and 3, so the factors need to start $(3x \quad)$ and $(x \quad)$. The second term in each bracket can be found by listing possibilities systematically. Look at the examples below and then attempt Reflect and Discuss 3.

Example 5

Factorize $3x^2 + 13x + 12$.

$3x^2 + 13x + 12 \equiv (3x \quad)(x \quad)$ — The numbers in the brackets have product 12.

Pairs of values with product 12 are:

1 and 12	2 and 6	3 and 4
12 and 1	6 and 2	4 and 3

$(3x + 1)(x + 12) \equiv 3x^2 + 37x + 12$ ✗

$(3x + 12)(x + 1) \equiv 3x^2 + 15x + 12$ ✗

$(3x + 2)(x + 6) \equiv 3x^2 + 20x + 12$ ✗

$(3x + 6)(x + 2) \equiv 3x^2 + 12x + 12$ ✗

$(3x + 3)(x + 4) \equiv 3x^2 + 15x + 12$ ✗

$(3x + 4)(x + 3) \equiv 3x^2 + 13x + 12$ ✓

Example 6

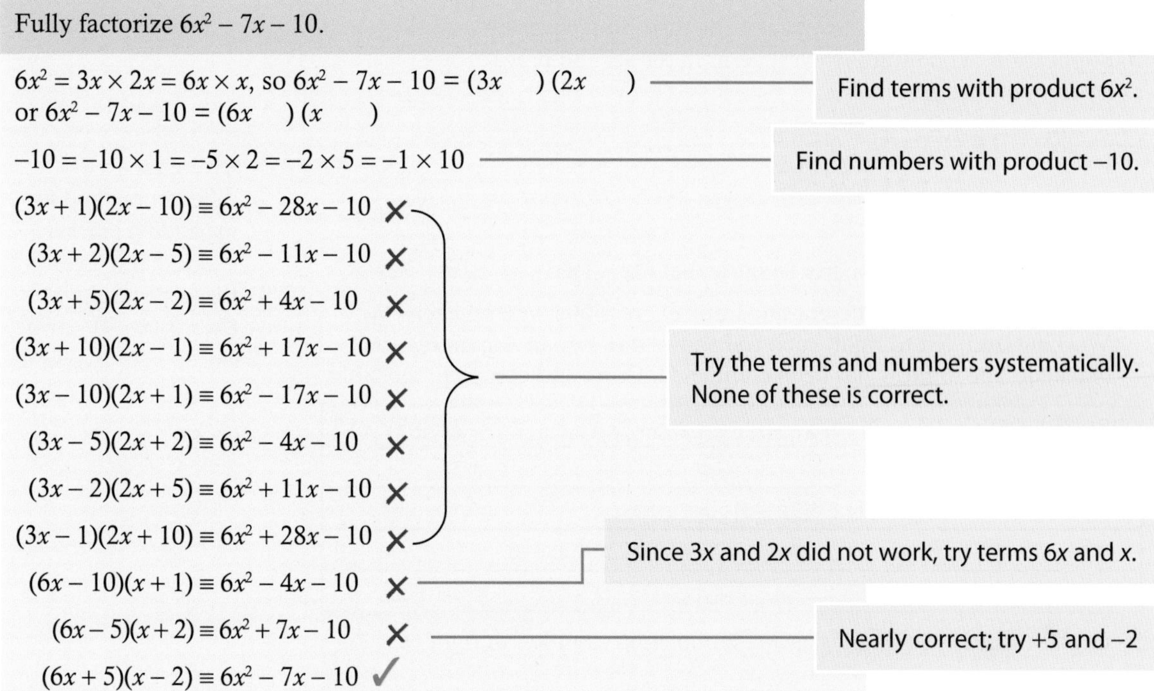

Fully factorize $6x^2 - 7x - 10$.

$6x^2 = 3x \times 2x = 6x \times x$, so $6x^2 - 7x - 10 = (3x \quad)(2x \quad)$ or $6x^2 - 7x - 10 = (6x \quad)(x \quad)$ — Find terms with product $6x^2$.

$-10 = -10 \times 1 = -5 \times 2 = -2 \times 5 = -1 \times 10$ — Find numbers with product -10.

$(3x + 1)(2x - 10) \equiv 6x^2 - 28x - 10$ ✗

$(3x + 2)(2x - 5) \equiv 6x^2 - 11x - 10$ ✗

$(3x + 5)(2x - 2) \equiv 6x^2 + 4x - 10$ ✗

$(3x + 10)(2x - 1) \equiv 6x^2 + 17x - 10$ ✗

$(3x - 10)(2x + 1) \equiv 6x^2 - 17x - 10$ ✗

$(3x - 5)(2x + 2) \equiv 6x^2 - 4x - 10$ ✗

$(3x - 2)(2x + 5) \equiv 6x^2 + 11x - 10$ ✗

$(3x - 1)(2x + 10) \equiv 6x^2 + 28x - 10$ ✗

Try the terms and numbers systematically. None of these is correct.

$(6x - 10)(x + 1) \equiv 6x^2 - 4x - 10$ ✗ — Since $3x$ and $2x$ did not work, try terms $6x$ and x.

$(6x - 5)(x + 2) \equiv 6x^2 + 7x - 10$ ✗ — Nearly correct; try +5 and −2

$(6x + 5)(x - 2) \equiv 6x^2 - 7x - 10$ ✓

You may not need to write down all the possible factorizations to test them. Each one gives the correct x^2 and constant terms, so you only need to check if expanding gives the correct middle term.

Reflect and discuss 3

- Which factorization was easier: the one in Example 5 or in Example 6? What made them easy or difficult?
- A student tries to factorize $3x^2 + 17x + 10$ and gives the answer $(3x + 5)(x + 2)$. Show that the middle term is incorrect.
- Why do you think listing the options systematically might be a time-consuming way to tackle this problem?

Practice 4

Fully factorize each quadratic expression.

1 $2x^2 + 5x + 3$ **2** $3x^2 + 7x + 2$ **3** $3x^2 + 7x - 6$

In Example 6 there are two possible pairs of factors for the x^2 term and the constant term, which gives a lot of possible factors to test. You will now explore another way to factorize quadratic expressions where it is not easy to identify the factors straight away.

Tip

Write down enough working to be sure you have the correct answer. Look for opportunities to be efficient: it's OK to do some working in your head.

Exploration 4

To expand a pair of brackets, you use the distributive property of multiplication. You multiply every term in the second bracket by every term in the first bracket.

$$(2x + 3)(x - 4) \equiv 2x(x - 4) + 3(x - 4)$$
$$\equiv 2x^2 - 8x + 3x - 12$$
$$\equiv 2x^2 - 5x - 12$$

If you use a different method for expanding brackets, check that it gives the same result for $(2x + 3)(x - 4)$.

1 Use the distributive property to show that $(2x - 1)(x + 5) \equiv 2x^2 + 9x - 5$.

2 The table shows five different quadratic expressions, one on each row. Working from left to right, each cell shows one step in expanding the brackets. The top row has been completed. Copy and complete the table.

$(2x + 3)(x - 4)$	$= 2x(x - 4) + 3(x - 4)$	$= 2x^2 - 8x + 3x - 12$	$= 2x^2 - 5x - 12$
$(2x - 3)(x + 7)$	$=$	$=$	$=$
	$= 3x(2x - 1) + 4(2x - 1)$	$=$	$=$
	$=$	$= 4x^2 + 12x - 2x - 6$	$=$
	$=$	$=$	$= 6x^2 + 5x - 4$

3 Explain which you think is easier: completing from left to right or right to left.

4 Comment on which of the steps is most difficult to complete. Explain why it is harder than the others.

5 Two students try to factorize the expression $6x^2 + 25x + 14$.

Carmen	Miranda
$6x^2 + 25x + 14 \equiv 6x^2 + 21x + 4x + 14$	$6x^2 + 25x + 14 \equiv 6x^2 + 20x + 5x + 14$

Explain which student's work helps to factorize the expression.

Exploration 4 shows that you can make a quadratic easier to factorize by splitting the middle term in the correct way. Exploration 5 investigates what happens when you expand some factorized expressions to find a pattern to help you factorize this kind of quadratic.

Objective: B. Investigating patterns
iii. prove, or verify and justify, general rules

In Exploration 5 you should gather enough information so you can conjecture a general rule. Then you should verify it, by showing that your conjecture holds for a few more examples.

ATL

Exploration 5

Investigate the relationship between the coefficients of the split middle term and the coefficients in $ax^2 + bx + c$.

To work out how to split the middle term of a quadratic by studying some examples, you could make a table like this:

▶ Continued on next page

Factorized expression	Expanded expression	Coefficients of the split middle term		Coefficients in $ax^2 + bx + c$		
				a	b	c
$(2x+3)(x+4)$	$2x^2+3x+8x+12$	3	8	2	11	12
$(2x+1)(x-2)$						
...						

Add more rows until you discover the pattern. You may find it useful to find the product and the sum of the coefficients of the split middle term.

You can gather more information by making up your own examples for the left-hand column. You can make new examples by changing the numbers in the examples you have already been given.

This is an important skill when investigating in mathematics.

Reflect and discuss 4

Compare what you found in Exploration 5 with others before moving on.

Example 7

Factorize $15x^2 - 2x - 24$.

$15x^2 - 2x - 24$ — Find two numbers with product $15 \times -24 = -360$ and sum -2.

$-360 = 18 \times -20$

$-2 = 18 + -20$

$15x^2 - 2x - 24 \equiv 15x^2 + 18x - 20x - 24$ — Split the middle term into $18x - 20x$. Take the common factors from each pair of terms.

$\equiv 3x(5x+6) - 4(5x+6)$ — $(5x+6)$ is a common factor.

$\equiv (3x-4)(5x+6)$

Does the order of the middle terms matter? Look at these two ways of factorizing the quadratic in Example 7:

$15x^2 - 2x - 24$

$\equiv 15x^2 + 18x - 20x - 24$	$\equiv 15x^2 - 20x + 18x - 24$
$\equiv 3x(5x+6) - 4(5x+6)$	$\equiv 5x(3x-4) + 6(3x-4)$
$\equiv (3x-4)(5x+6)$	$\equiv (5x+6)(3x-4)$

They both give the same factorization.

Quadratics where the coefficient of x^2 is not 1 may be factorized by splitting the middle term. To factorize $ax^2 + bx + c$, look for two numbers whose sum is b and whose product is ac.

Practice 5

Problem solving

1 Find pairs of numbers which have:

a product 240 and sum 32 **b** product −180 and sum 3

c product 96 and sum −35 **d** product −1000 and sum −117

2 Find values to fill these boxes and complete the identities.

a $\square x^2 - \square x - \square$

$= \square x^2 + \square x - 4x - 14$

$= 5x(2x + 7) - \square(2x + 7)$

$= (\square x - \square)(\square x + \square)$

b $4x^2 + 27x + \square$

$= \square x^2 + 3x + \square x + \square$

$= x(4x + 3) + \square(4x + 3)$

$= (x + \square)(\square x + \square)$

Use any of the methods you have learned.

3 Factorize each quadratic.

a $5x^2 - 9x + 4$ **b** $6x^2 + 7x + 2$ **c** $4x^2 + 8x + 3$

d $6x^2 - 11x + 4$ **e** $8x^2 - 14x - 15$ **f** $12x^2 - 17x + 6$

g $6x^2 + 5x + 1$ **h** $6x^2 + x - 2$ **i** $6x^2 - 5x - 1$

j $15x^2 - 7x - 2$ **k** $8x^2 + 10x - 3$ **l** $21x^2 - 23x + 6$

m $2x^2 - 3x + 1$ **n** $2x^2 - 9x + 10$ **o** $15x^2 - 40x - 15$

p $4x^2 + 7x - 15$ **q** $4x^2 + 8x - 21$ **r** $12x^2 + 4x - 5$

s $5x^2 - 2x - 3$ **t** $7x^2 - 16x + 4$ **u** $8x^2 + 24x + 16$

v $5x^2 - 14x - 3$ **w** $9x^2 + 11x + 2$ **x** $10x^2 - x - 21$

Problem solving

4 These expressions can be factorized. Determine the different values that could fill the empty boxes.

a $3x^2 + \square x + 2$ **b** $\square x^2 + 9x + 2$ **c** $6x^2 + 10x + \square$

D Non-factorizable expressions

- Can everything be written in a different form?

Exploration 6

1 Factorize the following expressions if possible. Verify your solution by expanding.

	factorized form	factorized form expanded
$x^2 - 16$	________	________
$49x^2 - 121$	________	________
$x^2 + 25$	________	________
$25x^2 + 81$	________	________
$4x^2 + 9$	________	________

▶ Continued on next page

2 Describe anything you notice about the expressions you could not factorize. Explain why they are not factorizable.

3 Factorize the following expressions if possible. Verify your solution by expanding.

	factorized form	factorized form expanded
$x^2 + 6x + 4$	________	________
$2x^2 - 3x - 1$	________	________
$x^2 - 10x - 24$	________	________
$x^2 + 10x + 24$	________	________
$4x^2 + x - 10$	________	________

4 For those expressions which you could not factorize, explain clearly why the methods you have already learnt did not work.

5 Expand these brackets: $(x+3+\sqrt{5})(x+3-\sqrt{5})$.
Describe anything you notice.

Tip

If you have not yet learned how to multiply out brackets like those in step **5**, either use a computer algebra system or complete it as a group with your teacher's assistance.

Reflect and discuss 5

- When we say that a quadratic expression is 'not factorizable', we usually mean that it cannot be factorized using integers. What other types of numbers could you use?

When quadratic expressions cannot be factorized using integers, there are other techniques that will help you work with quadratics.

The Fundamental theorem of algebra says that any quadratic (or higher power polynomial) can be factorized, although it might involve using numbers which are not in the set of real numbers. These numbers are known as complex numbers. You will learn about complex numbers if you take HL Mathematics as part of the Diploma. Proving the Fundamental theorem of algebra is university-level mathematics.

Summary

- The symbol ≡ means 'is identically equal to'. It shows that two expressions take the same value for any possible value of the unknown.
- To factorize a quadratic written in the form $x^2 + bx + c \equiv (x + p)(x + q)$, find two numbers p and q which have product c and sum b.
- Any expression of the form $a^2 - b^2$ is the **difference of two squares**, because it is one squared quantity subtracted from another.
  $a^2 - b^2 \equiv (a + b)(a - b)$
- Where the coefficient of x^2 does not equal 1, one method of factorizing involves listing all the pairs of terms with product ax^2 and all the pairs of numbers with product c. Then try the different combinations until you find pairs which work.
- Alternatively, it is often quicker to look for two numbers whose sum is b and whose product is ac.

Mixed practice

1 Factorize:

a x^2+4x+4 **b** $x^2-13x+36$

c $x^2+5x-14$ **d** $x^2+18x+81$

e x^2-7x-8 **f** $x^2-11x+24$

g $x^2-20x+100$ **h** $x^2-15x-100$

i x^2+4x+3 **j** $x^2+11x-42$

k $x^2+4x-96$ **l** x^2+7x+6

m $x^2-15x+56$ **n** $x^2-11x-60$

2 Factorize fully:

a x^2-49 **b** x^2-169

c $64-x^2$ **d** $25x^2-4$

e $144x^2-81$ **f** $256x^2-169$

g $4x^2-y^2$ **h** $16x^2-9y^2$

i $25x^2-289y^2$ **j** x^4-16

k $16x^4-1$ **l** $9x^4-729y^4$

3 Factorize completely:

a $2x^2+8x+6$ **b** $2x^2+14x+20$

c $2x^2-12x+10$ **d** $3x^2-9x+6$

e $2x^2+8x-24$ **f** $3x^2+21x-24$

g $4x^2-8x-32$ **h** $2x^2-2x-40$

4 Factorize:

a $2x^2+5x+2$ **b** $2x^2+13x+20$

c $2x^2-11x+12$ **d** $3x^2-8x-3$

e $5x^2-12x+4$ **f** $7x^2+38x-24$

g $6x^2+31x+5$ **h** $6x^2-17x+5$

i $8x^2+18x+9$ **j** $8x^2-41x+5$

k $9x^2+12x-32$ **l** $10x^2-13x-30$

5 Factorize:

a $4a^2-3a$ **b** $7b^3-35b^2-168b$

6 Factorize:

a $a^2-5a-36$ **b** $16b^2-9$

c $c^2+11c+24$ **d** $4d^2-17d-15$

e $4e^2-3e$ **f** $f^2+2f-48$

g $3g^2-23g+14$ **h** $16h^2-1$

Problem solving

7 Ornella thinks of a whole number, adds four to it, squares the result, and subtracts 9.

a By letting the original number be n, **write down** an expression for this process.

b Hence **show** the number that Ornella obtains can always be written as the product of two integers with a difference of 6.

8 Copy and complete:

a $x^2-\square x+15 \equiv (x-3)(x-\square)$

b $x^2\square\square x+20 \equiv (x+4)(x\square\square)$

c $x^2-8x\square\square \equiv (x+\square)(x-11)$

d $x^2\square 11x+\square \equiv (x\square 5)(x-\square)$

9 Dagmar thinks of an even number, squares it, and subtracts 1. By expressing his original number in the form $2n$, or otherwise, **show** that the result of this process can be written as the product of two consecutive odd integers.

10 You have two rectangular grids of 1 cm^3 cubes, one measuring $n+1$ by $n+6$ cm, and the other measuring $n+3$ by $n+3$ cm, where n is a positive integer.

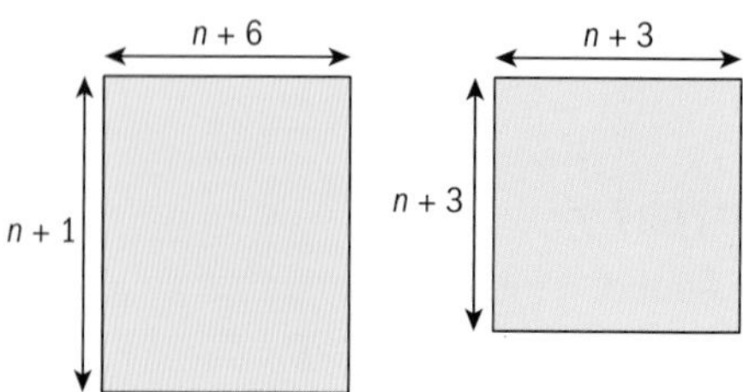

a **Find** and simplify an expression for the total number of 1 cm^3 cubes in the two rectangular grids combined.

b **Show** that if you take apart the rectangles and recombine the 1 cm^3 cubes you will always be able to form a rectangle with no cubes left over (where the rectangle will not simply be a straight line of cubes).

11 A rectangular sports pitch has a border of n meters on each side. The total area occupied by the pitch and its border is $4n^2 + 28n + 45$.

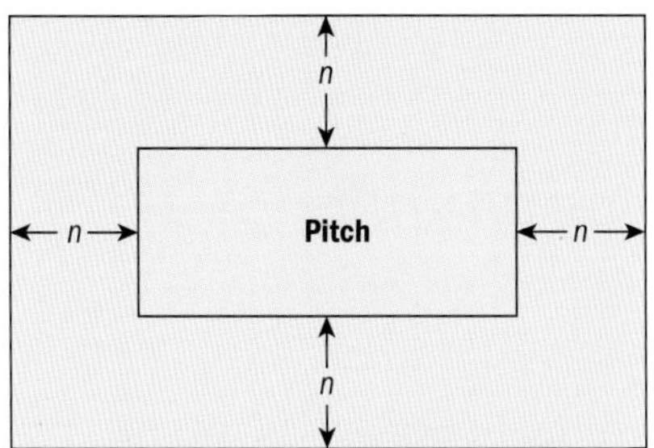

Find the dimensions of the pitch.

12 Start with a rectangular grid of 8 cubes:

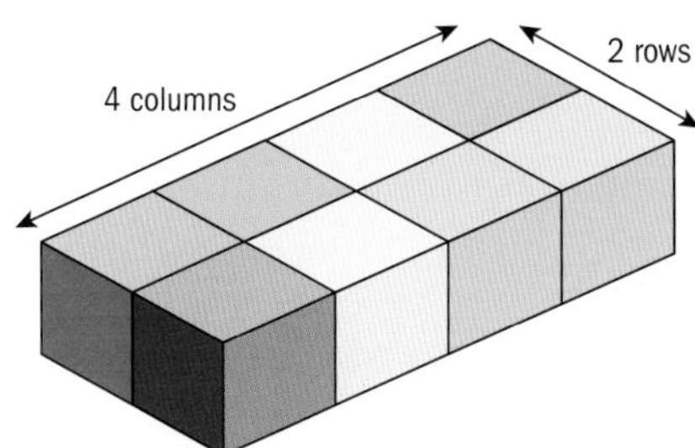

Somebody else then adds some extra rows and columns to the grid, but doesn't tell you how many they've added. They do tell you that the grid now contains 45 cubes.

They repeat the process, adding the same number of rows as before and the same number of columns as before. Now the grid contains 112 cubes.

They repeat the process a third time, and the resulting grid contains 209 cubes. After a final repetition, there are 336 cubes in the grid.

a **Construct** a difference diagram showing the sequence 8, 45, 112, 209, 336. Analyze the differences and **show** that these numbers follow a quadratic sequence.

b Letting u_n be the number of cubes in the nth grid (so $u_1 = 8$), **find** a formula for u_n in the form $an^2 + bn + c$.

c Factorize your formula for u_n.

d Hence **determine** the number of rows and columns being added each time.

Reflect and discuss

How have you explored the statement of conceptual understanding? Give specific examples.

Conceptual understanding:

Patterns can be represented in equivalent forms.

Space

The frame of geometrical dimensions describing an entity

Dimensions

We live in a three-dimensional world, meaning that everything in our world can be located using three aspects: latitude, longitude and altitude. Letting time be the 4th dimension, people and objects can be located at any particular place and moment in time.

We can imagine what a two-dimensional universe would be like, if we simply look into a mirror, or observe shadows on the ground. These images possess no thickness. You are familiar with using the 2D coordinate plane to create graphs. This representation of 2D space is called the Cartesian plane, named after the French philosopher, scientist and mathematician Rene Descartes.

In order to locate three-dimensional objects, we need to add a third axis to the Cartesian plane. A point in 3D would therefore have three coordinates instead of two.

There is a wonderful book written over a hundred years ago by Edwin Abbott called *Flatland, a Romance of Many Dimensions*, about a two-dimensional world where one of its inhabitants encounters a mysterious visitor from the third dimension and struggles to comprehend how such a being could exist and what it might look like.

Animation

Have you ever wondered how animated characters can look so fluid and real? An understanding of the three-dimensional space that we live in and move in is essential to achieve this effect. Therefore, a large part of this modern success in animations is due to the use of mathematics in creating the different animated figures. To create your favorite animations, mathematicians are using harmonic coordinates, which are a form of barycentric coordinates, or coordinates using a triple set of numbers. The advantage of this method is that it greatly simplifies the process of animating a character, and reduces the time and effort to complete the animation.

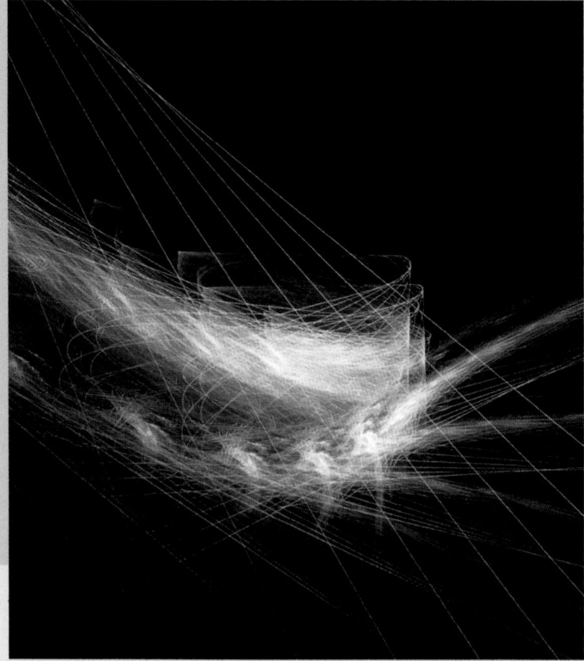

More than 3 dimensions

In the 1990s, mathematicians working on the string theories of physics discovered the likelihood of there being at least ten dimensions in existence. But what does this really mean? Mathematicians have always worked with any number of dimensional spaces, including infinite dimensional spaces. If mathematics is capable of explaining the inherent structure of our universe, then perhaps there are even more than ten dimensions waiting to be discovered.

Spacious interiors

Global context: Personal and cultural expression

RELATIONSHIPS

Objectives

- Finding the surface area of any 3D shape (including pyramids, cones and spheres)
- Finding the volume of any 3D shape (including pyramids, cones and spheres)

Inquiry questions

F
- What is the difference between area and surface area?
- What are some properties of prisms, cylinders, pyramids and cones?

C
- How are the surface areas of pyramids, cones and spheres related?

D
- Is there a best method for finding volume?

ATL Creative-thinking

Apply existing knowledge to generate new ideas, products or processes

Statement of Inquiry:

Generalizing relationships between measurements enables the construction and analysis of activities for ritual and play.

7.3

9.1

E9.1

You should already know how to:

• use Pythagoras' theorem	In each triangle, find the length of the missing side. If necessary, give your answer as a radical. **1** 5 m, 5 m, a **2** 9 cm, 41 cm, b
• find the volume and surface area of cuboids, prisms and cylinders	Find the volume and surface area of each solid. If necessary, round your answer to 3 significant figures. **3** 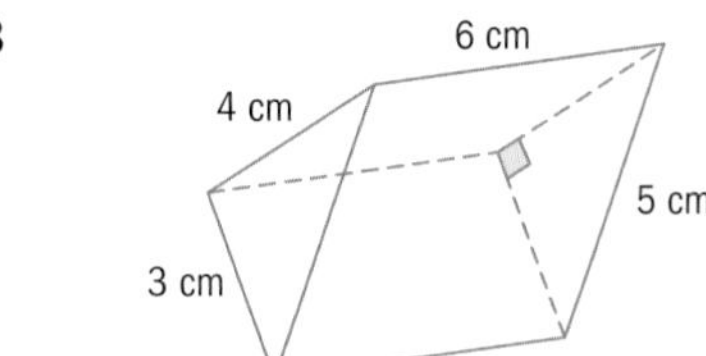 **4** 12 m, 8 m **5** 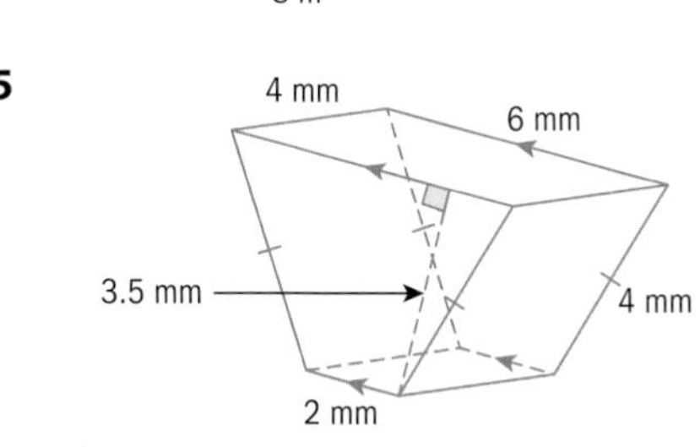
• find the area of a sector and length of an arc	**6** Find the area of the shaded sector and the length of its arc. 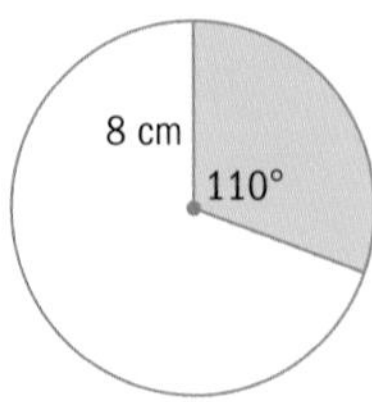

Area of a trapezium:
$A = \frac{(a+b)h}{2}$

F Properties of 3D solids

- What is the difference between area and surface area?
- What are some properties of prisms, cylinders, pyramids and cones?

Exploration 1

1. Make a list of shapes that have an area that you can calculate.
2. Make another list of shapes that have a surface area that you can calculate.
3. Next to each shape in step **2**, state the number of its faces.
4. Compare and contrast the shapes in each list. Identify similarities and differences between the ways of calculating area and surface area.

A **polyhedron** is a 3D solid that has only plane (flat) faces.

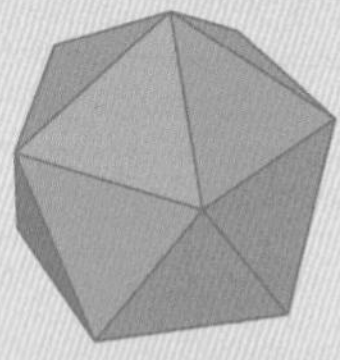

A **pyramid** is a 3D solid with a polygon base. The other faces are triangles that meet at a point called the **apex**. A pyramid is a polyhedron.

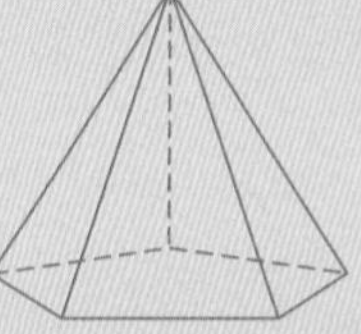

A **cone** is a 3D solid with a circular base and an apex or vertex. A cone is not a polyhedron.

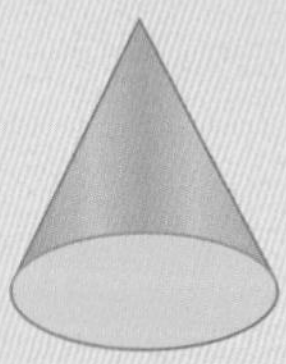

A **sphere** is a 3D solid with one curved face. All the points on the sphere's surface are equidistant (the same distance) from the center. A sphere is not a polyhedron.

You already know how to find the surface area and volume of cuboids, cylinders and prisms. Now you will look at finding the surface area and volume of pyramids, cones and spheres.

The **surface area** is the total area of all the faces of a 3-dimensional solid.

The **volume** of a 3-dimensional solid is the amount of space it occupies.

Exploration 2

Copy and complete this table with your observations about the different 3D solids.

	Prism	**Cylinder**	**Pyramid**	**Cone**
Base	• any shape, often a regular polygon • rectangular based prisms are called cuboids • two opposite faces could each be the base			
Cross-section parallel to the base		• always a circle • the same radius as the base all along the cylinder		
Sides			• triangles • if the base is a regular polygon, the triangles are all congruent • the triangles all meet at the apex	

Reflect and discuss 1

- How is a sphere similar to the 3D solids in the table in Exploration 2? How is it different?
- How do you think the volume and surface area of a sphere is found, given its differences compared to other 3D solids?

C Finding surface area

- How are the surface areas of pyramids, cones and spheres related?

Surface area

Pyramids, cones and spheres each require a different method for finding surface area.

The **slant height** of a pyramid is the distance from the center of one edge of the base to the apex of the pyramid.

The **vertical height** is the perpendicular distance from the apex to the base.

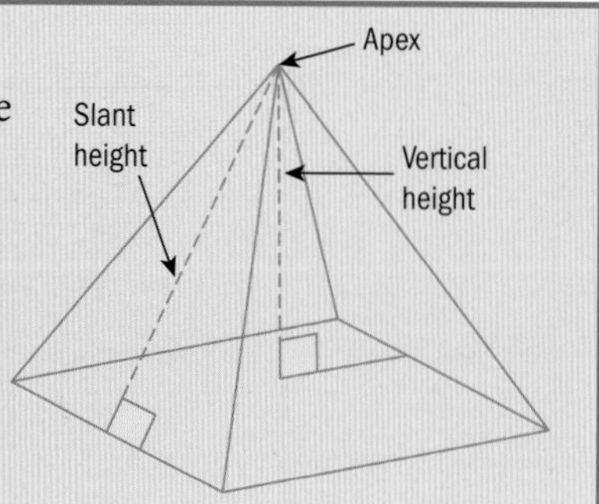

ATL

Exploration 3

1 a Determine which 2D (flat) shapes make up the different faces of this square-based pyramid.

b Identify the information you need to find the surface area of square-based pyramids.

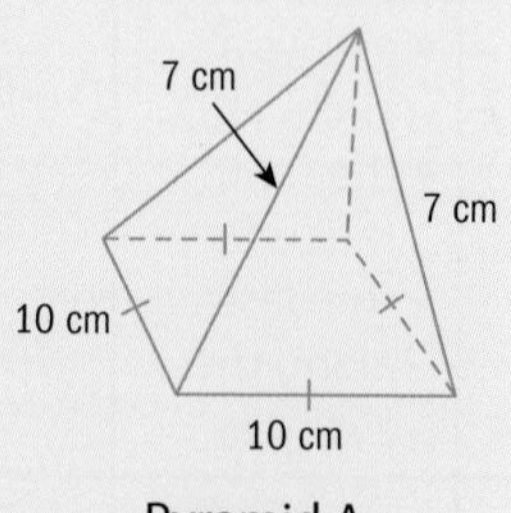

Pyramid A

2 a Use the information given in the diagram to find the area of the base of pyramid A.

b Justify whether the information in the diagram is enough to find the area of one of the triangles on the side of the pyramid.

c Calculate the surface area of pyramid A.

3 a In pyramid B, the side of the square base and the vertical height of the pyramid are given. Suggest how you could calculate the slant height of the pyramid using this information.

b Find the slant height of the pyramid.

c Use the slant height to calculate the area of one of the triangular faces of the pyramid.

d Hence find the total surface area of pyramid B.

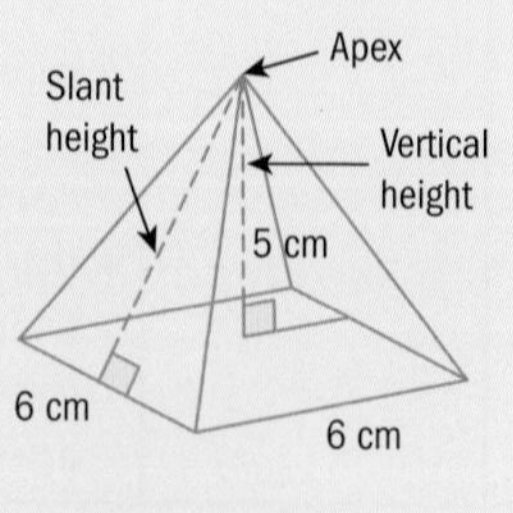

Pyramid B

Tip

The four triangles are isosceles and congruent.

Example 1

Find the surface area of this rectangular-based pyramid.

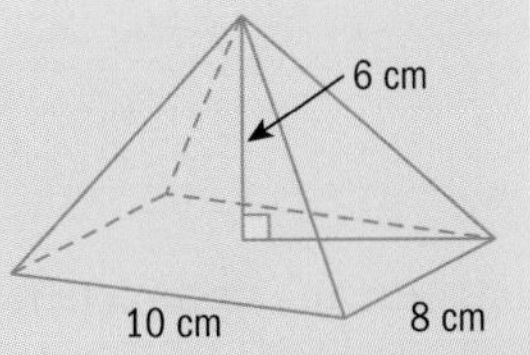

Area of base = $8 \times 10 = 80$ cm^2

There are three different side areas to find; the first one is the rectangular base.

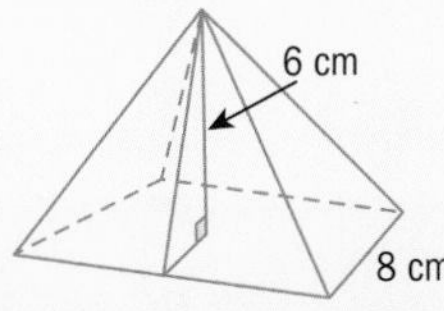

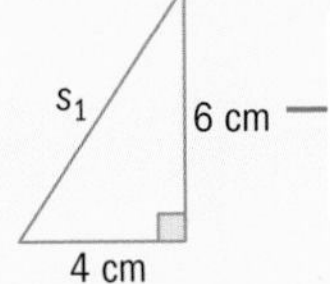

Use Pythagoras to calculate the slant height s_1 for the front and back triangular faces.

Use s for slant height and h for vertical height.

$s_1^2 = 6^2 + 4^2 = 52$

$\Rightarrow s_1 = 2\sqrt{13}$ cm

Area of front triangle $= \frac{1}{2} b \times s_1 = \frac{1}{2} \times 10 \times 2\sqrt{13}$

$= 10\sqrt{13}$ cm^2

The area of the back triangle is the same as the front triangle. Note: keep answers exact (using radicals) at each stage. Use your calculator only at the end to give your final answer correct to 3 s.f.

Similarly, calculate the slant height s_2 for the left and right triangular faces:

$s_2 = 6^2 + 5^2 = 61$

$\Rightarrow s_2 = \sqrt{61}$ cm

Area of left triangle $= \frac{1}{2} b \times s_2 = \frac{1}{2} \times 8 \times \sqrt{61} = 4\sqrt{61}$ cm^2

The area of the right side triangle is the same as the left side triangle.

Total area = area of base
+ 2 times the area of front triangle
+ 2 times the area of side triangle

Find the total area of all 5 faces.

$= 80 + 2 \times 10\sqrt{13} + 2 \times 4\sqrt{61}$

$= 215$ cm^2 (3 s.f.)

Reflect and discuss 2

- Compare and contrast the processes for finding the surface area of a square-based pyramid and a rectangular-based pyramid.
- How do you think this compares to finding the surface area of a pyramid with another regular polygon (for example, a pentagon or hexagon) as its base?
- How would you find the area of the base of a pyramid with a regular hexagon as its base? How would you find the slant height of this pyramid?

The surface area of a pyramid with an n-sided regular polygon base is

$S = A_{\text{base}} + nA_{\text{triangle}}$

Practice 1

In each question, copy the diagram and complete it as you solve the problem. You may need to draw more diagrams for different steps in the solution.

Find the surface area of each pyramid. If necessary, round your answer to 3 s.f.

1

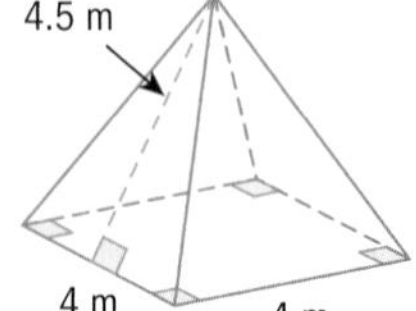

2

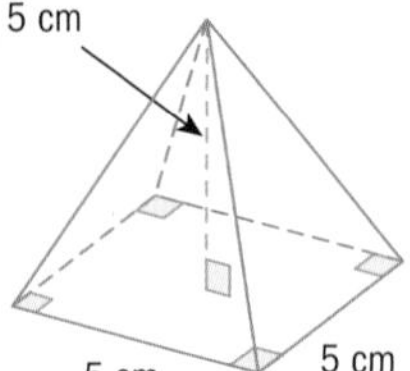

3

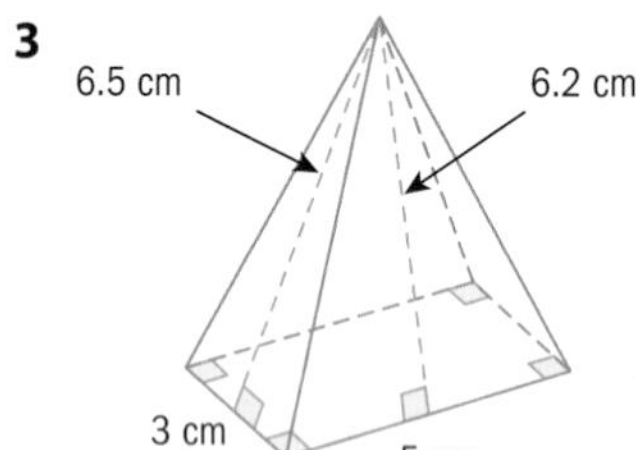

4 This square-based pyramid has four identical triangular faces. Each sloping edge measures 8 m.

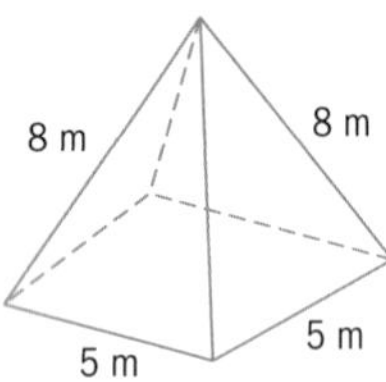

a Find the slant height.

b Hence find its surface area.

Problem solving

5 A tetrahedron is a 3D solid with all four faces equilateral triangles.
Find the surface area of a tetrahedron with side length 7 cm.

6 George bought a white plaster replica of the Great Pyramid on his trip to Egypt. Its height is 48 mm and the side length of its square base is 75 mm. George wants to paint the pyramid in stone-color to make it look more real.

a Explain which faces of the pyramid he needs to paint.

b Hence, find the area of the surface of the pyramid that he needs to paint.

7 This pyramid has a square base with side 30 cm and total surface area 2400 cm^2. Find its height.

30 cm

30 cm

8 These two square-based pyramids are mathematically similar.

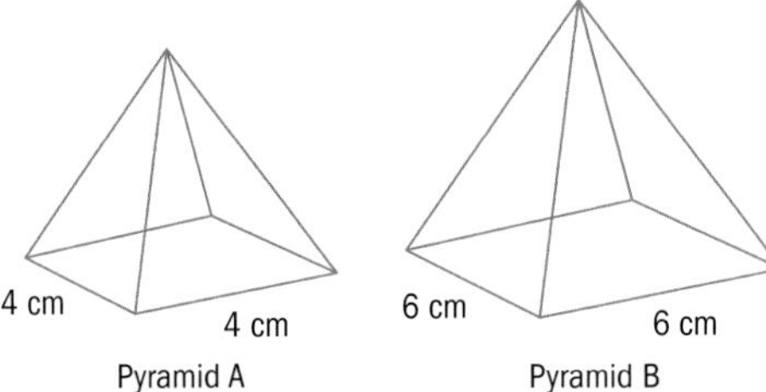

Pyramid A has surface area 64 cm^2.

Calculate the surface area of Pyramid B.

When the linear scale factor is r, the area scale factor is r^2.

The surface area of a cone with base radius r and slant height s can be unravelled into a circle and a sector:

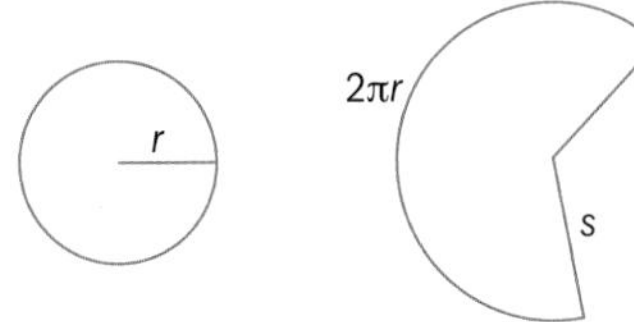

The arc length of the major sector is equal to the circumference of the circle, $2\pi r$. The radius of the sector is equal to the slant height s of the cone.

The formulae that show the relationships in a circle are:

Area of a circle $= \pi r^2$

Area of a sector $= \dfrac{\theta}{360}\pi s^2$

Arc length of a sector $= \dfrac{\theta}{360}2\pi s$

However, you know that arc length $l = 2\pi r$.

Therefore $2\pi r = \dfrac{\theta}{360}2\pi s$

$\Rightarrow \quad \theta = \dfrac{360r}{s}$

And the area of a sector $= \dfrac{\frac{360r}{s}}{360}\pi s^2$

$= \dfrac{r}{s}\pi s^2 = \pi rs$

Therefore the surface area of a cone $= \pi r^2 + \pi rs$.

The surface area of a cone $= \pi r^2 + \pi rs$ where r is the radius of the base and s is the slant height of the cone.

The curved surface area of the cone is πrs.

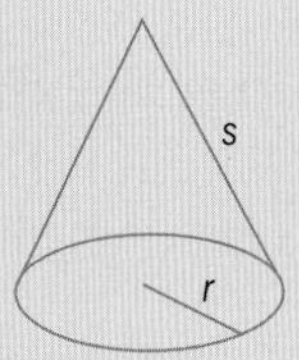

Practice 2

Find the surface area of each solid cone. If necessary, round your answer to 3 s.f.

1

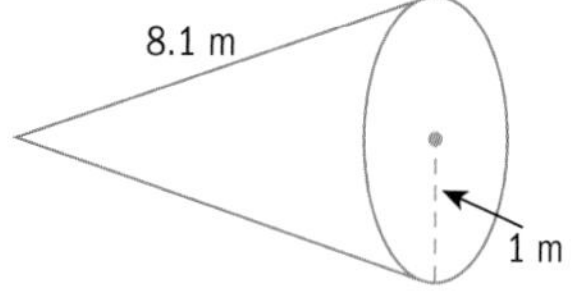

2

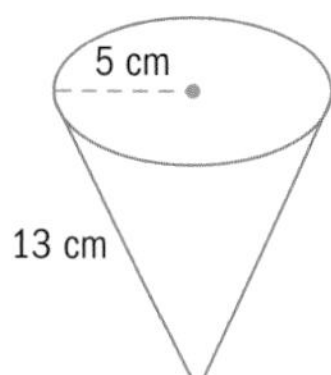

3 The diagram shows the dimensions of a paper cup. Calculate the area of paper used to make the cup.

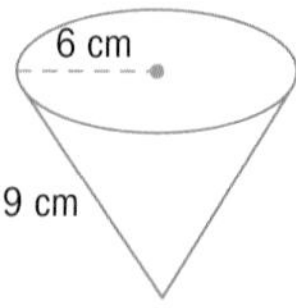

Problem solving

4 Belgium is famous for its French fries, which are often served in cones made from newspaper. Rene makes a paper cone from a semicircle cut from a page of his local newspaper. The page measures 410 mm by 315 mm.

a Draw a diagram to show the semicircle on the page of newspaper.

b Determine the largest possible diameter of the semicircle. Hence, find the radius.

c Find the area of the largest cone Rene can make from a page of this newspaper.

d Find the circumference of the circular base.

e Hence, find the radius of the base of the paper cone.

5 A solid cone has surface area 125.7 cm^2, and base radius 4 cm.

a Find the curved surface area of the cone.

b Find the slant height of the cone.

c Hence, find the vertical height of the cone.

As circles have no straight edges, you cannot use the same methods for finding the area as you would for polygons. Finding the surface area of a sphere (which has no edges at all) is not possible with methods that you know and requires more advanced mathematics.

Reflect and discuss 3

Suppose you traced around an orange several times, creating circles with the same radius as the orange. Suppose you then peeled the orange, breaking it into small pieces, and completely filled as many of the circles as possible.

- How many of the circles would you expect to fill with the orange peel pieces?
- How does this demonstrate the formula for the surface area of a sphere?

The surface area of a sphere = $4\pi r^2$

Practice 3

Find the surface area of each sphere. If necessary, round your answer to 3 s.f.

1

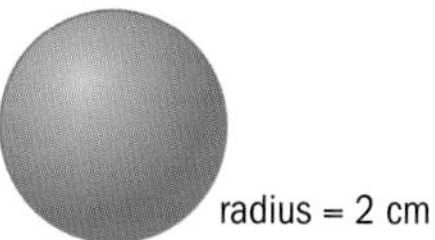

2

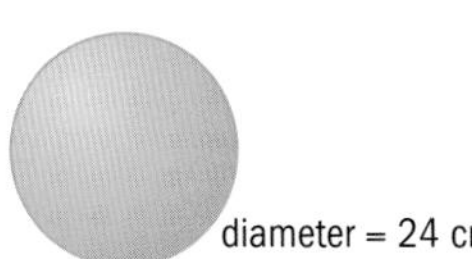

3 Calculate the surface area of these solids.

a

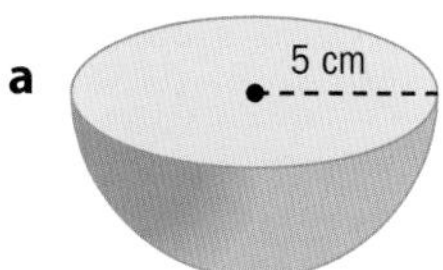

b

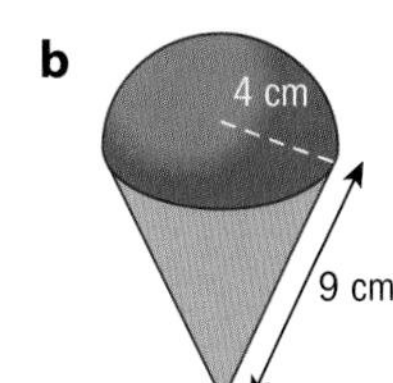

Problem solving

4 The surface area of a sphere is 400 cm^2. Calculate its radius.

5 How does the surface area of a sphere compare to the surface area of a cylinder in which it fits perfectly? (Diagram shown at right.)

D Finding volume

- Is there a best method for finding volume?

ATL

Exploration 4

Archimedes discovered the volume of a sphere in 225 BCE, and you will use a similar method here.

You will need a golf ball (or any ball that sinks in water), a measuring cylinder and water.

Part 1 – finding the volume of a golf ball

1 Put the golf ball in the measuring cylinder. Fill the cylinder with water so that the water covers the ball.

2 Record the water level and calculate the volume of the ball and the water together.

3 Remove the golf ball. Record the new water level, and calculate the volume of the water.

4 Use your results from steps **2** and **3** to find the volume of the golf ball.

1 ml = 1 cm^3

▶ Continued on next page

Part 2 – comparing spheres and cylinders

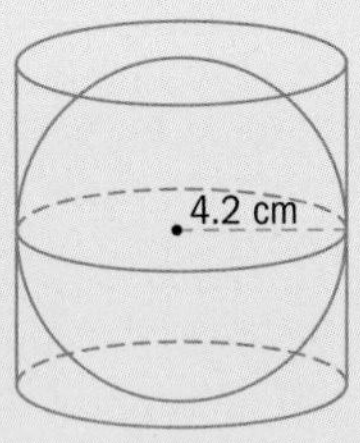

5 The diagram shows a golf ball in a cylinder. The radius of a golf ball is approximately 4.2 cm.

a State the radius of the cylinder

b State the height of the cylinder

6 Measure the diameter of the golf ball that you used in step **1**. Use this to find the volume of its circumscribed cylinder. Compare the volume of the golf ball with the volume of its circumscribed cylinder. Use your findings to suggest a formula for the volume of a sphere.

When a sphere fits perfectly inside the smallest possible cylinder, the cylinder **circumscribes** the sphere.

Archimedes of Syracuse (287 BCE – 212 BCE) was a Greek mathematician and inventor. It is said he had a sphere circumscribed by a cylinder on his tombstone, to represent his favourite mathematical proofs.

In Exploration 4, the volume of the sphere is two-thirds of the volume of its circumscribed cylinder, which has volume $2\pi r^3$. Two-thirds of this volume is $=\frac{2}{3}(2\pi r^3)=\frac{4}{3}\pi r^3$.

The volume of a sphere $=\frac{4}{3}\pi r^3$

You can use a similar method to verify the formulae for volume of a pyramid and volume of a cone.

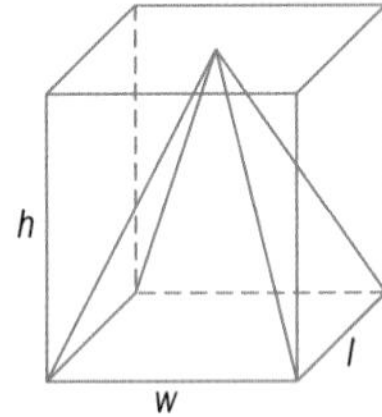

Volume of cuboid: $l \times w \times h =$ area of base $\times$ height

Volume of pyramid with same base and height: $\frac{1}{3} \times$ area of base $\times$ height

This formula can be generalized to a pyramid with any base.

Similarly, the volume of a cone is $\frac{1}{3}$ the volume of a cylinder with the same base radius and height.

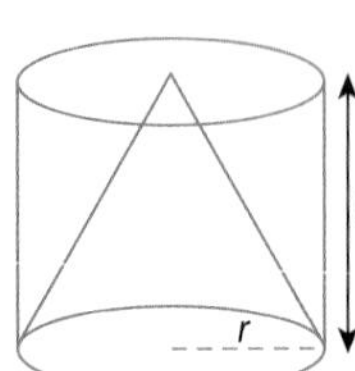

Volume of cone $=\frac{1}{3}\times$ volume of cylinder

$$V=\frac{1}{3}\pi r^2 h$$

Volume of pyramid $=\frac{1}{3}\times$ area of base $\times$ height

Volume of cone $=\frac{1}{3}\pi r^2 h$

Practice 4

Find the volume of each 3D solid.

1

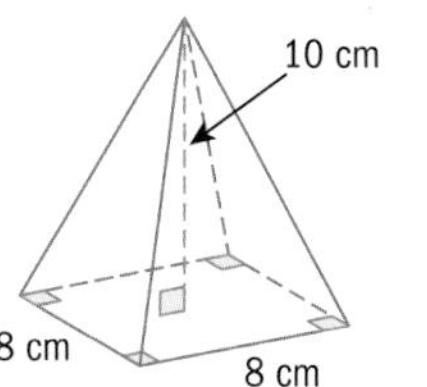

2

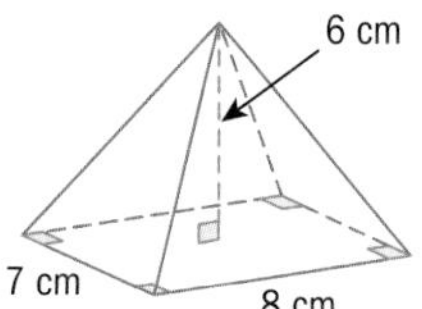

3

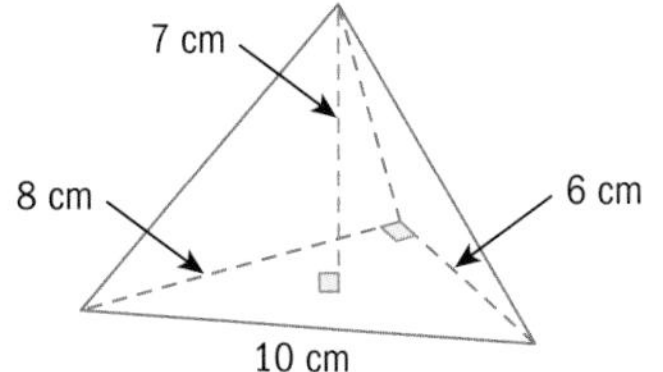

4

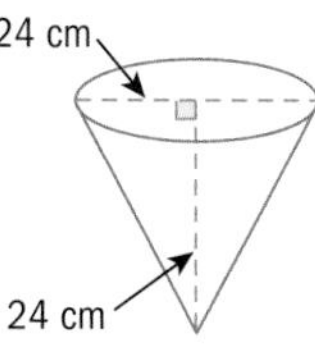

5

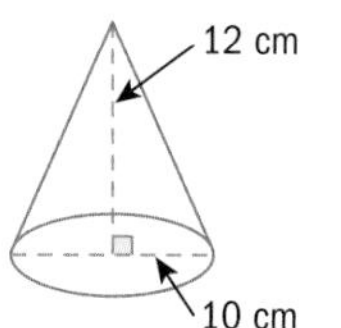

6

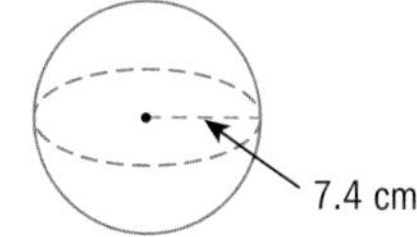

7

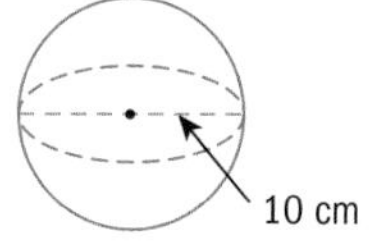

8

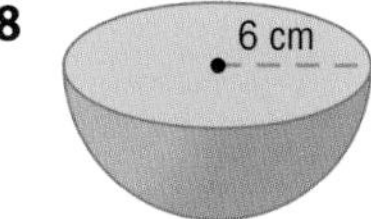

9

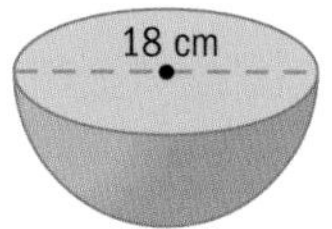

10 In Practice 2 Q4, you found the area of a newspaper cone made from a semicircle.

a Find the volume of this newspaper cone.

b Explain why the volume of potato in this cone of French Fries would be less than your answer to part **a**.

Problem solving

11 These two cones are mathematically similar.

Cone A has volume 2000 cm^3.

Calculate the volume of Cone B.

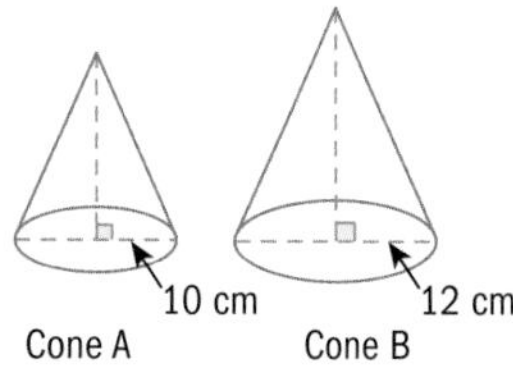

When the linear scale factor is r, the volume scale factor is r^3.

12 a Explain why any two spheres are mathematically similar.

b The volume of a sphere of radius 4 cm is 268 cm^3 (to 3 s.f.)
Find the volume of a sphere of

i radius 8 cm **ii** radius 2 cm

13 Find the height of a cone that has volume 270π mm^3 and base radius 9 mm.

14 Find the radius of a cone that has volume 8.38 cm^3 and height 2 cm.

15 Find the radius of a sphere with volume 523.6 cm^3.

16 Suppose you fill an empty sphere with water and then pour it into an empty cylinder that has the same height and diameter as the sphere. Find the height up the cylinder that the water would reach. Leave your answer as a fraction of the total height of the cylinder.

Summary

- A **polyhedron** is a 3D solid that has only plane (flat) faces.

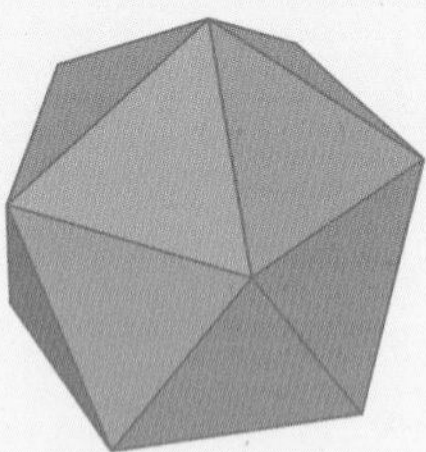

- A **pyramid** is a 3D solid with a polygon base. The other faces are triangles that meet at a point called the **apex**. A pyramid is a polyhedron.

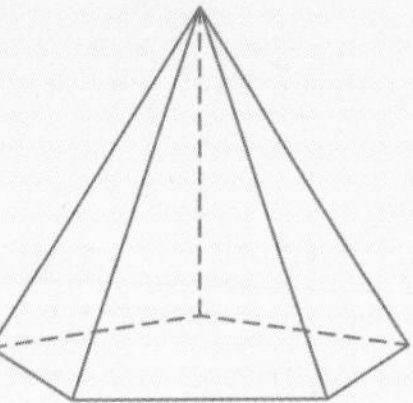

- A **cone** is a 3D solid with a circular base and an apex or vertex. A cone is not a polyhedron.

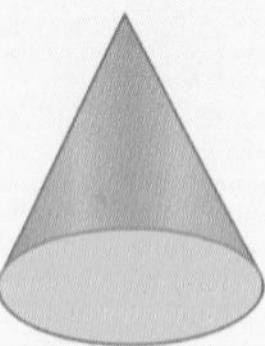

- A sphere is a 3D solid with one curved face. All the points on the sphere's surface are *equidistant* (the same distance) from the center. A sphere is not a polyhedron.

- The **slant height** of a pyramid is the distance from the center of one edge of the base to the apex of the pyramid.

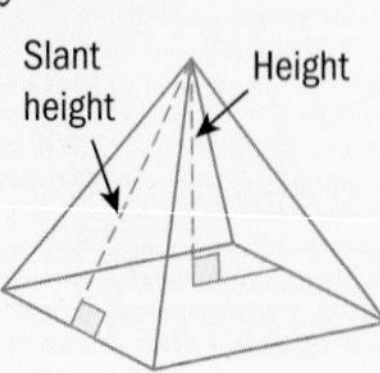

- The **surface area** is the total area of all the faces of a 3-dimensional solid.
- The **volume** of a 3-dimensional solid is the amount of space it occupies.
- The surface area of a pyramid with an n-sided regular polygon base is $S = A_{base} + nA_{triangle}$
- Volume of pyramid $= \frac{1}{3} \times$ area of base $\times$ height

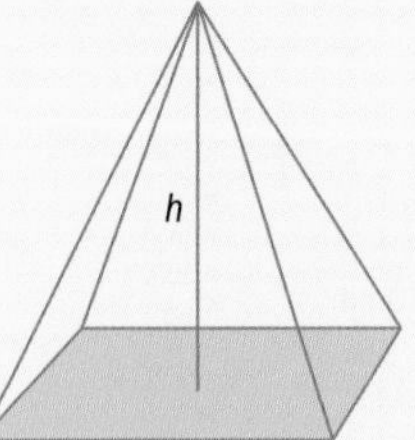

- Surface area of a cone $= \pi r^2 + \pi rs$ where r is the radius of the base and s is the slant height.

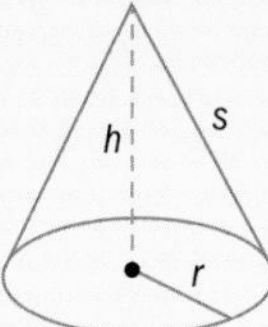

- Volume of a cone $= \frac{1}{3}\pi r^2 h$
- Surface area of a sphere $= 4\pi r^2$

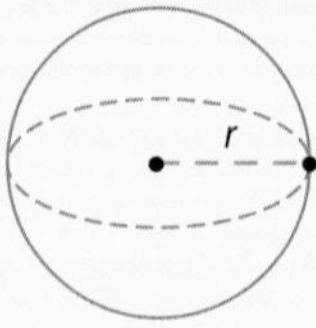

- Volume of a sphere $= \frac{4}{3}\pi r^3$

Mixed practice

1 Find the surface area and volume of each 3D solid.

a

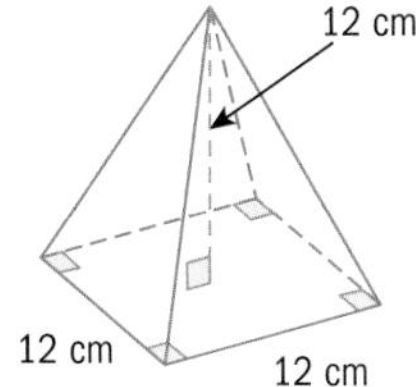

b

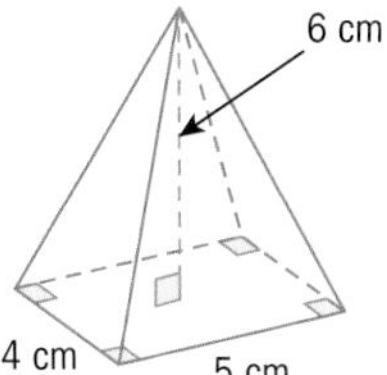

c

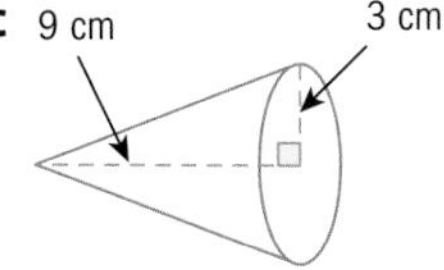

d

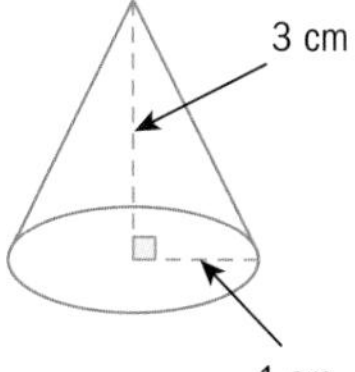

e

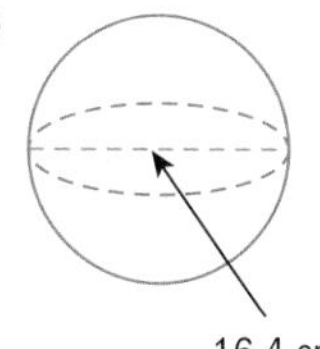

f

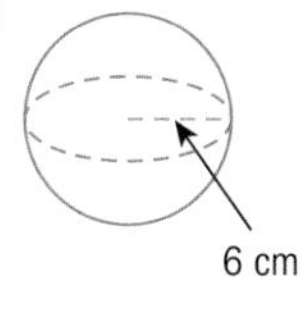

Problem solving

2 All eight edges of a square-based pyramid are 12 cm long.

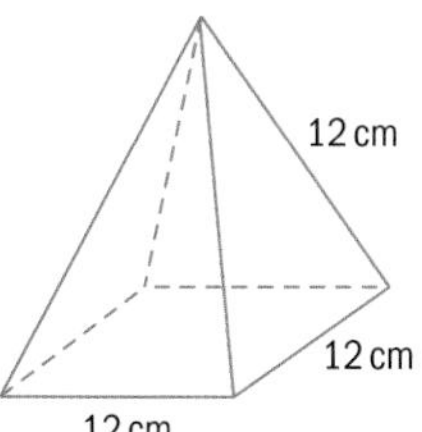

a Find its surface area and its volume.

b Find the surface area and volume of a mathematically similar pyramid, with base a square of side 9 cm.

3 A cone has surface area 204.2 cm^2, and base radius 5 cm.

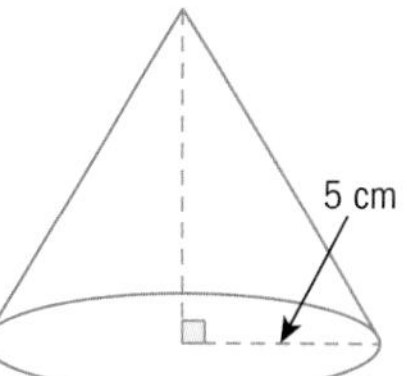

a Find the slant height of the cone.

b Find the volume of the cone.

4 The volume of a soccer ball is 5.6 dm^3.

According to the regulations, the circumference of the ball must be between 68 cm and 70 cm. **Determine** whether this soccer ball satisfies the regulations. **Justify** your answer.

5 A spherical scoop of ice cream is cut in half, and covered in chocolate (including the flat face). The total area of chocolate is 85 cm^2.

a Find the radius of the spherical scoop of ice cream.

b Hence, **find** the volume of the ice cream covered in chocolate.

6 Two paper cones are mathematically similar. The capacity of the larger cone is 8 times the capacity of the smaller cone. The surface area of the smaller cone is 50 cm^2.

Find the surface area of the larger cone.

Objective: D. Applying mathematics in real-life contexts
ii. select appropriate mathematical strategies when solving authentic real-life situations

In these real-life situations, select the strategies you have learned for finding the volume and surface area of 3D shapes to answer the questions.

Review in context

Personal and cultural expression

1 The Egyptians built square-based pyramids as part of the burial ritual for queens and pharaohs. The three largest and best preserved pyramids are located in the town of Giza, near Cairo. Many believe that the builders were influenced by the Golden Ratio (1.62).

a Each of the sides of the base of the Great Pyramid measures 230.4 meters while the height measures 145.5 meters. How close is the ratio of side length to height to the Golden Ratio?

b The entire pyramid was originally covered with a layer of white limestone. **Calculate** the area of limestone used.

c **Compare** the total surface area of the sides to the area of the square base. How does this relate to the Golden Ratio?

d **Calculate** the volume of the pyramid.

e **Calculate** the perimeter of the base of the pyramid as well as the circumference of a circle with a radius equal to the height of the pyramid. **Describe** what you notice about these results.

2 Near Mexico City is The Temple of the Feathered Serpent, one of hundreds of pyramids in the Mesoamerican city of Teotihuacan. Archaeologists used a robot and found hundreds of spheres with circumferences ranging from about 3.5 cm to 12.5 cm. The spheres were covered in a yellow material called jarosite.

a **Find** the amount of jarosite needed to paint:
i the smallest ball **ii** the largest ball.

b **Find** the volume of clay needed for each size ball.

3 Potentially the earliest team sport, ballgame, was played in Mesoamerica. A very large court is still visible at the archaeological site of Chichen Itza in Mexico. Players could end the game by getting a heavy ball through a ring mounted vertically on the wall of the court (no hands allowed). If the circumference of the ring was 95 cm and the surface area of the ball was 2830 cm^2, **justify** why putting the ball through the hoop resulted in ending the game.

4 The Christmas tradition of decorating evergreen trees began in Germany in the 16th century. With their roughly conical shape, it was possible to decorate them and enjoy their beauty from all sides. (No decorations were put underneath the circular face.)

a **Determine** which tree would have more area for decorations: one with a height of 215 cm and a maximum circumference of 200 cm or a tree with a height of 180 cm and a circumference of 210 cm? **Justify** your answer.

b A tree located in Rockefeller Center in New York City is lit every year in early December. The largest tree measured 100 feet tall and had a volume equivalent to 65 500 cubic feet. **Find** the area was available for decorations.

Reflect and discuss

How have you explored the statement of inquiry? Give specific examples.

Statement of Inquiry:

Generalizing relationships between measurements enables the construction and analysis of activities for ritual and play.

9.2 A parable about parabolas

Global context: Scientific and technical innovation

Objectives

- Finding the axis of symmetry and vertex of a quadratic function
- Expressing a quadratic function in three different forms: standard, factorized and vertex
- Finding a quadratic function given three distinct points on its graph
- Finding a function to model a real-life parabola
- Understanding how many unique points define an object in a given dimension of space

Inquiry questions

F
- What shape is the graph of a quadratic function?
- How do the parameters of a quadratic function affect the shape of its graph?

C
- How can you express a quadratic function in three different ways?
- What are the advantages and disadvantages of the different forms of a quadratic function?

D
- What makes one quadratic form better than another?

FORM

ATL Critical-thinking

Apply existing knowledge to generate new ideas or processes

8.2

9.2

11.2

Statement of Inquiry:

Representing patterns with equivalent forms can lead to better systems, models and methods.

You should already know how to:

• interpret graphs of linear functions	**1** In the linear function $y = mx + c$, what do m and c represent? **2** From the graph, find: **a** the x-intercept **b** the y-intercept **3 a** Determine if the gradient is positive or negative. **b** Find the equation of this line.
• factorize a quadratic expression	**4** Factorize: **a** $x^2 + 5x + 6$ **b** $2x^2 - 3x - 2$ **c** $x^2 - 49$ **d** $3 - 2x - x^2$

F Quadratic functions: standard form

- What shape is the graph of a quadratic function?
- How do the parameters of a quadratic function affect the shape of its graph?

You are surrounded by many different shapes which can be classified according to their mathematical properties. One of the most common shapes is an arch. Arches can be seen everywhere, from the shape of a banana to mouth guards, from water fountains to bridges and buildings, and in trajectories, such as the path of a basketball.

A **trajectory** is the flight path of a moving object.

Reflect and discuss 1

The photograph shows the trajectory of a basketball. The trajectory is a curve, called a parabola.

If you knew just one of the ball's positions in the air, could you tell whether or not the ball would go in the basket? What if you knew two of its positions, or three?

How many positions do you think you would need to know to be sure that the ball would go in the basket? Make a good guess. You will return to this question later.

Exploration 1

Graph the quadratic function $y = x^2$. The graph is a parabola.

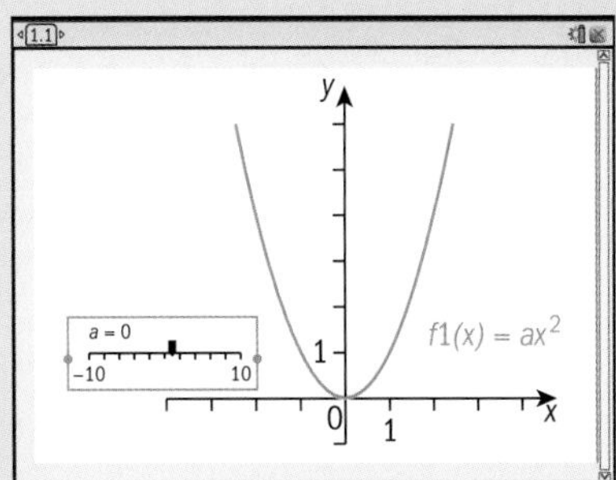

1 Graph $y = ax^2$ for a few values of a between -10 and $+10$. Include some non-integer values.

2 The parabola in the graph of $y = x^2$ is concave up ∨. The parabola of the basketball's trajectory is concave down ∧.

  a Find the values of a for which the parabola is concave up.

  b Find the values of a for which the parabola is concave down.

  c Find the values of a that make the parabola narrower than the graph of $y = x^2$.

  d Find the values of a that make the parabola wider than the graph of $y = x^2$.

3 Graph $y = x^2 + c$ for different values of c between -10 and $+10$. Include some non-integer values.

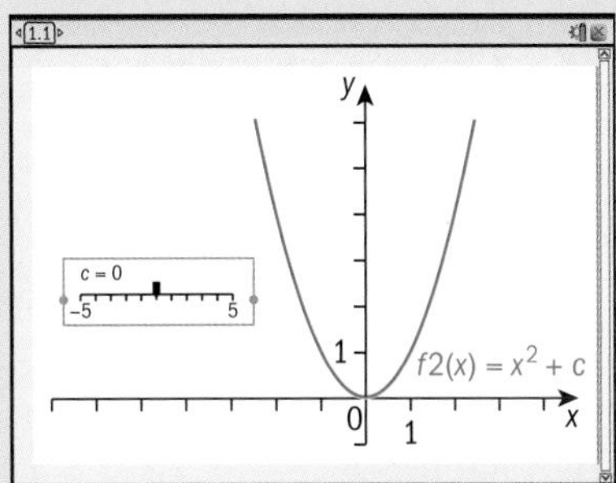

Describe the effect of changing the value of c.

4 Graph $y = ax^2 + bx$ for different values of b between -10 and $+10$. Include some non-integer values. Describe the effect of changing the value of b. (Leave $a = 1$ for the moment.)

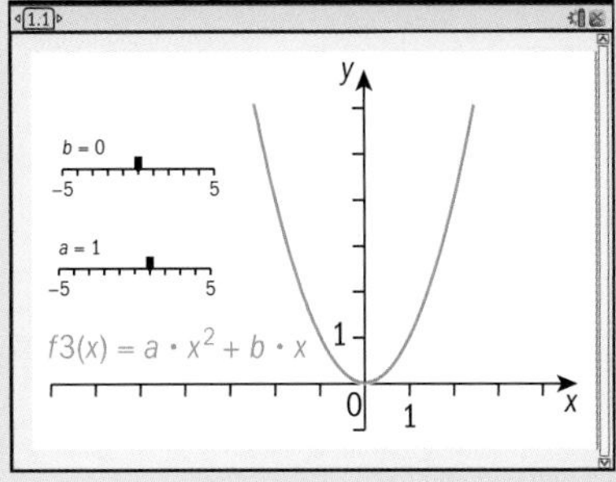

▶ Continued on next page

Tip

If your graphing program or GDC has a slider bar function, you can insert a slider to change the values in the functions easily.

Tip

A parabola can resemble a smile or a frown; could this help you remember the values of a for which the parabola will be concave down or concave up?

5 Graph the linear function $y = bx$ for different values of b. Describe how the graph changes for different values of b.

6 Go back to the parabola $y = ax^2 + bx$. Give a any value except 0. Keep a fixed and change b. Describe how the shape of the parabola changes for different values of b.

7 Use your findings to describe the similarities and differences in the shape of the graphs of each pair of quadratic functions.

a $y = 2x^2 + x + 1$; $y = 8x^2 + x + 1$ **b** $y = 3x^2 - 1$; $y = -3x^2 - 1$

c $y = 10x^2$; $y = \frac{1}{10}x^2$ **d** $y = x^2 + 3$; $y = x^2 - 1$

e $y = x^2 + x - 1$; $y = x^2 - x + 1$

Try to answer these questions without drawing the graphs. You can use technology to check your answers.

The **standard form** of a quadratic function is $y = ax^2 + bx + c$, where a, b and c are real numbers, and $a \neq 0$.

A **parameter** of a function defines the form of its graph.

- The graph of a linear function $y = mx + c$ is always a straight line. The parameter m defines its gradient; the parameter c defines its y-intercept.
- The graph of a quadratic function $y = ax^2 + bx + c$ is always a parabola. The parameters a, b and c define whether the parabola is slim or wide, concave up or concave down, and its y-intercept.

The parameters a, b and c are called the **coefficients** of the quadratic function. Parameter a, the coefficient of x^2, is called the leading coefficient. Parameter c is called the constant.

Quadratic and linear functions belong to the family of functions called **polynomial functions**. These are functions where the variable x has only positive integer exponents.

A polynomial is a mathematical expression involving a sum of powers of one or more variables multiplied by coefficients, for example, $x^4 - 3x^2 + 1$. The **degree** of a polynomial function is the value of its largest exponent of x. A linear function is a polynomial function of degree 1. A quadratic function is a polynomial function of degree 2.

A constant polynomial has degree 0. For example, the constant function $y = 2$ can be written $y = 2x^0$, since $x^0 = 1$.

ATL

Reflect and discuss 2

- Why is the parameter a in the quadratic function not allowed to equal 0? What type of function would result if $a = 0$?
- Why do you think parameter c is called the constant?
- One ordered pair of coordinates defines a 0-dimensional space, or point, for example: (2, 7). Two points define a 1-dimensional space, or straight line, for example: (1, 3) and (5, 2).

  How many points do you think are required to define a 2-dimensional space, such as the graph of a quadratic function?

A straight line has a constant gradient. A parabola does not. In fact, **each** point on the parabola has its own gradient. The point where its gradient changes from positive to negative, or vice versa, is called the **vertex** or turning point of the parabola.

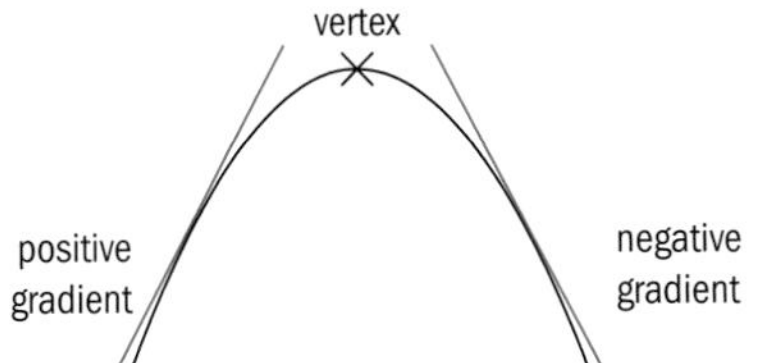

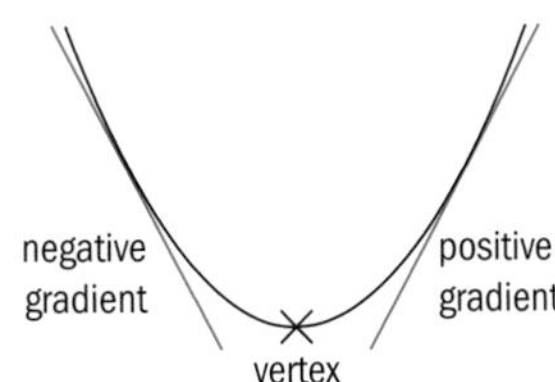

A concave down parabola has a maximum turning point.
$y = ax^2 + bx + c,\ a < 0$

maximum turning point

A concave up parabola has a minimum turning point.
$y = ax^2 + bx + c,\ a > 0$

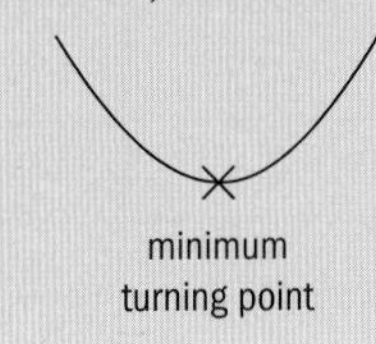

ATL

Exploration 2

Here is the graph of $y = x^2 - 2x - 3$.

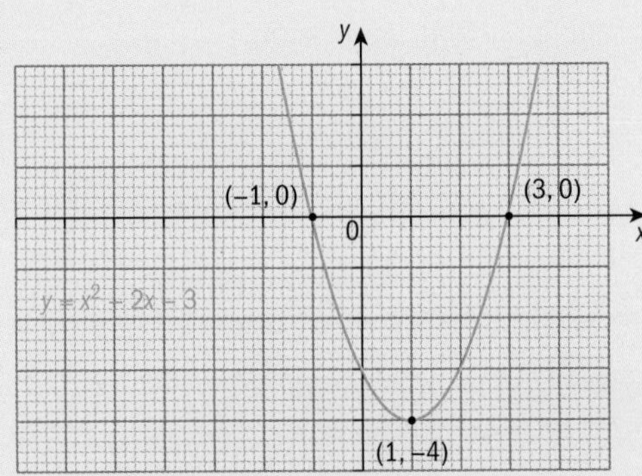

1 Describe the symmetry of the parabola.

2 State the equation of the axis of symmetry.

3 State the coordinates of the vertex.

4 Describe the relationship between the equation of the axis of symmetry and the x-coordinate of the vertex.

▶ Continued on next page

5 State the x-intercepts. Determine the relationship between the x-intercepts x_1 and x_2 of the function and the x-coordinate of the vertex x_v. Write this relationship as a formula: $x_v =$ ______.

6 Graph the following quadratic functions. Test that your formula from step **5** works for these graphs.

a $y = x^2 - 4x - 5$ **b** $y = x^2 + 6x - 7$ **c** $y = -x^2 + 2x + 10$

7 All the quadratic functions in step **6** have leading coefficient $a = 1$ or -1. Now test your formula from step **5** on the following quadratic functions whose leading coefficient $a \neq \pm 1$.

a $y = 2x^2 + 4x - 1$ **b** $y = \frac{1}{2}x^2 + 2x - 5$ **c** $2x^2 + 5x - 3$

8 Copy and complete this table with the parameters a and b and the x-coordinate of the vertex for all the quadratic functions in steps **6** and **7**.

Quadratic function	Parameters a and b	x-coordinate of the vertex, x_v
$y = x^2 - 2x - 3$	$a = 1; b = -2$	$x_v = 1$

Find a pattern relating x_v to a and b. Write this relationship as a formula.

9 Create your own quadratic functions, and test your formula in your examples.

10 Explain how to find the y-coordinate of the vertex, when you know its x-coordinate.

11 Factorize the quadratic expressions in step **6**. Describe the relationship between the x-intercepts and the factors of the quadratic expression. Explain why this relationship holds.

12 Determine how to find the y-intercept of any quadratic equation of the form $y = ax^2 + bx + c$.

The *x-intercepts* of a function are the x-coordinates of the points where the graph crosses the x-axis. They are also called the *zeros of the function*.

Reflect and discuss 3

- What do you think happens if a quadratic function isn't factorizable? Is it possible for it to still have x-intercepts? If so, how would you find them?
- Do you think it's possible for a quadratic function to have no x-intercepts? What would it look like? When do you think this would happen?

The main characteristics of the graph of a quadratic function are:

- x-intercepts and y-intercepts
- axis of symmetry
- vertex

Properties of quadratic functions

For a quadratic function $f(x) = ax^2 + bx + c$, $a \neq 0$:

- the x-coordinate of the vertex is $-\frac{b}{2a}$
- the equation of its axis of symmetry is $x = -\frac{b}{2a}$
- the coordinates of its vertex are $\left(-\frac{b}{2a}, f\left(-\frac{b}{2a}\right)\right)$
- the y-intercept is $(0, c)$

For a quadratic function $f(x)$ with x-intercepts x_1 and x_2:

- the x-coordinate of the vertex is $x_v = \frac{x_1 + x_2}{2}$
- the y-coordinate of the vertex is $f(x_v)$

Practice 1

For each quadratic function in questions **1** to **6**:

i find the coordinates of the vertex

ii find the equation of its axis of symmetry

iii determine whether the function is concave up or concave down

iv find the y-intercept

v if the function is factorizable, find the x-intercepts

vi draw a sketch of the quadratic function using your results from **i** to **v**.

Use your GDC to check your results only after you have worked them out.

1 $y = x^2 - x - 6$

2 $y = -x^2 + 2x - 4$

3 $y = x^2 - 4x - 2$

4 $y = -8x^2 + 16x - 11$

5 $y = -2x^2 + 20x - 51$

6 $y = 3x^2 - 6x + 1$

Problem solving

7 Write a concave up quadratic function with x-intercepts -2 and 3.

8 Write a quadratic function with axis of symmetry $x = 4$.

You can use graphs of quadratic functions to solve real-world problems that can be modelled as parabolas. For example, during a baseball match a player hits the ball at a height of 1 m.

The height h (meters) of the ball at time t (seconds) can be modelled by the quadratic function $h = -5t^2 + 14t + 1$. You can graph this function to find the maximum height the ball reaches, and how many seconds it takes before the ball hits the ground.

- Represent time on the x-axis, since time is the independent variable. Time cannot be a negative value, so $t \geq 0$. At $t = 0$, the height of the ball is 1 m.
- Represent height on the y-axis, since height is the dependent variable. The ball will always be above the ground, or on the ground, so $h \geq 0$.

The time does not depend on the height, so time is the independent variable.

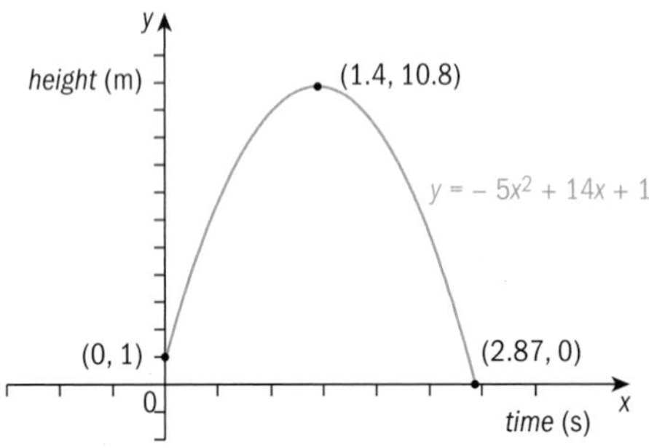

From the graph you can see the maximum height of the ball is 10.8 m and it hits the ground again after 2.87 seconds.

How could you find the times when the ball's height is 8 m? Add to your graph the function $y = 8$. Using a GDC, the points of intersection of the two graphs give the two times: 0.65 seconds and 2.2 seconds, both rounded to 1 d.p.

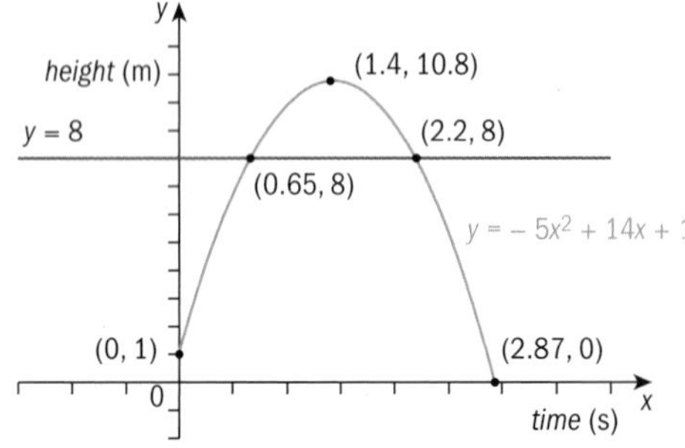

ATL

Practice 2

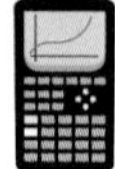

Graph the functions to answer these questions. When using a GDC, first set up a reasonable viewing window.

Problem solving

1 The US Food and Drug Administration uses this mathematical model for the number of bacteria B in food refrigerated at temperature T (°C):

$$B = 20T^2 - 20T + 120$$

The temperature in a particular refrigerator can be set anywhere between −2°C and 14°C. Determine the temperature where the number of bacteria will be at a minimum.

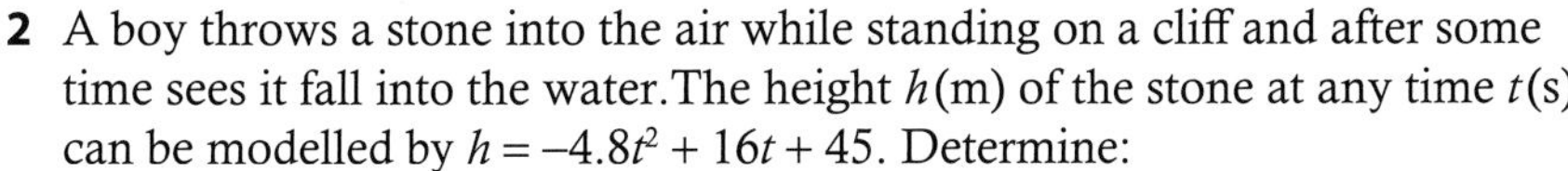

2 A boy throws a stone into the air while standing on a cliff and after some time sees it fall into the water.The height h(m) of the stone at any time t(s) can be modelled by $h = -4.8t^2 + 16t + 45$. Determine:

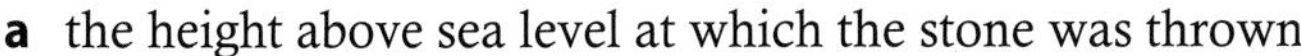

a the height above sea level at which the stone was thrown

b the time it takes the stone to reach its maximum height, and the maximum height it reaches

c the time it takes the stone to hit the water.

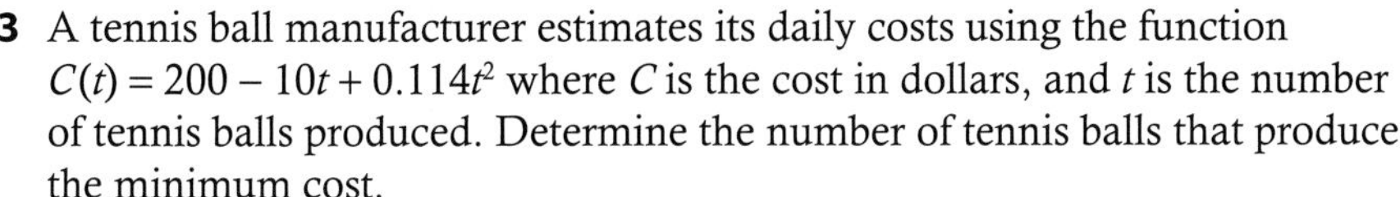

3 A tennis ball manufacturer estimates its daily costs using the function $C(t) = 200 - 10t + 0.114t^2$ where C is the cost in dollars, and t is the number of tennis balls produced. Determine the number of tennis balls that produce the minimum cost.

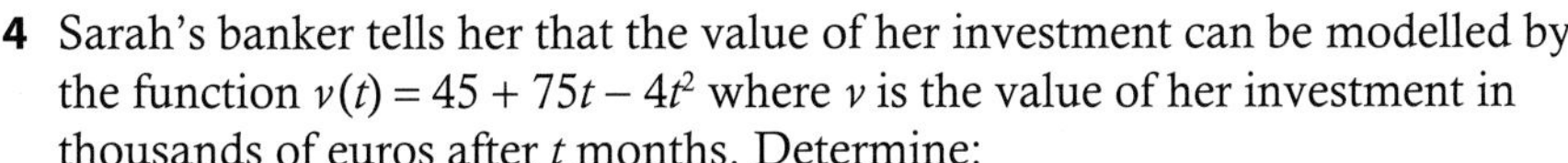

4 Sarah's banker tells her that the value of her investment can be modelled by the function $v(t) = 45 + 75t - 4t^2$ where v is the value of her investment in thousands of euros after t months. Determine:

a the initial amount that Sarah invested

b how many months it takes for Sarah's investment to reach maximum value.

5 A company's weekly profit P from selling x items can be modelled using the function $P(x) = -0.48x^2 + 38x - 295$. Determine the number of items the company needs to sell for maximum weekly profit.

C Algebraic forms of a quadratic function

- How can you express a quadratic function in three different ways?
- What are the advantages and disadvantages of the different forms of a quadratic function?

The quadratic function $y = x^2 - x - 2$ factorizes to $y = (x + 1)(x - 2)$. Graphing either of these functions gives the same parabola:

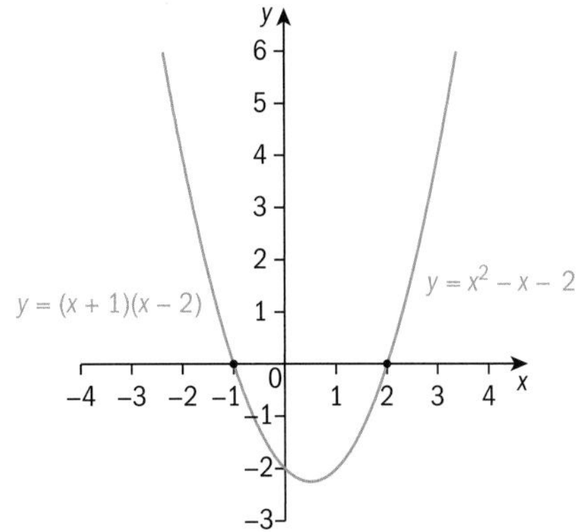

> The **factorized form** of a quadratic function is $y = a(x - p)(x - q)$, $a \neq 0$.

Reflect and discuss 4

- The x-intercepts of a function are also called its *zeros*. Why is this an appropriate name?
- How do you determine the coordinates of the vertex of a quadratic function when it is in factorized form?

Most of the quadratics you have considered so far have had two distinct zeros. Now consider the quadratic function $y = x^2 + 2x + 1$, which factorizes to $y = (x + 1)(x + 1) = (x + 1)^2$ and has graph:

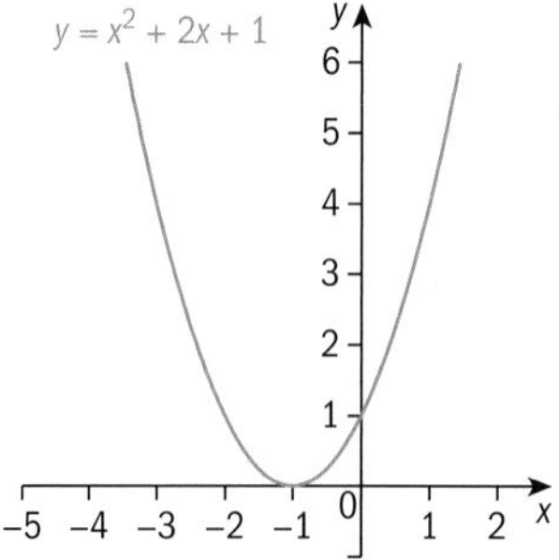

The function $y = x^2 + 2x + 1$ has only one unique factor, $(x + 1)$, and only one unique zero, $x = -1$. Does the formula you developed for the vertex of a quadratic also work when a quadratic function has only one unique factor?

Using the formula for finding the x-coordinate of the vertex using the zeros of this function, $x_v = \frac{x_1 + x_2}{2} = \frac{(-1)+(-1)}{2} = -1$

Using the formula $x_v = -\frac{b}{2a} = -\frac{2}{2(1)} = -1$

Both formulae work for this quadratic function.

Exploration 3

1 Graph these quadratic functions separately. For each, state the coordinates of the vertex. Verify that the x-coordinate of the vertex is equal to the x-intercept.

a $y = (x - 2)^2$ **b** $y = (x + 3)^2$ **c** $y = -(x - 1)^2$ **d** $y = -(x + 1)^2$

2 Write down the relationship between the x-coordinate of the vertex and the unique factor of the quadratic function.

3 Based on your findings in steps **1** and **2**, state the coordinates of the vertex of each function below. Verify your answers by graphing the functions.

a $y = (x + 2)^2$ **b** $y = (x - 4)^2$ **c** $y = -(x + 4)^2$

▶ Continued on next page

4 In all the quadratics in steps **1** to **3** the leading coefficient was either 1 or −1. Explore the relationship between the x-coordinate of the vertex and the x-intercept in the quadratic functions below with different leading coefficients. Write down your findings.

a $y = 2(x-3)^2$ **b** $y = -3(x+2)^2$ **c** $y = -2(x-2)^2$

5 The general form of a quadratic function with only one unique x-intercept is $y = a(x-h)^2$. Write down the coordinates of the vertex and the x-intercept in terms of h.

When a quadratic function has only one unique solution, its graph intercepts the x-axis at one point, the vertex of the graph. The function has a repeated factor and, therefore, a repeated zero at this point.

Graphing $y = (x-2)^2$ and $y = (x-2)^2 + 4$ on the same axes shows that adding 4 to the function translates the graph 4 units in the positive y direction. The graph of $y = (x-2)^2 + 4$ has vertex (2, 4) and no x-intercepts.

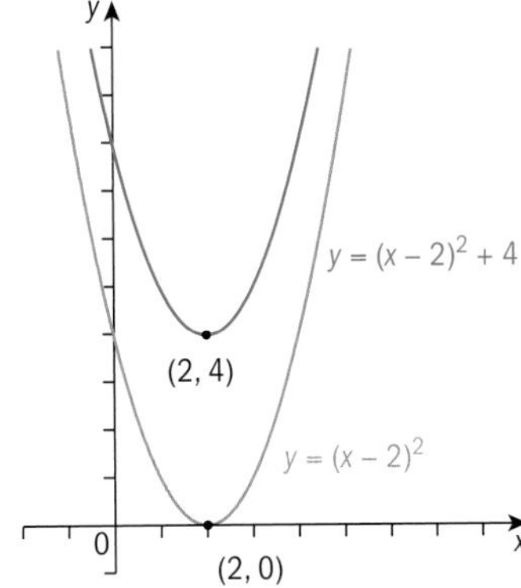

Graphing $y = (x-2)^2$ and $y = (x-2)^2 - 3$ on the same axes shows that subtracting 3 from the function translates the graph 3 units in the negative y direction. The graph of $y = (x-2)^2 - 3$ has vertex (2,−3) and two x-intercepts.

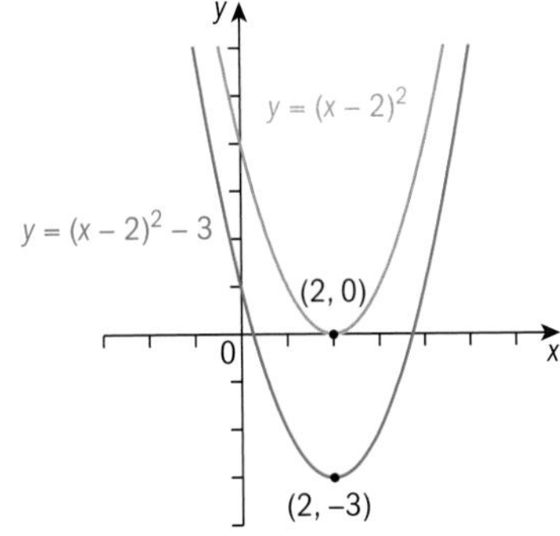

The quadratic functions $y = (x-2)^2 + 4$ and $y = (x-2)^2 - 3$ are written in vertex form.

The **vertex form** of a quadratic function is $y = a(x-h)^2 + k$, $a \neq 0$, where (h, k) is the vertex.

Practice 3

For **1** to **6**, state the coordinates of the vertex of each quadratic function and determine if the quadratic is concave up or concave down.

1 $y = (x - 3)^2 + 1$

2 $y = -(x - 4)^2 - 3$

3 $y = 2(x + 1)^2 - 1$

4 $y = -3(x + 2)^2 + 1$

5 $y = -x^2 + 2$

6 $y = 3x^2 - 1$

Problem solving

7 A quadratic function with leading coefficient 2 has a repeated zero at $x = -3$. Write the function in vertex form.

Converting quadratic forms

To convert from factorized form to standard form, expand the brackets:

$$y = (x + 3)(x - 2) \rightarrow y = x^2 + x - 6$$

If you can factorize a quadratic function, you can convert it from standard form to factorized form:

$$y = x^2 - x - 2 \rightarrow y = (x + 1)(x - 2)$$

To convert from vertex form to standard form, expand and simplify:

$$\begin{aligned} y &= (x - 1)^2 + 2 \\ &= (x - 1)(x - 1) + 2 \\ &= x^2 - 2x + 1 + 2 \end{aligned}$$

$$\Rightarrow \quad y = x^2 - 2x + 3$$

How do you convert from standard form to vertex form? (The next Exploration will help you answer this question.)

Exploration 4

1 Expand these expressions:

a $(x + 2)^2$ **b** $(x - 3)^2$ **c** $(x + 1)^2$

d $(x - 1)^2$ **e** $(x + 3)^2$ **f** $(x - 2)^2$

2 Look at your results in step **1**. In $(x + p)^2 = x^2 + bx + c$, find the relationships between p and b, and between p and c.

3 For these quadratic functions, choose values of c so that they factorize into two identical factors $(x + p)(x + p)$, in other words: the square of a linear factor. Write down the factorization for each one.

a $x^2 - 4x + c$ **b** $x^2 + 4x + c$

c $x^2 + x + c$ **d** $x^2 - x + c$

4 Write a general rule for what you did in step **3**. For a quadratic function $x^2 + bx + c$, express c in terms of b for the quadratic to be the square of a linear factor.

In steps **3** and **4** of Exploration 4 you were **completing the square**. Given the first two terms of a quadratic expression, you found c so that the quadratic factorizes into the square of a linear factor. For this, $c = \left(\frac{b}{2}\right)^2$, where b is the coefficient of the x term.

You can use completing the square to convert a quadratic function from standard form to vertex form.

Example 1

Write $y = x^2 + 2x - 2$ in vertex form.

$y = x^2 + bx + c$

$\left(\frac{b}{2}\right)^2 = \left(\frac{2}{2}\right)^2 = 1^2 = 1$ — Use $\left(\frac{b}{2}\right)^2$ to complete the square for the x^2 and x terms.

$x^2 + 2x + 1 = (x + 1)^2$

$y = x^2 + 2x - 2$ — Write the completed square in the right hand side. As this adds 1, subtract 1 at the end.

$\quad = (x^2 + 2x + 1) - 2 - 1$

$y = (x + 1)^2 - 3$ — You could check by expanding and simplifying.

Example 2

Sketch the graph of $y = (x + 1)^2 - 3$.

Vertex is $(-1, -3)$ — For $y = a(x - h)^2 + k$ the vertex is (h, k).

Axis of symmetry is $x = -1$

$y = (x + 1)^2 - 3 = x^2 + 2x - 2$

y-intercept $= -2$ — In standard form, y-intercept $= c$.

Graph is concave up. — Positive coefficient of x^2.

When $x = 1$, $y = 2^2 - 3 = 1$ — Find some other points on the curve by substituting a few values for x.

When $x = -3$, $y = (-2)^2 - 3 = 1$

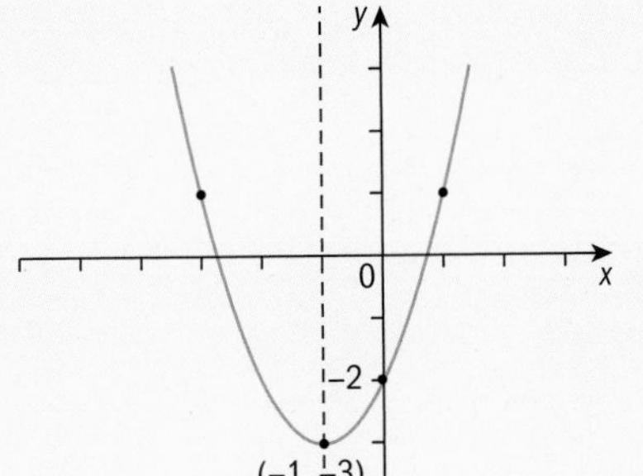

Connect the points with a smooth parabolic curve.

Reflect and discuss 5

Consider the two equivalent quadratic functions:

$y = x^2 + 2x - 2$ in standard form and $y = (x + 1)^2 - 3$ in vertex form.

- Which form of the quadratic function gives you more immediate information to use in sketching its graph?
- In Example 2, the y-intercept was found by expanding the original function. What is another way of finding the y-intercept? Which way is easier?
- How many points are necessary to determine a unique quadratic function or draw its graph? Explain.

Practice 4

1 The quadratic functions below are given in standard form.

  i Convert each function into vertex form.

  ii Sketch the graph of the quadratic by using the vertex and two other points on the quadratic.

  a $y = x^2 + 2x + 1$ **b** $y = x^2 - 4x - 2$ **c** $y = 3 + 6x + x^2$

  d $y = -x^2 - 6x + 1$ **e** $y = x^2 - 2x + 3$ **f** $y = 1 - 2x + x^2$

  g $y = x^2 + x + 2$ **h** $y = x^2 - x - 1$ **i** $y = 1 - 3x - x^2$

> **Tip**
>
> Find the y-intercept, and substitute an 'easy' value of x to find a third point.

2 For the quadratics below:

  i find the vertex using the formula for the x-coordinate of the vertex

  ii find two other points on the quadratic

  iii sketch the graph of the quadratic using the points you have found.

  a $y = 1 - 4x - x^2$ **b** $y = 2x^2 - 4x + 1$ **c** $y = 1 - 6x - 3x^2$

Problem solving

3 Match each quadratic function with its graph.

  a $y = (x + 1)^2 - 3$ **b** $y = x^2 + 2$ **c** $y = (x - 2)^2 + 1$

  d $y = -x^2 + 2$ **e** $y = x^2 - 3x - 1$ **f** $y = 2 - x - x^2$

A

B

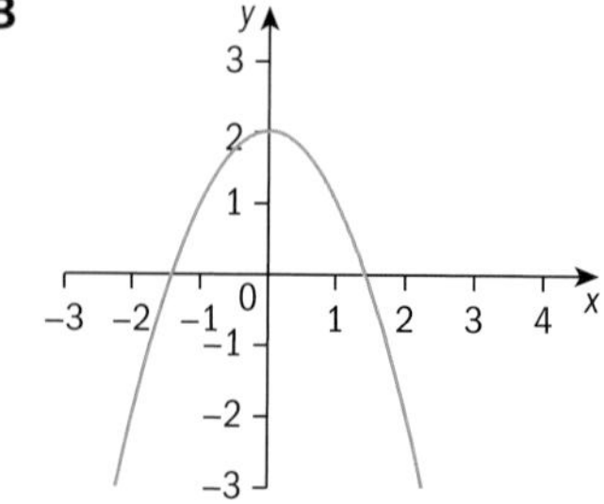

C

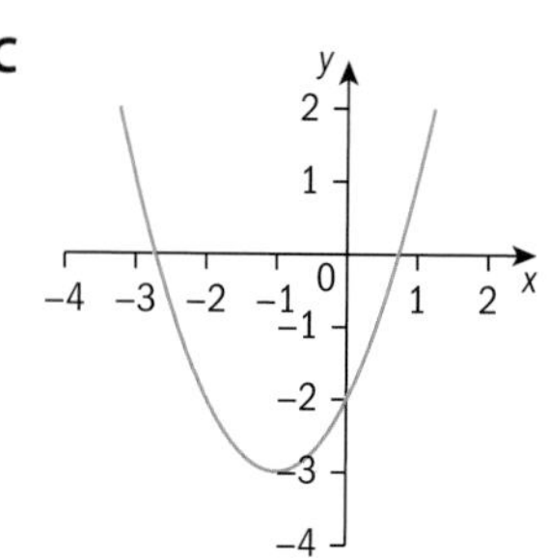

D
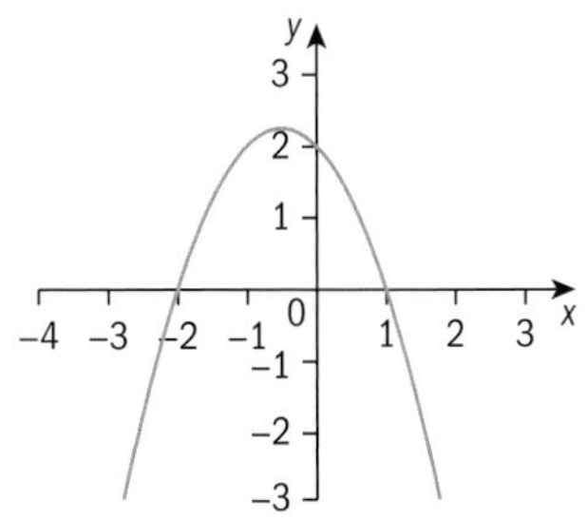

E
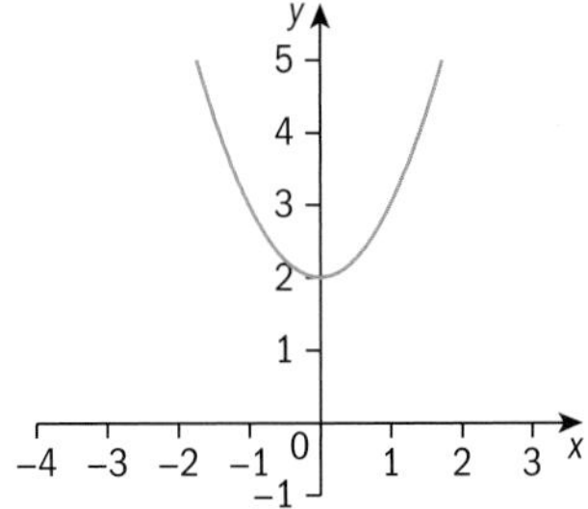

F
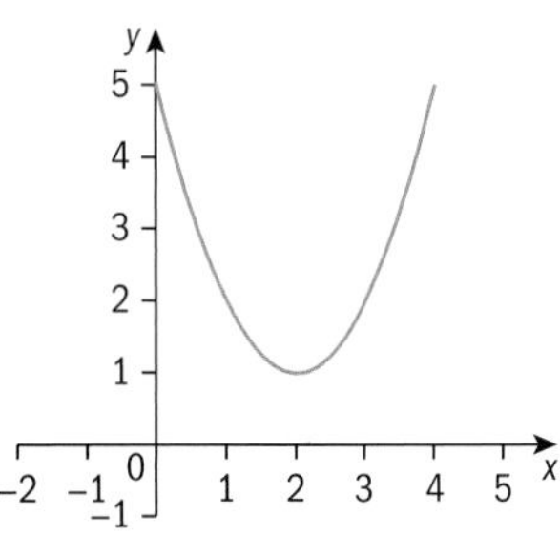

Reflect and discuss 6

- Find a quadratic function that goes through the point (0, 1).
- Find a quadratic function whose zeros are $x = 0$ and $x = 4$.
- Find a quadratic function with vertex (1, 2) and one of its zeros (−1, 3).
- Compare your answers with others. Did you all find the same functions? Explain.

Solving real-life problems

To solve real-life problems, you may have to derive the quadratic function from the given information, and then decide on the best form of the quadratic to use to answer the question.

Example 3

You have 100 meters of fencing to enclose a rectangular plot.

a Find the plot's maximum possible area.

b Determine the dimensions dimensions give the maximum area.

a $P = 2(l + w) = 100$ [1]

$A = lw$ [2]

Write equations for the perimeter and area of a rectangle.

From [1]:

$2(l + w) = 100$

$l + w = 50$

$l = 50 - w$

Write l in terms of w.

$A = (50 - w)w = 50w - w^2$

Substitute into [2], to get an equation linking A and w.

Area is given by the quadratic function $50w - w^2$, which has a maximum value at its vertex.

▶ Continued on next page

Method 1–finding the vertex using the factorized form

$50w - w^2 = w(50 - w) = 0$ — Find the zeros of the function.

$w = 0,\ w = 50$

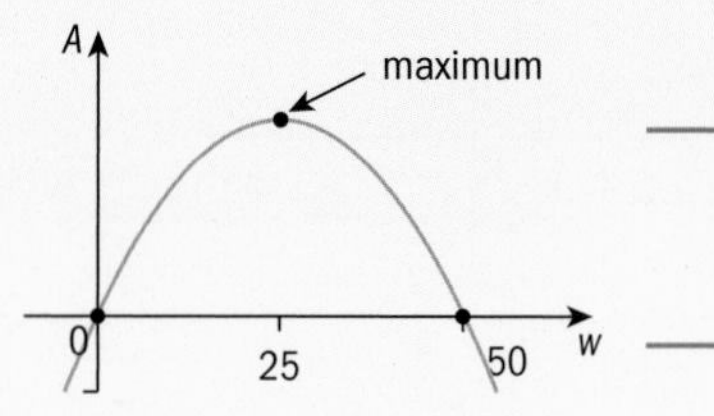

Sketch the graph of the function and identify its maximum point.

The maximum area of the rectangular plot occurs when $w = 25$.

When $w = 25$:

$A = 25(50 - 25) = 625$ — Calculate the maximum area.

Maximum area = 625 m^2

Method 2–finding the vertex using the standard form

$A = w(50 - w) = 50w - w^2$

$$x_v = -\frac{b}{2a} = -\frac{50}{2(-1)} = 25$$ Find the x-coordinate of the vertex.

When $w = 25$:

$A = 25(50 - 25) = 625$

Maximum area = 625 m^2

Method 3–finding the vertex using vertex form

$w^2 - 50w = (w - 25)^2 - 625$ — Complete the square for $w^2 - 50w$.

Hence $A = -(w^2 + 50w) = -(w - 25)^2 + 625$

Vertex = (25, 625)

Maximum area = 625 m^2

b When $w = 25$, $l = 50 - w = 25$. The dimensions that produce the maximum area are width = length = 25 m (a square).

Reflect and discuss 7

When sketching graphs of quadratic functions, what are the advantages and disadvantages of:

- standard form
- factorized form
- vertex form?

Practice 5

In these questions, derive the quadratic function that best fits the situation and use the most efficient method for answering the question.

Problem solving

1 You have 300 m of fencing to enclose a rectangular plot along the side of a river, as shown.

Determine the maximum area that you can enclose and the dimensions of the plot.

2 You have a total of 200 m of fencing to make two equal, adjacent, rectangular plots. Determine the dimensions of each plot, such that the total area enclosed is as big as possible.

3 Xixi wants to fence off a rectangular exercise area for her dog, using the house as one side of the area. She has 10 m of fencing. Find the maximum possible exercise area.

4 The perimeter of an athletics track is 0.4 km. The track has two parallel sides and a semicircle at each end. Determine the exact values of x and r that would maximize the area of the rectangular part of the track field, and use these dimensions to find the area of the entire track field to the nearest square meter.

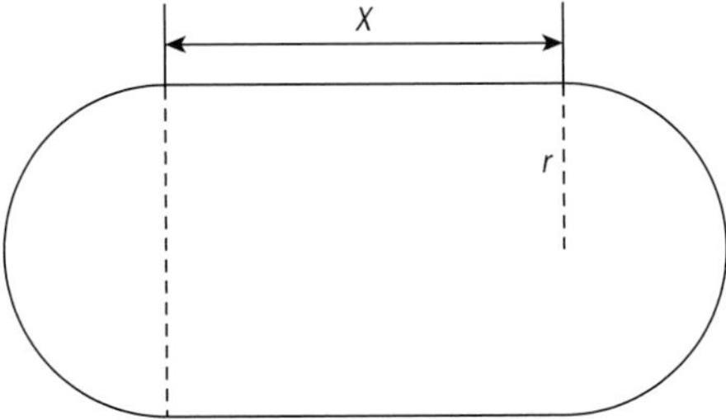

D Form and function

- What makes one quadratic form better than another?

Given a quadratic equation, you can sketch its graph by finding its x- and y-intercepts and vertex. If you start with a parabola, how can you find its quadratic function?

This photograph of water fountains is superimposed upon coordinate axes. Consider the parabolic shape of the 2nd fountain stream from left.

The x-intercepts and y-intercept, which is also the vertex, are labelled. Is this enough information to find the quadratic function modelling the parabolic fountain stream?

Tip

Some GDCs and graphing software have the functionality of superimposing a picture on a pair of coordinate axes.

The general form of a quadratic function with vertex is (h, k) is $y = a(x - h)^2 + k$.

Substituting the coordinates of the vertex (0, 7) for h and k gives $y = a(x - 0)^2 + 7$ or $y = ax^2 + 7$.

How can you find the leading coefficient a? It must have a negative value since the parabola is concave down. You can use one of the two other points given: (−4.4, 0) or (4.4, 0). Substituting the x- and y-values at point (4.4, 0) gives $a \times 4.4^2 + 7 = 0$. Solving for a gives:

$$a = \frac{-7}{4.4^2} = -0.361 \text{ (3 d.p.)}$$

The quadratic function that models the fountain stream is $y = -0.361x^2 + 7$. You can graph this quadratic on the same coordinate axes to check:

ATL

Reflect and discuss 8

- Again, find the general form of the quadratic that models the fountain stream, but this time use the other x-intercept to find the leading coefficient a. Is it the same function?
- Does your quadratic function exactly model the parabolic fountain, or is it an approximate model? Explain.
- What minimum information do you need to determine the equation of a unique quadratic function?
- When finding the equation of a quadratic function, under what circumstances would you use:

  i standard form

  ii vertex form

  iii factorized form?

Reflect and discuss 1 asked: How many positions do you think you would need to know to be sure that the ball would go into the basket? The correct answer is 'three'.

A point in space has no dimension, and needs one point to determine it.

A line has one dimension and needs two points to determine it.

A curve has two dimensions and needs three points to determine it.

Can you guess how many points determine a unique object in three-dimensional space? How many points would you need to uniquely define an object in a four-dimensional universe?

You can project 3D objects onto a 2D space, such as with television transmission. In the same way, 4D objects can be projected onto a 3D space.

Exploration 5

The Golden Gate Bridge in San Francisco, USA is a suspension bridge spanning the channel between San Francisco Bay and the Pacific Ocean. The diagram shows the central section of the bridge. The suspension cable is a parabola.

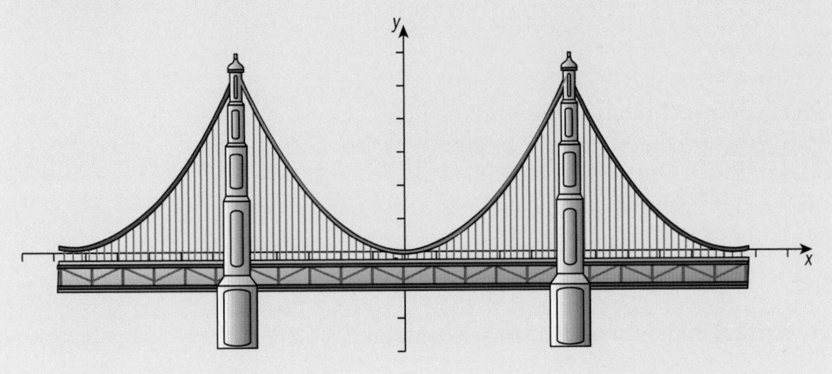

▶ Continued on next page

1 Research the heights of the towers and the length of the central section.

2 Trace the parabola. Draw a pair of coordinate axes, with the y axis as the axis of symmetry of the parabola.

3 Use the dimensions you found from step **1** to determine a suitable scale for the x and y axes.

4 Using your scale, find the coordinates of the vertex, and the x- and y-intercepts.

5 Using your quadratic function and another point on the curve, find the parameter a.

6 Compare your model with others. Discuss any similarities and differences.

Practice 6

1 Trace each parabola. Draw coordinate axes and decide on a suitable scale. Find a quadratic function to model the parabola.

a

b

c

d

Exploration 6

Take photos of parabolic objects you encounter in your everyday life. Use technology to find the quadratic function that defines your object, or use tracing paper over the photo.

Reflect and discuss 9

Why might it be that so many natural and built structures are parabolic?

Objective: D. Applying mathematics in real-life contexts
v. Justify whether a solution makes sense in the context of the authentic real-life situation.

In this activity you will write quadratic models that describe the given situation. You will then decide which one is better in describing the given scenarios, and explain your choice.

Activity

For a free throw shot in basketball (in US standard units):

- the free throw line is 15 ft away from the foot of the basket
- the basket is 10 ft above the ground
- the player releases the ball from an approximate height of 8 ft

1 Find **two different functions** that could model the shot into the basket. Make sure that each function satisfies the specifications.

2 Decide on the form of the quadratic function that is most appropriate for your models, and give reasons for your choice of form.

3 From your two models, select the one that you think best models a real-life shot. Explain why.

4 Your model does not take into account some factors that affect the path of the ball, or whether the shot goes into the basket. Suggest what some of these factors might be.

5 A player from the opposing team could intercept the ball on its way to the basket. Assume that an opposing player is 5 ft away from the free thrower and can jump to reach a maximum of 10 ft. Could this player intercept the ball in your model? If so, modify your function so that it models the ball reaching the basket.

Massachusetts physical education teacher James Naismith invented basketball in the late 1800s to keep his class occupied indoors in bad weather. He used a peach basket and a ball like a soccer ball.

Summary

- The **standard form** of a quadratic function is $y = ax^2 + bx + c$ where a, b and c are real numbers, and $a \neq 0$.
- The **factorized form** of a quadratic function is $y = a(x - p)(x - q)$, $a \neq 0$.
- The **vertex form** of a quadratic function is $y = a(x - h)^2 + k$, $a \neq 0$, where (h, k) is the vertex.
- The **degree** of a polynomial function is the value of its largest exponent of x. A linear function is a polynomial function of degree 1. A quadratic function is a polynomial function of degree 2. A constant function has degree 0.
- A concave down parabola has a maximum turning point.
  $y = ax^2 + bx + c$, $a < 0$

- A concave up parabola has a minimum turning point.
  $y = ax^2 + bx + c$, $a > 0$

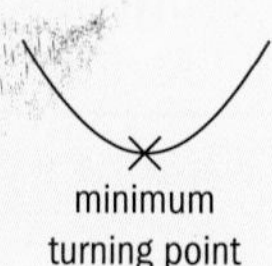

- For a quadratic function $f(x) = ax^2 + bx + c$, $a \neq 0$:
  - the x-coordinate of the vertex is $-\frac{b}{2a}$
  - the equation of its axis of symmetry is $x = -\frac{b}{2a}$
  - the coordinates of its vertex are $\left(-\frac{b}{2a}, f\left(-\frac{b}{2a}\right)\right)$
  - the y-intercept is c
- For a quadratic function $f(x)$ with x-intercepts x_1 and x_2:
  - the x-coordinate of the vertex is $x_v = \frac{x_2 + x_1}{2}$
  - the y-coordinate of the vertex is $f(x_v)$

Mixed practice

1 For each quadratic function:

i **find** the vertex, x-intercepts and y-intercept

ii **state** the axis of symmetry, and whether it is concave up or concave down

iii **sketch** the graph.

a $y = x^2 + x - 12$ **b** $y = x^2 + 7x + 12$

c $y = 2x^2 - x - 3$ **d** $y = 2 - x - 3x^2$

e $y = -6x^2 + 5x - 1$ **f** $y = 2x^2 - 9x - 5$

2 These quadratic functions are in standard form. Change each one to vertex form, and **state** the coordinates of the vertex.

a $y = x^2 - 4x + 6$ **b** $y = x^2 + 6x + 8$

c $y = x^2 + 2x - 9$ **d** $y = x^2 - 2x + 7$

e $y = x^2 + x - 5$ **f** $y = x^2 - x + 7$

Problem solving

3 Kanye has 600 m of fencing to make five adjacent pens like this:

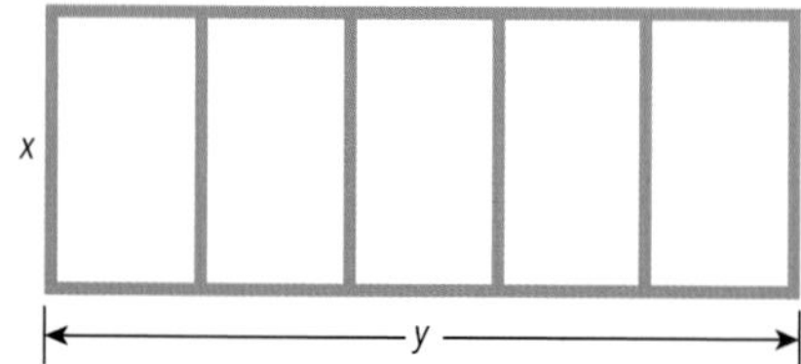

a Express the total area of the pens in terms of x.

b **Determine** the value of x that will maximize the total area.

c **Find** the maximum area.

4 Soraya has 40 m of fencing to make a rectangular play area. **Find** the dimensions for the maximum play area.

5 The population P of an animal species is modelled by the function $P(t) = -0.38t^2 + 134t + 1100$, where t is the time in months since the population was first observed. **Determine**:

a the number of months until the population is at its maximum

b the maximum population

c the number of months before the species disappears.

6 A ball is thrown up into the air, from 5 m above the ground. After 2 seconds, the ball reaches a maximum height of 9 m. It lands on the ground 5 seconds after it was thrown. **Find** a quadratic function that models this situation. Write it in standard and vertex form.

Review in context

Scientific and technical innovation

The area of a region formed by a parabola is $A = \frac{2}{3}bh$, where b is the length of the base of the parabola, and h is its height, i.e., the distance from its vertex to the base.

1 This rollercoaster is in Münich, Germany.

a Trace the outline of the parabolic section of the foreground section of track.

b **Draw** axes and use the map to determine a suitable scale.

c **Find** a suitable function to model the shape of section of track.

d **Use** the function to estimate the area under this section.

2 At right is the tunnel entrance that runs through Castle Hill in Budapest, Hungary.

a If the maximum height of the tunnel is the same as its maximum width, 9.8 meters, **determine** a quadratic model to represent the tunnel.

b **State** a geometric model that could be used to represent a truck passing through the tunnel.

c If a typical truck has a width of 2.62 m, determine the height limitation that should be put on these trucks if they are going to travel through the tunnel safely. **State** any assumptions that you made and be sure to show all of the steps in your solution.

d Because its length is almo… a truck stopped in the tunn… enough room for the passenger doo… fully. Suppose a truck with a width of 2.6… and height of 3.4 m is stuck in the right lane of the tunnel by the center line. The top of the driver's door is 0.2 m below the top of the truck, and the door opens 0.8 m to the side. **Show that** the door can open fully, showing all of your steps.

e **Determine** how close to the side of the tunnel the truck could be in **d** be and still be able to open the passenger door.

f **Describe** some issues/concerns that you think architects take into account when modelling a tunnel before its construction.

Reflect and discuss

How have you explored the statement of inquiry? Give specific examples.

Statement of Inquiry:

Representing patterns with equivalent forms can lead to better systems, models and methods.

10 Change

A variation in size, amount or behavior

Changing the problem

One strategy for solving mathematics problems is to change them from ones you can't solve into equivalent problems that you can solve.

For example, you cannot directly add fractions with different denominators, say:

$$\frac{3}{7}+\frac{2}{5}$$

But you can change this addition into an equivalent form which is possible to solve:

$$\frac{15}{35}+\frac{14}{35}=\frac{29}{35}$$

Changing between fractions, decimals and percentages changes the way you express a number.

How can you change the numbers in these calculations to make them easier to work out mentally?

0.75×24	65% of 20
$426 + 99$	$14 \div 0.07$

Changing the subject

A formula shows the relationship between quantities. You can calculate the speed of a car by substituting values into the formula: speed = distance/time or $s = d/t$.

- Calculate the speed when:

  a distance = 50km, time = 2 hours

  b distance = 80km, time = 2 hours

  c distance = 50km, time = 4 hours.

  What else changes when you change the distance and/or time?

- How can you change the formula to calculate distance, given the speed and time?

Evolutionary change

The theory of evolution describes how living species change over time. In *On the Origin of Species* (1859), Charles Darwin presented the evidence he had collected during a five-year voyage around the world, studying variation in plants and animals. He claimed that species have common origins but have adapted to different environments through natural selection.

We now know that there are differences in the genes of individuals within a species. In any environment, individuals most suited to that environment are most likely to survive and reproduce. They will then pass on their genes to their offspring. Individuals without these genes are less likely survive and reproduce in that environment.

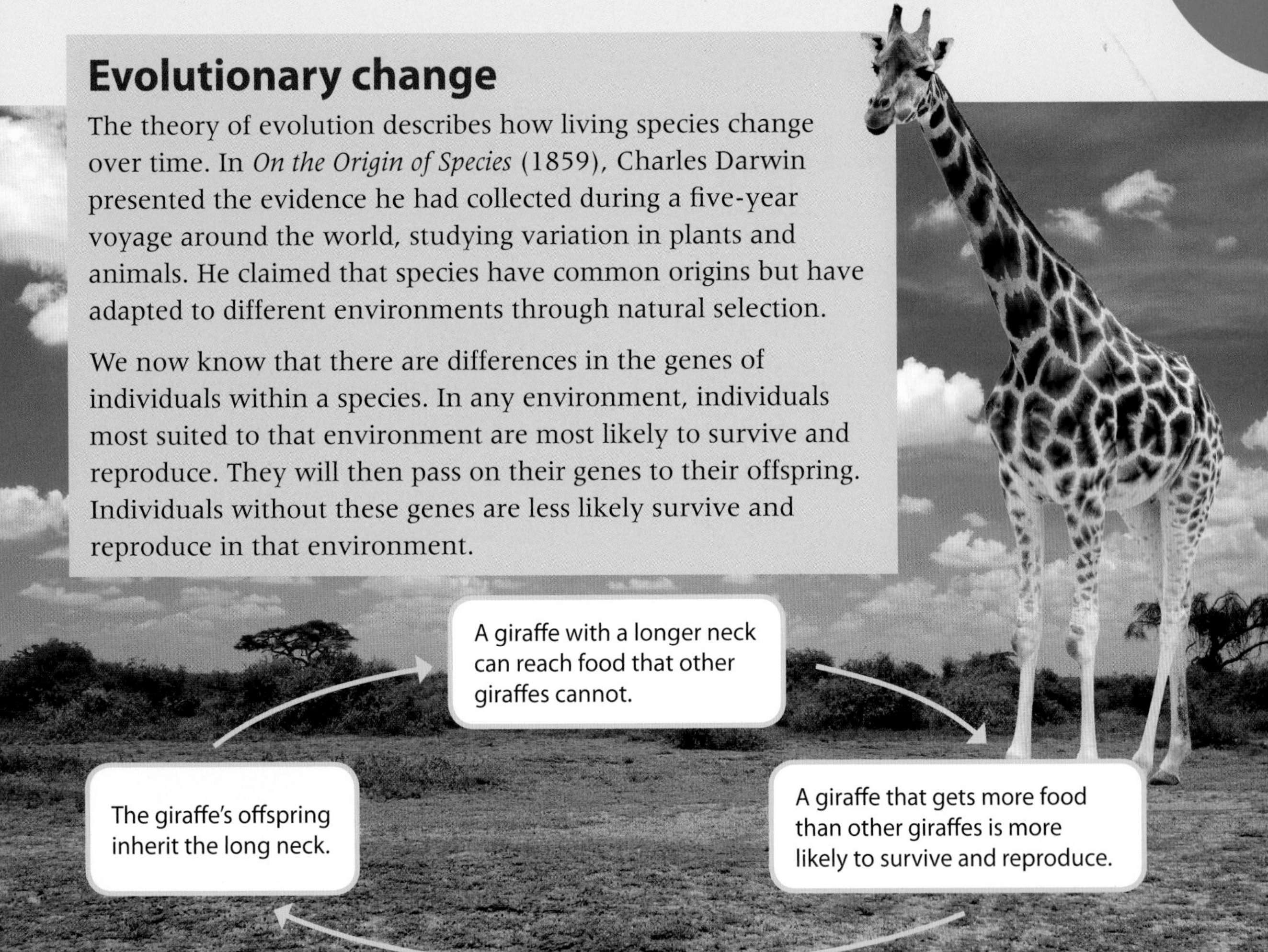

Climate change

From ice core sample data we know that the Earth's climate has changed over millions of years, between ice ages with warmer interglacial periods of about 10 000 years between them. Scientists debate the exact causes of glacial and interglacial periods, but almost all agree that the amount of carbon present in the Earth's atmosphere from burning fossil fuels – oil, coal and gas – is increasing global temperatures at a faster rate than ever before.

- Scientists analyze data to observe and predict change. How can predicting change also help us to make changes?

10.1 A frog into a prince

Global context: Orientation in space and time

FORM

Objectives

- Understanding how various parameters affect the shape and position of a graph
- Applying translations, reflections and dilations to graphs
- Describing the transformation of a function algebraically and graphically
- Describing combinations of transformations of a function algebraically and graphically
- Writing the equation of a graph following one or more transformations

Inquiry questions

F
- What transformations can be applied to lines and curves?

C
- What are the three transformations for linear and quadratic graphs?
- How can you use transformations to find the equation of a graph?

D
- If the graph of a transformed function is the same as the graph of the original function, has the function been transformed?
- If two different transformations give the graph of the same transformed function, are the transformations really different?

ATL Creative-thinking

Generating novel ideas and considering new perspectives

10.1

14.1

Statement of Inquiry:

Relationships model patterns of change that can help clarify and predict duration, frequency and variability.

You should already know how to:

• sketch graphs of linear and quadratic functions	**1** Sketch a graph of: **a** $f(x) = 2x + 3$ **b** $f(x) = x^2$
• write quadratic functions in vertex form	**2 a** Write the quadratic function $y = x^2 + 6x + 8$ in vertex form. **b** Write down the coordinates of the vertex of the graph of $y = x^2 + 6x + 8$.

F Changing graphs

- What transformations can be applied to lines and curves?

Changing the parameters of a line or quadratic curve changes its position on the coordinate plane.

For a straight line $y = mx + c$, the parameters are gradient m and y-intercept c.

For a quadratic function, $y = (x - h)^2 + k$, the parameters are h and k.

Exploration 1

1 Use your graphing software to draw the following pairs of curves on the given axes. In each case, describe briefly in words how the first curve must be transformed to match the second curve.

a $y = x^4 - 2x^3$ and $y = x^4 + 2x^3 - 2x - 1$.
Use x-values of −3 to 3 and y-values of −3 to 5.

b $y = x^3 - 13x^2 + 54x - 72$ and $y = x^3 - 4x^2 + 3x + 2$.
Use x-values of −2 to 8 and y-values of −4 to 4.

c $y = x^2 + x - 6$ and $y = 3(x + 3)(x - 2)$.
Use x-values of −4 to 3 and y-values of −20 to 6.

d $y = \sqrt{x+4}$ and $y = \sqrt{2x+4}$.
Use x-values of −4 to 4 and y-values of −1 to 4.

e $y = x^2 - \sqrt{8x}$ and $y = 2\sqrt{2x} - x^2$.
Use x-values of 0 to 4 and y-values of −6 to 6.

f $y = 0.5(x+5)\sqrt{4-x}$ and $y = \frac{1}{2}\sqrt{x+4}(5-x)$.
Use x-values of −10 to 10 and y-values of −6 to 6.

2 When describing the transformations in step **1**, you might have used everyday words like ‘shift’ and ‘stretch’. It is important to use the correct terminology to make sure your meaning is fully understood.

Research the meanings of the words *reflection, dilation* and *translation*. All three have non-mathematical meanings as well as specific mathematical meanings.

3 Describe each of the transformations from step **1** using one of the terms reflection, dilation or translation.

In this section, you will encounter only these three types of transformation. All the reflections will be reflections in either the x-axis or the y-axis. The dilations will be measured from the x- or y-axis.

> A **translation** occurs when every point on a graph moves by the same amount in the same direction.
>
> A **dilation** is a stretch or a compression. If a graph undergoes dilation parallel to the x-axis, all the x-values are increased by the same scale factor. Similarly, if it is dilated parallel to the y-axis, all the y-values are increased by the same scale factor.
>
> A **reflection** maps points on one side of a mirror line to points the same distance away but on the other side of the mirror line.

Practice 1

1 These diagrams show examples of graph transformations. In each case, describe the single transformation that has taken place to transform the red curve into the blue curve.

a

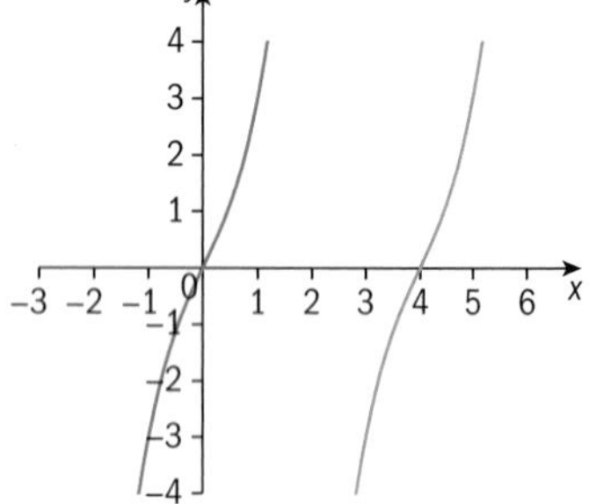

b

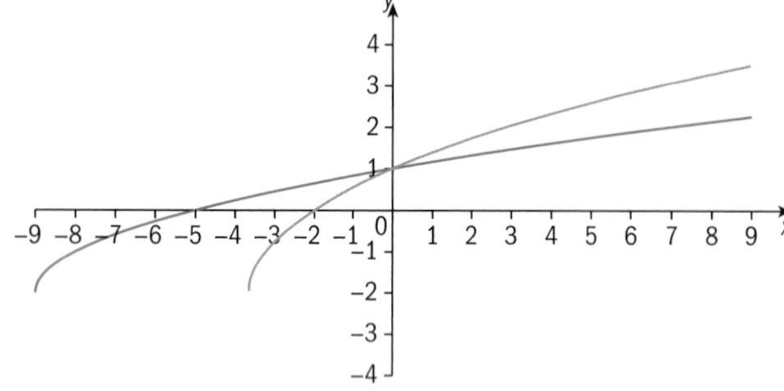

c

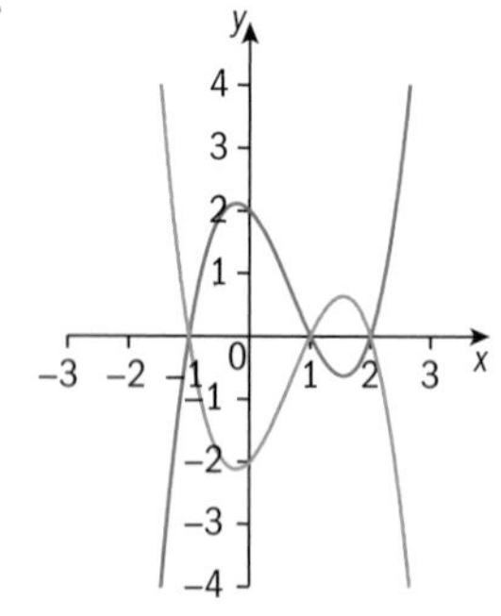

d

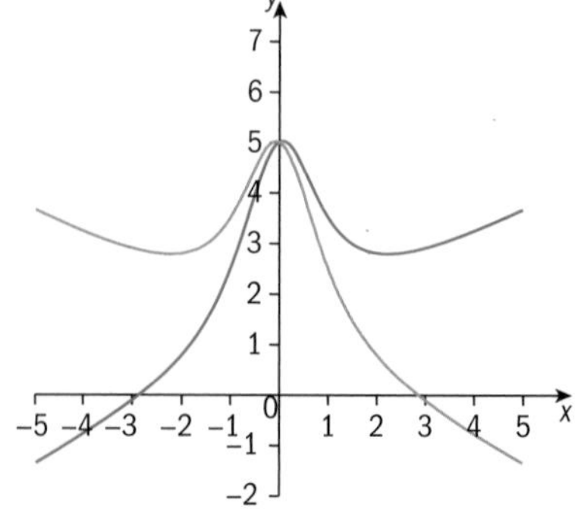

e

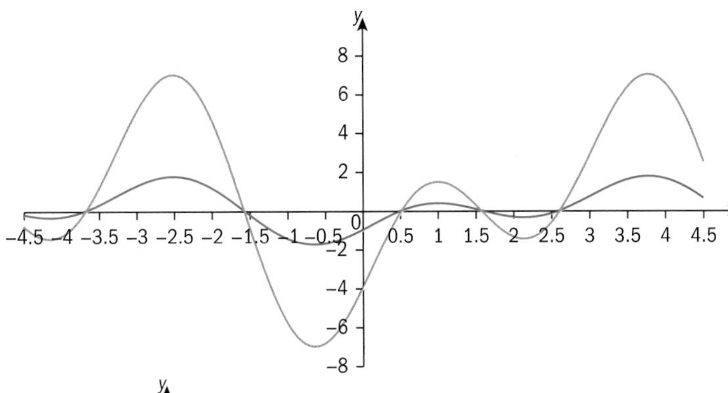

f

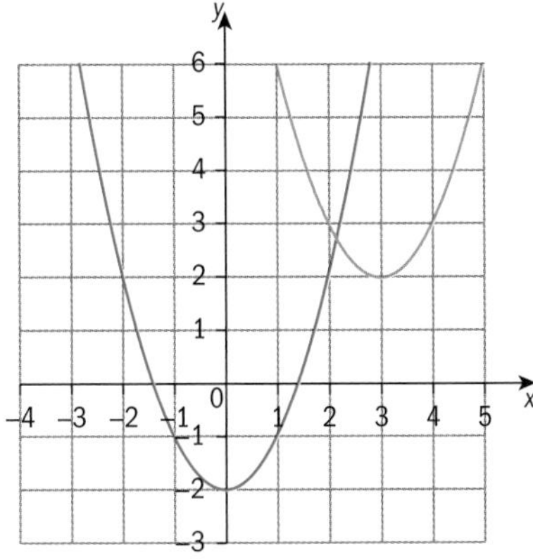

2 These diagrams show further examples of graph transformations. In each case, identify the two transformations which transform the red curve into the blue curve.

a

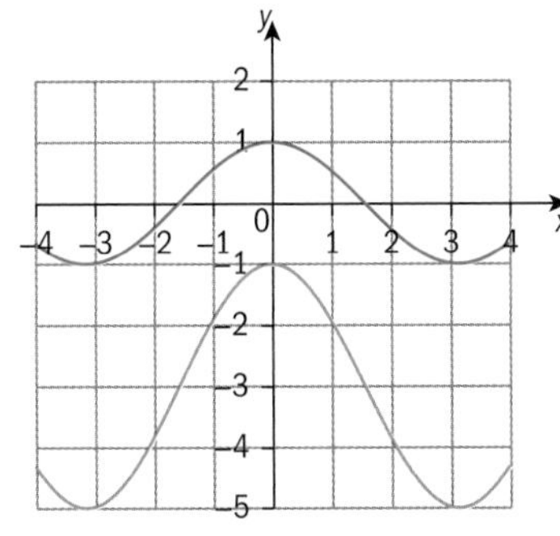

b

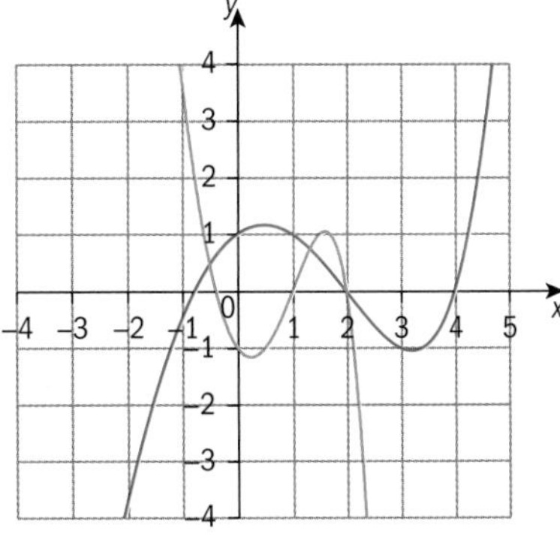

c

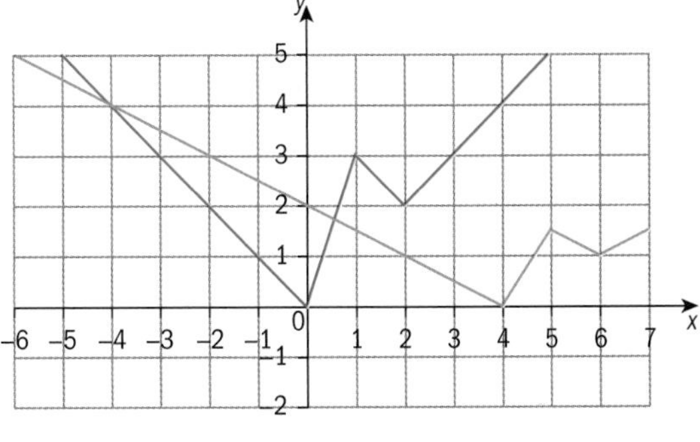

d

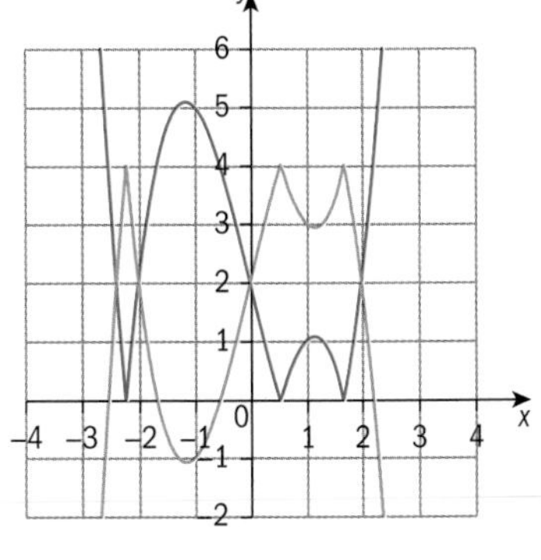

Problem solving

3 Use your GDC to draw the graphs of $y = (x - 2)^2$ and $y = (x + 2)^2$.

Ronnie says that one graph is a translation of the other.
Reggie says that one graph is a reflection of the other.
Decide whether their claims are correct.
Explain what you notice.

Reflect and discuss 1

Draw the graph of $y = f(x)$ where $f(x) = 4x$.
Using your answers to Exploration 1, predict what the graph of $y = f(x) + 4$ will look like.
Predict also what the graph of $y = f(x + 1)$ will look like.

Check using your graphical calculator. What do you notice?
Do $y = f(x + 1)$ and $y = f(x) + 4$ represent the same transformation or different transformations?

Linear and quadratic transformations

- How does transforming an equation affect its graph?

Objective: **B.** Investigating patterns
ii. describe patterns as general rules consistent with findings

In this exploration, you are asked to make observations about how varying a parameter affects the graph. Once you have observed the pattern, you should try to describe it as precisely as possible. Then you should check your rule by testing a few more examples. Sometimes it is a good idea to ask a friend if they think your rule makes sense. If it does not, perhaps you need to try rephrasing it.

Exploration 2

1 Graph the linear functions $y = x$ and $y = x + k$ on your graphing software, and insert a slider to change the value of k between −10 and 10. Start with $k = 0$.

Try different values of k to explore how changing $y = x$ to $y = x + k$ changes the graph.

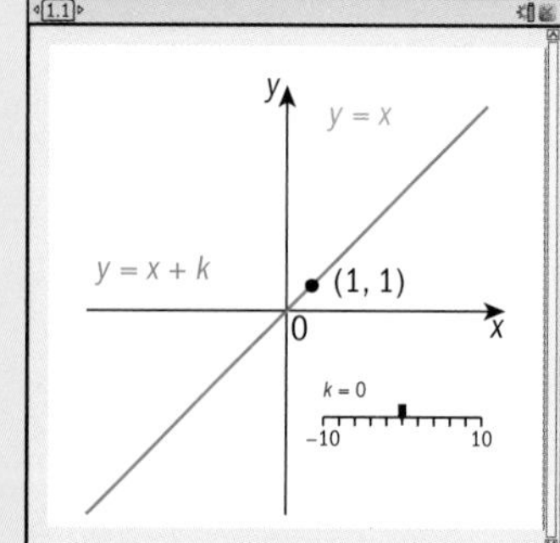

Try different values of k.

2 Repeat step **1** for three more linear functions: $y = 2x + k$, $y = -x + k$ and $y = 3x + k$

Describe how changing $y = f(x)$ to $y = f(x) + k$ changes the graph.

▶ Continued on next page

3 Graph the linear functions $y = 4x$ and $y = 4(x - h)$. Insert a slider to change the value of h between -10 and $+10$. Start with $h = 0$.

Try different values of h to explore how changing $y = 4x$ to $y = 4(x - h)$ changes the graph.

4 Repeat step **3** for three more linear functions: $y = 2(x - h) + 3$, $y = (x - h) - 2$ and $y = 3(x - h)$

Describe how changing $y = f(x)$ to $y = f(x - h)$ changes the graph.

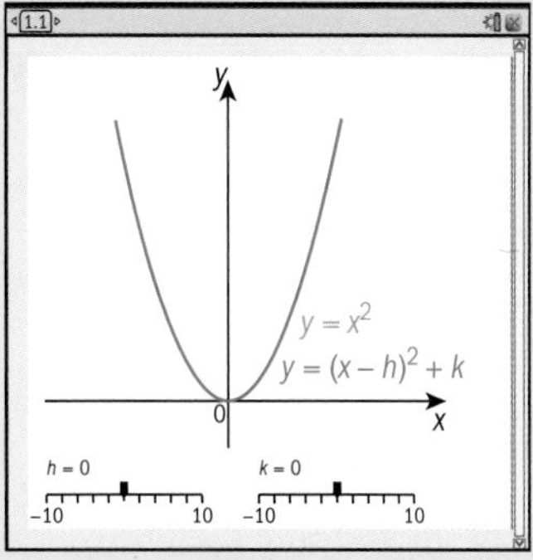

5 Graph the quadratic functions $y = x^2$ and $y = (x - h)^2 + k$ on your graphing software, and insert two sliders to change the values of h and k between -10 and 10. Start with $h = k = 0$, so $y = x^2$.

6 Find the values of h and k and write the equation of the parabola $y = (x - h)^2 + k$ that is:

a 4 units higher than $y = x^2$

b 4 units to the left of $y = x^2$

c both 4 units higher and 4 units to the left of $y = x^2$.

7 a Explain how changing the function $y = f(x)$ to $y = f(x - h)$ changes its graph.

b Explain how changing the function $y = f(x)$ to $y = f(x) + k$ changes its graph.

Vertical and horizontal shifts in the graph of a function are called **translations**.

- $y = f(x - h)$ translates $y = f(x)$ by h units in the x-direction.

  When $h > 0$, the graph moves in the positive x-direction, to the right.

  When $h < 0$, the graph moves in the negative x-direction, to the left.

- $y = f(x) + k$ translates $y = f(x)$ by k units in the y-direction.

  When $k > 0$, the graph moves in the positive y-direction, up.

  When $k < 0$, the graph moves in the negative y-direction, down.

- $y = f(x - h) + k$ translates $y = f(x)$ by h units in the x-direction and k units in the y-direction.

Reflect and discuss 2

When $h > 0$, why does $y = f(x - h)$ translate the function $y = f(x)$ by h units to the right and not to the left?

Example 1

The red, green and blue lines are parallel.
The red line has equation $y = f(x)$.
Find the equation of the other lines in terms of $f(x)$.

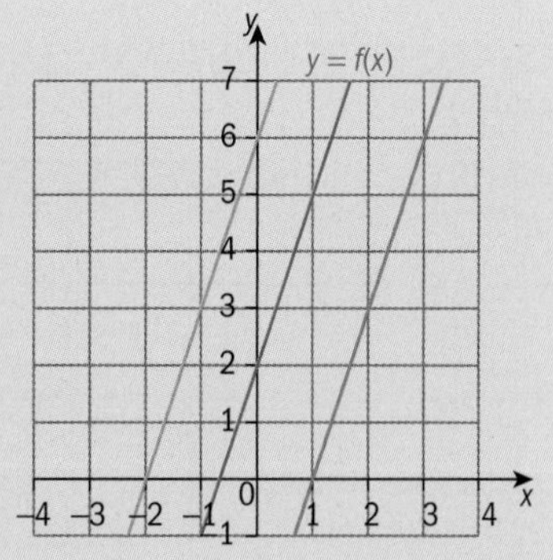

The blue line is a translation of the red line by 4 units in the y-direction.

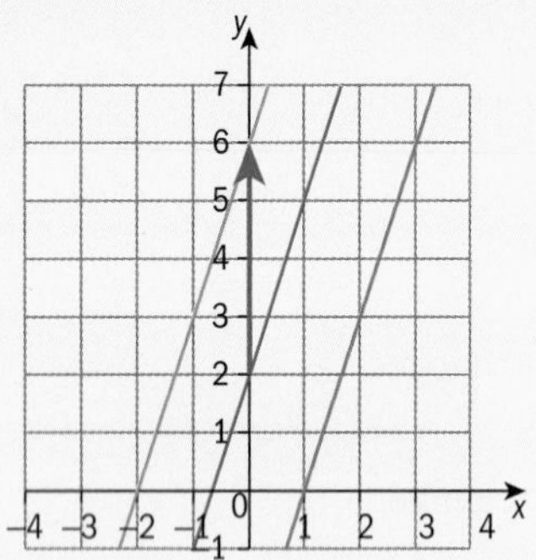

Find points on the two graphs that correspond to each other.

Hence the blue line has equation $y = f(x) + 4$.

The green line is a translation of the blue line to the right by 3 units.

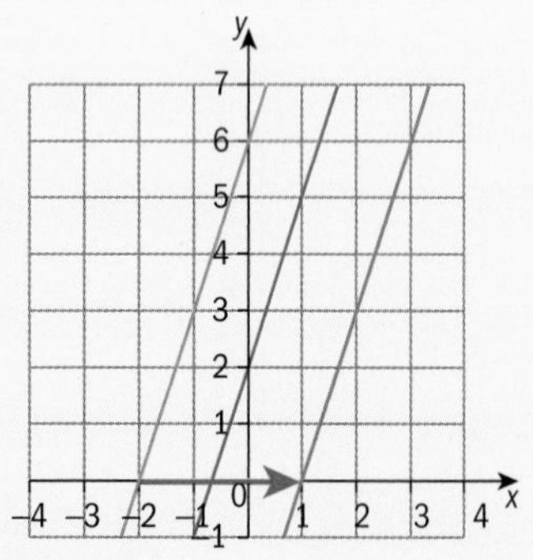

It is easiest to find corresponding points between the blue and green lines.

Hence the green line has equation $y = f(x - 3) + 4$.

Horizontal translations are of the form $f(x - h)$.

Example 2

The red curve is a translation of the blue curve, which has equation $y = f(x)$.

Find the equation of the red curve in terms of $f(x)$.

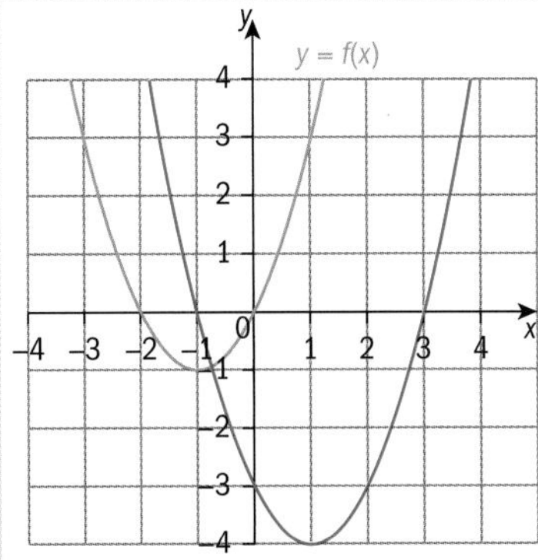

The vertex has moved from (–1, –1) to (1, –4). — Find corresponding points.

This is a translation of 2 units to the right and 3 units down. — Identify the translation.

The blue curve has equation $y = f(x - 2) - 3$.

Practice 2

1 In each graph you are given the equation of one of the functions. The other graphs are translations of that function. Find the equation of each function.

a

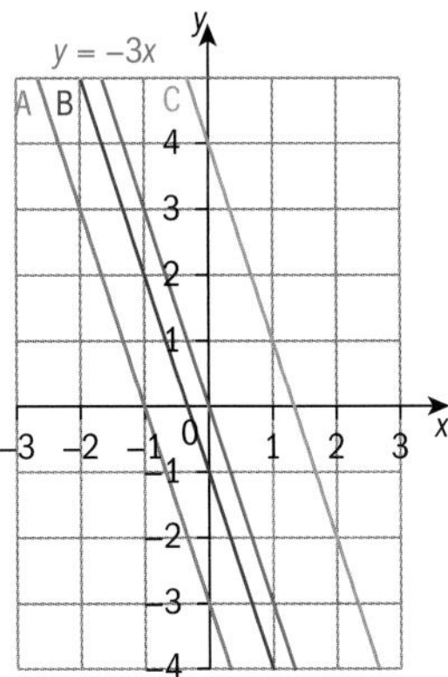

b

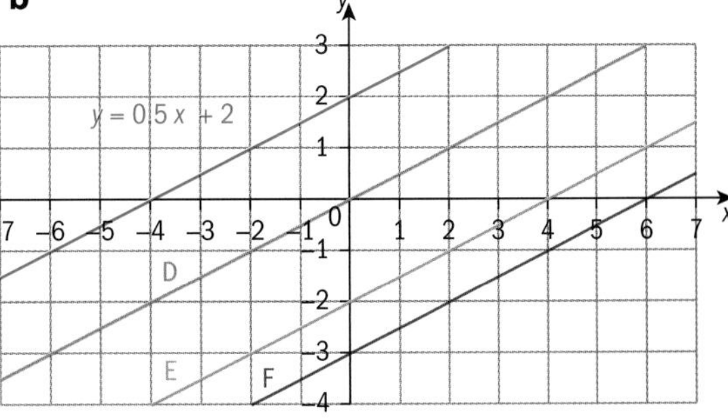

c

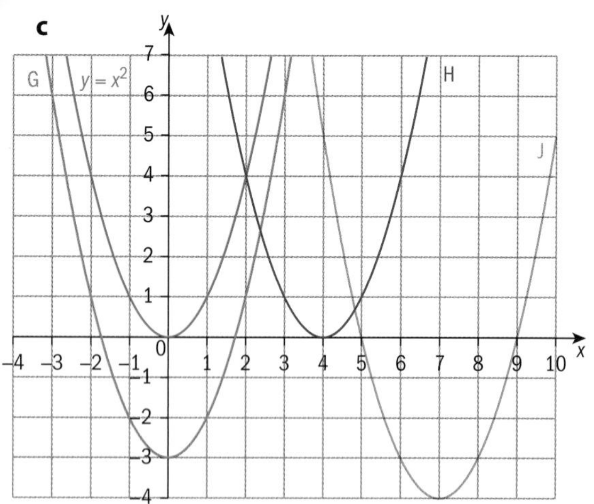

d

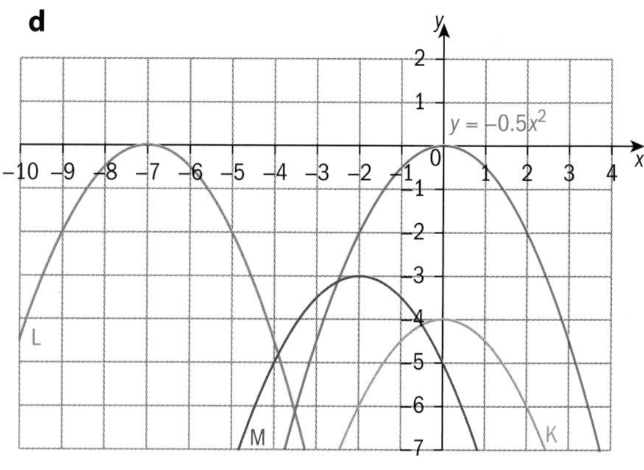

Problem solving

2 Without using graphing software or a calculator or finding any specific coordinates of points, sketch the graphs of these functions.

Label any intercepts with the axes.

a $y = x + 5$ **b** $y = 2x - 3$ **c** $y = -2x - 3$

d $y = x^2 + 5$ **e** $y = (x + 2)^2$ **f** $y = (x - 1)^2 - 1$

Start with $y = x$ or $y = x^2$ and translate the graph.

Exploration 3

1 Use graphing software to graph the linear function $f_1(x) = mx + c$ and insert a slider to change the values of m and c between -10 and 10. Start with $f_1(x) = x$, with $m = 1$ and $c = 0$.

2 Add the graphs of $y = -f_1(x)$ and $y = f_1(-x)$.

3 Change the values of m and c using the slider. Describe the relationship between the graphs $y = f_1(x)$ and $y = -f_1(x)$. Identify the transformation that takes the graph of $y = f_1(x)$ to $y = -f_1(x)$.

4 Continue to vary the values of m and c using the slider. By examining your graphs, describe the transformation that takes the graph of $y = f_1(x)$ to the graph of $y = f_1(-x)$.

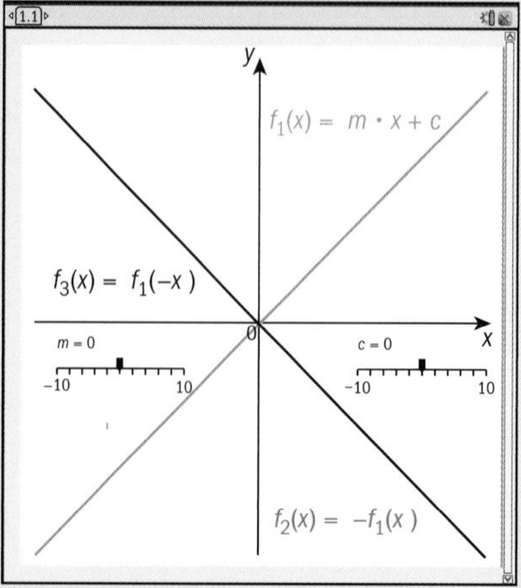

Continued on next page

5 On a clear screen, graph the quadratic function $y = f_1(x) = (x - h)^2 + k$. Insert sliders to change the values of h and k between –10 and 10.

6 Add the graphs $y = -f_1(x)$ and $y = f_1(-x)$. By changing the sliders as you did in steps **3** and **4**, find the relationship between the graph of $y = f_1(x)$ and the graphs of $y = -f_1(x)$ and $y = f_1(-x)$.

7 Explain whether the generalizations you made in steps **3** and **4** hold true for quadratic functions.

The graph of $y = -f(x)$ is a reflection of the graph of $y = f(x)$ in the x-axis.

The graph of $y = f(-x)$ is a reflection of the graph of $y = f(x)$ in the y-axis.

Example 3

Here is the graph of $y = (x - 2)^2 + 3$.

Find the equations of graphs A and B.

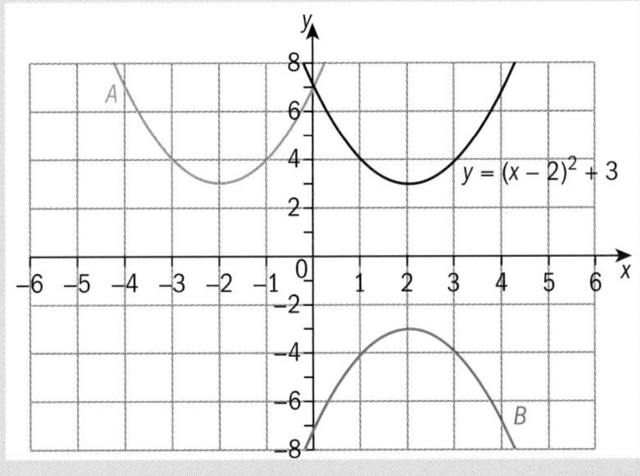

Graph A is a reflection of $y = (x - 2)^2 + 3$ in the y-axis.

Equation of $A = f(-x) = (-x - 2)^2 + 3$ — Reflection of $f(x)$ in the y-axis is $f(-x)$.

$= (-x - 2)(-x - 2) + 3$ — Substitute $-x$ into $f(x)$.

$= (x^2 + 4x + 4) + 3$

$= (x + 2)^2 + 3$ — Expand, then write in vertex form.

Graph B is a reflection of $y = (x - 2)^2 + 3$ in the x-axis. — Reflection of $f(x)$ in the x-axis is $-f(x)$.

Equation of $B = -f(x) = -(x - 2)^2 - 3$

Practice 3

1 Copy this sketch graph of $f(x) = 3x + 2$.

Sketch and label the graph of $f(-x)$ on the same axes.

Sketch and label the graph of $-f(x)$ on the same axes.

Label any intercepts with the y-axis.

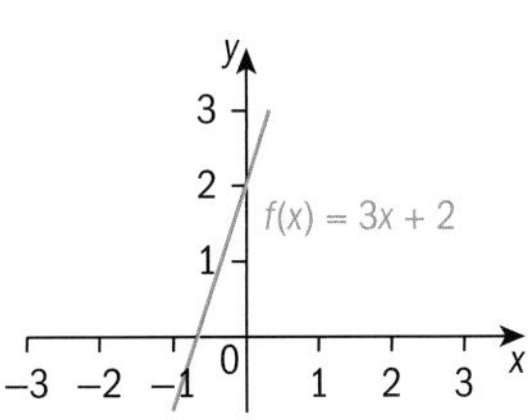

2 Copy this sketch graph of $f(x) = x^2 - 6x + 10$.

Sketch and label the graph of $f(-x)$ on the same axes.

Sketch and label the graph of $-f(x)$ on the same axes.

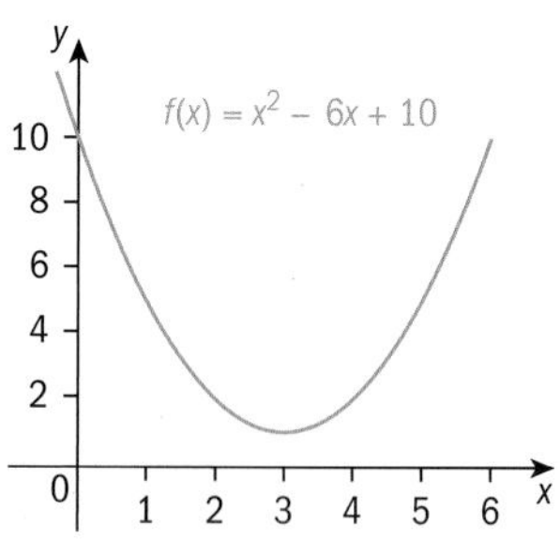

3 In each graph below you are given the equation of one of the functions. The other graphs are reflections of that function. Find the equation of each function.

a

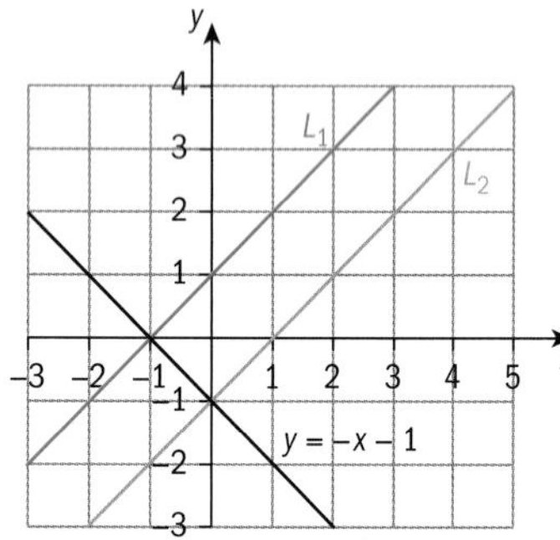

b

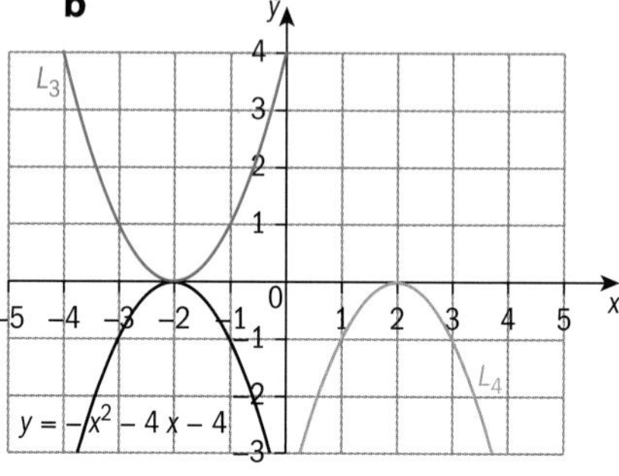

c

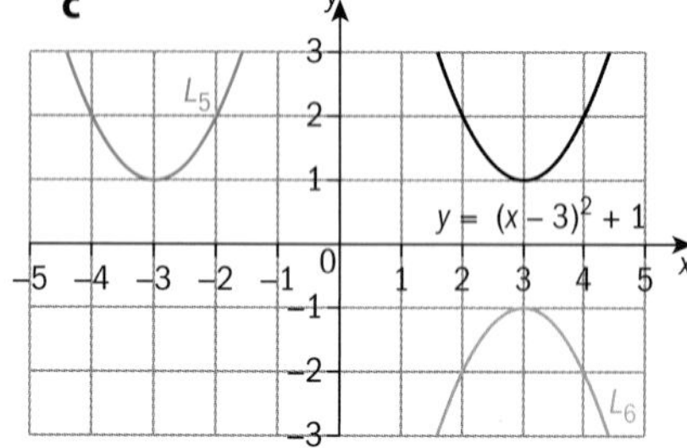

d

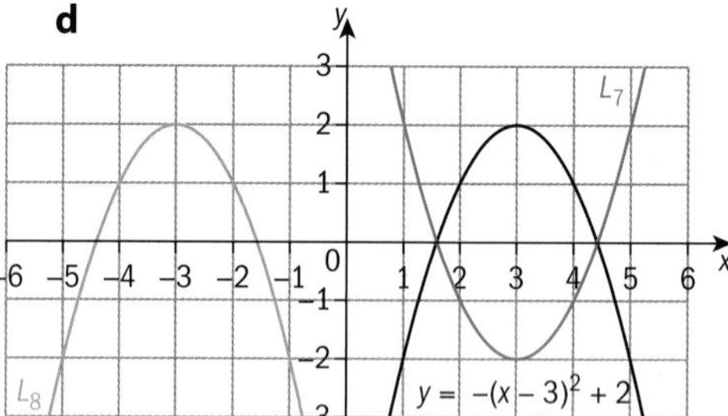

4 The graph of $y = x + 1$ is translated in the y-direction and then reflected in the y-axis to give line L.

Find the equation of line L.

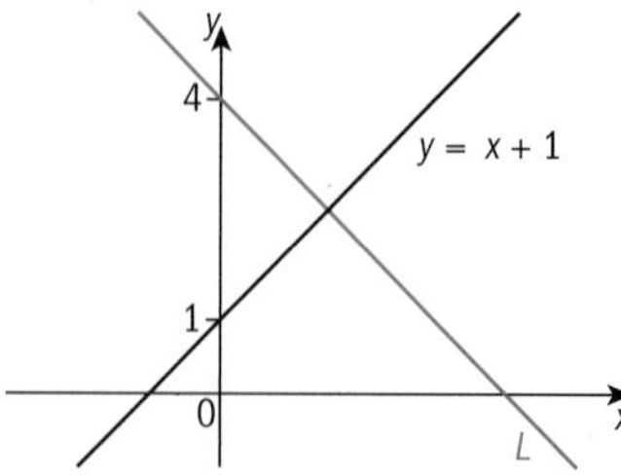

5 The graph of $y = x^2$ is translated in the x-direction and then reflected in the x-axis to give curve C. Find the equation of curve C.

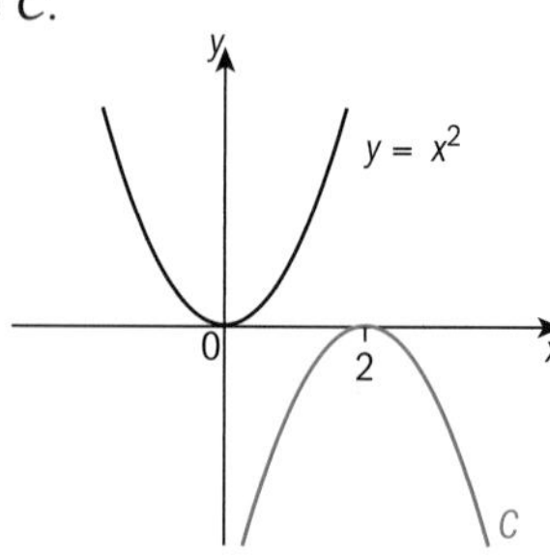

Exploration 4

1 Draw a coordinate plane with axes from –10 to +10.

Plot these points: (–4, – 5), (–3, – 2), (–2, 0), (–1, 1), (0, 0), (1, –1), (4, 1) and (5, 3).

This set of eight points defines the function $f(x)$. Join the points with a smooth curve.

2 A change is applied to $f(x)$ to give a new function $g_1(x)$. This change is defined as $g_1(x) = 2f(x)$.

a Write down the coordinates of the eight points of $g_1(x)$.

b Draw and label the graph of $g_1(x)$ on the same coordinate plane as $f(x)$.

c Describe the change between $f(x)$ and $g_1(x)$.

3 Repeat step **2** for a new change $g_2(x) = \frac{1}{2}f(x)$.

4 Make a generalization for what happens when the function $f(x)$ changes to $g(x) = af(x)$:

a when $0 < a < 1$ **b** when $a > 1$.

5 Draw the graph of $f(x)$ again on a new coordinate plane. A change is applied to $f(x)$ to give a new function $h_1(x)$. This change is defined as $h_1(x) = f(2x)$.

a Write down the coordinates of the eight points of $h_1(x)$, for example the point (–4, –5) will become (–2, –5).

b Draw and label the graph of $h_1(x)$ on the same coordinate plane as $f(x)$.

c Describe the change between $f(x)$ and $h_1(x)$.

6 Repeat step **5** for a new change $h_2(x) = f\left(\frac{1}{2}x\right)$.

7 Make a generalization for what happens when the function $f(x)$ changes to $h(x) = f(ax)$:

a when $0 < a < 1$ **b** when $a > 1$.

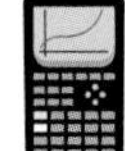

8 Use graphing software to see if your generalizations apply to linear and quadratic functions.

a Graph the linear functions $f_1(x) = x$, $f_2(x) = a \times f_1(x)$ and $f_3(x) = f_1(a \times x)$ and insert a slider to change the value of a between 0 and 3. Start with $a = 1$ and then explore what happens when $0 < a < 1$ and when $a > 1$.

b Repeat step **8a** for the quadratic function $f_1(x) = x^2$.

The mathematical term for stretch or compression is **dilation**.

$y = af(x)$ is a vertical dilation of $f(x)$, scale factor a, parallel to the y-axis.

$y = f(ax)$ is a horizontal dilation of $f(x)$, scale factor $\frac{1}{a}$, parallel to the x-axis.

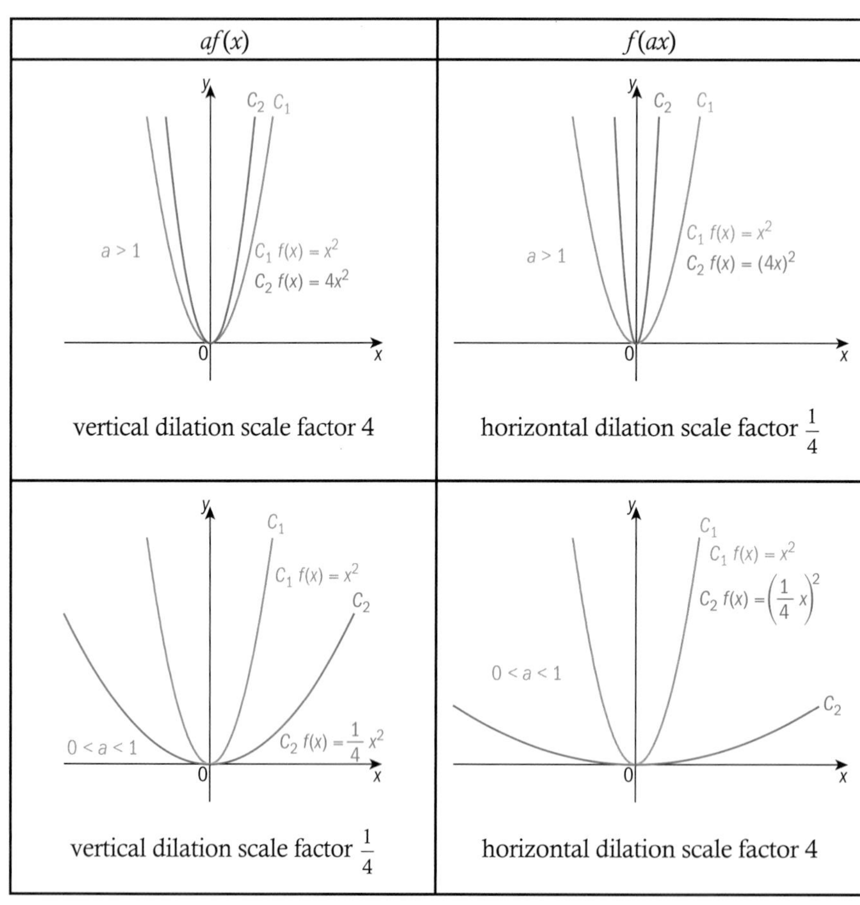

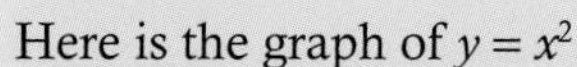

Example 4

Here is the graph of $y = x^2$.

Describe the effect of the transformation, giving examples of how individual points are transformed, and sketch the graphs of

a $2f(x)$

b $f(2x)$.

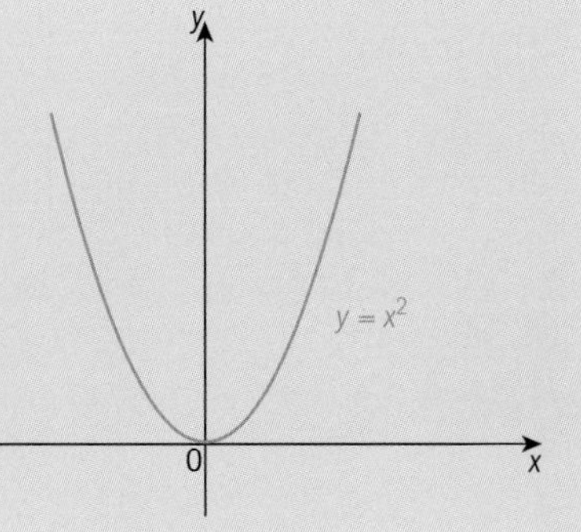

a The graph of $2f(x)$ is a dilation, scale factor 2, parallel to the y-axis.

$(0, 0) \to (0, 0)$

$(-2, 4) \to (-2, 8)$

$(1, 1) \to (1, 2)$

y-coordinates of points on the graph increase by scale factor 2.

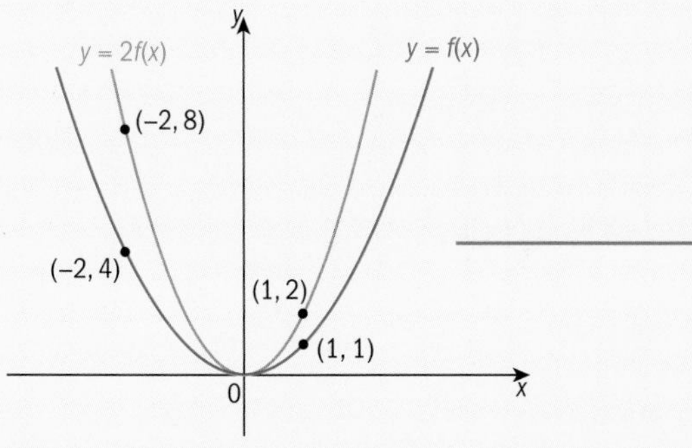

Sketch the graph. Label some points.

b Graph of $f(2x)$ is a dilation, scale factor $\frac{1}{2}$, parallel to the x-axis.

$(0, 0) \to (0, 0)$

$(2, 4) \to (1, 4)$

$(-2.5, 6.25) \to (-1.25, 6.25)$

x-coordinates of points on the graph increase by scale factor $\frac{1}{2}$.

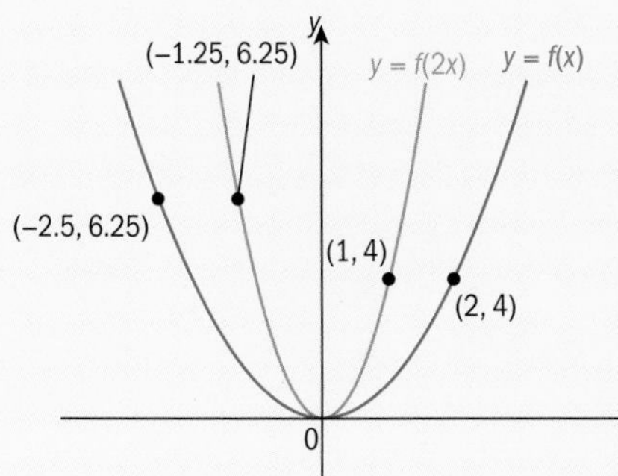

Practice 4

1 Copy this sketch of $f(x) = (x - 1)^2$.

Sketch the graph of $g(x) = 2f(x)$ on the same axes.

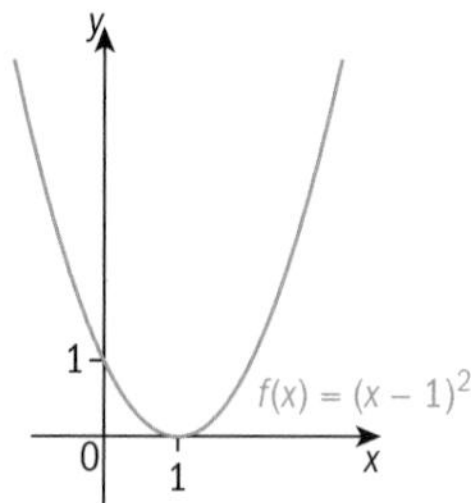

2 Copy this sketch of $f(x) = -x^2 + 4$.

Sketch the graph of $h(x) = f(2x)$ on the same axes.

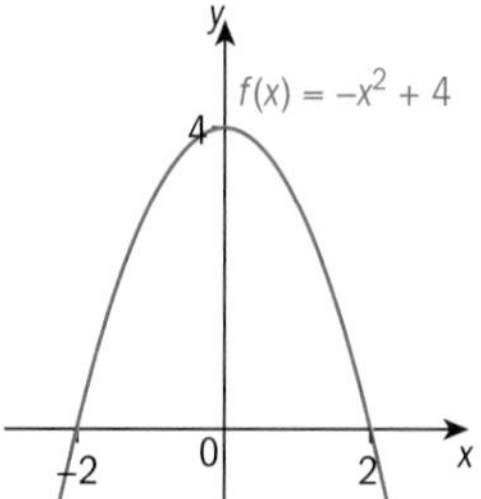

3 Copy this sketch of $f(x) = x^2 + 2x - 3$.

Sketch the graph of $g(x) = f(3x)$ on the same axes.

Sketch the graph of $h(x) = -f(3x)$ on the same axes.

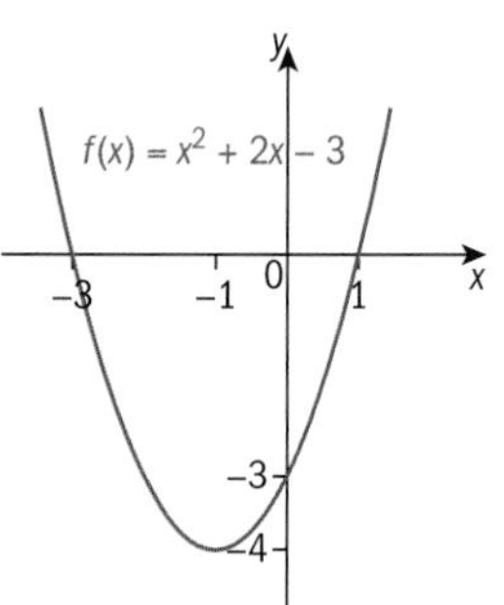

4 Draw the graph of $f(x) = x^2$.

Draw the graphs of $g(x) = f\left(\frac{1}{2}x\right)$ and $h(x) = f(2x)$ on the same axes.

5 Describe the transformation that maps the graph of $y = x^2$ to:

a $y = 3x^2$ **b** $y = \frac{1}{2}x^2$ **c** $y = (4x)^2$ **d** $y = \left(\frac{1}{3}x\right)^2$

6 Write down the equation of the graph which results when $y = x^2$ is transformed by a dilation:

a scale factor 5 parallel to the y-axis

b scale factor 4 parallel to the x-axis

c scale factor $\frac{1}{3}$ parallel to the y-axis.

7 Describe the transformation that takes the blue graph to the red graph:

a

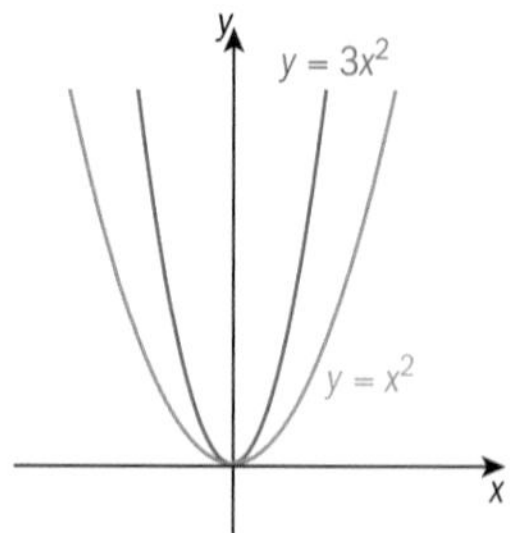

b

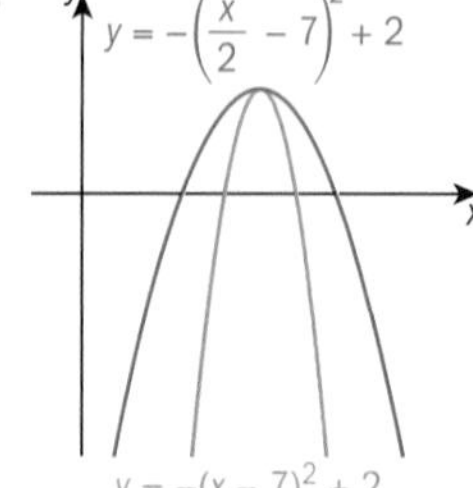

You need to be familiar with three types of transformation for linear and quadratic functions – translation, reflection and dilation.

Combinations of transformations:

- Any linear function can be defined by one or more transformations applied one after the other to the function $y = x$.
- Any quadratic function can be defined by one or more transformations applied one after the other to the function $y = x^2$.

Example 5

List the transformations applied to the function $y = x^2$ to give:

a $y = x^2 + 6x + 4$ **b** $y = -3x^2 - 12x - 11$

Sketch the transformed graphs.

a $x^2 + 6x + 4 = (x^2 + 6x) + 4$ — Write the function in vertex form.

$= [(x + 3)^2 - 9] + 4$

$= (x + 3)^2 - 5$

Start with $f(x) = x^2$. — Start with the original function $y = x^2$ and apply one transformation at a time.

$f(x + 3)$ is a horizontal translation of -3.

$f(x + 3) - 5$ is a vertical translation of -5 on $f(x + 3)$.

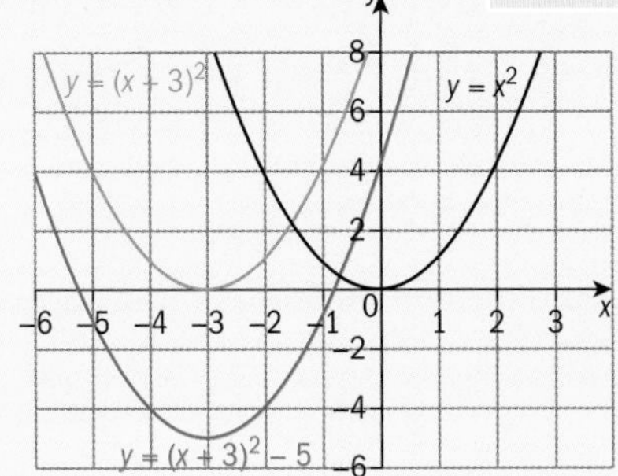

b $-3x^2 - 12x - 11 = -3(x^2 + 4x) - 11$

$= -3[(x + 2)^2 - 4] - 11$

$= -3(x + 2)^2 + 1$

Start with $f(x) = x^2$.

$y = 3x^2 = 3f(x)$ is a dilation parallel to the y-axis, scale factor 3.

$y = -3x^2 = -3f(x)$ is the reflection in the x-axis.

$y = -3(x + 2)^2 = -3f(x + 2)$ is a horizontal translation of -2.

$y = -3(x + 2)^2 + 1 = -3f(x + 2) + 1$ is a vertical translation of 1.

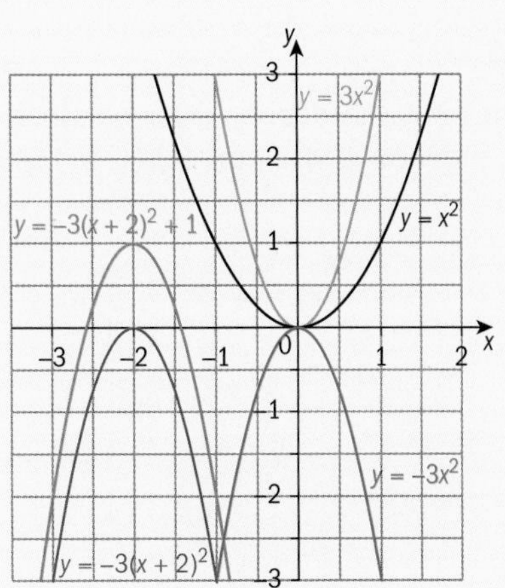

Reflect and discuss 3

- Why is the quadratic changed from standard form to vertex form in Example **5** in order to show the transformations that have been applied?
- In Example **5b**, does it matter in which order you apply the transformations?
- Why is it best to do the translations last?

Practice 5

1 List the transformations applied to the linear function $y = x$ to give:

a $y = -4x$ **b** $y = 2x - 9$ **c** $y = 0.5x + 4$ **d** $y = -1.2x - 2.5$

2 List the transformations applied to the quadratic function $y = x^2$ to give each function. Sketch the graphs of the transformations.

a $y = 25x^2$ **b** $y = 3(x - 9)^2$ **c** $y = -x^2 + 4$

d $y = x^2 + 8x + 16$ **e** $y = 2x^2 - 12x + 9$ **f** $y = -0.5x^2 - 4x + 1$

3 Express $g(x)$ in terms of $f(x)$ and state the transformation applied to $f(x)$ to get $g(x)$:

a $f(x) = x - 5$, $g(x) = 4x - 20$ **b** $f(x) = 2x + 5$, $g(x) = 6x + 5$

c $f(x) = -2x - 5$, $g(x) = 2x + 5$ **d** $f(x) = -2x - 5$, $g(x) = 2x - 5$

e $f(x) = 2x^2 + 3x$, $g(x) = 2x^2 + 3x - 12$ **f** $f(x) = -3x^2 + 6x - 7$, $g(x) = -3x^2 - 6x - 7$

g $f(x) = x^2 - 5x + 6$, $g(x) = (x - 2)^2 - 5(x - 2) + 6$

h $f(x) = 4x^2 - 8x + 12$, $g(x) = x^2 - 2x + 3$

4 Express $g(x)$ in terms of $f(x)$ and state the combination of transformations applied to $f(x)$ to get $g(x)$:

a $f(x) = 3x - 8$, $g(x) = -3x + 3$ **b** $f(x) = 2x + 6$, $g(x) = x - 3$

c $f(x) = x^2 + 4x$, $g(x) = x^2 - 4x + 4$ **d** $f(x) = 2x^2 - 2x - 3$, $g(x) = -x^2 + x + 1.5$

e $f(x) = 4x^2 - 5x + 7$, $g(x) = -4x^2 + 5x + 2$ **f** $f(x) = x^2 + 4x + 4$, $g(x) = x^2 - 6x + 9$

g $f(x) = x^2 - 2x - 3$, $g(x) = x^2 + 6x + 4$ **h** $f(x) = x^2 + x + 1$, $g(x) = 9x^2 - 3x + 1$

Problem solving

5 For each pair of graphs, the orange function $f_2(x)$ is obtained by applying one transformation to the blue function $f_1(x)$. Express $f_2(x)$ in terms of $f_1(x)$, and state the transformation.

a

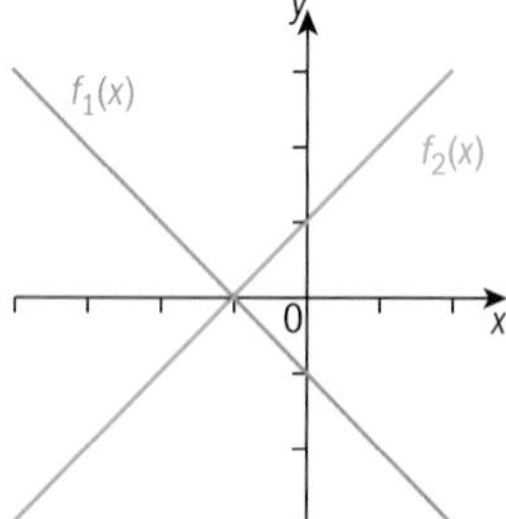

b

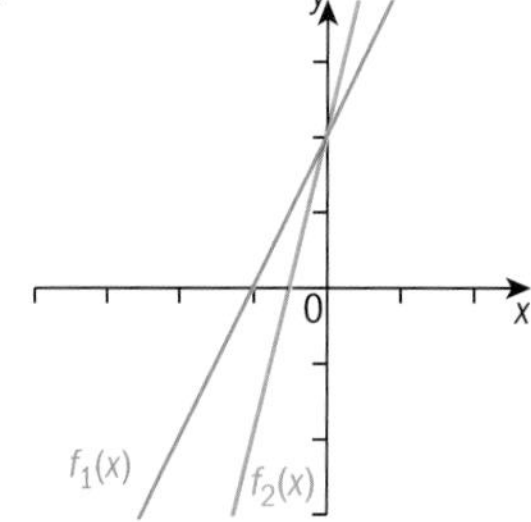

c

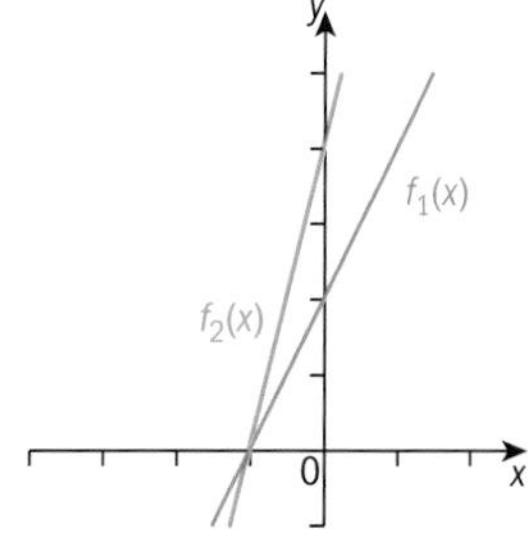

d

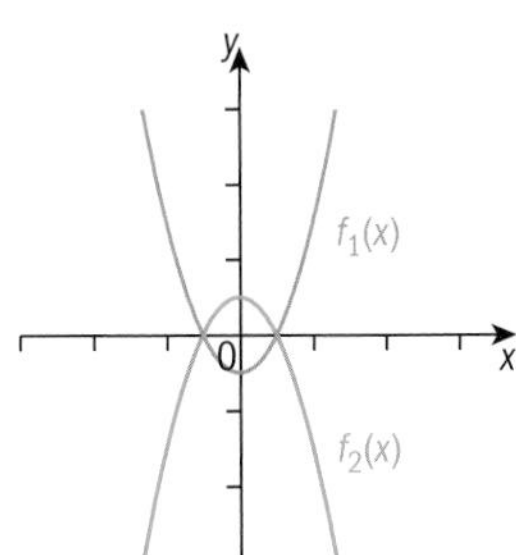

e

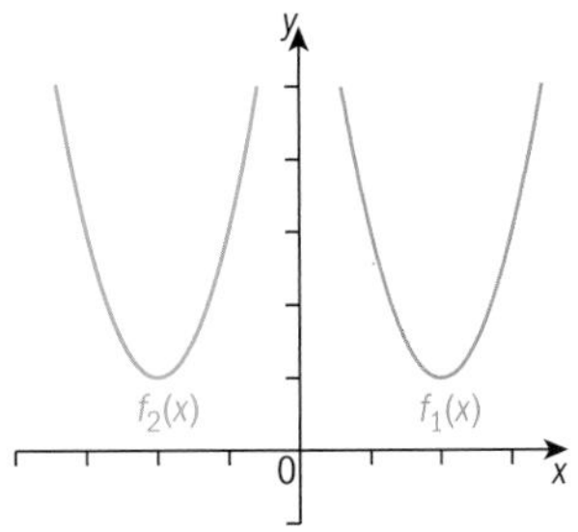

f

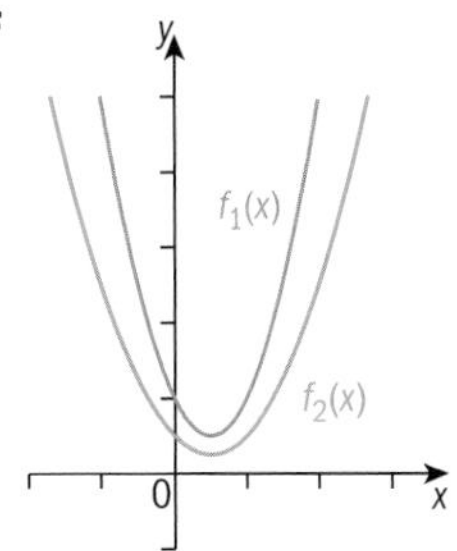

6 For each pair of graphs, the orange function $f_2(x)$ is obtained by applying two or more transformations to the blue function $f_1(x)$. Express $f_2(x)$ in terms of $f_1(x)$, and state the transformations.

a

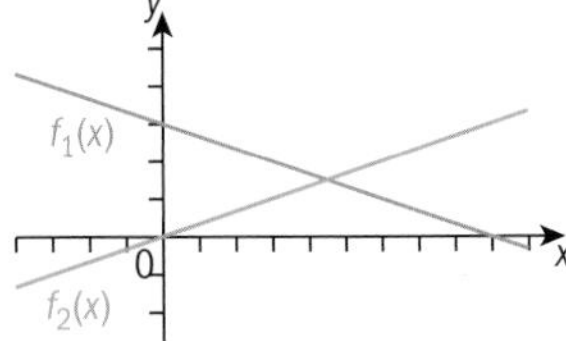

b

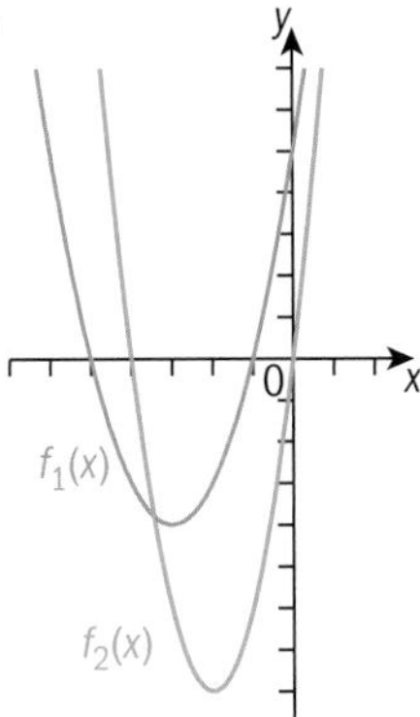

c

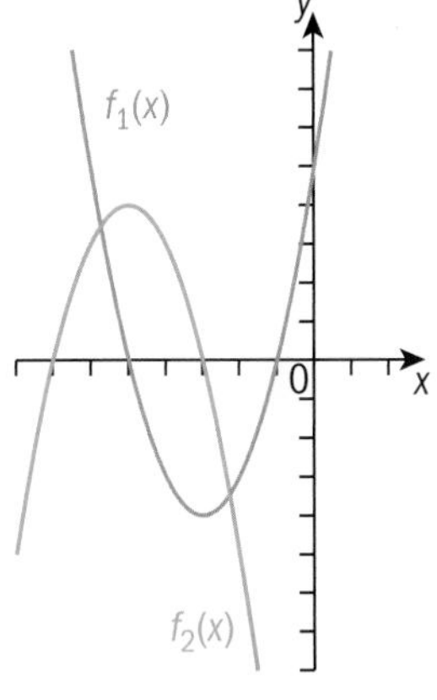

Translation, reflection and dilation transformations can be used on the graph of **any** function.

Example 6

Here is the graph of function $f_1(x)$ and the transformed function $f_2(x)$.

Identify the transformations that take $f_1(x)$ to $f_2(x)$.

Hence write the function $f_2(x)$ as a transformation of $f_1(x)$.

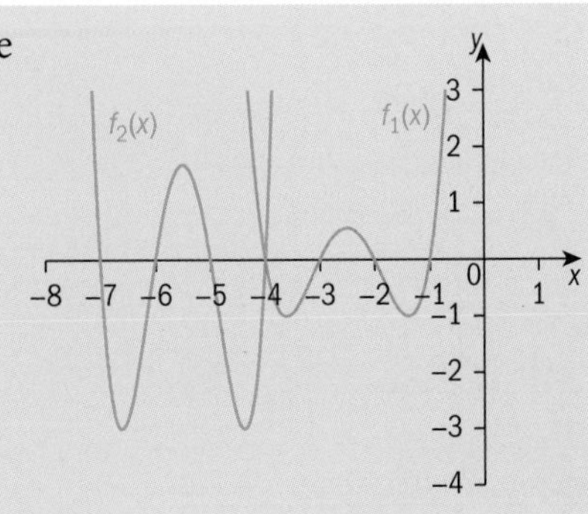

There has been a translation 3 units to the left.

There has been a dilation parallel to y-axis scale factor 3.

The minimum points at $y = -1$ have been transformed to minimum points at $y = -3$.

$f_1(x) \rightarrow f(x + 3)$ translation 3 units left

$\rightarrow 3f(x + 3)$ dilation scale factor 3 parallel to y-axis

Therefore, $f_2(x) = 3f(x + 3)$

Example 7

Given the graph of $f_1(x)$, draw the graph of $f_2(x) = 5f_1(-x)$.

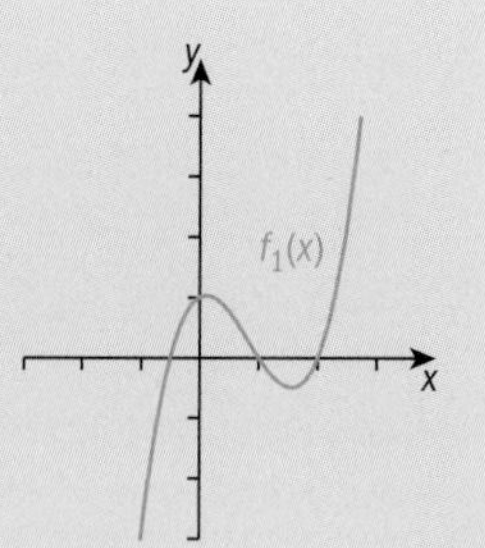

$f_1(-x)$: reflection in the y-axis. — Identify the transformations.

$5f_1(-x)$: vertical dilation of scale factor 5 — Multiply the y–value of all points by 5.

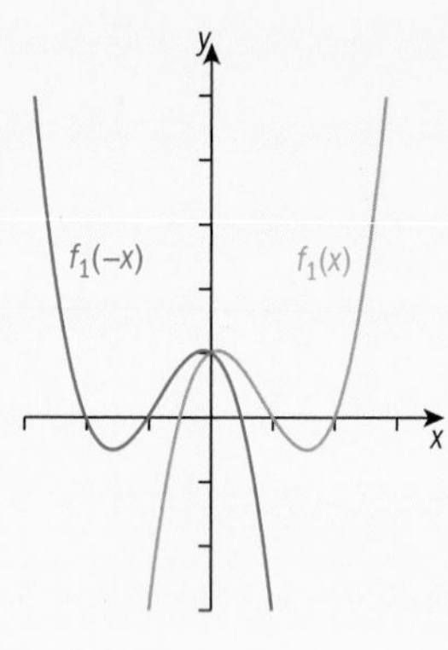

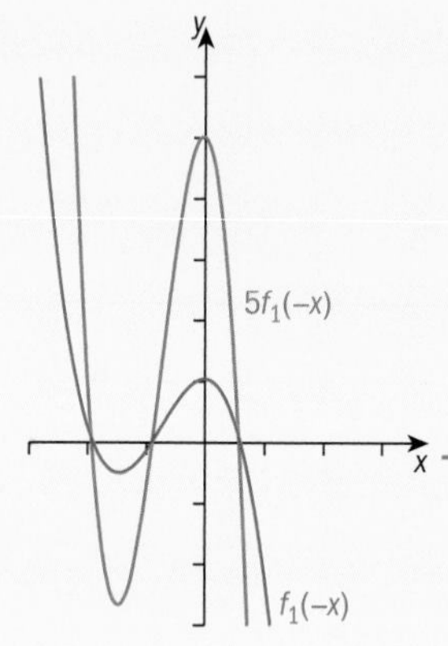

Draw the graph of each transformation.

Practice 6

1 For each pair of graphs, find the transformations applied to $f_1(x)$ to get $f_2(x)$. Hence write the function $f_2(x)$ as a transformation of $f_1(x)$.

a

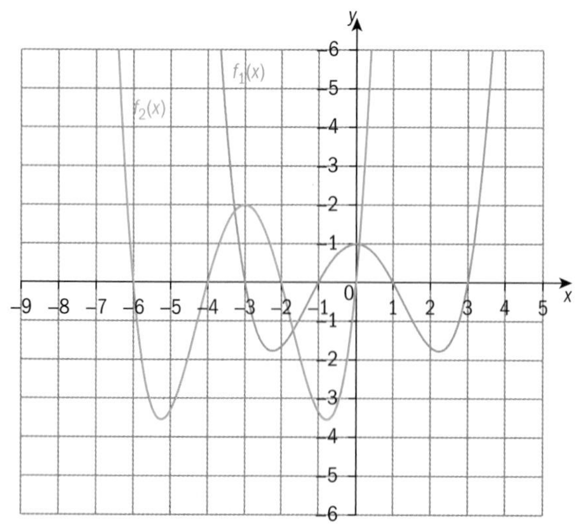

b

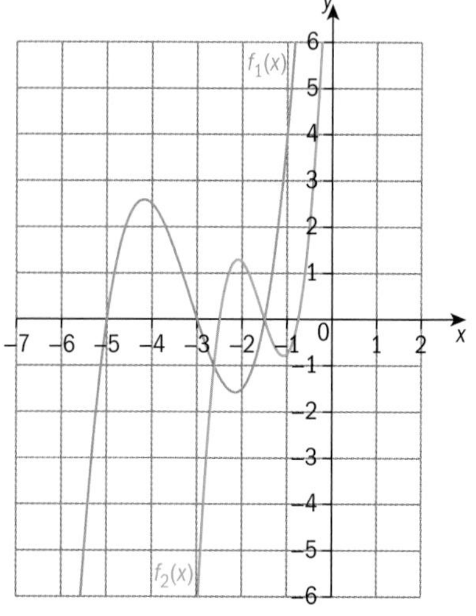

c

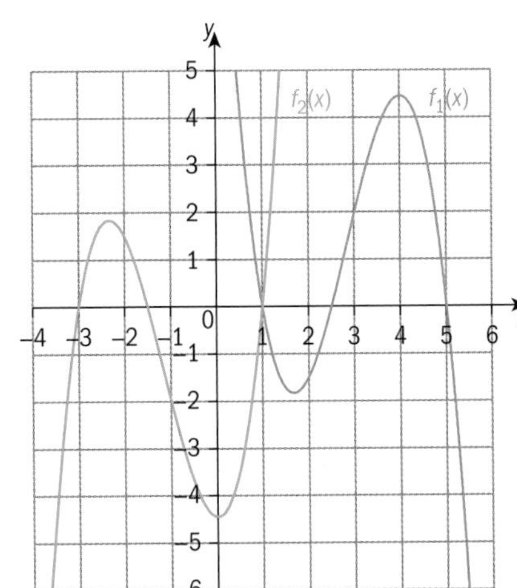

d

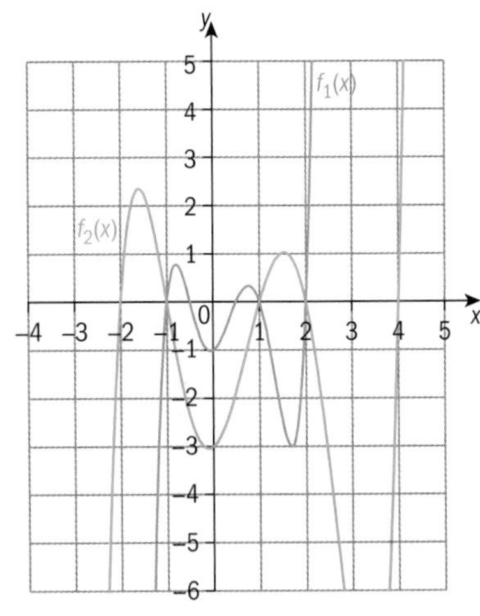

2 Given the graph of $f_1(x)$, draw the graph of $f_2(x)$.

a $f_2(x) = 2f_1(x - 3)$

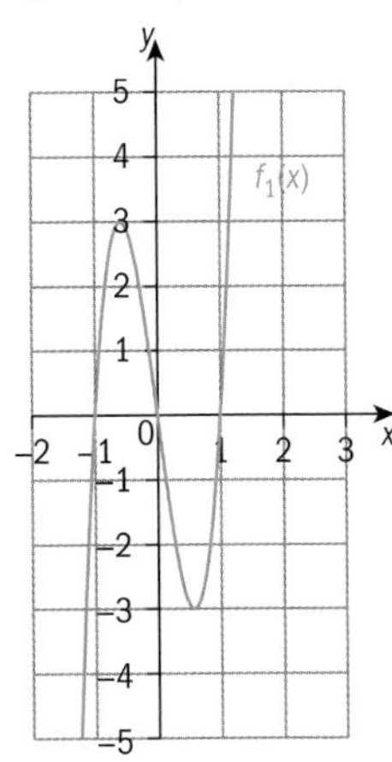

b $f_2(x) = f_1(-x + 5)$

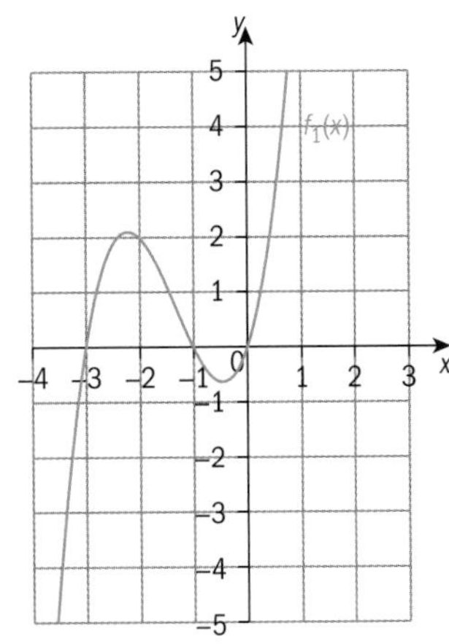

c $f_2(x) = 0.5f_1(-x)$

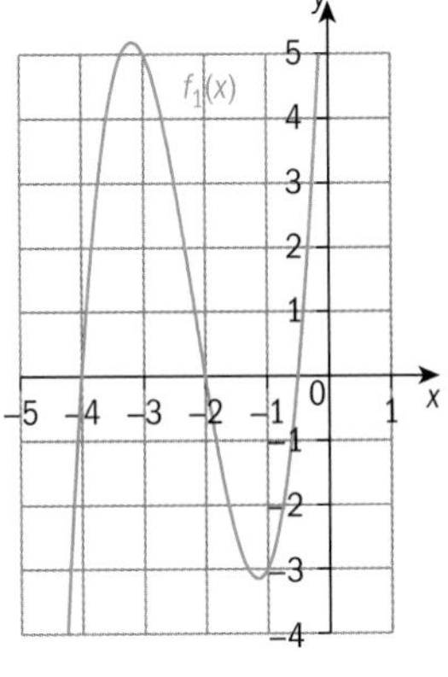

d $f_2(x) = 3f_1(x - 5)$

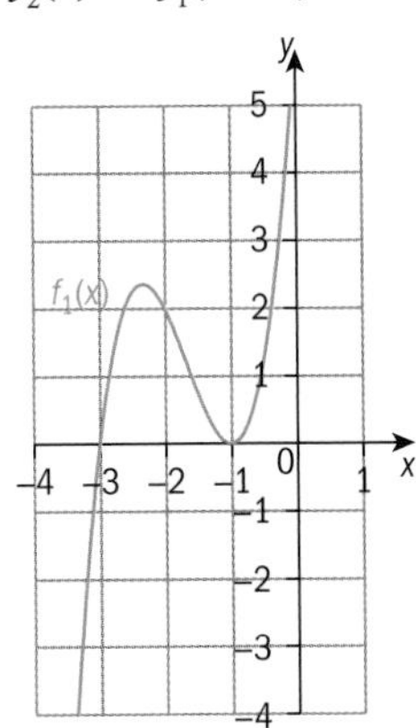

e $f_2(x) = -2f_1(x)$

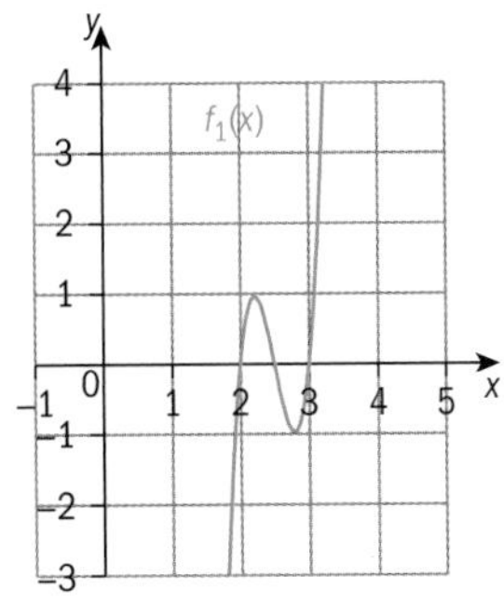

f $f_2(x) = f_1(-x) + 3$

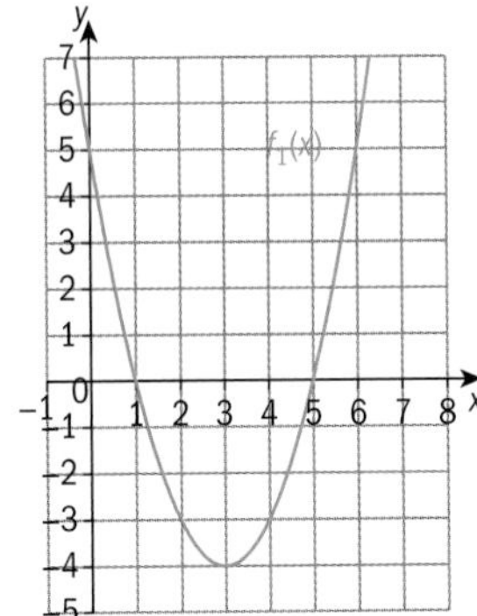

D Transformations or not?

- If the graph of a transformed function is the same as the graph of the original function, has the function been transformed?
- If two different transformations give the graph of the same transformed function, are the transformations really different?

Exploration 5

1 Use technology to graph different linear functions of the form $f(x) = mx + c$ and their reflections in the y-axis and the x-axis, $-f(-x)$.

2 Find functions where the graphs of $f(x)$ and $-f(-x)$ are the same. What is the value of c in these graphs?

In these graphs, are the original function and the transformed function identical?

3 For one of your graphs in step **2**, pick a point on the line. Apply the transformation $-f(-x)$ to that point by reflecting it in the x-axis and then the y-axis. Does the resulting point have the same coordinates as the original point? How does this affect your answer to step **2**?

4 What single transformation is equivalent to the two combined reflections?

Example 8

Consider the function $f(x) = 2x^2$. Show that the transformations $g(x) = 4f(x)$ and $h(x) = f(2x)$ both generate the same graph.

$g(x) = 4f(x)$ — Find the function generated by the first transformation.

$= 4(2x^2)$

$= 8x^2$

$h(x) = f(2x)$ — Find the function generated by the second transformation.

$= 2(2x)^2$

$= 8x^2$

$g(x) = h(x)$ — Both the functions generated by the two different transformations are equal.

The two transformations generate the same graph.

Reflect and discuss 4

- In Example 8 the function $g(x)$ is a vertical dilation by scale factor 4 of the function $f(x)$. The function $h(x)$ is a horizontal dilation by scale factor $\frac{1}{2}$. Do $g(x)$ and $h(x)$ represent the same transformation or different transformations of the function $f(x)$?
- Choose a point on the parabola $f(x) = 2x^2$. Apply each of the transformations to that point. Do the two new points have the same coordinates as each other? How does this affect your answer above?

Summary

Translations

- $y = f(x - h)$ translates $y = f(x)$ by h units in the x-direction.
- $y = f(x) + k$ translates $y = f(x)$ by k units in the y-direction.
- $y = f(x - h) + k$ translates $y = f(x)$ by h units in the x-direction and k units in the y-direction.

Reflections

- The graph of $y = -f(x)$ is a reflection of the graph of $y = f(x)$ in the x-axis.
- The graph of $y = f(-x)$ is a reflection of the graph of $y = f(x)$ in the y-axis.

Dilations

- $y = af(x)$ is a vertical dilation of $f(x)$, scale factor a, parallel to the y-axis.
- $y = f(ax)$ is a horizontal dilation of $f(x)$, scale factor $\frac{1}{a}$, parallel to the x-axis.

$af(x)$	$f(ax)$
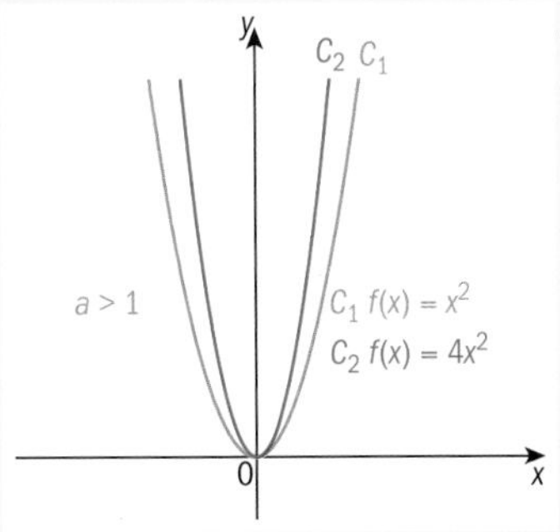 y, x, 0, C_2 C_1, $a > 1$, C_1 $f(x) = x^2$, C_2 $f(x) = 4x^2$ vertical dilation scale factor 4	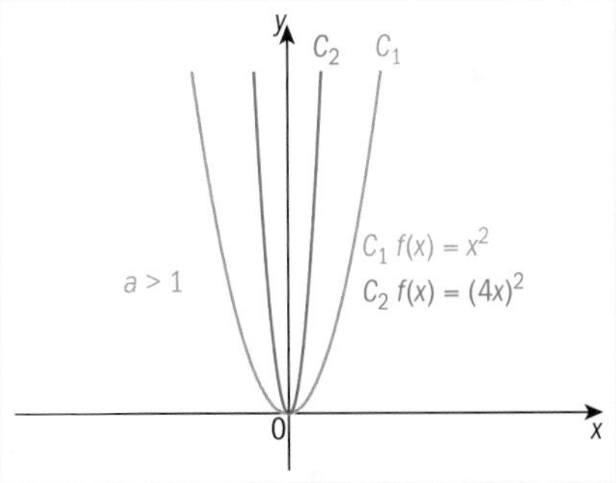 y, x, 0, C_2 C_1, $a > 1$, C_1 $f(x) = x^2$, C_2 $f(x) = (4x)^2$ horizontal dilation scale factor $\frac{1}{4}$
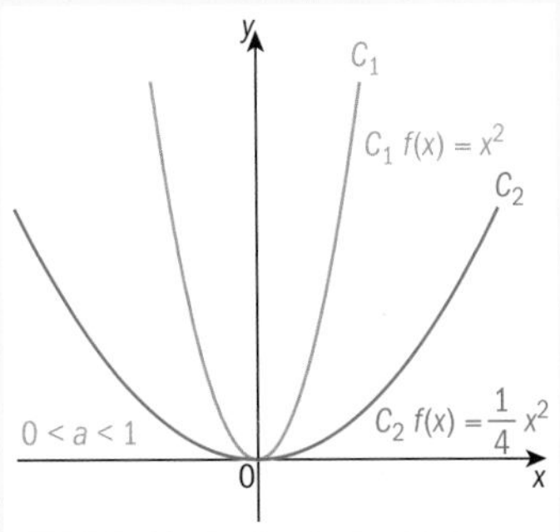 y, x, 0, C_1, C_1 $f(x) = x^2$, C_2, $0 < a < 1$, C_2 $f(x) = \frac{1}{4}x^2$ vertical dilation scale factor $\frac{1}{4}$	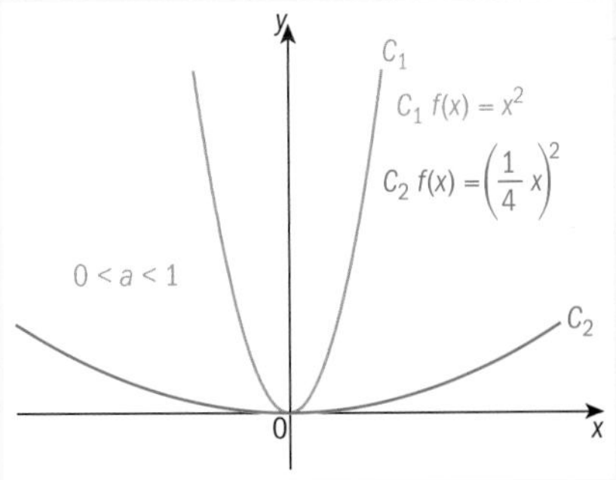 y, x, 0, C_1, C_1 $f(x) = x^2$, C_2 $f(x) = \left(\frac{1}{4}x\right)^2$, $0 < a < 1$, C_2 horizontal dilation scale factor 4

Combinations of transformations

- Any linear function can be defined by one or more transformations applied one after the other to the function $y = x$.
- Any quadratic function can be defined by one or more transformations applied one after the other to the function $y = x^2$.

Mixed practice

1 Here is the graph of $f(x) = x^2 - 2x - 1$.

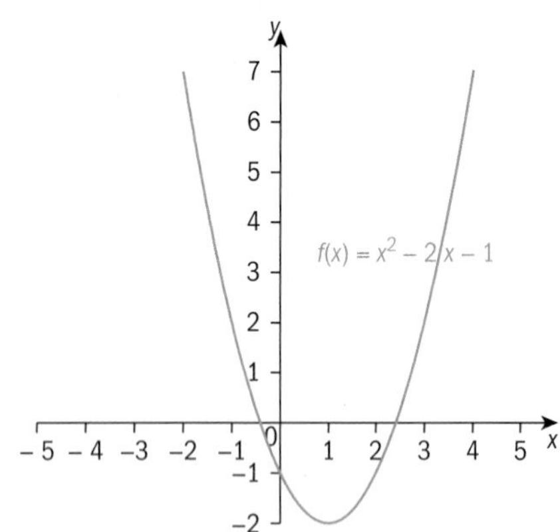

Use this graph to help you **sketch** the graph of each function and **write down** the coordinates of the new vertex.

a $f(x + 2)$ **b** $f(x) - 4$

c $-f(x)$ **d** $f(-x)$

e $2f(x)$ **f** $f(-3x)$

g $f(x + 1) - 3$ **h** $3 - f(x)$

i $-2f(x) - 1$ **j** $5f(-0.5x) + 4$

2 **State** which combination of transformations has been applied to $f(x)$ to get $g(x)$:

a $f(x) = x$, $g(x) = 2x - 8$

b $f(x) = -3x - 6$, $g(x) = x + 2$

c $f(x) = 12x + 5$, $g(x) = 6x + 6$

d $f(x) = x^2 + 4x + 4$, $g(x) = x^2 - 5$

e $f(x) = (x - 2)^2 + 1$, $g(x) = (x + 1)^2 - 2$

f $f(x) = 2x^2 + 8x + 4$, $g(x) = x^2 + 4x + 4$

3 For each pair of graphs, **find** the single transformation applied to $f_1(x)$ to get $f_2(x)$. **Write down** the function $f_2(x)$ as a transformation of $f_1(x)$.

a

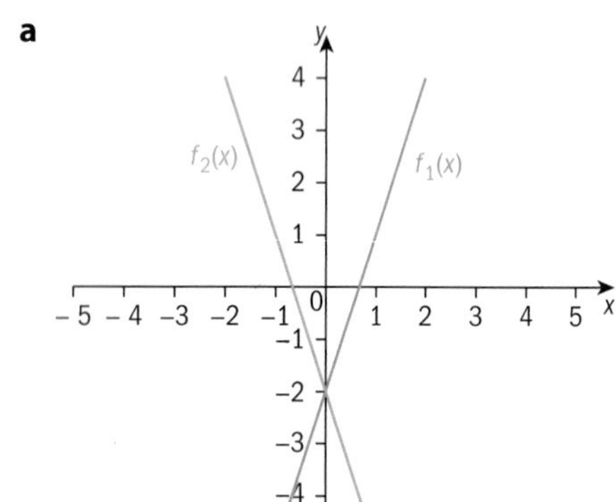

b

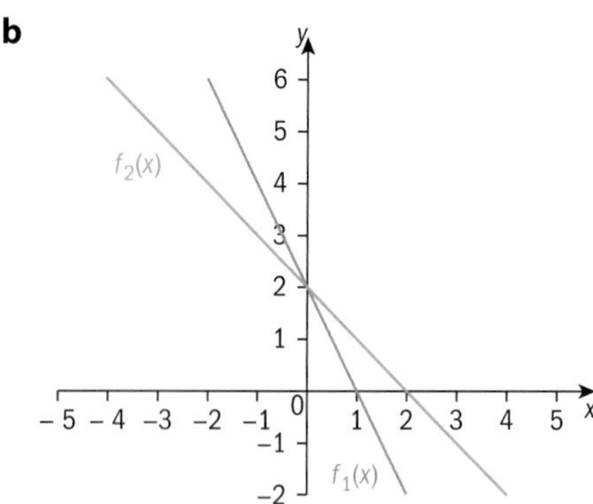

c

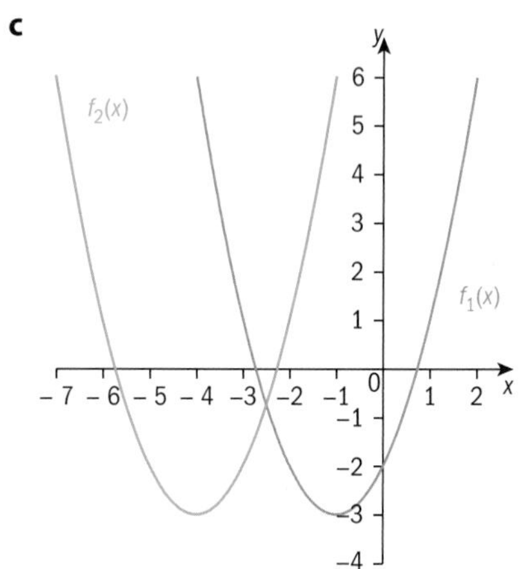

d

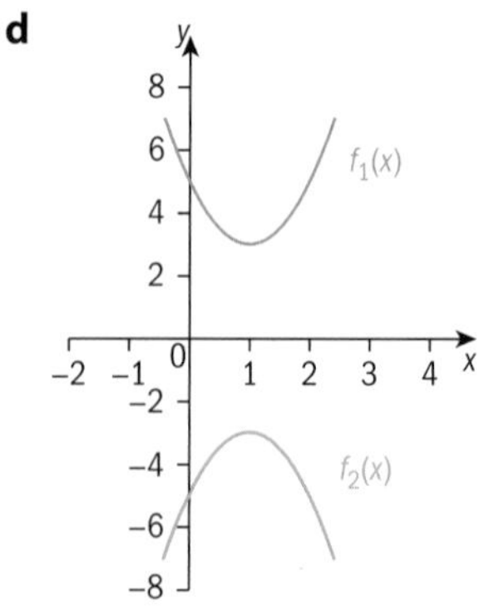

e

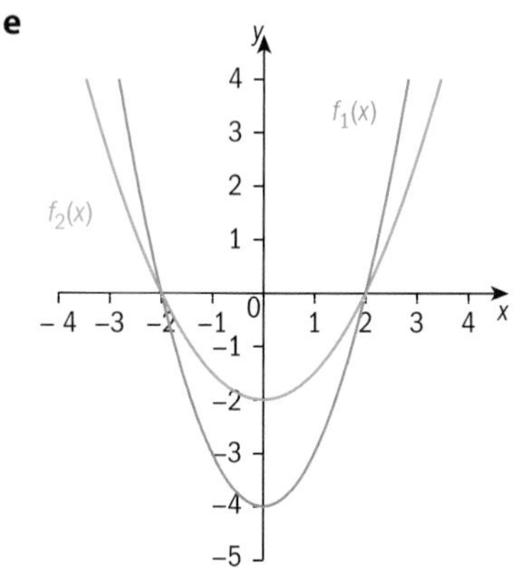

f

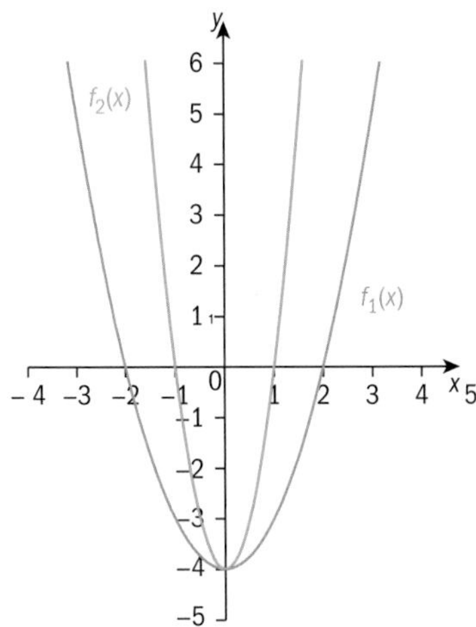

4 For each pair of graphs, **find** the transformations applied to $f_1(x)$ to get $f_2(x)$. **Write down** the function $f_2(x)$ as a transformation of $f_1(x)$.

a

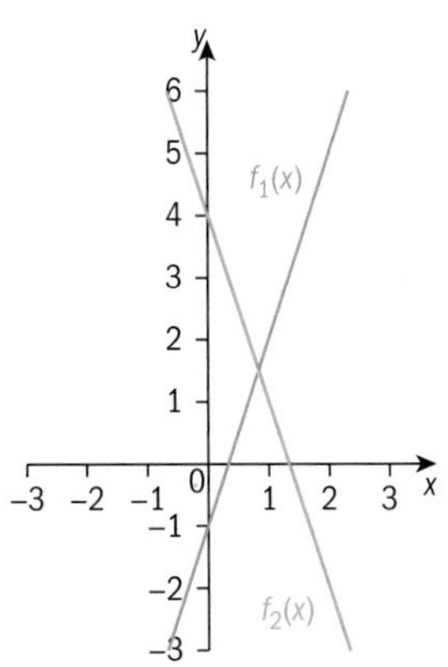

b

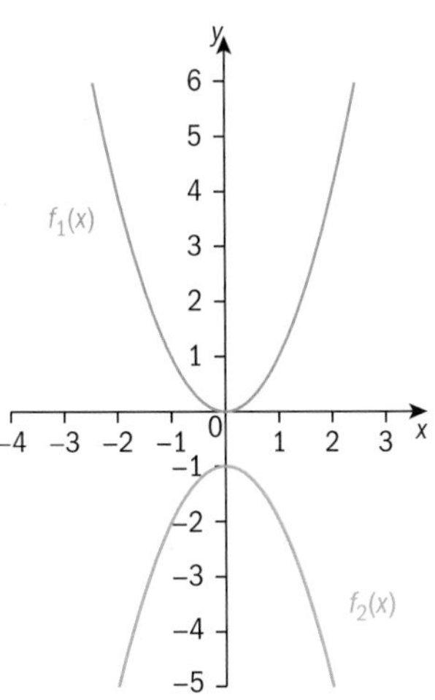

c

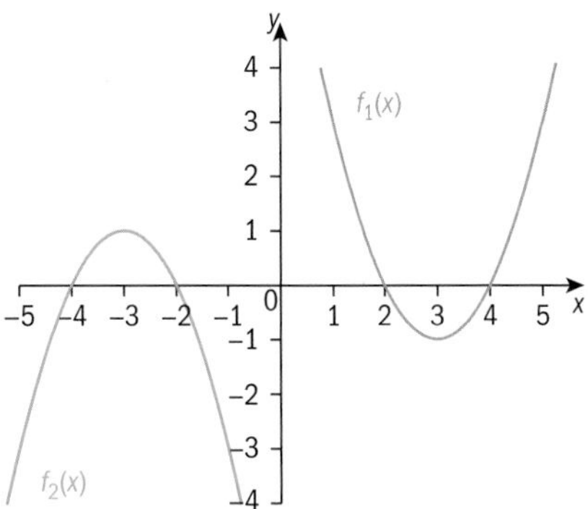

d

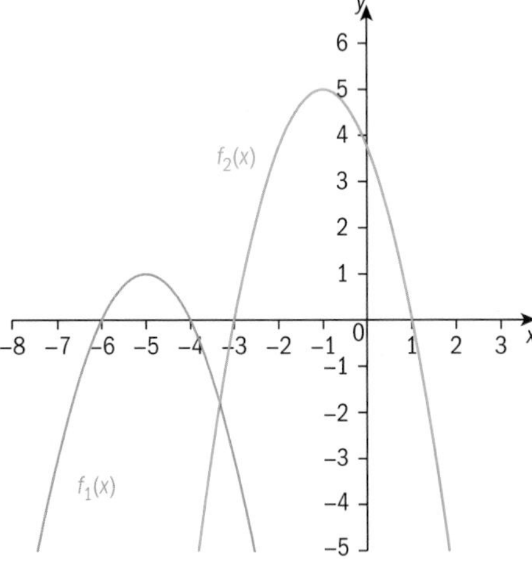

e

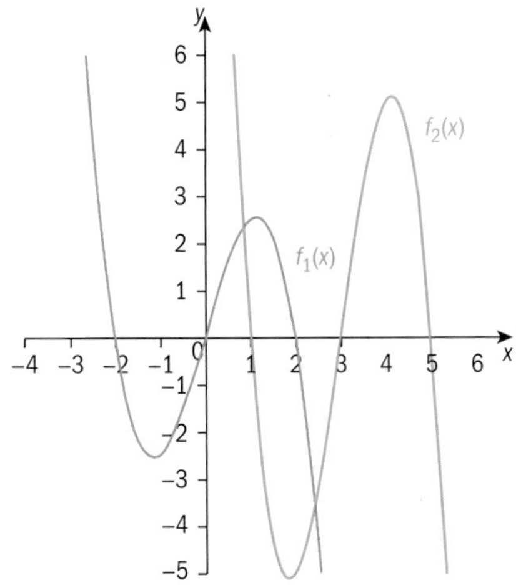

f

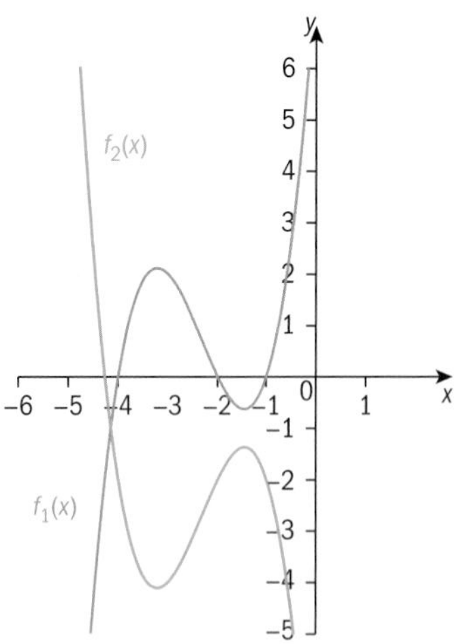

5 Given the graph of $f_1(x)$, **draw** the graph of $f_2(x)$.

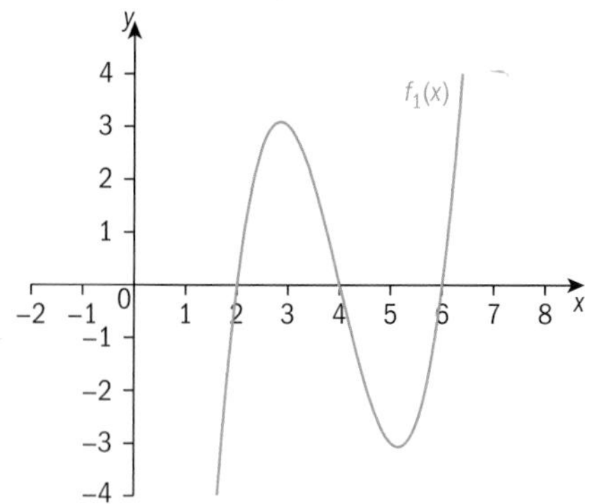

a $f_2(x) = 2f_1(x+4)$

b $f_2(x) = f_1(2x) + 4$

c $f_2(x) = -f_1(2x)$

d $f_2(x) = 2f_1(-x)$

Review in context

Orientation in space and time

Whenever an object such as a ball, arrow, or dart flies freely through the air, its motion is affected by gravity. Any object which isn't self-propelled (like a bird or a rocket would be) follows a path given by a parabola. This is because gravity always acts directly downward and exerts a constant force.

The path that an object takes in flight is known as its *trajectory*. The equation of a trajectory usually gives the height (y) in terms of the horizontal distance travelled (x).

1 A frog leaps from a lily pad. Viewed from the side, its initial position is given by (0, 0). The x-axis is horizontal and the y-axis is vertical; 1 unit represents 1 meter.

The trajectory of its flight is given by $y = f(x)$ where $f(x) = \frac{3}{4}x - \frac{1}{4}x^2$.

a **Draw** a graph of its trajectory.

b **Find** the distance it travels before returning to its starting height.

c The frog can change its jump by modifying the angle and speed with which it launches off. **Describe** clearly how its jump would vary if its new trajectory were given by

i $y = f(2x)$ **ii** $y = 1.2f(0.7x)$ **iii** $y = f(-x)$

d **Explain** clearly why $y = -0.8f(x)$ does not model a trajectory for the frog's jump.

2 A student throws a ball from a window in a tall building.

Its trajectory is given by $y = h - 5\left(\frac{x}{u}\right)^2$, where $(0, h)$ is the position from which it is thrown and u is its initial horizontal velocity in m/s.

a **Describe** the effect of increasing the value of h on the graph of the trajectory.

b **Describe** the effect of increasing the value of u on the graph of the trajectory.

3 A baseball player is practicing her batting on an indoor range. Let (0, 0) be the point at which she hits the ball. The indoor range is about 5 m higher than the point at which she hits the ball.

a Her first hit follows a trajectory given by $y = -\frac{1}{20}(x-10)^2 + 5$.

i **Describe** the transformations that map a curve with equation $y = x^2$ onto a curve with equation $y = -\frac{1}{20}(x-10)^2 + 5$.

ii **Show that** this curve passes through the point (0, 0) as described.

iii Use the graph transformations to determine whether the maximum height of this curve is more than 4 meters above her hitting point and hence would hit the roof.

b Her second shot follows a trajectory given by $y=-\frac{1}{40}\left[(x-10)^2-100\right]$.

By considering graph transformations, find the coordinates of the maximum point of the curve. Hence **determine** whether or not the ball will hit the ceiling.

4 Car loans use simple interest so that, every month, the amount you owe on the loan (the loan balance) decreases by the same amount. (House loans or mortgages do not work this way.) Suppose you have \$12 000 in car loans and you are making payments of \$150 per month.

a **Write** an equation to represent the loan balance (b) as a function of the number of months that have gone by (t).

b **Draw** a graph of this relationship, $b(t)$.

c **Find** $b(50)$. **State** what this amount represents.

d **Find** how long it will take you to pay off the loan.

e **Suggest** what change(s) to the scenario could produce a parallel graph with a higher y-intercept.

f **Suggest** what change(s) to the scenario could produce a graph with a lower y-intercept but the same x-intercept.

g **State** how the graph will be transformed if you increase your original payment to \$250 per month. **Write down** an equation for this new scenario.

Reflect and discuss

How have you explored the statement of inquiry? Give specific examples.

Statement of Inquiry:

Relationships model patterns of change that can help clarify and predict duration, frequency and variability.

10.2 A thin line divides us

Global context: Scientific and technical innovation

FORM

Objectives

- Changing the subject of a formula
- Simplifying rational algebraic expressions
- Performing mathematical operations on rational algebraic expressions

Inquiry questions

F
- What is the subject of a formula?
- How can you change the subject of a formula?
- How can you simplify a rational expression?

C
- How do operations on rational algebraic expressions compare to operations on rational numerical expressions?

D
- Does technology help or hinder understanding?

ATL **Organization skills**

Use appropriate strategies for organizing complex information

Statement of Inquiry:

Representing change and equivalence in a variety of forms has helped humans apply their understanding of scientific principles.

10.2

11.3

E10.2

You should already know how to:

• simplify fractions	**1** Simplify each fraction. **a** $\frac{4}{16}$ **b** $\frac{14}{35}$ **c** $\frac{x^2}{x^3}$ **d** $\frac{20x^2}{(20x)^2}$
• factorize quadratic expressions	**2** Factorize. **a** $x^2 + 5x + 6$ **b** $x^2 - 3x - 10$ **c** $x^2 - 9$
• solve a linear equation algebraically	**3** Solve. **a** $2x - 1 = 5$ **b** $5x - 7 = 8$ **c** $8x + 1 = 4x + 17$

Rearranging formulas

- What is the subject of a formula?
- How can you change the subject of a formula?
- How can you simplify a rational expression?

You already know how to solve linear equations like $5x - 7 = 8$, which has solution $x = 3$. This is sometimes called 'isolating the x' or 'rearranging the equation to solve for x'. You also know how to use substitution to solve equations that show relationships between variables, for example, solving $c^2 = a^2 + b^2$ when $a = 3$ and $b = 4$.

$5x - 7 = 8$ and $c^2 = a^2 + b^2$ are both equations, but $c^2 = a^2 + b^2$ is also a formula.

A **formula** is an equation that describes an algebraic relationship between two or more sets of values. Each variable in a formula can take different values, depending on the values of the other variables.

$c^2 = a^2 + b^2$ is valid for $a = 3$, $b = 4$, $c = 5$. It is also valid for $a = 5$, $b = 12$, $c = 13$ and an infinite number of other combinations of values a, b, and c.

By substituting values for a and b you can find a corresponding value for c. Similarly, by substituting values for b and c you can find a corresponding value for a. In other words, you can solve for any of the variables, given the values of the other two.

Exploration 1

The relationship between degrees Celsius and degrees Fahrenheit is given by the formula: $F = \frac{9}{5}C + 32$.

1 Find the temperature in °F when it is 35°C.

2 Find the temperature in Celsius when it is 77°F. To find a temperature in °C, you need to solve an equation. Carefully write down the steps you have used to solve the equation.

▶ Continued on next page

3 Using the same steps as you used in step **2**, begin with the formula for finding F and finish with a formula for C. C is now the 'subject' of the formula.

4 Use your new formula from step **3** to find the temperature in Celsius when it is 77°F. Compare your answer to that in step **2**.

5 Write down what you think it means to 'change the subject of the formula', and the necessary steps to do this.

The Fahrenheit scale was based on a mixture of salt and water freezing at 0°F. In this scale, water freezes at 32°F and boils at 212°F. The Celsius scale is based on water freezing at 0°C and boiling at 100°C.

Changing the subject of the formula is a way of rearranging algebraic relationships. When you change the subject of a formula, you solve a different variable in the formula.

Reflect and discuss 1

- In Exploration 1, does changing the subject of the formula change the relationship between F and C?
- What are similarities and differences between changing the subject of a formula and solving an equation?
- When could it be useful to change the subject of a formula?

Example 1

Pythagoras' theorem states that in a right-angled triangle, the relationship between the side lengths is $c^2 = a^2 + b^2$, where c is the hypotenuse and a and b are the other two sides.

Rearrange $c^2 = a^2 + b^2$ to make a the subject of the formula.

$c^2 = a^2 + b^2$

$c^2 - b^2 = a^2 + b^2 - b^2$ — Subtract b^2 from both sides.

$c^2 - b^2 = a^2$

$a^2 = c^2 - b^2$ — Rearrange the terms so that the variable a is on the left hand side.

$\sqrt{a^2} = \sqrt{c - b^2}$ — Take the square root of both sides. a is a length, so you can ignore the negative square root.

$a = \sqrt{c^2 - b^2}$ — Now a is the subject of the formula.

When the variable you are trying to isolate appears more than once in the formula, factorize so that the variable appears only once.

ATL

Example 2

Rearrange $F = \frac{mv - mu}{t}$ to make m the subject.

$F = \frac{mv - mu}{t}$ —— Multiply both sides by the denominator of the right hand side.

$Ft = mv - mu$ —— Factor out m to isolate it.

$Ft = m(v - u)$ —— Divide both sides by $v - u$.

$\frac{Ft}{v - u} = m$

Practice 1

1 Rearrange each formula to make the variable in brackets the subject.

a $y = 5x - 4$ [x]

b $y = \frac{1}{3}x^2 + 3$ [x]

c $p = \sqrt{m + 2}$ [m]

d $V = (2w - 5)^2$ [w]

e $V_f^2 = V_i^2 + 2ad$ [V_i]

f $V_f = V_i + at$ [a]

g $I = \frac{V}{R + r}$ [R]

h $F = \frac{mV^2}{r}$ [r]

i $F = \frac{mV^2}{r}$ [V]

j $F = \frac{1}{2}mV^2 + mgh$ [m]

k $\mu = \sqrt{\frac{3RT}{M}}$ [M]

l $c = (3a + 2b)^2 - 1$ [a]

You may have seen some of these in science.

2 Write down the formula for the area A of a circle with radius r. Make r the subject of the formula.

3 The formula for the volume V of a cone with height h and base of radius r is $V = \frac{1}{3}\pi r^2 h$. Make r the subject of the formula.

4 When driving a car, if you accelerate from rest at a constant rate a for an amount of time t, the distance d travelled is given by the formula $d = \frac{1}{2}at^2$. Rearrange the formula to make t the subject.

5 The time taken (T) for a pendulum to swing once each way and back to its starting position is given by the formula $T = 2\pi\sqrt{\frac{L}{g}}$, where L is the length of the pendulum.
Rearrange the formula to make L the subject.

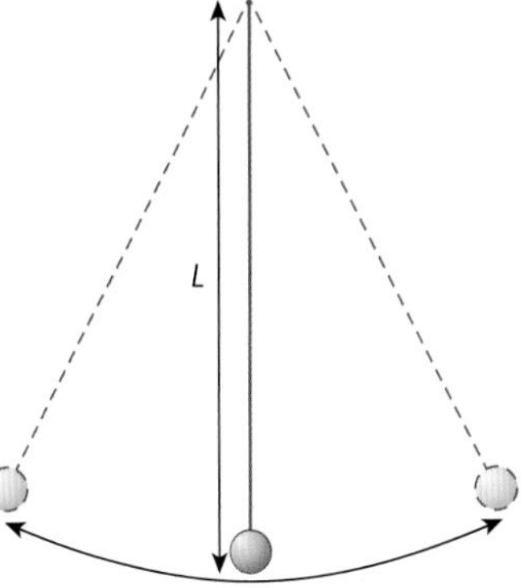

Problem solving

6 Between any two charged particles with charge q_1 and q_2, there is an electric force. You can calculate this force using the formula $F = \frac{kq_1q_2}{r^2}$, where r is the distance between the particles. Rearrange the formula to make r the subject.

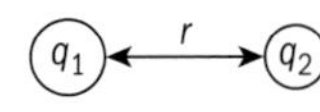

A **rational algebraic expression** is a fraction that contains variables.

Factorizing can change rational algebraic fractions to a simpler form. When the numerator and the denominator have a common factor, you can divide to simplify the fraction, as you would a numerical fraction.

For example $\frac{12}{20}=\frac{3}{5}$

$$\frac{x(x-2)}{(x-2)(x+5)}=\frac{x}{x+5}$$

ATL

Exploration 2

1 Simplify these fractions completely. Justify each step.

a $\frac{25}{100}$ **b** $\frac{45}{75}$ **c** $\frac{64}{80}$

d $\frac{91}{105}$ **e** $\frac{12a}{3a}$ **f** $\frac{15a^3}{60a}$

g $\frac{3(x+1)}{5(x+1)}$ **h** $\frac{(x+y)(x-y)}{4(x+y)}$ **i** $\frac{(a+b)(c-d)(e+f)(g-h)}{(a+b)(e+f)(g+h)}$

2 State the one value that you are not allowed to have in the denominator of a fraction. Explain why that is.

3 For each of these rational expressions, explain the steps needed to simplify it to the given expression.

a $\frac{x^2-9}{x-3}=x+3$ **b** $\frac{x^2-9}{x+3}=x-3$ **c** $\frac{x^2+3x}{x+3}=x$

4 For each expression in step **3**, state any value that x cannot take.

5 Based on your answers to steps **1** and **3**, write down the steps for simplifying a rational algebraic expression.

Reflect and discuss 2

- The values of a variable that are not allowed in the denominator are called the restrictions on the variable. Do you find these restrictions before or after you simplify? Justify your answer.
- Why do you need to find restrictions on the variable? What are they for?

To simplify a rational expression, you may need to factorize the numerator and/or denominator first.

The Ancient Egyptians used images to represent fractions in the same way they did to represent words. Horus was a sky god whose eye was shattered into six pieces. Each part of his shattered eye represented a different fraction. Interestingly if you add the six fractions together you get $\frac{63}{64}$ rather than 1. Some consider this to demonstrate that perfection is not possible.

Example 3

Simplify completely and state any restrictions on x.

$$\frac{x^2-3x-28}{x^3-9x^2+14x}$$

$$\frac{x^2-3x-28}{x^3-9x^2+14x}=\frac{(x+4)\cancel{(x-7)}}{x(x-2)\cancel{(x-7)}}$$

Factorize the numerator and denominator completely.

$$=\frac{(x+4)}{x(x-2)}$$

$x \notin \{0, 2, 7\}$

The restrictions are any values that make the expressions in the denominator 0, including any cancelled when simplifying.

Practice 2

Write these expressions in their simplest form. State any restriction(s) on the variable.

In simplest form the numerator and denominator have no common factors.

1 $\frac{10a^3}{2a}$

2 $\frac{5x^5+3x^4}{2x^4}$

3 $\frac{42x^6y^5}{35x^3y^7}$

4 $\frac{(x-7)(x+8)}{(x+7)(x+8)}$

5 $\frac{(a-b)(a+2b)}{(a-2b)(b-a)}$

6 $\frac{x^2+6x+8}{x^2+7x+10}$

7 $\frac{x^2+x-12}{x^2-x-6}$

8 $\frac{x^2-3x-10}{x^2+2x-35}$

9 $\frac{x^3-12x^2+32x}{x^2-7x+12}$

10 $\frac{x^2+4x-21}{x^2-3x-18}$

Problem solving

11 Write an unsimplified rational expression in x that has a quadratic expression in the numerator and the restriction on x that $x \notin \{-2, 5\}$.

12 Write a rational expression in x which simplifies to $\frac{x+3}{x-10}$, and where $x \notin \{6, 10\}$.

C Operations on rational algebraic expressions

- How do operations on rational algebraic expressions compare to operations on rational numerical expressions?

You can multiply and divide algebraic fractions in the same way as numeric fractions.

Exploration 3

1 Multiply these fractions by:

i simplifying first, and then multiplying,

ii multiplying first, and then simplifying.

▶ Continued on next page

Explain which method you found easier.

a $\frac{3}{10}\times\frac{2}{9}$ **b** $\frac{-4}{12}\times\frac{2}{6}$ **c** $\frac{12}{75}\times\frac{18}{60}$

2 Use your preferred method from step **1** to multiply these algebraic fractions.

a $\frac{x-5}{x+3}\times\frac{3x+9}{x+5}, x\neq -3, x\neq -5$

b $\frac{(x-2)(x-3)}{(x+2)(x+3)}\times\frac{(x+1)(x+2)}{(x-1)(x-2)}, x\notin\{-3, 1, \pm 2\}$

c $\frac{x^2+3x-4}{2x^2-14x+24}\times\frac{4x^2-16x+12}{2x^2+4x-16}, x\notin\{2, 3, \pm 4\}$

3 Divide these fractions, leaving your answers in simplest form. Explain your method.

a $\frac{5}{3}\div\frac{1}{6}$ **b** $\frac{4}{5}\div\frac{2}{10}$ **c** $\frac{12}{75}\div\frac{18}{60}$

4 Use your method in step **3** to divide these algebraic fractions. Explain your method.

a $\frac{x+5}{x+6}\div\frac{5x+25}{x+3}, x\neq -3, x\neq -5$

b $\frac{(x-8)(x-2)}{x(x+4)(x+3)}\div\frac{x(x+4)(x-2)}{(x+3)(x-5)}, x\notin\{\pm 3, 0, 5\}$

c $\frac{x^2+3x-4}{2x^2-14x+24}\div\frac{4x^2-16x+12}{2x^2+4x-16}, x\notin\{2, 3, \pm 4\}$

Reflect and discuss 3

- Write down two equivalent fractions. Take the reciprocal of each. Are the new fractions also equivalent?
- What are the differences between multiplying and dividing algebraic fractions and multiplying and dividing numeric fractions?

The *reciprocal* of $\frac{a}{b}$ is $\frac{b}{a}$. It is the multiplicative inverse.

Example 4

Simplify completely and state the restrictions on the variable.

$$\frac{x^2-9}{x^2-8x+16}\div\frac{x^2+6x+9}{x^2-7x+12}$$

$\frac{x^2-9}{x^2-8x+16}\div\frac{x^2+6x+9}{x^2-7x+12}$ — Factorize the quadratic expressions.

$=\frac{(x+3)(x-3)}{(x-4)(x-4)}\div\frac{(x+3)(x+3)}{(x-3)(x-4)}$ — To divide by a fraction, multiply by its reciprocal.

$=\frac{(x+3)(x-3)}{(x-4)(x-4)}\times\frac{(x-3)(x-4)}{(x+3)(x+3)}$ — Simplify.

$=\frac{(x+3)^2}{(x-4)(x+3)}, x\notin\{\pm 3, 4\}$ — Include values of x that make the denominators of the *original expressions* equal to 0.

Practice 3

Simplify each rational expression. State any restrictions on the variables.

1 $\frac{25p}{55k}\times\frac{66p}{5k}$

2 $\frac{3p^3c^4}{5c^5}\div\frac{24p^5c}{10c^3}$

3 $\frac{5x+10}{3x-9}\times\frac{6x-18}{4x+8}$

4 $\frac{6x-12y}{x+3y}\div\frac{3}{2x+9y}$

5 $\frac{x^2-x-2}{x^2-2x+1}\times\frac{x-1}{x+1}$

6 $\frac{a^3+ab^2}{a^3-2a^2b}\div\frac{a^2+b^2}{(a-2b)^2}$

7 $\frac{x^2-3x-10}{x^2+x-6}\times\frac{x^2+2x-3}{x^2+x-2}$

8 $\frac{x^2+5x-24}{x^2-9x+18}\div\frac{x^2+2x-48}{x^2+3x-18}$

9 $\frac{x^2-9}{x^2-4x+3}\times\frac{2x-2}{x^2+5x+6}\div\frac{6x}{2x^2+4x}$

10 $\frac{x^2-3x-10}{x^2-4}\div\frac{2x^2-2}{6x+12}\div\frac{x^2-6x+5}{2x^2+6x+4}$

Reflect and discuss 4

- Should you find the restrictions on the variable in a multiplication before or after multiplying? Explain your reasoning.
- In a division question, when would you find the restrictions – before or after finding the reciprocal? Explain your reasoning.

Exploration 4

There are two possible ways of connecting electrical components together: in series or in parallel. For components connected in parallel, the total resistance (R_{tot}) of the circuit is: $\frac{1}{R_1}+\frac{1}{R_2}+\frac{1}{R_3}+\cdots=\frac{1}{R_{tot}}$ (where R_1, R_2, ... are the resistances of the components).

1 Find the total resistance in a circuit with:

   a $R_1 = 4$ ohms, $R_2 = 8$ ohms

   b $R_1 = 2$ ohms, $R_2 = 6$ ohms

   c $R_1 = 3$ ohms, $R_2 = 5$ ohms

2 When two resistances are connected in parallel, determine if the combined resistance is greater or less than the individual resistances.

3 Write down a necessary step before adding fractions together, and explain.

4 Find an expression for the total resistance when one component has twice the resistance of the other.

5 Find an expression for the total resistance when one component has a resistance 2 ohms more than the other.

Example 5

Find each rational expression in simplest form.

a $\frac{1}{x+5}-\frac{2}{x-3}$ **b** $\frac{2}{x-2}+\frac{x}{x^2-5x+6}$

a $\frac{1}{x+5}-\frac{2}{x-3}=\frac{1}{x+5}\times\frac{x-3}{x-3}-\frac{2}{x-3}\times\frac{x+5}{x+5}$

If there is no common factor between the denominators then the LCM is simply the product of the denominators.

$=\frac{x-3}{(x+5)(x-3)}-\frac{2(x+5)}{(x-3)(x+5)}$

$=\frac{x-3-2x-10}{(x+5)(x-3)}$

Simplify your final answer completely.

$=\frac{-x-13}{(x+5)(x-3)}\ x\notin\{-5, 3\}$

b $\frac{2}{x-2}+\frac{x}{x^2-5x+6}=\frac{2}{x-2}+\frac{x}{(x-2)(x-3)}$

There is a common factor between the two denominators: $(x-2)$.

$=\frac{2}{x-2}\times\frac{x-3}{x-3}+\frac{x}{(x-2)(x-3)}$

With a common denominator, you can now add the terms.

$=\frac{2(x-3)+x}{(x-2)(x-3)}$

$=\frac{2x-6+x}{(x-2)(x-3)}$

$=\frac{x-6}{(x-2)(x-3)}\ x\notin\{2, 3\}$

Practice 4

Objective: **C.** Communicating
v. organize information using a logical structure.

In Practice 4, show the steps in your working in a logical order. Remember to include any restrictions on the variables.

Simplify each rational expression. State any restrictions on the variables.

1 $\frac{3}{x}+\frac{4}{y}$ **2** $\frac{3}{2x}-\frac{1}{4x}$ **3** $\frac{1}{3x}+\frac{1}{6x^2}$ **4** $\frac{2}{ab}+\frac{3}{ac}-\frac{5}{bc}$

5 $\frac{1}{x+2}+\frac{1}{x-2}$ **6** $\frac{x}{x+4}-\frac{5}{x-3}$ **7** $\frac{x-2}{x+4}+\frac{x+3}{x-2}$ **8** $\frac{x^2-2x+1}{x-3}-x$

9 $\frac{1}{x+5}-\frac{2x}{x^2+3x-10}$ **10** $\frac{5x}{x-8}+\frac{4}{x^2-5x-24}$ **11** $\frac{7}{x-1}+\frac{6}{x}-\frac{5}{x+1}$ **12** $\frac{3}{x+3}-\frac{x}{4}+\frac{5}{x-5}$

You have seen that algebraic fractions behave in much the same way as numerical fractions. Simplifying, multiplying, and dividing involve very similar processes:

- Simplify rational expressions by first factorizing the expression, and then dividing by common factors in the numerator and denominator.
- To multiply rational expressions, multiply the numerator and multiply the denominator. It is more efficient to simplify the rational expressions before multiplying them.
- To divide rational expressions, multiply by the reciprocal of the divisor.
- To add and subtract rational expressions, first rewrite the fractions with a common denominator.

D Finding restrictions on variables using technology

- Does technology help or hinder understanding?

Exploration 5

1 In Exploration 2 you saw that $\frac{x^2-9}{x+3}$ simplifies to $x-3$. Use your GDC or graphing software to draw the graphs of both expressions.

2 Describe any similarities or differences between the graphs.

3 State whether or not the GDC or graphing software has graphed both expressions correctly.

4 Explain whether or not you think both expressions are equal.

5 Repeat steps **2** to **4** for $\frac{x^2+3x}{x+3}$ which simplifies to $x+3$.

6 Explain whether you should always trust results you obtain using technology.

Summary

- A **formula** is an equation that describes an algebraic relationship between two or more sets of values. Each variable in a formula can take different values, depending on the values of the other variables.
- Changing the subject of the formula is a way of changing, or **rearranging** algebraic relationships. It usually involves isolating a different variable in the formula.
- A **rational algebraic expression** is a fraction that contains variables.

Rational expressions:

- must exclude all the values of x that make a denominator in the expression equal to 0
- can be simplified by dividing common factors
- can be multiplied, divided, added, or subtracted using similar rules as for numerical fractions.

Mixed practice

Rearrange the formulae to make the variable in brackets the subject.

1 $V=\frac{4}{3}\pi r^3$ [r]

2 $P=\frac{500T}{v}$ [T]

3 $F=G\frac{m_1m_2}{r^2}$ [r]

4 $W=\frac{1}{2}CU^2$ [C]

5 $E=mc^2$ [c]

6 $v=v_0\sqrt{\frac{c-v}{c+v}}$ [c]

Simplify each rational expression. State any restrictions on the variables.

7 $\frac{12x^4y}{36xy}$

8 $\frac{5x^7-6x^5}{4x^3}$

9 $\frac{x^2-4x-21}{x^2-8x-33}$

10 $\frac{x+4}{x^2-16}$

11 $\frac{2}{x}+\frac{y}{5}$

12 $\frac{16x^3y^2}{5z^3}\times\frac{15z}{8xy}$

13 $\frac{x^3y^2+x^2y^3}{x^3y^3}\div\frac{x^2-y^2}{x^4y^4}$

Simplify these expressions. **State** the restrictions on the variable.

14 $\frac{x-2y}{4}-\frac{1+2xy}{x}$

15 $\frac{7}{x-7}+\frac{5}{x+5}$

16 $\frac{-3}{x+9}+\frac{-2}{x-8}$

17 $\frac{x+1}{x-1}-\frac{x-4}{x+4}$

18 $\frac{9}{x-1}-\frac{9x-4}{x^2-1}$

19 $\frac{3}{a}+\frac{5}{b}-\frac{a+b}{4}$

20 $\frac{x+1}{6}+\frac{2}{x}-\frac{x}{x-2}$

Review in context

Scientific and technical innovation

1 An important relationship in the study of motion is $V=\frac{d}{t}$, where V is the velocity of an object, d is distance and t is the time taken. An Airbus A333 has a cruising speed of 870 km/h. In this exercise, consider this speed to be the average speed for the entire flight.

a On a flight from Brussels to Montreal, it is flying with a headwind of 40 km/h. **Find** the actual speed of the A333.

b On the return flight, it is flying with a tailwind of 30 km/h. **Find** the actual speed of the A333.

c **Write down** an expression for the actual speed of the A333 with a headwind of x km/h.

d **Write down** an expression for the actual speed of the A333 with a headwind of y km/h.

The distance between Brussels and Montreal is 5560 km.

e Make t the subject of the formula.

f **Find** the time it takes to travel without any wind.

g **Find** the time it took on the flight described in part **a**.

h **Find** the time it took on the return flight described in part **b**.

i **Find** the wind speed and direction (headwind or tailwind) if the flight took 6 hours and 57 minutes.

j **Find** the wind speed and direction (headwind or tailwind) if the flight took 6 hours 2 minutes and 30 seconds.

2 When resistors are connected in parallel, the total resistance R_{tot} is given by $\frac{1}{R_1}+\frac{1}{R_2}+\frac{1}{R_3}+\quad=\frac{1}{R_{tot}}$.

When resistors are connected in series, the total resistance is simply the sum of the individual resistances.

a **Write down** an expression for the total resistance of a circuit with 3 resistors, where one of them is 4 ohms larger than the smallest resistor and the largest resistor is 2 ohms larger than twice the smallest one. **Simplify** the expression into a single fraction.

b **Write down** an expression for the total resistance when a resistor the same as the largest one in **a** is connected in series to the circuit in **a**. Then **simplify** the expression into a single fraction.

3 Ohm's Law states that the voltage (V), in an electrical circuit with current (I) and two resistors, R and r connected in series, is given by $V = IR + Ir$.

a Make I the subject of the formula.

b Circuit A has a resistor (R) of 6 ohms and a smaller resistor with resistance r, connected in series. Circuit B has a resistor (R) of 5 ohms and a smaller resistor three times the smaller resistor in Circuit A. Circuit A and circuit B are connected in parallel, and have a voltage of 12 volts. **Find** an expression for the total current flowing in both circuits in terms of r.

Tip

The total current is divided between the circuits if they are connected in parallel.

Problem solving

4 Scientists calculate the gravitational force between *any* two masses (m_1 and m_2) using the formula $F = \frac{Gm_1m_2}{r^2}$, where r is the distance between them and G is the gravitational constant.

a **Determine** how many times larger the force of gravity is when the distance between the two masses is halved.

b One mass is doubled and the distance between the masses is increased by scale factor 4. **Determine** the effect on the gravitational force between them.

c One mass is doubled, the other is tripled and the distance between them is reduced by scale factor 4. **Determine** the effect on the gravitational force between them.

d One mass is halved and the other is made eight times larger. **Find** how the distance r needs to change to keep the gravitational force the same.

The minimum speed required for an object (usually a rocket) to break free of Earth's gravitational field is called 'escape velocity'. The equation for the escape velocity v_e of an object is given by: $v_e = \sqrt{\frac{2Gm_E}{r}}$ where G is the gravitational constant, m_E is Earth's mass, and r is the distance between Earth's center of mass and the object.

The escape velocity for objects leaving the Earth is approximately 25 000 miles per hour.

Now, rockets require a tremendous amount of fuel to break away from Earth's gravitational pull. This fuel adds considerable weight to the rocket, and thus it takes more thrust to lift it. But to create more thrust, you need more fuel. It's a vicious circle that scientists hope to overcome by building lighter vehicles, discovering more efficient fuels and new methods of propulsion.

Reflect and discuss

How have you explored the statement of inquiry? Give specific examples.

Statement of Inquiry:

Representing change and equivalence in a variety of forms has helped humans apply their understanding of scientific principles.

10.3 Getting more done in less time

Global context: Globalization and sustainability

FORM

Objectives

- Finding a constant of proportionality
- Setting up direct and indirect proportion equations to model a situation
- Graphing direct and indirect relationships
- Recognizing direct and inverse proportion from graphs
- Identifying direct and inverse proportion from tables of values

Inquiry questions

F
- How do you find a constant of proportionality?
- What does a proportional relationship look like?

C
- What does it mean to be proportional?
- How does changing one variable in a proportional relationship affect the other?

D
- Can situations seem proportional when they are not?
- Is simpler always better?

ATL **Transfer**

Combine knowledge, understanding and skills to create products or solutions

Statement of Inquiry:

Changing to simplified forms can help analyze the effects of consumption and conservation.

5.1

10.3

E5.1

You should already know how to:

• express a ratio as a fraction	**1** Express each ratio as a fraction in its simplest form. **a** 5 : 3 **b** 2 : 4 **c** 4.5 : 13.3 **d** 0.2 : 1.2
• write linear functions to model real-life situations	**2 a** John earns £10 per hour. Write a function for the amount he earns in h hours. **b** A taxi ride costs \$2 plus \$4.50 per mile. Write a function for the cost of any journey.
• draw graphs of linear functions	**3** Draw graphs of the two functions above.
• draw graphs of quadratic functions	**4** Draw the graph of $y = x^2$ for integer values of x from −3 to +3.

F Direct and inverse variation

- How do you find a constant of proportionality?
- What does a proportional relationship look like?

If you sell identical items, say mouse mats, the money you make varies with the number of mouse mats you sell. If you sell twice as many mouse mats, you will make twice as much money. If you sell half as many, you make half as much money. We say the two quantities are directly proportional.

Two variables are said to be in **direct proportion** if, and only if, their ratio is a constant for all values of each variable.

$y \propto x$ means 'y varies directly as x' or 'y is directly proportional to x'.

ATL

Exploration 1

A store sells regular light bulbs for 2 Euros per bulb.

1 Write down a function to represent the store's income from selling regular bulbs. Identify your variables and justify your choice of letters to represent each variable.

2 Determine which variable is dependent, and which one is independent.

3 Draw a graph representing the amount of money made from selling up to 10 light bulbs.

4 From your graph, determine what happens to the income when you multiply the number of light bulbs sold:

a by 2 **b** by 4 **c** by 5 **d** by 12 **e** by 14

To meet the growing demand for more efficient bulbs, the station also sells LED bulbs for 7 Euros per bulb.

▶ Continued on next page

5 Using the same variables as in step **1**, write down a new function to represent the amount of income from selling LED bulbs. Draw a line on your graph to represent this relationship.

6 Determine whether the amount of money made is in direct proportion to the number of light bulbs sold. Justify your answer.

Reflect and discuss 1

- Justify why the graphs you drew in Exploration 1 go through the origin.
- Describe a situation where, if one quantity increases, the other increases, but not in direct proportion.

$y \propto x$ means that $y = kx$ for a constant k, where $k \neq 0$.

k is called the **proportionality constant**, or **constant of variation**.

The function $y = kx$ is called a **linear variation function**.

In other words, if y is proportional to x, then the relationship between x and y is a function of the form $y = kx$.

Reflect and discuss 2

- In Exploration 1, identify the proportionality constant for the function for:
  - regular light bulbs
  - LED light bulbs.
- For linear relationships of the form $y = kx$, the proportionality constant k is also called the constant ratio of x and y. Explain why you think this is.

Example 1

A shop charges €1.50 for a liter of mineral water.

a Write down a function to represent the relationship between the number of liters of mineral water and the total price of mineral water purchased.

b Hence, determine the proportionality constant.

c Show that number of liters and price are in direct proportion by drawing a graph.

▶ Continued on next page

a P = price — Define the variables.

l = number of liters

$P(l) = 1.50l$

b $k = 1.50$ — For every extra liter the price increases by €1.50.

c The graph is a straight line that passes through the origin, so P and l are in direct proportion.

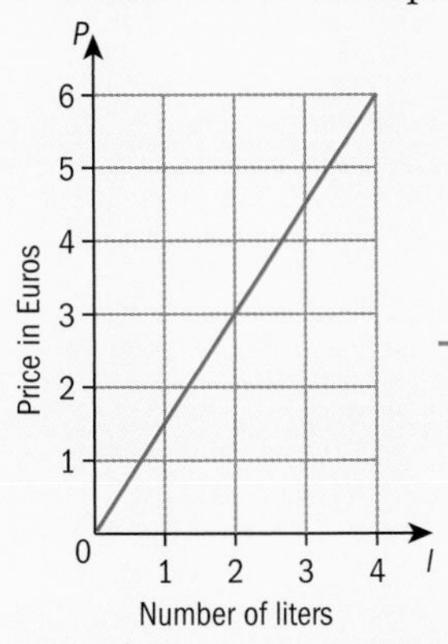

The total price depends on the number of liters, so P is the dependent variable and N is the independent variable.

If the graph is a straight line passing through the origin then the equation of the graph can be written as $y = kx$, and x is directly proportional to y.

Reflect and discuss 3

- How is the proportionality constant represented on the graph of a direct linear proportion?
- Explain why a linear graph that does not go through the origin does not represent a proportional relationship.

Example 2

The variable y varies as x. When $x = 4$, $y = 10$.

Find the value of y when $x = 7$.

'y varies as x' means 'y is directly proportional to x'.

$y \propto x$ means that $y = kx$

$10 = k \times 4$

$\Rightarrow k = \frac{10}{4} = 2.5$

When $x = 7$, $y = 2.5 \times 7 = 17.5$ — Use the proportionality constant to find y when $x = 7$.

Practice 1

1 The variable c is directly proportional to the variable b. Write an equation to represent this relationship. State what happens to:

a c if b is doubled

b b if c is tripled

c c if b is multiplied by $\frac{2}{3}$

d c if 4 is added to b.

2 The variable v is directly proportional to the variable t. When $t = 5$, $v = 40$.

a Write the function that represents the relationship of proportionality.

b Find the value of the constant of proportionality.

c Find v when $t = 25$.

3 P is directly proportional to V. When $V = 3.2$, $P = 8.64$.

Find V when $P = 5.4$.

4 Two variables x and Q are in direct proportion.

Find the missing values from this table:

x	0	4.7	6.1	
Q		6.58		13.72

Problem solving

5 In order to reduce trash in landfills, many people are beginning to compost food waste. Composting produces rich soil that can then be used in gardens or farms. In a science experiment, Kent records the number of containers of food waste dumped in a composter at one time and the mass of soil produced.

Number of containers	2	3	7	9
Mass of soil (kg)	1.6	2.4	5.6	7.2

a Verify that the number of containers and mass of soil are in direct proportion.

b Find the constant of proportionality.

c Calculate the amount of soil produced from 5 containers of food waste.

6 Determine which of these functions represent direct proportion, and find the constant of proportionality for those that do.

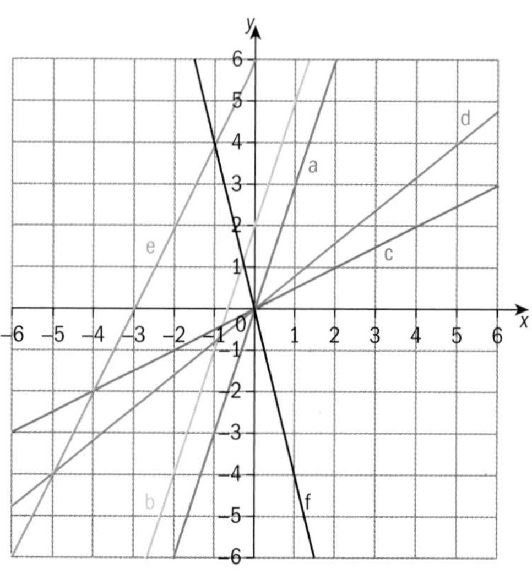

Objective: **B.** Investigating patterns
ii. describe patterns as general rules consistent with findings

In this Exploration, you will find a general rule in terms of a specific function that models the relationship between the time needed to plant trees and hourly income.

ATL

Exploration 2

In an effort to replace trees that have been harvested for consumer products, students can take a summer job that pays 8 cents per tree that they plant. One student can plant 300 trees in one hour.

Find the total income from planting 2400 trees.

1 Determine how long it would take one student to plant 2400 trees. Hence, find this student's income per hour.

2 Copy and complete the table to show the time taken and income per hour for the different groups of students.

	1 student	2 students	4 students	8 students
Time (t) needed to plant 2400 trees (h)	8	4		
Income (I) per hour ($/h)	24			

3 Predict the time needed and income per hour for 16 students. Do some calculations to check your predictions. Add the values for 16 students to your table.

4 Identify a pattern from this table. Generalize this pattern, and write down a function to represent the relationship between the number of students planting trees and the income per hour.

5 Describe the relationship between the number of students and the time needed to plant 2400 trees.

In Exploration 2, one relationship is directly proportional: doubling the number of students doubles the income per hour. But doubling the number of students halves the time taken to plant 2400 trees. The time taken to plant the trees is *inversely proportional* to the number of students doing the planting.

Two variables x and y are **inversely proportional** if multiplying one of them by a non-zero number results in the other variable being divided by the same non-zero number.

If x and y are in an inverse linear proportion, you can say 'y varies inversely as x' or 'y is inversely proportional to x', and you can write $y \propto \frac{1}{x}$ and $y = \frac{k}{x}$.

Reflect and discuss 4

- Write the relationship between the number of students (s) and time taken to plant 2400 trees (t) from Exploration 2 as a function.
- Does the function have a constant of proportionality? Explain.

If y is inversely proportional to x, then y is directly proportional to $\frac{1}{x}$.
An equation $y = k \times \frac{1}{x}$ or $y = \frac{k}{x}$ represents a relationship of inverse proportion or an **inverse relationship**.

Graphs of $y = \frac{k}{x}$ are called reciprocal graphs.

Example 3

It takes one person 2 days to prepare one hectare of field for planting and there are 16 hectares to be prepared.

a Find the time for one person to prepare the 16 hectares for planting.

b Determine if this relationship represents direct or inverse proportion.

c Write down a function to represent the relationship between the time to prepare the field and the number of workers.

d Draw the graph of this function.

a $2 \times 16 = 32$ days

One person, two days' prep time per person, total time = 32 days.

b

w: number of workers	1	2	4	8
t: time to prepare 16 hectares	32	16	8	4

The prep time halves as the number of workers doubles.

This describes an inverse proportion.

c $t \propto \frac{1}{w} \Rightarrow t(w) = \frac{k}{w}$

Substitute values for t and w, and solve for k.

$$32 = \frac{k}{1}$$

$$k = 32$$

$$t = \frac{32}{w}$$

d

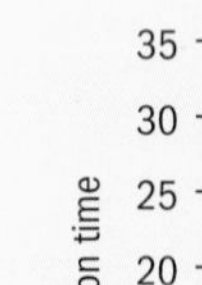

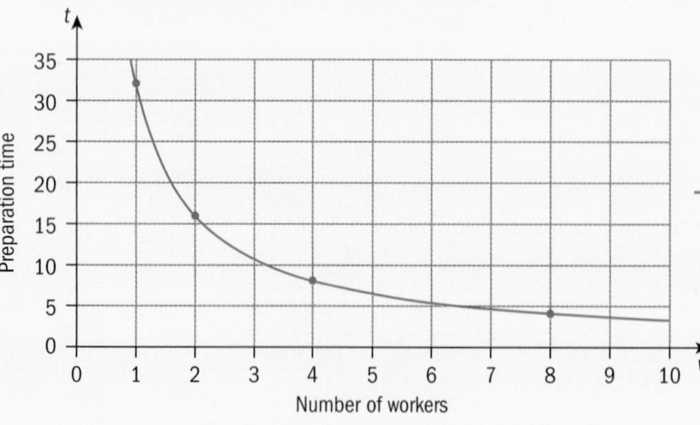

Plot the points from the table of values. Even though the function is in its own right a smooth curve, only the points with integer values on the n axis are meaningful since there can't be a non-integer number of workers.

Practice 2

1 For each part **a** to **e**:

i Determine whether there is a relationship of direct proportion, inverse proportion or neither. Justify your answer.

ii If there is a proportional relationship, find k.

a The perimeter p of an equilateral triangle with side length l.

b The number of liters n of paint needed to paint an area A, given that 3 liters will paint a 15 m^2 area.

c $y = 4x$ **d** $y = \frac{3}{x}$ **e** $y = \frac{x}{5}$

2 The variable f is inversely proportional to the variable v. When $v = 6$, $f = 4$.

a Write the function that represents the relationship of proportion.

b Find the value of the constant of proportionality.

c Find f when $v = 2$.

3 y is in inverse proportion to x. When $x = 5.2$, $y = 2$.

Find y when $x = 1.6$.

4 T is inversely proportional to m. When $m = 1.8$, $T = 4$.

Find T when $m = 3.1$.

5 Two variables s and t are in inverse proportion.

Find the missing values in this table:

s	4		4.8
t	0.8	2	

Problem solving

6 To balance a seesaw, Archimedes' Law of the Lever says that the distance d of each person from the pivot must be inversely proportional to their weight w.

a Write down a function to represent the Law of the Lever.

b Mark weighs 20 kg and is 1.5 m from the pivot. The seesaw is balanced when Emilie sits 1.25 m from the pivot. Use this information to find k.

c Hence, find Emilie's weight.

7 Milijenko has decided to encourage her class to participate in 'Adopt a Sumatran Tiger' which funds research and conservation of these critically endangered animals. The cost of adoption per person depends on the number of people participating as shown in the table below:

Number of people	1	4	8	10	20
Cost per person	\$300	\$75	\$37.50	\$30	\$15

a Verify that the number of people and the cost per person are inversely proportional.

b Find the constant of proportionality.

c Calculate the cost per person when 25 people participate.

8 Conservation efforts, like the adoption program, are trying to double the number of tigers in the wild. The amount spent on conservation is related to the increase in the number of wild tigers as shown in the table below:

Amount spent on conservation (millions of US $)	1.5	2.4	4.9	8.3
Increase in tiger population	390	624	1274	2158

a Determine whether or not this relation is in a proportion, and if so, state the type of proportion. Justify your answer.

b Write down an equation relating the two variables.

c Determine how much will need to be spent on conservation for the goal of adding 3200 tigers to the wild to be met. (This will double their population.)

9 a Determine whether the following relationships are proportional relationships. Justify your answer.

i The relationship between the distance and the time to see the lightning

ii The relationship between the distance and the time to hear the thunder

iii The relationship between the time to see the lightning and the time to hear the thunder

b Lightning strikes 5 km away.

i Find how long it takes to see the lightning.

ii Find how long it takes to hear the thunder.

iii Hence, verify your answer to **a iii**.

C Different types of variation

- What does it mean to be proportional?
- How does changing one variable in a proportional relationship affect the other?

You have seen examples of these proportional relationships:

- $y = kx$ is a relationship of direct proportion between x and y
- $y = \frac{k}{x}$ is a relationship of inverse proportion between x and y

In the next Exploration, you will see that y can vary, directly or inversely, with higher powers of x as well.

ATL

Exploration 3

This circle has radius r.

1 Write down the formula for the area A of the circle, in terms of r.

2 a Find the new area if you double the radius.

  b Find the new area if you triple the radius.

  c Find the new area if you halve the radius.

  d Multiply the radius by a few more factors, and find the new area each time.

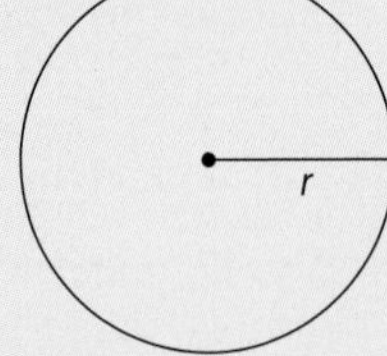

  e Suggest a general rule for what happens to the area when the radius of the circle is multiplied by a constant c, where $c \neq 0$.

3 Justify whether or not the formula represents a proportional relationship between the area and the radius of the circle.

4 Write a statement of proportionality linking area A and radius r. Find the constant of proportionality.

5 This cube has side length x.

  Write down the formula for the volume V of the cube. Find what happens to the volume if you double, halve or triple the side length. Suggest a general rule for what happens to the volume when the side length is multiplied by a constant c, where $c \neq 0$.

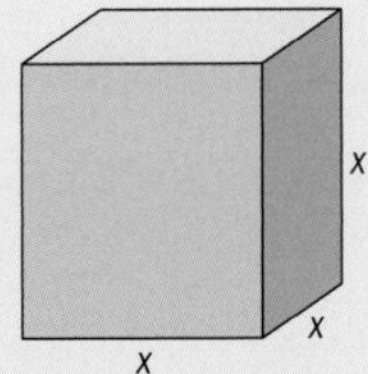

6 Justify whether or not the formula represents a proportional relationship between the volume and the side length of the cube.

7 Write a statement of proportionality linking volume V and side length x. Find the constant of proportionality.

Two variables x and y are in **direct non-linear proportion** if y is proportional to a power of x, or $y \propto x^n$, $n > 0$.

The variation function is $y = kx^n$, $k \neq 0$ and $n > 0$.

y varies directly as x^n or y is in direct proportion to x^n.

Exploration 4

Diego plans to tile a wall with square tiles of side length l.

1 Write down the formula for the area A_t of a square tile.

2 Let A_w represent the area of the wall.

  Let the number of tiles = n.

  a Write down an expression for n in terms of A_w and A_t.

  b Write down an expression for n in terms of A_w and l.

▶ Continued on next page

3 Determine whether the number of tiles is directly or inversely proportional to the area of the tile.

4 State the proportional relationship between the number of tiles and the side length of each tile. Find the constant of proportionality.

Diego changes his mind, and tiles the wall with square tiles of side length $\frac{2}{3}l$.

5 Write down the formula for the area A_s of this smaller tile.

6 Find the number m of new tiles he needs to cover A_w. Write down an expression for m:

  a in terms of A_w and A_s

  b in terms of A_w and l.

7 Compare m and n. How does changing to tiles with $\frac{2}{3}$ the side length affect:

  a the number of tiles needed

  b the proportionality constant?

Reflect and discuss 5

- Do you think that there has to be a whole number of tiles along the length and height of the wall?
- Does this affect the number of tiles you need to cover the wall entirely?

Two variables x and y are in an **inverse non-linear proportion** if y is proportional to a power of $\frac{1}{x}$, or $y \propto \frac{1}{x^n}, n > 0$. You can also write this as $y \propto x^{-n}, n > 0$.

The variation function is $y = \frac{k}{x^n}$, $k \neq 0$ and $n > 0$.

y varies inversely as x^n or y is inversely proportional to x^n

In Exploration 4 you should have found that:

- the smaller the tile, the more tiles you need for the given area
- the number of tiles is inversely proportional to the side length of each tile
- the proportionality constant k changes only when the initial conditions of the problem change.

Exploration 5

- Explain what happens to y when x is multiplied by a constant c in these variation functions:

  $y = kx$ $\quad$ $y = kx^2$ $\quad$ $y = kx^3$

  $y = \frac{k}{x}$ $\quad$ $y = \frac{k}{x^2}$

- Generalize your results first for $y = kx^n$, $n > 0$, and then for $y = \frac{k}{x^n}$, $n > 0$.

- $y = kx^n$ is a relationship of direct variation. When x is multiplied by a constant c, then y is multiplied by c^n.
- $y = \frac{k}{x^n}$ is a relationship of inverse variation. When x is multiplied by a constant c, then y is divided by c^n.

Example 4

The light intensity on a movie screen varies inversely with the square of the distance between the screen and the projector.

a When the screen is 3 m from the projector, the light intensity is 24 units.

Find the light intensity when the screen is 6 m away.

b Describe what happens to the light intensity when the distance between the projector and the screen is halved.

c The distance between the projector and the screen is increased by 25%. Find the effect on the light intensity on the screen.

d Draw a graph to represent the variation function.

a $I(d) = \frac{k}{d^2}$ — Choose sensible variable names and write the variation function.

$24 = \frac{k}{3^2}$ — Substitute the known values and solve for k.

$k = 24 \times 9 = 216$

$\Rightarrow I(6) = \frac{216}{6^2} = 6$ — Find I when $d = 6$.

The light intensity at 6 m is 6 units of illumination.

b $I(d) = \frac{k}{d^2}$

$I_{new}(d) = \frac{k}{\left(\frac{1}{2}d\right)^2} = \frac{k}{\frac{1}{4}d^2} = 4\frac{k}{d^2}$ — Replace d with $\frac{1}{2}d$ and substitute this into the variation function.

$= 4I_{original}(d)$

When the distance is halved, the light intensity is 4 times as great.

c $d_n = 1.25d = \frac{5}{4}d_o$ — Increasing d by 25% gives $1.25d$ or $\frac{5}{4}d$.

$I_n = \frac{k}{\left(\frac{5}{4}d_o\right)^2} = \frac{k}{\frac{25}{16}d_o^2} = \frac{16k}{25d_o^2} = \frac{64k}{100d_o^2}$ — $k \div \frac{25}{16} = \frac{16}{25}k$

$= 0.64$

When the distance is increased by 25% the light intensity is reduced to 64% of the original.

▶ Continued on next page

d $I = \frac{216}{d^2}$ — d is the independent variable, as the illumination I depends on the distance.

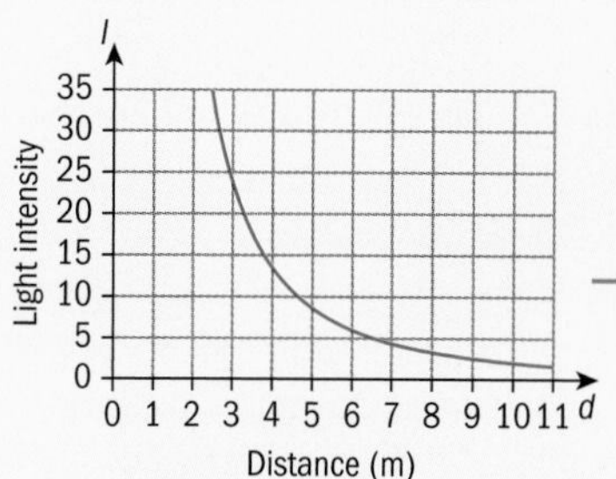

All points on the curve have meaning, as you can have any distance d on the horizontal axis.

Example 5

The table shows the resistances of four 1 m cables of the same material, but each a different radius.

Radius of cable (mm)	1	2	5	10
Resistance of 1 m of cable (ohms)	1.5	0.375	0.06	0.015

a Determine if this relationship between the radius and the resistance of the cable is one of direct variation, inverse variation or neither. Justify your answer.

b Hence, if the relationship is a variation function, write down the equation of the function.

a Let r = radius (mm) — Define the variables.

R = resistance (ohms)

As r increases, R decreases. There is either inverse variation, or no variation. — Compare the changes in the variables.

If $R = \frac{k}{r}$, then $k = r \times R$ — For $R = \frac{k}{r}$, k must be constant.

1st pair of values gives $k = 1 \times 1.5 = 1.5$

2nd pair of values gives $k = 2 \times 0.375 = 0.75$ — Test with pairs of values from the table.

k is not constant, so $R \neq \frac{k}{r}$

If $R = \frac{k}{r^2}$ then $k = r^2 \times R$ — Try $R = \frac{k}{r^2}$

1st pair of values gives $k = 1^2 \times 1.5 = 1.5$

2nd pair of values gives $k = 2^2 \times 0.375 = 1.5$ — Test with pairs of values from the table.

3rd pair of values gives $k = 5^2 \times 0.06 = 1.5$

4th pair of values gives $k = 10^2 \times 0.015 = 1.5$

Therefore, k is constant.

$R = \frac{k}{r^2}$, which is an inverse variation relationship, because k is constant.

b $R = \frac{1.5}{r^2}$ — Use the value of k from part **a**.

Reflect and discuss 6

- What conditions must always be met for direct or inverse variation?
- What do you think it means to be 'proportional'?

Practice 3

1 **i** For each table of values, determine whether the functions are a direct variation function, an inverse variation function or neither. Justify your answer.

ii If it is a variation function, write down the equation of the function.

a

x	1	2	3	4
y	1	8	27	64

b

x	1	2	3	4
y	24	12	8	6

c

x	−2	0	3	4
y	20	0	45	80

d

x	2	4	6	12
y	30	15	10	5

e

x	−2	0	2	5
y	−0.5	undefined	0.5	0.032

f

x	1	2	4	9
y	40	22	13	8

2 When $x = 2$, $y = 12$. Find the value of y when $x = 4$ if:

a y varies directly as x^2 **b** y varies directly as x^3

c y varies inversely as x **d** y varies inversely as x^2.

3 $y = 24$ when $x = 5$. Find the value of x when $y = 2$ if:

a y varies directly as x **b** y varies directly as x^3

c y varies inversely as x^2 **d** y varies inversely as x^3.

4 The surface area of a sphere varies directly as the square of its radius. The surface area of a sphere with radius 5 cm is 100π cm^2.

a Find the surface area of a sphere with radius 2 cm.

b From the information above, or otherwise, write down the formula for the surface area of a sphere.

c State what happens to the surface area of a sphere when the radius is enlarged by a factor of 5.

d Determine the effect on the radius of halving the surface area of a sphere.

Problem solving

5 The weight of an object in Newtons (N) varies inversely with the square of its distance in km from the center of the Earth. The radius of the Earth is approximately 3670 km. A certain astronaut weighs 850 N at the surface of the Earth. Find the weight of the same astronaut when she's on the International Space Station, which orbits at an average of 400 km from the surface of the Earth.

6 The rings of Saturn have been found to be made up of particles of a variety of sizes. Amazingly, the abundance is inversely proportional to the cube of the size of the particle. If a 2-meter size particle has an abundance of 10%, determine the abundance of a 3-meter size particle, to the nearest tenth of a percent.

7 Under the right conditions, pyrite (often called fool's gold) will form in the shape of a perfect cube. The cost of an ounce of this pyrite varies directly with its volume. If a cube with side length of 5 cm costs \$24 per ounce, determine the length of the side of a piece of pyrite that costs \$40 per ounce.

8 The use of low-flow shower heads has become an easy way to reduce our water consumption. Regular shower heads use 30 liters of water per minute while low-flow shower heads may use only 8 liters per minute. This volume of water is directly proportional to the square of the radius of the pipe in the shower head. If a regular shower head has a pipe with a radius of 5 cm what should be the radius of a pipe in a low-flow shower head?

Exploration 6

If your calculator/software has a slider bar or similar function you can change the parameters of the variation function and see how the shape of the graph changes.

Direct variation functions

Graph the direct variation function $y = kx^n$. Insert a slider so that you can change the value of k between -10 and 10, and the value of n between 1 and 3. If your calculator has a split-screen, you will be able to see the graph and table of values simultaneously.

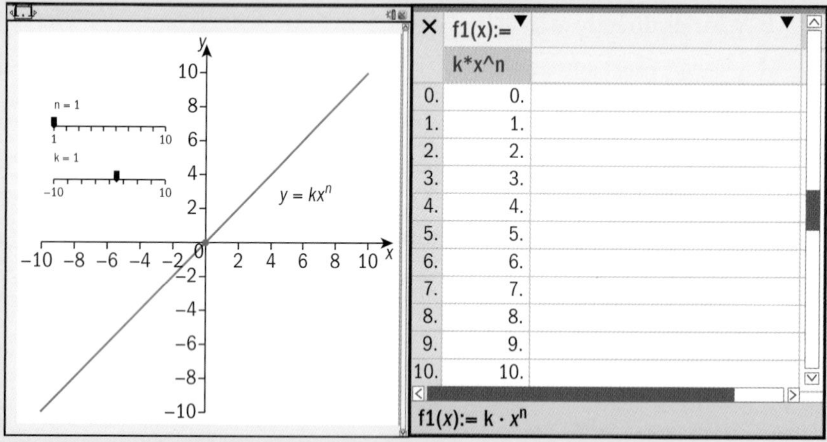

1 Start with the values $k = 1$ and $n = 1$. Change the value of k using the slider, and describe how the graph changes. State what does *not* change in the graph as you change the value of k.

2 State what k represents on the graph.

3 Repeat steps **1** and **2** for $n = 2$ and $n = 3$.

Continued on next page

Inverse variation functions

Graph the inverse variation function $y=\frac{k}{x^n}$.

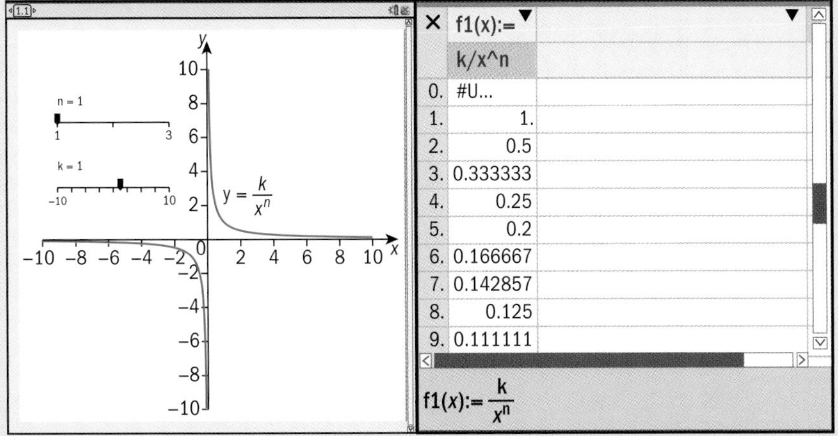

This reciprocal graph includes negative values of x.

4 Start with the values $k = 1$ and $n = 1$. Change the value of k using the slider, and describe how the graph changes.

5 State what happens as x gets closer to +10 and −10.

State what happens as x gets closer to 0 (from the positive side and from the negative side).

6 Explain how the proportionality constant affects the graph.

7 Repeat steps **4** to **6** for $n = 2$ and $n = 3$.

Reflect and discuss 7

- Consider the direct variation function $y = kx^n$.
  - Describe how n affects the shape of the graph.
  - Describe how k affects the shape of the graph.
  - Justify what happens when $x = 0$.
  - Justify what happens as x becomes extremely large in both the positive and the negative directions.
- Repeat these four steps for the inverse variation function $y=\frac{k}{x^n}$.
- Describe how you can recognize a direct variation function from a graph.
- Describe how you can recognize an inverse variation function from a graph.

Practice 4

1 Write down the variation function and sketch its graph.

a y varies directly as x, and when $y = 30$, $x = 5$.

b y varies inversely as x, and when $y = 2$, $x = 6$.

c y varies directly as the square of x, and when $x = 4$, $y = 120$.

d y varies inversely as the square of x, and when $x = 4$, $y = 5$.

e y varies directly as the cube of x, and when $x = 2$, $y = 62.5$.

f y varies inversely as the cube of x, and when $x = \frac{1}{2}$, $y = 4$.

Problem solving

2 Determine whether the following graphs show a direct variation function, an inverse variation function, or neither. Justify your answer.

a

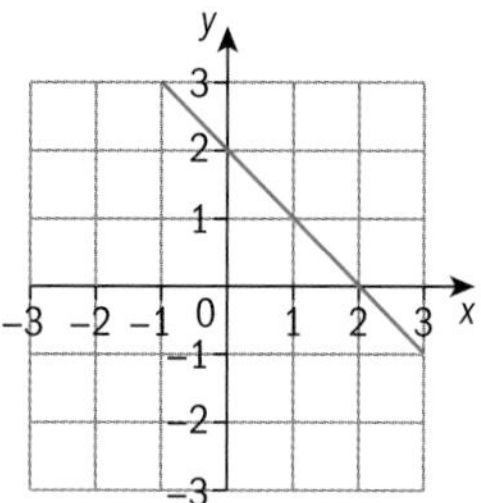

b

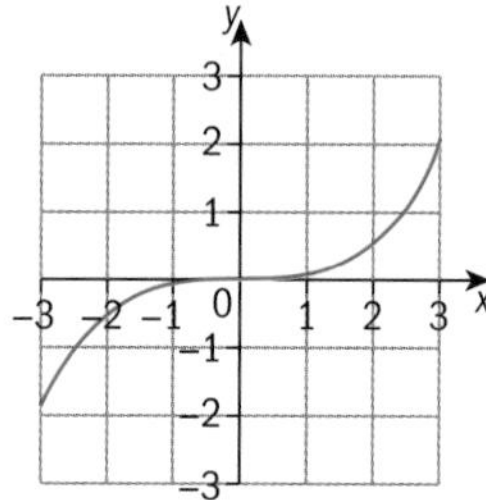

c

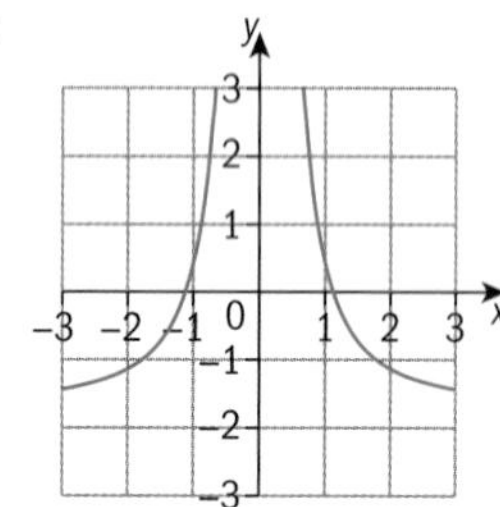

d

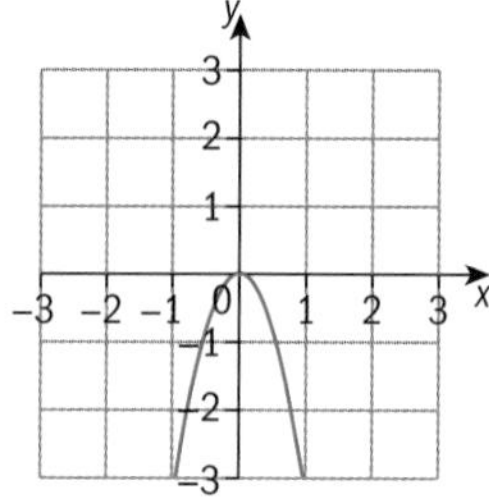

e

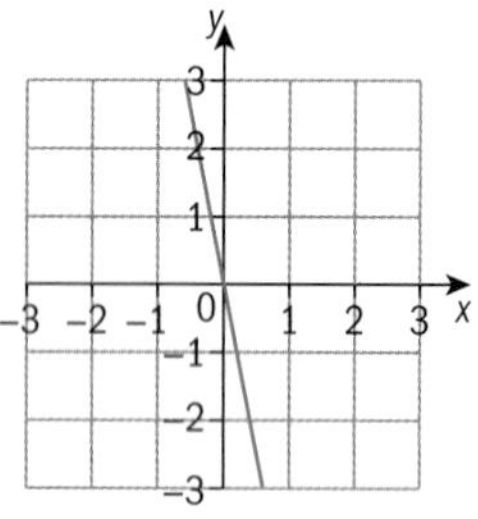

f

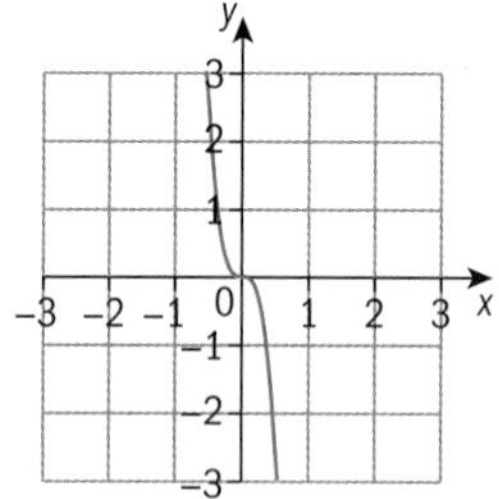

3 Sketch the graph of the variation function $y = \frac{2}{x^2}$.

a Determine whether or not y can have a negative value. Justify your answer.

b Explain what happens to y as x gets extremely large.

c Explain what happens to y as x gets closer and closer to 0.

d Hence, explain why this is an inverse variation function.

Modelling proportional relationships

- Can situations seem proportional when they are not?
- Is simpler always better?

Until now you have been dealing with variation functions that are either direct variation or inverse variation. Are there situations in which, even though a change in one variable causes a similar change in another, there is no direct or inverse variation?

Exploration 7

When you go to the coffee shop, you can either pay €4.50 for your favourite drink, or you can buy a cup for €5 and refill it every time, though each refill will cost you €3.50.

1 Write down an equation for each option and sketch a graph that represents the total cost as it relates to the number of drinks bought.

2 Determine which option represents a direct variation, and which one is not a direct variation. Justify your answers.

3 Suggest a rule for determining whether or not a function is a variation function:

a from its equation

b from its graph.

4 Describe a situation where one quantity increases as the other decreases, but that is not an inverse variation function. Explain how you know it is not a variation function.

In most of the examples and practice problems you have seen so far, the proportionality constant k has been positive. However, according to the definitions of direct and inverse proportion, k can be any number except 0, so technically, it could be negative.

Exploration 8

1 Sketch the graphs of these functions:

a $y = 3x$ **b** $y = -3x$ **c** $y = \frac{3}{x}$ **d** $y = -\frac{3}{x}$

2 Explain why they are all variation functions.

3 For each variation function, determine whether the following statements are true or false:

a As x increases, y increases

b As x decreases, y decreases

c As x increases, y decreases

d As x decreases, y increases

e Explain what happens as x gets closer and closer to 0.

f Explain what happens as x gets farther and farther away from 0, in the positive and negative directions.

Reflect and discuss 8

- Summarize your findings. What is the difference between negative proportionality and inverse proportionality?
- Describe a real-life situation in which there is a negative variation between two variables.

Summary

Proportional relationships	General shape of the graph
• Two variables are said to be in **direct proportion** if, and only if, their ratio is a constant for all values of each variable. • $y \propto x$ means 'y varies directly as x' or 'y is directly proportional to x'. • $y \propto x$ means that $y = kx$ for a constant k, where $k \neq 0$. k is called the **proportionality constant**, or **constant of variation**. • The function $y = kx$ is called a **linear variation function**.	$k > 0$ $k < 0$ $y = \frac{k}{x}$ direct linear proportion
• Two variables x and y are **inversely proportional** if multiplying one of them by a non-zero number results in the other variable being divided by the same non-zero number. • If x and y are in an inverse linear proportion, you can say 'y varies inversely as x' or 'y is inversely proportional to x', and you can write $y \propto \frac{1}{x}$ and $y = \frac{k}{x}$. • If y is inversely proportional to x, then y is directly proportional to $\frac{1}{x}$. • An equation $y = k \times \frac{1}{x}$ or $y = \frac{k}{x}$ represents a relationship of inverse proportion or a **reciprocal relationship**.	$k > 0$ $k < 0$ $y = \frac{k}{x}$ inverse linear proportion
• Two variables x and y are in **direct non-linear proportion** if y is proportional to a power of x, or $y \propto x^n$, $n > 0$. • The variation function is $y = kx^n$, $k \neq 0$ and $n > 0$. • y varies directly as x^n or y is in direct proportion to x^n. • $y = kx^n$ is a relationship of direct variation. When x is multiplied by a constant c, then y is multiplied by c^n.	$k > 0$ $k < 0$ $y = kx^n$ direct non-linear proportion, when n is even $k > 0$ $k < 0$ $y = kx^n$ direct non-linear proportion, when n is odd

- Two variables x and y are in an **inverse non-linear proportion** if y is proportional to a power of $\frac{1}{x}$, or $y \propto \frac{1}{x^n}, n>0$. You can also write this as $y \propto x^{-n}, n>0$.
- The variation function is $y=\frac{k}{x^n}$, $k\neq 0$ and $n>0$.

  y varies inversely as x^n or y is inversely proportional to x^n
- $y=\frac{k}{x^n}$ is a relationship of inverse variation. When x is multiplied by a constant c, then y is divided by c^n.

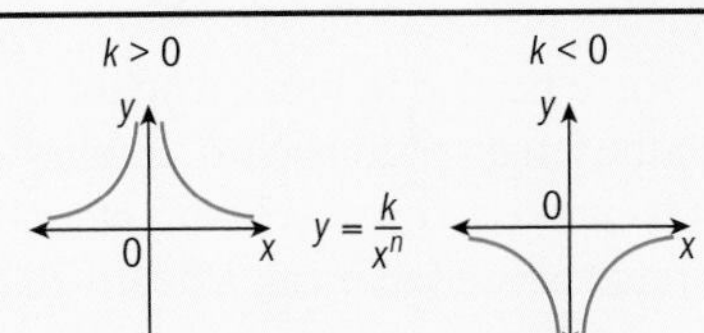

indirect non-linear proportion, when n is even

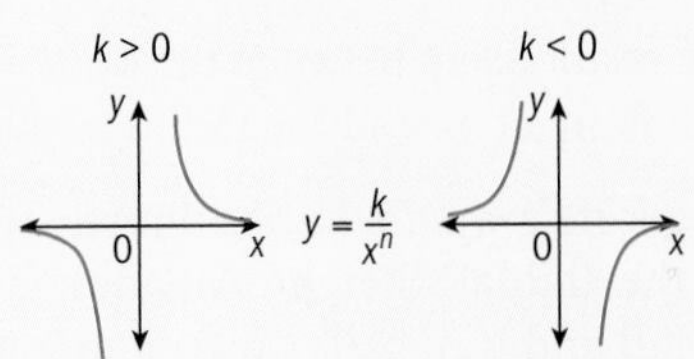

indirect non-linear proportion, when n is odd

Mixed practice

1 **Sketch** the graph of each variation function and **determine** whether it is a direct or inverse relationship.

a $y=5x$ **b** $y=\frac{x}{3}$

c $y=1.4x^2$ **d** $y=\frac{2}{x^2}$

e $y=\frac{7}{x}$ **f** $y=\frac{1}{3x}$

g $y=\frac{3x^2}{5}$ **h** $y=\frac{3}{5x^2}$

2 The variable x is directly proportional to the variable t. When $t = 4.6$, $x = 3.45$

a **Find** the value of the constant of proportionality.

b **Write down** the function relating t and x.

c Find x when $t = 8.2$

3 P is inversely proportional to y. When $P = 15$, $y = 0.2$. **Find** P when $y = 1.5$

4 M varies directly as the square of c. When $M = 12.6$, $c = 3$. **Find** c when $M = 17.15$

5 p is inversely proportional to the square of q. **Find** the missing values in this table:

p	100		1
q		0.4	2

6 **Write down** the function connecting x and y for each table of values:

a

x	−1	3	5	8
y	0.25	2.25	6.25	16

b

x	−2	−1	0.5	4
y	2.5	5	−10	−1.25

7 A supermarket sells 300ml bottles of ketchup for £1.30 a bottle.

a **Explain** why total cost of ketchup is in direct proportion to the number of bottles of ketchup bought.

b In a special offer, if you buy two bottles you get one free.

Is the total cost still in direct proportion to the number of bottles bought? **Justify** your answer.

Problem solving

8 The force needed to break a board varies inversely with the length of the board. It takes a force of 120 Newtons to break a board 60 cm long. **Find** the force needed to break a board 20 cm long.

9 The number of tennis balls you can pack in a box varies inversely with the volume of the balls. **Write down** the variation function that expresses the relationship between the number of balls that can fit in a box and the radius of the balls.

10 A giraffe's weight varies directly with the cube of the animal's height. An adult giraffe is 5 m tall and weighs 1.1 metric tonnes. **Find** the weight of a 2 m tall baby giraffe.

11 During freefall, the distance an object falls is directly proportional to the square of the time spent falling. An object falls 40.8 m in 20 s. **Find** how far it falls in 1 minute.

12 The shutter speed of a camera varies inversely with the square of the aperture setting. When the aperture setting is 8, the shutter speed is 125. **Find** the shutter speed when the aperture setting is 4.

13 The volume of a cylinder is given by the formula $V = \pi r^2 h$, where r is the radius of the base and h is the height of the cylinder.

a If the cylinder has a fixed height, **write down** the variation function between V and r. Hence, **write down** the proportionality constant in this case.

b If the cylinder has a fixed radius, **write down** the variation function between V and h. Hence, **write down** the proportionality constant in this case.

Review in context

Globalization and sustainability

Problem solving

1 In order to reduce overcrowding in their cities, Boomcity and Alphatown have begun pricing office real estate in a way that they hope will encourage developers to construct buildings outside the city center.

In Boomcity, the price of a building plot varies inversely with the distance of the plot from the city's center.

In Alphatown, the price of a plot varies inversely with the square of the distance from the center.

In both Alphatown and Boomcity, a plot 5 km from the center costs $250 000.

Find the cost of a plot 10 km from the center of each city.

2 The amount of water that flows through a water pipe is directly proportional to the square of the diameter of the pipe. A pipe of diameter 10 cm can serve 50 houses.

a Make a table of the number of houses served by water pipes of diameter 10 cm, 20 cm, 30 cm, 40 cm and 50 cm.

b **Use** your table to **estimate** the number of houses that can be served by a water pipe with diameter 25 cm.

c **Draw** a graph of the variation function that represents the relationship between the diameter of the water pipe and the number of houses served.

d Now **use** your graph to **estimate** the number of houses that can be served by a water pipe with diameter 25 cm.

e **Use** the equation of the variation function to **find** the actual number of houses that can be served by a water pipe with diameter 25 cm.

f **Discuss** how far off your estimations were from the actual number of houses.

g Water conservation is a priority in many parts of the world, yet humans want and need water for so many of our activities. **Discuss** what the trade-offs might be between using larger diameter of pipes and our need to conserve water.

> The average water consumption per head varies widely between the northern and southern hemispheres. But there is also great variation between European countries. Luxembourg's average of 80 cubic meters per person per year is 5 times less than Germany's average of 400 cubic meters per person.

Problem solving

3 Generating electricity, whether through hydroelectricity, fossil fuels, wind or nuclear energy, can have serious negative effects on the environment. We should all do our part to decrease the amount of energy we consume; something that will also save money.

In the construction of an office building, a heating engineer tries to place a heater in the spot where it will be most efficient. The heat received at X is inversely proportional to the square of the distance from the heat source. The engineer plans to put the heater at position A, however, the company's manager prefers position B.

Determine how much moving the heater to B will reduce the heat received at X. Give your answer as a percentage.

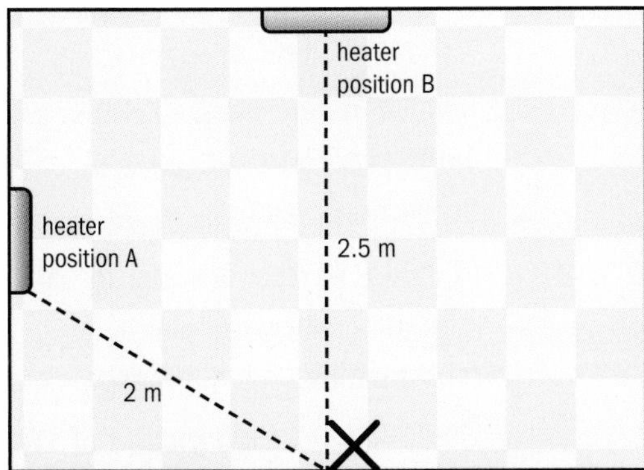

4 The 'urban heat island effect' occurs when buildings and parking lots are constructed, covering grass and dirt with concrete. Temperatures in these areas tend to be higher than normal, leading to health issues. To combat this, some cities now have ordinances that require the planting of trees to provide shade and decrease these temperature changes.

Suppose the temperature change varies directly with the cube of the diameter of the trees planted. Trees with a diameter of 1.5 meters produce an average temperature change of –0.3°C.

a **Find** the temperature change that would be expected for a tree with a diameter of 2 meters.

b If a temperature change of –1°C was desired, **determine** what size trees should be used.

5 Companies use a variety of ways to try to entice consumers to buy their products. One of those is by setting a price that will encourage people to purchase their goods. A clothing company makes jeans in a variety of price ranges. The Bootcut jeans sell for \$25 while the Classic jeans sell for \$160. The demand for this company's jeans is inversely proportional to the square of the price.

a If 50 000 pairs of the Bootcut jeans were bought, **find** how many of the Classic jeans were bought.

b If the company wanted to double the number of pairs of Classic jeans bought, **determine** at what price they should be sold. Your answer should be to the nearest dollar.

6 Since the volume of a sphere is given by $V = \frac{4}{3}\pi r^3$, you would think that the price of ice cream should vary directly with the cube of the radius of the scoop. However, the price of one small scoop of ice cream at Dave's Ice Cream Shop (radius = 3 cm) is \$2.50 and the price of a large scoop of ice cream (radius = 5 cm) is \$5.00.

a **Show that** this does not represent a direct cube variation.

b If this were a direct cube variation, **determine** the price of one large scoop.

c **Suggest** reasons why the price does not seem to vary based on the volume of the ice cream.

Reflect and discuss

How have you explored the statement of inquiry? Give specific examples.

Statement of Inquiry:

Changing to simplified forms can help analyze the effects of consumption and conservation.

11 Equivalence

The state of being identically equal to or interchangeable, applied to statements, quantities or expressions

Did you know that 0.999... = 1?

0.999... is not *almost* 1, but is actually *identical* to 1. You can prove this result using the equivalence transformations that you will study in this chapter. Read the following and make sure that you understand each step.

Let $n = 0.999...$

Then $10n = 9.999...$

Subtracting the first line from the second gives

$9n = 9$

Dividing both sides by 9 gives
$n = 1$

0.999... is a recurring decimal, meaning that the number keeps repeating forever after the decimal point.

In the same way, you can also show that any infinite repeating decimal is the same as its fractional equivalent. Use the same procedure as above to prove that 0.333... is equal to $\frac{1}{3}$, or that 0.2424.... is equal to $\frac{8}{33}$ (Hint: when two places are repeating, multiply both sides of the equation by 100 instead of 10).

In 1847, Englishman Oliver Byrne published an edition of *Euclid's Elements* where he used color in diagrams to prove theorems, instead of text and labels.

- Do you think a visual 'proof' of a theorem can be equivalent to a written one?

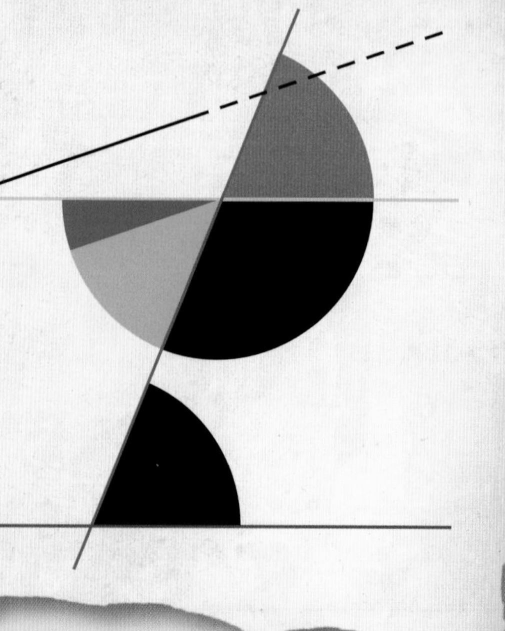

Logical equivalence

In mathematics, two statements can be proved to be (or not to be) logically equivalent.

The following statements are logically equivalent:

- If Ellie travels from Madagascar to Iceland then she has crossed the equator.
- If Ellie has not crossed the equator, then she can't have travelled from Madagascar to Iceland.

The second statement is called the contrapositive of the first statement.

For each of the following statements, write the contrapositive and state whether or not it is logically equivalent with its partner.

- If Ellie speaks Portuguese then she speaks a romance language.
- If Ellie studies hard then she will do well on her algebra test.
- If Ellie passes her algebra test then she will take her brother to the zoo.
- If Ellie's bus is on time then she will arrive at school on time.

A model of equality

Global context: Identities and relationships

FORM

Objectives

- Solving linear equations and systems of linear equations algebraically and graphically
- Using equivalence transformations to solve linear equations and systems of equations
- Creating a mathematical model to solve real-life problems
- Determining if a model solution is equivalent to the real-life solution
- Evaluating and interpreting your solutions in light of the real-life problems

Inquiry questions

F
- What is an equivalence transformation?
- How can you solve linear equations using equivalence transformations?

C
- Are all solution methods for systems of equations equivalent?
- How do the graphs of systems of equations relate to the types of solutions they may have?

D
- Can good decisions be calculated?

ATL **Transfer**

Apply skills and knowledge in unfamiliar situations

Statement of Inquiry:

Modelling with equivalent forms of representation can improve decision making.

11.1

14.3

E14.1

You should already know how to:

• expand and factorize algebraic expressions to obtain equivalent expressions	**1** Expand: **a** $4(x+3)$ **b** $5(2x-1)$ **c** $-3(8-6x)$ **d** $4x(7+x-x^2)$ **2** Factorize: **a** $3x+6$ **b** $5x-15$ **c** $14+35x$ **d** $34-85x$
• solve linear equations	**3** Solve these equations. **a** $\frac{x}{5}=3$ **b** $4x=52$ **c** $5x-7=8$ **d** $\frac{x}{2}+4=32$
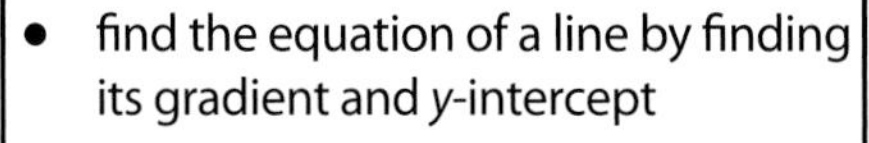 • find the equation of a line by finding its gradient and y-intercept	**4** Find the equations of these two lines. 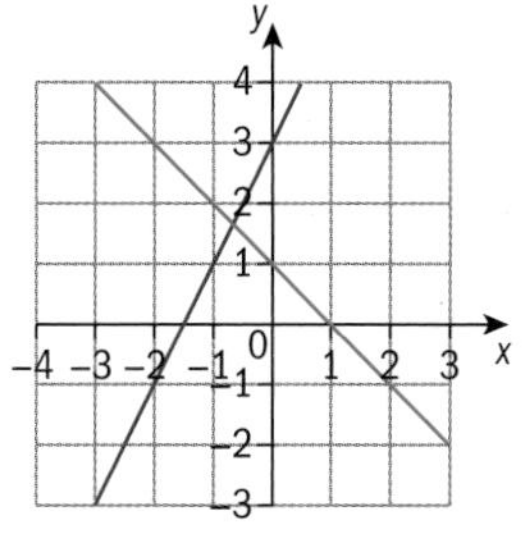

F Linear equations

- What is an equivalence transformation?
- How can you solve linear equations using equivalence transformations?

You know how to solve linear equations, but do you know the mathematical principles that you use to do this?

Exploration 1

1 The examples below show steps that are typically used to solve linear equations. Decide whether you think the example demonstrates 'the addition principle' or 'the multiplication principle'.

a $3x-5=2$
$3x=7$

b $6x=12$
$x=2$

c $\frac{1}{4}x=3$
$x=12$

d $6-2x=11$
$-2x=5$

e $4n+2=9n$
$2=5n$

f $7x-1=2x+8$
$5x=9$

g $\frac{2}{3}x+\frac{1}{6}=1$
$4x+1=6$

h $16x+4=4x$
$4x+1=x$

i $5x=2x+9$
$3x=9$

2 Label the stages in the solution below as either 'the addition principle' or 'the multiplication principle'. If it's neither of these, then state the mathematical property that is being used.

$2(3x-4)=3x+7$
$6x-8=3x+7$ ____________
$3x=15$ ____________
$x=5$ ____________

Reflect and discuss 1

- These principles are called 'equivalence transformations'. Why is this an appropriate name?
- Why don't we need a subtraction principle or a division principle?

When solving linear equations in the past, you may have used very informal language to describe your working steps. In this section, you will learn the formal terms for the mathematical principles used in solving linear equations.

Equivalence transformations

To solve an equation you can use these mathematical principles:

Principle	Example
Addition Principle: Add the same value or variable to both sides of an equation.	Add 3 to both sides of $2x - 3 = 5$ to get the equivalent equation $2x = 8$.
Multiplication Principle: Multiply by the same non-zero value or variable on both sides of an equation.	Multiply both sides of $11 = 5x$ by $\frac{1}{5}$ to get the equivalent equation $\frac{11}{5} = x$.

An **equivalence transformation** uses mathematical principles to transform an equation into an equivalent equation.

Example 1

Solve the equation $\frac{1}{4}(x-2)=\frac{1}{2}(3x+4)$. Show the equivalence transformation used at each step. Remember to check your solution.

$\frac{1}{4}(x-2)=\frac{1}{2}(3x+4)$

$4\cdot\frac{1}{4}(x-2)=4\cdot\frac{1}{2}(3x+4)$ — Multiply both sides by 4.

$x - 2 = 2(3x + 4)$

$x - 2 = 6x + 8$

$x - 2 - 8 = 6x + 8 - 8$ — Add −8 to both sides (or subtract 8 from both sides).

$x - 10 = 6x$

$-x + x - 10 = -x + 6x$ — Add $-x$ to both sides (or subtract x from both sides).

$-10 = 5x$

$\frac{1}{5}\cdot -10=\frac{1}{5}\cdot 5x$ — Multiply both sides by 5.

$-2 = x$

Check: LHS: $\frac{1}{4}(-2-2)=-\frac{4}{4}=-1$

RHS: $\frac{1}{2}(3\times(-2)+4)=\frac{1}{2}(-2)=-1$

LHS = RHS ✓

Practice 1

1 Solve these equations. Show the equivalence transformation that you use at each step. Remember to check your solutions.

a $2x+3=x-7$

b $5x-4=2x+6$

c $-(x+2)-3x=2(x+1)$

d $1-3(x+2)=\frac{1}{2}(2x-8)+3$

e $\frac{x}{3}+2=\frac{1}{2}x-4$

f $\frac{x+2}{5}=\frac{2x-4}{2}$

g $\frac{1}{3}(6x-3)=\frac{1}{4}(8-4x)$

h $2x=\frac{1}{5}(9-8x)$

Problem solving

2 For the equation $x=3$, use equivalence transformations to write an equivalent equation with brackets and the variable x on both sides of the equals sign.

3 Apply equivalence transformations to the equation $5(x+4)-8+x=6(x+2)$.

Describe what happens, and explain what you think this means.

Reflect and discuss 2

- 'Multiplying both sides of an equation by zero' is not an equivalence transformation. Explain why.
- Can you think of operations that are not equivalence transformations? Explain why these operations do not result in equivalent equations.

Systems of linear equations

- Are all solution methods for systems of equations equivalent?
- How do the graphs of systems of equations relate to the types of solutions they may have?

A **system of equations** is two or more equations with the same unknowns. Solving a system of equations means finding values for each unknown that satisfy every equation in the system.

You have seen how you can solve an equation like $3(x+2)-6=4(2x-3)+1$ using an algebraic method (equivalence transformations), but did you know that you can also solve it using a graphical method?

You can consider the two sides of the equation as two separate linear equations, each equal to the variable y. You can then graph each equation on the same set of axes.

Here is a graph of $y=3(x+2)-6$

and $y=4(2x-3)+1$

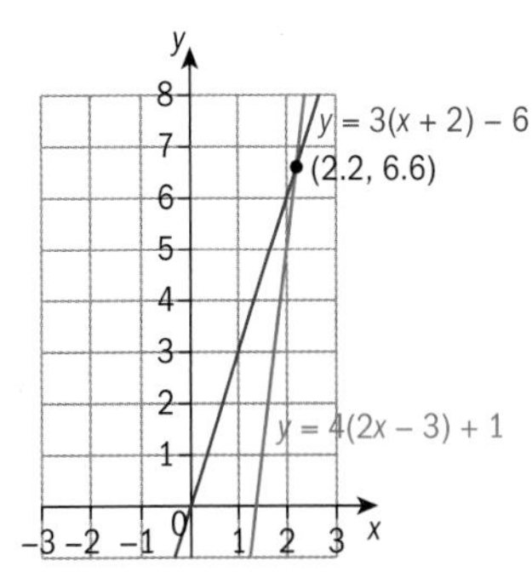

The graph shows that the value of $x = 2.2$ makes both equations equal to the same value, $y = 6.6$. So the solution that satisfies both equations simultaneously is $x = 2.2$.

Reflect and discuss 3

- Why is drawing a graph by hand not the *best* method to solve this equation?
- When is using technology not an appropriate method to finding the solution of an equation? (Hint: Can you always find exact solutions using technology?)

You can solve the two equations $y = 3(x + 2) - 6$ and $y = 4(2x - 3) + 1$ algebraically.

Expanding and simplifying:

$$y = 3(x + 2) - 6 \Rightarrow y = 3x$$

$$y = 4(2x - 3) + 1 \Rightarrow y = 8x - 11$$

So, we have two equations in two unknowns, x and y, to solve.

[1] $y = 3x$

[2] $y = 8x - 11$

There is a third equivalence transformation you can use here:

Substitution Principle: Replace part of an equation by an equivalent expression.	For example, for the equations $y = 3x$ and $y = 8x - 11$, substitute the value of y from [1] into equation [2] to give $3x = 8x - 11$.

[3] $3x = 8x - 11$

[4] $11 = 5x$

[5] $x = 2.2$

Substituting this value of x into both equations gives $y = 6.6$.

For a system of two equations in two unknowns the solution is an ordered pair (x, y).

Check the solution pair satisfies both original equations:

[1] $y = 3x$

LHS: 6.6

RHS: $3 \times 2.2 = 6.6$

LHS = RHS ✓

[2] $3x = 8x - 11$

LHS: $3 \times 2.2 = 6.6$

RHS: $8 \times 2.2 - 11 = 17.6 - 11 = 6.6$

LHS = RHS ✓

The *substitution method* reduces two separate equations in two unknowns into one single equation in one unknown.

Example 2

Use the method of substitution to solve the system of equations $7x + 2y = 19$ and $x - y = 4$, and check your solution.

$x - y = 4 \Rightarrow y = x - 4$ — Choose one of the equations and solve for *y*.

$7x + 2(x - 4) = 19$ — Substitute the expression for *y* into the other equation, and solve for *x*.

$$7x + 2x - 8 = 19$$
$$9x - 8 = 19$$
$$9x = 27$$
$$x = 3$$

$y = x - 4 \Rightarrow y = 3 - 4 = -1$ — Substitute the value for *x* into one of the equations to find the value of *y*.

The solution is (3, –1). — Write the solution as an ordered pair (*x*, *y*).

$7x + 2y \Rightarrow 7(3) + 2(-1) = 21 - 2 = 19$ ✓

$x - y \Rightarrow 3 - (-1) = 4$ ✓

— Check algebraically by substituting the *x* and *y* values into both original equations.

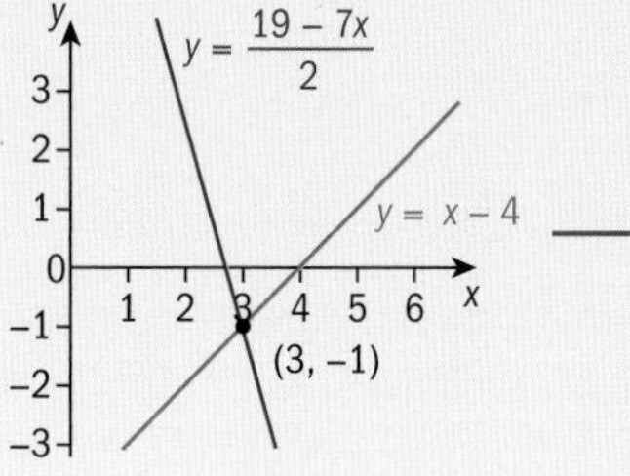

— Check graphically by graphing the two equations and finding their intersection.

Practice 2

Solve these systems of equations using the substitution method.

1 $3x - 2y = 0$ and $y = 7 - 2x$

2 $y = 9 - 2x$ and $3x - 4y = 8$

3 $x = 25 + 9y$ and $6x - 5y = 3$

4 $3x = 12 - 5y$ and $x + 4y = 11$

5 $2x + 3y = -6$ and $3x + 2y = 25$

Problem solving

6 Rupa and George solve this system of equations: $5x + y = 8$ and $x + 3y = 10$

Rupa	**George**
[1] $5x + y = 8 \Rightarrow y = 8 - 5x$	[1] $5x + y = 8$
[2] $x + 3y = 10$	[2] $x + 3y = 10 \Rightarrow x = 10 - 3y$
$x + 3(8 - 5x) = 10$	$5(10 - 3y) + y = 8$

Copy and complete their working to find their solutions.

Reflect and discuss 4

- Does your final solution for the system of equations depend on which variable you choose to solve for and substitute? Hint: look back at Practice 2, question **6**.
- When is it easier to substitute **i** for x **ii** for y?

Exploration 2

1 Find a solution for the following system of equations:

$x + y = 4$

$2x - y = 5$

2 Write the equations one above the other (as above) and then add them together, one term at a time.

3 Explain what happens to the y's in the equation.

4 Find the value of x from this new equation.

5 Starting with the original system,

$x + y = 4$

$2x - y = 5$

multiply each term in the first equation by –2.

6 Rewrite this new equation and the second equation directly above one another, as you did in step **2**.

7 Add the two equations together term by term. Explain what happens to the x's.

8 Find the value of y from this new equation.

Tip

If the question doesn't tell you which method to use, you can choose either the algebraic or graphical method.

Sometimes the substitution method can be a bit tricky, because it is difficult to solve for either x or y, or because equations with fractions result. There is another method that uses the ideas from Exploration 2. Let's use it to solve the system of equations $3y - x = -3$ and $y - x = 1$:

[1] $3y - x = -3$

[2] $y - x = 1$

Subtracting [2] from [1] eliminates the variable x.

$$\begin{array}{r} 3y - x = -3 \\ -\quad y - x = \ \ 1 \\ \hline 2y = -4 \\ y = -2 \\ \hline \end{array}$$

Tip

Recall that subtraction is the addition of the opposite.

Substituting $y = -2$ into either [1] or [2] gives $x = -3$. The solution to this system of equations is (–3, –2).

Checking the solution in both original equations:

(1) $y - x = 1$: (2) $3y - x = -3$

$-2 - (-3) = 1$ ✓ $3(-2) - (-3) = -6 + 3 = -3$ ✓

This method is called the *elimination method* for solving systems of equations. Can you see why?

To solve systems of equations using the elimination method, add or subtract the equations to eliminate one of the variables. You may need to use an equivalence transformation on one or both equations first.

Practice 3

Solve each system of equations using the elimination method.
Check your solutions algebraically or graphically.

1 [1] $5x + y = 27$
[2] $2x + y = 12$

2 [1] $2y - x = 7$
[2] $4y + x = 11$

3 [1] $2x - 3y = -16$
[2] $2x + y = 0$

4 [1] $3x + 2y = 11$
[2] $5x - 2y = 13$

Reflect and discuss 5

- How do you decide which variable to eliminate?
- How do you decide what to multiply one of the equations by in order to eliminate a variable?

Exploration 3

In Practice 2, you solved the following system of equations by substitution:

$2x + 3y = -6$

$3x + 2y = 25$

1 Determine if it is possible to eliminate a variable by multiplying just one of the equations by an integer and then adding. Explain.

2 Determine if it is possible to eliminate x by multiplying each equation by a different integer and then adding. Explain.

3 Solve the system by eliminating x.

4 Solve the system by using a similar method to eliminate y instead.

5 Check your answer by substituting it into both of the original equations.

6 Explain how you would decide which variable to eliminate first.

As early as 200 BC the Chinese had invented a method of solving a system of two equations in two unknowns. The method is described in the book *Jiuzhang suanshu* (Nine Chapters in the Mathematical Art), written during the Han Dynasty.

Example 3

Solve the system of equations $5x - 7y = 27$ and $3x - 16 = 4y$, using the elimination method. Check your result algebraically or graphically.

[1] $5x - 7y = 27$ — Rearrange the equations to line up like terms.

[2] $3x - 4y = 16$

Multiply both sides of [1] by 4, and both sides of [2] by −7. — Choose equivalence transformations that give the same coefficient for one variable.

[3] $20x - 28y = 108$

[4] $-21x + 28y = -112$

$-x = -4 \Rightarrow x = 4$ — Add [3] and [4] to eliminate y.

$5(4) - 7y = 27$ — Substitute into one of the original equations to find the value of the 2nd variable.

$20 - 7y = 27$

$-7y = 7$

$y = -1$

Solution is $(4,-1)$ — Write the solution as an ordered pair.

$5(4) - 7(-1) = 20 + 7 = 27$ ✓ — Check algebraically by substituting the solution into both original equations.

$3(4) - 4(-1) = 12 + 4 = 16$ ✓

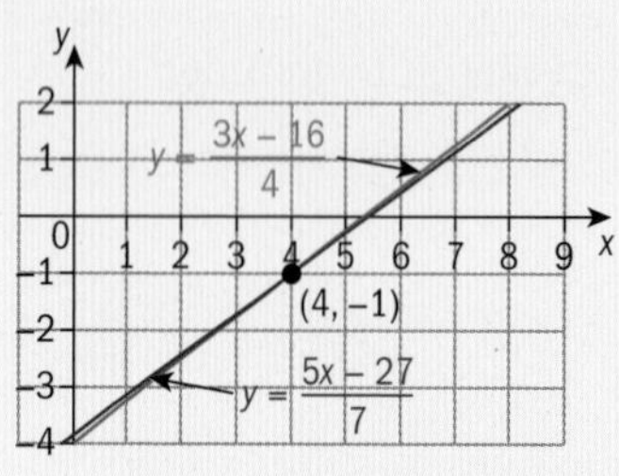

Check graphically to find the point of intersection.

Practice 4

Solve each system of equations using the elimination method.
Check your answers either algebraically or graphically.

1 $2x + 5y = 24$ and $4x + 3y = 20$

2 $2a + 3b = -9$ and $4a + b = 13$

3 $x + 3y - 7 = 0$ and $2y - x = 3$

4 $5r = 23 - 3s$ and $4s = 12 - 2r$

5 $3x = 17 + y$ and $\frac{x}{5} + \frac{y}{2} = 0$

6 $4x - 0.5y = 12.5$ and $3x = 8.2 - 0.8y$

Reflect and discuss 6

- Would it be more efficient to use the substitution method on any of the systems of equations in Practice 4? If so, explain why.
- When solving a system of linear equations, how do you select the most efficient algebraic method? Write yourself a set of guidelines on how to solve a system of linear equations efficiently.

Practice 5

Use the most efficient algebraic method for solving each system of equations. Explain why you selected your chosen method each time. Check your solutions algebraically or graphically.

1 $3x + y = 9$ and $5x + 4y = 22$

2 $y = 2x + 4$ and $3x + y = 9$

3 $3y - 2x = 11$ and $y + 2x = 9$

4 $3x + 2y = 16$ and $7x + y = 19$

5 $4x + 3y = -2$ and $4x - y = 6$

6 $3x + 2y = 19$ and $x + y = 8$

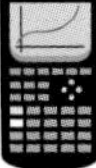

Exploration 4

1 Draw the systems of equations on separate graphs:

a $-x + 2y = 16$

$4x - 8y = -64$

b $-x + 2y = 16$

$4x - 8y = 24$

2 Generalize your results from your work so far, and from step **1**. Write down the conditions necessary for a system of two linear equations to have:

- a unique solution (an ordered pair)
- no solution
- a non-unique solution (an infinite number of solutions).

3 Describe how the graphs of systems of equations relate to the number of solutions they may have. Include a sketch for each case.

4 Without actually solving these systems of equations, state, giving reasons, whether or not the system has a solution, and explain why.

a $2x - y = 5$

$x + 2y = 0$

b $-2x + 3y = 12$

$2x - 3y = 6$

c $2x - y = 5$

$-4x + 2y = -10$

5 Explain why it is not possible for a system of two linear equations to have two unique solutions.

6 Explain why you should test your solution in the *original* equations, and not in any of the equivalent equations you get from using equivalence transformations.

D Applications of systems of equations

- Can good decisions be calculated?

You are going to take a written driver's license exam that contains 80 questions. The scoring is such that:

- each correct answer scores 1 point
- each incorrect answer is a $\frac{1}{4}$ point deduction.

If you answer all of the questions, how many do you need to get correct in order to earn 70 points (the minimum to pass)?

Step 1: Identify the variables:

Let x represent the number of correct answers.

Let y represent the number of incorrect answers.

Step 2: Identify the constraints:

Conditions on the values of the variables are called constraints.

Both x and y must be positive numbers or zero, since you cannot answer a negative number of questions.

There are 80 questions, so the maximum for x or for y is 80.

Expressed mathematically: $0 \le x \le 80$; $0 \le y \le 80$

Step 3: Create the model:

Write equations to represent the situation.

The 80 questions are correct or incorrect, so:

number of correct answers + number of incorrect answers = 80 $\Rightarrow x + y = 80$

Total score = (number of correct answers) $-\frac{1}{4}\times$ (number of incorrect answers) = 70

$$\Rightarrow x - \frac{1}{4}y = 70$$

The model is a system of two equations:

[1] $x + y = 80$

[2] $x - \frac{1}{4}y = 70$

Solve your model:

Use an efficient method to solve the system of equations.

Check your solution:

Check your solution algebraically, or by using a GDC or graphing software.

Finish working out the problem and then fill in the answers in the last sentence.

Interpret your solution in the context of the problem:

To earn 70 points, you need to answer ☐ questions correctly, and you can answer ☐ questions incorrectly.

Example 4

From 1996 to 2012, the total imports and exports from a particular company can be modelled by the following system of equations, where y represents the total exports or imports in millions of dollars, and x represents the time in years. The year 2000 is represented by $x = 0$.

Exports: $y = 1310x + 5165$

Imports: $y = 725x + 7430$

By analyzing the system of equations, describe the company's pattern of imports and exports between 1996 and 2012.

$x = 0$ represents the year 2000, so the range of x (from 1996 to 2012) is $-4 \leq x \leq 12$.

y represents the total exports or imports, thus $y > 0$.

Identify the variables and constraints.

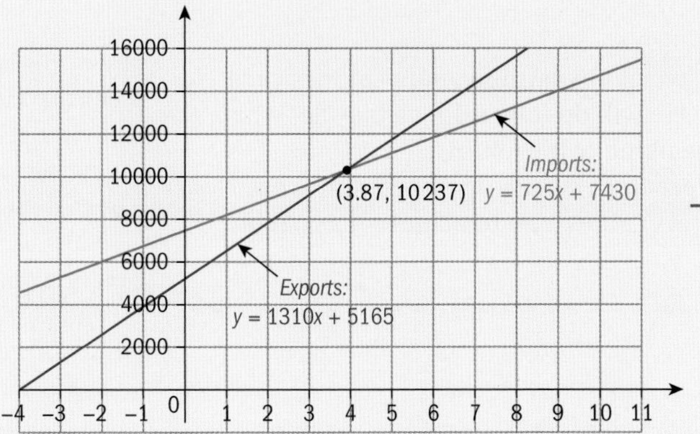

Graph the system of equations.

The intersection is at $x = 3.87 \approx 4$, which is around the year 2004.

The company had no exports in 1996. Then, until about 2004, it spent more on imports than it made on exports. After 2004 the company's export income exceeded its import expenses.

Interpret the information from the graph in your own words, and in the context of the problem.

ATL

Practice 6

Problem solving

In the following problems, create or use the given mathematical model to answer the question. Choose the most efficient method for solving the system of equations, and check your solution.

1 You open a small business with an initial investment of \$90 000. The weekly running costs for the business are \$7800. The weekly revenue (income) from your business is \$8800. Determine how long will it be before you break even (when your total profit matches the amount you have invested).

2 Between 2000 and 2010, coal and petroleum production from the fuel-producing countries can be modeled by the following system of equations, where $x = 0$ represents the year 2005:

Coal production: $y = 93.2x + 3100$

Petroleum production: $y = -29.1x + 2942$

Calculate the year in which production from both forms of fuel was equal.

3 A phone company has two weekly plans to choose from. Plan A charges an operating fee of \$31.45, plus 15.5 cents per minute. Plan B charges an operating fee of \$5.20, plus 37 cents per minute. Determine how you would decide on which plan would be best for you, and explain your reasoning.

4 You gave \$4500 to an investment broker, who put part of the money in a fund paying 2% annual interest and the rest in a fund paying 5% annual interest. At the end of the year you receive \$210 in interest. Determine how much of your initial investment was placed into each fund.

Reflect and discuss 7

- Why would someone want to place their money in two different funds, when one of them offers a lower return than the other one?
- What kind of advantages might the bank offer in the fund with the lower interest rate?

5 Your uncle Karl wants to lose some weight by burning extra calories each week (2 kg = approx. 3000 calories). He will do it by riding a bicycle, which burns 350 calories an hour, and by walking, which burns 200 calories an hour. If uncle Karl has 10 hours per week to devote to exercise, what combination of walking and cycling will allow him to achieve his goal of burning exactly 3000 calories?

During the Tour de France race, a cyclist will burn between 4000 and 5000 calories per stage. That is around 123 900 calories for the entire race, equivalent to 252 double cheeseburgers.

ATL

Reflect and discuss 8

- Is it possible to create more than one correct mathematical model to solve a particular problem?
- Are all methods for solving the model equal or equivalent?
- Explain in your own words the similarities and differences between the concepts of equality and equivalence.

Summary

1 An **equivalence transformation** is the application of one or more of the following principles in solving an equation so that all resulting equations are equivalent:

Addition Principle: Adding the same value or values to both sides of an equation.

Multiplication Principle: Multiplying by the same non-zero value or values on both sides of an equation.

Substitution Principle: Replacing part of an equation by an equivalent expression.

2 Methods of solution for systems of two linear equations:

Substitution: Solve one of the equations for one of the variables, and substitute this expression into the other equation, resulting in one equation in one unknown.

Elimination: By applying equivalence transformation(s), reduce the two equations to one equation in one unknown, and solve. Substitute this value into one of the original equations to solve for the other variable.

Graphical: Solve both equations for y and enter the equations into a GDC or graphing software. The point of intersection is the solution of the system.

3 Solution scenarios for a system of two linear equations:

One solution: The algebraic solution to the system of equations is an ordered pair. The graphs of the equations will intersect at one point.

No solution: There is no ordered pair that will satisfy both equations. The graphs of the two equations are parallel lines.

Infinitely many solutions: There is an infinite number of ordered pairs that will satisfy the system of equations.The graphs of the equations are coincident, which means the two lines are actually the same line.

4 Creating a mathematical model using **systems of equations**:

- Identify the variables and constraints
- Translate the real-life problem into a system of equations
- Solve the system of equations
- Check the solution in the original equations
- Interpret the solution in the context of the real-world problem

Mixed practice

1 **Solve** these equations by identifying the equivalence transformations you use at each step:

a $3x+4=19$

b $4-3x=5x-3$

c $3(3x-1)=4(2x-3)$

d $\frac{2}{3}x+2=4-\frac{1}{2}x$

2 **Use** either elimination or substitution to solve each system of equations. **Justify** your choice of method. Check your solutions graphically.

a $y=4x+3$ and $y=-x-2$

b $x-7y=19$ and $5x-8y=-13$

c $\frac{1}{4}x+\frac{1}{6}y=1$ and $x-y=3$

d $\frac{1}{3}x+\frac{1}{2}y=-\frac{1}{4}$ and $\frac{1}{6}x-\frac{5}{6}y=\frac{11}{16}$

e $3.5x+2.5y=17$ and $-1.5x-7.5y=-33$

Objective: **A.** Knowing and understanding
iii. solve problems correctly in a variety of contexts

For the mathematical model that describes each real-world problem, select the most efficient solution method for solving systems of equations.

Review in context

Identities and relationships

1 You decide that instead of taking vitamin and mineral supplements you will get your calcium and vitamin A by drinking milk and orange juice. An ounce of milk contains 38 mg of calcium and 56 µg (micrograms) of vitamin A; an ounce of orange juice contains 5 mg of calcium and 60 µg of vitamin A. **Determine** how many ounces of milk and orange juice you would need to drink daily in order to meet your minimum requirement of 550 mg of calcium and 1200 µg of vitamin A. Decide whether or not this a realistic amount for you to drink.

2 A chemist has been asked to make a solution of 8 liters containing 20% acid. The problem is that he has only two acidic solutions, one containing 12% acid and the other containing 32% acid. **Determine** how many liters of each he should use in order to create the solution he has been asked to make.

3 A coffee distributor has two types of coffee. The premium blend sells for \$10.50 per kilogram, and the standard blend sells for \$8.25 per kilogram. The distributor wishes to create 20 kilograms of a mixture containing these two blends to sell at \$9 per kilogram. **Determine** how many kilograms of each blend should be in the mixture.

4 You've decided to buy a printer and have narrowed it down to two choices. The laser printer costs \$150 but the average cost of each page is just 1.5 cents. The other option is an inkjet printer, which costs \$30, but each page has an average cost of 6 cents. **Determine** the conditions (e.g. the number of pages) where each printer is the better buy.

Coffee is the world's second most valuable traded commodity, surpassed only by petroleum.

Since Brazil produces around 40% of the world's coffee, the single most influential factor in world coffee prices is the weather in Brazil.

In 1991, a group of Cambridge University scientists aimed a fixed camera on their department's coffee pot, streaming the live footage on the web so that they could tell if the pot was empty or not. This made it the world's first live webcam.

Before coffee caught on in the US in the 1700s, the preferred breakfast drink of Americans was beer.

Reflect and discuss

How have you explored the statement of inquiry? Give specific examples.

Statement of Inquiry:

Modelling with equivalent forms of representation can improve decision making.

11.2 More than one way to solve a problem

Global context: Scientific and technical innovation

Objectives

- Solving quadratic equations algebraically and graphically
- Solving real-life problems by creating and using quadratic models

Inquiry questions

F
- What is the null factor law?
- How do you solve a quadratic equation in factorized form?

C
- How can you use equivalence transformations to solve quadratic equations?
- How are the three methods for solving quadratic equations equivalent?

D
- How do you determine a 'best method' among equivalent methods?
- Do systems, models and methods solve problems or create them?

FORM

ATL **Critical-thinking**

Propose and evaluate a variety of solutions

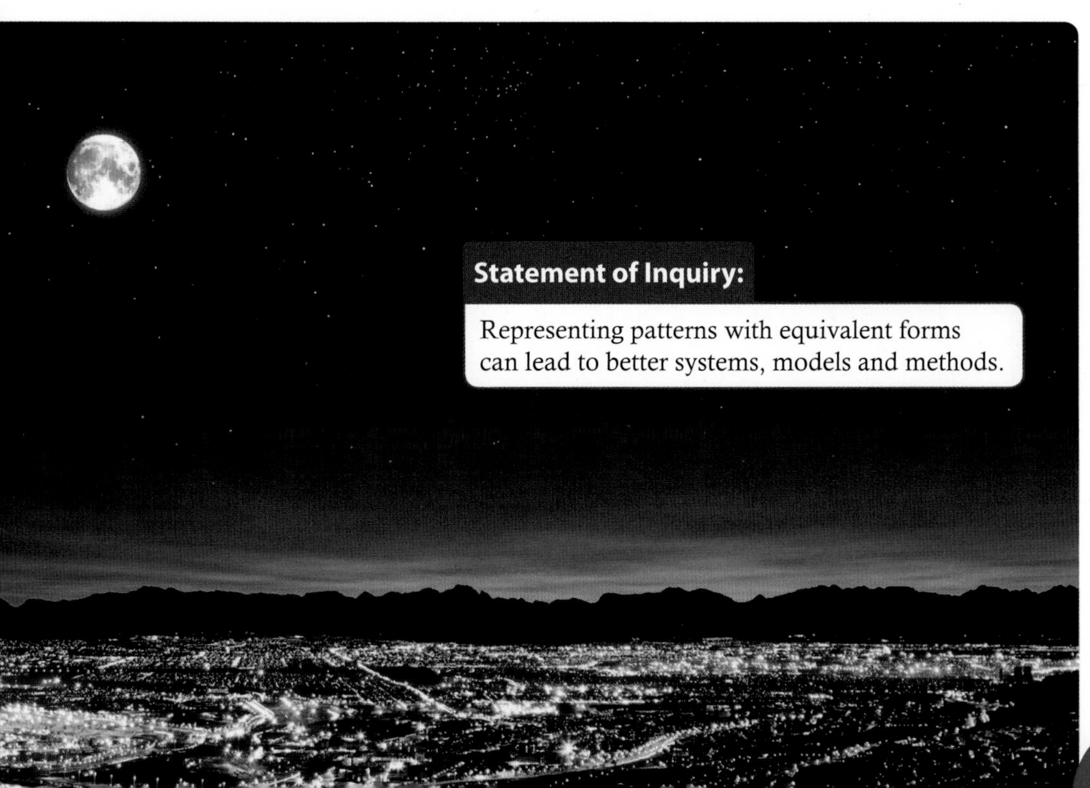

Statement of Inquiry:

Representing patterns with equivalent forms can lead to better systems, models and methods.

8.2

9.2

11.2

You should already know how to:

• convert quadratic expressions into standard form	**1** Rearrange into standard form: **a** $(x+2)(3x-1)$ **b** $(x-1)^2-16$
• convert quadratic expressions into factorized form	**2** Factorize: **a** x^2+x-6 **b** $2x^2+5x-3$ **c** $4x^2-81$
• convert quadratic functions into vertex form	**3** Write in the form $y=a(x-h)^2+k$, where (h, k) is the vertex. **a** $y=x^2-6x+14$ **b** $y=3x^2-6x+7$
• use equivalence transformations in solving linear equations	**4** Solve $3(x+2)-6=4(2x-3)+1$. State the equivalence transformation you use in each step.

F Quadratic and linear equations

- What is the null factor law?
- How do you solve a quadratic equation in factorized form?

The graph shows the two linear functions $f_1(x) = x + 1$ and $f_2(x) = 2 - x$, and the function which is the product of these two linear functions, $f_3(x) = (x + 1)(2 - x)$.

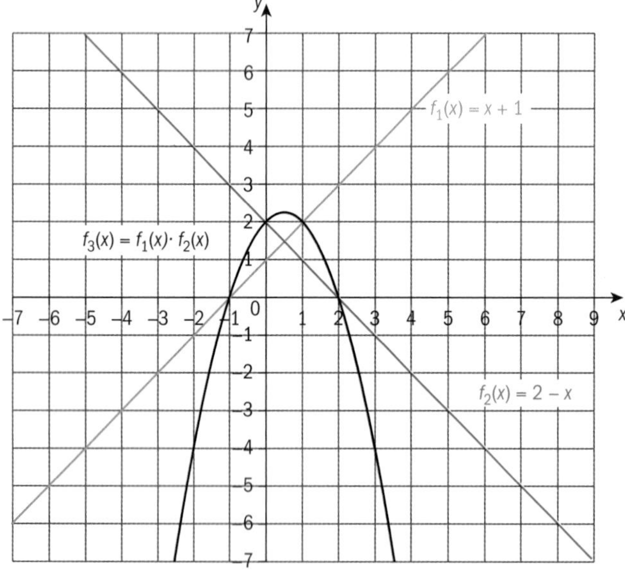

The graph of the product of the two linear functions is a parabola, because $f_3(x) = (x + 1)(2 - x)$ is a quadratic function.

Expanding $(x + 1)(2 - x)$ gives a quadratic expression.

Exploration 1

1 Write these quadratic functions as the product of two linear functions:

a $f(x) = x^2 + 3x + 2$ **b** $f(x) = -x^2 - x + 2$

c $f(x) = 2x^2 - 5x - 3$ **d** $f(x) = -3x^2 + 4x - 1$

2 Graph each quadratic function with its two related linear functions.

3 Write the function $f(x) = 2x^2 - 4x + 1$ as a product of two linear factors. Graph the quadratic function and its two related linear functions.

Reflect and discuss 1

- What is the relationship between the x-intercepts of the original linear functions and the x-intercepts of the parabola?
- Which parameter of the linear functions is responsible for the parabola being concave up or concave down?

The zeros of a function are the x-coordinates of the points where the graph crosses the x-axis. So, to find the x-intercepts of any function algebraically, you need to find the x-value(s) such that $f(x) = 0$.

Exploration 2

1 **a** Plot the graph of $f(x) = x^2 + 2x - 3$, and find the zeros of this function.

b Factorize $x^2 + 2x - 3$ to find two linear functions g and h whose product is $f(x)$.

c Solve $g(x) = 0$ and $h(x) = 0$.

d Check your solutions algebraically by substituting them into $f(x)$ and verifying that they satisfy $f(x) = 0$.

2 Plot the graph of $x^2 - 3x - 4 = 0$ and repeat parts **a** to **d** in step **1**.

3 Explain the relationship between the solutions of the linear equations, the zeros of the function and the solutions of the quadratic equation.

You can use the null factor law to solve quadratic equations.

The **null factor law** states that if the product of two or more numbers is zero, then at least one of the numbers must be zero.

For two numbers a and b, if $ab = 0$ then $a = 0$ or $b = 0$, or both.

When $f(x) = g(x) \times h(x)$, where g and h are functions, if $f(x) = 0$ then $g(x) = 0$ or $h(x) = 0$, or both.

Example 1

Solve $2x^2 - 6x = 0$.

$2x^2 - 6x = 0 \Rightarrow 2x(x-3) = 0$ — Factorize the quadratic.

$2x = 0 \Rightarrow x = 0$ — Use the null factor law.

$x - 3 = 0 \Rightarrow x = 3$

$x_1 = 0;\ x_2 = 3$

$x_1 = 0$: $2(0) - 6(0) = 0$, LHS = RHS ✓ — Check both solutions in the original equation.

$x_2 = 3$: $2(3)^2 - 6(3) = 0$, LHS = RHS ✓

Example 2

Solve $2x^2 - 5x - 3 = 0$.

$(2x+1)(x-3) = 0$ — Factorize the quadratic.

$2x + 1 = 0;\ x - 3 = 0$

$x_1 = -\frac{1}{2};\ x_2 = 3$ — Set both linear functions equal to 0 and solve.

$x_1 = -\frac{1}{2}$: $2\left(-\frac{1}{2}\right)^2 - 5\left(-\frac{1}{2}\right) - 3 = 0$ ✓ — Check the solutions in the original quadratic.

$x_2 = 3$: $2(3)^2 - 5(3) - 3 = 0$ ✓

Practice 1

1 Solve each equation, leaving your answers exact.

a $x^2 - 8x = 0$ **b** $x^2 - 5x + 6 = 0$ **c** $6x^2 + 4x - 16 = 0$

d $x^2 - 9x = 0$ **e** $4x^2 = 1$ **f** $3x^2 - 48 = 0$

g $10 - 3x = x^2$ **h** $x = 6x^2 - 2$ **i** $9x^2 = 10x$

j $3x^2 - 4 = 23$ **k** $13 - 2x^2 = 5$ **l** $18 = 9x + 2x^2$

> **Tip**
>
> To use the null factor law, first rearrange the quadratic so it is equal to zero.

Problem solving

2 Write a quadratic equation in standard form that has solutions:

a –2 and 4 **b** $\frac{1}{2}$ and 6 **c** $-\frac{2}{5}$ and 3

3 Find a quadratic equation with roots $\frac{3}{4}$ and –5:

a whose leading coefficient is 1

b with only integer coefficients.

You have now seen quadratic equations that have two unique solutions. How many solutions can quadratic equations have?

Reflect and discuss 2

- Graph $f(x) = 2x^2 - 6x$ and $g(x) = 4x^2 - 4x + 1$.
- Explain why they have a different number of unique solutions.
- Can a quadratic equation have more than two unique solutions? Explain.

Practice 2

1 Solve each equation, if possible, leaving your answers exact, or to 3 s.f.

a $x^2 + 2x + 1 = 0$ **b** $x^2 - 6x + 9 = 0$ **c** $x^2 - 10x = -25$

d $9x^2 - 6x + 1 = 0$ **e** $4x^2 + 9 = 12x$ **f** $9x^2 = -4 - 12x$

Problem solving

2 Write a quadratic equation in standard form whose only solution is

a $x = -2$ **b** $x = \frac{1}{5}$ **c** $x = -\frac{3}{4}$

Example 3

The length of a rectangle is 4 cm more than its width. Its area is 21 cm^2.

a Write an equation that represents the area A of the rectangle.

b Solve the equation.

c Write down the dimensions of the rectangle.

a Let x = width. Then length $= x + 4$. — State your variables and write the equation.

$A = x(x + 4) = 21$

b $x^2 + 4x = 21$ — Rearrange, factorize, and solve.

$x^2 + 4x - 21 = 0$

$(x + 7)(x - 3) = 0$

$x + 7 = 0 \Rightarrow x_1 = -7$

$x - 3 = 0 \Rightarrow x_2 = 3$

c width = 3 cm, length = width + 4 cm = 7 cm — x is a length and cannot be negative hence x_1 is not a solution.

Practice 3

1 The width of a rectangle is 3 cm less than its length. The area of the rectangle is 18 cm^2. Find the length and width of the rectangle.

2 The length of a rectangle is 1 cm less than 2 times its width. Its area is 45 cm^2. Find its dimensions.

3 A rectangular shaped garden's length is 2 m less than twice its width. If the area of the garden is 420 m^2, find the dimensions of the garden.

4 The base of a triangle exceeds its height by 17 cm. If its area is 55 cm^2, find the base and height of the triangle.

Problem solving

5 A square and a rectangle have the same area. The length of the rectangle is 5 cm more than twice the length of the side of the square. The width of the rectangle is 6 cm less than the side of the square. Find the length of the side of the square.

6 Find the lengths of the sides of this triangle.

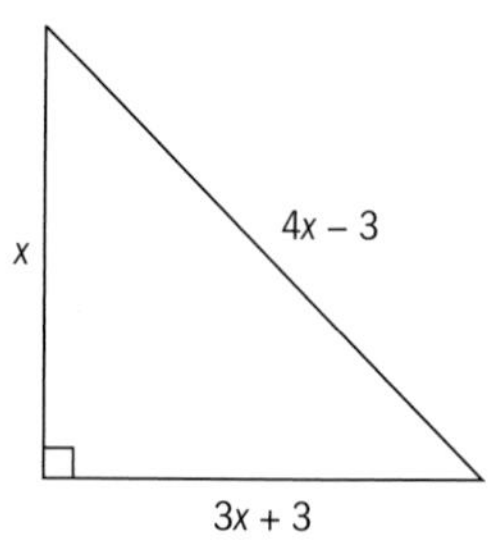

7 A pole leans against a vertical wall. The top of the pole touches the wall at a height of 15 m. The length of the pole is 1 m more than twice its distance from the wall. Find the distance from the wall to the bottom of the pole.

C Solving quadratic equations

- How can you use equivalence transformations to solve quadratic equations?
- How are the three methods for solving quadratic equations equivalent?

ATL

Reflect and discuss 3

You have seen that quadratic equations can have one or two solutions. Do you think it is possible for a quadratic equation to have no solutions? What would its graph look like?

Exploration 3

1 Using a GDC or graphing software, graph the function $f(x) = x^2 - 2x - 3$.

a Find the points where the parabola intersects the x-axis.

b Construct lines through the intersection points that are parallel to the y-axis.

c Grab the parabola and move it around the coordinate plane. Describe what happens to the two vertical lines as you move the curve so that its vertex is to the left and the right of the y-axis.

d The vertex of the original parabola is below the x-axis. Describe what happens to the vertical lines as you move the vertex so that it is above the x-axis, and when it is exactly on the x-axis.

2 Graph the function $f(x) = 3 + 2x - x^2$. Repeat parts **a** to **d** in step **1**.

▶ Continued on next page

3 Based on your answers to steps **1** and **2**, determine:

a whether or not a quadratic function is always the product of two linear functions.

b whether or not a quadratic function has zeros if the function is always above or below the x-axis.

c the number of zeros of the quadratic, if its vertex lies on the x-axis.

ATL

Reflect and discuss 4

$f(x) = x^2 - 2$ cannot be factorized using only integers.

- Solve the quadratic equation $x^2 - 2 = 0$ by factorizing using the difference of two squares.
- Solve the quadratic equation $x^2 - 2 = 0$ by first using the addition principle.
- Do your two methods give the same solution?

When a quadratic is of the form $ax^2 - c = 0$, $c > 0$, you can use the addition principle to rearrange to $ax^2 = c$, and the multiplication principle to get $x^2 = \frac{c}{a}$. The solution is $x = \pm\sqrt{\frac{c}{a}}$.

Taking the square root of both sides of an equation does not always give an equivalent equation. For example $(-1)^2 = 1^2$, but taking the square root of both sides leaves $-1 = 1$, which is obviously false. Therefore 'taking the square root' is not an equivalence transformation.

Reflect and discuss 5

- Discuss the reasons for each step in solving this equation.

  $x = \sqrt{2x+3}$

  [1] $x^2 = 2x + 3$

  [2] $x^2 - 2x - 3 = 0$

  [3] $(x - 3)(x + 1) = 0$

  [4] $x - 3 = 0$ or $x + 1 = 0$

  [5] $x_1 = 3, x_2 = -1$

- Is 'squaring both sides' an equivalence transformation? Explain.

When a quadratic function can be written as the product of linear factors, the quadratic equation has solutions. When quadratics do not have factors with integer coefficients, it may be difficult to express them as the product of linear factors. Instead of factorizing, another method for solving quadratic equations is completing the square.

Example 4

Solve $x^2 - 2x - 4 = 0$.

$x^2 - 2x - 4$ is not factorizable.

$x^2 - 2x - 4 = 0$

$x^2 - 2x = 4$ — 'Complete the square' to write the quadratic in vertex form.

$x^2 - 2x + 1 = 4 + 1$

$(x - 1)^2 = 5$ — Take the square root of both sides. Don't forget the positive and negative square roots.

$x - 1 = \pm\sqrt{5}$

$x = 1 \pm \sqrt{5}$

$x_1 = 1 + \sqrt{5}\,;\, x_2 = 1 - \sqrt{5}$

Check both solutions in the original equation.

when $x_1 = 1 + \sqrt{5}$: $x^2 - 2x - 4 = \left(1+\sqrt{5}\right)^2 - 2\left(1+\sqrt{5}\right) - 4$

$= (1 + 2\sqrt{5} + 5) - 2 - 2\sqrt{5} - 4$

$= 0$ ✓

when $x_2 = 1 - \sqrt{5}$: $x^2 - 2x - 4 = \left(1-\sqrt{5}\right)^2 - 2\left(1-\sqrt{5}\right) - 4$

$= (1 - 2\sqrt{5} + 5) - 2 + 2\sqrt{5} - 4$

$= 0$ ✓

Example 5

Solve the quadratic equation $2x^2 = 4x + 3$. Rationalize the denominator in your answer.

$2x^2 - 4x = 3$ — Rearrange the equation and factorize so that x^2 has coefficient 1.

$2(x^2 - 2x) = 3$

$x^2 - 2x = \frac{3}{2}$

$x^2 - 2x + 1 = \frac{3}{2} + 1$ — Complete the square.

$(x - 1)^2 = \frac{5}{2}$ — Solve for x.

$x - 1 = \pm\sqrt{\frac{5}{2}}$

$x = 1 \pm \sqrt{\frac{5}{2}} = 1 \pm \frac{\sqrt{10}}{2}$

Practice 4

Solve, if possible, by completing the square. Leave your answers exact.

1 $x^2 - 4x = -1$ **2** $x^2 + 2x = 1$ **3** $x^2 - 2x - 2 = 0$

4 $x^2 - 2x + 2 = 0$ **5** $x^2 - 4x + 6 = 0$ **6** $x^2 + 6x + 1 = 0$

7 $x^2 + 20x + 40 = 0$ **8** $2x^2 + 4x = 1$ **9** $3x^2 + 6x = 2$

10 $2x^2 + 20x + 3 = 0$ **11** $4x^2 + 4x = 3$ **12** $2x^2 + 12x = -9$

When a quadratic equation cannot be factorized, you can use the method of 'completing the square' to solve it. You can also use a method that involves the 'quadratic formula'. Just like with factorized form, the equation should first be set equal to zero before using the formula.

> The quadratic formula to solve $ax^2 + bx + c = 0$, $a \neq 0$, is $x = \frac{-b \pm \sqrt{b^2 - 4ac}}{2a}$.

Example 6

Solve $x^2 + 3x - 1 = 0$, giving your solutions to 1 d.p.

$x = \frac{-b \pm \sqrt{b^2 - 4ac}}{2a}$ — Write the quadratic formula and the values of *a*, *b* and *c*.

$a = 1,\ b = 3,\ c = -1$

$x = \frac{-3 \pm \sqrt{3^2 - 4(1)(-1)}}{2} = \frac{-3 \pm \sqrt{9+4}}{2} = \frac{-3 \pm \sqrt{13}}{2}$ — Substitute the values into the formula and simplify.

$x_2 = \frac{-3 + \sqrt{13}}{2} = 0.3$ — Separate the two solutions.

$x_2 = \frac{-3 - \sqrt{13}}{2} = -3.3$

Practice 5

1 Solve for the variable, if possible, using the quadratic formula, leaving your answers exact. Remember to set the equation equal to 0, when necessary.

a $4p^2 + 8p - 1 = 0$ **b** $5a^2 + 3a + 1 = 0$ **c** $-r^2 + 3r - 6 = 0$

d $-q^2 + 2q = -5$ **e** $-3c^2 = -6c + 3$ **f** $2m^2 + 23 = 14m$

g $9x^2 - 11 = 6x$ **h** $4b^2 + 4b - 8 = 1$ **i** $-4t^2 + 8 = -t$

2 Solve, if possible, the quadratic equations giving your answers to 1 d.p. Check your solutions using a GDC or graphing software.

a $x^2 + x - 3 = 0$ **b** $x^2 - 2x - 4 = 0$ **c** $4p^2 + 8p - 1 = 0$

d $5a^2 + 3a + 1 = 0$ **e** $-r^2 + 3r - 6 = 0$ **f** $-x^2 + 4x + 7 = 0$

3 Solve these quadratic equations giving your answers in radical form.

a $x^2 = x + 5$ **b** $x^2 + 5x = 1$ **c** $2x - x^2 = 1$

d $x^2 - 7x = 3$ **e** $-2z^2 + 6z = -95$

4 Solve, giving your answer to 3 s.f.

a $x^2 + 0.6x - 2.6 = 0$ **b** $0.2x^2 - 3x + 0.3 = 0$ **c** $4 - 2x - 3x^2 = 0$

d $3.2 + x - 1.2x^2 = 0$ **e** $2.2x^2 = 3x + 1.1$

Tip

For the equations in **3**, rearrange so that the right hand side is equal to 0.

Exploration 4

1 Create a table with the same column headings as the one below, and give it 8 rows. Copy the first two rows as shown here, then enter the six equations below in the first column. Solve each equation, if possible, using one method from the three equivalent solution methods: factorization (if possible), completing the square, or the quadratic formula. In the fourth column, substitute the parameters a, b and c of the quadratic into $b^2 - 4ac$.

$2x^2 + 5x - 3 = 0$ $-x^2 + 4x - 4 = 0$ $2x^2 + x + 1 = 0$

$x^2 + 2x - 2 = 0$ $3x^2 + 4x + 1 = 0$ $2x^2 - 6x + 5 = 0$

Equation	Solutions	Number of distinct solutions	$b^2 - 4ac$	Can the quadratic be factorized?	Sketch of graphical solution (include x-axis, but not y-axis)
$x^2 + 2x + 1 = 0$	$x_1 = -1; x_2 = -1$	1	$2^2 - 4(2)(1) = 1$	Yes	
$x^2 + 1 = 0$	No solutions	0	$0^2 - 4(1)(1) = -4$	No	

2 Use the answers in your table to complete these sentences.

a When $b^2 - 4ac = 0$, the quadratic equation has _____ distinct solution(s).

b When $b^2 - 4ac > 0$, the quadratic equation has _____ distinct solution(s).

c When $b^2 - 4ac < 0$, the quadratic equation has _____ distinct solution(s).

3 State at how many points the quadratic expression intersects the x-axis when:

a $b^2 - 4ac = 0$ **b** $b^2 - 4ac > 0$ **c** $b^2 - 4ac < 0$

4 Explain how the expression $b^2 - 4ac$ helps you determine whether or not a quadratic expression can be factorized.

You may need to look back at other quadratics that are factorizable in Practice 1 and evaluate $b^2 - 4ac$.

The part of the quadratic formula, $b^2 - 4ac$, is called the **discriminant**, and the symbol we use for it is the Greek letter delta Δ. We write $\Delta = b^2 - 4ac$.

In 820 CE, mathematician Mohammad bin Musa Al-Khwarismi derived the quadratic formula for positive solutions. His method was brought to Europe by the mathematician Abraham bar Hiyya, who lived in Spain around 1100. In 1545 Girolamo Cardano compiled the existing work on quadratic equations. The quadratic formula first appeared in the form we know today in Rene Descartes' *La Géométrie* in 1637.

Practice 6

1 By considering the discriminant of each quadratic, state:

i the number of distinct solutions of the quadratic equation

ii whether or not the quadratic is factorizable.

a $2x^2 - 3x + 2 = 0$ **b** $x^2 - 7x + 6 = 0$ **c** $-3x^2 + 17x - 2 = 3$

d $3x + 7 = -5x^2 - 4$ **e** $4x^2 - 28x + 40 = 0$

Problem solving

2 Find the values of k such that the quadratic has the given number of solutions.

a $x^2 + kx + 16 = 0$ (1 solution) **b** $3x^2 + kx + 12 = 0$ (1 solution)

c $-4x^2 + 8x + k = 0$ (2 solutions) **d** $kx^2 + 3x + 5 = 0$ (no solutions)

D Solutions of quadratic equations

- How do you determine a 'best method' among equivalent methods?
- Do systems, models and methods solve problems or create them?

When solving quadratic equations, particularly for real-life problems, how do you decide which of the three methods of solution (factorizing, completing the square, or the quadratic formula) is the best method?

ATL

Reflect and discuss 6

- Which of the three methods would you try first? Explain why. If this method did not work, which method would you try next?
- How could you find out if the equation is factorizable?
- Which method would you try first if the question asked you to give your solutions to 2 decimal places? Explain why.

Example 7

A gardener has 16 m of fencing to section off a rectangular area of 15 m^2. Determine the constraints on the dimensions of the rectangle, and find the dimensions of the rectangle.

Tip

'Determine the constraints' means 'write an inequality for the values'.

Perimeter = 2(length + width) = 16

$\Rightarrow$ length + width = 8

Let width = x. Then length = $8 - x$.

Constraint 1: $x > 0$ — Length and width can be only positive values. Write inequality statements to reflect this.

Constraint 2: $8 - x > 0$, so $x < 8$

$\Rightarrow 0 < x < 8$

Area = $x(8 - x) = 15$

$8x - x^2 = 15$

$x^2 - 8x + 15 = 0$ — Calculate the discriminant to see if the equation is factorizable.

Here, $a = 1$, $b = -8$ and $c = 15$

$\Rightarrow b^2 - 4ac = (-8)^2 - (4 \times 1 \times 15) = 4$ — The discriminant is positive, so there will be two unique solutions.

$x^2 - 8x + 15 = (x - 3)(x - 5)$

$x_1 = 3$; $x_2 = 5$ — Both values 3 and 5 satisfy $0 < x < 8$.

When $x = 3$: width = 3, length = 8 − 3 = 5

When $x = 5$: width = 5, length = 8 − 5 = 3

The dimensions of the rectangle are 3 m by 5 m.

Practice 7

In these problems, create a mathematical model to solve the problem. Select the most efficient method to solve the quadratic equation.

With geometric models, draw a sketch to help you visualize the problem.

Problem solving

1 A gardener has 40 m of fencing to make three equal rectangular plots. The total area of the three plots is 30 m^2. Determine the constraints on the dimensions of the rectangles, and find the dimensions of each rectangle.

2 50 m of fencing is used to make three sides of a rectangular area, using an existing wall as the fourth side. The area of the rectangle is 150 m^2. Determine the dimensions of the rectangular area.

3 A rectangular garden measures 24 m by 32 m. A path of uniform width is built all around the outside of the garden. The total area of the path and garden is 1540 m^2. Find the width of the path.

4 A rectangular swimming pool measures 10 m by 6 m. A paved area on two sides of the pool has width x meters. The total area including the paved area is $\frac{4}{3}$ of the pool area on its own. Find x.

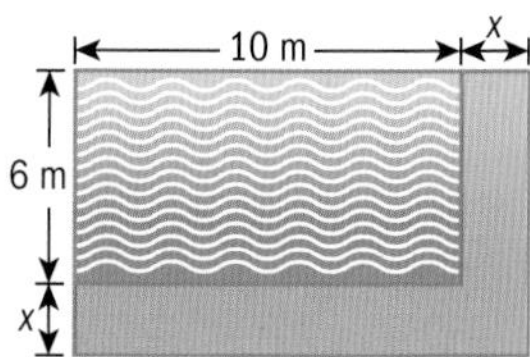

5 The perimeter of a rectangle is 42 cm. Its diagonal is 15 cm. Find the width of the rectangle.

6 A student cycles from home to school, due north for a distance x km, then the same distance plus an additional 7 km due east. If the direct distance from home to school is 17 km, find the distances that the student cycles north and east.

7 A box has a square base of side x cm, height 4 cm, and volume 289 cm^3. Find x.

8 A box with an open top is made from a rectangular piece of cardboard by cutting squares of equal size from each corner. The cardboard measures 50 cm by 40 cm. The area of the base is 875 cm^2.

a Determine the side length of the squares cut from each corner.

b Find the volume of the box.

9 Two numbers have a difference of 3 and a product of 88. Find the numbers.

10 The product of two consecutive odd numbers is 143. Find the numbers.

11 The sum of two numbers is 9 and the sum of their squares is 153. Find the two numbers.

12 When the square of a number is decreased by 1, the result is 4 times the original number. Find the number.

Example 8

When the brakes are applied in a certain make of car, the length L of the skid in meters is given by $L = 0.168s^2 - 0.8s + 0.3$, where s is the speed of the car in km/h at the time when the brakes were first applied. A car involved in an accident leaves skid marks measuring 50 m. Determine the car's speed at the time of braking.

$0.168s^2 - 0.8s + 0.3 = 50$ — Write the equation.

$0.168s^2 - 0.8s - 49.7 = 0$

Using the quadratic formula: — *a*, *b* and *c* are decimals, so the quadratic cannot be factorized.

$a = 0.168$, $b = -0.8$, $c = -49.7$

$$s = \frac{-(-0.8) \pm \sqrt{(-0.8)^2 - 4(0.168)(-49.7)}}{2(0.168)}$$

$s_1 = -15.0$; $s_2 = 19.7$ — Only the positive value makes sense in the context of the question.

The car was travelling at approximately 20 km/h.

The distance a car skids depends on several factors, including the speed of the car and weather conditions. Car accident investigators use skid distances to calculate travelling speeds, to help discover the cause of an accident.

Practice 8

Problem solving

1 The amount of money in bank account A is given by the formula $A = P(1 + r)^t$ where P is the initial investment, r is the interest rate compounded annually and t is the time in years. Your mother wants to invest \$5000 and would like it to be worth \$6500 after two years, to help you with your first year of university costs. Determine the interest rate she needs.

2 An internet company was listed on a small stock exchange for several months before going bankrupt. The price P of the company's stock in terms of the number of months x that the stock traded on the exchange is given by $P(x) = -2.25x^2 + 40.5x + 42.75$. Determine the number of months it took for the company to go bankrupt, in other words when $P(x) = 0$.

3 A charity organizes a soccer game to raise money. There are 15 000 seats in the stadium. From past experience they know that when the tickets cost \$15, they sold 9500 tickets. For every \$1 decrease in ticket price, the number of tickets sold increases by 1000.

a Copy and complete the table.

Price (\$)	Decrease in ticket price	Increase in attendance	Total attendance	Total sales (\$)
15	$15 - 15 = 0$	0	9500	$15 \times 9500 = 142\,500$
14	$15 - 14 = 1$	$1000 \times 1 = 1000$	$9500 + 1000 = 10\,500$	$14 \times 10\,500 = 147\,000$
13				
12				
p	$15 - p$			

b From the last row in the table, find a function for sales generated in terms of ticket price p.

c Write an equation to find the ticket price that would be so expensive that nobody would buy a ticket. Solve it using the most efficient method.

d Determine the ticket price that would maximize sales. How many tickets would you sell at that price?

e Determine the ticket price needed to fill all 15 000 seats.

f Consider your answers for **d** and **e**. Decide what you could do with the unsold seats if you sold tickets at the price that gives maximum sales.

4 A swimming team sells t-shirts with their logo to raise money for an international competition. A survey into prices and predicted sales is shown in this table.

Price of t-shirt in Euros (p)	Predicted number of t-shirts sold (s)
5	165
10	150
15	130
20	118
25	97
30	85
35	65
40	56
45	42
50	28

a Find the line of best fit that best describes the relationship between price and number of t-shirts sold. This is the demand function.

Lines of best fit are covered in topic 4.3.

b Determine the revenue function.

c Determine the t-shirt price that would maximize revenues.

d Determine the t-shirt price that would guarantee that no t-shirts are sold.

5 A ski rental shop charges a rental fee of \$50 per pair of skis and averages 36 rentals per day. A ski journal article states that for every \$5 increase in rental price, the average ski rental business can expect to lose two rentals per day, and vice versa.

a Determine the revenue function.

b Determine the ski rental price that would maximize revenues.

c Determine the ski rental price that would guarantee the shop could not rent out any skis.

6 You have designed a new kind of exercise bike which you want to sell to retailers. Your advertising and start-up costs will be \$800 000. Each exercise bike will cost \$120 to make. Sales of such bikes (demand) tend to follow a linear function modeled by $s = 80\,000 - 180p$, where s is the number of bikes sold and p is the price of the bike in dollars.

a Determine how many bikes you could sell at the price of
i \$100 **ii** \$300 **iii** \$500.

b Use the linear function to write the revenue function for the sale of bikes.

c Write a function for the total costs of producing the bikes.

d Write a function for the profit you would make on producing the bikes.

e Calculate what the price of the bike would be if you were to make no profit on the bikes, and use this to calculate the price that would ensure a maximum profit.

Objective: D. Applying mathematics in real-life contexts
ii. select appropriate mathematical strategies when solving authentic real-life situations

Draw and label suitable diagrams, create equations and select the most efficient method to solve them.

Activity

Do the open box problem in Practice 7 question 8, before this activity.

1 Suitcases are made from rectangular sheets of leather 60 cm by 92 cm.
Each sheet is folded in half, squares are cut from the corners and the sheet is opened up again.

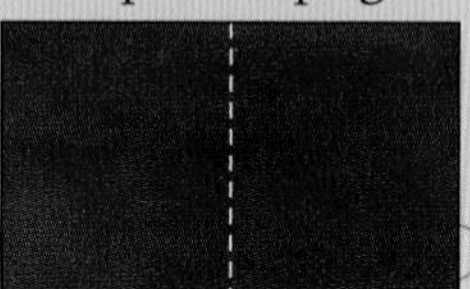 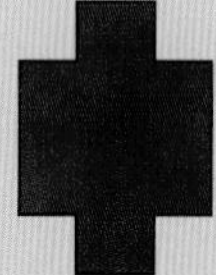 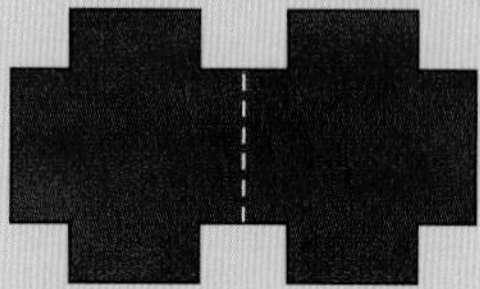

a Draw a diagram and label the variable for the side of each square cut out.
b Express all the dimensions in terms of your variable.
c Express the volume of the case using these dimensions.
d Define the constraints.
e Use your GDC or software to determine the maximum volume of the case and the size of the cut-out squares that give this maximum volume.

2 Determine the size of the squares to cut out to give a case with volume 18 350 cm^3.

3 Create a mathematical model which enables you to decide on the sizes of the squares to be cut out from any size rectangle.
Use suitable software to explore the size of the square giving the maximum volume for several different starting rectangles.

4 Research suitable size cases for a travel set of 3 cases. Use your model to calculate the size of the rectangle and the size of squares to cut out to give maximum volume for your set of cases.

Summary

The **null factor law** states that if the product of two or more numbers is zero, then at least one of the factors must be zero.

For two factors a and b, if $ab = 0$ then $a = 0$ or $b = 0$, or both.

When $f(x) = g(x) \times h(x)$, where g and h are functions, if $f(x) = 0$ then $g(x) = 0$ or $h(x) = 0$, or both.

When a quadratic is of the form $ax^2 - c = 0$, $c > 0$, you can use the addition principle to rearrange to $ax^2 - c$, and the multiplication principle to get $x^2 = \frac{c}{a}$. The solution is $x = \pm\sqrt{\frac{c}{a}}$.

The quadratic formula to solve $ax^2 + bx + c = 0$, $a \neq 0$, is $x = \frac{-b \pm \sqrt{b^2 - 4ac}}{2a}$.

The part of the quadratic formula, $b^2 - 4ac$, is called the **discriminant**, and the symbol we use for it is the Greek letter delta Δ. We write $\Delta = b^2 - 4ac$.

- If $\Delta < 0$, the quadratic has no real roots.
- If $\Delta = 0$, the quadratic has one real root (a repeated root).
- If $\Delta > 0$, the quadratic has two real roots.

When the discriminant is a perfect square (e.g. 1, 4, 9, 16, ...), then the quadratic expression is factorizable.

Mixed practice

1 **Solve** each quadratic by factorizing, and check your answers graphically. Leave your answers exact.

a $x^2 - 2x - 24 = 0$ b $x^2 - 6x = 27$

c $6x^2 = 5x + 4$ d $17x - 2x^2 = 21$

e $-5x^2 + 7x = 2$ f $2x^2 = 8x - 6$

2 **Solve**, if possible, by completing the square, and check your answers graphically. Leave your answers exact.

a $x^2 + 6x - 59 = 0$ b $x^2 + 12x = -23$

c $x^2 - 10x + 26 = 8$ d $3x^2 + 6x = 7$

e $2x^2 = 8x - 3$ f $2x^2 - 5x - 3 = 0$

3 **Solve**, if possible, using the quadratic formula, giving answers to 2 d.p., and check your answers graphically.

a $x^2 + 10x + 13 = 0$ b $x^2 - 5x = 7$

c $3x^2 = 1 + 3x$ d $x^2 - \frac{1}{3}x = 2$

e $4x^2 + 8x = 1$ f $-3x^2 + 2x = 12$

4 **Solve**, if possible, using the most appropriate method. Leave your answer exact and completely simplified. Check your answers graphically.

a $(x+3)^2 + x^2 - 9x = 8$ b $\frac{x^2}{2} + \frac{5}{2} = -x$

c $(x-2)^2 - 11 = 0$ d $2x(x-1) - 5 = -x^2$

e $(2x-6)^2 = 12$

f $(2x+5)(x-1) = (x-3)(x+8)$

Solve these next problems using the most efficient method. Round your answers appropriately.

Problem solving

5 The length of a rectangle is 3 cm greater than its width. Its area is 108 cm^2. **Find** the dimensions of the rectangle.

6 The length of a rectangle is 2 cm more than 3 times its width. Its area is 85 cm^2. **Find** the dimensions of the rectangle.

7 The length of a rectangle is 1 cm more than its width. If the length of the rectangle is doubled, the area of the rectangle increases by 30 cm^2. **Find** the dimensions of the original rectangle.

8 A rectangular lawn measures 8 m by 4 m and is surrounded by a border of uniform width. The combined area of the lawn and border is 165 m^2. **Find** the width of the border.

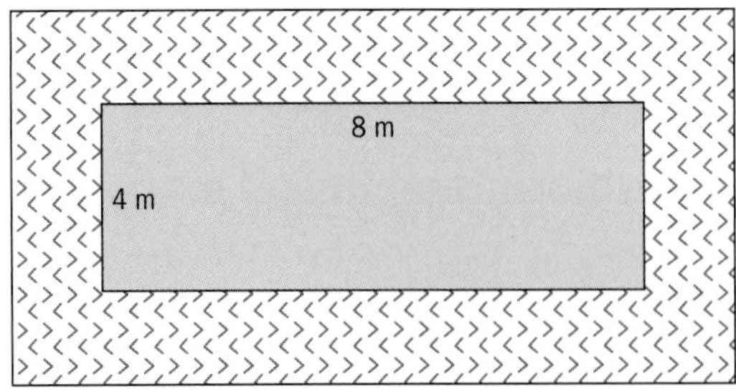

9 The height of a right-angled triangle is 5 cm less than its base. The area of the triangle is 42 cm^2. **Find** its base and height.

10 One side of a triangle is 2 cm shorter than the hypotenuse and 7 cm longer than the third side. **Find** the side lengths of the triangle.

11 **Find** two consecutive odd integers whose product is 99.

12 The square of a number exceeds the number itself by 72. **Find** the number.

13 **Find** two consecutive positive integers such that the square of the first less 17 equals 4 times the second.

14 The height h in meters of a football kicked into the air can be modeled by the function $h(t) = -4.9t^2 + 24.5t + 1$ where t is in seconds.

a **Find** how long after being kicked it takes the ball to hit the ground.

b **Determine** how long it takes the object to reach a height of 20 meters.

c For how long was it above this height?

15 The profits of an international ticket agency can be modeled by the function $P(t) = -37t^2 + 1258t - 7700$, where t is the number of tickets sold and P is in dollars. **Determine** the ticket price that would leave the agency no profit or loss.

16 A square piece of cardboard is to be formed into a box. After 5 cm squares are cut from each corner and the sides are folded up, the box will have a volume of 400 cm^3. **Find** the length of a side of the original piece of cardboard.

Objective: D. Applying mathematics in real-life contexts
ii. select appropriate mathematical strategies when solving authentic real-life situations

In the Review in context section, draw and label suitable diagrams, create equations, and select the most effective method to solve them.

Review in context

Scientific and technical innovation

Quadratic equations can model the path of a projectile through space, and its height above the ground at different times. A projectile is any object that is propelled with force through the air, such as kicking a soccer ball, doing the high jump or even launching fireworks.

Quadratic models of projectile motion take into account three factors:

- the initial height off the ground
- the initial velocity with which the object moves
- the acceleration due to gravity that affects all falling objects.

The formula for calculating the height above ground is:

$h(t)=\frac{1}{2}(-9.8)t^2+V_it+h_i$

where the acceleration due to gravity (on Earth) is -9.8 m/s², V_i is the initial velocity, and h_i is the initial height.

Problem solving

1 A model rocket is launched from 2.5 m above the ground with initial velocity 49 m/s. **Use** the quadratic model $h(t)=\frac{1}{2}(-9.8)t^2+49t+2.5$ to **determine** how long it takes for it to land.

2 An object is launched directly upward at 24 m/s from a platform 30 m high.

a Write the quadratic model for this object.

b Determine:

i the maximum height the object reaches

ii how long the object takes to reach this height

iii how long the object takes to strike the ground again.

3 A projectile is launched from ground level directly upward at 39.2 m/s. **Determine** how long the projectile's altitude is 34.3 m or above.

4 Suppose NASA wants to launch a probe on the surface of the moon, which has one sixth the gravitational pull of the Earth (and therefore one sixth the acceleration due to gravity). The height of the launcher is 0.5 m.

a On the moon, how long would it take the probe to reach a height of 60 m if its initial velocity is 15 m/s?

b If the initial velocity is 24.5 m/s, how much longer will it be in the air on the moon compared to a similar launch on Earth? (It is launched from the surface in both places.)

c Explain how you think scientists test their hypotheses about the behavior of objects on the moon (or Mars) when the force of gravity there is so different than on Earth.

Reflect and discuss

How have you explored the statement of inquiry? Give specific examples.

Statement of Inquiry:

Representing patterns with equivalent forms can lead to better systems, models and methods.

11.3 Seems rational to me

Global context: Scientific and technical innovation

Objectives

- Solving linear and quadratic rational equations algebraically and graphically
- Using equivalence transformations to solve rational equations
- Using rational equations to model situations and solve problems

Inquiry questions

F
- What is a rational algebraic equation?
- How do you solve equations with fractions?

C
- How are the different methods of solving rational algebraic equations related?
- What makes different methods equivalent?

D
- Is there a 'best method' for solving rational equations?
- Does science solve problems or create them?

FORM

ATL **Transfer**

Apply skills and knowledge in unfamiliar situations

Statement of Inquiry:

Representing change and equivalence in a variety of forms has helped humans apply their understanding of scientific principles.

10.2

11.3

E10.2

You should already know how to:

• apply equivalence transformations to solve linear equations	**1** Solve using equivalence transformations: **a** $2x+3=5x-9$ **b** $4(x-1)=2x+7$ **c** $\frac{3x}{2}=10$
• carry out the four arithmetic operations on rational algebraic expressions	**2** Simplify: **a** $\frac{x}{3}+\frac{3x}{4}$ **b** $\frac{7}{x+2}-\frac{5}{x}$ **c** $\frac{2x-6}{x^2-9}\times\frac{x^2+4x+3}{5x+5}$ **d** $\frac{16x^2}{4x-12}\div\frac{x^2-14x}{x^2-49}$
• factorize quadratic expressions	**3** Factorize: **a** x^2-16 **b** $3x^2-6x$ **c** x^2-5x-6 **d** $3x^2-5x-12$
• solve quadratic equations	**4** Solve: **a** $2x^2-18=0$ **b** $x^2+2x=3$ **c** $2x^2+7x+3=0$ **d** $3x^2-4x-1=0$

F Rational algebraic equations

- What is a rational algebraic equation?
- How do you solve equations with fractions?

A rational number is of the form $\frac{p}{q}$ where p and q are integers, and $q \neq 0$. Similarly, a rational algebraic expression is of the form $\frac{P(x)}{Q(x)}$ where $Q(x) \neq 0$. A rational algebraic equation is an equation containing rational algebraic expressions.

The rules and procedures you use with rational numbers also apply to rational algebraic expressions and equations.

Example 1

The total electrical current (16 amps) used by two appliances is given by the equation $\frac{x}{10}+\frac{x}{30}=16$, where x volts is the voltage used by each appliance. Find the voltage used by each appliance.

$\frac{x}{10}+\frac{x}{30}=16$

$30\times\frac{x}{10}+30\times\frac{x}{30}=30\times16$

Use the multiplication principle. Multiply each term by 30 to simplify.

$3x+x=480$

$4x=480$

$x=120$

Each appliance uses 120 volts.

$\frac{120}{10}+\frac{120}{30}=12+4=16$ ✓

Check by substitution.

There is no standard mains voltage used by all countries. In most countries it is between 200 and 240 volts while in Japan and most of the Americas it is between 100 and 127 volts.

Multiplying each fraction in the equation by the lowest common denominator eliminates one of the denominators so it is no longer a rational equation.

Practice 1

Solve these rational equations, leaving your answers exact, or to 3 s. f.

1 $\frac{x}{4}+\frac{3x}{5}=2$

2 $\frac{5x-9}{6}+\frac{2x-4}{7}=1$

3 $\frac{2x}{7}-\frac{1}{3}=\frac{x}{2}$

4 $\frac{x+3}{6}+\frac{2x}{5}=\frac{5x}{2}$

5 $\frac{2x-1}{4}-\frac{x+4}{3}=\frac{5x}{6}$

6 $\frac{4x+1}{2}+\frac{3x-5}{5}=\frac{2x-2}{4}$

7 $\frac{x-3}{2}+\frac{3x-1}{4}=\frac{4x+7}{6}+\frac{1-5x}{8}$

8 $\frac{5x}{3}-\frac{4x-1}{8}=\frac{x+2}{6}-\frac{x+1}{12}$

Problem solving

9 $\frac{2}{x}-\frac{3x}{4}=\frac{1}{3}$

Reflect and discuss 1

- How is question **9** in Practice 1 similar to and different from the other questions?
- How did that affect the process you used to solve the equation?

When different is the same

- How are the different methods of solving rational algebraic equations related?
- What makes different methods equivalent?

Multiplying by the lowest common denominator is a useful method for solving simple rational algebraic equations. The LCM is easy to find when the denominators are rational numbers. What happens when rational expressions or equations have variables in the denominator?

Tip

Finding the lowest common denominator is the same as finding the lowest common multiple, LCM, of two or more numbers or two or more algebraic expressions.

Exploration 1

Look at these pairs of rational expressions and their lowest common denominators.

Rational expressions	Lowest common denominator
$\frac{2}{x}, \frac{3}{x+2}$	$x(x+2)$
$\frac{-1}{x+1}, \frac{7}{x-3}$	$(x+1)(x-3)$
$\frac{5}{x}, \frac{1}{x^2-7x}$	$x(x-7)$

▶ Continued on next page

$\frac{10}{x^2-4}, \frac{-6}{x-2}$	$(x+2)(x-2)$
$\frac{8}{(x+1)(x-5)}, \frac{-3}{(x-1)(x+1)}$	$(x+1)(x-5)(x-1)$
$\frac{2x-5}{x^2+x-2}, \frac{3x}{x^2-1}$	$(x+2)(x+1)(x-1)$
$\frac{x+2}{2x^2+x}, \frac{3x-1}{2x^2-x-1}$	$x(2x+1)(x-1)$
$\frac{x-1}{x^2-9}, \frac{4}{x^2+6x+9}$	$(x-3)(x+3)^2$

1 Determine the first step in finding the LCM of two algebraic rational expressions.

2 After that first step, explain how you would find the LCM of two rational algebraic expressions.

3 Find the LCM of these three expressions: $\frac{7}{x}, \frac{3}{x^2-6x}$ and $\frac{x+1}{x^2-7x+6}$.

Practice 2

1 The denominators of two or more algebraic fractions are given. Find the lowest common denominator of the given denominators.

a x, x **b** $2x, 3x$ **c** $x, x^2, 3x^4$

d $x, x-2$ **e** $x, 2, x$ **f** $x-2, x+2$

g $x, x-2, x+2$ **h** x, x^2+1 **i** $x-1, x^2-1$

j $x^2-x, x+1$ **k** $x+4, 1, 2x+8$ **l** $x+1, x-1, x^2-1$

m $x-a, x+1, x^2-1$ **n** $x, x+2, x^2-x-6$ **o** $2x, 2x-3, x+1$

p $3x^2, 2x^2-8, x+2$

Reflect and discuss 2

- Do you see any potential issues with having variables in the denominator? Explain.
- Find any value(s) of x that are not allowed in each expression in Exploration 1.

To solve rational algebraic equations with variables in the denominator you use the same procedure as for rational equations with numerical denominators.

Example 2

Solve the equation $\frac{3}{x}=\frac{8}{x-2}$ and check your answer algebraically.

$\text{LCM}(x, x-2) = x(x-2)$ — Find the lowest common denominator.

$x(x-2)\times\frac{3}{x}=x(x-2)\times\frac{8}{(x-2)}$ — Rearrange to place each equal algebraic expression in a fraction.

$\frac{\cancel{x}}{\cancel{x}}\times 3(x-2)=\frac{\cancel{(x-2)}}{\cancel{(x-2)}}\times 8x$ — $\frac{x}{x}=1$ for $x\neq 1$, $\frac{x-2}{x-2}=1$ for $x\neq 2$.

$3(x-2) = 8x$

$3x-6 = 8x$

$-5x = 6$

$x=-\frac{6}{5}=-1.2$ — This solution is valid since the only excluded values are 0 and 2.

LHS: $\frac{3}{x}=\frac{3}{-1.2}=-2.5$

RHS: $\frac{8}{x-2}=\frac{8}{-1.2-2}=-2.5$ — Check your answer.

LHS = RHS = −2.5 ✓

Reflect and discuss 3

- Why are you allowed to put lines through the factors that are common to both numerator and denominator?
- What number are you really multiplying by when you do this?

Practice 3

Solve these rational equations using appropriate equivalence transformations, and check your answers algebraically. Make sure you check for extraneous solutions.

1 $\frac{4}{x}=\frac{9}{x-2}$

2 $\frac{10}{x+4}=\frac{15}{4(x+1)}$

3 $\frac{6}{3x}+\frac{5}{4}=\frac{3}{x}$

4 $\frac{4}{3x}+\frac{5}{4}=\frac{3}{x}$

5 $\frac{10}{x(x-2)}+\frac{4}{x}=\frac{5}{x-2}$

6 $\frac{5}{x-2}=7+\frac{10}{x-2}$

7 $\frac{3}{x-2}+\frac{2}{2-x}=2-x$

8 $x+1=\frac{72}{x}$

9 $x+\frac{3}{x+1}=4$

In rational equations, you must exclude any values of the variable that make the denominator equal to 0.

For example, in the rational equation $\frac{3}{x}=\frac{8}{x-2}$ you need to exclude the values $x = 0$ and $x = 2$ because they make one of the denominators equal to zero. If the solution to the equation includes either 0 or 2, you need to eliminate that answer.

An extraneous solution is one that you find algebraically, but does not satisfy the original equation. First determine any values of x that are not allowed in the equation because they make the denominator zero. When you have the solutions, reject any that are not allowed.

Reflect and discuss 4

For the equation: $\frac{7}{x^2-4}+\frac{2}{x^2+2x}=\frac{3}{x}$

- How easy is it to determine the LCM of the algebraic fractions?
- What could you do first, to make it easier to determine the LCM?
- What is the LCM?
- What values of x are not allowed?

Example 3

Solve $\frac{1}{x-6}+\frac{x}{x-2}=\frac{4}{x^2-8x+12}$ and check your answers algebraically.

$x^2 - 8x + 12 = (x - 6)(x - 2)$ — Factorize the denominator on the right side.

$\frac{1}{x-6}+\frac{x}{x-2}=\frac{4}{(x-6)(x-2)}$; $x \neq 6, x \neq 2$ — Identify any values of x that are not allowed.

$(x-6)(x-2)\left[\frac{1}{x-6}+\frac{x}{x-2}\right]=(x-6)(x-2)\left[\frac{4}{(x-6)(x-2)}\right]$ — Use the multiplication principle with the LCM of all denominators.

$(x - 2) + x(x - 6) = 4$

$x^2 - 5x - 2 = 4$

$x^2 - 5x - 6 = 0$ — Set the quadratic equal to 0.

$(x - 6)(x + 1) = 0 \Rightarrow x = 6, x = -1$ — Factorize and solve.

Since $x \neq 6$, the only possible solution is $x = -1$. — Reject any invalid solutions.

LHS: $\frac{1}{-1-6}+\frac{-1}{-1-2}=-\frac{1}{7}+\frac{1}{3}=\frac{4}{21}$

RHS: $\frac{4}{(-1)^2-8(-1)+12}=\frac{4}{1+8+12}=\frac{4}{21}$

Substitute the solution found into the original equation.

LHS = RHS ✓

Practice 4

Solve these equations using appropriate equivalence transformations.

Make sure you check for extraneous solutions.

1 $\frac{7}{x+2}+\frac{5}{x-2}=\frac{10x-2}{x^2-4}$

2 $\frac{x}{x-2}+\frac{1}{x-4}=\frac{2}{x^2-6x+8}$

3 $\frac{2x-4}{x^2-10x+16}=\frac{2}{x+2}$

4 $\frac{2}{x^2-x}=\frac{1}{x-1}$

5 $\frac{2x}{x-3}=\frac{3x}{x^2-9}+2$

Are there any other methods for solving rational equations?

Exploration 2

1 For the fractions $\frac{a}{b}$ and $\frac{c}{d}$ to be equal, determine what conditions must hold for a, b, c, and d.

2 If $b = d$, determine what must be true about a and c in order for the fractions to be equal.

3 Select any pair of equivalent fractions, for example $\frac{3}{6}=\frac{12}{24}$, and multiply diagonally across the equal sign, so 3×24 and 6×12. Suggest what is true about these products. Explain if you think that it is always true.

4 Generalize your result for any pair of equivalent fractions $\frac{a}{b}=\frac{c}{d}$.

5 Prove your result using equivalence transformations.

6 State in which examples from Practice 3 and Practice 4 this result would be useful.

7 State the condition necessary to use this result in solving rational equations.

Consider again the rational equation in Example 3: $\frac{1}{x-6}+\frac{x}{x-2}=\frac{4}{x^2-8x+12}$.

Instead of multiplying both sides by the LCM, what happens if you write all the terms in the equation with the LCM as denominator? The LCM is $(x-2)(x-6)$, so you get an equivalent equation:

$$\frac{1(x-2)}{(x-6)(x-2)}+\frac{x(x-6)}{(x-2)(x-6)}=\frac{4}{(x-6)(x-2)}$$

which simplifies to:

$$\frac{1(x-2)+x(x-6)}{(x-6)(x-2)}=\frac{4}{(x-6)(x-2)}$$

$$\frac{x^2-5x-2}{(x-6)(x-2)}=\frac{4}{(x-6)(x-2)}$$

Since the denominators are equal, the numerators must be as well, so:

$$x^2-5x-2=4$$

Tip

Try using this method in Practice 5, before you decide which method you prefer for solving a rational equation.

Set the quadratic equal to 0 and solve:

$$x^2 - 5x - 6 = 0$$
$$(x-6)(x+1) = 0$$
$$x = 6 \text{ or } x = -1$$

From the original equation $x \neq 2$ and $x \neq 6$, since these values would make denominators equal to zero.

So $x = 6$ is an extraneous solution. The only solution is $x = -1$.

Practice 5

Solve these rational equations by first obtaining equal denominators on both sides of the equation equals sign. Make sure that you check your answers in the original equation to exclude any extraneous answers.

1 $\frac{5}{x-2} - \frac{4}{x} = \frac{10}{x(x-2)}$ **2** $\frac{4}{x} = \frac{3}{x-2}$ **3** $\frac{10}{x^2-2x} = \frac{5}{x-2} - \frac{4}{x}$

4 $\frac{2x^2-5}{x^2-4} + \frac{6}{x+2} = \frac{4x-7}{x-2}$ **5** $\frac{6}{x-1} + \frac{2x}{x-2} = 2$ **6** $\frac{x}{x^2-8} = \frac{2}{x}$

Reflect and discuss 5

What makes different methods equivalent?

D Modelling real-life problems using rational equations

- Is there a 'best method' for solving rational equations?
- Does science solve problems or create them?

Exploration 3

1 Choose two questions from Practice 5 and solve them again, this time multiplying by the LCD.

2 State any differences between the equations from the two methods.

3 Which method do you think is better? Explain your choice.

4 Solve these equations using the method you prefer:

a $\frac{1}{a+1} + \frac{1}{a-1} = \frac{2}{a^2-1}$ **b** $3 - x + \frac{1}{x-2} = \frac{x-1}{x-2}$ **c** $\frac{1}{x+1} = \frac{1}{x^2-4x-5}$

5 State what you notice about your answers, and explain whether or not the results would be the same independent of the method you use. Justify your answer.

Objective: A. Knowing and understanding
i. solve problems correctly in a variety of contexts

In this practice set you will translate the various problems into appropriate rational equations in order to find the correct solutions.

Practice 6

Problem solving

Translate each situation described below into a rational equation. Make sure that you identify all variables. Select the most efficient method for solving the rational equation and use appropriate equivalence transformations. Interpret your answer in the given context.

1 A motorist makes a 1080-mile journey on brand A of gasoline and averages x miles per gallon. For the return trip, she uses the less expensive brand B. On brand B, she travels 3 miles fewer per gallon, and uses 4 gallons more for the same journey.

a Determine the number of miles per gallon for the initial journey.

b Brand A costs 35 cents per gallon, and brand B costs 32 cents per gallon. Calculate the difference in the cost for the 1080-mile journey.

2 The concentration of a medication in a patient's bloodstream C (in mg/l), can be modelled by:

$C(t) = \frac{6t}{t^2 - 4}$, where t is the time (in hours) after taking the medication.

a Determine how many hours after taking the medication the concentration will be 2 mg/l.

b How many solutions did you find? Explain whether or not all the solutions you found made sense.

c One dose equals 5 mg/l and the concentration should never be more than $6\frac{1}{8}$ mg/l. Determine a safe time interval between doses.

Doctors and nurses need to know the concentration of medication in a patient's bloodstream, so they can decide on a safe time interval between doses.

3 On a mountain trail, a group of hikers walked for 7 miles on a level path, and then hiked 12 miles uphill. They generally walked 3 miles per hour faster than they hiked. The entire excursion lasted 4 hours. Determine the rate at which they hiked uphill.

4 To stay underwater, divers need air that contains enough oxygen. However, oxygen under high pressure can act like poison in the body, so the recommended percentage of oxygen in the air changes as you dive deeper. The percentage of oxygen (P) recommended at depth d meters is calculated using the formula $P = \frac{1980}{d + 99}$.

a At sea level the percentage of oxygen in the air is approximately 20%. Explain how this compares to the value calculated using the formula.

b Find the depth at which the recommended amount of oxygen is 10%.

5 Juan drives 20 km/h per hour faster than Maria. Juan drives 100 km in the same time that Maria drives 75 km. Find their driving speeds.

6 Carl's water pump removes 10 000 l of water in 15 minutes. His neighbor Sareeta's pump removes the same amount in 20 minutes. Determine how long it would take to remove 10 000 l of water using both pumps.

7 A scientist studying fish wants to find out how fast a salmon swims in still water. When the river speed is 0.5 m/s she records that a salmon swims 20 m upstream (against the current) and 45 m downstream (with the current) in 70 seconds. Determine how fast the salmon swims in still water.

Summary

A rational number is of the form $\frac{p}{q}$ where p and q are integers, and $q \neq 0$. Similarly, a rational algebraic expression is of the form $\frac{P(x)}{Q(x)}$, where $Q(x) \neq 0$. A rational algebraic equation is an equation containing rational algebraic expressions.

To solve a rational equation, first find the lowest common denominator (LCM) of all the denominators.

Either: Multiply both sides of the equation by the LCM to eliminate all the fractions.

Or: Rewrite the equation so that all terms have the common denominator. The two numerators are then equal.

An **extraneous solution** to an equation is one that you find algebraically, but does not satisfy the original equation. First determine any values of x that are not allowed in the equation because they make the denominator zero. When you have the solutions, reject any that are not allowed.

Mixed practice

Solve these equations using appropriate equivalence transformations and the most efficient method. Make sure you check for extraneous solutions. Leave all answers exact.

1 $\frac{2x-3}{x+3} = \frac{3x}{x-4}$

2 $\frac{4x-3}{x-4} = \frac{x}{x-3}$

3 $\frac{2}{x(x-2)} + \frac{3}{x} = \frac{4}{x-2}$

4 $\frac{10}{x^2+2x} + \frac{4}{x} = \frac{5}{x+2}$

5 $\frac{2(x+7)}{x-4} - 2 = \frac{2x+20}{2x+8}$

Problem solving

6 a **Determine** if a pair of consecutive integers exists whose reciprocals add up to $\frac{5}{6}$.

b **Determine** if a pair of consecutive integers exists whose reciprocals add up to $\frac{3}{4}$.

c **Determine** if a pair of consecutive even integers exists whose reciprocals add up to $\frac{3}{4}$.

d **Determine** if a pair of consecutive odd integers exists whose reciprocals add up to $\frac{11}{60}$.

ATL

Review in context

Scientific and technical innovation

1 The water in the Danube Delta flows at about 2 km/h. On a kayaking trip Max paddles upstream for 15 km, takes a half-hour lunch break, and then returns to his original starting point. His entire trip takes 3.5 hours (including his lunch break).

Let x km/h represent the kayak's average speed. Then the kayak's average speed upstream is $(x-2)$ km/h and downstream is $(x+2)$ km/h.

$$\text{Average speed} = \frac{\text{total distance}}{\text{total time}}$$

a **Justify** the expressions for the average speed of the kayak going upstream and downstream.

b **Write** an expression for the time taken for the kayak to travel:

i upstream

ii downstream.

c **Use** your expressions from part **b** to write an equation representing the whole journey.

d **Find** the average speed of the kayak in still water.

2 When resistors are connected in parallel, the total resistance R_{tot} is given by:

$$\frac{1}{R_1}+\frac{1}{R_2}+\frac{1}{R_3}+\cdots=\frac{1}{R_{tot}}.$$

Find the resistances of two resistors connected in parallel where:

a One resistance is three times as large as the other. Total resistance is 12 ohms.

b One resistance is 3 ohms greater than the other. Total resistance is 2 ohms.

c **Find** the resistances of three resistors connected in parallel where the resistance of one is 2 ohms greater than the smallest one, and the other has resistance twice as large as the smallest one.

The total resistance is $\frac{10}{7}$ ohms.

d An engineer wants to build a circuit with total resistance of 3 ohms. **Show** how he can do this with two resistors, where the resistance of one is 8 ohms greater than the other.

3 Lucy makes an 8-hour round trip of 45 km upstream and 45 km back downstream on a motorboat travelling at an average speed of 12 km/h relative to the river. **Determine** the speed of the current.

'speed 12 km/h relative to the river' means the speed travelling upstream is $12 - x$, where x is the speed of the current.

4 Lenses are used to look at objects, and to project images on to a screen.

The focal length of a lens is the distance between the lens and the point where parallel rays of light passing through the lens would converge.

The distance between a lens and the image produced d_i is related to the focal length f and the distance between the lens and the object d_o by the formula $\frac{1}{d_o}+\frac{1}{d_i}=\frac{1}{f}$.

a The focal length of a lens is 15 cm. The image appears to be twice as far away as the original object. **Find** how far the object is from the lens.

b An object is placed in front of a lens, at a distance 2 cm more than the focal length. The image appears to be 4 cm from the lens. **Find** the focal length of the lens.

c The human eye contains a lens that focuses an image on the retina for it to be seen clearly.

An object is at a distance from the eye that is 20 cm longer than the focal length. The distance from the eye to the retina is –20 mm (on the other side of the lens from the object). **Find** the focal length of this eye.

5 An airplane travels 910 miles with a tailwind in the same time that it travels 660 miles with headwind. The speed of the airplane is 305 m/h in still air. **Determine** the wind's speed.

Reflect and discuss

How have you explored the statement of inquiry? Give specific examples.

Statement of Inquiry:

Representing change and equivalence in a variety of forms has helped humans apply their understanding of scientific principles.

12 Generalization

A general statement made on the basis of specific examples

In mathematics we talk about two different types of generalization. The first involves looking at specific information and trying to find out about something less specific that might underlie it. The other is to tackle specific problems by looking at more general ones.

'All generalizations are dangerous, even this one.'

Alexandre Dumas

Generalization in language - Hypernyms and Hyponyms

A **hypernym** is a word with a broad meaning that includes other, more specific words. So, *vegetable* is a hypernym for *onion, carrot, potato* and *cabbage*, since it is a general term which includes them all.

The specific words are called **hyponyms**. It would be appropriate to say that we use hypernyms to generalize individual ideas: rather than ask someone if they'd like broccoli, carrots, cauliflower and beans with their meal, we might generalize and ask if they would like vegetables.

Horse sense

Your friend says she passes a field every day on the way to school, and each day, without exception, she sees a farmer wearing a straw hat feeding a carrot to a brown horse.

Look at the following general statements and rank them in order from most likely to be true to least likely to be true. Can you think of any other general statements that could be made based on your friend's specific observations?

- The horse never eats anything but carrots
- The farmer feeds the horse only on school days
- The farmer is the only person who feeds the horse
- Your friend never walks to school
- The farmer always wears a hat
- The horse is not a racing horse
- Your friend travels to school at the same time every day

Is 3843 evenly divisible by 7?

To see if a number is evenly divisible by 7, remove the units digit of the number, double the removed digit and subtract the doubled amount from the remaining number. If the result is evenly divisible by 7, then the original number is also divisible by 7. The method may need to be repeated several times.

3843	remove the units digit, 3, leaving 384:
384	
− 6	subtract the doubled number 3
378	remove the units digit, 8, leaving 37:
37	
− 16	subtract the doubled number 8
21	this is a multiple of 7, so 3843 is evenly divisible by 7.

This is a specific example that works with the method described.

- Is this enough to say that the general method always works? Try some other numbers for yourself.

12.1 Seeing the forest *and* the trees

FORM

Objectives

- Identifying patterns in number problems
- Solving complicated problems by looking at a more general case
- Making generalizations from a given pattern

Inquiry questions

F
- What is a conjecture?
- What is a generalization?

C
- How can generalizations be used to solve specific problems?

D
- What are the risks of making generalizations?

ATL **Critical thinking**

Draw reasonable conclusions and generalizations

Conceptual understanding:

Different forms can be used to generalize and justify patterns.

8.1

12.1

12.2

E8.1

You should already know how to:

• expand brackets and simplify algebraic expressions	**1** Expand the brackets in each expression. **a** $2x(3-4x)$ **b** $(4+x)(1-3x)$ **2** Simplify these expressions. **a** $\frac{6x^3y^2}{2x^2y}$ **b** $\frac{12x^3y}{6y}-\frac{12x^5}{3x^4}$
• understand the terms LCM (lowest common multiple) and GCD (greatest common divisor) or HCF (highest common factor)	**3** Find the LCM of: **a** 3 and 8 **b** 6 and 9 **4** Find the GCD (HCF) of: **a** 12 and 54 **b** 10 and 75

Generalization in mathematics

- What is a conjecture?
- What is a generalization?

In mathematics, one of the meanings of generalization relates to reasoning from the specific to the general. Often, it means looking at a limited amount of information and trying to make a statement (a conjecture) about an underlying trend, rule or pattern.

Reflect and discuss 1

Look at the following powers of 4.

$4^1 = 4$	$4^7 = 16\,384$
$4^2 = 16$	$4^8 = 65\,536$
$4^3 = 64$	$4^9 = 262\,144$
$4^4 = 256$	$4^{10} = 1\,048\,576$
$4^5 = 1024$	$4^{11} = 4\,194\,304$
$4^6 = 4096$	$4^{12} = 16\,777\,216$

Make a general statement about the powers of 4.

Typically, forming a generalization involves collecting some information and trying to describe any patterns you notice. If you don't find a pattern, try looking at more information.

There are several things you might notice about the powers of 4. For example, they are all even. You might also notice something about the final digits. Suitable generalizations might include:

- Every power of 4 ends with a 4 or a 6.
- Every odd power of 4 ends with a 4.
- Every even power of 4 ends with a 6.

It's important to bear in mind that you don't yet know whether these generalizations will always hold true. Proving that a particular conjecture is *always* true is discussed in the chapter on justification.

A **conjecture** is a mathematical statement which has not yet been proved. Usually it is consistent with some known information. Once a conjecture has been proven to be true, we usually refer to it as a **theorem**.

A conjecture is similar to the idea of a hypothesis in science.

In 1994 Andrew Wiles found the first successful proof to Fermat's conjecture, which stated that no three positive integers a, b and c can satisfy the equation $a^n + b^n = c^n$ for any integer value of n greater than 2, a conjecture that mathematicians had been trying to prove for 358 years.

ATL

Practice 1

1 Write down the first few powers of 5.
Suggest a general statement about the last digits of the powers of 5.

2 For any number, the *digit sum* of the number is found by adding together the values of its digits. For example, the digit sum of 842 is $8 + 4 + 2 = 14$.

Write down some multiples of the number 9. List their digit sums.
Suggest a general statement about the digit sums of multiples of 9.

3 Look at the multiplication square below.

x	1	2	3	4	5	6	7	8	9	10
1	1	2	3	4	5	6	7	8	9	10
2	2	4	6	8	10	12	14	16	18	20
3	3	6	9	12	15	18	21	24	27	30
4	4	8	12	16	20	24	28	32	36	40
5	5	10	15	20	25	30	35	40	45	50
6	6	12	18	24	30	36	42	48	54	60
7	7	14	21	28	35	42	49	56	63	70
8	8	16	24	32	40	48	56	64	72	80
9	9	18	27	36	45	54	63	72	81	90
10	10	20	30	40	50	60	70	80	90	100

From within the square, select any 2×2 square, for example:

35	42
40	48

Find the difference between the sums of the entries in the two diagonal cells, in this case:

$(35 + 48) - (40 + 42)$.

Describe what you notice. Suggest a suitable general statement.

4 The *arithmetic mean* of two numbers a and b is given by $\frac{1}{2}(a + b)$.

The *geometric mean* of two numbers a and b is given by $\sqrt{ab}$.

Choose five pairs of different numbers. For each pair, calculate the arithmetic mean and the geometric mean. Form a general statement about the arithmetic mean and the geometric mean of a pair of numbers.

5 Consecutive integers are whole numbers which differ by one.

For example, 56 and 57 are a pair of consecutive integers.

a Choose four pairs of consecutive integers.

b For each pair, double the smaller number and add it to the larger number. Write down the total each time.

c Divide each total by 3 and describe anything you notice. Suggest a suitable general statement.

Problem solving

6 a Choose a few pairs of numbers (a, b).

b For each pair, find:

i their product, ab

ii their lowest common multiple, LCM(a, b)

iii their greatest common divisor, GCD(a, b).

c Organize your results.

d Describe what you notice. Suggest a suitable general statement.

7 Observe that $4 = 2 + 2$, $6 = 3 + 3$, $8 = 5 + 3$, $10 = 7 + 3$ and $12 = 7 + 5$. All these even numbers have been written as the sum of two prime numbers.

a Show that 14, 16, 18 and 20 can also be written as the sum of exactly two prime numbers.

b Determine whether or not the following even numbers can be written as the sum of exactly two primes.

i 30 **ii** 98

iii 128 **iv** 2

c Form a general statement based on your observations.

Tip

You can use GCD(a, b) to mean greatest common divisor, or HCF(a, b) to mean the highest common factor of a and b. The two terms mean the same thing.

The general statement formed in question **7** is known as the Goldbach Conjecture, after Christian Goldbach, who first posed it in 1742. Although his claim has been verified for numbers as high as 4×10^{18}, it remains unproven to this day.

Conjectures are usually based on a significant amount of evidence that we have collected and studied; the more evidence we have collected, the more likely we are to believe that the conjecture will prove to be true. Whenever we generalize in this way, we take a series of observations, find something that they have in common, and suggest that it will always prove to be true.

Objective: **B.** Investigating patterns
i. select and apply mathematical problem-solving techniques to discover complex patterns

In this practice you are not told what constitutes enough information to form a generalization, or even what information you should collect. Choosing how you start the problem is important – maybe some of the techniques you used in Practice 1 will help.

ATL

Practice 2

Problem solving

1 Investigate the value of $7^n - 3^n$ for different values of n, where $n \in \mathbb{N}$. Generalize and suggest a conjecture regarding the value of $7^n - 3^n$.

2 Investigate the value of $n^3 - n + 3$ for different values of n, where $n \in \mathbb{N}$. Generalize and suggest a conjecture regarding the value of $n^3 - n + 3$.

3 Investigate the value of $p^2 - 1$, where p is a prime greater than 3. Generalize and suggest a conjecture regarding the value of $p^2 - 1$.

Remember that $n \in \mathbb{N}$ means 'n is a natural number'.

Sometimes a general approach can help us to solve numerical problems. For example:

Simplify $2014^2 - 2012^2$.

This is a very specific problem and it has a specific answer. You could solve it by finding 2014^2 and 2012^2 and then finding the difference, but if you were working without a calculator then this would prove challenging.

Look at the approach taken in Example 1.

Example 1

Simplify $2014^2 - 2012^2$.

Let $n = 2013$

Then $2014^2 - 2012^2 = (n+1)^2 - (n-1)^2$ — Since n 2013, then 2014 n 1, and 2012 n 1.

$= (n^2 + 2n + 1) - (n^2 - 2n + 1)$ — Expand the brackets.

$= 4n$ — Simplify.

$= 4 \times 2013 = 8052$ — Substitute $n = 2013$.

Reflect and discuss 2

In Example 1:

- What was the specific problem?
- What was the general problem that was solved instead?
- How did solving the general problem make it easier to solve the specific problem?

Generalization in mathematics

- How can generalizations be used to solve specific problems?

Exploration 1

1 By letting $n = 2000$, show that you can write $\frac{2003^2 - 2000^2}{2002 + 2001}$ as $\frac{6n+9}{2n+3}$.

Simplify this expression and hence find the value of $\frac{2003^2 - 2000^2}{2002 + 2001}$.

2 By letting $n = 312$, show that you can write $\frac{1}{2}(314 \times 315 - 312 \times 313)$ as $2n + 3$.

Hence find the value of $\frac{1}{2}(314 \times 315 - 312 \times 313)$.

3 Find a suitable value of n to simplify this expression:

$\left(\frac{1}{507} - \frac{1}{508}\right)(507^2 + 508^2 - 1)$.

Use this to evaluate the expression.

4 Find a suitable value of n to simplify the expression $\frac{6000^2 \times 6001 - 5999^2 \times 6000}{17999}$.
Use this to evaluate the expression.

> **Tip**
>
> Remember that 'Find' means that you must show working.

Reflect and discuss 3

- In Exploration 1, how did using n make it easier to evaluate the expressions?
- Explain how your work in Exploration 1 has also made it easier to evaluate $\frac{3003^2 - 3000^2}{3002 + 3001}$.
- Did it matter that we chose n 2000 in step **1**? Could we have chosen n 2003 to find the value of $\frac{2003^2 - 2000^2}{2002 + 2001}$?
What difference would it have made?
- Is there always only one way to generalize a problem? If there is more than one way, is there a best way?

Although we could have easily evaluated each of the expressions in Exploration 1 with the aid of a calculator, using algebraic expressions in this way can help to generalize and provide better understanding of a problem.

In each of the previous number puzzles we picked a value of n which made it easy to simplify the structure of the problem. To do this for yourself, look for numbers which are similar in value (e.g. 5999 and 6001) or numbers which are close to multiples of one another (e.g. 6000 and 17999). Then pick n so that the expression simplifies nicely.

For example, if you were attempting a question containing the numbers 999, 1001, 1500 and 1501, you might pick $n = 500$, because 999, 1001, 1500 and 1501 are all close to multiples of 500.

Practice 3

Without using a calculator, find the value of each of these expressions.

1 $500 \times 502 - 501^2$

2 $\dfrac{5002 \times 5006 - 5003 \times 5004}{5}$

3 $\dfrac{2001}{2000} - \dfrac{2000}{2001} + \dfrac{1}{2000 \times 2001}$

4 $1001 \times 2000 - 1000 \times 2001$

5 $\dfrac{3001 + 3002}{3001 \times 3003 - 3000 \times 3002}$

Generalization in mathematics

- What are the risks of making generalizations?

Consider the expression $n^2 + n + 41$. It is quite well known among mathematicians as an example of one of the potential pitfalls of careless generalization.

The table here shows the value of $n^2 + n + 41$ for some values of n.

n	$n^2 + n + 41$
1	43
2	47
3	53
4	61
5	71
6	83
7	97
8	113
9	131
10	151

Verify that each of the numbers in the right-hand column is prime. This leads us to wonder whether the expression $n^2 + n + 41$ always generates a prime number. On the evidence here, it seems like a sensible generalization.

Reflect and discuss 4

- Read about RSA encryption to the right. How many examples of $n^2 + n + 41$ being prime would convince you that it could be used for generating primes for use in secure data transmission?
- What is the smallest positive integer n for which the rule fails?
- Investigate the expression $n^2 + n + 11$ in the same way.
- Can you find any other expressions with similar properties?

Prime numbers are very important for a method of cryptography known as RSA encryption.

This type of encryption, which relies on prime factorization of some truly enormous numbers, underpins our modern methods of secure data transmission.

Given that in this case generalization has created a false claim, you might wonder whether generalization is a useful skill at all.

Importantly, generalization is usually only one part of a mathematical process. Once you have formed a conjecture, it is important that you then proceed to justify your claim. In this case, you would not be able to justify the claim that '$n^2 + n + 41$ is prime for all n', because it is not true. Forming a conjecture is essential if we are to discover new mathematics, such as an efficient way of generating prime numbers, but you must always bear in mind that a conjecture is not reliable until it has been proved.

Summary

- A **conjecture** is a mathematical statement which has not yet been proved.
- Once a conjecture has been proved to be true, we usually refer to it as a **theorem**.
- Generalization takes two forms:
  - the creation of a conjecture based on a collection of pieces of evidence
  - the creation of a more general problem in the hope of simplifying a problem
- If you are trying to generalize to form a conjecture, gather your information in a table or other logical layout to help you identify patterns.
- If you are trying to generalize a number problem, look at numbers which are similar, or are close to being factors of other numbers in the problem.

Mixed practice

1 Consider the expression $n(n + 1)(2n + 1)$ for different positive integer values of n.

Suggest a suitable general statement about the value of $n(n + 1)(2n + 1)$.

2 Consider the expression $4^{n+1} + 3^{2n}$ for different positive integer values of n.

Suggest a suitable general statement about the value of $4^{n+1} + 3^{2n}$.

3 Mersenne numbers are numbers of the form $2^n - 1$ where $n > 2$.

Marin Mersenne, a French mathematician, observed that when $n = 2, 3, 5$ or 7, then $2^n - 1$ is prime.

a **Verify** this observation.

b Based on this observation, **explain** why a reasonable generalization might be '$2^n - 1$ is prime whenever n is prime.'

c **Show** that this generalization does not hold true for all prime values of n.

4 Without the aid of a calculator, **find** the value of:

a $2002 \times 3001 - 2001 \times 3002$

b $1000 \times 1003^2 - 3000\,(1001 + 1002)$

c $\dfrac{1005 \times 1995 - 995 \times 2005}{1005 + 1995 + 995 + 2005}$

d $\dfrac{542 \times 536 + 8}{540}$

5 A 10×10 grid contains the numbers 1 to 100:

1	2	3	4	5	6	7	8	9	10
11	12	13	14	15	16	17	18	19	20
21	22	23	24	25	26	27	28	29	30
31	32	33	34	35	36	37	38	39	40
41	42	43	44	45	46	47	48	49	50
51	52	53	54	55	56	57	58	59	60
61	62	63	64	65	66	67	68	69	70
71	72	73	74	75	76	77	78	79	80
81	82	83	84	85	86	87	88	89	90
91	92	93	94	95	96	97	98	99	100

Pick any set of five boxes forming a cross shape, for example:

	74	
83	84	85
	94	

Suggest a general rule linking the number in the centre of the cross to the total of all five of the boxes added together.

Reflect and discuss

How have you explored the statement of conceptual understanding? Give specific examples.

Conceptual understanding:

Different forms can be used to generalize and justify patterns.

12.2 Growing predictably

Global context: Scientific and technical innovation

FORM

Objectives

- Finding and justifying (or proving) general rules and formulae for sequences
- Using explicit and recursive formulae to describe arithmetic sequences and geometric sequences
- Recognizing arithmetic and geometric sequences in context

Inquiry questions

F
- What is an arithmetic sequence?
- What is a geometric sequence?
- What are the recursive and explicit formulae for arithmetic and geometric sequences?

C
- How can you recognize arithmetic and geometric sequences in real-life problems?
- How can you solve problems involving arithmetic and geometric sequences?

D
- How does the behavior of a geometric sequence vary depending on the value of the common ratio?
- How can you use the general formulae for arithmetic and geometric sequences to predict future terms?

ATL Critical-thinking

Identify trends and forecast possibilities

Statement of Inquiry:

Using different forms to generalize and justify patterns can help improve products, processes and solutions.

8.1

12.1

12.2

E8.1

You should already know how to:

• use notation to describe sequences	**1** Write down the first three terms of each sequence. **a** $u_{n+1} = u_n + 1, u_1 = 5$ **b** $v_n = 2n - 1$ **c** $w_{n+1} = 2w_n, w_1 = 2$
• write recursive and explicit formulae	**2** Write a recursive and an explicit formula for the sequence that begins 1, 4, 9, 16, …
• use the laws of indices for positive integer powers	**3** Evaluate: **a** $3^5 \times 3^2$ **b** $6^5 \div 6^3$ **c** $2 \times 5^1 \times 5^{n-1}$ **d** $\frac{4^7}{4^2}$
• solve simultaneous equations	**4** By hand, or using a GDC, solve the simultaneous equations: $y = x + 6$ and $y = 3x$
• calculate compound interest	**5** Find the value of \$150 invested at a compound interest rate of 5% per year for 3 years.

F General formulae for arithmetic and geometric sequences

- What is an arithmetic sequence?
- What is a geometric sequence?
- What are the recursive and explicit formulae for arithmetic and geometric sequences?

ATL

Exploration 1

1 Here are four different sequences.

4, 7, 10, 13, 16, 19, …

23, 30, 37, 44, 51, 58, …

48, 42, 36, 30, 24, …

11, 42, 73, 104, 135, …

Copy each sequence and write down anything you notice about it.

Describe any similarities about the way you think the sequences could have been be generated.

Suggest ways in which the sequences might continue.

▶ Continued on next page

2 Here are four more sequences.

4, 12, 36, 108, 324, ...

5, −10, 20, −40, 80, ...

−96, −48, −24, −12, ...

100, 25, 6.25, 1.5625, ...

Write down each sequence and anything you notice about it.

Describe any similarities about the way you think the sequences have been generated.

Suggest ways in which the sequences might continue.

3 Discuss your conclusions with others.

4 The sequences in step **1** are **arithmetic** sequences.

The sequences in step **2** are **geometric** sequences.

Consider each of the following sequences and decide if it is an arithmetic sequence, a geometric sequence, or neither.

a 17, 19, 21, 23, 25, ... **b** 18, 9, 4.5, 2.25, ...

c 1, 4, 9, 16, 25, ... **d** 5, 10, 20, 40, 80, ...

e $1, \frac{1}{2}, \frac{1}{3}, \frac{1}{4}, \ldots$ **f** 13, 39, 117, 351, ...

g 64, 56, 48, 40, 32, ... **h** 1, 5.5, 10, 14.5, 19, ...

i 3, 3, 3, 3, 3, ... **j** $\sqrt{1}, \sqrt{2}, \sqrt{3}, \sqrt{4}, \ldots$

5 The following recursive formulae each define a sequence.

In each case, list the first few terms of the sequence and determine if it is an arithmetic sequence, a geometric sequence, or neither.

a $a_{n+1} = a_n + 3, a_1 = 4$ **b** $b_{n+1} = b_n - 2, b_1 = 5$

c $c_{n+1} = 2c_n, c_1 = 3$ **d** $d_{n+1} = \frac{1}{4}d_n, d_1 = 1024$

e $e_{n+1} = 3e_n - 1, e_1 = 1$ **f** $f_{n+1} = (f_n + 2)(f_n - 3), f_1 = 5$

g $g_{n+1} = 4 - g_n, g_1 = 1$ **h** $h_{n+1} = h_n - 6.4, h_1 = 13.5$

A recursive formula is sometimes called a **term-to-term** formula.

In an **arithmetic** sequence the *difference* between consecutive terms is constant.

In a **geometric** sequence the *ratio* between consecutive terms is constant.

When working with arithmetic sequences, we typically use a for the first term, and n for the term number. The constant difference between consecutive terms is d. Therefore you could write:

$$u_1 = a \text{ and } u_{n+1} - u_n = d$$

or, rearranging the second equation:

$$u_1 = a \text{ and } u_{n+1} = u_n + d.$$

The difference d is usually called the **common difference**.

If d is positive the series is **increasing**, and if d is negative it is **decreasing**.

ATL

Exploration 2

1 Use the recursive formula to write each of the terms $u_1, u_2, \ldots u_6$ of an arithmetic sequence with first term a and common difference d.

2 Look for patterns as the sequence moves from one term to the next.

Describe briefly how these patterns relate to what you know about arithmetic sequences.

3 Now look for patterns linking the term number to the right-hand side of the formula. Generalizing from the patterns you observe, write down a conjecture for a formula for u_n in terms of n. This is an **explicit** formula.

4 Verify that your explicit formula gives $u_7 = a + 6d$.

5 Use your conjecture to find explicit formulae for these recursive formulae:

a $s_{n+1} = s_n + 3,\ s_1 = 4$

b $t_{n+1} = t_n - 2,\ t_1 = 5$

An explicit formula is sometimes called a **position-to-term** formula.

Mathematicians often use a to represent the first term of a geometric sequence and r to represent the constant ratio between consecutive terms. So when n is the term number, you can write:

$$u_1 = a \text{ and } \frac{u_{n+1}}{u_n} = r$$

or, rearranging the second expression:

$$u_1 = a \text{ and } u_{n+1} = ru_n$$

r is usually called the **common ratio**.

Looking again at Exploration 1, step **5**, which of the formulae given are of the same form as the formula above?

ATL

Exploration 3

1 Use the recursive formula to write each of the terms $u_1, u_2, \ldots u_6$ of a geometric sequence with first term a and common ratio r.

2 By considering the pattern you obtain, and generalizing from it, write down a conjecture for an explicit formula for u_n, the nth term of a geometric sequence with first term a and common ratio r.

3 Use your conjecture to find explicit formulae for the geometric sequences described by these recursive formulae:

a $c_{n+1} = 2c_n,\ c_1 = 3$

b $d_{n+1} = \frac{1}{4}d_n,\ d_1 = 1024$

An **arithmetic sequence** with first term a and common difference d has recursive formula $u_{n+1} = u_n + d$, with $u_1 = a$ and explicit formula $u_n = a + (n-1)d$ for the nth term.

A **geometric sequence** with first term a and common ratio r has recursive formula $u_{n+1} = ru_n$, with $u_1 = a$ and explicit formula $u_n = ar^{n-1}$ for the nth term.

Sometimes arithmetic sequences are called **arithmetic progressions**, and geometric sequences are called **geometric progressions**. The names are interchangeable.

Example 1

An arithmetic sequence has first term 7 and common difference −3. Find a general formula for u_n, the nth term. Simplify your answer.

$a = 7,\ d = -3$

$u_n = 7 + (n-1)(-3)$ — Substitute the given information into $u_n = a + (n-1)d$.

$u_n = 7 - 3n + 3$ — Expand the brackets.

$u_n = 10 - 3n$ — Simplify.

Example 2

A geometric sequence has first term 2 and common ratio 5. Find a general formula for u_n, the nth term.

$a = 2,\ r = 5$

$u_n = ar^{n-1}$

$u_n = 2 \times 5^{n-1}$ — Substitute the given information into $u_n = ar^{n-1}$.

Example 3

A geometric sequence has the general formula $u_n = 3 \times 6^n$.

a Express the formula in the form $u_n = ar^{n-1}$.

b Hence write down its first term (u_1) and common ratio.

a $u_n = 3 \times 6^n$ — You are given the formula, but it is not in the form $u_n = ar^{n-1}$.

$u_n = 3 \times 6^1 \times 6^{n-1}$ — Use the laws of indices.

$u_n = 18 \times 6^{n-1}$ — This is now in the form $u_n = ar^{n-1}$.

b The first term is 18 and the common ratio is 6. — The general form is $u_n = ar^{n-1}$, so $a = 18$ and $r = 6$.

Example 4

A geometric sequence has the general formula $u_n = 2 \times (-4)^n$.
Find its first term and common ratio.

$u_1 = 2 \times (-4)^1 = -8$ — Using the given formula, you can find the first term directly.

The first term is -8.

$u_2 = 2 \times (-4)^2 = 32$ — Using the formula again gives u_2. The common ratio r is equal to $u_2 \div u_1$.

$r = u_2 \div u_1 = 32 \div -8 = -4$

The common ratio is -4.

Practice 1

1 Find explicit formulae for the nth term of:

a an arithmetic sequence with first term 5 and common difference 7
b an arithmetic sequence with first term 4 and second term 7
c a geometric sequence with first term 10 and common ratio 4
d a geometric sequence with first term 3 and second term 1
e a geometric sequence with first term 20 and common ratio $\frac{1}{4}$

2 Find the first term and common difference of each arithmetic progression.

a $u_n = 3 + 5(n - 1)$ **b** $u_n = 2n - 4$
c $u_{n+1} = u_n - 4,\ u_1 = 11$ **d** $u_n = -8 - 5n$

A sequence is often referred to as a progression.

3 Find the first term and common ratio of each geometric sequence.

a $u_n = 4 \times 5^{n-1}$ **b** $u_n = \frac{1}{2} \times 7^{n-1}$
c $u_n = 3 \times 5^n$ **d** $u_n = \frac{8}{3^n}$
e $u_n = 0.4 \times 10^{n+3}$ **f** $u_1 = 7,\ u_{n+1} = \frac{1}{2} u_n$

4 A Norwegian glacier was 6.4 km long in 1995. Since then it has been retreating at a constant rate of 80 m per year. Write an explicit formula for the length l_n of the glacier n years after 1995. Use this formula to predict the length of the glacier in 2025 if the retreat continues at the same rate.

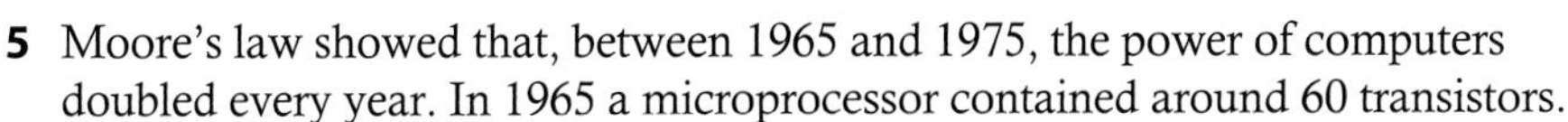

5 Moore's law showed that, between 1965 and 1975, the power of computers doubled every year. In 1965 a microprocessor contained around 60 transistors.

a Write an explicit formula for the number of transistors t_n in a microprocessor n years after 1965.
b Moore later revised his law, saying that after 1971 doubling would occur every two years. How many transistors did the law predict that there would be in a processor in 1993?

C Using arithmetic and geometric sequences in context

- How can you recognize arithmetic and geometric sequences in real-life problems?
- How can you solve problems involving arithmetic and geometric sequences?

In this section you will explore how the general formulae for arithmetic and geometric progressions can help solve problems involving sequences.

Example 5

An arithmetic sequence has terms $u_1 = 7$ and $u_2 = 18$. Find u_{10}.

$a = u_1 = 7$

$d = u_2 - u_1 = 18 - 7 = 11$ — The common difference is the difference between any two consecutive terms.

$u_n = 7 + 11(n - 1)$ — Although the question did not ask you to, it is useful to find the explicit formula for the *n*th term.

$u_{10} = 7 + 11(10 - 1)$ — For the 10th term, let $n = 10$ in the general formula.

$u_{10} = 7 + 99 = 106$

Example 6

An arithmetic sequence has terms $u_3 = 14.2$ and $u_5 = 19.6$. Find u_{11}.

$14.2 = u_1 + 2d \quad (1)$

$19.6 = u_1 + 4d \quad (2)$ — Use $u_n = u_1 + (n - 1)d$ to write two simultaneous equations.

$2d = 5.4$ — Subtract (1) from (2) and rearrange.

$d = 2.7$

$u_1 = 14.2 - 2 \times 2.7 = 8.8$ — Solve the simultaneous equations to find u_1.

$u_n = 8.8 + 2.7(n - 1)$ — Write a general formula for u_n.

$u_{11} = 8.8 + 2.7(11 - 1)$ — Use $n = 11$ to find u_{11}.

$u_{11} = 8.8 + 27 = 35.8$

Example 7

The second term of a geometric progression is 4 more than the first term. The common ratio is 3. Find the first term.

$u_2 = u_1 + 4$ — u_2 is 4 greater than u_1.

$u_2 = 3u_1$ — Since the common ratio is 3, u_2 is 3 times u_1.

$\Rightarrow 3u_1 = u_1 + 4$

$u_1 = 2$

Practice 2

1 An arithmetic sequence has first term 4 and second term 7. Find its tenth term.

2 An arithmetic sequence has second term 19 and common difference 4. Find its eighth term.

3 An arithmetic sequence has third term 11 and fifth term 6. Find its first term, tenth term and common difference.

4 An arithmetic sequence has first term 3. The third term is three times the second term. Find the value of the second and third terms.

5 An arithmetic sequence has sixth term 15 and ninth term 16. Find its common difference. Determine which term has value 22.

6 A geometric sequence has first term 8 and second term 24. Find its common ratio and fifth term.

7 A geometric sequence has third term 48 and fourth term 96. Find its common ratio and fifth term.

You could solve simultaneous equations by hand, or use your GDC.

Problem solving

8 A geometric sequence has second term 30 and third term 45. Find its first term, and write down an expression for the nth term. By using trial and improvement, or otherwise, find the first term in the sequence that is greater than 1000.

9 Two arithmetic sequences have the same first term. The 5th term of the first sequence and the 4th term of the second are both equal to 16. The 9th term of the first sequence and the 7th term of the second are both equal to 28. Find the first term of each sequence and their common differences.

Objective: D. Applying mathematics in real-life contexts
i. identify relevant elements of authentic real-life situations

To identify whether each of the following should be modelled with an arithmetic or a geometric sequence, you must first decide whether the real life factors lead to growth at a constant rate or by a constant scale factor.

ATL

Exploration 4

Sometimes a real-life problem can be modelled using an arithmetic or geometric sequence. Consider each of the following scenarios and decide if it could be described using an arithmetic or a geometric sequence, and explain why. In each case, write down what the terms of the sequence would represent.

1 You are saving up for a summer trip. Your parents give you \$10 to get you started, and then you save half your pocket money each month. (Your pocket money is the same amount every month.)

2 The population of the Earth is increasing by 1.13% per year. This means that next year it will have increased by a factor of 1.0113.

3 The developers of a new website are trying to predict how its number of active users will grow. The developers believe that every month, each existing user will introduce 3 new users to the site.

▶ Continued on next page

4 The owners of a new coffee shop are trying to work out how their customer numbers will grow in the first year. They think that each month, 20 new customers will find their business.

5 A population of Chloroflexi bacteria doubles in number every 27 minutes. A researcher records the size of the population every hour after the start of the experiment.

6 Outside my house the snow is 15 cm deep one morning. As the snow melts, the total depth of snow decreases by 15% every hour.

Reflect and discuss 1

- Explain to others why you chose either arithmetic or geometric sequences to describe each of the scenarios in Exploration 4.
- Assumptions sometimes need to be made in order to make real-life problems simpler so that they can be successfully modelled. How has this happened in the examples in Exploration 4?
- What is it about a real-life problem that means **i** an arithmetic sequence will be appropriate, or **ii** a geometric sequence will be appropriate?

Example 8

At a certain ski resort, some guests stay overnight and all the other skiers arrive in full coaches of equal size during the day. One coach arrives every 10 minutes, starting at 08:10, ten minutes after the resort opens. After the 7th coach has arrived, there are 785 skiers at the resort. After the 10th coach has arrived, there are 920 skiers at the resort.

a Find the number of skiers that will be at the resort once the 12th coach has arrived.

b Find the number of skiers that stayed overnight.

a Let u_n be the number of skiers in the resort after the nth coach has arrived. — An arithmetic sequence models this problem because each coach brings the same number of people.

$u_n = a + (n-1)d$

$u_7 = 785 = a + 6d$

$u_{10} = 920 = a + 9d$ — Form two equations.

so $135 = 3d$

$\Rightarrow d = 45$, and $a = 515$ — Solve to find the values of *a* and *d*.

$u_{12} = 515 + 11 \times 45 = 1010$ — Use the formula to find the twelfth term.

b The number of skiers before the first coach is given by $u_0 = 515 - 45 = 470$. — Since the number of overnight guests is the number of guests when 0 coaches have arrived, finding u_0 tells you the number of overnight guests.

ATL

Practice 3

1 Some money is invested in an account which pays compound interest annually. At the end of 20 years the account is worth \$29 136.22. At the end of 21 years it is worth \$29 718.95. Find the amount of money in the account at the end of the 30th year.

2 When standing vertically upright, a ladder's 11th rung is 170 cm above the ground and its 14th rung is 215 cm above the ground.

- **a** Explain why the heights of the rungs could be expected to form an arithmetic sequence.
- **b** Find the height of the first rung above the ground and the distance from one rung to the next.
- **c** Find the height of the 20th rung.
- **d** Given that the top rung is 350 cm above the ground, determine the total number of rungs that the ladder has.

3 After being cleaned, a swimming pool is being refilled with water. The water level is rising at a constant rate. After three hours the water is 16 cm deep and after five hours it is 19.5 cm deep.

- **a** Determine how deep the water is after only one hour.
- **b** Find the number of hours it will take for the pool to fill to a depth of 25 cm.

4 A computer virus spreads via emails from an infected computer. Each infected computer successfully manages to infect five new computers every day.

- **a** If u_n represents the number of infected computers at the start of day n, explain why $u_{n+1} = 6u_n$.
- **b** Given that there are 2592 computers infected at the start of day 5, find the number of computers from which the virus was initially launched at the start of day 1.
- **c** Determine on what day the virus will infect the millionth computer.

5 The seats in an arena are arranged so that each row is longer than the row in front of it by the same number of seats. The 13th row has 71 seats and the 19th row has 95 seats. Determine the total number of seats in the front three rows combined.

6 In the 2014 Ebola outbreak, the number of deaths from the disease was doubling every month from April. In June there were 252 deaths.

- **a** If u_n represents the number of deaths n months after April explain why $u_{n+1} = 2u_n$.
- **b** Find the number of deaths in April and write an explicit formula for finding the number of deaths in any month after April.
- **c** Find the number of deaths from Ebola in September.
- **d** Explain why this model does not fully explain the outbreak and the continued number of deaths over a longer period of time.

D When do geometric sequences defy expectations?

- How does the behavior of a geometric sequence vary depending on the value of the common ratio?
- How can you use the general formulae for arithmetic and geometric sequences to predict future terms?

ATL

Exploration 5

For each of the sequences described in steps **1** to **4**:

a find the first ten terms of the sequence

b describe what you notice about the terms

c predict what will happen to the value of u_n, the nth term, as n becomes very large.

1 A geometric sequence, first term 10, common ratio 0.8

2 A geometric sequence, first term 10, common ratio −0.8

3 A geometric sequence, first term 10, common ratio 1.2

4 A geometric sequence, first term 10, common ratio −1.2

Reflect and discuss 2

- Compare and contrast the geometric sequences in steps **1** and **2**.
- Do the same for those in steps **3** and **4**.

Exploration 6

A sequence has terms 10, 10, 10, 10, 10, 10, ...

Another sequence has terms 10, −10, 10, −10, 10, −10, ...

For each sequence:

a determine whether or not there is a common ratio between successive terms

b explain whether or not it is a geometric sequence.

Reflect and discuss 3

- How would you best describe the behavior of the sequences?
- Is the sequence 0, 0, 0, 0, 0, ... a geometric sequence? What happens if you try to find the common ratio by dividing any term by the term before it?

Geometric sequences either become larger and larger or smaller and smaller depending on the common ratio. If r is between −1 and 1, the values will get closer to zero. If r is greater than 1 or less than −1, the terms will become infinitely large. When r is negative, the terms of a geometric sequence alternate between being positive or negative.

The sequence 0, 0, 0, 0, ... has a lot in common with geometric sequences but is not normally regarded as geometric because of the problems involved with dividing by zero.

Summary

- In an **arithmetic sequence** the difference between consecutive terms is constant.
- An arithmetic sequence with first term a and common difference d has recursive formula $u_{n+1} = u_n + d$, $u_1 = a$, and explicit formula $u_n = a + (n - 1)d$ for the nth term.
- In a **geometric sequence** the ratio between consecutive terms is constant.
- A geometric sequence with first term a and common ratio r has recursive formula $u_{n+1} = ru_n$, with $u_1 = a$ and explicit formula $u_n = ar^{n-1}$ for the nth term.
- In a geometric sequence, the common ratio, r, cannot be equal to 0.

Mixed practice

1 An arithmetic sequence begins 7, 25, 43, …

a **Find** the common difference.

b **Write down** a recursive formula linking the nth term to the $(n + 1)$th term.

c **Write down** an explicit formula for the nth term.

d **Find** the value of the 15th term.

e **Find** the term number of the term with value 367.

2 An arithmetic sequence has fourth term 253 and fifth term 291.

a **Write down** the value of the common difference.

b **Hence find** the first term.

c **Find** the sum of the first three terms.

3 An arithmetic sequence has third term 31 and sixth term 52. Let the first term be a and the common difference be d.

a Form two simultaneous equations in terms of a and d.

b **Hence find** a, d and the value of the 10th term.

4 An arithmetic sequence has common difference 6. The product of the first two terms is 91. **Find** the value of the first term, given that it is negative.

5 The third term of an arithmetic sequence is twice the first term. The second term is 45.

a **Find** the value of the first term and the common difference.

b **Determine** whether or not 310 is a term of the sequence.

6 A geometric sequence begins 6, 42, 294, …

a **Write down** the value of the first term and the common ratio.

b **Write down** a recursive formula linking the nth term to the $(n + 1)$th term.

c **Write down** an explicit formula for the nth term.

d **Find** the value of the fifth term.

e **Find** the value of the first term to exceed one million.

7 A geometric sequence begins 6, 3, 1.5, …

a **Write down** the value of the first term and the common difference.

b **Write down** a recursive formula linking the nth term to the $(n + 1)$th term.

c **Write down** an explicit formula for the nth term.

d **Find** the value of the tenth term, correct to three significant figures.

8 A geometric sequence has common ratio 4. The second term is 9 more than the first term.

a **Write down** two different expressions for the second term in terms of a, the first term.

b **Hence** write an equation using this information and **find** the value of a.

c **Find** how many terms there are in the sequence that are less than 1000.

9 A geometric sequence has first term 12 and third term 48. The common ratio is r.

a **Show that** $r^2 = 4$.

b **Hence find** two possible values for the common ratio.

c **Find** the possible values of the sixth term.

Review in context

Scientific and technical innovation

1 Consider the following diagram, which illustrates the chemical structure of methane, ethane and propane, three examples of chemicals known as alkanes.

methane ethane propane

a **Show that** the number of hydrogen atoms (represented by an H) in the three alkanes pictured forms an arithmetic sequence.

b Let u_n be the number of hydrogen atoms in an alkane with n carbon (C) atoms. **Write down** an explicit formula for u_n.

c **Find** the number of hydrogen atoms in an alkane with 20 carbon atoms.

d **Find** the number of carbon atoms in an alkane with 142 hydrogen atoms.

2 The developers of a new social media website think that its membership will grow by the same scale factor every month. At the end of the first month it has 20 000 members, and at the end of the second month it has 25 000 members.

a **Explain** which information in the statement above suggests that this can be modelled with a geometric sequence.

b **Determine** how many members this model would predict for the website to have at the end of the 12th month.

c **Determine** how many members this model predicts for the end of the second year.

3 A design company produces business cards. They charge a set fee for design, and then sell the cards in boxes of 100 cards. Each box costs the same amount.

The total cost (including the design fee) for 400 cards is \$108.

The total cost (including the design fee) for 600 cards is \$133.

Find the cost for 1000 cards, and **calculate** the design fee.

4 A machine produces a constant number of components per hour, except in the first hour of operation.

By the end of the second hour of operation, it has made 6000 components. By the end of the seventh hour, it has made 24 000 components.

a **Explain** why the number of components made by the nth hour forms an arithmetic sequence.

b **Find** the number of components made in the first hour.

c **Find** the number of components made in total over a nine-hour working day.

Reflect and discuss

How have you explored the statement of inquiry? Give specific examples.

Statement of Inquiry:

Using different forms to generalize and justify patterns can help improve products, processes and solutions.

So, what do *you* think?

Global context: Identities and relationships

Objectives

- Selecting samples and making inferences about populations
- Understanding the purpose of taking a sample
- Using different sampling techniques
- Understanding when it is appropriate to generalize from a sample to a population
- Understanding the effect of sample size on the reliability of your generalizations

Inquiry questions

F
- What are the different sampling methods?

C
- How are conclusions drawn from experimental data?

D
- How do we know when to say 'when'?
- Do you want to follow the crowd?
- What effect does sample size have on the reliability of our generalizations?

RELATIONSHIPS

ATL Critical-thinking

Recognize unstated assumptions and bias

Statement of Inquiry:

Generalizing and representing relationships can help to clarify trends among individuals.

4.3

12.3

E4.1

You should already know how to:

<table>
<tr>
<td>• represent data using both stem-and-leaf diagrams and box-and-whisker diagrams</td>
<td>1 Below are the number of customers per day at a noodle bar over the course of 20 days. Present this data in a stem-and-leaf diagram.

65 54 77 71 73 89 55
19 17 53 70 75 88 29
34 33 67 71 55 50

Represent the same data on a box-and-whisker diagram.</td>
</tr>
<tr>
<td>• represent data using cumulative frequency graphs</td>
<td>2 35 people made note of how many radio ads they had heard on their commute to work, shown here. Create a cumulative frequency graph from this data.
<table>
<tr><th>Ads</th><th>Frequency</th></tr>
<tr><td>$1 < t \leq 4$</td><td>7</td></tr>
<tr><td>$5 < t \leq 9$</td><td>14</td></tr>
<tr><td>$10 < t \leq 14$</td><td>10</td></tr>
<tr><td>$15 < t \leq 20$</td><td>3</td></tr>
<tr><td>$21 < t \leq 25$</td><td>1</td></tr>
</table>
</td>
</tr>
<tr>
<td>• calculate the mean and the median for a set of data</td>
<td rowspan="2">3 The heights of boys (in meters) in a school football team were measured as follows:
1.55, 1.64, 1.49, 1.72, 1.81, 1.58, 1.73, 1.68, 1.75, 1.70, 1.69
Calculate the mean and median heights.
Calculate the range.
Calculate the lower and upper quartiles and hence the interquartile range.</td>
</tr>
<tr>
<td>• calculate the range and interquartile range for a set of data</td>
</tr>
</table>

The need for sampling

- What are the different sampling methods?

Some things we learn from first-hand experience, while others are based on deductions we draw from the evidence we have. Using observation and background knowledge to draw a conclusion is known as *inference*. For example, if you walk into your classroom and your teacher has laid out a sheet of paper on each desk, you might infer that you were having an assessment.

To *infer* means to draw conclusions based on evidence or reasoning.

Inference in statistics

The mathematics eAssessment results for two schools where the same teacher taught in consecutive years are shown in the stem-and-leaf diagram here.

Crescent Moon High School		Oyster Bay Academy
4	4	
7 5 4 3	5	
9 8 6 4 3 1	6	
8 7 5 3 3 2 0 0	7	2 7
8 4 2 1	8	0 3 5 5 8 9
5 3	9	1 2 4 4 5 8
8 1	10	4 4 6 7
1	11	3

Key: 1 | 8 | 5 means 81 for Crescent Moon and 85 for Oyster Bay

Reflect and discuss 1

Look at the stem-and-leaf diagram and compare the performance of the two schools.

- What conclusions can you draw based on this data?
- How do your conclusions differ from those of others?
- Did you come up with any answers that they did not?
- Did you come up with any conflicting conclusions?

Inference is the process of drawing conclusions about general characteristics from specific pieces of information. In the eAssessment scenario, it's difficult to make inferences from the data because you are given so little information about it. Various factors, such as time spent on revision, personal circumstances, and even class size might explain the difference in performance between one school and the other.

If you don't know anything about the quality or the nature of the data, it's difficult to draw conclusions about any underlying trends. In this case, you may have reached conclusions about the school, or the students, which were unfair or unrepresentative.

Reflect and discuss 2

Calculate the median score for the classes from the two schools. Which of the following statements do you think are justified?

- The students at Oyster Bay Academy are more intelligent than the students at Crescent Moon High.
- The teacher did a better job of teaching the class at Oyster Bay than at Crescent Moon.
- The class at Crescent Moon scored lower on average than the Oyster Bay class.

Again, it's difficult to compare the schools, or to compare the teacher's performance, because we don't know enough about the factors surrounding the data. The only statement that we can be really confident of is saying that on average, the Crescent Moon class scored lower than the Oyster Bay class, but even this is problematic, since they sat the eAssessment in different years and you don't know if the standard of the tests was the same.

Even though you might not agree with the first statement: 'The students at Oyster Bay Academy are more intelligent than the students at Crescent Moon High', it is an example of *inference*. Specifically, *statistical inference* is the action of drawing conclusions about a population based on a sample from that population.

A **population** is a set of objects under consideration in a statistical context. It could be a group of people, or a collection of objects.

A **sample** is a subset of a population.

Inference is the act of drawing conclusions about a population based on a sample.

Any set of objects could form a population, for example the bags of sugar filled by a machine could be a population, as could all the cars sold in Japan in 2014.

Reflect and discuss 3

Suppose you believe that the students in one school are better at mathematics than the students at another school. What sort of evidence would you need to collect to justify such a claim?

A **census** is a collection of data from every member of a population.

A sample is **representative** if its properties reflect those of the population that it comes from.

Often a sample is taken because it is not practical to conduct a census. It is important to ensure that a sample is representative of the population. In the example concerning the two schools, the sample would be representative only if it covered the whole ability range within the school.

Objective: D. Applying mathematics in real-life contexts
i. identify relevant elements of authentic real-life situations

You will need to draw on your general knowledge to help answer the questions in Exploration 1. Trying to find underlying assumptions even when you don't know everything about the context is a key mathematical skill.

Exploration 1

The managers of a large factory, which has around 10 000 workers, would like to know more about the commuting times of their staff.

1 Explain why it might not be sensible to ask every single employee.

▶ Continued on next page

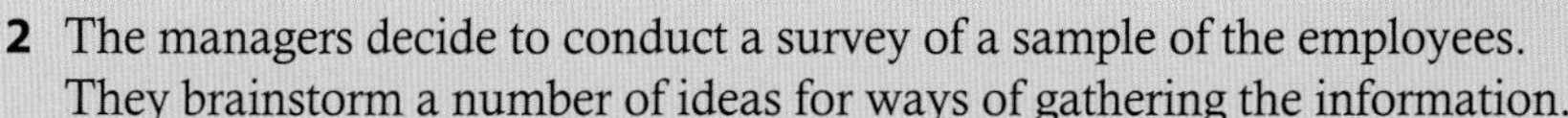

2 The managers decide to conduct a survey of a sample of the employees. They brainstorm a number of ideas for ways of gathering the information.

a They could ask each employee individually as they arrived at work.

b They could put up a notice asking employees to respond by email.

c They could ask one person in each department to estimate an average commuting time for that department.

d They could ask every 200th person on the company's staff list, having sorted it alphabetically.

e They could ask every 200th person on the company's staff list, having sorted it by salary.

f They could put every employee's name into a bag and draw out 50 names at random.

Comment on each of these ideas. Explain how useful you think the surveys would be for finding out representative information about the commute times of the factory's employees.

There are good reasons why a company might conduct a survey of a sample, rather than the whole workforce, two of which are cost and practicality. There are, as you have seen in Exploration 1, problems that can arise when taking a sample, however. Most of these are to do with the idea of bias. A sample is biased if it is not a fair representation of the population.

You will notice that some of the samples above required having a list of all the employees, whereas others did not. This list is known as a *sampling frame*. A sampling frame is any list that identifies every member of the population. In a case like this one, where we are surveying employees, it is quite likely that we would be able to obtain such a list. If, however, the company wished to run a customer satisfaction survey, they would probably not be able to obtain a list of everybody who had bought their products.

A **sampling frame** is a list of every member of the population.

Different sampling methods are available to us depending on whether it is possible, or practical, to obtain a sampling frame. The following three sampling methods all require a sampling frame.

- A **simple random sample** is a sample such that every member of the population is equally likely to be included in the sample, independent of any other member of the population.
- A **systematic sample** takes members of the population at regular intervals from an ordered sampling frame – e.g. taking every 10th person from a list sorted alphabetically.
- A **stratified sample** identifies different sections of the population and ensures that those groups are represented in appropriate proportions in the sample.

Simple random samples

When creating a simple random sample it is important that the selections you make are genuinely random. This could be done by pulling names from a hat or by using the random number generator on a calculator.

Systematic sampling

A systematic sample is formed by taking an ordered sampling frame and then – having first picked a random starting point – taking items from the sampling frame at regular intervals.

Stratified sampling

To form a stratified sample, you first have to divide the population into groups that you want to represent. For example, if you were surveying a group of people, you might want to make sure that men and women were fairly represented, or that different age ranges were fairly represented.

In this case, 'fairly represented' means that the proportion of men in the sample should match the proportion of men in the population, and likewise for women; or that the proportions of people in different age ranges in the sample reflected the proportions in the population.

Surveying 10 students from a year group of 80 after an IB exam to see how difficult they thought it was. The students were taught in five different classes of 16 students.	You are sampling $\frac{10}{80}=\frac{1}{8}$ of the students. From each class of 16, you need $\frac{1}{8}$ of $16 = 2$ students. From each class, pick a simple random sample of 2 students.
Checking 20 of the cars in a full lot of 500 cars to see whether their tires meet regulation requirements. 100 cars were compacts, 250 cars were estates and 150 cars were saloons.	You need $\frac{20}{500}=\frac{1}{25}$ of the cars. So you need $\frac{1}{25}$ of $100 = 4$ compacts; $\frac{1}{25}$ of $250 = 10$ estates; $\frac{1}{25}$ of $150 = 6$ saloons. Pick a simple random sample of 4 of the compacts, 10 of the estates and 6 of the saloons.

Reflect and discuss 4

- What are the advantages and disadvantages of the three types of sampling described above?
- Suggest some situations in which you would use each of the sampling techniques.
- Describe the difficulties you might encounter when trying to apply them.

Example 1

Javier is going to sample 50 students at his school.

The school secretary provides a list of the 500 total students in the school. There are five year groups, each with 100 students, and each year group has 60 boys and 40 girls.

Explain how Javier could create:

a a simple random sample

b a systematic sample

c a stratified sample.

a Method 1

Cut up a printed list to separate all the names and place them in a bag. Shuffle its contents and then randomly draw out 50 names.

Method 2

Number each student from 1 to 500 in alphabetical order. Use a random number generator to pick 50 numbers between 1 and 500 (ignoring any duplicates).

b List the students in alphabetical order. Use a random number generator to pick a random starting point. From that point, take every 10th name until you get back to the starting point.

50 students out of a total of 500 means taking every 10th name.

c Each year group is $\frac{100}{500} = \frac{1}{5}$ of the school, so each group should be a sample of 10 students. Within each group 60% are boys and 40% girls, so from each year group take a simple random sample of 6 boys and 4 girls.

Practice 1

1 You are concerned that the eggs in your kitchen are rotting. You can tell if an egg is rotten by breaking it open and smelling the contents. If the egg is rotten, there will be a pungent smell.

a Explain why you might wish to sample the eggs, rather than conduct a census.

b Explain how you could conduct a simple random sample of the eggs.

ATL

2 Alex lives in a high-rise apartment building containing 80 apartments, numbered 1–80. He wants to conduct a survey to find out how many people on average live in each apartment.

a Explain how he could create a simple random sample of 20 apartments.

b Explain why a systematic sample of 20 apartments starting with apartment 3 and then selecting every fourth apartment might lead to bias.

c Describe any factors that would be relevant if Alex wished to take a stratified sample.

Tip

There is information relevant to the situation that is not contained within the question. Could the design of the building be relevant here?

3 A teenager wants to know how many of his friends watch cartoons. He sends a message to every 20th person in the contact list on his phone. Determine what sort of sample this is. Comment on the suitability of this sampling method.

4 A company employs 900 people, employed on different pay grades. The number of men and women at each pay grade is as follows:

Pay grade	Men	Women
A	1	2
B	12	15
C	40	30
D	80	320
E	240	160

A full list of the employees has been made available for a survey on childcare provision at work.

a Describe how the company could construct a simple random sample of 40 employees.

b Describe how the company could construct a systematic sample of 45 employees.

c Describe how the company could construct a stratified sample of 90 employees. Explain clearly why it would be necessary to combine pay grades A and B.

5 Explain in which of the following scenarios it would be possible to conduct a random sample by selecting people at random from a sampling frame.

a A survey of the students in a school to find out the proportion of students who are vegetarian.

b A survey of the supporters of a particular ice hockey team to find out whether or not they thought the manager was doing a good job.

c A survey of the voters in a particular town to see if they would vote in favour of re-electing the town mayor.

d A survey of the civil servants in an American city to see if they attended fire safety training sessions regularly.

In situations where a sampling frame cannot be obtained, or it is not possible to guarantee responses, *quota sampling* is often used. To create a quota sample, first decide how big your sample is to be. If desired, the sample can still be split into different subgroups as in stratified sampling. Then collect data by any means possible until you have as many as needed to fill your quota.

Suppose you wish to know what local shoppers think of public transport in your town. You could decide to survey 30 people. You want your sample to be representative of the people who live in your town, so you are going to ask questions only of people who live locally, and want to represent men's and women's views equally.

An appropriate way to conduct the survey might be to stand in the downtown area and ask the first people that were willing to talk to you. When you have received responses from 15 local men and 15 local women, you can stop.

ATL

Reflect and discuss 5

- Why might the choice of where you stand to conduct the survey be important (and thus have unwanted influence on the survey)?
- What problems might you encounter when taking a quota sample?

A **quota sample** is used when the sampling frame is unknown, or does not exist. It involves collecting data until enough pieces of data are found.

Practice 2

1 Identify a suitable sampling method for each of the following scenarios. Give as much detail as possible.

a A botanist wants to find the mean length of the leaves on a type of tree.

b You want to find out how students at your school travel to school each day.

c You want to find the average mark obtained by students in your year group in a recent test, but don't want to collect everybody's scores.

d A school principal wishes to sample 20 employees in a school which employs 150 teachers, 20 administrative staff and 30 facilities staff.

e An event organizer wishes to find the average time taken by runners to complete a 10 km charity race.

2 A company wants to assess customer satisfaction with their latest style of coffee machine. Some different plans are proposed.
One plan is to put a feedback form in the box for every product that is sold and ask customers to return the form by post.
Another plan is to call a random sample of customers based on the telephone numbers filled in on the guarantee registration forms.
Describe the advantages and disadvantages of the plans. Determine whether they are random samples.

3 **a** A student wishes to estimate the average height of students in his year group of 150 students. He uses the 25 members of his school athletics team as a sample and records their heights correct to the nearest cm. Comment on the suitability of his sample.

b Another student wishes to estimate the average height of the 150 students in her year group. She uses her MYP History class of 25 students as a sample and records their heights correct to the nearest cm. Comment on the suitability of her sample.

c Explain how a simple random sample of 25 students could be taken from this year group of 150 students.

d Explain how a stratified sample of 25 students could be taken from this year group.

C Generalization and inference

- How are conclusions drawn from experimental data?

If a sample is representative of the population from which it is drawn, its properties should reflect the properties of the population. This means that you can use the properties of the sample to make deductions about the whole population. In doing so, you are generalizing – taking specific knowledge from the sample data you have obtained and using it to make broader conclusions about the whole population.

Exploration 2

Antonia thinks that boys who make the school rowing team are taller on average than the rest of their year group.

She measures the height of each of the 8 boys on the rowing team, and takes a random sample of 16 other boys from the year group.

The rowers have the following heights, each measured to the nearest whole centimeter: 173, 192, 181, 192, 174, 184, 182, 179.

The other students have the following heights: 170, 154, 192, 165, 168, 190, 188, 167, 187, 181, 194, 180, 167, 157, 193, 165.

1. Calculate the mean height of the rowers.
2. Calculate the mean height of the non-rowers.
3. Comment on her claim that boys who make the school rowing team are taller on average than the rest of their year group.

ATL

Reflect and discuss 6

- Where have you generalized in Exploration 2?
- Is the generalization reasonable?
- In this problem you have used the mean of a sample as a way of estimating the mean of the population. Explain why it might not be sensible to use the range of the sample to estimate the range of the population.

In this context, you should have made two generalizations: first, that the information gained from the sample told you something about the population from which it came; second, that by comparing the means of two groups you could draw conclusions about which group contained taller students. Both of these generalizations are types of inference.

Practice 3

1 Your tutor says that your test scores in Physics are better than your test scores in Chemistry. Your results (out of 100) for the last six tests in each subject are presented in this table:

Chemistry	56	64	66	58	72	81
Physics	64	62	72	73	60	70

a Calculate your mean test score in Chemistry and in Physics.

b Comment on your tutor's claim.

c Explain how you have used generalization in this problem.

ATL

2 You suspect that boys are spending more money than girls in your school canteen. You stand by the cashier one lunch time and record the amount spent by the first ten girls and by the first ten boys. The amounts you record are as follows:

Boys		Girls	
\$3.12	\$1.00	\$4.00	\$2.51
\$4.00	\$6.37	\$1.37	\$0.88
\$4.00	\$4.00	\$4.00	\$4.00
\$5.16	\$3.82	\$4.00	\$4.80
\$2.13	\$3.65	\$2.32	\$3.55

a Describe the type of sampling used.

b Outline one way in which this manner of sampling might have introduced bias.

c Comment on the claim that boys spend more than girls in the canteen.

d Explain where you have generalized in your answer.

Tip

Why does \$4.00 appear so frequently in the table? Something that stands out like this should cause you to reflect on what other information you can infer from the data you have, and then question whether it is likely to have an impact on the validity of the sample.

ATL

Exploration 3

GDP per capita is a measure of the average income per person in a country. You can find this data in annual tables in the World Bank database (data.worldbank.org). Ask your teacher if you need help with this.

1 There are a large number of countries in this table so it will be necessary to select a sample. Make a simple random sample of 10 countries with the GDP per capita for the years 2000 and 2010.

2 Calculate the mean GDP per capita for each year and determine whether it rose over the period from 2000 to 2010.

3 Determine if the selection of countries you made is fair for you to generalize about the GDP per capita for the whole world. Explain whether it would be better to use a stratified sample. Plan how to do this and see how your results compare using this method.

GDP stands for Gross Domestic Product.

Exploration 4

A factory owner wishes to compare the output of two employees. The number of components they each produce per day is recorded for ten days:

Employee A	108	101	100	99	93	108	100	106	90	95
Employee B	116	102	100	98	86	116	100	112	80	90

1 Present each employee's production using two side-by-side box-and-whisker diagrams.

2 Compare the performance of the two employees.

3 If the factory owner was able to keep only one of the two employees, suggest which you would recommend keeping. Justify your decision.

Reflect and discuss 7

- The factory owner asks you to comment on the consistency of his employees. What does 'consistent' mean in this context?
- Why might consistency be considered a positive attribute?
- What statistical measures are you able to calculate that describe how consistent a set of data is?

Measures of spread such as the range and interquartile range are appropriate ways of comparing how densely a set of data is packed together – how consistent it is.

Look back at the eAssessment data for Crescent Moon High School and Oyster Bay Academy (page 477).

You have already looked at how you might compare the location of the data by examining the median. What do the range and interquartile range tell you about the spread of the data?

Crescent Moon High School has range 67 and interquartile range 19.5. Oyster Bay Academy has range 41 and interquartile range 16. Both pairs of statistics (the range and the IQR) suggest that the marks at Oyster Bay Academy are grouped together within a smaller interval. You might say that the performance of the students at Oyster Bay Academy was more consistent, or that the results at Crescent Moon High School showed greater variety.

In general, you will be more interested in comparing the IQR than the range; this is because the range is easily affected by outliers, whereas the IQR is not. In general, however, if your measures of spread have smaller values, then the data is more densely packed together. You should always try to interpret this in the context of the data as has been done here – so rather than saying simply that the IQR is smaller, you should observe that the marks at Oyster Bay Academy were more consistent.

An outlier is a piece of data that lies far outside the overall distribution pattern of the rest of the data.

Practice 4

1 Joe can travel to work by bus or by train. The total journey times for 40 bus journeys and 40 train journeys are illustrated in the box-and-whisker diagrams below:

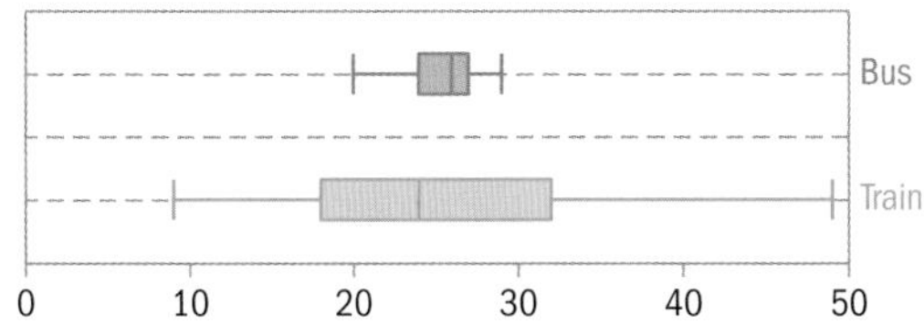

a Determine which mode of transport is faster on average.

b Determine which mode of transport has a more predictable journey time.

c Explain, with reference to these statistics, why Joe might prefer to take the bus than the train.

2 Two computer scientists write programs designed to crack encoded messages. The two programs (A and B) work in different ways, and so their speed can be assessed. Both programs are timed for decoding 11 messages. The times taken (in seconds) are as follows:

A	12.1	11.8	12.4	12.3	12.5	12.1	12.1	12.5	12.1	12.3	12.5
B	10.9	12.4	14.2	13.2	13.7	14.4	13.3	11.8	14.2	13.8	10.8

a Determine which program is faster on average.

b The program is to be used in a coding competition. The program which decodes a new message the fastest will win a prize. Select which program you would choose to use under the following scenarios. Justify your answer.

i You are the last programmer to compete and the current best time is 12.5 seconds.

ii You are the last programmer to compete and the current best time is 11.5 seconds.

iii You are the first programmer to compete.

3 In a factory, two machines are used to fill sugar bags. They are expected to fill them to a weight of 1000 g. To compare the two machines, 12 bags are taken from each and their contents weighed. The results found, in grams, are as follows:

Machine A	995	999.6	1005.3	996.7	997.1	999.5	1005.4	1002	1005.2	1001.1	994.7	1000.7
Machine B	1009.5	991.3	1022.3	1005.9	1000	994.7	1002.4	1016.2	996.2	1010.1	1009.6	1005.3

a Find the mean weight for each machine.

b Find the interquartile range for each sample.

c Compare the output of the two machines.

D The effect of sample size

- How do we know when to say 'when'?
- Do you want to follow the crowd?
- What effect does sample size have on the reliability of our generalizations?

In spite of using sampling techniques with great care, the sample may still, by chance, end up being unrepresentative. Another factor that affects the reliability of a sample is its size.

Exploration 5

1 Use a spreadsheet or your GDC to generate a list of 400 randomly selected integers between 1 and 20.

> In Microsoft Excel or Google Sheets the command to generate a random integer between 1 and 20 is '=RANDBETWEEN(1, 20)'.

2 Select a random sample of 5 numbers from this list.

3 Calculate the mean of your sample.

4 Collect all the sample means that have been calculated in your class and organize them in a stem-and-leaf diagram and then present the data using a box-and-whisker diagram.

5 Repeat the experiment using a sample size of 20 instead of 5, gathering your class data as before.

6 Explain why different members of the class have found different values for the sample mean.

7 Compare the box-and-whisker diagrams for the two sample sizes. Suggest what effect varying the sample size would have on the consistency of the estimates found.

8 Predict what the mean value would be for the whole set of data. Suggest whether the mean of a large sample or the mean of a small sample is a better estimate for the mean of the population.

Reflect and discuss 8

Given that larger sample sizes produce less variation in the estimated mean value, why would we ever use small samples? Consider the following scenarios and explain why taking large samples might be problematic:

- A manufacturer wants to know long their light bulbs last before failing.
- A marine biologist wants to study the mass of a particular species of crab in different areas of the Mediterranean Sea.

Summary

- A **population** is a set of objects under consideration in a statistical context. It could be a group of people, or a collection of objects.
- A **census** is a collection of data from every member of a population.
- A **sampling frame** is a list of every member of the population.
- A **sample** is a subset of a population.
  - A sample is **representative** if its properties reflect those of the population from which it comes.
  - A **simple random sample** is a sample such that every member of the population is equally likely to be included in the sample, independent of any other member of the population.
  - A **systematic sample** takes members of the population at regular intervals from the sampling frame.
  - A **stratified sample** identifies different sections of the population and ensures that those groups are represented in appropriate proportions in the sample.
  - A **quota sample** is used when the sampling frame is unknown, or does not exist. It involves collecting data until enough pieces of data are found.
- A sample is **biased** if it is not a fair representation of the population.
- **Inference** is the act of drawing conclusions about a population based on a sample.
  - You can use the sample mean to estimate the mean of a population.
  - Larger samples provide more reliable estimates.

Mixed practice

1 Tom and Jo are comparing their marks in recent English assessments. The IQR for Tom's marks is 13 and the IQR for Jo's marks is 18. Tom says that because his IQR is smaller he has performed better in the assessments. Jo says that because her IQR is larger she has performed better in the assessments.

Comment on their claims.

2 A school's ballet class has 11 students, with a mean height of 1.51 m and a range of heights of 21 cm. The school's chess club also has 11 students, with a mean height of 1.73 m and a range of heights of 43 cm.

a **Comment** on the differences between the heights of the members of the ballet class and the chess club.

b **Explain** why it would be preferable to know the interquartile range rather than the range of the data sets.

c Claire proposes to use the data to make predictions concerning adult ballet dancers and adult chess players. **Explain** why these clubs may not be a suitable sample.

3 Paul wants to estimate the average length of the fish in a lake.

He knows that roughly 40% of the fish in the lake are salmon and 60% of the fish are trout. To measure the fish, he will catch them with a large net and measure them on the shore before returning them to the water. He plans to take a sample of size 50.

a **Suggest** a suitable sampling method for Paul to use.

b **Explain** how he would take such a sample.

4 A polling company would like to conduct a survey to determine the extent to which the residents of a city feel that they are subject to unwanted advertising. The company proposes to conduct a simple random sample using the local phone directory as a sampling frame.

Evaluate the company's plan.

Review in context

Identities and relationships

1 The average household income in two neighboring towns is assessed by taking a random sample of residents using a recent census as a sampling frame. The data obtained is as follows:

Income range ($)	Town A	Town B
$20000 \leq x < 30000$	4	1
$30000 \leq x < 40000$	7	4
$40000 \leq x < 50000$	11	7
$50000 \leq x < 60000$	15	6
$60000 \leq x < 70000$	5	12
$70000 \leq x < 80000$	4	9
$80000 \leq x < 90000$	4	7
$90000 \leq x < 100000$	0	3
$100000 \leq x < 110000$	0	1

a **Present** this information using a pair of cumulative frequency graphs on the same axes.

b **Comment** on the differences between the earnings of households in towns A and B.

c The statistician conducting the survey claims that the samples are representative of the towns they came from. **Explain** what is meant by the word *representative* in this context.

2 To assess the impact of a new literacy strategy in 50 primary schools, a local education authority assesses the reading age of all 11-year-olds in three of the schools before and after implementation. The following table summarizes the data.

	Before	After
Minimum	8.4	8.2
Q_1	9.2	10.3
Q_2	10.1	10.8
Q_3	10.5	11.1
Maximum	13.2	12.8

a **Present** the data with over-and-under box-and-whisker diagrams.

b **Describe** the similarities and differences between the two sets of data.

c **Comment** on the effectiveness of the education authority's intervention.

d **Evaluate** the authority's sampling method.

3 You will be able to find data about the representation of women in national parliaments online from the Inter-Parliamentary Union (www.ipu.org) or from the World Bank (data.worldbank.org). You are to investigate the proposition that the percentage of women's representation varies in different continents.

a Take a random stratified sample of 5 countries from Africa, Asia and Europe. **Compare** the means of these samples.

b **Comment** on whether the samples are representative of the data in this case.

c **Describe** how you would take a *census* from this data to compare the average representation from the three continents.

4 Elderly people were surveyed about the number of hours, t, of social contact they have with volunteers after the introduction of a new assistance scheme.

Hours	Before	After
$0 \leq t < 2$	4	0
$2 \leq t < 4$	10	8
$4 \leq t < 5$	7	9
$5 \leq t < 6$	5	8
$6 \leq t < 8$	6	5
$8 \leq t < 10$	4	5
$10 \leq t < 12$	3	2
$12 \leq t < 16$	0	2
$16 \leq t < 20$	1	1

a **Interpret** the data and **suggest** whether or not the scheme was successful.

The sample was obtained by collecting contact details for 40 elderly people. The researcher waited in the local bus depot and asked people who he thought were over 75 whether they would be interested in such a scheme and whether he could have their contact details to evaluate the success of the scheme.

b Determine the type of sampling used.

c Evaluate the researcher's methods.

d Suggest a more appropriate way of constructing the sample.

5 You want to find out if students in your school are more likely to donate to local, national or international causes.

a Outline difficulties you expect to encounter in obtaining data relevant to your inquiry.

You decide that the easiest way to collect reliable data is to give people forms to keep for a whole year, which they can fill in when they donate money to charity.

You give a form to every student in your school to take home.

You believe that donation trends will vary depending on whether the parents in the family grew up locally, grew up elsewhere in the country, or grew up in a different country.

Separate research suggests that 30% of families had parents that grew up locally, 45% grew up elsewhere in the country and 25% grew up in a different country.

b Explain how you would use this information to try to make your sample representative.

After a year, you receive 40 responses from families with parents that grew up locally, 75 from those that grew up elsewhere in the country and 50 from those that grew up in a different country.

c Calculate the number of responses you should use from each category of respondent in order to obtain the largest representative sample possible.

Reflect and discuss

How have you explored the statement of inquiry? Give specific examples.

Statement of Inquiry:

Generalizing and representing relationships can help to clarify trends among individuals.

13 Justification

Valid reasons or evidence used to support a statement

Verifying and justifying

To **verify** a statement (in other words, prove the truth of it), you need to check that it works for some examples.

You could verify the statement 'the difference between consecutive square numbers is odd' by giving these examples: 144 − 121 = 23 and 49 − 36 = 13

- Give some more examples to verify the statement.

To **justify** the statement, you need to give reasons why it is true.

- Show that every even square number is followed by an odd square number and vice versa, so the difference between consecutive square numbers must always be an odd number.

You can **verify** that an airplane flies, by observing it flying.

You can **justify** why it flies by explaining about the difference in air pressure above and below its wings, and how this gives it lift.

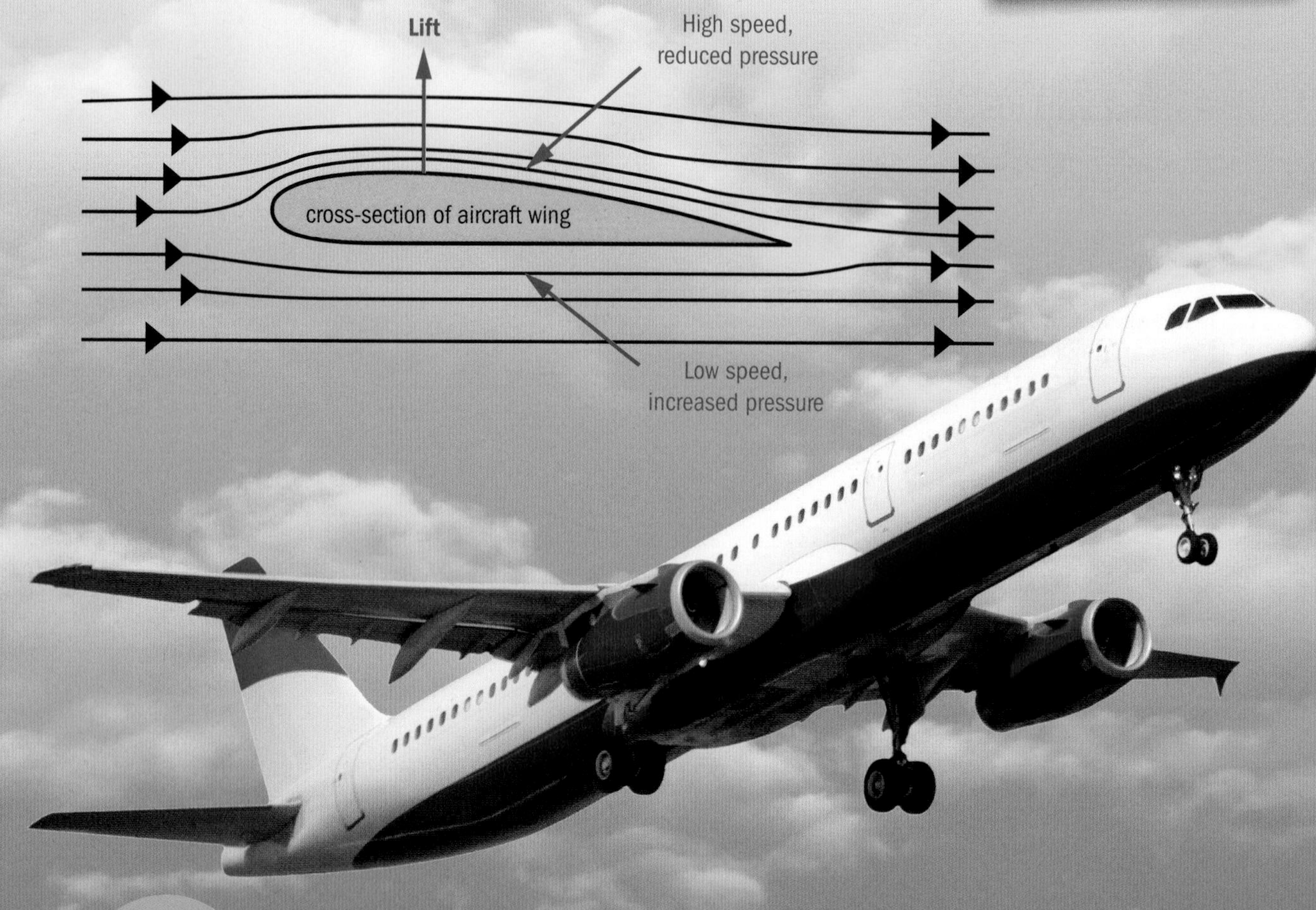

Justification using evidence

The word *justify* comes from the 12th-century French verb 'justifier', which meant 'to submit to court proceedings'. The standard in most courtsis that the prosecution must prove guilt beyond reasonable doubt.

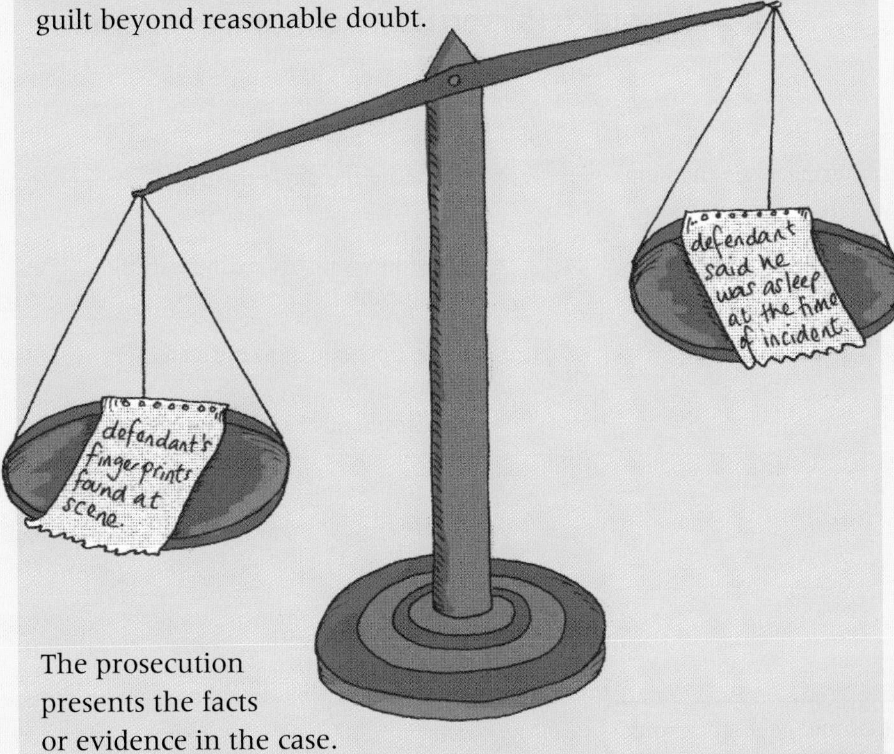

The prosecution presents the facts or evidence in the case.

If the only logical explanation of these facts is that the defendant committed the crime, then the defendant is found guilty. If there is no such evidence, then the defendant is found not guilty.

Justification in medical research

Pharmaceutical companies are continually trying to produce new medicines to treat as-yet incurable conditions, or to treat curable conditions more effectively. Before a new drug can go on sale, the company needs to justify that the drug is safe and effective. To do this, they run clinical trials to test the drugs on humans and collect evidence on positive results as well as negative side effects.

All clinical trials have to be approved by a scientific and an ethical committee. The researchers must justify that the volunteers in the trial will not be exposed to unnecessary risk, and that the new drug can reasonably be expected to improve patient care.

13.1 Well-rounded ideas

Global context: Personal and cultural expression

LOGIC

Objectives

- Finding angles and lengths using circle theorems
- Proving results using circle theorems
- Examining 'If ... then ...' statements and testing the truth of their converses

Inquiry questions

F
- What are the circle theorems?

C
- How do we justify mathematical conclusions?

D
- Is the opposite of a true statement always false?
- Can aesthetics be calculated?

ATL Critical-thinking

Draw reasonable conclusions and generalizations

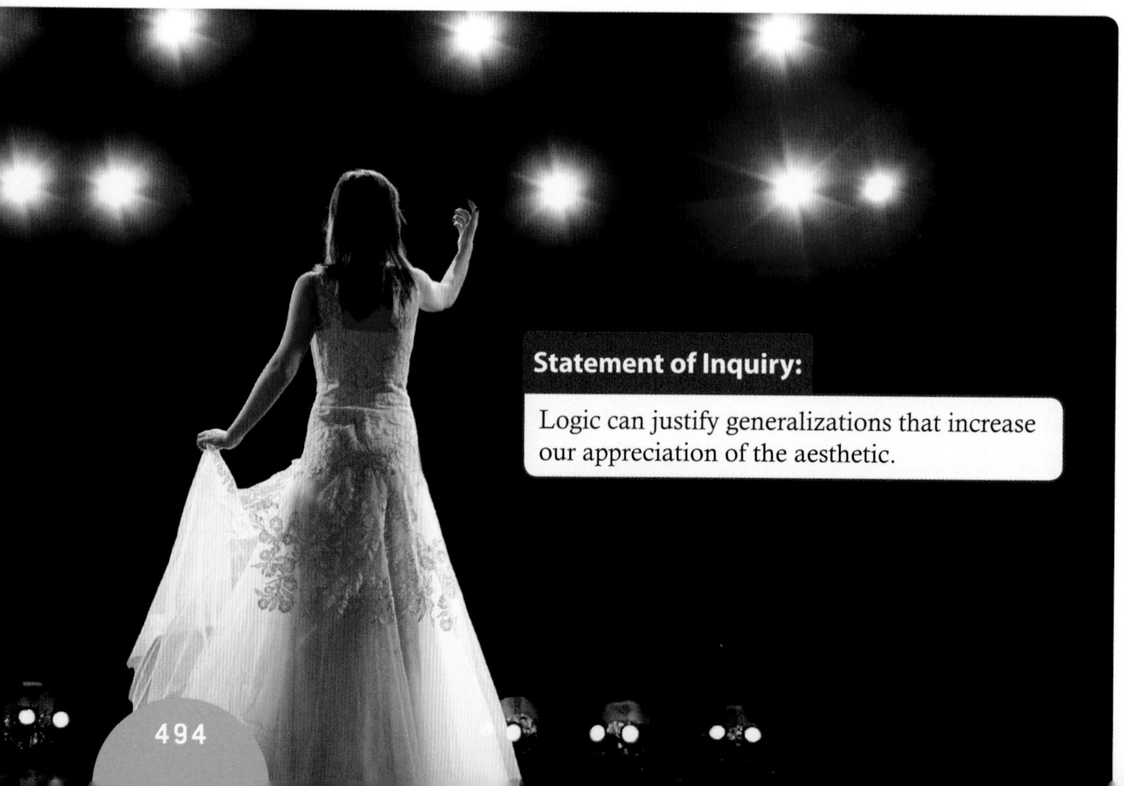

Statement of Inquiry:

Logic can justify generalizations that increase our appreciation of the aesthetic.

13.1

13.2

E13.1

You should already know how to:

• name the parts of a circle	**1** Draw a circle, then draw and label: **a** a radius **b** a diameter **c** a chord **d** a segment **e** an arc **f** the center **g** the circumference

Discovering the basic circle theorems

- What are the circle theorems?

These diagrams all illustrate tangent lines.

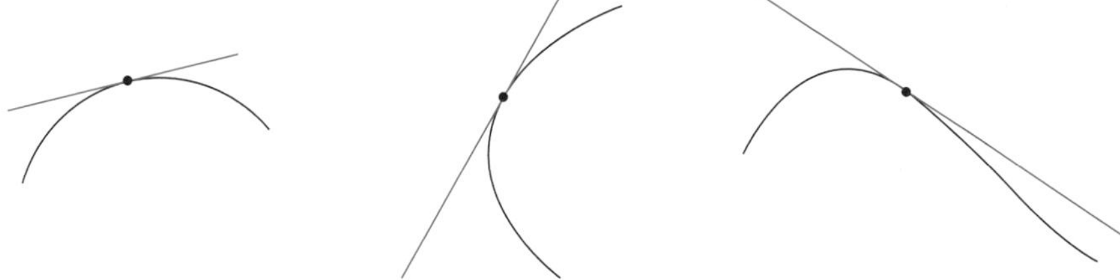

In each case, the tangent line touches the curve at the point of contact and does not pass through the curve.

A tangent line has the same gradient (slope) as the curve at the point where the tangent touches the curve.

Line L is the **tangent** to a circle at point P if L intersects the circle at only one point, P.

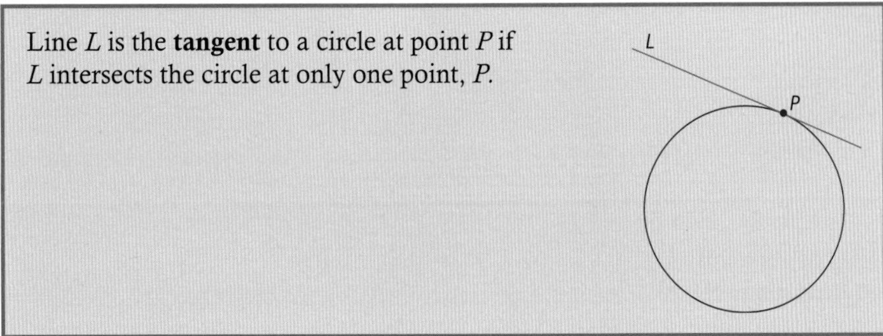

ATL

Exploration 1

1 Using a GDC or dynamic geometry software, construct a circle with center O which passes through a second point, P. Add a third point Q, not on the circle. Draw a line segment joining O to P, and add the line PQ.

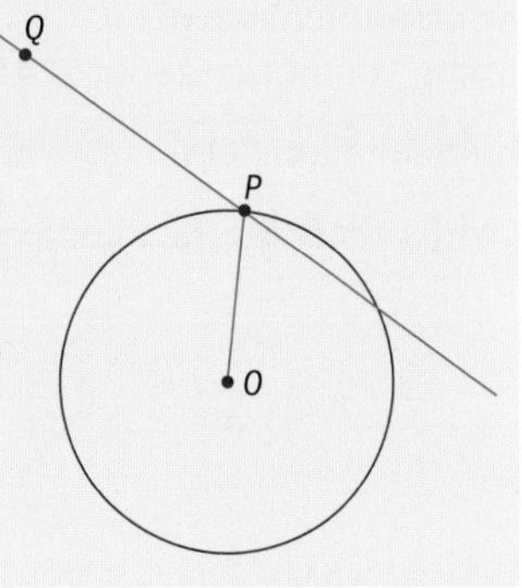

A line **segment** connects two points, but a line extends in both directions.

Continued on next page

2 Move the point Q freely until the line PQ is a tangent to the circle at P. Repeat a few times by repositioning P (and hence changing the circle) and finding a new position for Q so that PQ is a tangent to the circle.

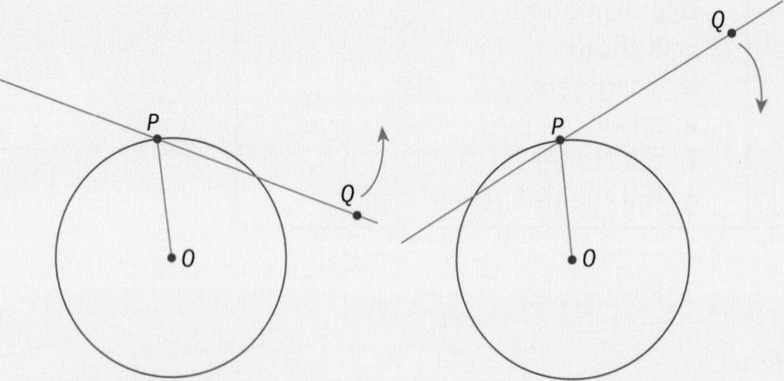

3 Now use your GDC or dynamic geometry software to measure $\angle OPQ$. Describe anything you notice about angle OPQ when PQ is a tangent to the circle. Compare your findings with others. Form a conjecture about the relationship between a tangent and the radius at a point P on the circumference of a circle.

If P is a point on the circumference of a circle, the tangent to the circle at P is perpendicular to the radius OP.

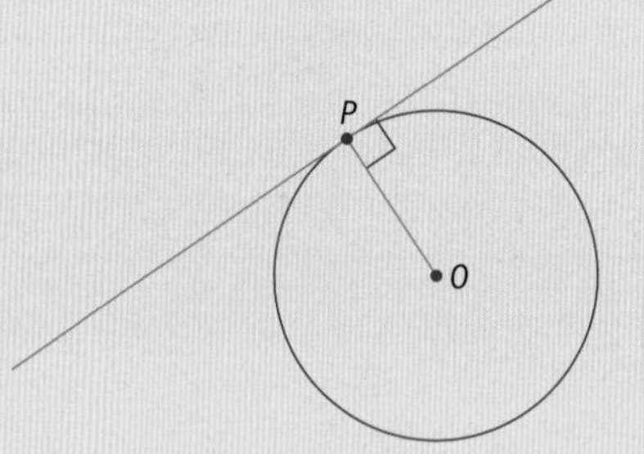

ATL

Exploration 2

1 Using a GDC or dynamic geometry software, draw a circle with center O and diameter AB. Draw another point, P, on the circumference of the circle. Draw the line segments AP and BP.

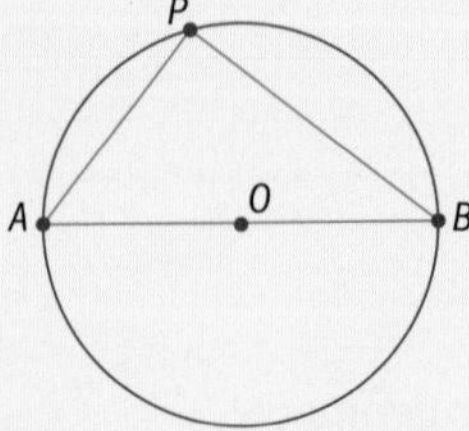

2 Use the software to measure $\angle APB$.

3 Move point P around the circumference of the circle. Describe anything you notice about $\angle APB$.

4 Repeat with circles of different diameters.

5 Compare your findings with others. Form a conjecture based on your findings.

Exploration 3

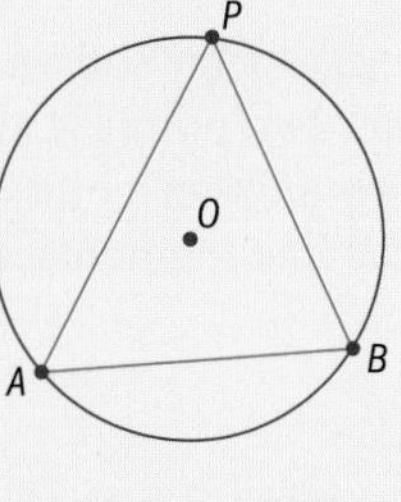

1 Using a GDC or dynamic geometry software, draw a circle with center O and chord AB. Draw another point, P, on the circumference of the circle. Draw the line segments AP and BP.
2 Use the software to measure $\angle APB$.
3 Without moving A or B, move point P around the circumference of the circle. Describe anything you notice about $\angle APB$.
4 Now reposition A or B (or both). Describe the effect of this on $\angle APB$.

5 Move point P around the circumference of the circle, without moving A and B. Describe anything you notice about $\angle APB$.
6 Compare your findings with others. Form a conjecture based on your findings.

A **cyclic quadrilateral** is a quadrilateral whose vertices all lie on the circumference of a circle.

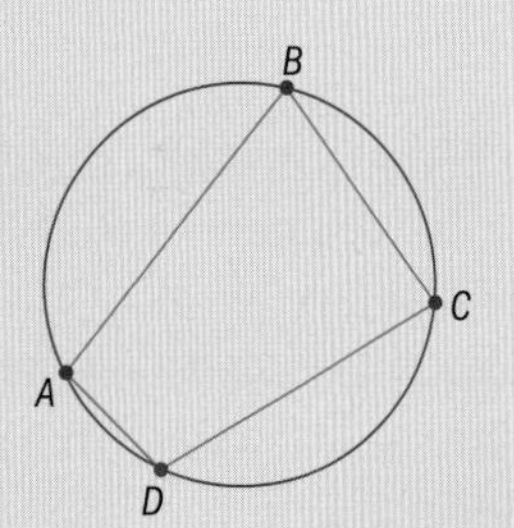

Exploration 4

1 Using a GDC or dynamic geometry software, draw a circle and mark four points P, Q, R and S on the circumference. Join the four points with straight line segments so that $PQRS$ is a cyclic quadrilateral.
2 Use the software to measure $\angle PQR$ and $\angle RSP$.
3 Now reposition any or all of the points. Describe anything you notice about the relationship between $\angle PQR$ and $\angle RSP$ as the cyclic quadrilateral changes.
4 Compare your findings with others. Form a conjecture based on your findings.

Tip

Make sure the points P, Q, R and S are in order, so your quadrilateral does not cross itself.

The **major arc** is the long way around and the **minor arc** is the short way around a circle. To show the direction you can include a third point on the circumference of the circle, e.g. ADB.

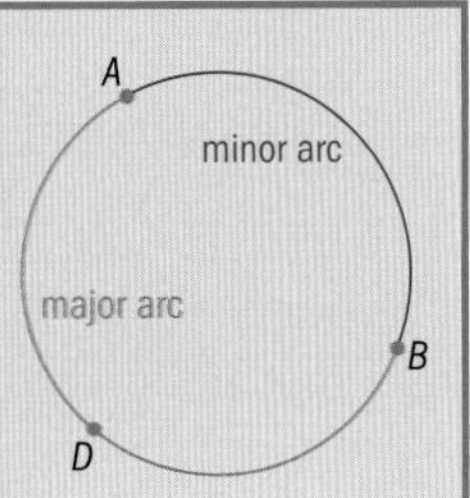

ATL

Exploration 5

1 Using a GDC or dynamic geometry software, draw a circle with center O. Mark two points A and B on the circumference. Draw another point, P, on the major arc AB. Draw the line segments AP, BP, AO and BO.

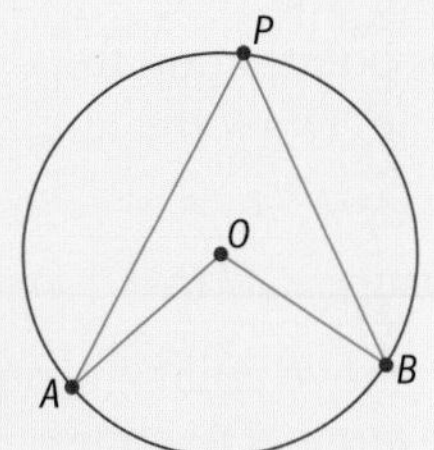

2 Measure $\angle AOB$ and $\angle APB$.

3 Move points A, B and P, keeping P in the major arc. Describe anything you notice about the relationship between $\angle AOB$ and $\angle APB$ as you vary the positions of A, B and P.

4 Repeat the exploration with a circle of a different diameter.

5 Compare your findings with others. Form a conjecture based on your findings.

ATL

Exploration 6

1 Using a GDC or dynamic geometry software, draw a circle with center O and points A, B and C on its circumference. Draw the line segments AB, BC, and CA. Also draw the tangent to the circle at A.

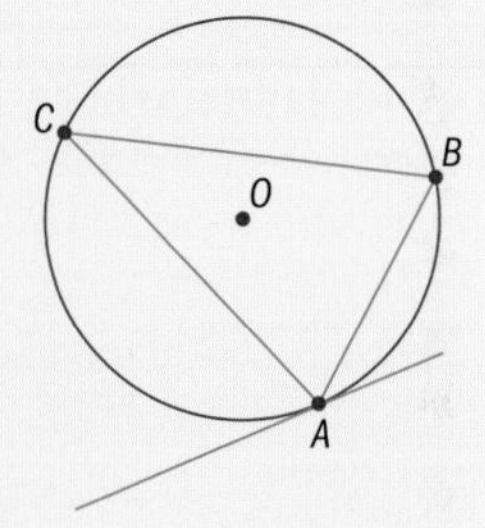

2 Measure $\angle ABC$ and $\angle ACB$.
Measure the acute angles between AB and the tangent, and AC and the tangent.

3 Move points A, B and C. Describe anything you notice.

4 Repeat the exploration with a circle of a different diameter.

5 Compare your findings with others. Form a conjecture based on your findings.

Tip

Your software should be able to construct a tangent to a circle automatically.

Activity

Did you find it difficult to explain the rules you discovered in Explorations 2 to 6? Below are five ways of writing these rules and five diagrams to illustrate them. Read the five rules, look at the diagrams, and then answer the questions below the diagrams.

Five rules:

i *Thales' theorem*: An angle inscribed in a semicircle is a right angle.

ii Opposite angles in a cyclic quadrilateral are supplementary.

iii Angles in the same segment subtended by equal chords are equal in size.

iv *Alternate segment theorem*: the angle between a chord and tangent at a point is equal to the angle subtended by the chord in the alternate segment.

Continued on next page

v The angle subtended by a chord at the center (central angle) is twice the angle at the circumference (inscribed angle) in the same segment subtended by the same chord.

Five diagrams:

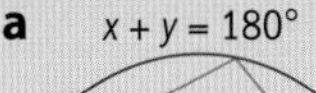

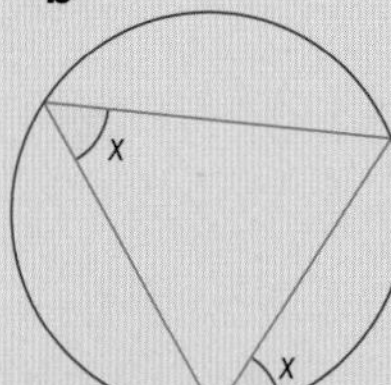

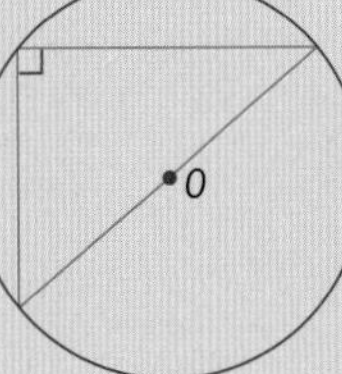

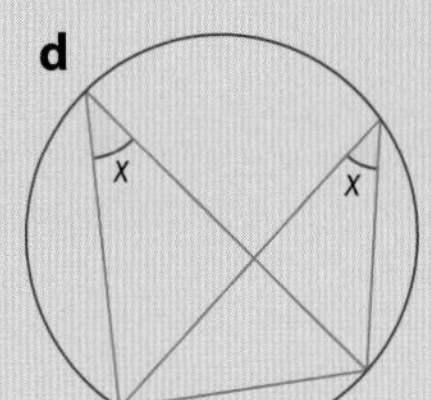

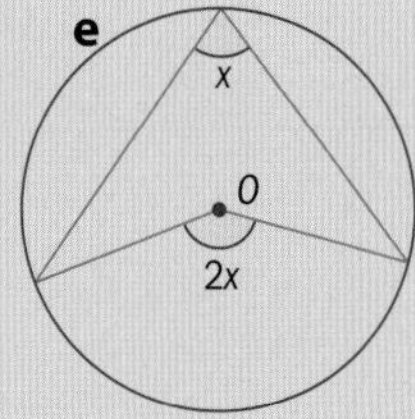

- What do these terms mean?: *subtended, supplementary, opposite angle, inscribed, same segment, alternate segment*. Look them up if you need to.
- Match each rule to its diagram.
- Match each rule to one of the Explorations 2 to 6.

Tip

Make sure you understand all these terms, so that you can understand the information given in a question.

You can use these five results to find angles in geometric problems.

Example 1

Find the value of angle t, and justify your reasoning.

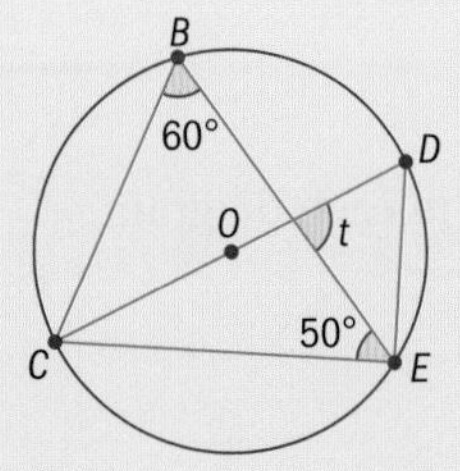

$\angle CDE = 60°$ because $\angle CDE$ and $\angle CBE$ are angles in the same segment.

'Justify' means 'give clear reasons for your answers'.

$\angle CED = 90°$ because the angle inscribed in a semicircle is a right angle.

CD passes through O, so it is a diameter.

$\angle BED = 90° - 50°$
$= 40°$ because $\angle CED = 90°$

$\angle t = 180° - 60° - 40°$
$= 80°$ because the interior angles in a triangle sum to 180°.

Example 2

Calculate $\angle BOD$.

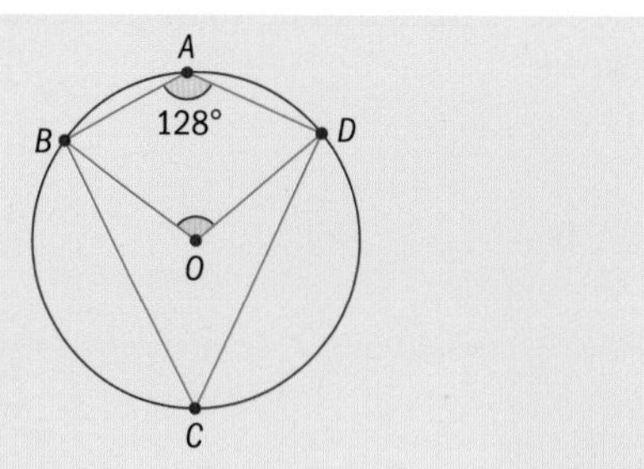

$\angle BCD = 180° - 128°$

$= 52°$

Opposite angles in a cyclic quadrilateral are supplementary.

$\angle BOD = 2 \times 52°$

$= 104°$

The central angle is twice the inscribed angle.

Example 3

Prove that $\angle BOD = 64°$.

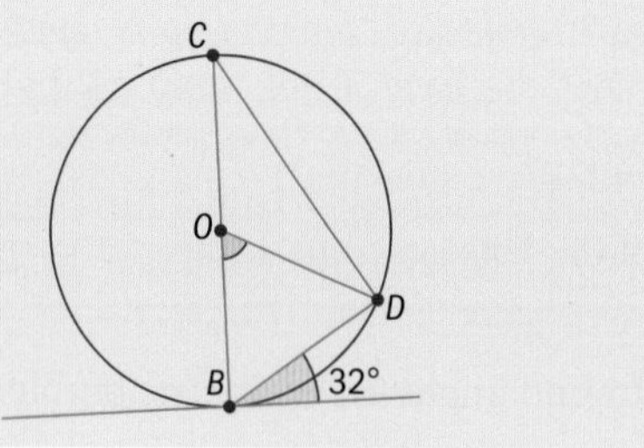

$\angle BCD = 32°$ by the alternate segment theorem.

$\angle BOD = 64°$ because the central angle is twice the inscribed angle.

Practice 1

1 Calculate the size of the marked angles in each diagram.

a

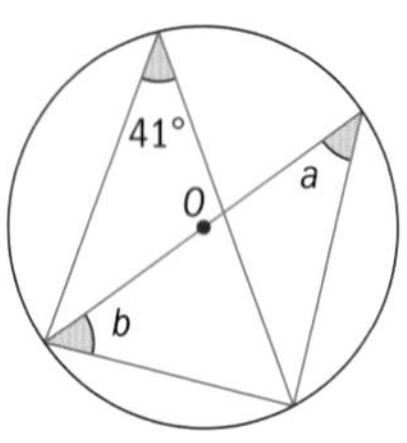

b

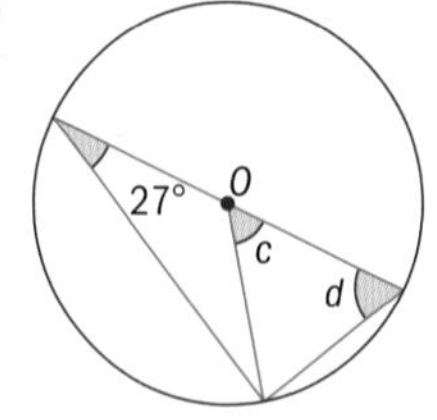

c

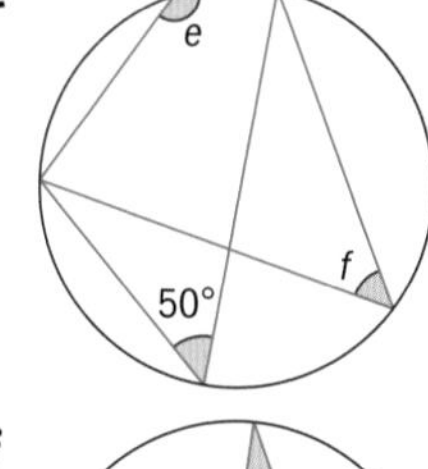

d

e

f

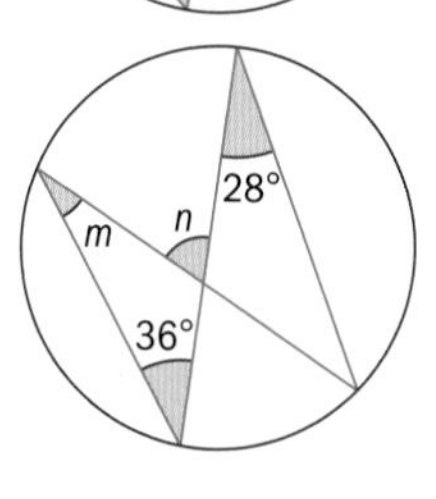

Tip

Calculate and **find** both mean that you should show relevant stages in your working, but you do not need to justify all your reasoning.

Problem solving

2 Find the size of the marked angle in each diagram.
In each case, justify your answer.

a

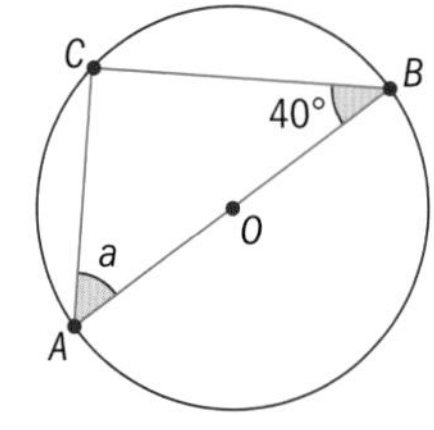

b

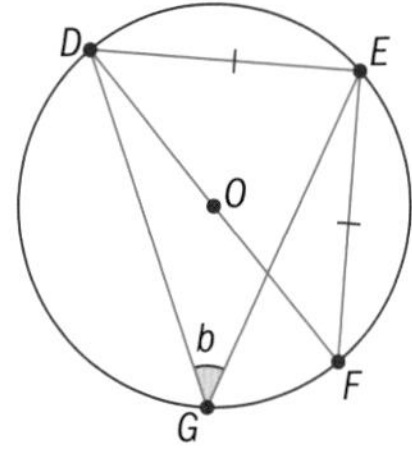

c

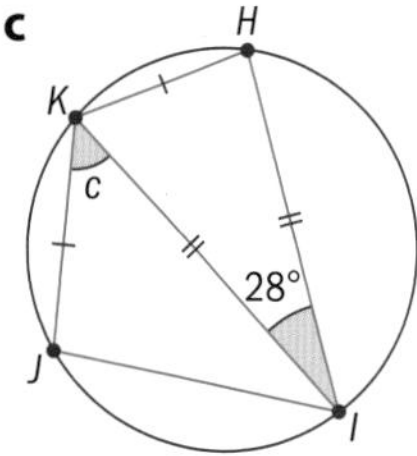

d

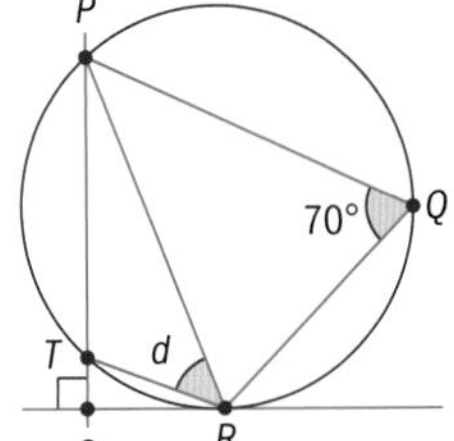

e

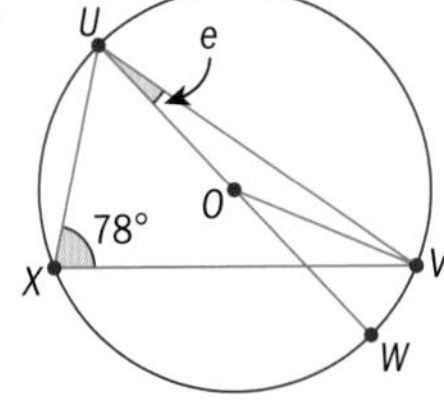

f

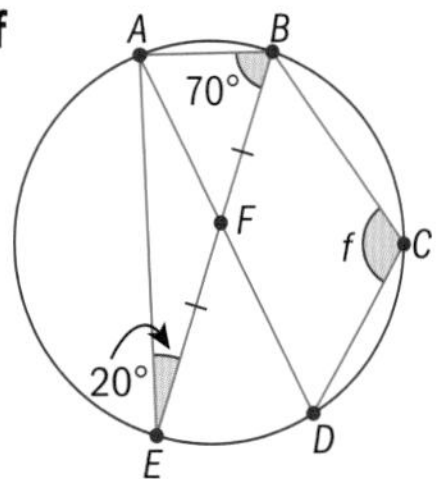

Justifying the circle theorems

- How do we justify mathematical conclusions?

In this section you will use direct (deductive) proof to justify the circle theorems.

Objective: **C.** Communicating **iv.** communicate complete, coherent and concise mathematical lines of reasoning
The following two proofs complete the proof of the cyclic quadrilateral theorem. Make sure your proofs are concise and that they make sense.

Exploration 7

1 Copy and complete this skeleton proof.

Theorem

Opposite angles in a cyclic quadrilateral are supplementary.

Proof

Consider a cyclic quadrilateral *ABCD* inscribed in a circle with center *O*.

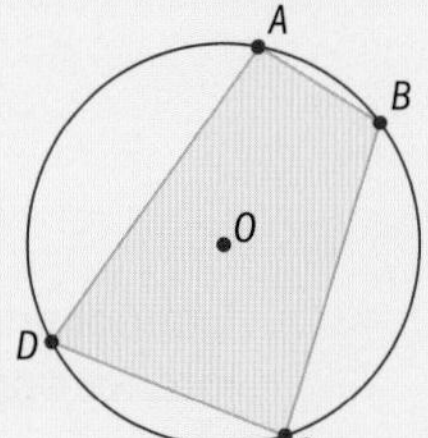

A diagram makes it easier for readers to understand your proof.

Continued on next page

$OA = OB = OC = OD$ because ___________

Therefore ΔOAB, _______, _______, and _______ are all _______ triangles.

What type of triangle is ΔOAB?

Therefore $\angle OAB = \angle OBA$, _______ = _______, _______ = _______, and _______ = _______

Use the 'triangle fact' from the line above.

because ___________

$\angle OAB + \angle OBA +$ ____ + ____ + ____ + ____ + ____ + ____ $= 360°$

$\Rightarrow$ means 'implies'

$\Rightarrow 2(\angle OAB + \angle OAD +$ _______ + _______$) = 360°$

$\Rightarrow \quad 2(\angle DAB +$ _______$) = 360°$

Notice how $\angle OAB + \angle OAD = \angle DAB$.

$\Rightarrow \quad$ _______ + _______ $= 180°$

Therefore the angles at A and C are supplementary, and the angles at B and D are supplementary.

2 The proof is incomplete as you need to consider two other cases. By drawing suitable diagrams and modifying the proof, prove that the opposite angles are supplementary in cyclic quadrilaterals where:

a one edge of the quadrilateral passes through the center of the circle
b the center of the circle is not inside the quadrilateral.

Exploration 8

Copy and complete this skeleton proof of Thales' theorem.

Theorem

Any angle inscribed in a semicircle is a right angle.

Proof

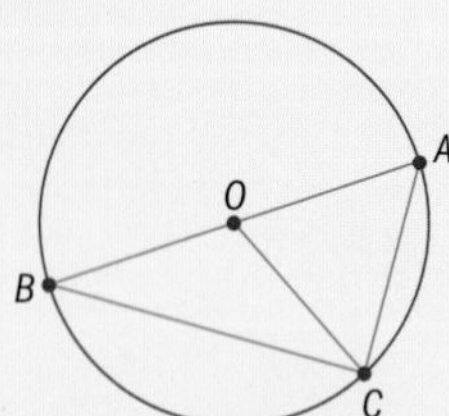

Consider a circle with center O where AB is a diameter. Let C be any point on the circumference of the circle.

ΔOAC and ΔOCB are _______ triangles.

Hence $\angle OCA =$ _______ and $\angle OCB =$ _______

$\angle ACB = \angle OCA + \angle OCB$

$\angle AOC =$ _______ + _______ and $\angle BOC =$ _______ + _______

Exterior angle = sum of interior opposite angles.

▶ Continued on next page

$\Rightarrow \angle AOC = 2\angle OCB$ and $\angle BOC = 2\angle OCA$

$\Rightarrow \angle AOC + \angle BOC = 2(\_\_\_\_\_\_ + \_\_\_\_\_\_)$

$\Rightarrow \angle AOC + \angle BOC = 2\ \_\_\_\_\_\_$

$\Rightarrow 2\_\_\_\_\_\_ = 180°$

$\Rightarrow \_\_\_\_\_\_ = 90°$

$\angle OCA + \angle OCB = \angle ACB$

Exploration 9

Copy and complete this skeleton proof.

Theorem

Angles in the same segment are equal.

Proof

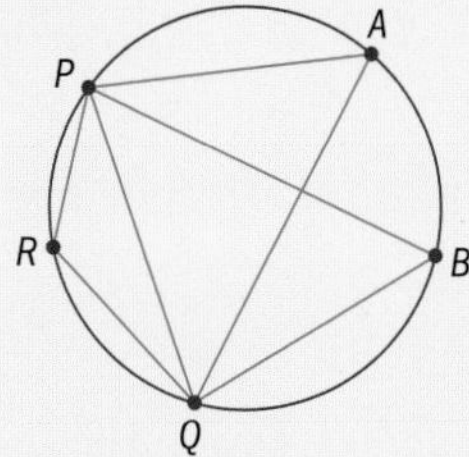

Consider points A and B that lie in the same segment of a circle divided by the chord through P and Q. Point R lies in the opposite segment.

$\angle PAQ + \angle PRQ = \_\_\_\_\_\_$

$\angle PBQ + \angle PRQ = \_\_\_\_\_\_$

because ________________________

Hence $\angle PAQ = \_\_\_\_\_\_$

Example 4

A circle, center O, has points S and T on its circumference. RST is a straight line segment. Prove that $\angle SOT = 2 \times \angle RSO - 180°$.

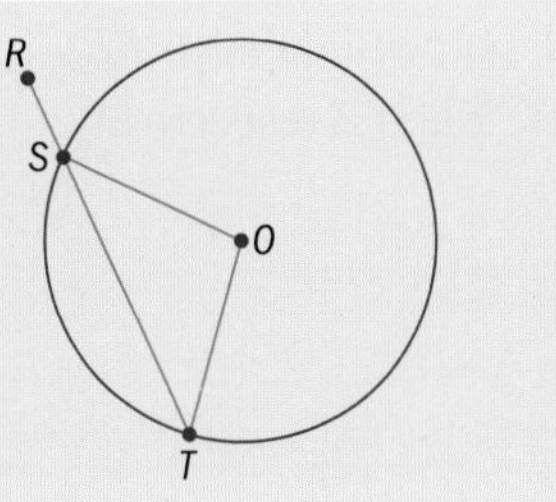

$\angle OST = 180° - \angle RSO$ (angles on a straight line)

$\angle OST = \angle OTS$ (base angles in an isosceles triangle)

$\angle SOT = 180° - \angle OTS - \angle OST$ (angles in a triangle sum to 180°)

$\Rightarrow \angle SOT = 180° - 2\angle OST$

$\Rightarrow \angle SOT = 180° - 2(180° - \angle RSO)$

$\Rightarrow \angle SOT = 2 \times \angle RSO - 180°$

Practice 2

Problem solving

1 Points A, B and C lie on a circle with center O such that A, B and C all fall within the same semicircle. Prove that $\angle ABC = 180° - \frac{1}{2}\angle AOC$.

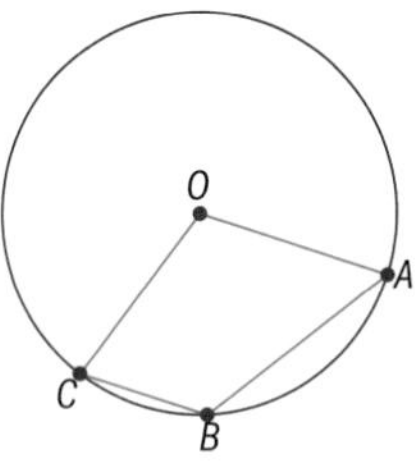

2 Points P, X and Y lie on the circumference of a circle. The tangents to the circle at X and Y meet at Q. Let O be the center of the circle. Prove that

$\angle XPY = \frac{1}{2}(180° - \angle XQY)$.

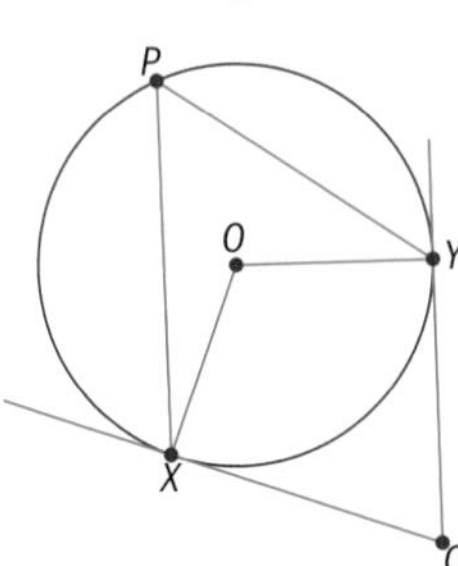

3 Points B, D and E lie on a circle. ABC is a tangent to the circle at B. DE is parallel to ABC. Prove that ΔBDE is isosceles.

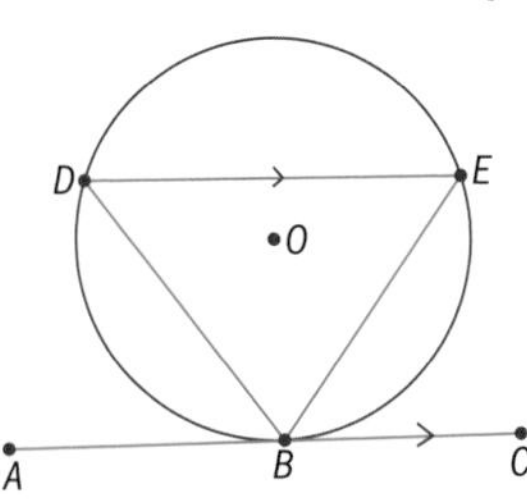

4 A, B, C and D lie on the circumference of a circle, and when the line segments AB and DC are extended they meet at X, outside the circle.

Prove that ΔACX is similar to ΔDBX.

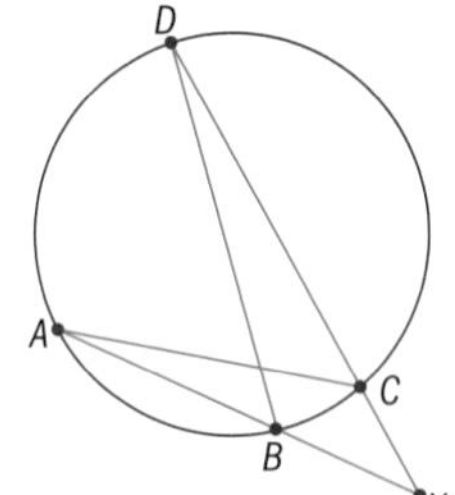

5 A, B, C, D, E and F lie on the circumference of a circle, with $AB = BC$ and DF perpendicular to BE. AE, BE and CE meet DF at X, Y and Z respectively.

Prove that $\angle EXY = \angle EZY$.

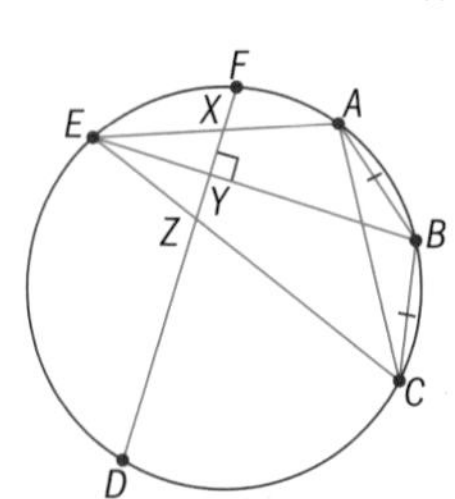

Tip

Remember that two triangles are similar if their angles are the same.

'Bisect' means to cut in half exactly.

D The converse of the circle theorems

- Is the opposite of a true statement always false?
- Can aesthetics be calculated?

The theorems you have proved are all of the form 'If p, then q', where p and q are statements. For example, 'If points A, B and C lie on the circumference of a circle and AC is a diameter, then $\angle ABC$ is a right angle'. You can see that the statement can be split into two parts, the premise (the 'if' part) and the conclusion (the 'then' part).

- If ABC is a triangle, then $\angle ABC + \angle BCA + \angle CAB = 180°$.
- If x is a whole number which ends in a zero, then x is not prime.
- If $x^2 = 4$, then $x = 2$.
- If $x + 3 = 7$, then $x = 4$.

Tip

You can also write the statement 'If p then q' as $p \Rightarrow q$. This is read as 'p implies q'.

Reflect and discuss 2

Which of the 'If … then …' statements above are true?

The converse of a statement takes its two constituent parts and places them in the opposite order. So the converse of 'If p then q' is 'If q then p'.

These statements are the converses of the four previous statements.

- If $\angle ABC + \angle BCA + \angle CAB = 180°$, then ABC is a triangle.
- If x is not prime, then x is a whole number which ends in a zero.
- If $x = 2$, then $x^2 = 4$.
- If $x = 4$, then $x + 3 = 7$.

Reflect and discuss 3

- Which of the preceding list of 'If … then …' statements are true?
- If the original statement is true, then is the converse also true?
- If the original statement is false, then is the converse also false?

You should have seen that statements and their converses can be true or false independently of one another. This means that you have to be careful when working with proof and justification; proving an 'If … then …' statement does not necessarily mean that its converse is true.

Consider Thales' theorem once more: 'If points A, B and C lie on the circumference of a circle and AC is a diameter, then $\angle ABC$ is a right angle'. The converse of the theorem would be: 'If $\angle ABC$ is a right angle then points A, B and C lie on the circumference of a circle, and AC is a diameter'. It's a little harder to see exactly what the converse means – and certainly it is not obvious whether or not it is true.

Exploration 10

1 Use a GDC or dynamic geometry software to draw two points A and C.

2 Draw a third point D, and the line segment AD.

3 Draw a line perpendicular to AD through C. Let point B be the intersection of AD and the perpendicular through C.

You have created a set of three points, A, B and C that satisfy the 'if' part of the converse of Thales' theorem: three points such that $\angle ABC$ is a right angle. By moving point D, you should see that as B moves $\angle ABC$ stays the same size.

4 Use the Trace facility of your GDC or software to track the position of point B. Vary the position of B by moving D.

5 Explain if your diagram supports the converse of Thales' theorem.

Tip

Your software should be able to construct a perpendicular to a line through a point.

Reflect and discuss 4

Is Exploration 10 a proof of the converse of Thales' theorem? If not, is it sufficiently convincing that you believe the converse to be true anyway?

Exploration 11

Use a GDC or dynamic geometry software to verify that the converses of the other four circle theorems also hold. Some are significantly easier than others, provided you know how to use the software to produce an angle that is the same size as another. Exploring the converse of the theorem that the opposite angles in a cyclic quadrilateral are supplementary is quite tricky – so tackle this one last!

Summary

Line L is the **tangent** to a circle at point P if L intersects the circle at only one point, P.

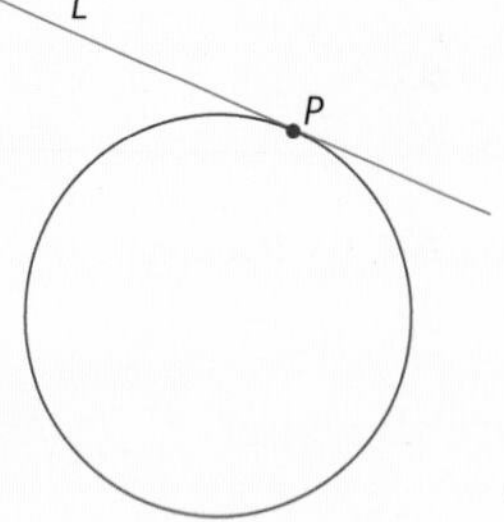

A **cyclic quadrilateral** is a quadrilateral whose vertices all lie on the circumference of a circle.

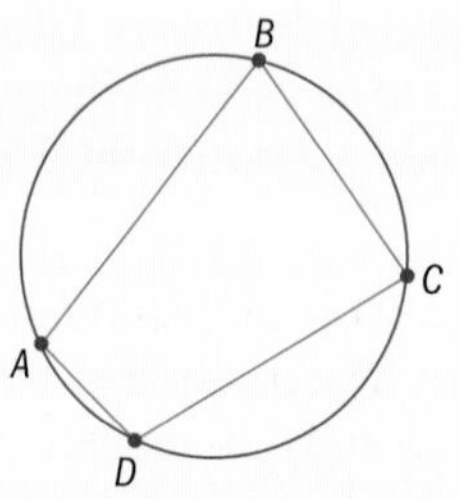

The **major arc** is the long way around the circle; the **minor arc** is the short way around. To indicate a major arc you can include a third point on the circumference of the circle, e.g. ADB.

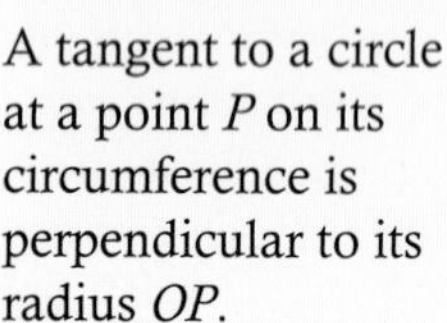

A tangent to a circle at a point P on its circumference is perpendicular to its radius OP.

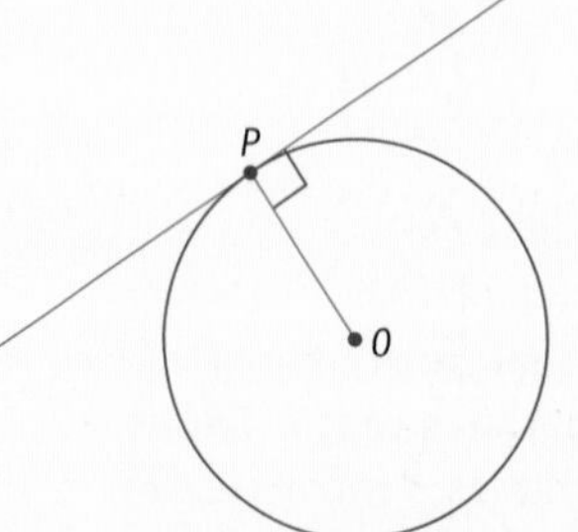

- **Thales' theorem:** An angle inscribed in a semicircle is a right angle.

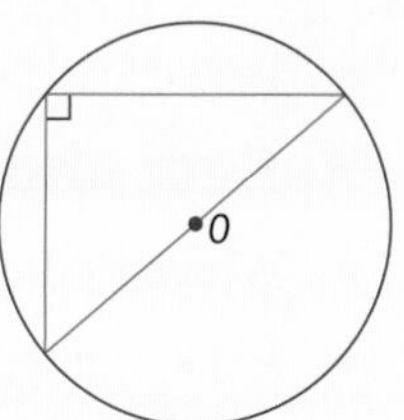

- Opposite angles in a cyclic quadrilateral are supplementary.

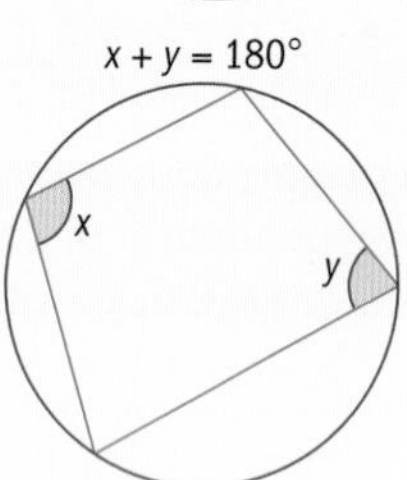

- Angles in the same segment subtended by equal chords are equal in size.

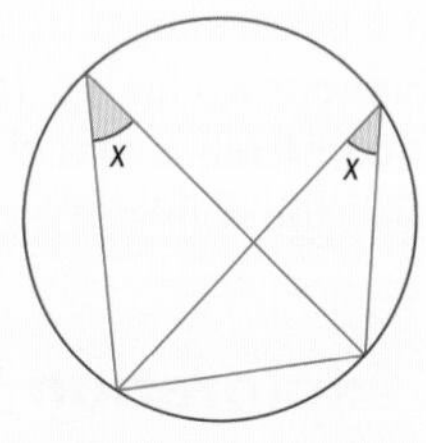

- **Alternate segment theorem:** The angle between a chord and tangent at a point is equal to the angle subtended by the chord in the alternate segment.

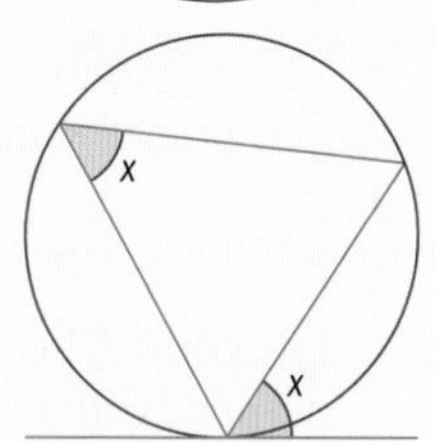

- The angle subtended by a chord at the center is twice the angle at the circumference in the same segment subtended by the same chord.

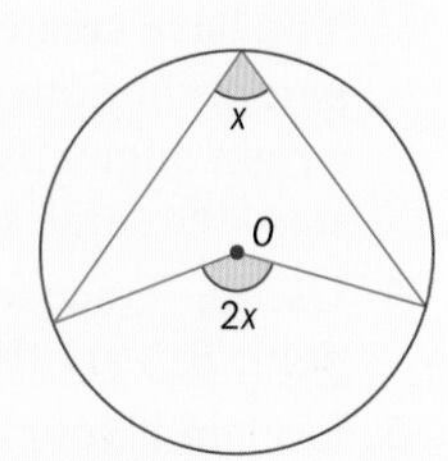

Mixed practice

1 **Find** the size of the marked angles.

a

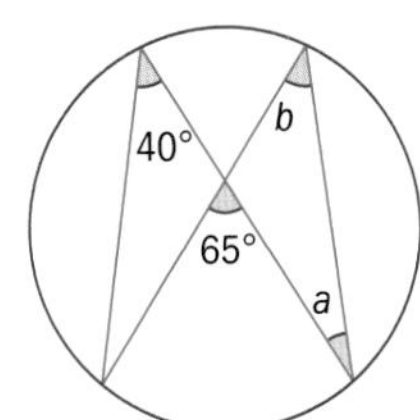

b

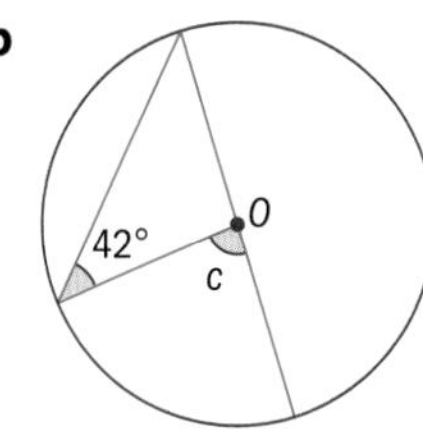

2 **Find** the size of the marked angle. **Justify** your answers.

a

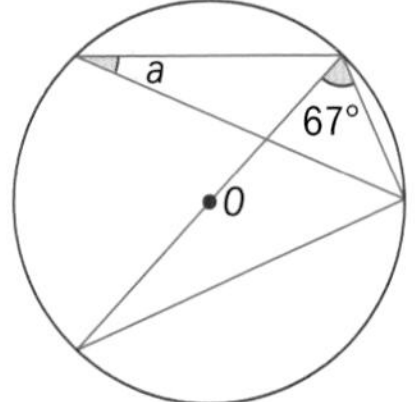

b

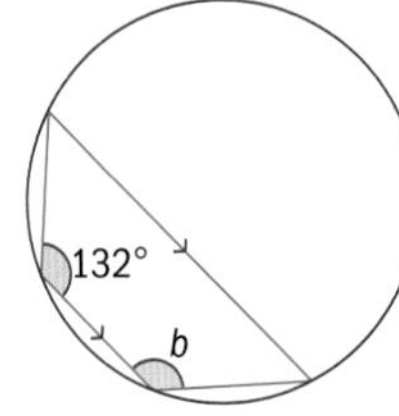

3 For each statement, **write down** the converse. Determine which statements, and which of their converses, are true.

a If $a = 7$ then $3a - 2 = 19$.

b If $b = a$ then $a - b = 0$.

c If it is a bird then it has wings.

d If a polygon has four sides, then it is a square.

e If $c = 9$ then $c^2 = 81$.

f If a right-angled triangle is drawn with its vertices on the circumference of a circle then its hypotenuse is a diameter of the circle.

4 In the cyclic quadrilateral $ABCD$, $\angle DAB \cong \angle ABC$.

Prove that $AD = BC$.

5 A, B, C and D lie on a circle center O, and DOB bisects $\angle ABC$.

Prove that ΔABC is isosceles.

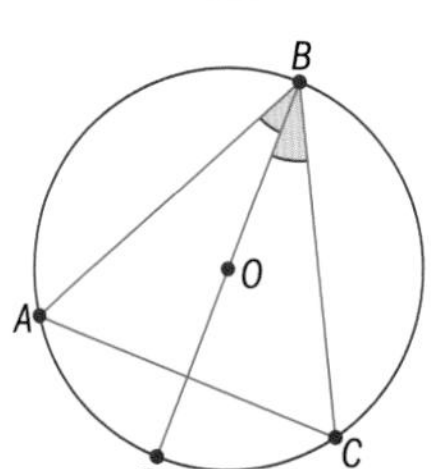

Review in context

Personal and cultural expression

1 A circular auditorium has a stage across the end of it. When a patron sits at O, the center of the auditorium, the stage occupies 88° of the patron's field of vision.

Find the amount of the field of vision that the stage would occupy when viewed from P, at the very edge of the auditorium.

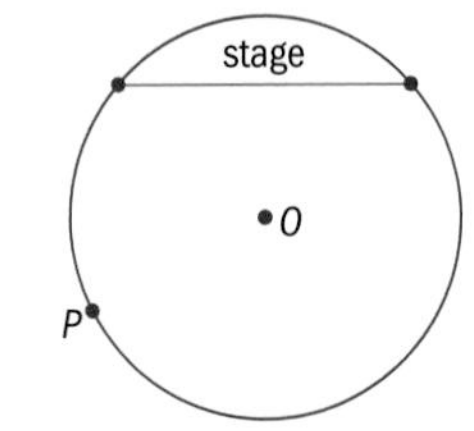

2 Three projectors, used to display images on the wall of a circular performance space, are mounted on the opposite wall. The projectors each have a beam width of 17°, and are placed one above another at a single point P.

The circle has diameter 6 m.

Find the total length of the wall projected on when the projectors are positioned so their beams overlap by 20 cm.

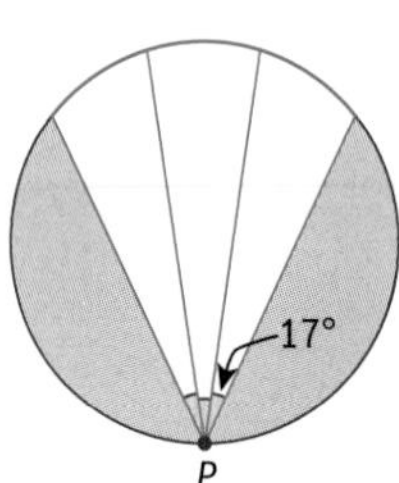

Reflect and discuss

How have you explored the statement of inquiry? Give specific examples.

Statement of Inquiry:

Logic can justify generalizations that increase our appreciation of the aesthetic.

13.2 It strikes a chord

Global context: Personal and cultural expression

LOGIC

Objectives

- Using circle theorems to find lengths of chords
- Finding lengths using the intersecting chord theorem

Inquiry questions

F
- What is the intersecting chords theorem?
- How can you apply the theorem?

C
- How can a theorem have different cases?

D
- Can aesthetics be calculated?

ATL Critical-thinking

Test generalizations and conclusions

13.1

13.2

E13.1

Statement of Inquiry:

Logic can justify generalizations that increase our appreciation of the aesthetic.

You should already know how to:

• solve quadratic equations	**1** Solve the equation $2x^2 - x - 3 = 0$ for x by factorizing.
• apply the circle theorems	**2** A circle of radius 5 cm has a tangent at a point on the circumference. If a point on the tangent is 13 cm from the center of the circle, how far is this point from the point where the tangent touches the circle?

F Intersecting chords

- What is the intersecting chords theorem?
- How can you apply the theorem?

ATL

Exploration 1

1 Draw a circle of radius 5 using a GDC or dynamic geometry software.
Draw four points A, B, C and D on the circumference of the circle.

Draw line segments AB and CD. You should have a diagram similar to this one.

2 Note that chords AB and CD do not meet in this diagram. By varying the position of the points, modify your diagram so that chords AB and CD intersect.

State a suitable condition on the points A, B, C and D such that chords AB and CD will intersect. Label the point of intersection of AB and CD as X.

3 Use the software to measure the lengths AX, BX, CX and DX.

4 Find the values of $AX \times BX$ and $CX \times DX$.
Vary the position of the points. Describe anything you notice, and suggest a suitable general rule.

The rule you have discovered is known as the **intersecting chords theorem**. It has a number of cases, of which this is the first.

5 Verify that the theorem holds true as you continue to vary the positions of points A, B, C and D. Determine whether it continues to hold when you change the radius of the circle. Describe any limitations to the rule you have observed.

Reflect and discuss 1

The result that you have just found is proved in the Extended chapter on circle theorems. In the Exploration you have just done, you have justified the intersecting chords theorem by demonstrating that it works in all the situations you have seen. A mathematical proof must show the result to be true beyond any doubt. Do you think that the dynamic geometry approach does this?

Theorem: If A, B, C and D are points on the circumference of a circle such that AB meets the chord CD at a point X interior to the circle then $AX \times BX = CX \times DX$.

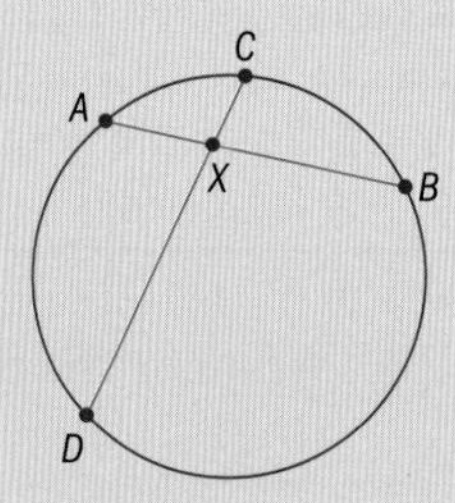

Example 1

In the circle shown here, chords AB and CD meet at P. Find the value of x.

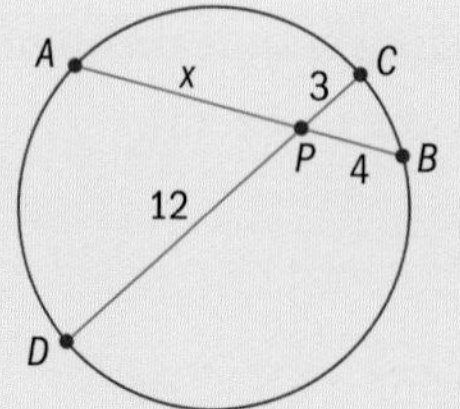

$AP \times BP = CP \times DP$ — Make sure that you correctly identify the sides involved.

$\Rightarrow x \times 4 = 3 \times 12$

$x = \frac{36}{4} = 9$

Example 2

In this circle, chords PR and QS meet at T. Find the value of x and hence find the possible values of the lengths of PT, RT and QT.

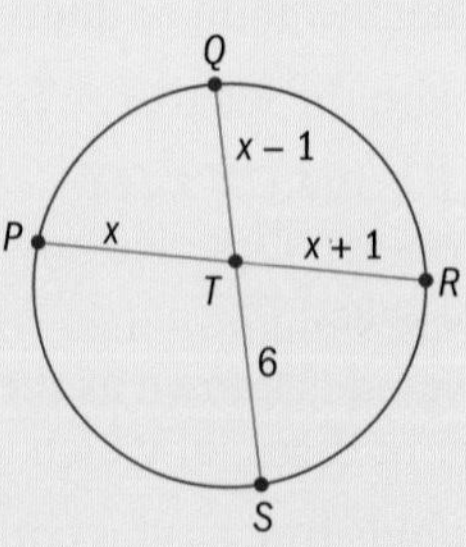

$PT \times RT = QT \times ST$

$x \times (x+1) = (x-1) \times 6$ — Substitute in the known values.

$x^2 + x = 6x - 6$

$x^2 - 5x + 6 = 0$

$(x-2)(x-3) = 0$ — Factorize.

So $x = 2$ or $x = 3$. — There are two possible values of x.

$PT = 2$ or 3, $RT = 3$ or 4 and $QT = 1$ or 2.

Practice 1

1 Find the value of x in each diagram. Hence find the missing lengths.

a

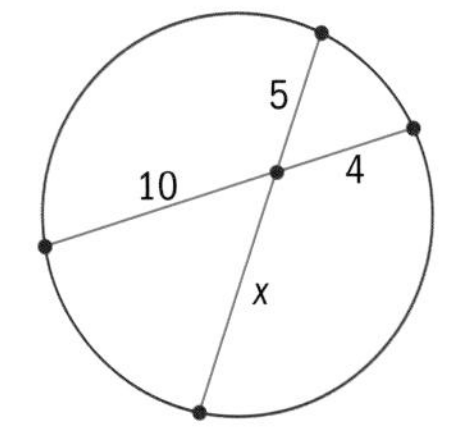

b

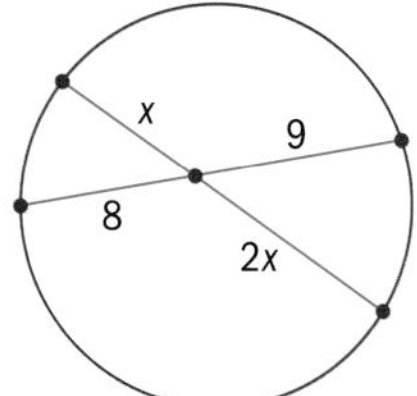

c

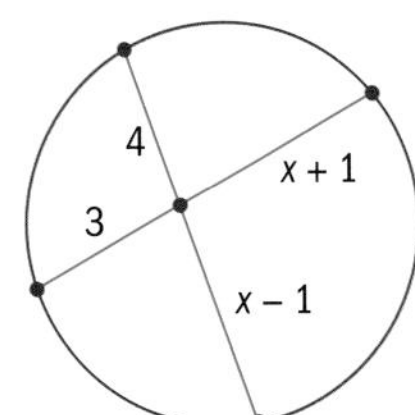

d

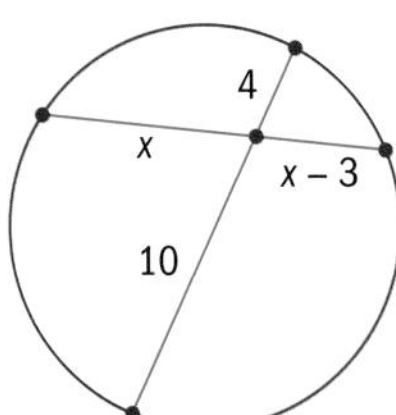

e

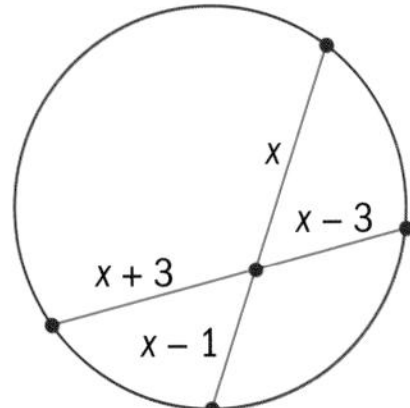

f

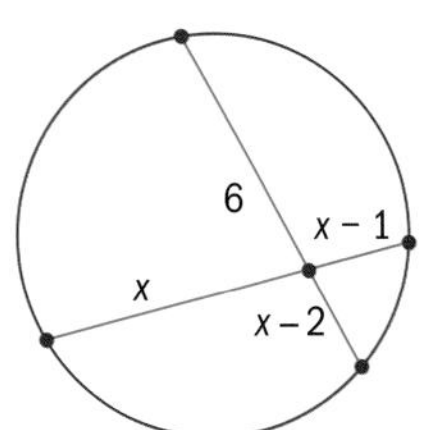

C Chords that intersect outside the circle

- How can a theorem have different cases?

ATL

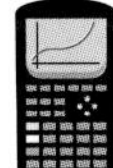

Exploration 2

1 Draw a circle using a GDC or dynamic geometry software.

Draw four points A, B, C and D on the circumference of the circle.

Draw lines AB and CD.

Draw point X, where AB and CD and intersect.

You should have a diagram similar to this.

2 Move the points so that lines AB and CD meet exterior to the circle. This is the second case of the intersecting chords theorem.

Using the software, measure the lengths AX, BX, CX and DX. Verify that the relationship $AX \times BX = CX \times DX$ still holds when the lines AB and CD intersect exterior to the circle.

Can you describe any conditions under which it would not be possible to say that $AX \times BX = CX \times DX$?

> Note that a **line** continues in both directions, unlike a **line segment** which stops at its endpoints.

The intersecting chords theorem referred to chords that intersect at an interior point. You have now seen that the theorem is identical if point P is exterior to the circle.

Example 3

Points J, K, L and M lie on the circumference of a circle. The lines that pass through JK and LM intersect at P. Find the value of x.

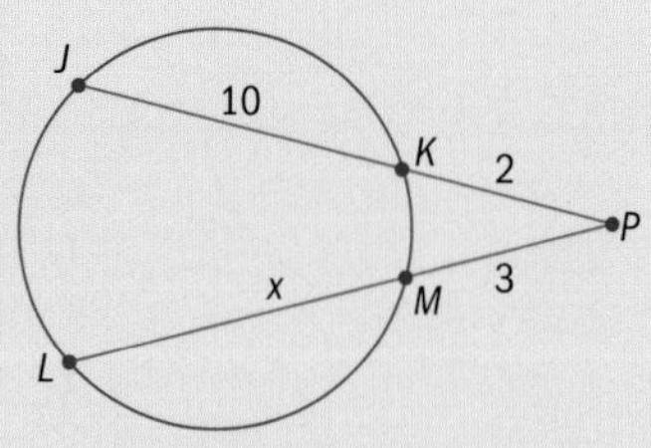

$JP \times KP = LP \times MP$

$\Rightarrow 12 \times 2 = (x+3) \times 3$ — Substitute in the known values.

$\Rightarrow 24 = 3(x+3)$

$\Rightarrow 8 = x+3$

$\Rightarrow x = 5$

Practice 2

1 Use the intersecting chords theorem to find the value of x in each diagram. Hence find the missing lengths in the diagrams.

a

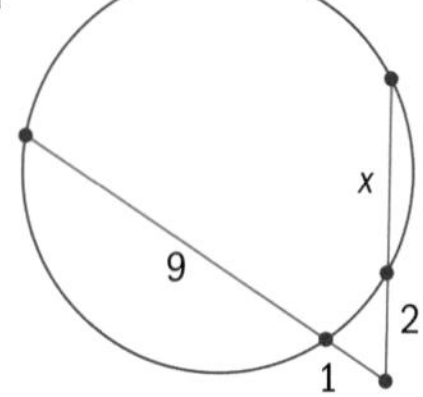

b

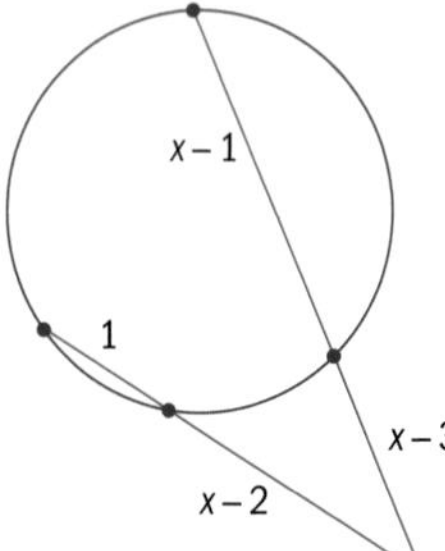

c

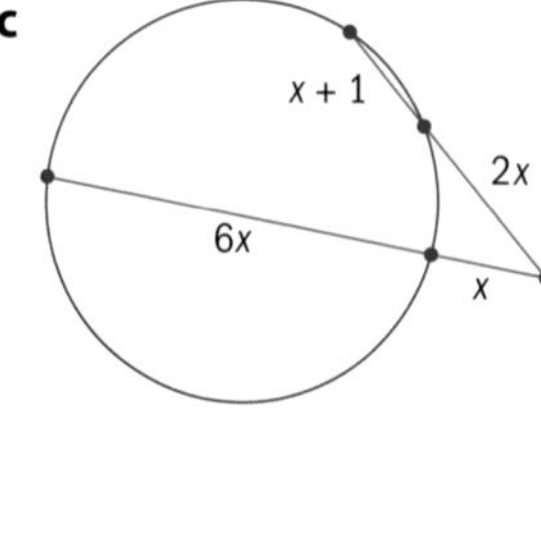

D When a chord becomes a tangent

- Can aesthetics be calculated?

ATL

Exploration 3

1 Recall the diagram from Exploration 2 step **2** where the lines AB and CD meet externally at X.

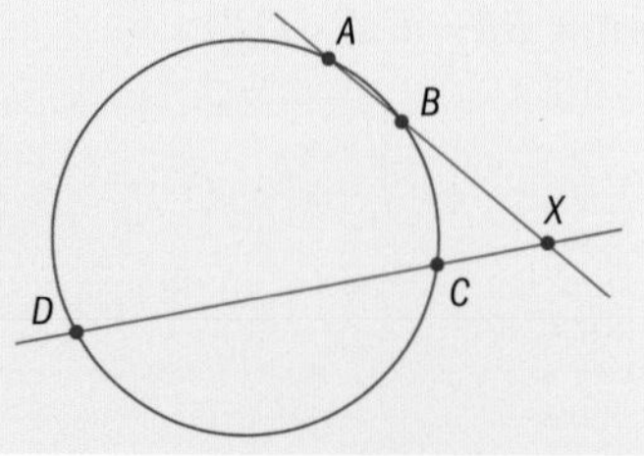

You may need to move A to make sure that X remains on the page.

▶ Continued on next page

2 Use the software to measure the lengths of AX, BX, CX, and DX and confirm that $AX \times BX = CX \times DX$.

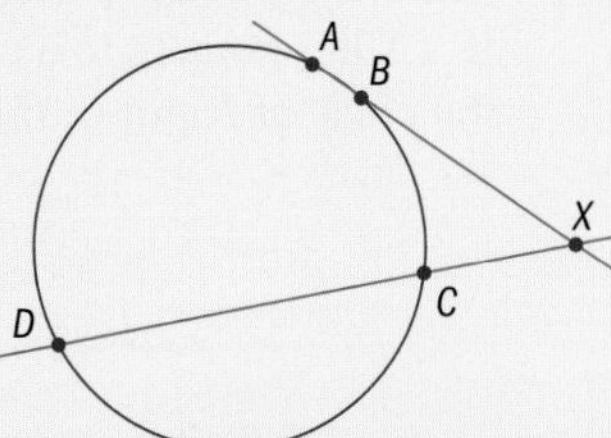

3 Move the point B so that it gets closer to A. You should make sure that the other three points remain stationary. Re-measure the four lengths and check that the intersecting chords theorem still applies.

4 What do you notice about the two lengths AX and BX as B approaches A?

5 Repeat steps **3** and **4** a few more times, allowing B and A to get as close as possible.

6 From your observations, make a conjecture about the limiting case as B approaches A. Test your conjecture.

Reflect and discuss 2

- In Exploration 3, what happened to the line AX as the point B approached the point A? What other line does this resemble?
- Make a conjecture for a *tangent-chord theorem* that is the limiting case of the intersecting chord theorem.
- Use the software to construct a tangent to a circle and confirm that your conjecture is true.

The tangent-chord theorem:

If TX is a tangent to a circle at T and PQX meets the circle at P and Q, then $PX \times QX = TX^2$.

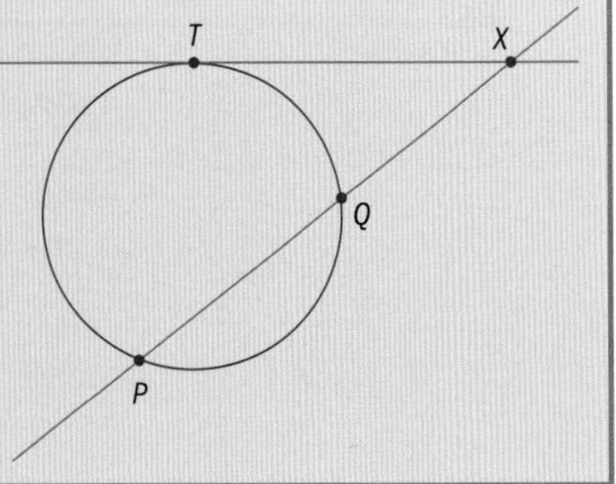

A soccer player running down the sideline wants to take a shot with the best chance of scoring a goal. It turns out that the best chance for a successful shot happens when the player is at the point of tangency of the circle that passes through the goal posts and is tangent to the sideline.

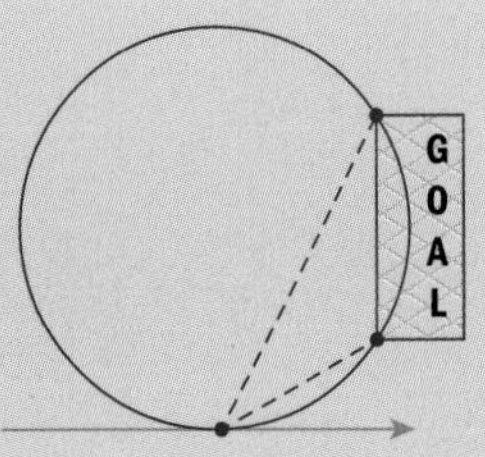

Just as in Explorations 1 and 2, you have used software to form and test a conjecture. To fully justify the conjectures that you have made, a proof is needed.

If you have already written proofs using similar triangles, you might be able to prove the theorems in this section for yourself. To get started, see if you can explain why triangles PTX and TQX are similar.

Summary

- The intersecting chords theorem:
  If A, B, C and D are points on the circumference of a circle such that AB meets the chord CD at a point X interior to the circle then $AX \times BX = CX \times DX$.

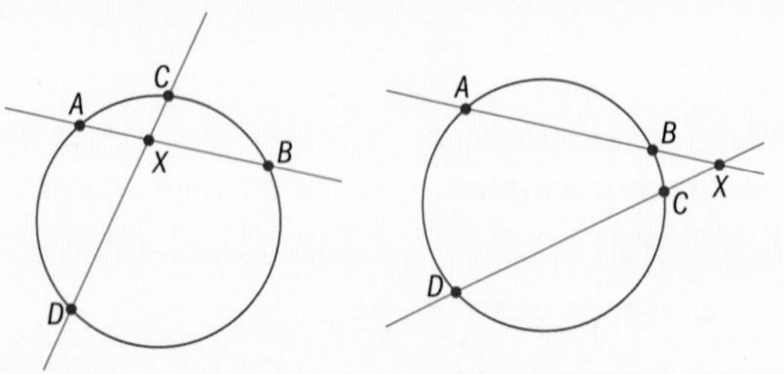

- The tangent-chord theorem:
  If TX lies tangent to a circle at T and PQX meets the circle at P and Q, then $PX \times QX = TX^2$.

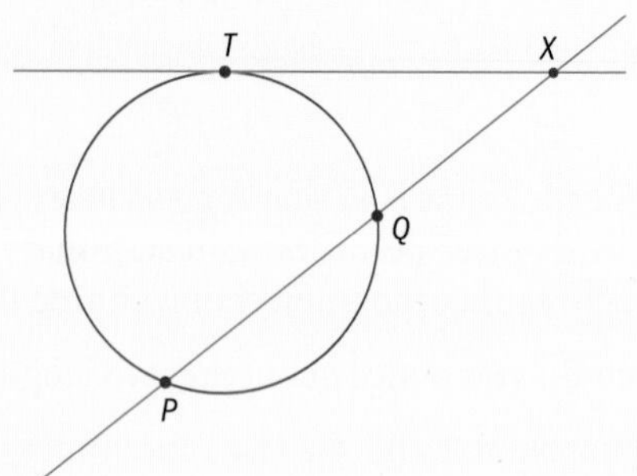

Mixed practice

1 A circle has points A, B, C and D on its circumference. Chords AB and CD have length 11 and 7 respectively, and meet interior to the circle at point X.

a **Sketch** the circle and indicate points A, B, C, D and X on the sketch.

b Given that AX, BX, CX and DX are integers, **find** the possible values that they can take.

2 **Find** the value of the unknown(s) in each diagram.

a **b** **c** **d** **e** **f**

3 **Find** the values of x, y and z.

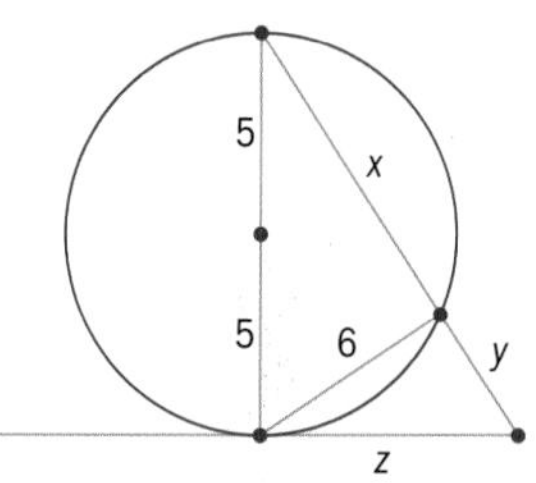

4 In the diagram, BC is a chord and meets a diameter AB at B. $AO = BO = 3$ and $BC = 8$. **Find** x and y.

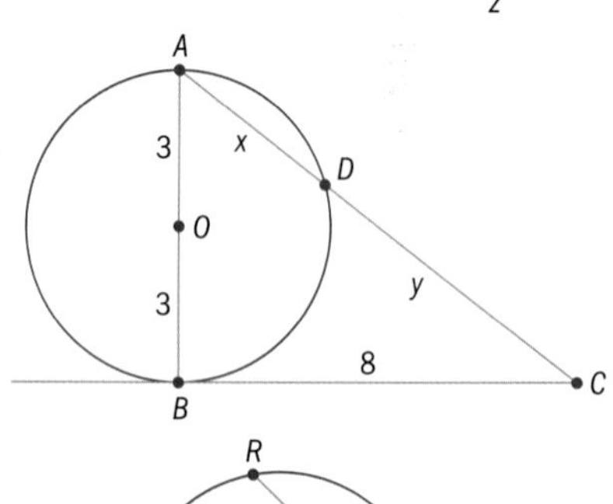

5 **Find** the values of x and y.

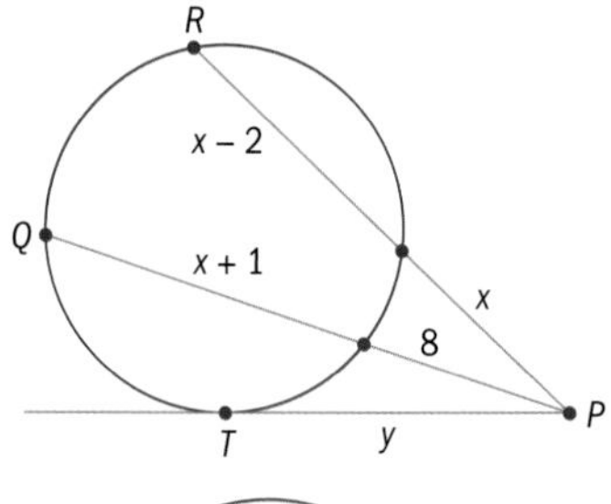

6 In this diagram:

a **show** that

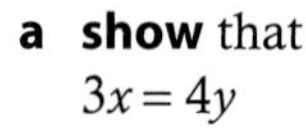

$3x = 4y$

b hence **find** the values of x and y.

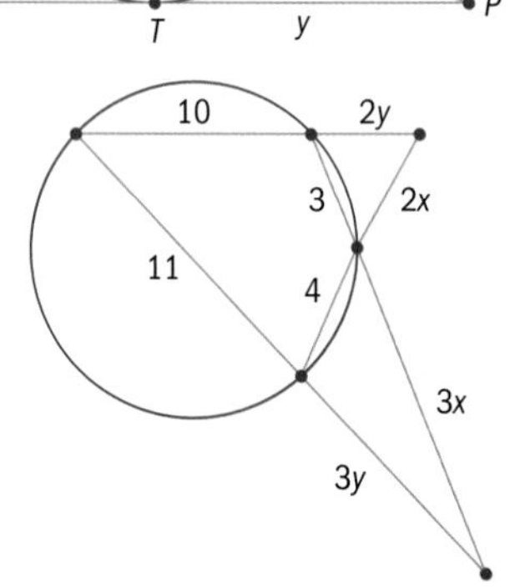

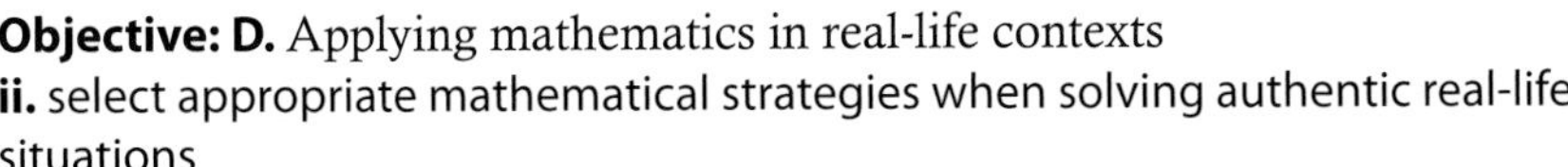

Objective: D. Applying mathematics in real-life contexts
ii. select appropriate mathematical strategies when solving authentic real-life situations

For each of the questions below, consider the geometry of the real-life situation and determine which version of the intersecting chords theorem you need to use.

Review in context

Personal and cultural expression

1 A fragment of a bowl, bearing an ancient inscription, dates from the 7th century BCE. Archaeologists believe the original bowl was circular.
Points A and B are chosen on the edge of the bowl and AB is measured to be 5 cm. M is the midpoint of AB, and MC is perpendicular to AB where C lies on the circumference of the bowl. MC is measured to be 0.25 cm.

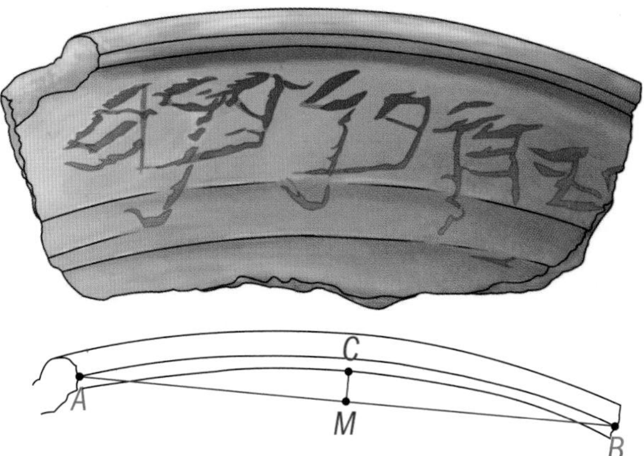

a **Sketch** the complete bowl, including the line segments AB and MC.

b **Show** that the line segment MC passes through the center of the bowl when extended.

c **Use** the intersecting chords theorem to **find** the diameter of the bowl.

2 Huge stone circles have been discovered in Jordan in the Middle East. Archaeologists have been attempting to find out more about these circles. Measuring hundreds of meters across, one of the challenges has been to measure them accurately, especially as some of them are incomplete.

Two points, A and B are marked on the circle and the distance between them is measured. The mid-point of AB is found and the distance from this point to the nearest point on the circle is measured.

Surveyors found that AB was 114 m and the distance from the midpoint was 9 m.

a **Use** this information to **calculate** the diameter of the circle.

A second group of archaeologists use a different technique. From a point outside the circle they make a straight line which is a tangent to the circle and measure the distance to the point of contact. Then they measure the shortest distance from the original point to the circle.

The distances are 150 m and 52 m respectively.

b **Calculate** the diameter from these measurements and then **compare** the two results.

Reflect and discuss

How have you explored the statement of inquiry? Give specific examples.

Statement of Inquiry:

Logic can justify generalizations that increase our appreciation of the aesthetic.

14 Models

Depictions of real-life events using expressions, equations or graphs

Modelling a problem

The length of a rectangular garden is twice as long as its width.

The area of the garden is 72 m^2.

Find the dimensions of the garden.

Drawing a diagram can help you write an equation to represent a situation.

$2x$

x

- For the rectangular garden, write and solve an equation to find x, and then find the length and width.

Translating a word problem into a mathematical expression or equation creates a model of the problem. Sometimes you can adapt a model to solve similar problems.

- Adapt your garden model to find the dimensions of a rectangular field whose length is three times as long as its width, and has an area of 7500 m^2.

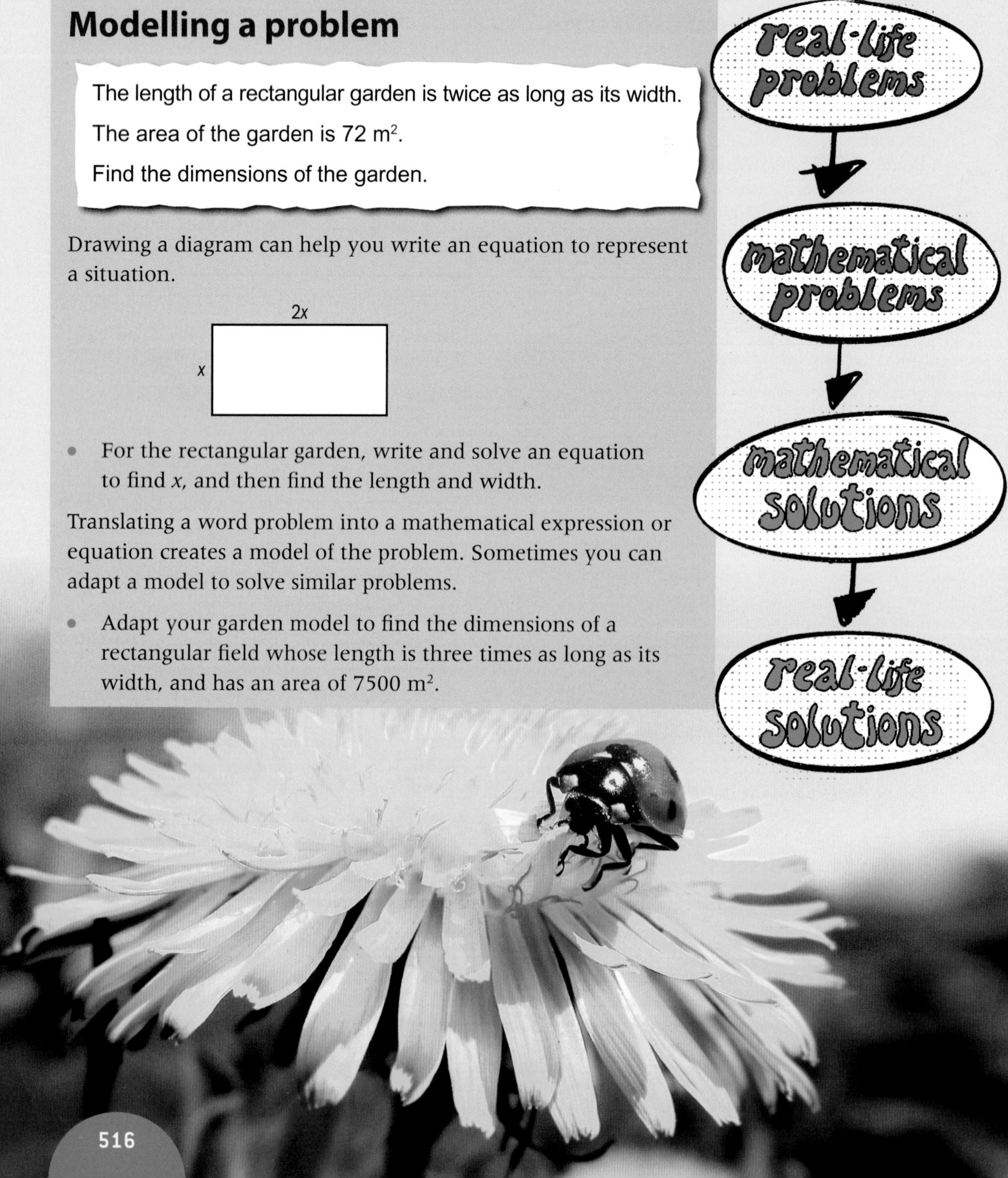

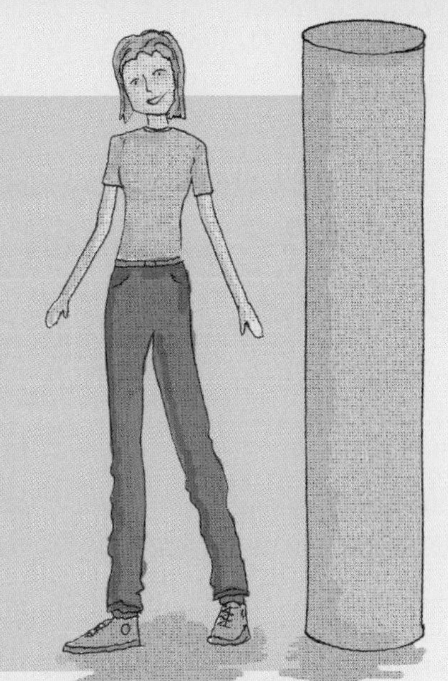

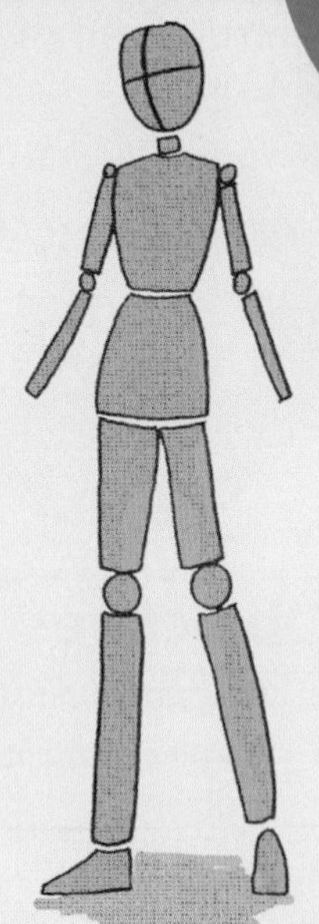

Estimating

Using a model can help you calculate reasonable estimates.

The average human body is about 60% water.

- Modelling your body as a cylinder, calculate an estimate for the volume of water in your body.
- How could you improve your model to make your estimate more accurate?

Should you take an umbrella?

How reliable are weather forecasts? Data shows that on the whole they are around seven times more accurate now than they were 20 years ago, because computers are getting faster, and the mathematics more sophisticated. The mathematical models take account of the formation of the atmosphere, equations of motion and thermodynamics and the uncertainty in a chaotic system.

14.1 The power of exponentials

Global context: Orientation in space and time

FORM

Objectives

- Recognizing exponential functions
- Using exponential functions to model real-life problems
- Identifying and using translations, reflections and dilations with exponential functions

Inquiry questions

F
- What are exponential functions?

C
- How do you transform exponential functions?
- How do you recognize transformations of exponential functions?

D
- Do exponential models have limitations?
- Do patterns lead to accurate predictions?

ATL **Critical-thinking**

Draw reasonable conclusions and generalizations

Statement of Inquiry:

Relationships model patterns of change that can help clarify and predict duration, frequency and variability.

10.1

14.1

You should already know how to:

• evaluate expressions with exponents	**1** Write down the value of: **a** 2^0 **b** 2^4 **c** 4^{-1} **d** 3^{-2}
• identify and apply translations, reflections and dilations to functions	**2** Here is the graph of $y = f(x)$. 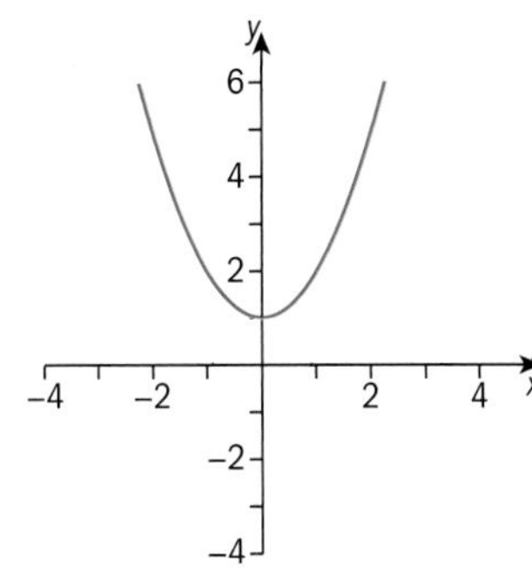 Sketch the graph of: **a** $y = f(x) + 3$ **b** $y = f(x - 2)$ **c** $y = -f(x)$ **d** $y = f(-x)$ **e** $y = 2f(x)$ **f** $y = f(2x)$

F Exponential functions

- What are exponential functions?

Exploration 1

1 A wealthy donor offers money to a charity in two ways:

Option 1: A one-time donation of \$5000.

Option 2: A donation of one cent on day 1, two cents on day 2, four cents on day 3, and so on, doubling each day, for 20 days.

Without doing any calculations, predict which option gives the most money.

2 Now find the total amount donated under Option 2.

a Copy and complete the table:

Day	1	2	3	4	5
Amount (cents)	1	2	4	8	16
Amount as a power of 2	$2^{\square}$	$2^{\square}$			

b Write a function for the amount on day x.

3 Use your GDC or a spreadsheet to work out the amounts for days 1 to 20. Find the total in US dollars for all 20 days.

ATL **4** Graph your function from step **2b** for $1 \le x \le 20$. Explain which option gives the most money.

In a function $y = b^x$, where b is a positive integer greater than 1, as the x values increase, the y values increase very rapidly.

Exploration 2

1 Graph the function $y = b^x$ and insert a slider for the parameter b. Move the slider to explore what happens to the graph for values of b greater than 1.

  a State what happens to the value of y as b increases.

  b Determine if the y-intercept remains the same for all values of b. Explain.

  c Determine if the function has an x-intercept, that is, whether the function will ever be equal to 0.

  d Explain what happens to the value of y for $x < 0$, as the values of x get less and less.

ATL

  e Suggest why an exponential function with $b > 1$ is called an exponential growth function.

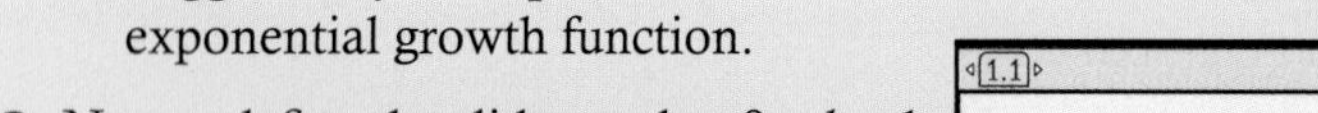

2 Now redefine the slider so that $0 < b < 1$.

  Answer parts **a**, **b** and **c** in step **1** for this function.

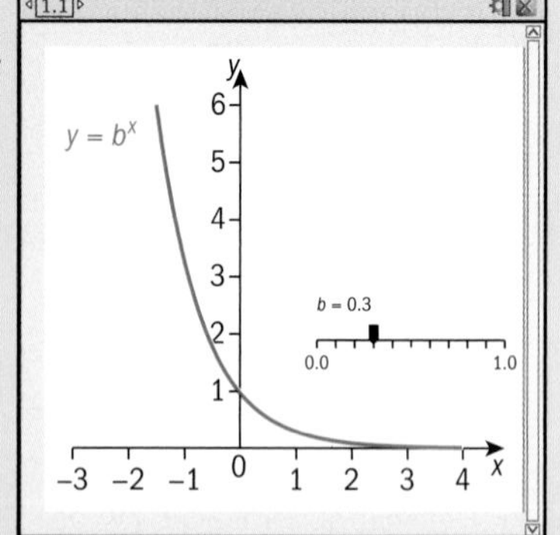

  a Explain what happens to the value of y for $x > 0$, as the values of x get larger and larger.

  b Suggest why an exponential function with $0 < b < 1$ is called an exponential decay function.

3 Explain what happens to the graph if $b < 0$, and justify your answer.

An **exponential function** is of the form $f(x) = a \times b^x$, where $a \neq 0$, $b > 0$, $b \neq 1$.

The independent variable x is the exponent.

Exploration 3

1 By choosing a fixed value for b where $b > 1$, graph the function $y = a \times b^x$ and insert a slider for the parameter a. Describe how the y-intercept changes.

2 Repeat step **1**, but this time choose a fixed value of b such that $0 < b < 1$.

3 Explain how the value of a is related to the y-intercept for any exponential function.

The exponential function $y = b^x$ does not have any x-intercepts.

For $b > 1$, as x gets less and less, the function approaches 0, but never equals 0.

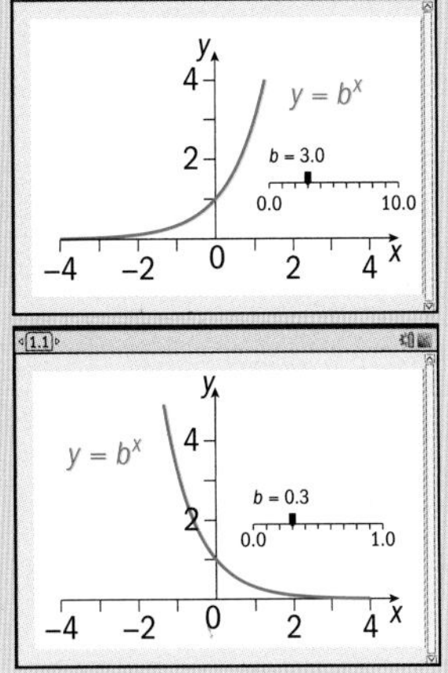

For $0 < b < 1$, as x gets larger and larger, the function approaches 0, but never equals 0.

For any value of $b > 0$ the function $y = b^x$ has a horizontal asymptote at $y = 0$.

A horizontal asymptote is the line that the graph of $f(x)$ approaches as x gets larger and larger.

Reflect and discuss 1

- Here is the graph of exponential function $y = b^x$.

  Deduce the value of b from the graph. Check your answer by graphing the exponential function with the value of b you chose. Explain how you found b.

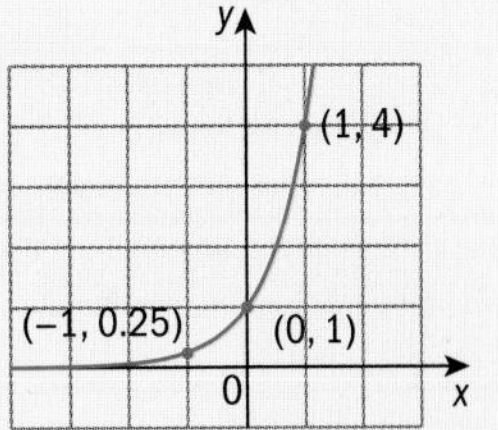

- Identify the value of b from the second graph of $y = b^x$. Explain how you found b.
- Given the standard form of the exponential function is $y = a \times b^x$, for these two graphs the value of a is 1. Explain how you could identify the value of a from a graph of $y = a \times b^x$ where $a \neq 1$.

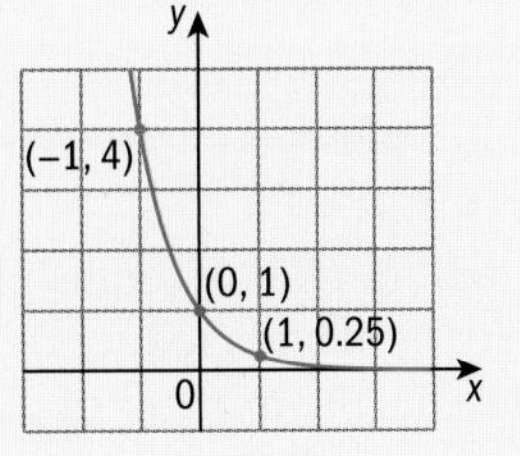

Tip

Look at any graphs you drew in Practice 1.

Population growth, bacterial and viral growth, spread of epidemics, decay of radioactive substances, and finances such as interest and credit card payments can all be modelled with exponential functions.

Example 1

There are 128 players in a tennis tournament. Half of the players are eliminated after each round.

a Write a table of values, and a function for the number of players after round x.

b Determine the number of rounds needed to declare a winner.

a

Round number (x)	Number of players at end of round (y)
0	128
1	$64\left(=128\times\frac{1}{2}\right)$
2	$32\left(=64\times\frac{1}{2}=128\times\left(\frac{1}{2}\right)^2\right)$
3	$16\left(=128\times\left(\frac{1}{2}\right)^3\right)$
4	$8\left(=128\times\left(\frac{1}{2}\right)^4\right)$
5	$4\left(=128\times\left(\frac{1}{2}\right)^5\right)$
6	$2\left(=128\times\left(\frac{1}{2}\right)^6\right)$
7	$1\left(=128\times\left(\frac{1}{2}\right)^7\right)$

Look for a pattern in the y values.

$y=128\times\left(\frac{1}{2}\right)^x$

b 7 rounds are needed.

After 7 rounds only one player, the winner, is left.

In Example 1, $a = 128$ (the initial number of tennis players) and $b=\frac{1}{2}$ (the factor by which the number of players reduces). The variable x represents the round number, and y represents the number of players left after round x.

In the standard form of the exponential function $y = a \times b^x$:

- the parameter a represents the initial amount
- the parameter b represents the growth or reduction (decay) factor
- x is the independent variable (for example: time)
- y is the dependent variable (for example: population)

Practice 1

1 One bacterium cell divides to produce two cells every minute.

a Write down a function for the number of cells after x minutes.

b Use your function to calculate the number of cells after 15 minutes.

c Graph your formula to determine:

i the number of cells after half an hour

ii when there will be 1 million cells (after how many minutes).

2 The half-life of carbon-11 is 20 minutes.

a A sample of carbon-11 has 10 000 nuclei. Make a table of values for five 20-minute intervals, and write down a function for the number of nuclei after x number of 20 minute intervals.

b Find the number of nuclei after 3 hours.

c Determine how long it takes for the number of nuclei to reduce to less than 10% of the original number.

The half-life of a radioactive material is the time taken for the number of nuclei to halve.

Problem solving

3 Drug testing trials show that the amount of a pain-relieving medicine in a person's body reduces by one quarter every hour. It is safe to take another dose of the medicine when there is less than 200 mg in the body.

Determine how long after a dose of 400 mg it will be safe to take another dose.

Exploration 4

1 By choosing a fixed value for b such that $b > 1$, graph the function $y = a \times b^x$ and insert a slider for the parameter a. Describe how the horizontal asymptote changes, and write an equation for the horizontal asymptote of any exponential equation $y = a \times b^x$ when $b > 1$.

2 Do the same as in step **1**, but this time select a fixed value of b such that $0 < b < 1$. Write an equation for the horizontal asymptote of any exponential equation $y = a \times b^x$, for $0 < b < 1$.

The parameter b, the growth factor, may be a percentage instead of an integer or fraction. To increase an initial amount by 10%, you multiply by $1 + 0.1 = 1.1$. So if an initial amount y increases by 10% each day, then you have:

End of day 1	End of day 2	End of day 3	End of day x
$1.1y$	1.1^2y	1.1^3y	1.1^xy

To decrease an initial amount by 5%, you multiply by $1 - 0.05 = 0.95$. So if an initial amount y decreases by 5% each day, then you have:

End of day 1	End of day 2	End of day 3	End of day x
$0.95y$	0.95^2y	0.95^3y	0.95^xy

Exponential growth is modelled by $y = a(1 + r)^x$ where r is the growth rate in decimal form (the percentage expressed as a decimal).

Exponential decay is modelled by $y = a(1 - r)^x$ where r is the decay rate in decimal form.

Example 2

The population in a town was estimated to be 38 000 in 2015. Since then the population has increased each year by approximately 2.4%.

a Determine the growth factor b of the population in the town.

b Write a function to model future population growth.

c Use your function to estimate the population (to the nearest hundred people) in the year 2030.

a Growth factor $b = 1 + r = 1 + 0.024 = 1.024$ — Change 2.4% to a decimal and add 1.

b $y = 38\,000 \times 1.024^x$, where x = number of years, and y = total population after x years. — a = initial population, $y = a(1 + r)^x$

c $x = 2030 - 2015 = 15$

$y = 38\,000 \times 1.024^{15} = 54\,234.4$ — Find x and substitute into the model.

$y \approx 54\,200$

Example 3

Sarah bought a car for \$15 000. Its value depreciates (loses value) by approximately 15% each year.

a Find the depreciation (decay) factor.

b Write a function to model the future value of the car.

c Use your function to estimate the value of Sarah's car after three years, to the nearest dollar.

a Decay factor $b = 1 - r = 1 - 0.15 = 0.85$ — Change 15% to a decimal and subtract from 1.

b $y = 15\,000 \times 0.85^x$, where x = number of years, and y = value of the car after x years. — Define the variables x and y.

c $y = 15\,000 \times 0.85^3 = 9211.88$

$y \approx \$9212$ in three years.

Practice 2

1 For each exponential function, identify **i** the growth/decay factor b, **ii** the growth/decay rate r, and **iii** the initial value a.

a $y = 1.43^x$ **b** $y = 3 \times 0.62^x$ **c** $y = 4 \times \left(\frac{2}{3}\right)^x$ **d** $y = 0.8 \times 1.2^x$

2 A jeweler estimates that the value of gold is increasing by 12% per year.

a Determine the growth factor of the value of gold.

b Write a function to model the future value of a gold bracelet valued at £300 in 2015.

c Use your function to estimate the value of the bracelet in 2030.

3 An electrical store calculates that the number of DVD players sold is decreasing by 22% per year. In 2010 they sold 470 DVD players.

 a Write a function to model future DVD sales.

 b Use your function to estimate the number of DVD players the store will sell in 2018.

4 In 2015, there were 3014 registered cell phone users in a small town. The number of cell phone users is estimated to increase by 42% per year. Estimate the number of cell phone users in 2025.

5 The initial population of an ant colony was approximately 600. The population grows at a rate of 12% per week.

 a Find:

  i the growth factor of the population of ants

  ii a function to model the population growth of the ants

  iii the approximate number of ants in the colony after 15 weeks and after 30 weeks.

 b Graph your function. Use the graph to estimate how many weeks the ant population takes to double in size.

In **5b**, find the point on the graph where $y = 1200$, and read off the value of x.

6 A population of 10 000 insects decreases by 9% every year.

 a Write down a formula and use it to calculate the number of insects left

  i after 3 years

  ii after 10 years.

 b Use your formula to determine how long it would take for the insect population to reduce to less than half its present size.

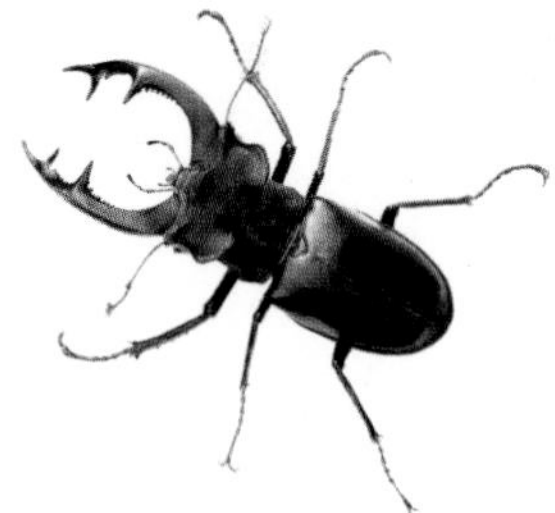

Problem solving

7 Sahil's parents invest $5000 in a long-term money fund offering 4% interest compounded annually. Determine how many years it will take for this amount to double.

8 Conservationists estimate that there are 80 wolves in a forest and that the population is decreasing at a rate of 3.5% per year. Estimate how long it will take for the present population to be halved.

9 An antibiotic destroys 10% of bacteria present in one hour. Determine how long it would take for the antibiotic to reduce 50 million bacteria down to less than 1 million bacteria.

C Transformations of exponential functions

- How do you transform exponential functions?
- How do you recognize transformations of exponential functions?

Objective: A. Knowing and understanding **i.** select appropriate mathematics when solving problems in both familiar and unfamiliar situations
In Exploration 5, use the rules for transformations of graphs to justify your conjectures.

Exploration 5

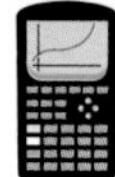

1 Graph $y = 3^x$ and $y = \left(\frac{1}{3}\right)^x$ on the same set of axes. Describe and justify the symmetry between the two graphs. Make and justify a conjecture about the symmetry of all pairs of graphs of the form $y = b^x$ and $y = \left(\frac{1}{b}\right)^x$.

2 Graph $y = 3^x$ and $y = -3^x$ on the same axes. Describe and justify the symmetry between the two graphs. Make and justify a conjecture about the symmetry of all pairs of graphs of the form $y = b^x$ and $y = -b^x$.

3 Graph the functions below and insert a slider for the parameter a, such that $a > 0$. Describe what happens to the shape of the graph as the value of a varies, and state the type of transformation this represents.

a $y = a \times b^x$, for $b = 2$ **b** $y = a \times b^x$, for $b = 0.5$

4 Graph the function $y = 2^x + k$, and insert a slider for the parameter k. By changing the value of k, state the transformation that takes the function $y = 2^x$ to $y = 2^x + k$. Describe the transformation from any exponential function $y = ab^x$ to $ab^x + k$.

5 Graph the function $y = 2^{(x-h)}$, and insert a slider for the parameter h. By changing the value of h, state the transformation that takes the function $y = 2^x$ to $y = 2^{(x-h)}$. Describe the transformation from any exponential function $y = ab^x$ to $y = ab^{(x-h)}$.

6 Explain how to transform the function $y = ab^x$ to $y = ab^{(x-h)} + k$.

Tip

You may get a better idea of what the transformation is by looking at $y = 2^{(x-h)} - 2$ instead of $y = 2^{(x-h)}$.

Transformations of exponential functions

Reflection: For the exponential function $f(x) = a \times b^x$:

- the graph of $y = -f(x)$ is a reflection of the graph of $y = f(x)$ in the x-axis.
- the graph of $y = f(-x)$ is a reflection of the graph of $y = f(x)$ in the y-axis.

Translation: For the exponential function $f(x) = a \times b^x$:

- $y = f(x - h)$ translates $y = f(x)$ by h units in the x-direction.
  When $h > 0$, the graph moves in the positive x-direction, to the right.
  When $h < 0$, the graph moves in the negative x-direction, to the left.
- $y = f(x) + k$ translates $y = f(x)$ by k units in the y-direction.
  When $k > 0$, the graph moves in the positive y-direction, up.
  When $k < 0$, the graph moves in the negative y-direction, down.
- $y = f(x - h) + k$ translates $y = f(x)$ by h units in the x-direction and k units in the y-direction.

Continued on next page

Dilation: For the exponential function $f(x) = a \times b^x$:

- $y = af(x)$ is a vertical dilation of $f(x)$, scale factor a, parallel to the y-axis.
- $y = f(ax)$ is a horizontal dilation of $f(x)$, scale factor $\frac{1}{a}$, parallel to the x-axis.

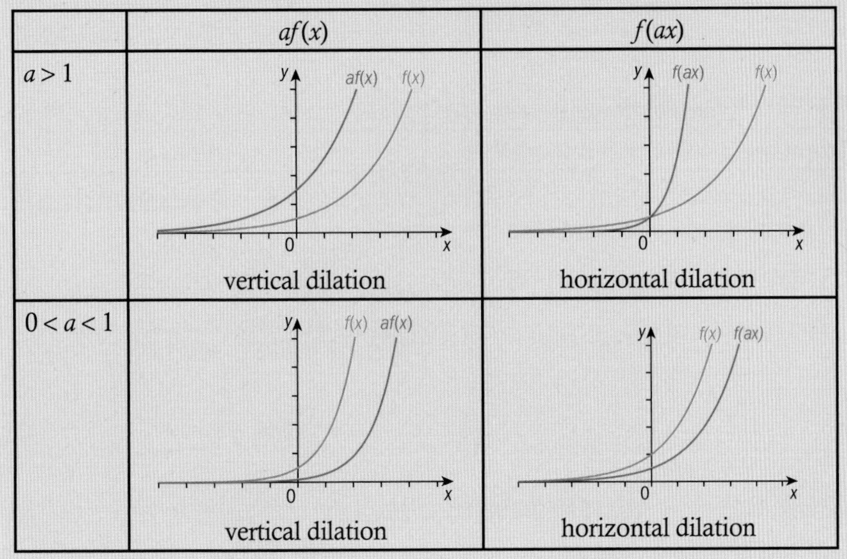

	$af(x)$	$f(ax)$
$a > 1$	vertical dilation	horizontal dilation
$0 < a < 1$	vertical dilation	horizontal dilation

Example 4

Start with the function $f(x) = 2^x$.

a In words, describe the transformation **i** $0.5f(x)$, and **ii** $f(0.5x)$.

b Write the transformed functions as **i** $g(x) = 0.5f(x)$, and **ii** $h(x) = f(0.5x)$

c Without using a graphing program or GDC, sketch both the function and the transformed function on the same set of axes.

a i The transformation $0.5f(x)$ is a vertical dilation of the graph of $f(x)$, scale factor 0.5, parallel to the y-axis.

ii The transformation $f(0.5x)$ is a horizontal dilation of the graph of $f(x)$, scale factor $\frac{1}{0.5} = 2$, parallel to the x-axis.

b i $g(x) = 0.5 \times 2^x$

ii $h(x) = 2^{0.5x}$

c i Graph of $f(x) \to$ Graph of $g(x)$

$(0, 1) \to (0, 0.5)$

$(1, 2) \to (1, 1)$

$(2, 4) \to (2, 2)$

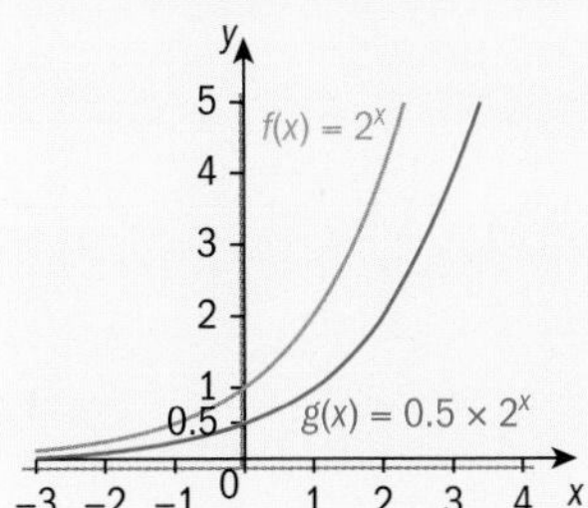

ii Graph of $f(x) \to$ Graph of $h(x)$

$(0, 1) \to (0, 1)$

$(1, 2) \to (1, \sqrt{2})$

$(2, 4) \to (4, 4)$

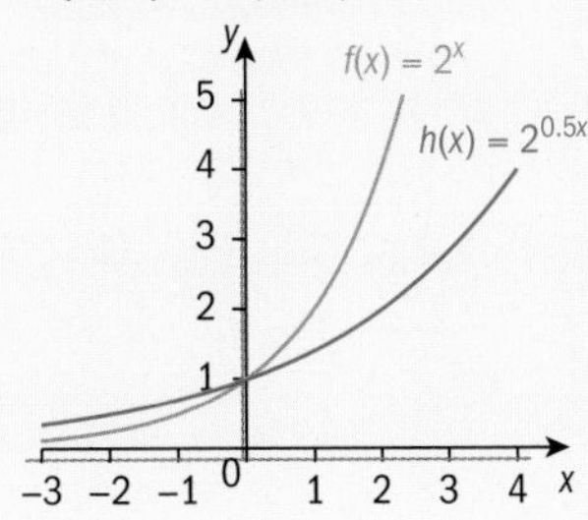

Example 5

Given the functions $f(x)=\left(\frac{1}{2}\right)^x$, $g(x)=\left(\frac{1}{2}\right)^{x+4}$ and $h(x)=0.5\left(\frac{1}{2}\right)^x$:

a Describe the transformations **i** g, and **ii** h of the function f.

b Write **i** g, and **ii** h in terms of f.

c Sketch on the same set of axes **i** f and g, and **ii** f and h.

a i g is a horizontal translation of f by 4 units in the negative x direction.

ii h is a vertical dilation of f, scale factor 0.5, parallel to the y-axis.

b i $g(x) = f(x + 4)$ **ii** $h(x) = 0.5f(x)$

c i Graph of $f(x) \rightarrow$ Graph of $g(x)$

$(0, 1) \rightarrow \left(0, \frac{1}{16}\right)$

$\left(1, \frac{1}{2}\right) \rightarrow \left(1, \frac{1}{32}\right)$

$(-1, 2) \rightarrow \left(-1, \frac{1}{8}\right)$

ii Graph of $f(x) \rightarrow$ Graph of $h(x)$

$(-2, 4) \rightarrow (-2, 2)$

$(-1, 2) \rightarrow (-1, 1)$

$(0, 1) \rightarrow (0, 0.5)$

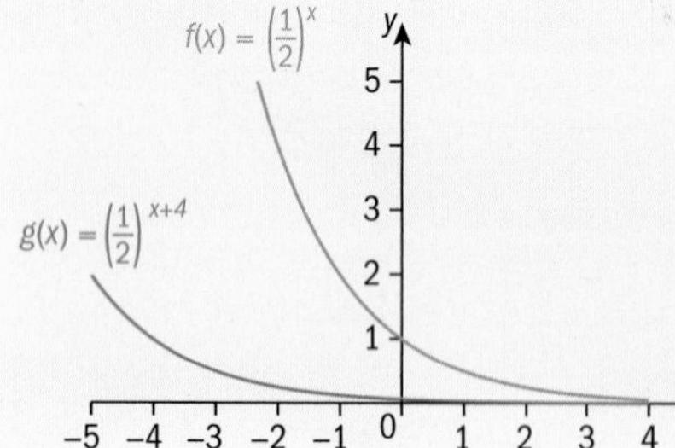

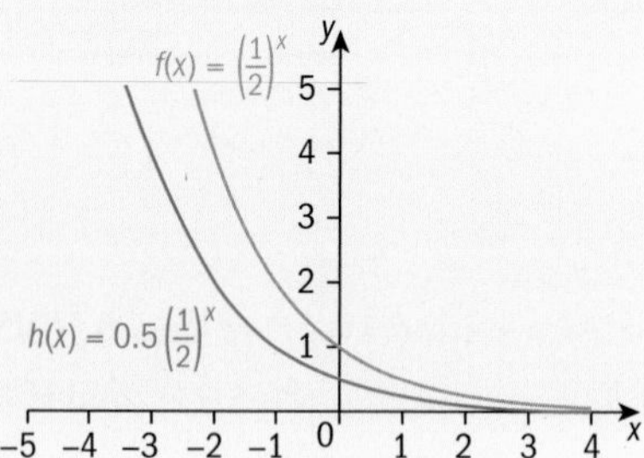

Practice 3

1 You are given the function $f(x) = 3^x$.

i Describe each transformation below in words.

ii Write the transformed function as $g(x) =$ ___.

iii Sketch both the function and transformed function on the same set of axes.

a $f(x + 2)$ **b** $-f(x)$ **c** $f(x) - 2$ **d** $f(-x)$

e $3 + f(x - 1)$ **f** $f(2x)$ **g** $2f(x)$ **h** $f(0.5x)$

Problem solving

i $-2f(x)$ **j** $f(-2x)$

2 For each pair of functions f and g:

i Describe the transformation(s) of function f to function g.

ii Write g in terms of f.

iii Sketch both functions on the same set of axes.

a $f(x)=1.5^x,\ g(x)=-1.5^x$

b $f(x)=\left(\frac{1}{2}\right)^x,\ g(x)=2^x$

c $f(x)=3^x,\ g(x)=3^{x+2}-3$

d $f(x)=0.5^x,\ g(x)=0.5^{2x}$

e $f(x)=2^x,\ g(x)=2^{0.5x}$

f $f(x)=\left(\frac{1}{3}\right)^x,\ g(x)=3\times\left(\frac{1}{3}\right)^x$

Problem solving

g $f(x)=\left(\frac{1}{2}\right)^x,\ g(x)=\left(\frac{1}{2}\right)^{-2x}$

3 On the same set of axes, sketch the graph of the function f and its three transformations. Label each graph and any intercepts with the axes.

$f(x)=2^x;\ f_1(x)=2^{2x};\ f_2(x)=-2\times 2^{0.5x};\ f_3(x)=\frac{1}{2}\times 2^{-2x}$

D Modelling with exponential functions

- Do exponential models have limitations?
- Do patterns lead to accurate predictions?

Exploration 6

In this exploration you will investigate how much \$1 can grow at an interest rate of 100%, when the interest is compounded at different intervals.

1 Find the amount \$1 is worth at the end of one year at 100% interest.

2 Find the amount \$1 is worth at the end of one year at 100% interest compounded half-yearly. In the function $y=a(1+r)^x$, the initial investment is \$1, the growth rate is $\frac{100}{2}=50\%=0.5$, and $x=2$, as interest is now compounded two times per year. Therefore, $y=(1+0.5)^2$ or $y=1.5^2$.

3 Find the amount \$1 is worth at the end of one year at 100% interest compounded quarterly (every 3 months). In this case, $y=(1+0.25)^4$ or $y=1.25^4$.

4 Write down an exponential function representing the investment of \$1 at 100% interest compounded x times per year.

5 Use your function to find the amount your investment of \$1 is worth at the end of one year at 100% interest if it is compounded:

a monthly **b** weekly **c** daily

d hourly **e** each minute **f** each second.

6 Graph your function. Determine if there is a limit to the amount the \$1 investment can grow in one year if it is compounded an infinite number of times at 100% interest. Justify your answer.

Bank accounts usually pay compound interest per annum, which means the interest is calculated and added to the account once a year.

The limit to the amount that the \$1 could grow in one year is \$2.72, to two decimal places. It is actually the number e, which is a special number just like π, and is an infinite non-repeating number. It is called 'e' after Leonhard Euler, the mathematician who first discovered it.

You have seen problems that model the growth of populations of people, bacteria, and insects. Is such a model appropriate for all phases of population growth?

Reflect and discuss 2

An exponential model for population growth can look like this one.

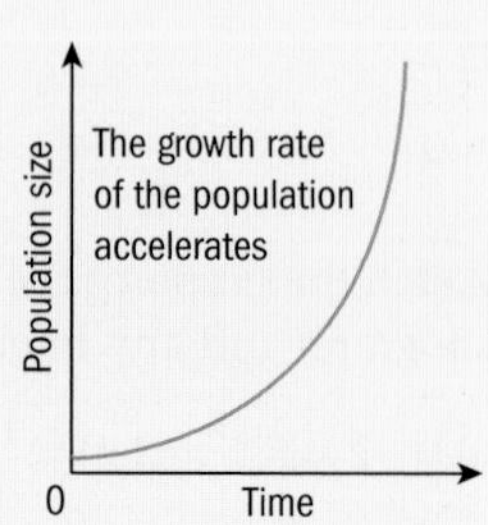

- What environmental factors are necessary to sustain growth of a bacteria population? What factors will limit this growth?

  Sketch a graph that more accurately reflects population growth of bacteria over a long period of time.

- What environmental factors are necessary to sustain growth of a human population? What environmental factors limit human population growth?

  Sketch a graph that more accurately reflects population growth of humans over a long period of time.

- Is your human growth graph different from your bacteria growth graph? Why, or why not?
- Research the types of functions used to more accurately model and predict the growth of human populations over an extended period of time.

Summary

An exponential function is of the form $f(x) = a \times b^x$, where $a \neq 0$, $b > 0$, $b \neq 1$.

The independent variable x is the exponent.

In the standard form of the exponential function $y = a \times b^x$:

- the parameter a represents the initial amount
- the parameter b represents the growth or reduction (decay) factor.
- x is the independent variable (e.g. time)
- y is the dependent variable (e.g. amount at time x)

The exponential function $y = b^x$ does not have any x-intercepts.

For $b > 1$, as x gets less and less, the function approaches 0, but never equals 0.

For $0 < b < 1$, as x gets larger and larger, the function approaches 0, but never equals 0.

For any value of $b > 0$ the function $y = b^x$ has a horizontal asymptote at $y = 0$.

A horizontal asymptote is the line that the graph of $f(x)$ approaches as x gets larger and larger.

Exponential growth is modeled by $y = a(1 + r)^x$ where r is the growth rate in decimal form (the percentage expressed as a decimal). Exponential decay is modeled by $y = a(1 - r)^x$ where r is the decay rate in decimal form.

Transformations of exponential functions

Reflection: For the exponential function $f(x) = a \times b^x$:

The graph of $y = -f(x)$ is a reflection of the graph of $y = f(x)$ in the x-axis.

The graph of $y = f(-x)$ is a reflection of the graph of $y = f(x)$ in the y-axis.

Translation: For the exponential function $f(x) = a \times b^x$:

- $y = f(x - h)$ translates $y = f(x)$ by moving it h units in the x-direction.

  When $h > 0$, the graph moves in the positive x-direction, to the right.

  When $h < 0$, the graph moves in the negative x-direction, to the left.

- $y = f(x) + k$ translates $y = f(x)$ by moving it k units in the y-direction.

  When $k > 0$, the graph moves in the positive y-direction, up.

  When $k < 0$, the graph moves in the negative y-direction, down.

- $y = f(x - h) + k$ translates $y = f(x)$ by h units in the x-direction and k units in the y-direction.

Dilation: For the exponential function $f(x) = a \times b^x$:

$y = af(x)$ is a vertical dilation of $f(x)$, scale factor a, parallel to the y-axis.

$y = f(ax)$ is a horizontal dilation of $f(x)$, scale factor $\frac{1}{a}$, parallel to the x-axis.

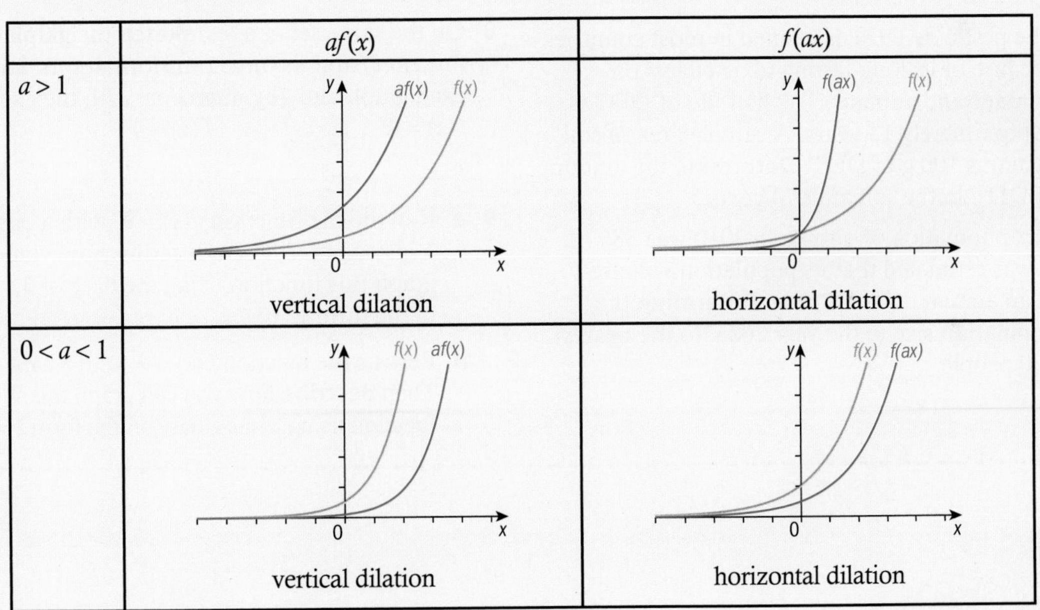

	$af(x)$	$f(ax)$
$a > 1$	vertical dilation	horizontal dilation
$0 < a < 1$	vertical dilation	horizontal dilation

Mixed practice

1 Xixi bought a car five years ago for \$18 000. It depreciates approximately 15% every year.

  a **Write down** the depreciation (decay) factor for this problem.

  b **Write down** a function to model the depreciation of the car over time.

  c **Find** the value of her car now, to the nearest \$100.

Problem solving

2 A strain of bacteria is doubling in a laboratory every 5 minutes. At midday there are 20 bacteria. **Determine** the number of bacteria at the end of one hour.

3 The isotope radium-226 has a half-life of approximately 1600 years. A sample has 10 000 nuclei. **Determine** how many nuclei will be left after 150 years.

4 The pesticide DDT is banned in most countries because of its long-lasting toxic effects on humans and animals. The half-life of DDT is approximately 15 years. A sample area of soil contains 100 g of DDT. **Determine** the amount of DDT in the soil after 70 years.

5 The population of a town in 2010 was 38 720. It was estimated that its population will grow at an annual rate of 2.68%. **Determine** the population size in the year 2025, to the nearest 100 people.

6 Gina invests \$400 in a savings account that pays 1.4% interest compounded annually.

  a **State** the growth factor on this investment.

  b **Write down** a function to model the growth of her savings.

  c **Determine** how many years it will take for her savings to double.

7 Write these transformations in terms of the function $f(x) = 2^x$ when f is:

  a reflected in the y-axis

  b translated 3 units to the left and 4 units down

  c reflected in the x-axis

  d dilated by scale factor 3 parallel to the x-axis

  e dilated by scale factor 4 parallel to the y-axis.

Problem solving

8 On the same set of axes, **sketch** the graph of the function f and its three transformations. **Label** each graph and any intercepts with the axes.

$f(x) = 4^x$ $\quad$ $f_1(x) = 4^{2x}$

$f_2(x) = -2 \times 4^{0.5x}$ $\quad$ $f_3(x) = \frac{1}{2} \times 4^{-2x}$

9 a Rewrite the function $f(x) = 9^x$ with a base of 3 instead of 9. Then **describe** how you can graph this function using the base of 3, in the form $y = 3^{ax}$.

  b Rewrite the function $f(x) = 9^x$ with a base of 81. Then **describe** how you can graph this function using 81 as a base, in the form $y = 81^{ax}$.

Objective: D. Applying mathematics in real-life contexts
iii. apply the selected mathematical strategies successfully to reach a solution

In this Review in context you will apply exponential functions to determine the age of fossils using carbon dating.

Review in context

Orientation in space and time

Ötzi was found on 19 September 1991 by two German tourists, Helmut and Erika Simon at an elevation of 3210 meters (10 530 ft) on the Austrian–Italian border. Because the body, clothing and tools were so well preserved, the Simons thought the man had died recently. Scientists at the University of Innsbruck in Austria used carbon dating to estimate that Ötzi died about 5300 years ago, making him the oldest, best preserved mummy in the world.

Carbon dating of fossils is based upon the decay of ^{14}C, a radioactive isotope of carbon with a relatively long half-life of about 5700 years.

All living organisms get ^{14}C from the atmosphere. When an organism dies, it stops absorbing ^{14}C, which begins to decay exponentially. Carbon dating compares the amount of ^{14}C in fossil remains with the amount in the atmosphere, to work out how much has decayed, and therefore how long ago the organism died.

1 Let N_0 = the initial amount of ^{14}C at the time of death.

Let x = the number of half-lives, where each half-life is 5700 years.

Let N = the amount present after x number of half-lives.

Write an exponential function relating N_0, x, and N.

You can use your function from **1** to help you solve the following problems.

Problem solving

2 When it dies, an organism contains 30 000 nuclei of ^{14}C. **Calculate** the number of ^{14}C nuclei in the organism after 11 400 years.

3 The amount of ^{14}C in a fossil is calculated to be 0.25 times the amount when the organism died. **Calculate** the approximate age of the fossil.

4 A fossil bone is approximately 16 500 years old. **Estimate** the fraction of ^{14}C still in the fossil.

5 Only 6% of the original amount of ^{14}C remains in a fossil bone. **Estimate** how many years ago it died.

6 **Calculate** the approximate percentage of ^{14}C left in a fossil bone sample after 35 000 years.

7 Analysis on an animal bone fossil at an archeological site reveals that the bone has lost between 90% and 95% of its ^{14}C. **Determine** an approximate interval for the possible ages of the bone.

8 Using your function, **explain** why fossils older than 50 000 years may have an undetectable amount of ^{14}C.

Reflect and discuss

How have you explored the statement of inquiry? Give specific examples.

Statement of Inquiry:

Relationships model patterns of change that can help clarify and predict duration, frequency and variability.

14.2 Like gentle ocean waves

Global context: Scientific and technical innovation

RELATIONSHIPS

Objectives

- Graphing sine and cosine functions
- Understanding periodicity
- Transforming sine and cosine functions using translations, reflections and dilations
- Recognizing transformations of sine and cosine graphs, and finding equations of graphs
- Modelling real-life problems using sine and cosine functions

Inquiry questions

F
- What is a periodic function?
- What are the main characteristics of the graphs of $y = a\sin bx$ and $y = a\cos bx$?

C
- How do you translate, dilate and reflect periodic functions?
- How does the equation of a sinusoidal function represent the transformations performed on it?
- Does the order in which transformations are performed matter?

D
- What real-life phenomena are periodic?
- Have scientific models and methods provided more answers or questions?

ATL **Critical-thinking**

Draw reasonable conclusions and generalizations

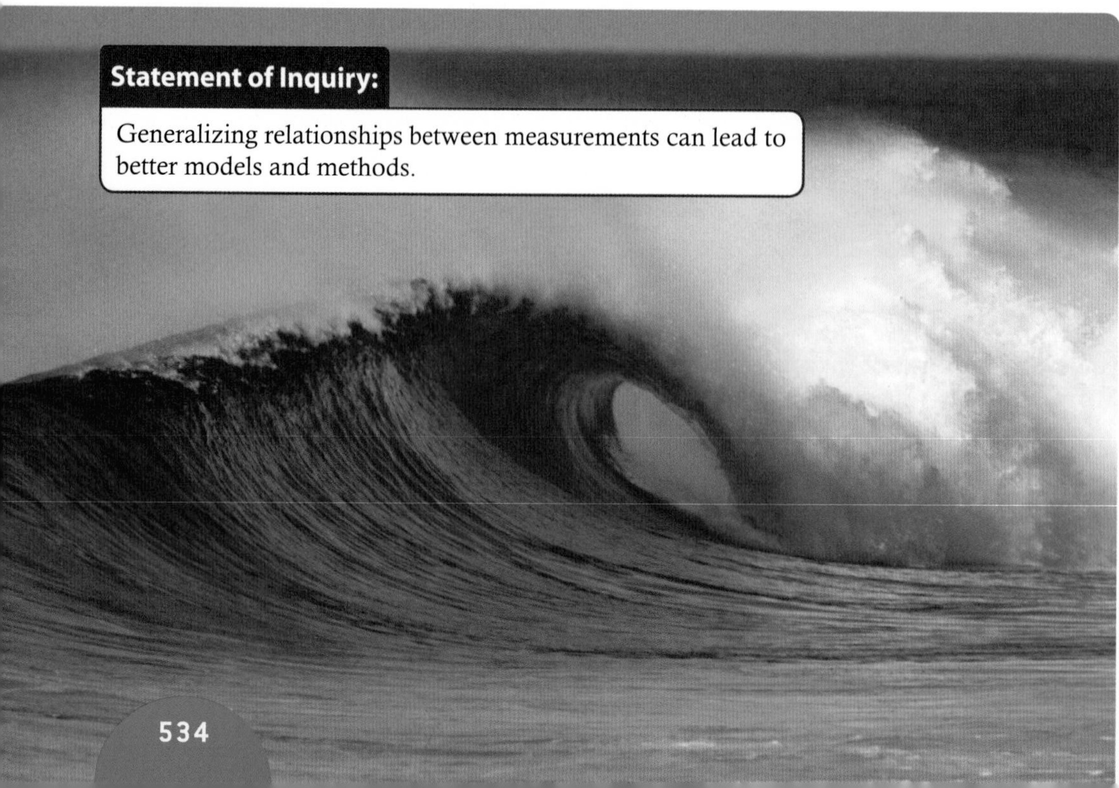

Statement of Inquiry:

Generalizing relationships between measurements can lead to better models and methods.

7.4

14.2

You should already know how to:

• use the trigonometric ratios for sine and cosine	**1** Here is a right-angled triangle. Write down **a** sin x **b** cos x
• find values of sine and cosine	**2** Use a calculator to find the value of **a** sin 45° **b** cos 37
• transform functions using translations, reflections and dilations	**3** Here is the graph of $f(x)$. On graph paper, sketch $f(x)$ and **a** $f(x)$ 2**d** $f(x)$ **b** $3f(x)$**e** $f(\frac{1}{2}x)$ **c** $f(x)$**f** $f(2x)$ Make sure your GDC is in degree mode.

The sine and cosine curves

- What is a periodic function?
- What are the main characteristics of the graphs of $y = a \sin bx$ and $y = a \cos bx$?

You have used the trigonometric ratios sine, cosine and tangent in right-angled triangles. You can use your GDC to find these ratios for different angles.

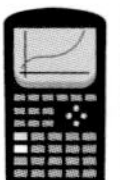

Exploration 1

1 a Draw a table of values for angle $\theta = 0°, 10°, 20°, \ldots, 360°$, and $\sin\theta$.

θ	0°	10°	20°	30°	40°	50°	60°	...	360°
$\sin\theta$									

b Find sin0°, sin10°, sin20°, etc. using your GDC. Round the values to 2 decimal places and write them in your table.

c On graph paper, plot the graph of sin θ, with θ from 0° to 360° on the x-axis and the sine values on the y-axis. Join your points with a smooth curve.

Tip

Make sure your GDC is on degree mode.

▶ Continued on next page

2 Follow the instructions in step **1** to draw a table of values and plot the graph of $\cos\theta$.

3 Check your graphs by graphing $\sin\theta$ and $\cos\theta$ on your GDC.

4 For the domain $0° \le \theta \le 360°$, write down the range of

a $\sin\theta$

b $\cos\theta$.

Reflect and discuss 1

- What is similar about the graphs for sine and cosine? What is different?
- Predict how you think each curve will continue from:
  - 360° to 720°
  - 0° to −360°
- Describe the symmetry of each curve.
  Use your GDC to draw graphs to check your predictions.

The sine and cosine curves have the same shape. Translating the sine graph 90° to the left gives the cosine graph. Translating the cosine graph 90° to the right gives the sine graph.

The **amplitude** is the height from the mean value of the function to its maximum or minimum value. The graphs of $y = \sin x$ and $y = \cos x$ have amplitude 1.

The **period** is the horizontal length of one complete cycle. The graphs of $y = \sin x$ and $y = \cos x$ have period 360°.

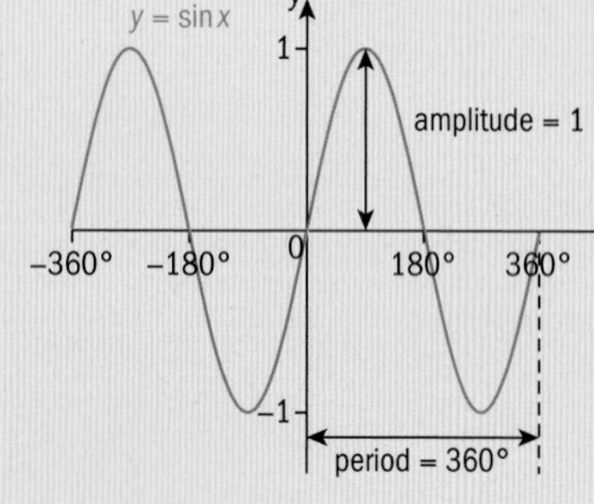

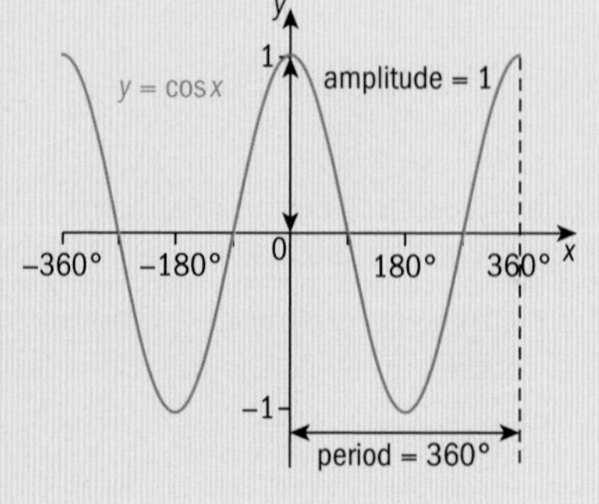

Sine and cosine graphs are also called sine and cosine waves. Waves have an amplitude and a period.

The graph of $y = \sin x$ passes through (0, 0) and has a maximum at (90°, 1).

The graph of $y = \cos x$ has a maximum at (0, 1).

Exploration 2

1 Use what you know about transformations of functions from $f(x)$ to $af(x)$ to predict the shape of the graph of $y = 2\sin x$. Graph the function on your GDC to check your prediction.

2 Predict the shape of the graph of $y = \frac{1}{2}\sin x$. Graph the function on your GDC to check your prediction.

▶ Continued on next page

3 Find the amplitude of the graph of:

a $y = 2\sin x$

b $y = \frac{1}{2}\sin x$

c $y = a\sin x$.

4 Predict the shapes of $y = 2\cos x$ and $y = \frac{1}{2}\cos x$. Graph the functions to check your predictions.

5 Find the amplitude of the graph of:

a $y = 2\cos x$

b $y = \frac{1}{2}\cos x$

c $y = a\cos x$.

6 Make generalizations about the effect of parameter 'a' on the graphs of $y = a\sin x$ and $y = a\cos x$.

The graph of $y = a\sin x$ and the graph of $y = a\cos x$ both have amplitude a.

The amplitude of a curve is half the distance between the maximum and minimum values.

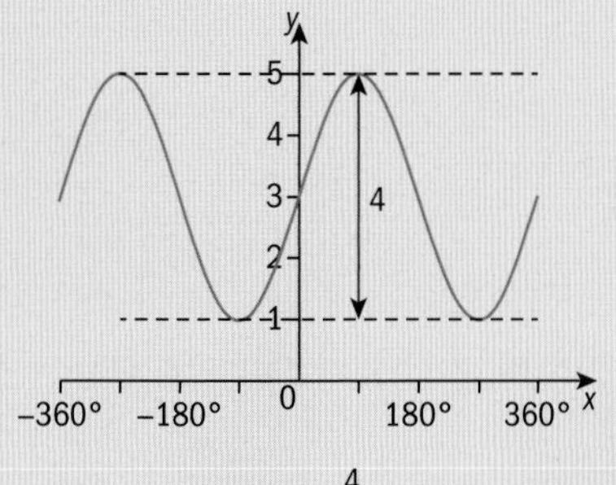

Amplitude = $\frac{4}{2}$ = 2

Example 1

Write down the function represented by this graph:

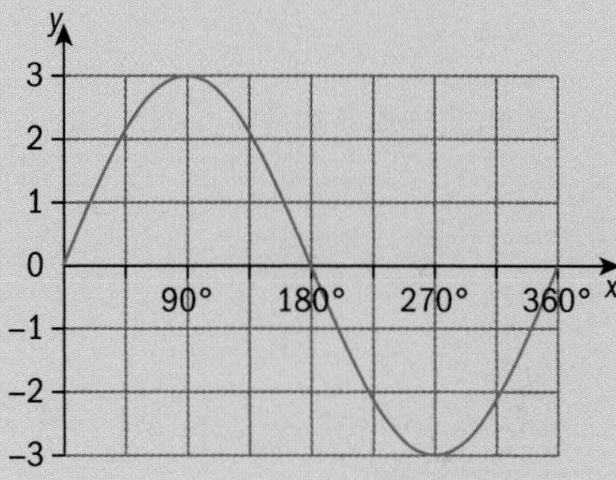

This is a sine graph. — Compare to the sine and cosine graphs. This passes through (0, 0).

The amplitude is 3. — $a = \frac{3-(-3)}{2} = 3$

$\Rightarrow y = 3\sin x$

Practice 1

1 Sketch the graph of these functions. Label all the x-intercepts on your sketch graphs.

a $y = 4\sin x$ **b** $y = 2\cos x$

Problem solving

2 Write down the function shown in each graph.

a

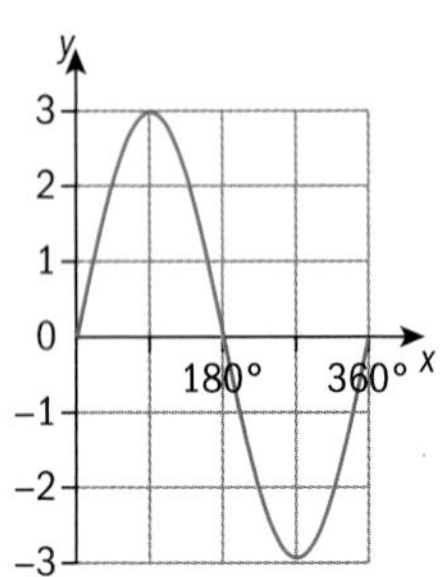

b

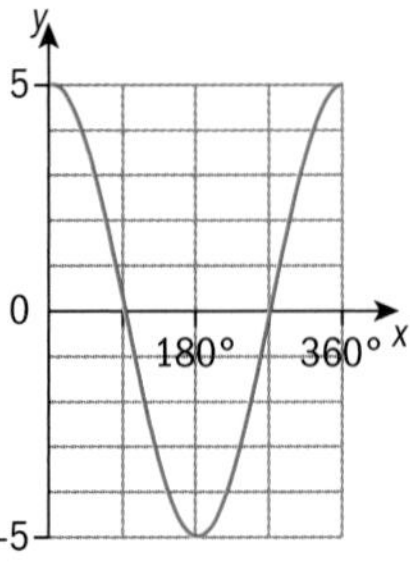

c

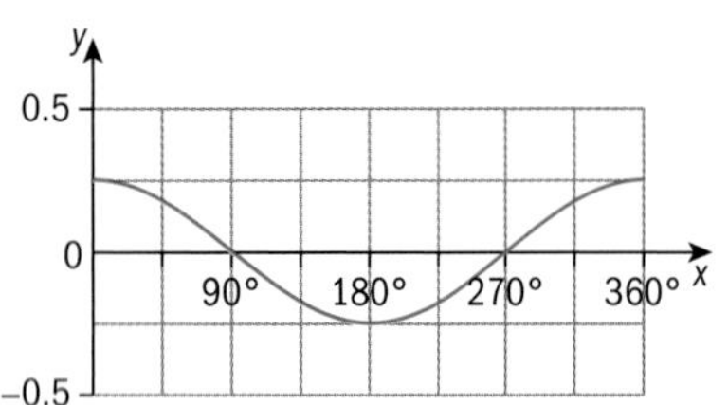

d

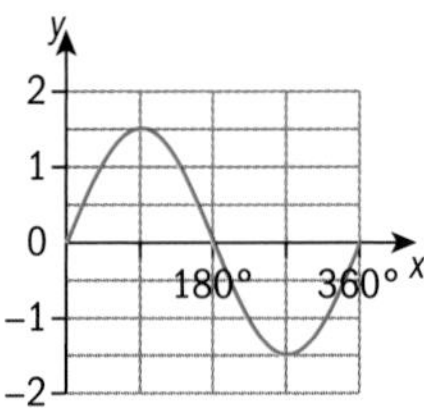

3 Four functions are shown at the right. When graphed, choose the function(s) that

a intersect the y-axis at (0, 2)
b intersect the x-axis at (180, 0) and (360, 0)
c intersect the x-axis at (90, 0) and (270, 0)
d has the smallest amplitude.

$y = 2\cos x$

$y = 0.5\sin x$

$y = 4\cos x$

$y = 2\sin x$

Exploration 3

1 Describe the effect on the graph of $y = f(x)$ when it is transformed to $y = f(2x)$.

2 On the same axes, sketch the graphs of the functions $y = \sin x$ and $y = \sin 2x$ for $0 \le x \le 360°$. Check with your GDC or with dynamic geometry software.

3 Add the graph of $y = 2\sin 2x$ to your sketch.

4 Write down the amplitude and period of each graph, in a table like this:

Graph	$y = \sin x$	$y = \sin 2x$	$y = 2\sin 2x$
Period			

5 Predict how the graph of $y = \sin x$ changes when it is transformed to $y = \sin \frac{1}{2}x$. Predict the period of the graph of $y = \sin \frac{1}{2}x$. Check with your GDC or geometry software.

Continued on next page

6 Add a column for $y = \sin\frac{1}{2}x$ to your table.

7 Repeat steps **1–6** for the graphs of $y = \cos x$, $y = \cos 2x$ and $y = \cos\frac{1}{2}x$.

8 Conjecture a rule for finding the period of a function $y = \sin bx$, $y = \cos bx$, $y = a\sin bx$ and $y = a\cos bx$, and explain why your rule makes sense.

Sine and cosine functions are periodic functions.

A **periodic function** repeats a pattern of y-values at regular intervals.

One complete repetition of the pattern is called a **cycle**.

The period of the functions $y = a\sin bx$ and $y = a\cos bx$ is $\frac{360°}{b}$.

The parameter b is the **frequency**, or number of cycles between 0° and 360°.

The graph $y = 3\sin 2x$ has period $\frac{360°}{2} = 180$

The frequency is 2 and the amplitude is 3.

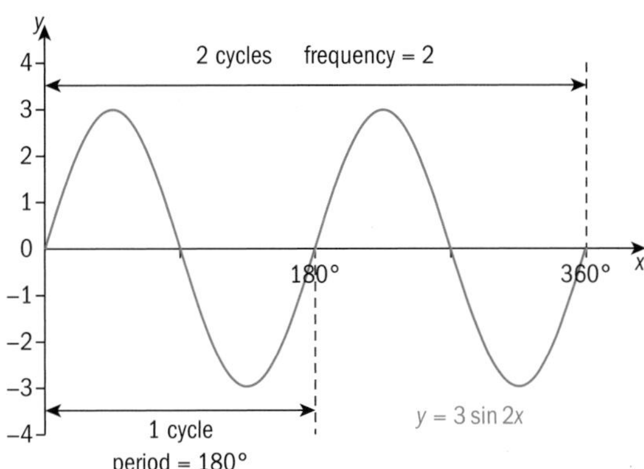

Practice 2

1 State the period and frequency of the graph of each function. Sketch the graph for the domain $0° \le x \le 360°$. Then check your graphs with a GDC or graphing software.

a $y = \sin 3x$ **b** $y = \cos 4x$

Problem solving

2 State the period and frequency of the graph of each function. Sketch the graph for the domain $0° \le x \le 720°$. Then check your graphs with a GDC or graphing software.

a $y = \sin\frac{1}{2}x$ **b** $y = \cos\frac{1}{3}x$

3 Write the equation of each graph.

a

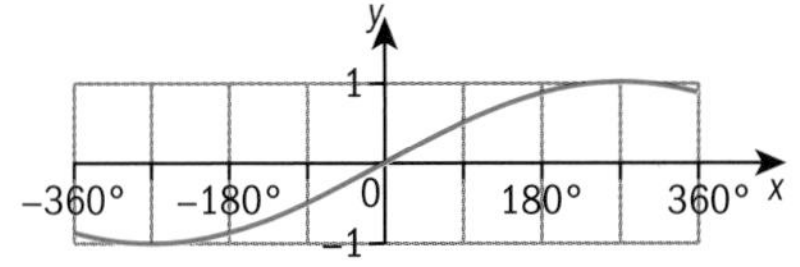

b

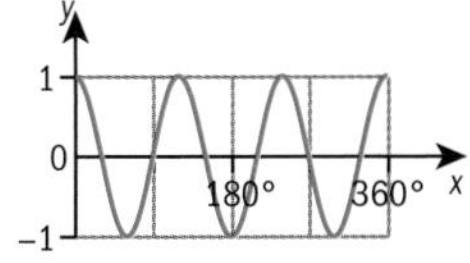

c

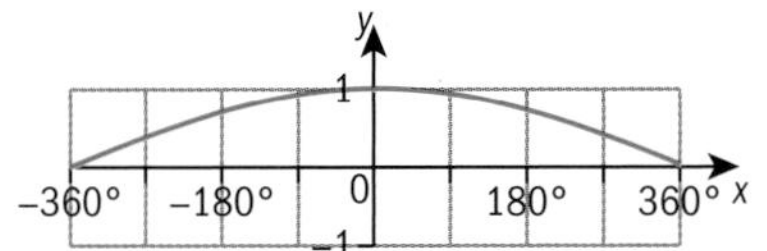

d

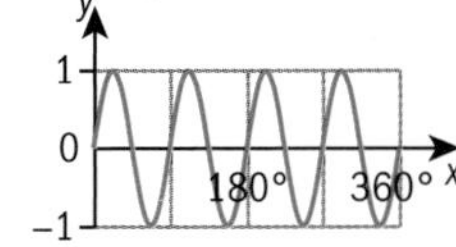

Problem solving

4 Match each equation to its graph.

$y\ 0.5\cos 3x$ $\quad$ $y\ 3\sin\frac{x}{2}$ $\quad$ $y\ 3\sin 2x$ $\quad$ $y\ 0.5\cos 0.5x$

a

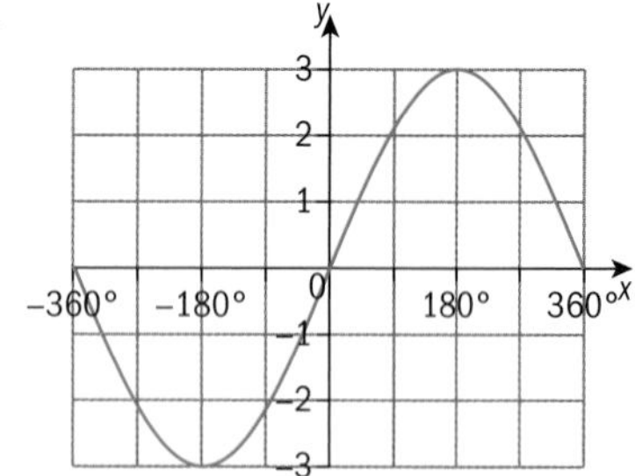

b

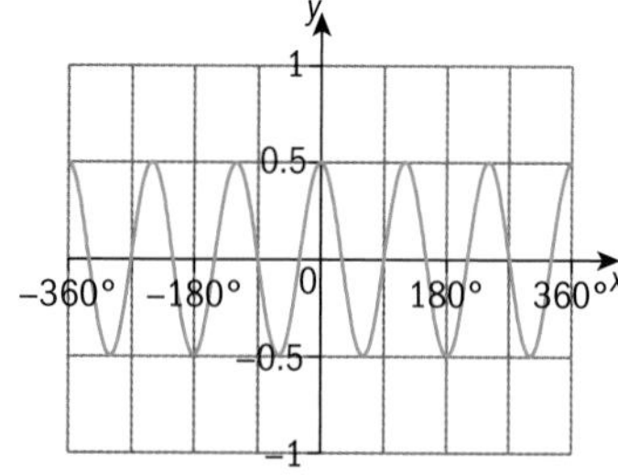

c

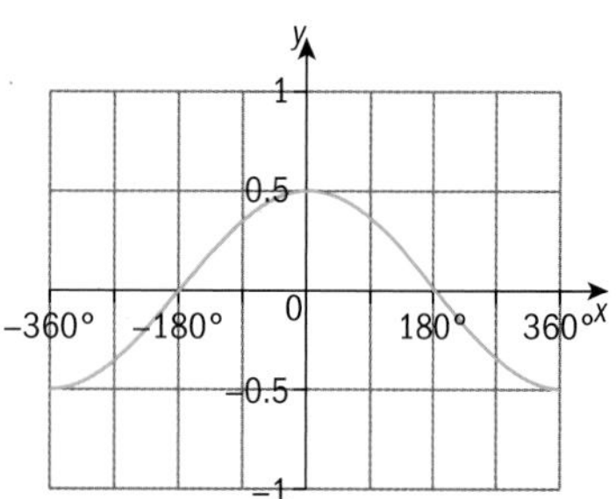

d

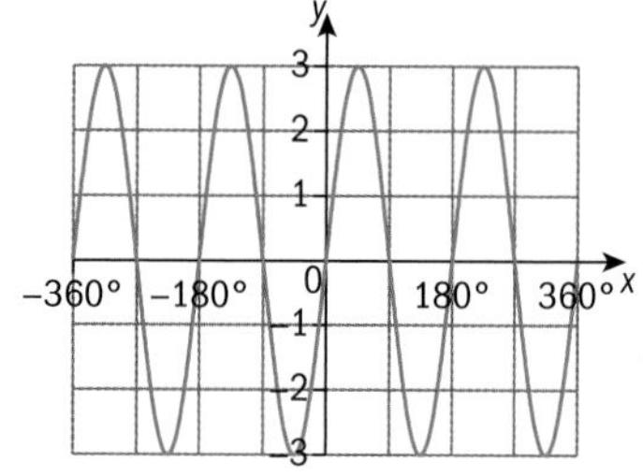

Example 2

Sketch the graph of each function on the domain given.

a $y = 0.5\sin 3x$, for $0° \leq x \leq 360°$

b $y = 3\cos 0.5x$, for $0° \leq x \leq 720°$

a amplitude $= 0.5$

frequency $= 3$

period $= \frac{360°}{3} = 120°$

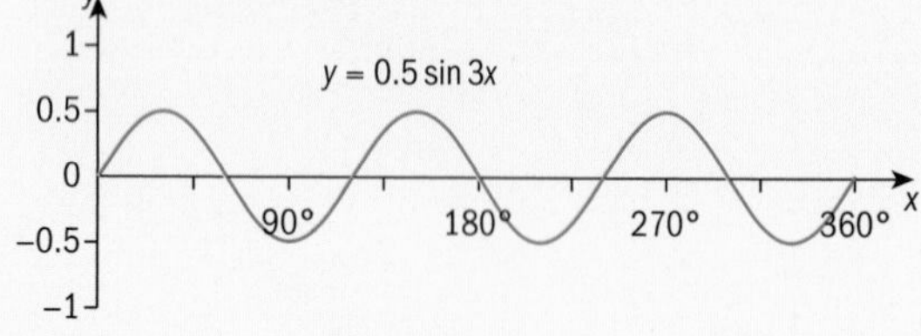

State the amplitude, frequency and period.

Draw axes for the domain.

The maximum and minimum *y* values are 0.5 and −0.5 respectively.

Frequency = 3, so there are 3 complete sine curves between 0° and 360°.

Period = 120°, so there is a complete sine curve between 0° and 120°.

Sine curve passes through (0, 0).

▶ Continued on next page

b amplitude = 3

frequency = 0.5

period = $\frac{360°}{0.5} = 720°$

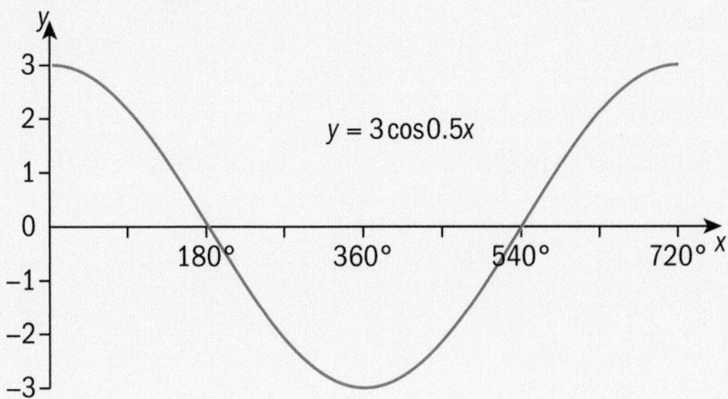

Frequency = 0.5, so there is half a complete cosine curve between 0° and 360°.

Period = 720° so there is a complete cosine curve between 0° and 720°.

Cosine curve has a maximum at $x = 0°$ and at 720°.

Practice 3

1 Write down the amplitude, frequency and period of each function, then sketch the graph for the given domain. Check your graphs on a GDC or graphing software.

a $y = 2\sin 4x$, for $0° \le x \le 360°$ **b** $y = 4\cos 3x$, for $0° \le x \le 360°$

c $y = \frac{1}{3}\sin\frac{x}{3}$, for $0° \le x \le 1080°$ **d** $y = 0.7\cos\frac{x}{4}$, for $0° \le x \le 1440°$

C Transformations of periodic functions

- How do you translate, dilate and reflect periodic functions?
- How does the equation of a sinusoidal function represent the transformations performed on it?
- Does the order in which transformations are performed matter?

A **sinusoidal function** is any function that can be obtained from a transformation of the sine function. For example, the graph of $y = \cos x$ can be obtained by translating the graph of $y = \sin x$ by 90° to the left. So $y = \cos x$ is a sinusoidal function.

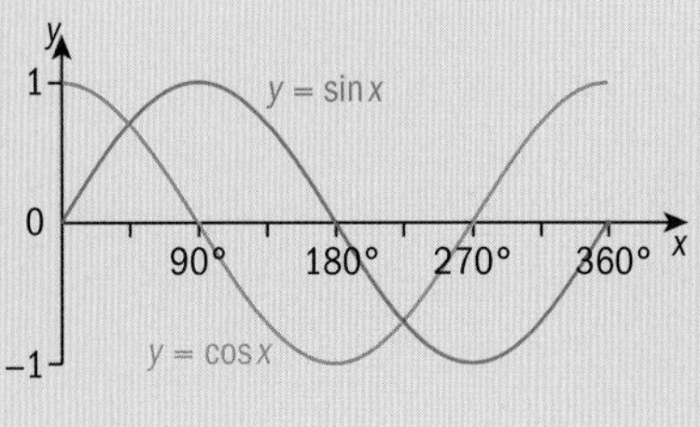

Exploration 4

Transforming $f(x)$ to $f(x) + d$

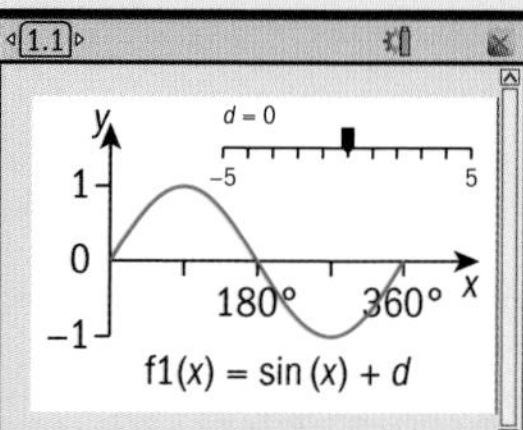

1. Use what you know about transformations of graphs from $f(x)$ to $f(x) + d$ to predict how the transformation of $y = \sin x$ to $y = \sin x + d$ will affect the graph.
2. Graph the function $y = \sin x + d$ on the domain $0° \le x \le 360°$ for different values of d. Investigate how changing the value of d affects the graph. Was your prediction from step **1** correct?
3. Verify that the amplitude and period of the function $y = a\sin bx + d$ is not affected by the value of d.
4. Repeat steps **1** to **3** for the graph of $y = \cos x + d$.

Transforming $f(x)$ to $-f(x)$

5. Use what you know about transformations of graphs from $f(x)$ to $-f(x)$ to predict how the transformation of $y = \sin x$ to $y = -\sin x$ will affect the graph. Graph the functions $y = \sin x$ and $y = -\sin x$ on the domain $0° \le x \le 360°$ to check your prediction.
6. Verify that $y = a\sin bx$ and $y = -a\sin bx$ have the same amplitude and period.
7. Repeat steps **5** and **6** for the graphs of $y = \cos x$ and $y = -\cos x$.

Transforming $f(x)$ to $f(-x)$

8. Use what you know about transformations of graphs from $f(x)$ to $f(-x)$ to predict how the transformation of $y = \sin x$ to $y = \sin(-x)$ will affect the graph. Graph the functions $y = \sin x$ and $y = \sin(-x)$ on the domain $-360° \le x \le 360°$ to check your prediction.
9. Verify that $y = a\sin bx$ and $y = a\sin(-bx)$ have the same amplitude and period.
10. Repeat steps **8** and **9** for the graphs of $y = \cos x$ and $y = \cos(-x)$.

Transformations of sine and cosine functions

Reflection: For the sinusoidal functions $f(x) = \sin x$ and $f(x) = \cos x$:

The graph of $y = -f(x)$ is a reflection of the graph of $y = f(x)$ in the x-axis.

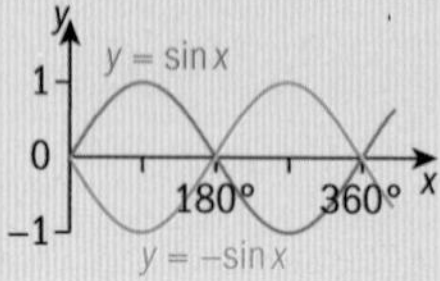

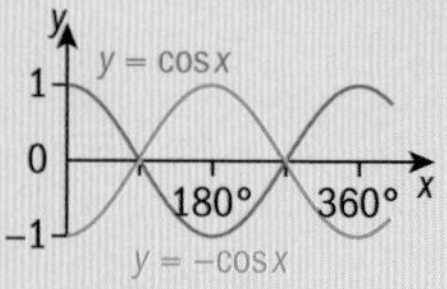

The graph of $y = f(-x)$ is a reflection of the graph of $y = f(x)$ in the y-axis.

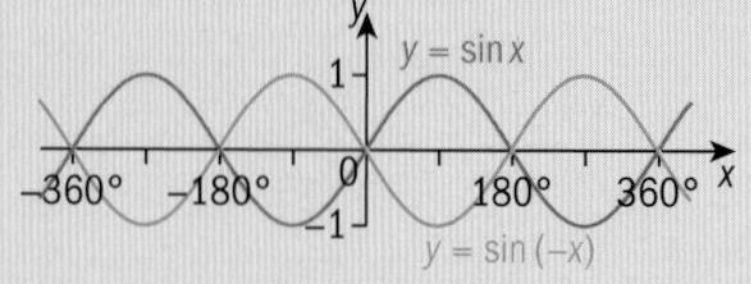

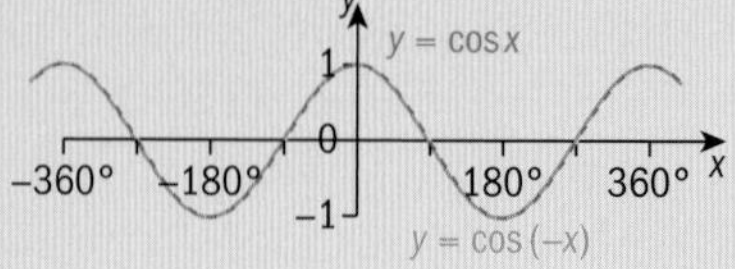

▶ Continued on next page

Translation: For the sinusoidal functions $f(x) = \sin x$ and $f(x) = \cos x$, $y = f(x) + d$ translates $y = f(x)$ by d units in the y-direction.

When $d > 0$, the graph moves in the positive y-direction (up).

When $d < 0$, the graph moves in the negative y-direction (down).

Dilation: For the sinusoidal functions $f(x) = a \sin bx$ and $f(x) = a \cos bx$:

$y = af(x)$ is a vertical dilation of $f(x)$, scale factor a, parallel to the y-axis. a is the amplitude.

$y = f(bx)$ is a horizontal dilation of $f(x)$, scale factor $\frac{1}{b}$, parallel to the x-axis. b is the frequency.

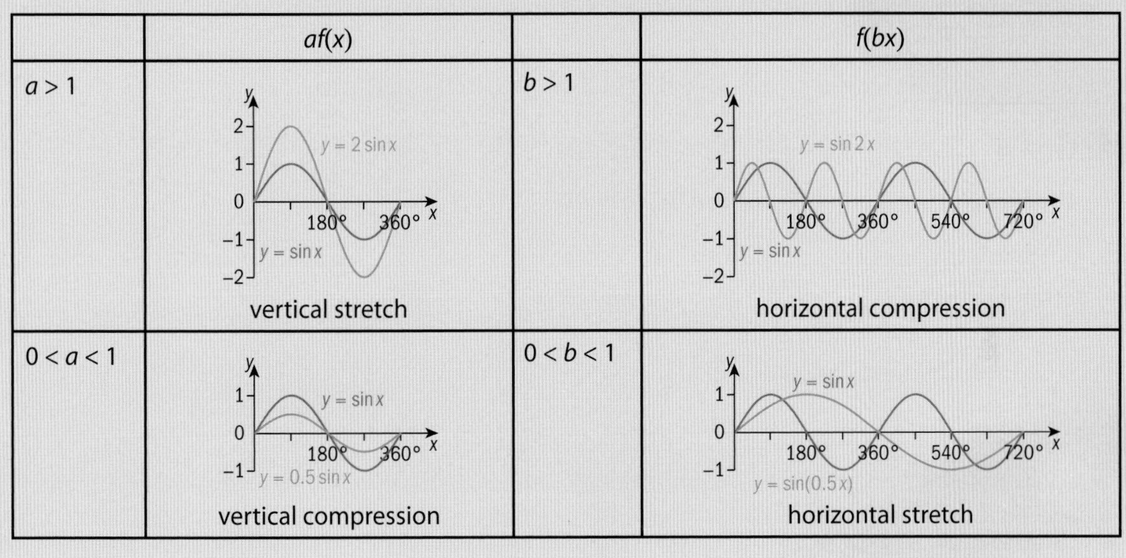

	$af(x)$		$f(bx)$
$a > 1$	vertical stretch	$b > 1$	horizontal compression
$0 < a < 1$	vertical compression	$0 < b < 1$	horizontal stretch

While the graphs of $y = \sin x$ and $y = \cos x$ share many similarities, they are easy to recognize by looking at a specific point on the graph. Exploration 5 will help you find that point.

ATL

Exploration 5

1 Sketch the graphs of $y = \sin x$ and $y = \cos x$.

2 In Reflect and Discuss 1 you informally noted differences between the sine and cosine graphs. Now, using proper mathematical terminology, state the similarities and differences between the sine and cosine graphs.

You can recognize the graphs of $y = \sin x$ and $y = \cos x$ from their values at $x = 0$.

3 Use your GDC to graph several cosine functions with equations of the form $y = a \cos(bx) + d$. Generalize how to find its y-intercept from the equation.

4 Repeat step **3** with the function $y = a \sin(bx) + d$.

The y-intercept of a function is the y-value when $x = 0$. Specifically, for $y = a\sin bx + d$, the y-intercept is the average value of the function: $\frac{y_{max} + y_{min}}{2}$

For $y = a\cos(bx) + d$, the y-intercept is the maximum value of y.

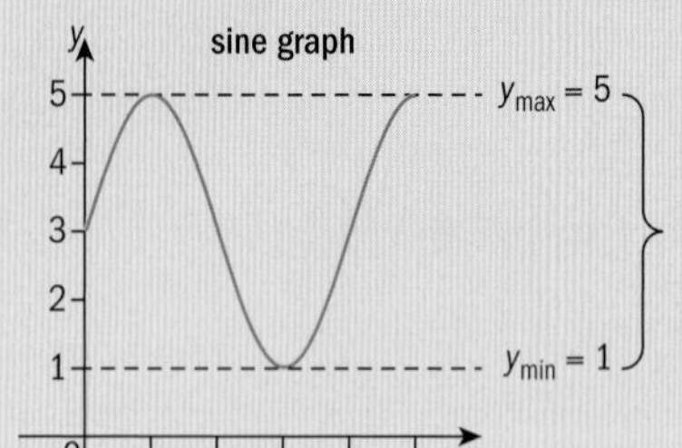

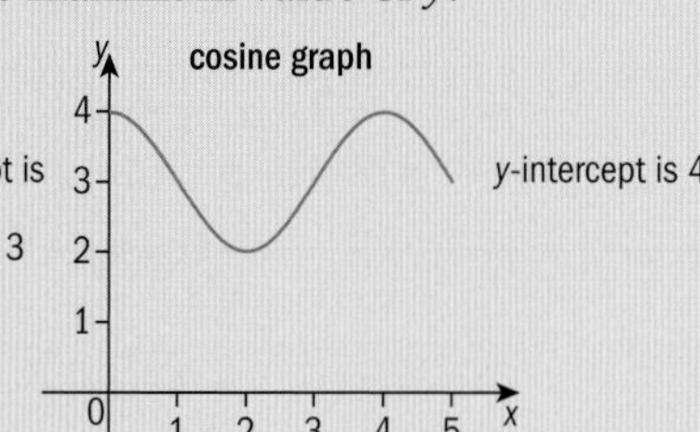

Practice 4

1 Describe how to transform the graph of $y = \cos x$ to:

a $y = \cos x + 7$ **b** $y = -\cos x$

c $y = \cos x - 2$ **d** $y = \cos(-x)$

2 Identify the transformation of each graph from either $y = \sin x$ or $y = \cos x$. Then, write down the equation of each graph.

a

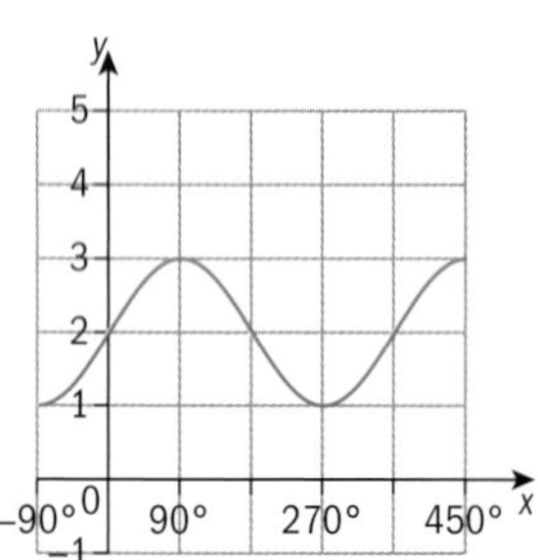

b

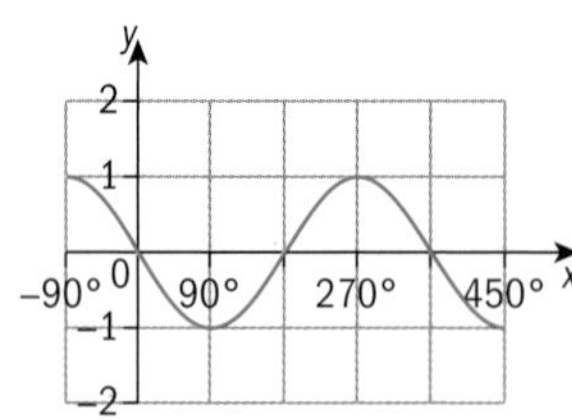

c

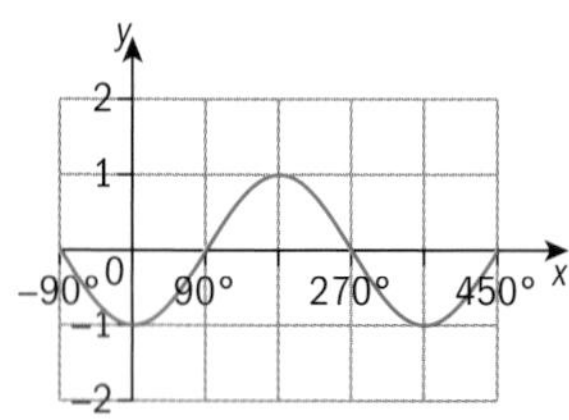

d

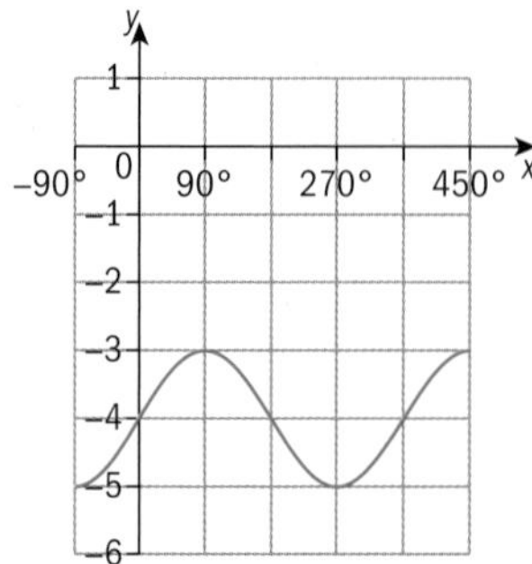

3 Draw the graph of each function for $0° < x \leq 360°$. Be sure to clearly indicate on your graph the amplitude, the y-intercept and the period.

a $y = \sin x - 2$ **b** $y = -\cos x$

c $y = \cos x + 11$ **d** $y = -\sin x$

e $y = \cos x - 8$ **f** $y = \sin x + 1$

4 The graph of $y = \sin x$ was transformed to produce the following graph:

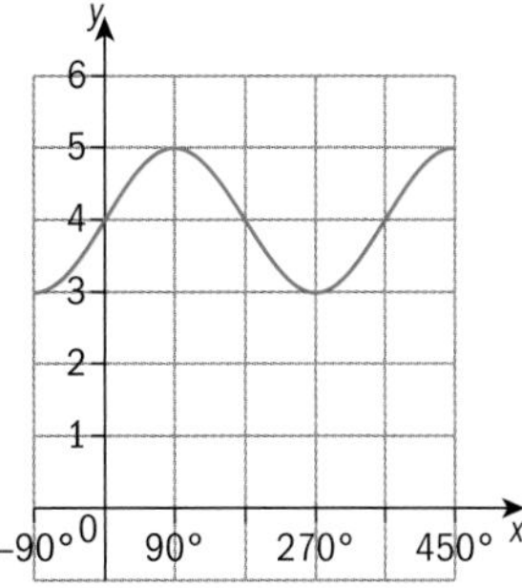

a Find the amplitude, frequency and period of the function.

b Write down the equation of the graph.

c Dennis wants to transform the graph so that its minimum value is $y = 0$. Describe the transformation that will allow him to accomplish this.

d Write down the equation for Dennis' graph.

Problem solving

5 Jawad and Aaron disagree about the equation of the following graph:

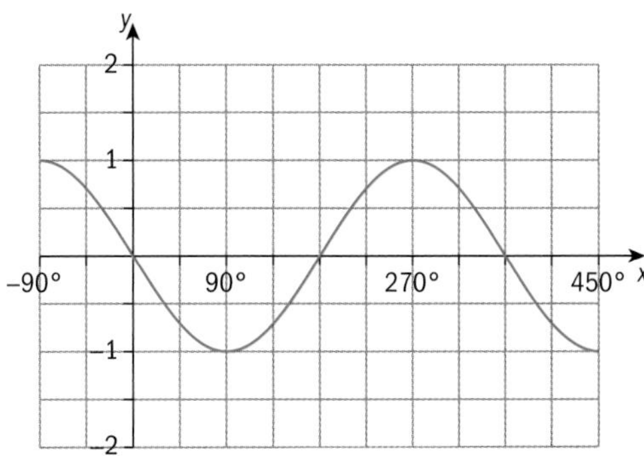

Aaron says the equation is $y = -\sin x$ while Jawad thinks it is $y = \sin(-x)$. State who is correct, and justify your answer.

Example 3

By describing the transformations on the sinusoidal graph, determine the function for the graph.

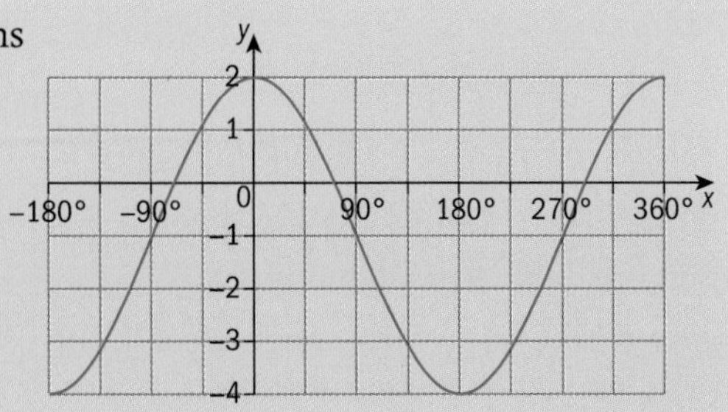

This is a cosine curve, form: $y = a\cos(bx) + d$ — Cosine curve has maximum at $x = 0$.

The amplitude, a, is half the distance between the max and min points: $a = \frac{1}{2} \times 6 = 3$ — Find the amplitude a and frequency b.

The frequency, b, is the number of cycles between 0° and 360° = 1

The amplitude is 3 but the maximum value is 2, so the graph is translated down 1 unit. Hence $d = -1$ — Since the amplitude is 3, the graph has shifted 1 unit downward, hence $d = -1$.

$y = 3\cos x - 1$ — Substitute your values for a, b and d into $y = a\cos(bx) + d$.

Example 4

Write a function that describes this sinusoidal graph.

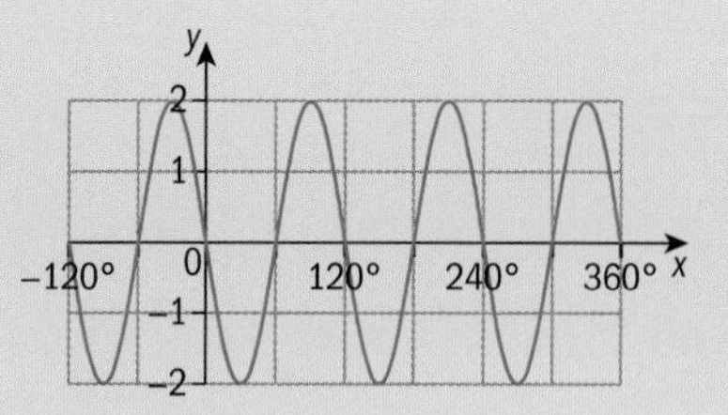

> For the sine function, a reflection in the x-axis gives the same graph as a reflection in the y-axis.

This is a sine curve, reflected in either the x-axis or y-axis.

> This is a sine curve because at $x = 0$ it has average value.

The amplitude, $a = \frac{4}{2} = 2$

There are 3 cycles from 0° to 360°, so $b = 3$.

y-intercept $c = 0$

$\Rightarrow y = -2\sin(3x)$ or $y = 2\sin(-3x)$

> Graph of $y = \sin(x)$ increases from 0° to 90° but this graph decreases over that range.

ATL

Exploration 6

1 a Determine which two transformations transform $y = \sin x$ to $y = 3\sin 2x$.

  b Start with the graph of $y = \sin x$ and apply the two transformations.

  c Start with the graph of $y = \sin x$ and apply the two transformations in the opposite order.

  d Do you get the same graph each time?

2 Repeat step **1** for the graphs of:

  a $y = -3\sin 2x$

  b $y = 3\sin(-2x)$

  c $y = -3\sin x$.

Does the order you carry out these transformations affect the graph?

3 a Repeat step **1** for the graph of $y = 3\cos x + 1$ and $y = 3\sin x + 1$.

  b Write down the equation of all the graphs you produce.

4 Summarize your findings by stating when the order you carry out transformations is important, and when it doesn't matter.

Practice 5

1 Describe how to transform the graph of $y \ \sin x$ to:

a $y = -4\sin x$	**b** $y = 2\sin x - 9$	**c** $y = 0.5\sin x + 4$
d $y = \sin\left(\frac{1}{2}x\right) + 3$	**e** $y = -\sin\left(\frac{1}{3}x\right)$	**f** $y = \sin(-2x) + 1$

2 Describe how to transform the graph of $y \ \cos x$ to:

a $y = -3\cos x$	**b** $y = 2\cos x + 1$	**c** $y = \cos\left(\frac{1}{2}x\right) + 2$
d $y = 0.5\cos x - 1$	**e** $y = -\cos 3x$	**f** $y = \cos\left(-\frac{1}{2}x\right) - 3$

Problem solving

3 Each graph is a transformation of either $y = \sin x$ or $y = \cos x$.
For each graph, identify the transformations on the graph and write the function that describes the graph.

a

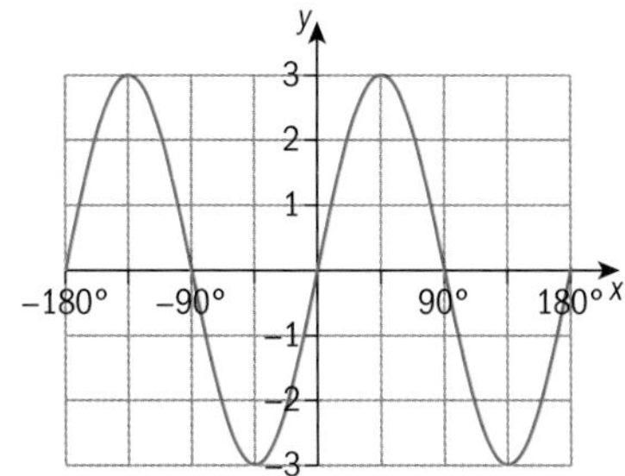

b

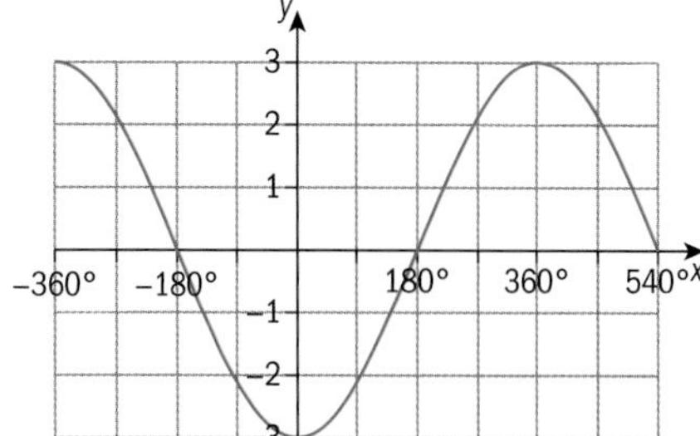

c

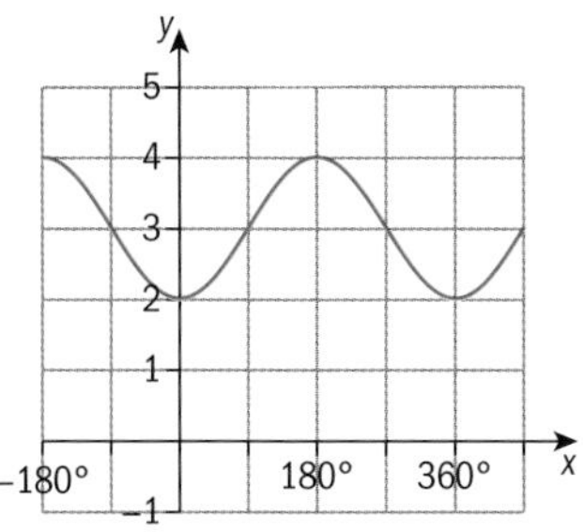

d

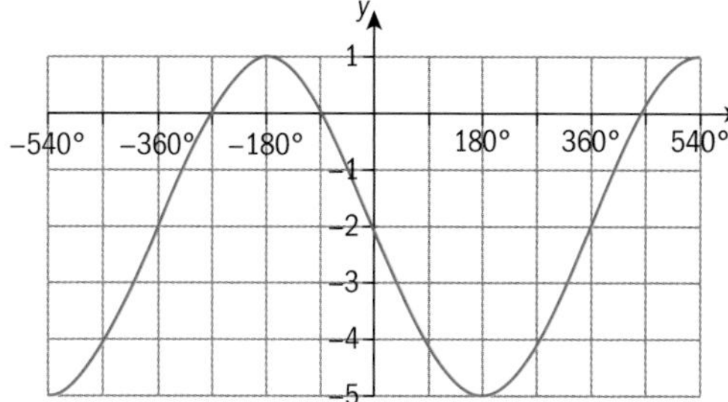

e

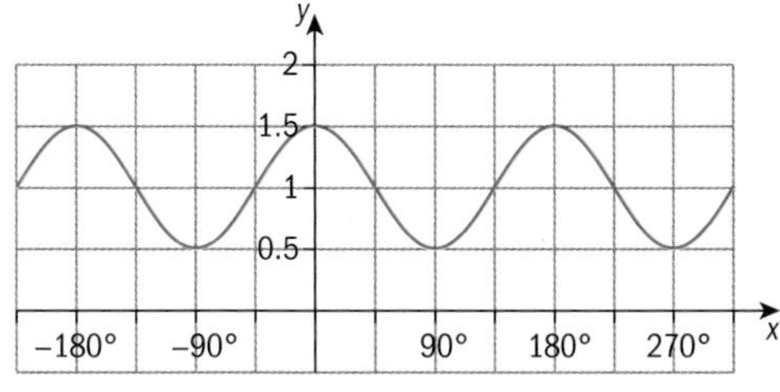

f

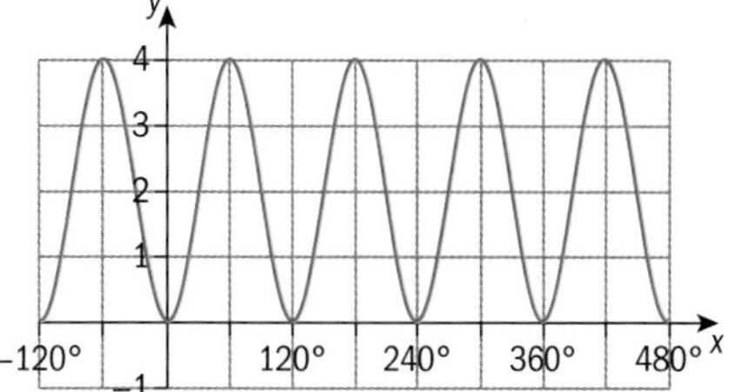

g

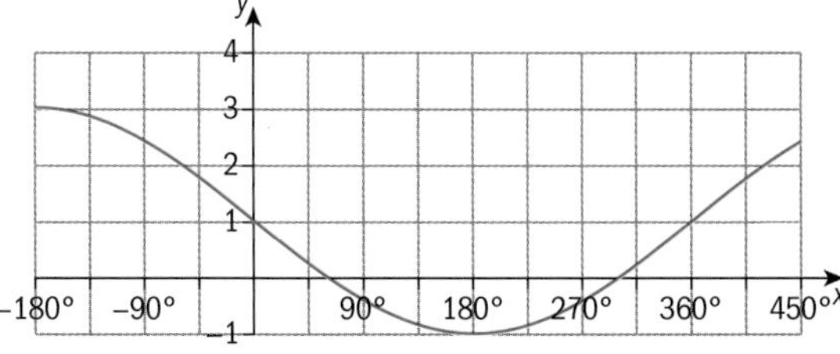

h

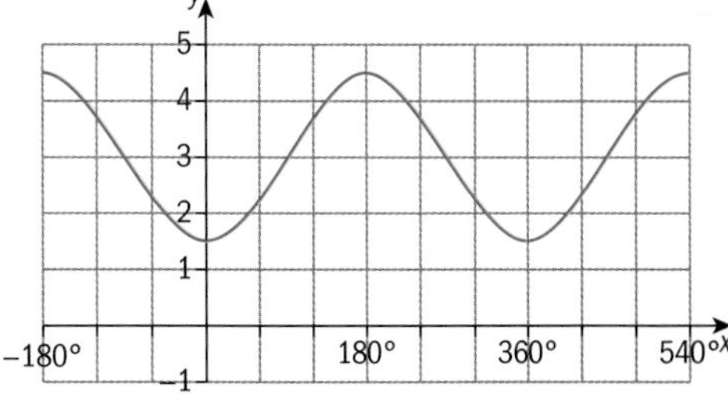

D Modelling real-life situations

- What real-life phenomena are periodic?
- Have scientific models and methods provided more answers or questions?

In the sinusoidal graphs you have worked with so far, the x-axis has shown degrees, and the y-axis shows the sine or cosine function for each angle. In sinusoidal curves representing real-life situations, the x and y axes may show other units, such as time on the x-axis and height on the y-axis.

The graph shows the distance (in meters) of a wrecking ball from its original position as it swings left and then right.

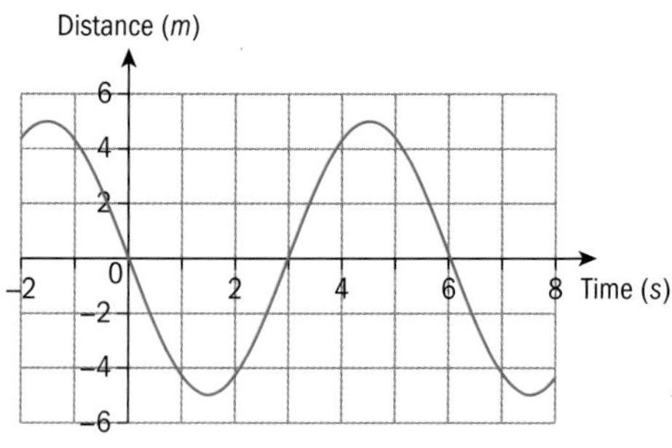

From the graph:

- The amplitude is 5, so the maximum distance the wrecking ball reaches from its original position is 5 m.
- The negative y-values represent the distance when it swings to the left, and the positive values represent the distance when it swings to the right.
- The period is 6 seconds (the time needed for one cycle).

To find the equation of the graph:

The graph is a sine curve. It has been reflected in the x- or y-axis.

Frequency $b = \frac{360°}{\text{period}} = \frac{360°}{6} = 60°$

Amplitude $a = 5$

$c = 0$

So, $y = 5\sin(-60°x)$ or $y = -5\sin 60°x$

Tip

When graphing sinusoidal curves with degrees in the function, make sure to type the degree symbol on the $y =$ line.

Exploration 7

A Ferris wheel has 18 seats arranged every 20°. Its diameter is 75 m and the lowest seat is 2 m above the ground. As you ride the Ferris wheel your height above the ground changes continuously. Draw an accurate scale drawing of the wheel. Determine the kind of function that describes its motion.

1 Suppose you get on at the bottom and the wheel rotates so that it takes 5 seconds between positions of the seats. Use a ruler to determine your height above the ground as the wheel passes through each position. Record your results in a table. Graph height h (m) above the ground against time t (sec) for $0 \le t \le 180$.

2 Find the amplitude, period and frequency of your graph.

3 Use your results to write down the equation of your graph.

4 Identify the transformations of the graph $y = \cos x$.

5 Determine the changes that would need to be made to the Ferris wheel in order for the relationship to be described by transformations of $y = \sin x$.

Practice 6

1 After you exercise, the velocity of air flow (in liters per second) into your lungs can be modelled by the equation $y = 2\sin 90°t$.

a Find the velocity of air flow at:

i 1 s ii 2.5 s iii 4 s

b Explain what you think negative velocities represent.

c Find the period of the function, and what it represents.

d Find the amplitude and maximum velocity.

e Draw the graph of the function in the first 8 seconds.

2 The graph here shows how a buoy's distance from the ocean floor changes with the waves. Find a sinusoidal function that models the graph.

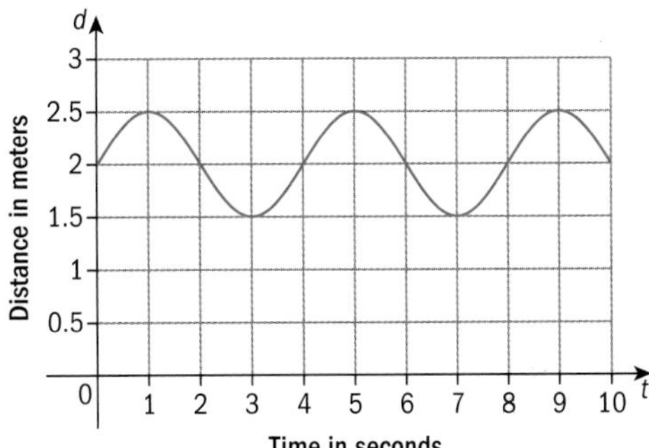

Problem solving

3 Find a sinusoidal function that models the relationship shown in each graph.

a

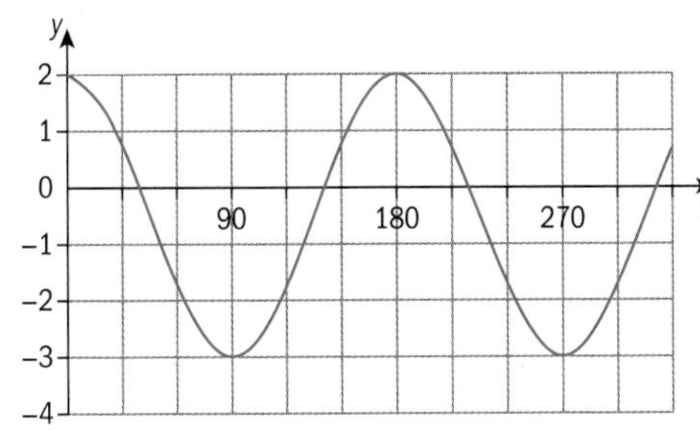

b

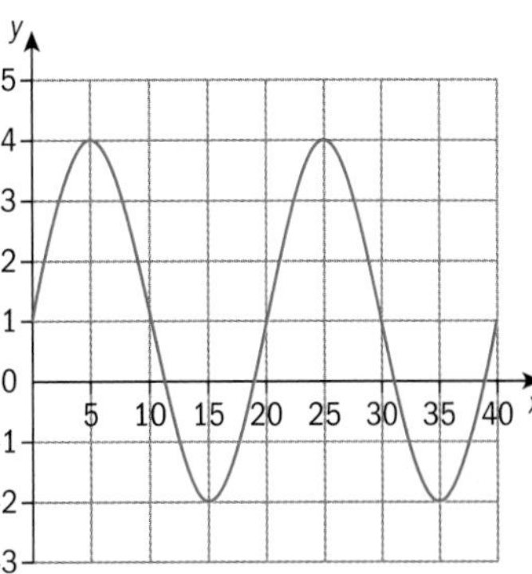

c

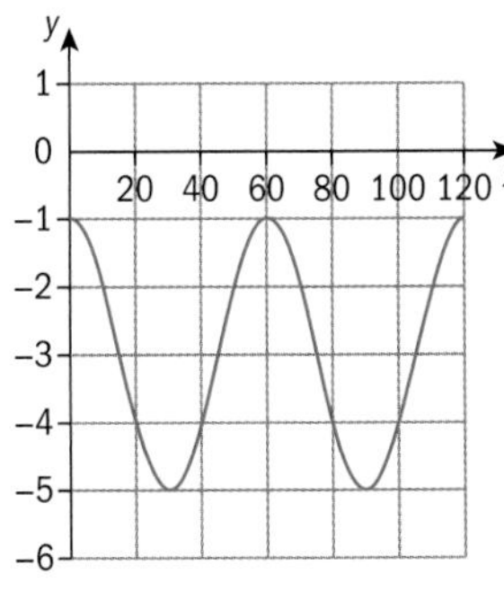

d

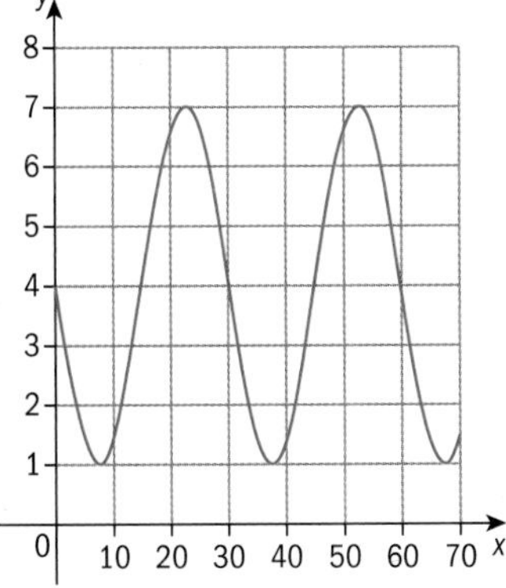

4 The height in meters (from its position at rest) of a spring as it bounces up and down can be modelled by the function $h(t) = 2\sin(60°t) + 1$, where t is measured in seconds.

a Find the amplitude of $h(t)$.

b Sketch the graph of the function.

c Determine if the amplitude is the same as the maximum height from its position at rest.

5 When you ride a Ferris wheel, your vertical height above the ground changes. The relationship between your height above the ground and the time for a complete revolution of the wheel can be modeled with a sinusoidal graph like this:

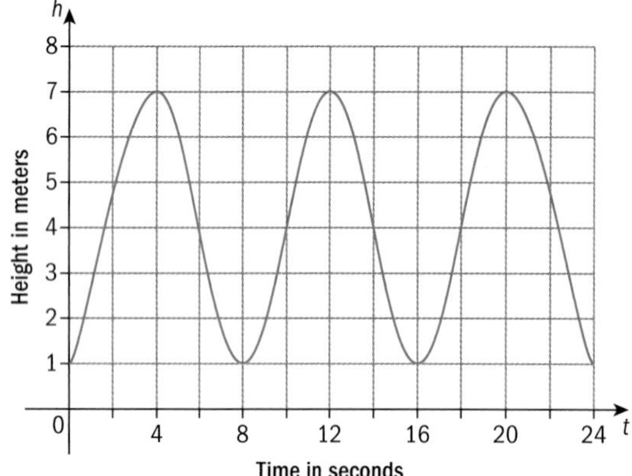

a State the height at which you board the Ferris wheel.

b State the maximum height you reach on the Ferris wheel.

c Determine how long it takes to rise to approximately 5 m above the ground for the first time.

d Determine your height above the ground after 8 seconds.

e Determine how long it takes for one complete revolution of the wheel.

f The Ferris wheel is circular. Determine its radius.

g Find a function that models this graph.

6 The height in meters of the tide above the mean sea level one day at Bal Harbor can be modeled by the function $h(t) = 3\sin(30°t)$, where t is the number of hours after midnight.

a Find h when $t = 0$, 6, 12, 18 and 24 hours.

b Draw the graph of the function for a 24 hour cycle.

c Determine the time of high tides, and their maximum heights.

d Determine the height of the tide at 4 o'clock in the afternoon.

e A ship can cross the harbor if the tide is at least 2 m above the average sea level. Determine the times when it can cross the harbor.

Problem solving

7 One day, high tide in Venice, Italy was at midnight. The water level at high tide was 3.3 m, and later at low tide the water level was 0.1 m. Assume that the next high tide is 12 hours later and that the height of the water level can be modelled with a sinusoidal function.

a Find a function that models this situation.

b Draw a graph of your function.

Summary

The **amplitude** is the height from the mean value of the function to its maximum or minimum value. The graphs of $y = \sin x$ and $y = \cos x$ have amplitude 1.

The **period** is the horizontal length of one complete cycle. The graphs of $y = \sin x$ and $y = \cos x$ have period 360°.

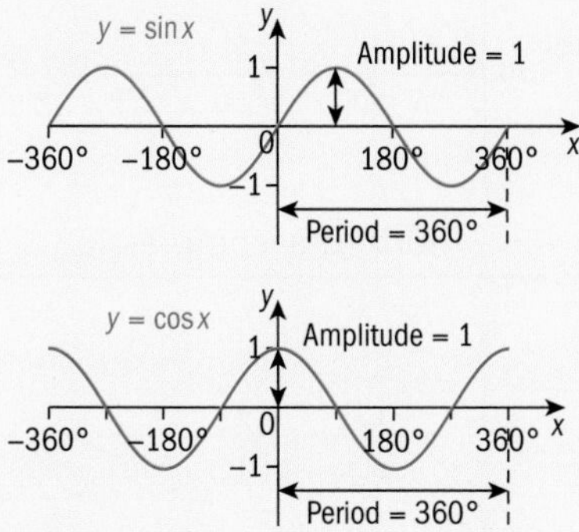

The graph of $y = a \sin x$ has amplitude a.
The graph of $y = a \cos x$ has amplitude a.

A periodic function repeats a pattern of y-values at regular intervals.

The period of the functions $y = a \sin bx$ and $y = a \cos bx$ is $\frac{360°}{b}$

$$\text{Period} = \frac{360°}{\text{frequency}}$$

One complete repetition of the pattern is called a **cycle**. The parameter b is the **frequency**, or number of cycles between 0° and 360°.

The amplitude is $(y_{max} - y_{min}) / 2$.

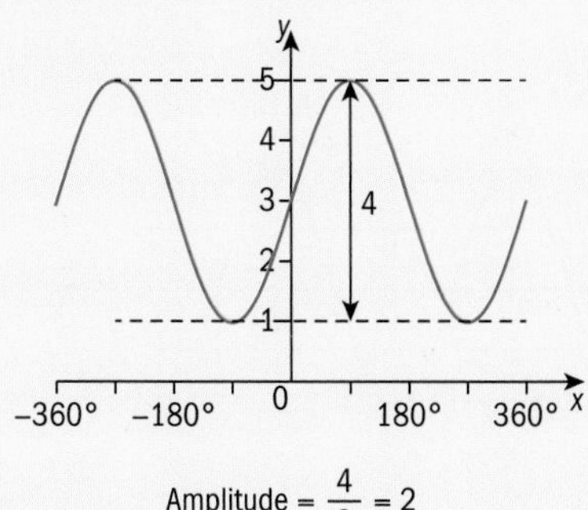

A **sinusoidal function** is any function that can be obtained from a transformation of the sine function. The graph of $y = \cos x$ can be obtained by translating the graph of $y = \sin x$ 90° to the left. So $y = \cos x$ is a sinusoidal function.

Reflection: For the sinusoidal functions $f(x) = \sin x$ and $f(x) = \cos x$.

The graph of $y = -f(x)$ is a reflection of the graph of $y = f(x)$ in the x-axis.

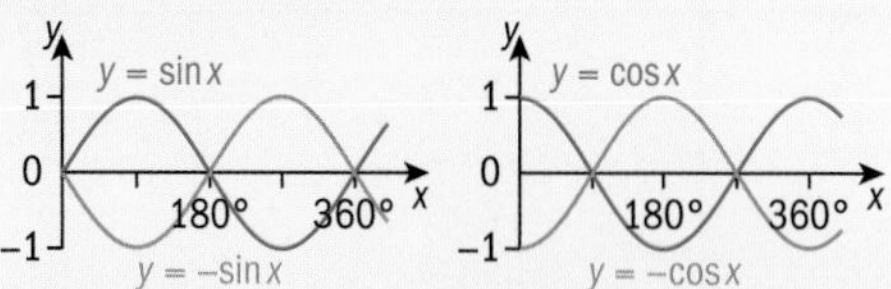

The graph of $y = f(-x)$ is a reflection of the graph of $y = f(x)$ in the y-axis.

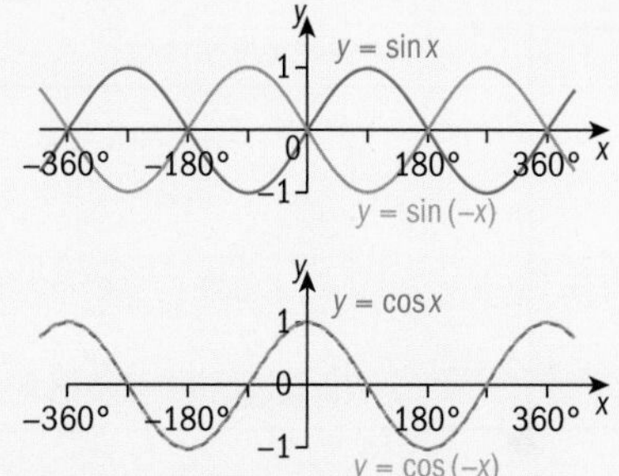

Translation: For the sinusoidal functions $f(x) = \sin x$ and $f(x) = \cos x$:

- $y = f(x) + d$ translates $y = f(x)$ by d units in the y-direction.
  When $d > 0$, the graph moves in the positive y-direction, up
  When $d < 0$, the graph moves in the negative y-direction, down.

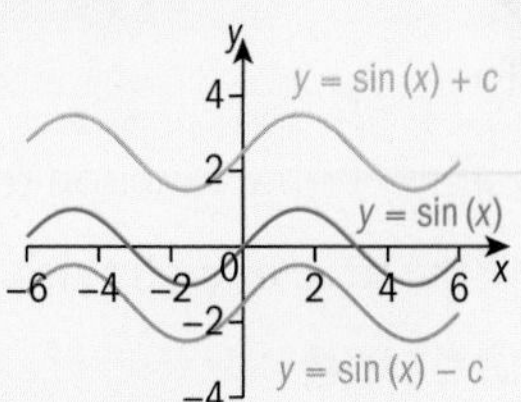

- $y = f(x - h)$ translates $y = f(x)$ by h units in the x-direction.
  When $h > 0$, the graph moves in the positive x-direction, to the right.
  When $h < 0$, the graph moves in the negative x-direction, to the left.

> Horizontal translation of cosine and sine graphs is an extended topic, but here you can see that the rules are the same as for translations of all functions $f(x)$.

Dilation: For the sinusoidal functions $f(x) = a \sin bx$ and $f(x) = a \cos bx$:

$y = af(x)$ is a vertical dilation of $f(x)$, scale factor a, parallel to the y-axis. a is the amplitude.

$y = f(bx)$ is a horizontal dilation of $f(x)$, scale factor $\frac{1}{b}$, parallel to the x-axis. b is the frequency.

	$af(x)$		$f(bx)$
$a > 1$	$y = 2\sin x$, $y = \sin x$ vertical stretch	$b > 1$	$y = \sin 2x$, $y = \sin x$ horizontal compression
$0 < a < 1$	$y = \sin x$, $y = 0.5\sin x$ vertical compression	$0 < b < 1$	$y = \sin x$, $y = \sin(0.5x)$ horizontal stretch

For $y = a\sin bx + d$, the y-intercept is the average value of the function: $\frac{y_{max} + y_{max}}{2}$

sine graph

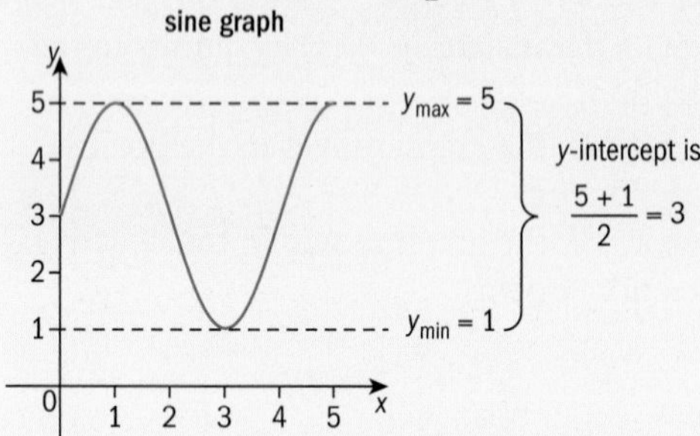

For $y = a\cos(bx) + d$, the y-intercept is the maximum value of y.

cosine graph

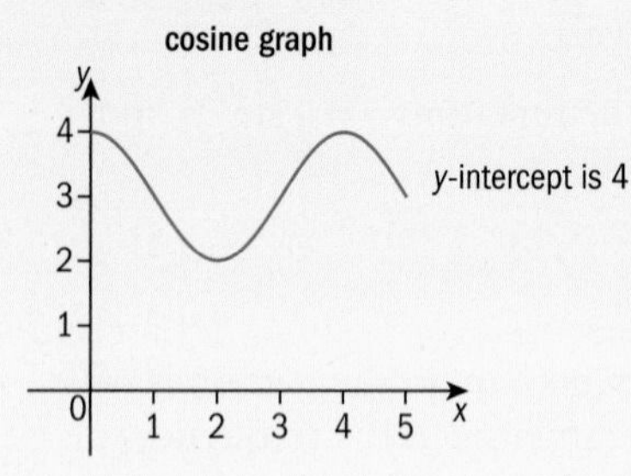

Mixed practice

1 For each of the following graphs, **find** the amplitude, frequency and period. Then, **write down** the equation the represents the function.

a

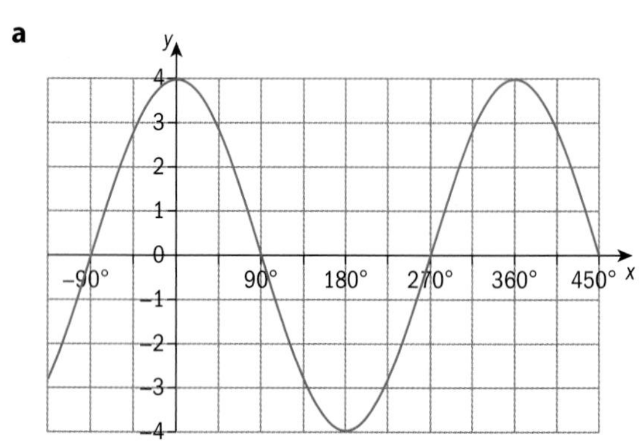

b

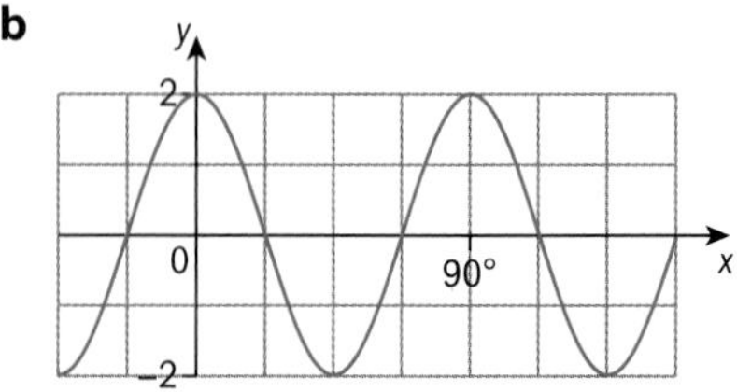

c

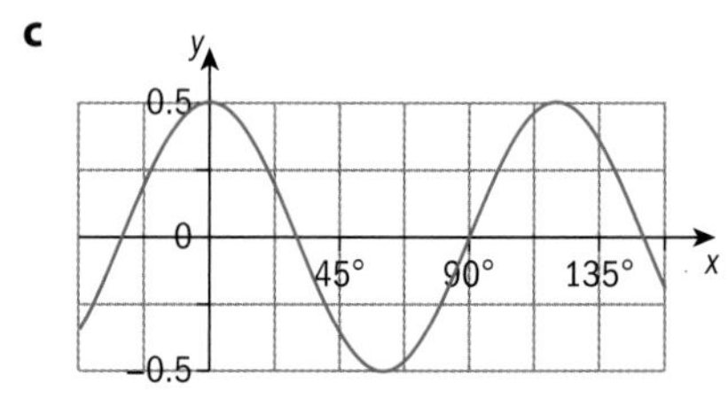

2 **Draw** the graph of each function showing at least two cycles. Be sure to clearly indicate the amplitude and period.

a $y = 0.4\sin(2x)$ **b** $y = 7\cos(5x)$

c $y = \frac{2}{3}\cos(8x)$ **d** $y = \frac{5}{2}\sin\left(\frac{-x}{3}\right)$

3 **State** the amplitude, frequency and period of each graph. **Sketch** the graph for $-360° \leq x \leq 360°$.

a $y = 4\cos(3x)$

b $y = -\sin(0.5x)$

c $y = 0.5\sin(2x) + 3$

d $y = 3\cos(0.5x) - 1$

4 This graph is a result of a transformation on $y = \sin x$ or $y = \cos x$.

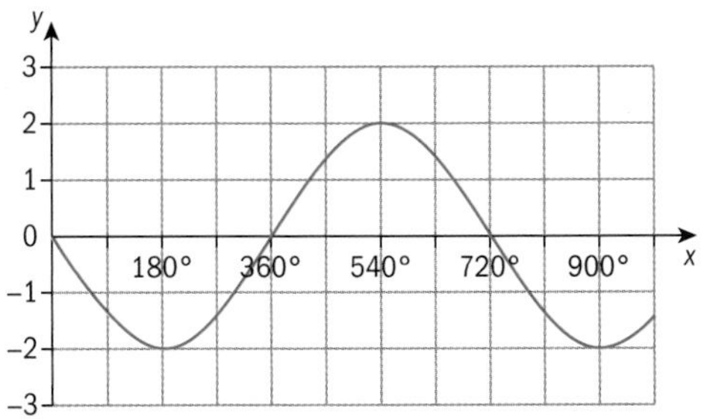

Identify the transformation, and determine the function of the transformed graph.

5 **Sketch** the graph of each of the following, indicating the amplitude and period, and showing at least two cycles:

a $y = \sin x + 12$ **b** $y = \sin(-x)$

c $y = \cos x - 1$ **d** $y = \cos x + 6$

6 **Find** the amplitude, period and equation of the following graphs:

a

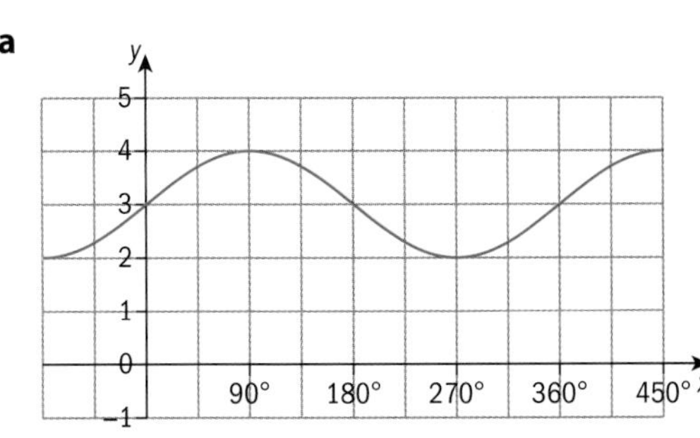
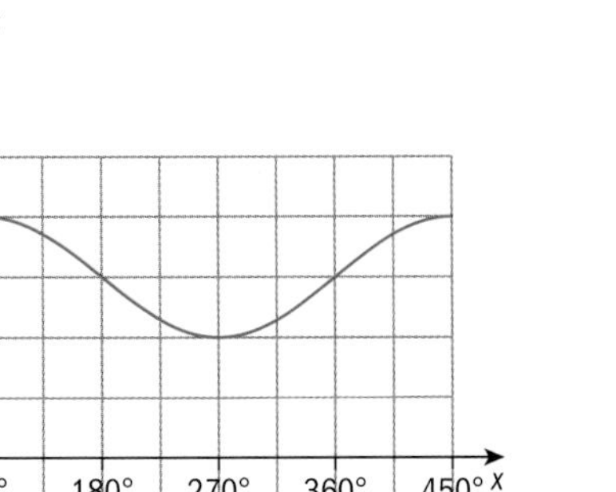

b

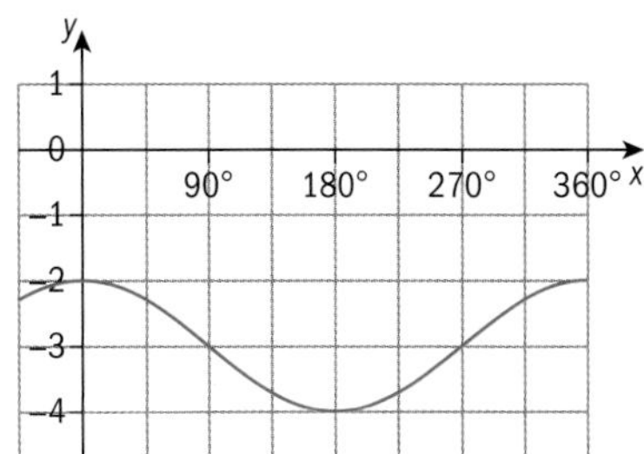

c

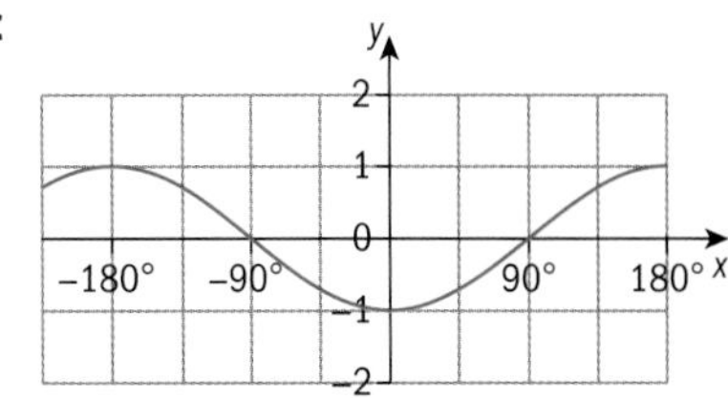

d

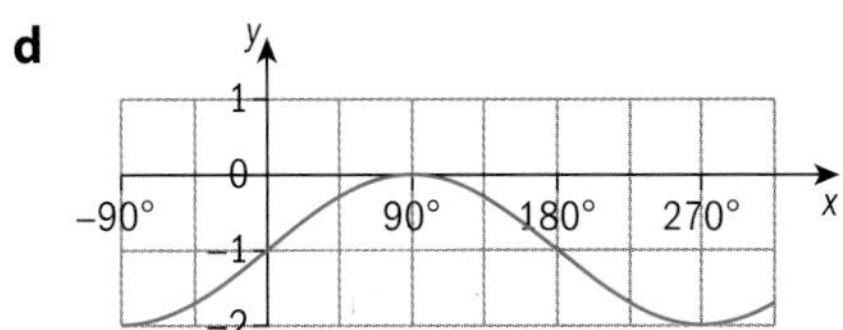

7 **Describe** in words how to transform the graph from $y = \sin x$ to:

a $y = 3\sin x + 2$ **b** $y = -2\sin(4x)$

c $y = -\sin(2x) - 3$ **d** $y = \frac{1}{2}\sin\left(\frac{x}{3}\right) - 4$

e $y = -3\sin(3x) + 3$ **f** $y = \frac{3}{4}\sin\left(\frac{3x}{4}\right)$

8 Each graph is a result of a combination of transformations on $y = \sin x$ or $y = \cos x$.

Identify the transformations, and **determine** the function of the transformed graph.

a

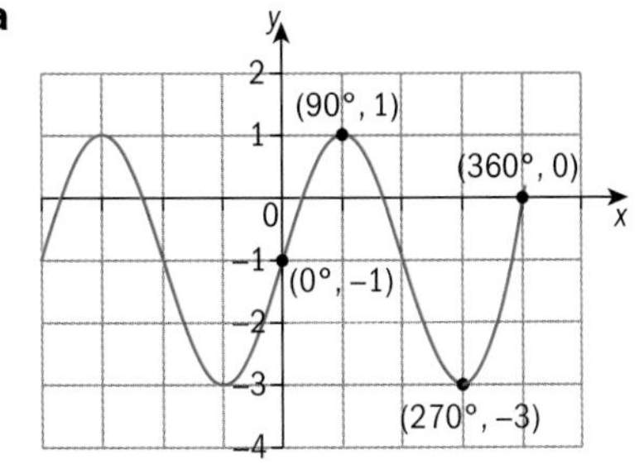

b

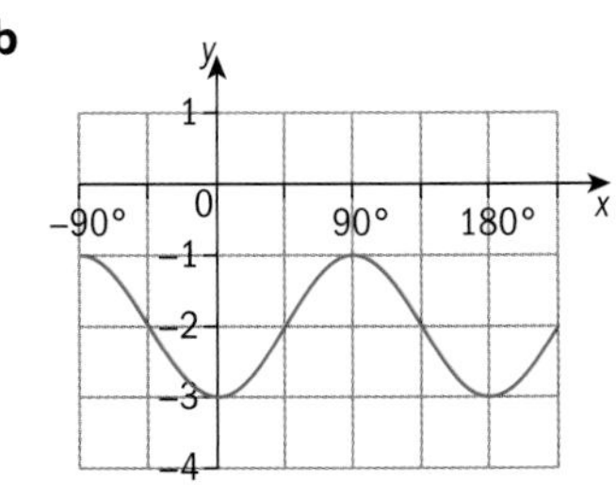

c

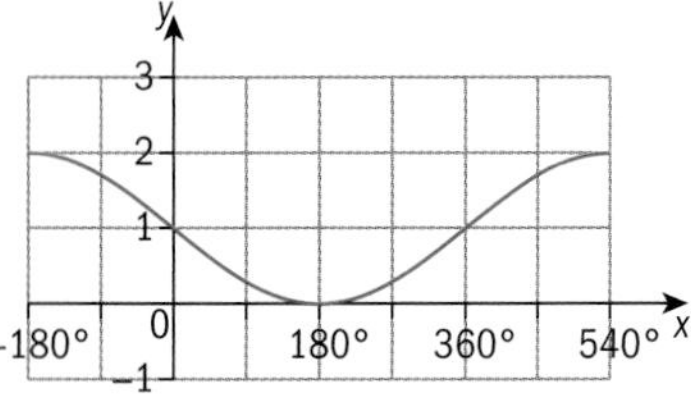

d

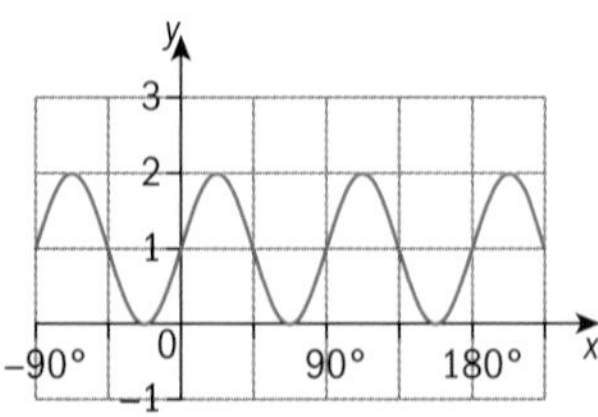

e

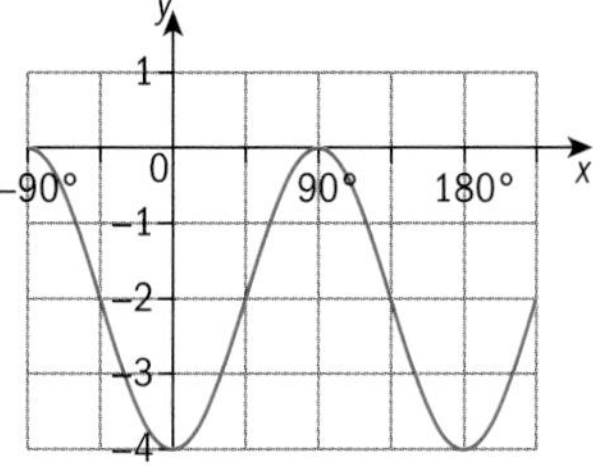

Problem solving

9 a A Ferris wheel has a radius of 10 meters. The bottom of the wheel is 2 meters above the ground. The wheel, rotating at a constant speed, takes 100 seconds to complete one revolution.

i Find a function that models this Ferris wheel.

ii Draw a graph of your function.

b Another Ferris wheel has a radius of 5 meters and is 1 meter above the ground. It takes 2 minutes to make one complete revolution.

i Find a function that models this Ferris wheel.

ii Draw a graph your function.

iii For which age group do you think this Ferris wheel has been designed?

10 The graph shows the motion of a tall building as it sways to and fro in the wind.

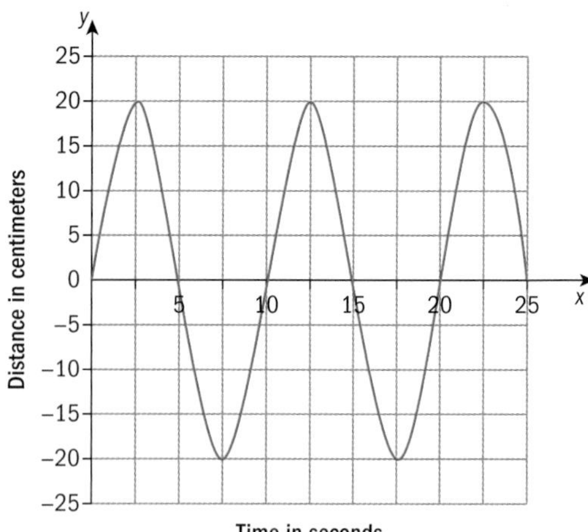

a Determine the period, and explain what it means in this problem.

b State the maximum number of centimeters that the tall building sways from its vertical position.

c Find a sinusoidal function to model this situation.

11 A nail is stuck in a car tire. The height of the nail above the ground varies as the wheel turns and can be modeled by this graph.

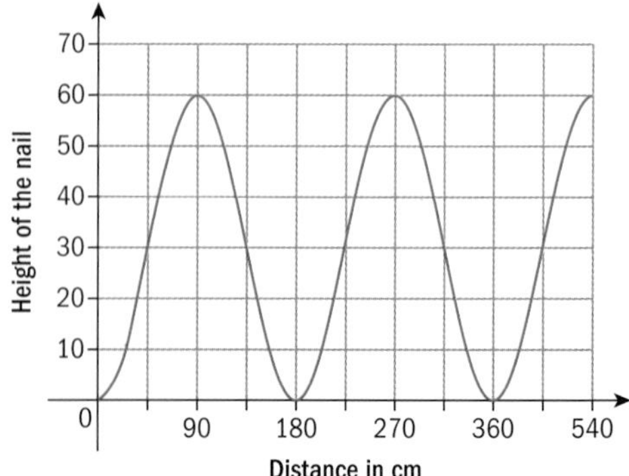

Find:

a the period of the graph, and explain what it means in this situation.

b the amplitude, and explain what it means in this situation.

c the radius of the wheel.

d a sinusoidal function to describe the relationship between the distance the tire travels and its height above the ground.

Problem solving

12 When the sun first rises, the angle of elevation increases rapidly at first, then more slowly, until the maximum angle is reached at about noon. The angle then decreases until sunset.

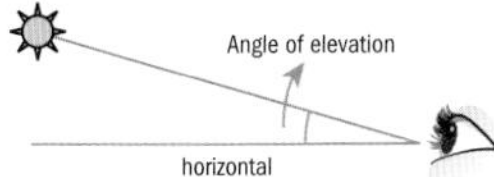

Assume that the relationship between the time of the day and the angle of elevation is sinusoidal.

- Let t be the number of hours since midnight.
- Let the amplitude be 60 degrees.
- The maximum angle of elevation occurs at noon.
- The period is 24 hours.
- The angle of elevation at midnight is –65 degrees.

a **Sketch** a graph of hours after midnight against angle of elevation.

b **Explain** the significance of the t-intercepts.

c **Explain** what the values below the t-axis mean.

d **Predict** the angle of elevation at 9:00 in the morning and 2:00 in the afternoon.

e **Predict** the time of sunrise.

Objective: **D.** Applying mathematics in real-life contexts
iii. apply the selected mathematical strategies successfully to reach a solution

In this Review you will apply the strategies of interpreting amplitude, period and dilation to sketch the graphs and reach appropriate solutions.

Review in context

Scientific and technical innovation

1 A tsunami, or monster tidal wave, can have a period anywhere between ten minutes and 2 hours, with a wavelength well over 500 km. Because of its incredible destructive powers, warning systems have been developed in regions where tsunamis are most likely to happen. A particular tsunami to hit a coastal town in Japan had a period of about a quarter of an hour. The normal ocean depth in this town was 9 m, and a tsunami of amplitude 10 m hit the coast.

a Write a sinusoidal function to describe the relationship of the depth of water and time.

b Use your function to predict the depth of the water after **i** 1 minute, **ii** 5 minutes, and **iii** 10 minutes.

c Use your function to determine the minimum depth of the water.

d Interpret your answer to **c** in terms of the real world.

2 In 2005, the United States had one of the worst hurricane seasons on record. Computer models were continuously generated and updated in order to try to predict the path of oncoming storms. For almost five days, a portion of Hurricane Franklin's path could be approximated by a sinusoidal function, as seen in the following data:

Time (hours)	0	20	40	60	80	100	120
Longitude (degrees)	–78	–77	–74.5	–71.5	–69	–68	–69

a **Draw** a graph of the data and clearly indicate the amplitude and period.

b **Write down** the function representing this model.

c At what longitude was the hurricane at $t = 50$ hours?

d **Describe** how the graph of $y = \cos x$ was transformed to obtain this graph.

e Do you think the latitude of the hurricane followed a similar model? **Explain**.

3 The Bay of Fundy and Ungava Bay, both in Canada, have some of the highest tides in the world. Tides show periodic behavior, with their constant shift from high to low tide and back again. Suppose a low tide of 1.5 m occurs at 6 am and a high tide of 18.5 m occurs twelve hours later.

a **Draw** a graph to represent the height of the tide as a function of the time of day.

b **Write down** the equation for this model.

c **Find** the height of the tide at 3 am and at 3 pm.

d **Find** the time(s) at which the height of the tide is 10 m.

e **Explain** why harnessing electricity from this tidal energy is considered more reliable than wind energy.

4 A chemotherapy treatment is designed to kill cancer cells but it also kills red blood cells which are vital in the transport of oxygen in the body. The amount of red blood cells can be modelled by a sinusoidal function since it decreases after the treatment and then steadily increases until the next treatment. Suppose a patient's red blood cell count is 5 million cells per microliter on the day of treatment and hits a low of 2 million cells per microliter 10 days after treatment. If treatments occur every 20 days:

a **Draw** a graph to represent the patient's red blood cell count over the course of 30 days.

b From the graph, **state** the amplitude, period and frequency, and **write down** the function that models the situation.

c On day 15, does the patient have more or less than half of their red blood cell count back? **Justify** your answer.

d If the 20th day were a holiday, would you rather go in for the treatment before the holiday or after? **Explain**.

Problem solving

5 The electricity delivered to your home is called 'alternating current' because it alternates back and forth between positive and negative voltage. It does so with regularity and so it can be described by a sinusoidal function.

a The voltage in many European countries is 220 V with a frequency of 50 Hz (which means that 50 cycles occur in one second). **Draw** a graph of the amount of voltage over time, assuming at $t = 0$ you have 220 V.

> Draw the graph between $t = 0$ seconds and $t = 0.02$ seconds.

b **Write down** the period of your graph.

c **Write down** the equation of your graph.

d In North America, the typical voltage is 120 V at 60 Hz. Draw a graph of voltage as a function of time, assuming at $t = 0$ you have 120 V.

e **Write down** the period, amplitude and equation for your graph.

f **Describe** the transformations that occurred from the European model to the North American model.

Reflect and discuss

How have you explored the statement of inquiry? Give specific examples.

Statement of Inquiry:

Generalizing relationships between measurements can lead to better models and methods.

14.3 Decisions, decisions

Global context: Identities and relationships

Objectives

- Solving systems of inequalities algebraically and graphically
- Modelling real-life problems with linear programming

Inquiry questions

F
- Can you use all the equivalence transformations to solve inequalities algebraically?
- How can you solve inequalities graphically?

C
- What does it mean to be linear?
- How do the gradients of different lines demonstrate the relationship between them?

D
- What are the advantages and limitations of linear programming?
- Can good decisions be calculated?

FORM

ATL **Communication**

Organize and depict information logically

11.1

14.3

E14.1

Statement of Inquiry:

Modelling with equivalent forms of representation can improve decision making.

You should already know how to:

• work with inequalities	**1** Start with the inequality $4 < 6$. **a** Add 2 to each side. **b** Subtract 4 from each side. **c** Multiply each side by 3. **d** Divide each side by -2. Determine if you still have a valid inequality after each step.
• solve equations using equivalence transformations	**2** Solve these equations. Describe the equivalence transformations you use. **a** $3x + 7 = 13$ **b** $\frac{x}{4} + 5 = 8$
• solve linear inequalities algebraically	**3** Solve these inequalities. **a** $4x > 12$ **b** $2x - 1 < 3$ **c** $5x + 7 \geq 22$ **d** $-7 < 3x - 1 < 5$
• graph linear equations	**4** Draw the graph of **a** $y = 3x - 2$ **b** $x + 2y = 6$

Solving inequalities algebraically

- Can you use all the equivalence transformations to solve inequalities algebraically?
- How can you solve inequalities graphically?

In solving inequalities, you learned to use the equivalence transformations that are used in solving equalities. For example:

$3x - 2 > 1$ —— Addition principle: add 2 to each side

$3x > 3$ —— Multiplication principle: multiply each side by $\frac{1}{3}$

$x > 1$

Solution set is $\{x | x \in \mathbb{R}, x > 1\}$

In this section you will investigate why you can use these equivalence transformations for inequalities as well.

Exploration 1

1 Choose two numbers a and b, so that $a > b$.

Fill in the missing sign: $a - b \square 0$

Determine if this is true for all a and b, where $a > b$. Try with positive and negative values of a and b.

2 Choose two numbers a and b, so that $a < b$.

Fill in the missing sign: $a - b \square 0$

Determine if this is true for all a and b, where $a > b$. Try with positive and negative values of a and b.

Definitions of inequality:

$a > b$ if and only if $a - b > 0$

and

$a < b$ if and only if $a - b < 0$

You can use these definitions to prove some theorems about inequalities and equivalence transformations.

Theorem 1: Adding the same number to both sides of an inequality gives another valid inequality.

Using algebra:

i if $a > b$ then $a + c > b + c$

ii if $a < b$ then $a + c < b + c$

Choose some numbers for a, b and c and check these statements. Make sure to include some negative numbers.

The theorem states that if you add the same number to both sides of an inequality, the inequality still holds. The proof shows that this is true for all real numbers.

Proof of Theorem 1i:

When $a > b$:

$a - b > 0 \Rightarrow (a - b) \in \mathbb{R}^+$ — Using the definition of inequality.

$a - b + c - c > 0$ — $c - c = 0$, and adding 0 does not change the value.

$a + c - b - c > 0$

$a + c - (b + c) > 0$ — Rearrange.

So $a + c > b + c$ — Using the definition of inequality.

Reflect and discuss 1

- Prove Theorem **1ii** for $a < b$, to show that when $a < b$, $a + c < b + c$.
- Does it matter whether c is a positive or negative number? Explain.
- When working with equations or inequalities, do we need a subtraction principle, or is the addition principle enough? Explain.

Theorem 2: Multiplying both sides of an inequality by the same *positive* real number c gives another valid inequality.

Using algebra:

i if $a > b$ then $ac > bc$

ii if $a < b$ then $ac < bc$

Exploration 2

1 Prove Theorem 2i and ii, making sure that you justify all your steps.

2 Determine if the theorem holds when you divide both sides by a positive, non-zero real number.

3 Use Theorem 2 to prove that if $a < 1$ then $a^2 < a$.

4 Explain if it is possible that $-a > 0$.

Try multiplying both sides of inequalities like $2 < 5$ or $4 > -1$ by negative numbers, to see what happens to the inequality sign.

Theorem 3: If both sides of an inequality are multiplied by a negative real number, the direction of the inequality changes. This means that if $a, b \in \mathbb{R}$, and $c \in \mathbb{R}^-$ then:

i if $a > b$ then $ac < bc$ and

ii if $a < b$ then $ac > bc$

Proof of Theorem 3i:

$a > b$

$a - b > 0 \Rightarrow (a - b) \in \mathbb{R}^+$ — Using the definition of inequality.

If $c < 0$, then $c(a - b) < 0$ — The product of numbers with different signs is a negative number.

$ca - cb < 0$ — Expand the brackets.

$ca < cb$ — Using the definition of inequality.

Exploration 3

1 Prove Theorem 3ii.

2 Determine if Theorem 3 holds when you divide both sides by a negative real number.

The definition of inequality and theorems 1, 2 and 3 all hold for $\geq$ and $\leq$.

When you multiply both sides of an inequality by a negative real number, the direction of the inequality changes. Theorem 3 shows this is true for **all** real numbers, and not just the ones you have used.

Solve the inequalities in Practice 1 using Theorems 1, 2 and 3, which you have just proved.

Practice 1

Solve:

1 $3x - 7 < 5$

2 $\frac{x}{6} - 7 > 2$

3 $3(4 - m) > 9$

4 $5x - 7 \geq 3x + 9$

5 $1 - \frac{3}{2}x \leq x - 4$

6 $-\frac{2x+1}{3} > 5x + 2$

7 $-2(a - 3) < 5(a - 2) - 12$

8 $2(1 - b) + 5 \geq 3(2b - 1)$

9 $4k - 11 \leq \frac{3k}{2} + 5$

Exploration 4

One way to solve the double inequality $-4 < 3x + 2 < 5$ is:

Add -2 to both sides: $-6 < 3x < 3$

Multiply both sides by $\frac{1}{3}$: $-2 < x < 1$

Another way is to rewrite the double inequality as two distinct inequalities:
$-4 < 3x + 2$ and $3x + 2 < 5$

Solve them separately:
$-6 < 3x$ and $3x < 3$
$-2 < x$ and $x < 1$

Combine the solution sets: $-2 < x < 1$

1 Solve the double inequality $-9 < 5 - 7x \leq 12$ in two ways algebraically.

2 Solve $-14 < -7(3x + 2) < 21$

- **a** by first using the equivalence transformation 'multiply by $-\frac{1}{7}$'
- **b** by expanding the brackets first.

Explain which method you prefer.

3 Does the pair of inequalities $4x < 4$ and $3x - 5 \geq 1$ have a common solution set? Explain.

4 Given that $-1 < x < 4$, find two values m and n such that $m < 2x + 3 < n$.

> In step **3**, is there one set of values that satisfies both inequalities?

You can solve inequalities graphically in a similar way to solving equations graphically.

For the inequality $3x - 7 < 5$, draw the lines $f_1(x) = 3x - 7$ and $f_2(x) = 5$ and find where they intersect.

The inequality $3x - 7 < 5$ means that $3x - 7$ cannot take the value 5. Use a dashed line for $f_2(x) = 5$ to show that points on the line $f_2(x) = 5$ are not possible solutions.

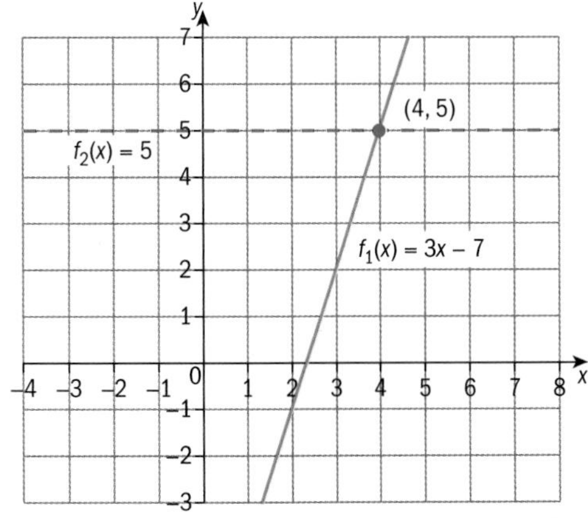

The lines intersect at $x = 4$. This is the value of x that makes both sides equal.

For $3x - 7 < 5$ you need values of x such that $f_1(x) < f_2(x)$. From the graph, the line for $f_1(x)$ is below $f_2(x)$ when $x < 4$, showing that the values of $f_1(x)$ are less than the values of $f_2(x)$ when $x < 4$.

The solution is $x < 4$.

Check your solution by choosing a point with $x < 4$, for example: (2, 3).

Substitute $x = 2$ into the inequality:
$3x - 7 = 6 - 7 = -1 < 5$ ✓

Now choose a point with $x > 4$, say: (5, 7).

Substitute $x = 5$ into the inequality: $3x - 7 = 15 - 7 = 8$

8 is not less than 5, so the point does not satisfy the inequality.

Example 1

Solve the inequality $5x - 7 \geq 3x + 9$ graphically.

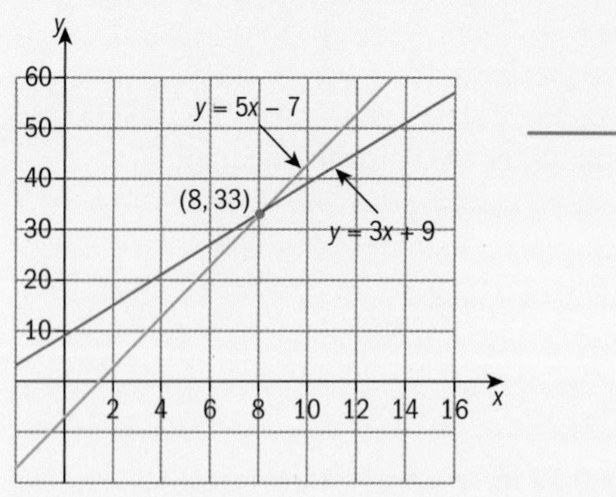

Plot the graphs of $y = 5x - 7$ and $y = 3x + 9$. The inequality sign is $\geq$, so use solid lines to show that the values on the lines are included.

The point of intersection is where the two functions are equal.

$x \geq 8$ — The line $y = 5x - 7$ is above the line $y = 3x + 9$ when $x \geq 8$.

When $x = 9$,

$5x - 7 = 45 - 7 = 38$ $\quad 3x + 9 = 27 + 9 = 36$ — Check by substituting an x-value into the inequality.

$38 \geq 36$ ✓

Practice 2

1 Solve the inequalities in Practice 1 graphically. Check against your algebraic solutions.

Problem solving

2 Write down the two inequalities that could be represented by the non-shaded region of this graph.

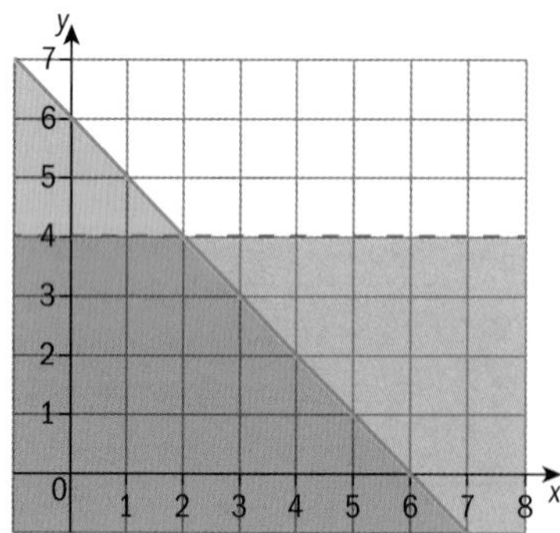

3 Verify that the inequality $3x + 2 < 9x + 6$ has an infinite number of solutions.

C Systems of linear inequalities

- What does it mean to be linear?
- How do the gradients of different lines demonstrate the relationship between them?

A system of equations has more than one variable and more than one equation that are all true simultaneously.

A system of inequalities has more than one variable and more than one inequality that are all true simultaneously.

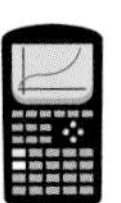

Exploration 5

1 Graph the system of equations $y = x + 1$ and $y = 4 - 2x$ and label their point of intersection clearly.

2 Choose a point on your coordinate grid. Substitute the x and y values for this point into the two equations. Determine if they satisfy the inequality $y \le x + 1$. Choose more points until you have identified the region of the graph that contains the set of points that satisfy $y \le x + 1$. Shade this region of the graph.

3 Repeat step **2** to find and shade the region of the graph where the points satisfy the inequality $y \ge 4 - 2x$.

4 On your graph, find the region that satisfies both inequalities $y \le x + 1$ and $y \ge 4 - 2x$. Test some points in the region.

5 Determine how you would modify your graph if you had $y < x + 1$ and $y > 4 - 2x$ instead of $y \le x + 1$ and $y \ge 4 - 2x$. State whether or not the solution set would be the same.

Reflect and discuss 2

- Why do you think that you have learned how to find the solution set for a system of linear inequalities graphically rather than algebraically?
- Try to find the solution to the system of linear inequalities $y \le x + 1$ and $y \ge 4 - 2x$ algebraically. What difficulties do you experience?

Example 2

Solve this system of inequalities graphically, showing clearly the region that satisfies both $y < x + 2$ and $y \ge 3 - 2x$.

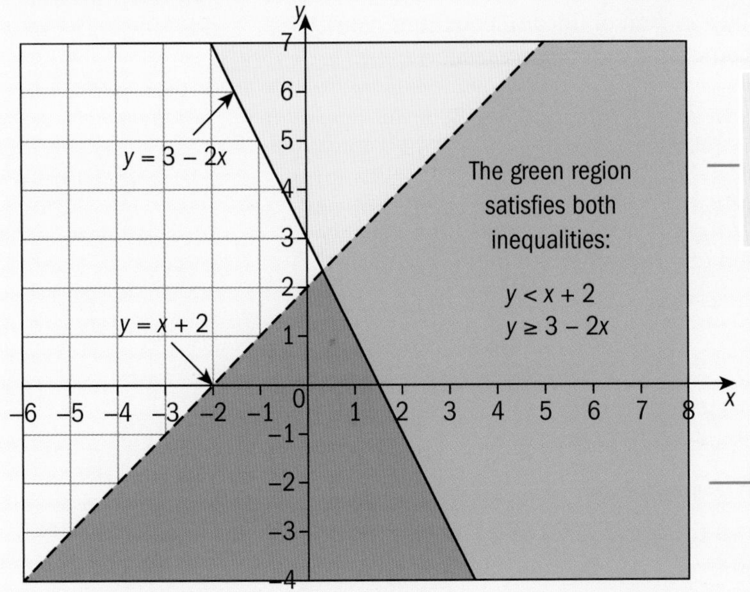

Graph $y = x + 2$. Use a dashed line.

Test a point on one side of the line. Shade the region that satisfies $y < x + 2$.

Graph $y = 3 - 2x$. Use a solid line. Test points, then shade the region that satisfies $y \ge 3 - 2x$.

All points that satisfy the system of linear inequalities are in the overlap of the two shaded regions. Points on dashed line are not in the solution set.

Practice 3

ATL

1 Graph these systems of inequalities. Shade the region of the solution set.

a $y < 4\ ;\ y \leq 2x - 6$

b $x \geq 3\ ;\ y > x$

c $y \leq -x - 2\ ;\ y \geq 2 - 5x$

d $2y - x < 4\ ;\ y + 2x < 3$

e $x \geq -3\ ;\ 3y > 5x + 6$

f $3x + 2y > -2\ ;\ x + 2y < 2$

g $4x + 2y \leq 2\ ;\ y > -2$

h $y > 23x + 3\ ;\ y \geq -43x - 3$

Problem solving

2 Write the system of inequalities satisfied by the region shaded in the graph.

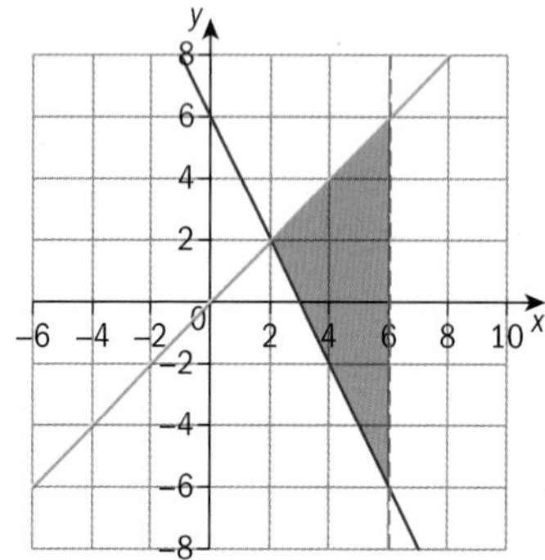

3 Graph these systems of inequalities. Shade the region of the solution set.

a $y < 5\ ;\ x \geq -3\ ;\ y < x$

b $y > x - 2\ ;\ y > 0\ ;\ x < 0$

4 Graph these systems of inequalities.

a $x \geq 0\ ;\ x + y < 7,\ y < 6$

b $-2 \leq x < 7\ ;\ y > 3x + 2$

c $x \geq -1\ ;\ 2x - y < 12;\ x > y$

d $y \leq 2x\ ;\ 3x + y > 2;\ x \leq 8$

When a real-life problem is modelled by a system of linear inequalities, you can solve the system graphically. This process is called *linear programming*.

Linear programming was developed during World War II to help optimize the allocation of labor and materials for the war effort. It is sometimes called linear optimization. Nowadays it is used as a tool in the decision-making process by a wide range of groups. For example, industries use it to allocate labor, transport, materials and other resources to maximize profit and minimize cost.

ATL

Exploration 6

1 Graph the inequalities $y \leq -\frac{1}{2}x + 7$, $y \leq 3x$ and $y \geq x - 2$.

2 Shade the region which contains the solution set.

3 The region is a triangle. Write down the coordinates of the vertices of this triangle.

4 If the vertices give the maximum and minimum values of *z*, explain why you think this happens.

In Exploration 6, you optimized (found the maximum/minimum) of a function z in x and y, where x and y satisfy a system of inequalities.

> The region of the graph that contains the solution set is called the **feasible region**. This region might be a polygon.
>
> A system of linear equalities is optimized on one of the vertices of the feasible region polygon.

Explaining why the system of linear inequalities is optimized on one of the vertices of the feasibility region polygon uses mathematics beyond the level of this course.

You can use this method to solve real-world problems, such as this one:

A furniture manufacturing company makes desks and chairs.

It takes 8 hours to make a desk, and 2 hours to varnish it.

It take 2 hours to make a chair, and 1 hour to varnish it.

One week there are 400 hours of worker time available for making desks and chairs, and 120 hours of worker time for varnishing them.

The company sells each desk for \$50 and each chair for \$20. Determine the number of desks and chairs the company should produce that week to maximize their revenue.

Revenue is the amount of money they will be paid for the desks and chairs.

1 Identify the variables.

Let x be the number of desks made and y be the number of chairs.

It takes $8x$ hours to make x desks, and $2y$ hours to make y chairs.

It takes $2x$ hours to varnish x desks, and y hours to varnish y chairs.

2 Identify the constraints.

There are 400 hours available for making, so one constraint is 'the total time for making desks and chairs is less than or equal to 400'.

This is expressed mathematically as $8x + 2y \leq 400$.

There are 120 hours available for varnishing, so another constraint can be expressed as $2x + y \leq 120$.

Also, both x and y are greater than or equal to 0, so $x \geq 0$ and $y \geq 0$.

This is a system of four inequalities.

3 Graph the system of inequalities.

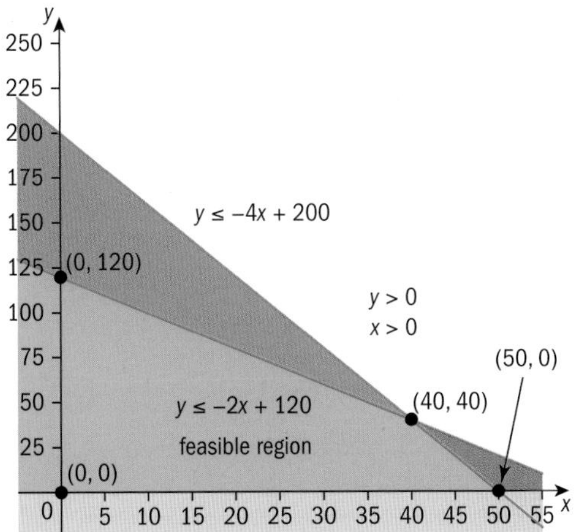

4 **Shade the feasible region**.
The numbers of desks and chairs that can be manufactured given the company's constraints all fall within this region. Write down the coordinates of its vertices: (0, 0), (0, 120), (50, 0) and (40, 40).

5 **Determine the function to be maximized**.
Desks sell for \$50 and chairs for \$20, so for selling x desks and y chairs gives total revenue $R = 50x + 20y$.

$R = 50x + 20y$ is the **objective function** – the one to be optimized using values that satisfy the constraints.

6 **Test the coordinates of the vertices with the constraints and objective function:**

	x	y	$8x + 2y$	$2x + y$	$50x + 20y$
(0, 0)	0	0	0	0	\$0
(50, 0)	50	0	400	100	\$100
(0, 120)	0	120	240	120	\$2400
(40, 40)	40	40	400	120	\$2800

You can see from the table that the maximum revenue is \$2800 from making and selling 40 desks and 40 chairs.

Linear programming

Linear programming is a method of maximizing or minimizing a linear function subject to linear constraints within the problem.

The objective function is the linear function representing cost, profit, or some other quantity to be optimized (maximized or minimized) subject to the constraints.

The constraints form a system of linear inequalities which model the real-life problem.

The feasible region is the area of the graph that contains the solutions to the system of linear inequalities.

If there is an optimal solution to a linear programming problem, it will occur at one or more vertices of the feasibility region polygon.

Example 3 is a simplified version of a real-life problem at the end of World War II, where British and US planes were flying supplies into the Berlin airport.

Example 3

Up to 10 planes can fly into a particular airport each day. Each US plane can carry 30 tons of supplies and requires two crew members. Each British plane can carry 20 tons and requires one crew member. There are 14 crew members in total (they can fly on either US or British planes). Find the maximum weight of supplies that can be transported each day.

x = number of US planes

y = number of British planes

Identify the variables.

▶ Continued on next page

$0 \le x \le 10$; $0 \le y \le 10$

$x + y \le 10$ — The maximum number of planes that can fly in is 10.

x US planes need $2x$ crew members,
y British planes need y crew members.

$2x + y \le 14$ — The total number of crew members is 14.

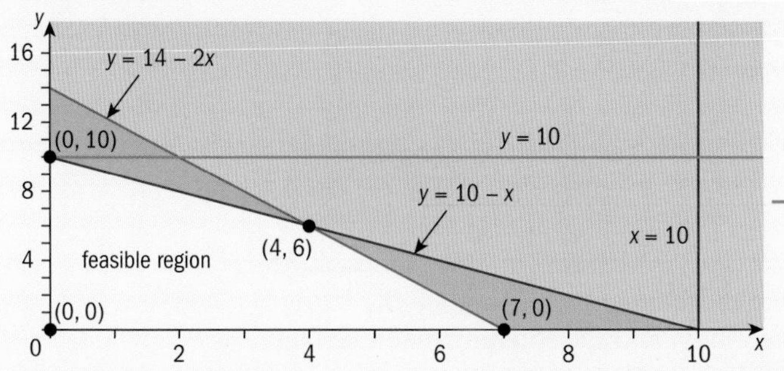

Another way of showing the solution is to shade the regions that *do not satisfy* the inequalities. The feasible region is left unshaded.

Vertices: (0, 0), (0, 10), (7, 0), (4, 6) — Identify the vertices of the unshaded polygon which is the feasible region.

Objective Function:

Total weight of supplies in x US planes and y British planes is $W = 30x + 20y$. — Determine the objective function.

	x	y	$x + y$	$2x + y$	$30x + 20y$
(0, 0)	0	0	0	0	0
(0, 10)	0	10	10	10	200
(7, 0)	7	0	7	14	210
(4, 6)	4	6	10	14	240

Test the coordinates of the vertices with the constraints and objective function.

Find the one that optimizes the objective function.

The maximum weight of supplies that could be transported each day is 240 tons.

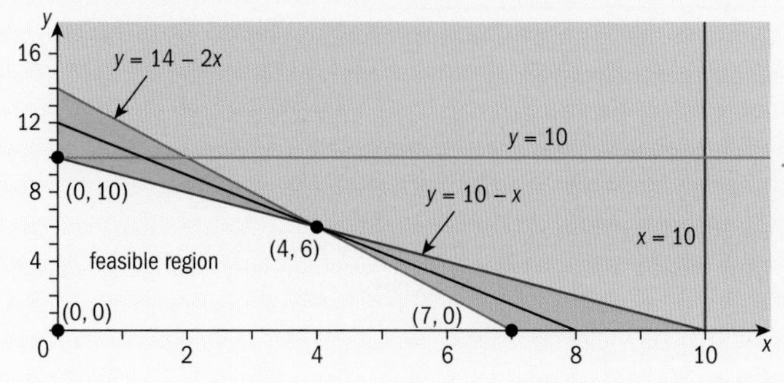

As a check, you can graph the objective function $30x + 20y = 240$ (here in black) to check that it intersects the other lines at (4, 6).

Practice 4

1 Optimize the objective function $P = 50x + 20y$, given the four constraints:

$x + 4y \le 21 \qquad -4x + 6y \le 0$

$0 \le x \le 15 \qquad y \ge 0$

2 Optimize the objective function $z = 0.4x + 3.2y$ given the four constraints:

$x \ge 0,\ y \ge 0 \qquad x + y \le 7$

$x + 2y \ge 4 \qquad y \le x^2 + 5$

3 Optimize the objective function $R = 5y - 2x$, given the three constraints:

$99 \le x < 201 \qquad 79 < y \le 181 \qquad y + x \le 300$

4 A shoe manufacturer produces mid-top and high-top athletics shoes. Cutting machines cut the fabric for the shoes. Stitching machines stitch the fabric. The cutting machines run 4 hours per day, and the stitching machines run 5 hours per day.

A mid-top shoe takes 1 minute to cut, and 2 minutes to stitch.

A high-top shoe takes 3 minutes to cut, and 2 minutes to stitch.

The profit is \$13 on a mid-top shoe and \$16 on a high-top shoe.

Determine how many shoes of each kind should be made in order to maximize profits, and what the maximum possible profit is.

Tip

Graph the system of inequalities, and test the vertices of the feasible region polygon in the objective function.

Problem solving

5 A school wants to buy small and large minibuses to transport students to sports activities. It has \$220 000 to buy the minibuses, and \$1200 to insure them.

The table gives some information about different sizes of minibuses.

Size	Maximum number of people	Cost	Insurance
Small	8	\$11 000	\$120
Large	16	\$24 000	\$80

Determine how many of each size minibus the school should buy to maximize the number of students they can transport.

6 An educational software company produces two software packages: an algebraic solver and a graphing program. They project a demand for at least 100 algebraic solvers and 80 graphing programs each day. They can produce up to 200 algebraic solvers and 170 graphing programs per day. They need to produce at least 200 software packages each day to satisfy existing orders. Each algebraic solver makes a loss of \$1.50. Each graphing program makes a profit of \$4.50.

Determine how many of each software package the company should produce each day to maximize its profits.

7 World Polar Products makes skis for the Alpine Skiing World Championship competitions.

To make one downhill ski takes an average of 2 hours to cut the material needed, 1 hour to shape the ski, and 3 hours for the finishing. To make one cross-country ski takes 2 hours to cut the material, 2 hours for shaping, and 1 hour for finishing. Each week the company has 140 staff and machine hours for cutting, 120 hours for shaping and 150 hours for finishing. They make a profit of $10 for each downhill ski and $8 for each cross-country ski. Find the maximum profit they can make in one week.

8 Amelie received $120 000 from a trust fund on her 25th birthday. She wants to invest it in different funds to maximize the interest she receives.

She can invest in three types of investment:

Type of investment	Interest per annum
Municipal bond	2%
Bank Mutual fund	3.2%
Speculative money market fund	5%

To minimize her risk, she decides to invest only $20 000 in the speculative money market fund. Her tax adviser says she has to invest at least three times as much in the municipal bond as in the bank's mutual fund. Determine the optimum investment amounts for each type of investment.

Linear programming

- What are the advantages and limitations of linear programming?
- Can good decisions be calculated?

In Practice 4 you used linear programming to solve problems in different real-life situations.

Activity

Research the use of linear programming in industry. Which types of industry use it, and for what applications?

Reflect and discuss 3

- What type of problems can you solve using linear programming? What kinds of functions are used?
- All the examples and problems you have seen deal with only positive numbers. Why do you think this is?
- What kinds of real-world problems would not be suitable for linear programming?
- What are the advantages of using linear programming?
- What are the limitations of mathematical modelling using linear programming?

Summary

Definitions of inequality

$a > b$ if and only if $a - b > 0$

and

$a < b$ if and only if $a - b < 0$

Theorem 1: Adding the same number to both sides of an inequality gives another valid inequality.

Using algebra:

i if $a > b$ then $a + c > b + c$

ii if $a < b$ then $a + c < b + c$

Theorem 2: Multiplying both sides of an inequality by the same positive real number gives another valid inequality.

Using algebra:

i if $a > b$ then $ac > bc$

ii if $a < b$ then $ac < bc$

Theorem 3: If both sides of an inequality are multiplied by a negative real number, the direction of the inequality changes. This means that if $a, b \in \mathbb{R}$, and $c \in \mathbb{R}^-$ then:

i if $a > b$ then $ac < bc$ and

ii if $a < b$ then $ac > bc$

Linear programming

Linear programming is a method of maximizing or minimizing a linear function subject to linear constraints within the problem.

The **objective function** is the linear function representing cost, profit, or some other quantity to be optimized (maximized or minimized) subject to the constraints.

The constraints form a system of linear inequalities which model a real-life problem.

The feasible region is the area of the graph that contains the solutions to the system of linear inequalities.

If there is an optimal solution to a linear programming problem, it will occur at one or more vertices of the feasibility region polygon.

Using linear programming to solve real-life problems:

- Identify all variables and parameters.
- Identity all constraints. This is your system of inequalities.
- Graph your system of inequalities, and identify the polygon that makes up the feasibility region.
- Substitute the values of the vertices of the polygon into the system of inequalities and the objective function. Identify the ordered pair that optimizes the function.

Mixed practice

1 **Solve** these inequalities algebraically:

a $4 - 2x \leq 2$ **b** $\frac{x}{3} + 5 < 11$

c $2(x + 1) > x - 7$ **d** $3(x + 1) < 2(1 - x)$

e $3x + 1 \geq 2x + 5$ **f** $10 \leq 2x \leq x + 9$

g $x < 3x - 1 < 2x + 7$

2 **Solve** the inequalities in **1a–d** graphically.

3 **Graph** these inequalities. **Show** clearly the region that satisfies the system of inequalities.

a $0 < x < 4$; $y < 2$

b $x > 0$; $x + y < 10$; $y > x$

c $2y > x$; $y < 2x$; $x + y < 8$

d $y \geq 2x - 3$; $y \geq -3$; $y \leq -0.8x + 2.5$

Problem solving

4 A company hires out buses:

Bus	Hire cost
40-seater	£80
24-seater	£50

A school can spend up to £400 to hire buses to take 120 students on an outing.

Determine how many of each bus the school should hire.

5 A factory manager has to decide which machines to install to manufacture a product.

Machine A

Output is 300 units per week. Needs 500 m^2 of space, and 10 operators.

Machine B

Output is 200 units per week. Needs 600 m^2 of space, and 6 operators.

The factory has 5050 m^2 of space available and 80 operators. Its production target is 2400 units per week. **Determine** if the factory can meet its production target. If it can, **determine** how many of each type of machine it should buy.

6 Your company has a maximum of $2200 to spend on advertising. You plan to run at most 20 advertisements because each one needs to be different and creative. An advertisement in the daily newspaper costs $50, and in the weekend newspaper costs $200. **Find** the maximum number of weekday and weekend newspaper advertisements you can buy.

Objective: D. Applying mathematics in real-life contexts
ii. select appropriate mathematical strategies when solving authentic real-life situations

Make sure you identify variables and constraints in each problem in order to solve the real-world problems.

Review in context

Identities and relationships

1 Trees are planted in urban areas for beauty but also to remove air pollution. Trees with a large diameter, which cost £100 to plant, need 80 m^2 of space to grow and remove 1.5 kg/year of pollutants. Smaller trees, costing £30, need 50 m^2 of space to grow but remove only 0.02 kg/year of air pollutants. If your school has a budget of £1200 and 1500 m^2 of space, **determine** how many of each type of tree should be planted in order to maximize the amount of air pollutants removed.

2 A nutritionist prescribes this diet to a patient:

- 400 units of carbohydrates
- 500 units of fat
- 300 units of protein

The nutritionist has combined these into two types of protein shake. Type A costs $3 per pack and type B costs $2.50 per pack. One pack of type A contains 10 units of carbohydrates, 20 units of fat and 15 units of protein. One pack of type B contains 25 units of carbohydrates, 10 units of fat and 20 units of protein. **Find** the minimum cost for a diet that consists of a mixture of these two shakes and meets the minimum requirements.

Reflect and discuss

How have you explored the statement of inquiry? Give specific examples.

Statement of Inquiry:

Modelling with equivalent forms of representation can improve decision making.

15 Systems

Groups of interrelated elements

Axiomatic systems

An axiom is a starting point of reasoning that is so evident it can be accepted as true. Mathematics is an axiomatic system – a set of axioms that can be used to derive theorems. The axiomatic system for mathematics is similar to grammatical rules for a language – once you know the rules, you can interpret the meaning.

Gödel's incompleteness theorems

Mathematician Kurt Gödel (1906–1978) proved that no matter what axioms you choose to start with, any axiomatic system will eventually run into statements that can not be proven true or false – they just 'are'.

Gödel's revolutionary theorems turned mathematical and philosophical thinking on its head, and paved the way for new ideas in the field of logic and the foundation of mathematics.

Ecosystems

When an ecosystem is in balance, all the living organisms within it are healthy and capable of reproducing themselves. If one part of the ecosystem is damaged – either by natural events or human activity – every other element of the ecosystem is affected.

Forest fires cause widespread destruction, but fire is a natural and vital element in some forest ecosystems. After a fire, more sunlight can reach the forest floor, which is free from dead wood and leaf litter and has increased nutrient levels from the ash. This creates perfect conditions for some trees and plants to thrive. Some species have evolved to withstand fire (or sometimes even rely on it) as part of their life cycle. For example, sand pine cones open to disperse their seeds only in intense heat.

Axiomatic system for number

You have been working with the real number system since you first learned to count. The number system is based on a set of axioms, which are assumed to be true for real numbers.

Verify the axioms below for real numbers x, y and z.

Addition and multiplication are:

Commutative: $x + y = y + x$
$xy = yx$

Associative: $(x + y) + z = x + (y + z)$
$(xy)z = x(yz)$

Distributive: $x(y + z) = xy + xz$

These axioms form the basis for all of your number skills.

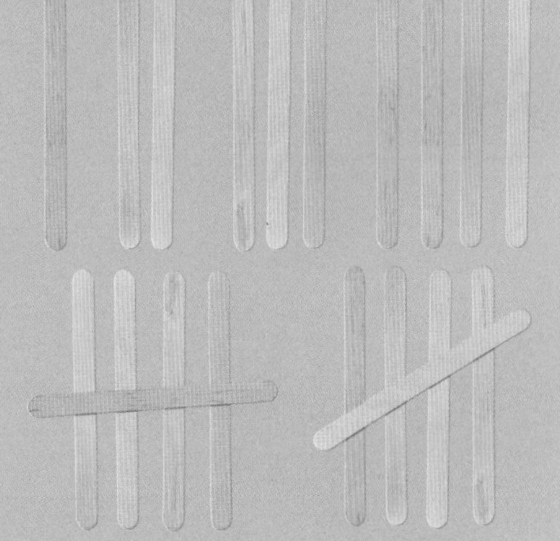

15.1 More than likely, less than certain

Global context: Identities and relationships

LOGIC

Objectives

- Understanding and using formal probability notation
- Calculating probabilities of independent, mutually exclusive and combined events
- Proving probability theorems
- Determining whether or not events are mutually exclusive and/or independent

Inquiry questions

F
- What are the axioms of probability?
- How do you calculate the probability of mutually exclusive events and of independent events?

C
- How can an axiomatic system be developed?

D
- Does a system help solve a problem?
- What affects the decisions we make?

ATL Communication

Use and interpret a range of discipline-specific terms and symbols

4.4

15.1

E15.1

Statement of Inquiry:

Understanding health and making healthier choices result from using logical representations and systems.

You should already know how to:

- use basic probabilities and sample spaces

1 a What is the sample space for the event 'rolling a single die'?

b For an ordinary, fair die, find the probability of:

i rolling a 3

ii rolling an odd number.

- find probabilities from Venn diagrams, tree diagrams and two-way tables.

2 Find P(A) and P($A \cup B$) from the Venn diagram.

U
A
B
4
8
7
3

3 Find P(C), P(D') and P(C and D) from the tree diagram.

0.2
C
0.4
D
0.6
D'
0.8
C'
0.4
D
0.6
D'

4 Using the table below, find the probability that a person chosen at random is a female teacher.

	Teacher	Not a teacher
Male	13	12
Female	15	10

The axiomatic system for probability

- What are the axioms of probability?
- How do you calculate the probability of mutually exclusive events and of independent events?

Probability theory determines the likelihood of an event happening. For an event A, probability assigns a numerical value P(A), called the probability of the event A. P(A) is a measure of the likelihood or chance that A occurs.

Probability is a system which satisfies a set of axioms.

A *system* is a group of interrelated elements. An axiomatic system is governed by a set of axioms or rules. Everything in the system obeys the rules.

Exploration 1

1 Use these probability terms to describe the probability of each event listed below.

certain | very likely | likely | even chance | unlikely | very unlikely | impossible

a You will get full marks on your next English test.

b You will be given homework tonight.

c The last meal you ate was breakfast.

d It will snow tomorrow.

e When you add 2 and 2 you will get 5.

f Brazil will win the next World Cup.

g The Queen of Denmark will come to your mathematics class tomorrow.

h The sun will come up in the morning.

2 Suggest approximate numerical values for the probability of each event in step **1**.

3 In a raffle, a total of 500 tickets are sold. Find the probability that you win the prize if you buy:

a 10 tickets **b** 50 tickets

c 0 tickets **d** 500 tickets

4 Describe each event in step **3** with a probability term (like those in step **1**).

We use probability in many applications, including weather forecasting, science, social science, philosophy and psychology.

In Exploration 1 you should have seen that that the smallest value a probability can ever take is zero (an impossible event). This is the first axiom of probability.

Axiom 1

For any event A, $P(A) \geq 0$. This means that all probabilities have a value greater than or equal to zero.

An event that is certain to occur has probability 1. The whole sample space S includes all the possible events, so the probability that one of those events occurs is 1. This is the second axiom of probability.

Axiom 2

For a sample space S, $P(S) = 1$. In other words, the probability of all occurrences is equal to 1.

Tip

If you ever calculate a probability which is less than 0 or greater than 1, you have made a mistake.

Exploration 2

1 Listed below are events that could occur when rolling a fair, six-sided die. Select and write down pairs of events that could **not** occur at the same time when rolling the die just once.

rolling a 4	rolling an odd number	rolling a 2
rolling a multiple of 2	rolling a 3	rolling a factor of 6
rolling an even number	rolling a factor of 18	rolling a 5

2 Listed below are some of the possible events that could happen when drawing a card from a standard deck of 52 cards.

drawing a red card	drawing a face card	drawing a factor of 8
drawing a heart	drawing a black 6	drawing an odd-numbered card
drawing a red ace	drawing a multiple of 3	drawing a ten

a Select and write down pairs of events that could **not** occur at the same time.

b Invent a pair of events that could not happen at the same time if you drew a card from the deck.

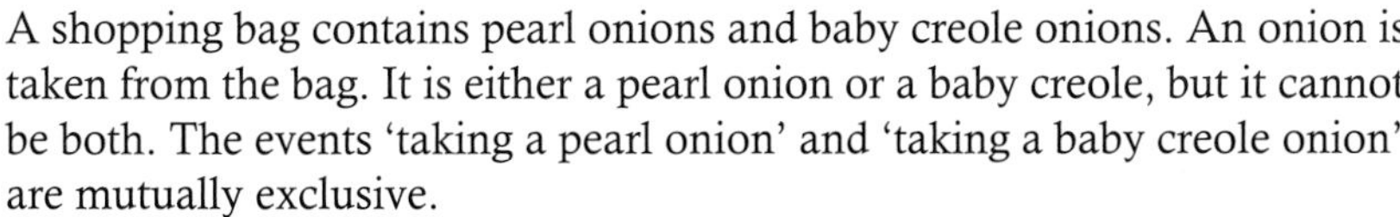

A shopping bag contains pearl onions and baby creole onions. An onion is taken from the bag. It is either a pearl onion or a baby creole, but it cannot be both. The events 'taking a pearl onion' and 'taking a baby creole onion' are mutually exclusive.

In probability, two events are **mutually exclusive** if only one can happen in any given experiment.

Formally, two events are mutually exclusive when the sets representing the events are disjoint, in other words: $A \cap B = \varnothing$ (the empty set).

For example, when a die is rolled once:

- the sample space $S = \{1, 2, 3, 4, 5, 6\}$
- Event A = 'odd number' = $\{1, 3, 5\}$
- Event B = 'even number' = $\{2, 4, 6\}$
- $A \cap B = \varnothing$, so A and B are mutually exclusive.

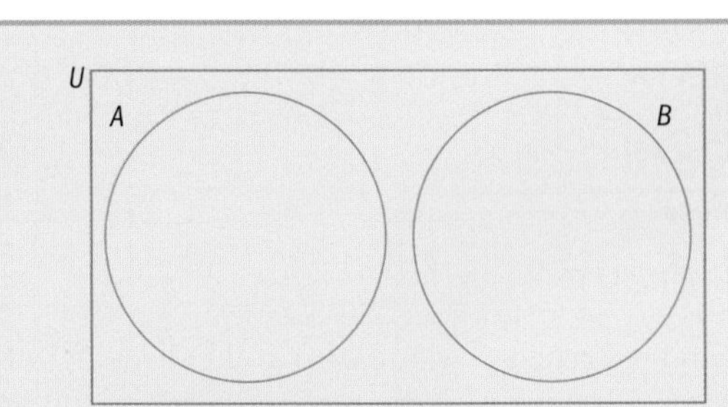

For mutually exclusive events A and B, $A \cap B = \varnothing$.

$$\text{Probability of event } A = \frac{\text{number of ways event } A \text{ can occur}}{\text{total number of possible outcomes}}$$

or:

$$P(A) = \frac{n(A)}{n(S)}$$

So

$$P(A \cap B) = \frac{n(A \cap B)}{n(S)}$$

For the die example, $P(A \cap B) = \frac{0}{6} = 0$. This means the probability that event A and event B both happen is impossible, and therefore equal to zero.

> Two events A and B are mutually exclusive if it is impossible for them to happen together. In mathematical language, if $A \cap B = \varnothing$, then $P(A \cap B) = 0$.

ATL

Exploration 3

1 Construct a Venn diagram to illustrate this information:

- 30 students study one or more of three languages: French, German and Mandarin
- 5 study all three languages
- 6 study Mandarin only
- 2 study French and Mandarin but not German
- 15 students in total study Mandarin
- 2 study only French
- 3 study only German.

2 From your diagram, find n(French) and $n(S)$. Use these to find the probability that a student picked at random studies French.

3 Find n(French and Mandarin) and P(French and Mandarin) from your diagram. State what each of these numbers represent.

4 Explain how you can find P(French $\cap$ German $\cap$ Mandarin) from your Venn diagram.

5 Summarize your findings:

$P(F) =$	$P(F \cap G) =$
$P(G) =$	$P(G \cap M) =$
$P(M) =$	$P(F \cap G \cap M) =$
$P(F \cap M) =$	

6 Find n(French $\cup$ Mandarin) and P(French $\cup$ Mandarin) from your Venn diagram. Use the probabilities you found in step **5** to find and verify a formula for P(French $\cup$ Mandarin). Do the same for P(French $\cup$ German) and P(Mandarin $\cup$ German).

Tip

$\cup$ represents union or 'OR'

$\cap$ represents intersection or 'AND'

Your results from Exploration 3 lead to the addition rule, used to determine the probability that event A **or** event B occurs, **or** both occur.

Addition rule

$P(A \cup B) = P(A) + P(B) - P(A \cap B)$, where:

$P(A)$ is the probability that event A occurs

$P(B)$ is the probability that event B occurs

$P(A \cup B)$ is the probability that event A **or** event B occurs, **or** both occur

$P(A \cap B)$ is the probability that event A **and** event B both occur

Tip

Calculating $P(A) + P(B)$ adds the intersection $P(A \cap B)$ twice, so you need to subtract it once.

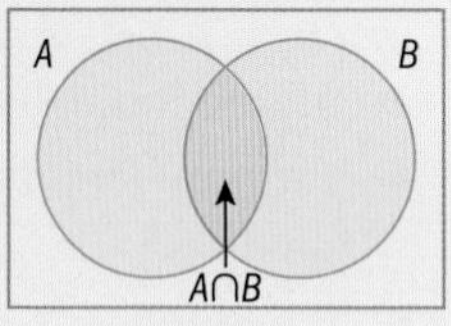

A fair die is rolled once.

Let A be the event 'rolling a 1'.

Let B be the event 'rolling a 2', and so on.

$P(A) = \frac{1}{6}$ $\quad P(B) = \frac{1}{6}$ $\quad P(C) = \frac{1}{6}$ $\quad P(D) = \frac{1}{6}$ $\quad P(E) = \frac{1}{6}$ $\quad P(F) = \frac{1}{6}$

The events are mutually exclusive as it is impossible, for example, to roll a 1 and a 2 at the same time, so $P(A \cap B) = 0$.

$$P(A \cup B) = P(A) + P(B) - P(A \cap B)$$
$$= \frac{1}{6} + \frac{1}{6} - 0 = \frac{2}{6}$$

$$P(A \cup B \cup C) = P(A) + P(B) + P(C) - P(A \cap B \cap C)$$
$$= \frac{1}{6} + \frac{1}{6} + \frac{1}{6} - 0 = \frac{3}{6}$$

Tip

Venn diagrams – if the circles are intersecting they are not mutually exclusive.

Two-way tables – sections in two-way tables are mutually exclusive unless you are told otherwise.

Reflect and discuss 1

Why is the term 'mutually exclusive' an appropriate term for events that cannot happen at the same time?

For mutually exclusive events $P(A \cap B) = 0$, so the addition rule is

$$P(A \cup B) = P(A) + P(B)$$

Axiom 3

If $\{A_1, A_2, A_3, \dots\}$ is a set of mutually exclusive events then $P(A_1 \cup A_2 \cup A_3 \dots) = P(A_1) + P(A_2) + P(A_3) + \dots$

When the occurrences do not coincide, their associated probabilities can be added together.

Axioms 1 to 3 give a good starting point to the structure of probability and start to define its system. This system will help to explain other areas of probability.

Practice 1

> Write down the possible outcomes for event A and event B. If necessary draw a Venn diagram.

1 A fair, six-sided die is rolled once. State whether or not the following events are mutually exclusive (i.e. they cannot happen together).

a 'rolling a 6' and 'rolling a 3'

b 'rolling a multiple of 2' and 'rolling a multiple of 3'

c 'rolling a 4' and 'rolling a factor of 42'

2 The numbers 1, 2, 3, 4, 5, 6, 7, 8 are written on table tennis balls and placed in a bag. A ball is selected at random. State whether or not the following events are mutually exclusive.

a 'the number on the ball is even' and 'the number on the ball is a square number'

b 'the number on the ball is a multiple of 5' and 'the number on the ball is a prime number'

c 'the number on the ball is a factor of 8' and 'the number on the ball is an odd number'

3 Events A and B have probabilities $P(A) = 0.4$, $P(B) = 0.65$ and $P(A \cup B) = 0.85$.

a Calculate $P(A \cap B)$.

b State with a reason whether events A and B are mutually exclusive.

4 A group of 30 students were asked if they'd ever been stung by a bee or a wasp. Of these, 18 students said they'd been stung by a bee, 10 said they'd been stung by a wasp, and 6 said they'd never been stung by either insect.

a Show this information on a Venn diagram.

b Find the number of students who had been stung by both insects.

c Explain why being stung by a bee and being stung by a wasp are not mutually exclusive events.

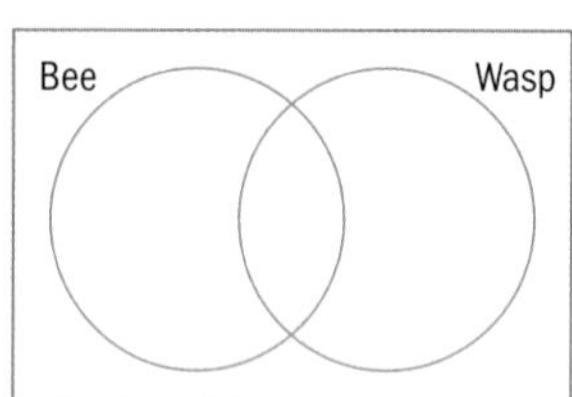

5 Children in a class of 30 students are asked whether they like strawberries (S) or bananas (B).

There are 12 girls in the class: 8 girls like strawberries, 6 girls like bananas and 4 girls like both.

16 boys like strawberries, 13 boys like bananas and 12 boys like both.

This information is shown in the Venn diagram. The teacher chooses a student at random.

U
Strawberry
Girls
4
p
12
4
q
2
Boys
Banana

a Use the information given to find p and q.

b Find P(boy) and P(girl). State whether these events mutually exclusive.

c Find $P(S)$ and $P(B)$. State whether these events mutually exclusive.

Problem solving

6 For the experiment 'Pick a card from a standard, 52 card deck', write:

a two mutually exclusive events

b two events that are not mutually exclusive.

Italian mathematician Gerolamo Cardano (1501–76) analyzed the likelihood of events with both cards and dice. His *Liber de Ludo Aleae* (Book on Games of Chance) discussed in detail many of the basic concepts of probability theory, but received little attention. Cardano often found himself short of money, so he would gamble as a means to make some. His *Ludo Aleae* even contains tips on how to cheat by using false dice and marked cards!

Independent events

If you roll a die twice, the outcome of the first roll does not affect the probabilities of the outcomes of the second roll. The results of the two rolls are independent of each other.

If the outcome of an event in one experiment does not affect the probability of the outcomes of the event in the second experiment, then the events are **independent**.

In a tree diagram representing successive trials with replacement, the trials are independent. In a tree diagram represents successive trial **without** replacement the trials are **not** independent.

Activity

A magician's trick bag contains three rubber snakes: one black, one yellow and one green. A snake is chosen at random, its color is noted and then it is replaced in the bag.

Again, a snake is chosen at random and the color is noted.

Let A be the event 'at least one snake is black'.

Let B be the event 'both snakes are the same color'.

1 List the outcomes from the trials.

2 List the elements of the sets A, B, $A \cup B$, and $A \cap B$.

3 Draw a Venn Diagram to represent the events.

4 Find P(A) from your Venn diagram.

If A occurs, what is the probability that B then occurs? You need to look only at the events for B that are also in A.

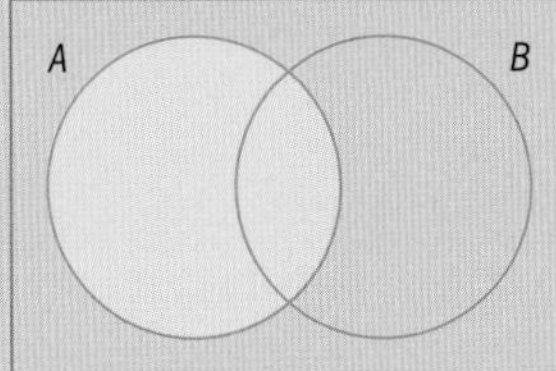

Once A has occurred, the probability of event B is now $\frac{1}{5}$.

This is not the same as P(B). Therefore the outcome of event A affects the probability of event B, so events A and B are not independent events.

5 Find P(B) from your Venn diagram.

Use your Venn diagram to check this.

Tree diagrams can either represent successive trials (such as pulling a snake out of a bag, noting its color, then repeating) or multiple events (such as pulling a numbered ball out of a hat and noting whether or not it is an even number or a square number).

Successive trials **with** replacement mean that events are independent and thus probability remains the same. Successive trials **without** replacement mean events are not independent.

Multiple events require more calculations to determine independence and this is explored further in the Extended book.

Tip

In the Activity, pulling the snakes out of the bag were independent events. However, the events 'at least one snake is black' and 'both snakes are the same color' are **not** independent events.

Exploration 4

1 A regular pack of playing cards is shuffled and one card is drawn. A is the event of drawing a red card and B is the event of drawing a king.

a Find P(A) and P(B).

b Draw a tree diagram to represent these two events.

c If you choose a red card, the sample space is only the red cards. What is the probability of drawing a king?

d Is the probability of getting a king affected by the color of the card?

e Draw a tree diagram for these events.

If A and B are two independent events with probabilities P(A) and P(B) respectively, then $P(A \cap B)$, the probability that A and B will both happen, is found by multiplying the two probabilities together.

Multiplication rule

If A and B are independent events, then $P(A \cap B) = P(A) \times P(B)$.

Example 1

ATL

Tickets numbered 1, 2, 3, 4, 5, 6, 7, 8 and 9 are placed in a bag.

One ticket is then taken out of the bag at random.

Let A be the event 'the ticket's number is even', and let B be the event 'the ticket's number is a square number'.

a Represent the information on a Venn diagram.

b Calculate P(A), P(B) and $P(A \cap B)$.

c Explain whether or not the events are independent.

d Verify your answer to **c** using the multiplication rule.

e Verify that $P(A \cup B) = P(A) + P(B) - P(A \cap B)$. (the addition rule)

▶ Continued on next page

a

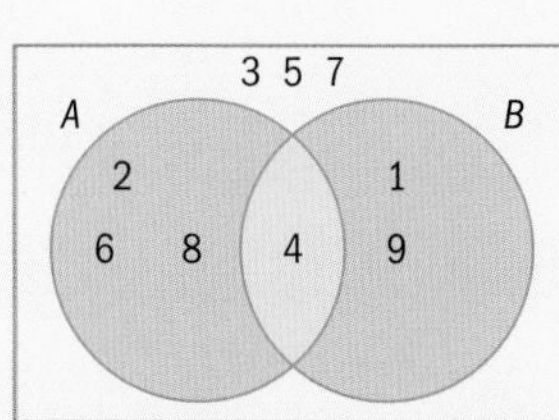

b $P(A) = \frac{n(A)}{n(S)} = \frac{4}{9}$

$P(B) = \frac{n(B)}{n(S)} = \frac{3}{9}$

$P(A \cap B) = \frac{n(A \cap B)}{n(S)} = \frac{1}{9}$

c The events are not independent because whether or not the ticket is an even number affects the probability of it being a square number.

d $\frac{1}{9} \neq \frac{4}{9} \times \frac{3}{9}$

$P(A \cap B) \neq P(A) \times P(B)$ — If two events are independent then $P(A \cap B) = P(A) \times P(B)$.

e $P(A \cup B) = \frac{6}{9}$

$P(A) + P(B) - P(A \cap B) = \frac{4}{9} + \frac{3}{9} - \frac{1}{9} = \frac{6}{9}$ — Verify that $P(A \cup B) = P(A) + P(B) - P(A \cap B)$.

Practice 2

1 A fair six-sided die has the digits 1, 2, 3, 4, 5, 6 on its faces.

A fair four-sided die has the digits 1, 2, 3, 4 on its faces.

The two dice are rolled simultaneously.

The diagram represents the sample space of the possible outcomes.

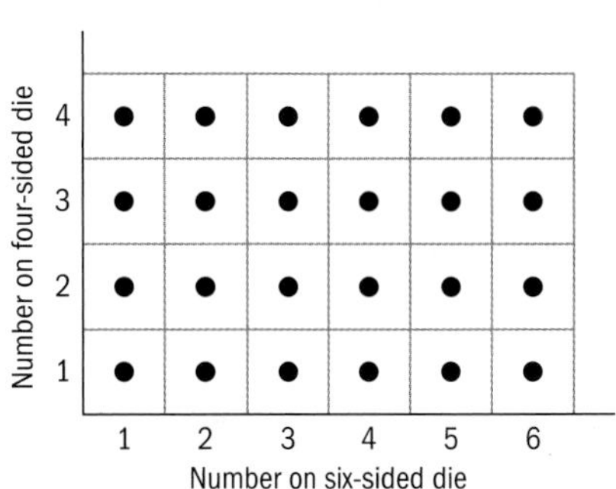

a Let A be the event 'rolling a 1 on the four-sided die', and B be the event 'rolling a 1 on the six-sided die'.

Find $P(A \cap B)$, the probability that you roll a 1 on the four-sided die and a 1 on the six-sided die.

b Explain whether or not the events are independent.

c Verify your result using the multiplication rule.

2 A red die has faces numbered 1, 2, 3, 4, 5, 6 and a green die has faces numbered 0, 0, 1, 1, 2 and 2.

Let A be the event 'rolling a 2 on the red die'.

Let B be the event 'rolling a 2 on the green die'.

The two dice are rolled together.

a Calculate the probability of rolling a 2 on both dice.

b Explain whether or not these events are independent.

c Verify your result using the multiplication rule.

3 Here is a standard set of dominos.

One domino is selected at random. It is then replaced and a second domino is drawn at random.

a Write down $n(S)$.

b Let A be the event 'at least one of the numbers on the domino is a six' and B be the event 'the domino is a double' (i.e. both numbers are the same). Explain whether or not these events are independent.

c Verify your answer to **b** using the multiplication rule.

Problem solving

4 The Venn diagram shows the number of students in a class taking Mathematics (M) and Science (S). Use it to determine whether or not taking Mathematics and Science are independent events.

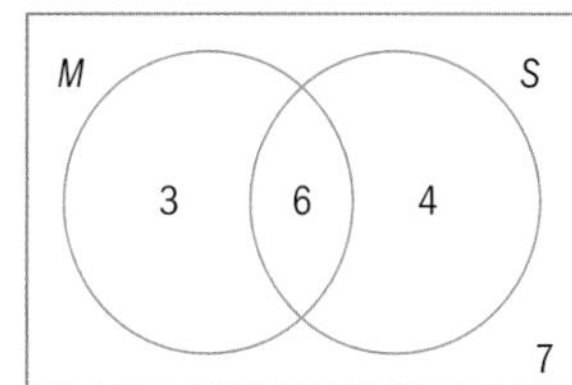

5 In a different class, four students take Mathematics only, two take both Mathematics and Science, six take Science only and 12 take neither subject.

Determine whether the choice of taking Mathematics and taking Science are independent events for this class.

6 Students in an after school activity program register for trampolining or table tennis. The table shows students' choices by gender:

	Trampolining	Table tennis	Total
Male	39	16	55
Female	21	14	35

A student is selected at random from the group. Find:

a P(male trampoliner)

b P(trampoliner)

c P(female)

d Determine whether or not the events trampolining and table tennis are independent events.

C Developing an axiomatic system

- How can an axiomatic system be developed?

Here are three important set definitions.

1 There is a universal set S, containing all other sets.

For any subset $A \subset S$:

$A \cup S = S$ and $A \cap S = A$

2 There is an empty set $\varnothing$.

For any subset $A \subset S$:

$A \cup \varnothing = A$ and $A \cap \varnothing = \varnothing$

3 For any subset $A \subset S$ there exists a unique complementary set.

$A \cup A' = S$ and $A \cap A' = \varnothing$

In probability, the sample space S is equivalent to the Universal set.

The symbol for the complement of set A is either A' or A^c.

Reflect and discuss 2

- Use the set definitions to help you explain why these probability definitions are true.

$P(A \cup S) = P(S)$	$P(A \cup A') = P(S)$
$P(A \cap S) = P(A)$	$P(A \cap A') = P(\varnothing)$
$P(A \cup \varnothing) = P(A)$	$P(S) = 1$
$P(A \cap \varnothing) = P(\varnothing)$	

- Explain what each definition means, in your own words.

Drawing a Venn diagram may help.

Starting from these rules and Axioms 1, 2 and 3, we can deduce new rules, which will hopefully agree with your intuition about probability.

Exploration 5

1 Suppose you roll a fair four-sided die. Find the probability of:

a rolling a 3

b not rolling a 3

c rolling an even number

d not rolling an even number

e rolling a number less than 3

f not rolling a number less than 3

2 Suppose there are 5 choices on a multiple choice test.

a What is the probability that you get the answer correct by guessing?

b What is the probability of not getting the answer correct by guessing?

▶ Continued on next page

c Repeat steps **a** and **b** for a multiple choice question with just 4 answer choices.

d Repeat steps **a** and **b** for a True/False question.

3 How do your answers for P(A) and P(not A) relate to each other?

4 Generalize a rule for calculating the probability of an event's complement P(A') if you know P(A).

A' means the same thing as not A.

If A is an event then $P(A') = 1 - P(A)$

Proof of $P(A') = 1 - P(A)$:

By Axiom 2, $P(S) = 1$. And by definition, $P(A \cup A') = P(S)$. Therefore, $P(A \cup A') = 1$.

A and A' are mutually exclusive, so $P(A \cup A') = P(A) + P(A')$, by the addition rule. Hence $P(A) + P(A') = 1$.

This rearranges to $P(A') = 1 - P(A)$.

Proof of $P(\varnothing) = 0$:

$S' = \varnothing$	by definition
$P(S) = 1$	Axiom 2
$P(A') = 1 - P(A)$	complementary events
$P(S') = 1 - P(S)$	substituting S for A
$P(\varnothing) = 1 - P(S)$	because $S' = \varnothing$
$P(\varnothing) = 1 - 1$	because $P(S) = 1$
$P(\varnothing) = 0$	

ATL

Exploration 6

1 Draw three Venn diagrams showing overlapping sets A and B and shade them to illustrate these three relationships:

$A = (A \cap B) \cup (A \cap B')$

$B = (B \cap A) \cup (B \cap A')$

$A \cup B = (A \cap B) \cup (A \cap B') \cup (A' \cap B)$

2 Hence complete these probability statements:

$P(A) = P((A \cap B) \cup (A \cap B'))$

$P(B) = P(\qquad)$

$P(A \cup B) = P(\qquad)$

▶ Continued on next page

3 From your Venn diagrams you can see that the events $A \cap B$ and $A \cap B'$ are mutually exclusive and hence you can apply Axiom 3. Hence, complete these statements.

$\text{P}(A) = \text{P}(A \cap B) + \text{P}(A \cap B')$

$\text{P}(B) =$

$\text{P}(A \cup B) =$

4 Add together the statements for P(A) and P(B).

5 Rearrange to produce the proof of the addition rule.

Definitions and axioms are the building blocks of any mathematical system. An axiom is a statement whose truth is assumed without proof because it is self-evident.

Theorems are established and proven using axioms.

A corollary is a theorem that is a direct result of a given theorem. Usually, a theorem is a larger, more important statement, and a corollary is a statement that follows simply from a theorem.

A proposition is any statement whose truth can be ascertained.

Practice 3

Axioms, propositions, corollaries and theorems for Practice 3

Axiom 1: For any event A, $\text{P}(A) \geq 0$

Axiom 2: $\text{P}(S) = 1$

Axiom 3: If $\{A_1, A_2, A_3, \ldots\}$ is a set of mutually exclusive events then:

$\text{P}(A_1 \cup A_2, \cup \ldots) = \text{P}(A_1) + \text{P}(A_2) + \ldots$

Proposition 1: Complementary events: if A is an event then $\text{P}(A') = 1 - \text{P}(A)$

Corollary 1: $\text{P}(\varnothing) = 0$

Corollary 2: $\text{P}(A) \leq 1$

Theorem 1: For any events A and B, $\text{P}(A \cup B) = \text{P}(A) + \text{P}(B) - \text{P}(A \cap B)$

Theorem 2: For independent events A and B, $\text{P}(A \cap B) = \text{P}(A) \times \text{P}(B)$

1 A group of 50 students were asked if they liked canoeing and camping. There were 12 students who liked canoeing, 42 students who liked camping, and 8 students who liked both.

- **a** Draw a Venn diagram showing this information.
- **b** Determine how many students did not like either recreation.
- **c** Find the probability that a student chosen at random likes canoeing.
- **d** Determine if liking canoeing and liking camping are independent events.
- **e** State the axioms, propositions and corollaries you used in answering each of the questions **a** through **d**.

2 A group of 100 people were asked if they own a pair of sandals and if they own a pair of slippers. Of these, 48 people own sandals, 14 people own both, 18 people own neither.

a Draw a Venn diagram to show this information.

b Determine how many people own sandals.

c Find the probability that a person selected at random owns a pair of sandals, given that the person owns a pair of slippers.

d Determine if the events are **i** mutually exclusive **ii** independent.

e State the axioms, propositions and corollaries you used in answering each of the questions **a** through **d**.

3 A survey asked if people were right- or left-handed. There were 30 women and 70 men in the survey; 27 of the women were right-handed, 12 of the men were left-handed. Draw a two-way table showing this information.

a Find the probability of being left-handed, given the person is female.

b Determine whether being left- or right-handed are mutually exclusive events.

c Determine whether being left- or right-handed are independent events.

d State the axioms, propositions and corollaries you used in answering each of the questions **a** through **c**.

D Do you need a formal system?

- Does a system help solve a problem?
- What affects the decisions we make?

You have discovered and used the axioms that lead to important probability theorems. How does this system make it easier to answer probability questions?

Example 2

Jenna and Dan are playing volleyball. The probability that Jenna serves the ball to the back of the court is $\frac{1}{5}$.

If Jenna's serve goes to the back of the court, the probability that Dan returns her serve is $\frac{1}{4}$.

If Jenna's serve does not go to the back of the court then the probability that Dan returns it is $\frac{5}{8}$.

The probability that Jenna's serve goes to the back of the court or that Dan returns it is $\frac{7}{10}$.

Let A be the event 'Jenna serves the ball to the back of the court'.

Let B be the event 'Dan returns Jenna's serve'.

a Find $\mathrm{P}(A \cap B)$.

b Find $\mathrm{P}(B)$.

c Show that A and B are **not** independent.

▶ Continued on next page

a

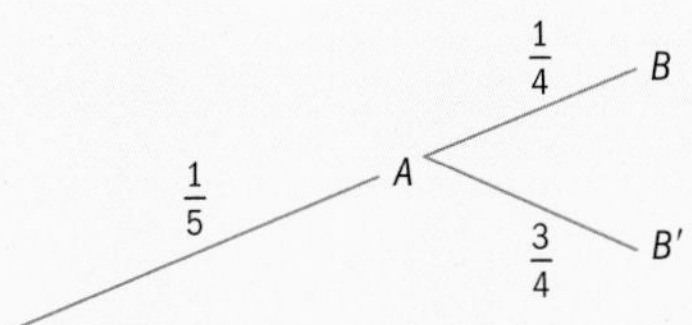

Draw just the part of the tree diagram you need to find P(A ∩ B).

$$P(A \cap B) = \frac{1}{5} \times \frac{1}{4} = \frac{1}{20}$$

b $P(A \cup B) = P(A) + P(B) - P(A \cap B)$ — Use Theorem 1

$$\frac{7}{10} = \frac{1}{5} + P(B) - \frac{1}{20}$$

Use the information given in the question, plus P(A ∩ B) which you found from the tree diagram.

$$P(B) = \frac{7}{10} - \frac{1}{5} + \frac{1}{20} = \frac{11}{20}$$

c

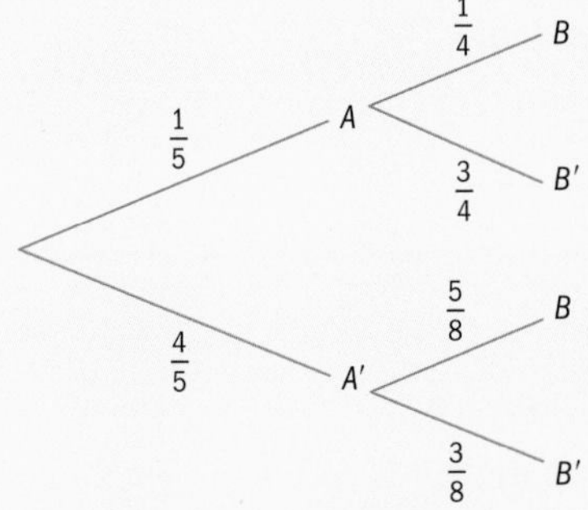

Draw a full tree diagram.

In the tree diagram, the probabilities are different for B depending on where the original serve was. Therefore events A and B are not independent.

Objective: C. Communicating
v. Organize information using a logical structure

In Practice 4, you will use the axioms of probability to put in place a structure to answer the problems.

Practice 4

1 In a class of 20 students, 12 of them study History, 15 study Geography and 2 students study neither History nor Geography.

Let A be the event 'number of students who study History'.

Let B be the event 'number of students who study Geography'.

a Write down:

i $n(S)$

ii $n(A)$

iii $n(B)$

iv $n(A \cup B)$

v $P(A)$ and $P(B)$

vi $P(A \cup B)$

Tip

For **1 a iv**, look at the number of students who study neither History nor Geography and use Proposition 1.

b **i** Let $P(A \cap B) = x$. Use theorem 1 to calculate x.

ii Represent the information in a Venn diagram.

c Given that a student picked at random studies History, find the probability that this student:

i also studies Geography

ii does not study Geography.

d Given that a student picked at random does **not** study History, find the probability that this student:

i studies Geography

ii does not study Geography.

e Draw a tree diagram to represent this information.

2 Events A and B are such that $P(A) = 0.3$ and $P(B) = 0.4$.

a **i** Use Axiom 3 to find $P(A \cup B)$ given that the events are mutually exclusive.

ii Use Theorem 2 to find $P(A \cap B)$ given that A and B are independent.

b Draw a tree diagram to represent this information.

Summary

- Axiom 1: For any event A, $P(A) \geq 0$
- Axiom 2: $P(S) = 1$
- Two events A and B are mutually exclusive if it is impossible for them to happen together $A \cap B = \varnothing$ and so $P(A \cap B) = 0$
- Axiom 3: If $\{A_1, A_2, A_3, \ldots\}$ is a set of mutually exclusive events then:
  $P(A_1 \cup A_2 \cup A_3 \cup \ldots) = P(A_1) + P(A_2) + P(A_3) + \ldots$
- Proposition 1: Complementary events: if A is an event then $P(A') = 1 - P(A)$
- Corollary 1: $P(\varnothing) = 0$
- Corollary 2: $P(A) \leq 1$
- If the outcome of an event in one experiment does not affect the probability of the outcomes of the event in the second experiment, then the events are **independent**.
- Theorem 1: For any events A and B, $P(A \cup B) = P(A) + P(B) - P(A \cap B)$
- Theorem 2: For independent events A and B, $P(A \cap B) = P(A) \times P(B)$

Mixed practice

1 A group of 30 students were asked about their favourite way of eating eggs.

- 18 liked boiled eggs (B)
- 10 liked fried eggs (F)
- 6 liked neither

a **Find**

i $P(B)$ **ii** $P(B \cup F)$
iii $P(F)$ **iv** $P(B' \cup F')$

b Use Theorem 1 to **find** $P(B \cap F)$.

c Represent the information on a Venn diagram.

d A student chosen at random likes boiled eggs. **Find** the probability that the student also likes fried eggs.

Problem solving

2 A survey was carried out at an international airport. Travelers were asked their flight destinations and results are shown in the table.

Destination	Geneva	Vienna	Brussels
Number of males	45	62	37
Number of females	35	46	25

a **i** **Determine** whether or not the destinations are mutually exclusive.

ii **Determine** whether or not gender is mutually exclusive.

b One traveller is chosen at random. **Find** the probability that this traveller is going to Vienna.

c One female traveller is chosen at random. **Find** the probability that she is going to Geneva.

d One traveller is chosen at random from those **not** going to Vienna. **Find** the probability that the chosen traveller is female.

e Copy and complete the tree diagram for the data in the table.

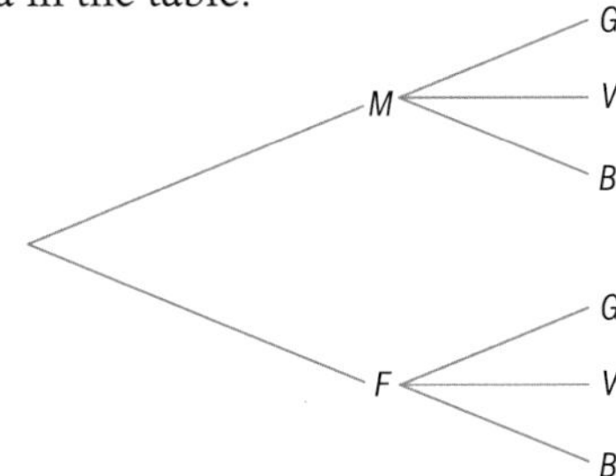

For this data, test whether or not destination and gender are independent.

3 Let $P(A) = 0.5$, $P(B) = 0.6$ and $P(A \cup B) = 0.8$

a **Find** $P(A \cap B)$.

b **Draw** a tree diagram. **Find** the probability of getting B, given that A has occurred.

c **Determine** whether or not A and B are independent events. Give a reason for your answer.

4 100 students were asked if they liked various toast toppings.

- 56 like avocado
- 38 like marmalade
- 22 like soft cheese
- 16 like avocado and marmalade, but not soft cheese
- 8 like soft cheese and marmalade, but not avocado
- 3 like avocado and soft cheese, but not marmalade
- 4 like all three toppings

a **Draw** a Venn diagram to represent this information.

b **Find** the number of students who didn't like any topping.

c **Determine** if the toppings are mutually exclusive.

d A student is chosen at random from the group who like soft cheese. **Find** the probability that they also like marmalade.

5 Events A and B have probabilities $P(A) = 0.4$, $P(B) = 0.65$, and $P(A \cup B) = 0.85$.

a **Calculate** $P(A \cap B)$.

b **Determine** whether or not events A and B are independent.

c **Determine** whether or not events A and B are mutually exclusive.

6 Stefan rolls two 6-sided dice at the same time. One die has three green sides and three black sides. The other die has sides numbered from 1 to 6.

a **Draw** a tree diagram to represent this sequence of events.

b **Find** the probability that Stefan rolls a 5.

c **Find** the probability that he rolls a number less than 3.

d **Find** the probability that he gets green on the one die and an even number on the other.

e **Determine** if the events 'rolling green' and 'rolling an even number' are independent, and explain your answer.

7 The table below shows the number of left- and right-handed golf players in a sample of 50 males and females.

	Left-handed	Right-handed	Total
Male	3	29	32
Female	2	16	18
Total	5	45	50

A golf player is selected at random.

a **Find** the probability that the player is:

i male and left-handed

ii right-handed

iii right-handed, given that the player selected is female.

b **Determine** whether the events 'male' and 'left-handed' are:

i independent

ii mutually exclusive.

Review in context

Identities and relationships

Remember to use the probability axioms when answering these questions.

Problem solving

1 The table shows 60 students' choices of yoghurt.

Sugar levels	Yoghurt			
–	Strawberry	Chocolate	Vanilla	Total
Low sugar	3	8	14	25
High sugar	8	9	18	35
Total	11	17	32	60

a One student is selected at random.

i **Find** the probability that the student chose vanilla yoghurt.

ii **Find** the probability that the student chose a yoghurt that was not vanilla.

iii The student chose chocolate yoghurt. **Find** the probability that the student chose the low sugar one.

iv **Find** the probability that the student chose a high sugar or vanilla yoghurt.

v The student chose a low sugar yoghurt. **Find** the probability that the student chose strawberry.

b The 60 yoghurts were then classified according to fat content type: 15 of the yoghurts had high fat, 37 had medium fat and 8 had low fat content. Two yoghurts were randomly selected.
Find the probability that:

i both yoghurts had low fat content

ii neither of the yoghurts had medium fat content.

2 All human blood can be categorized into one of four types: A, B, AB or O. The distribution of the blood groups varies between races and genders. The table shows the distribution of the blood types for the US population.

Blood type	O	A	B	AB
Probability	0.42	0.43	0.11	x

a **Find** the probability that a person chosen at random:
  i has blood type AB
  ii does **not** have blood type AB.

b **Determine** if the two events 'have blood type AB' and 'do not have blood type AB' are mutually exclusive.

c **Find** the probability that a randomly selected person in the US has blood type A or B.

d Damon has blood type B, so he can safely receive blood from people with blood types O and B. **Find** the probability that a randomly chosen person can donate blood to Damon.

e Given that Damon received blood, **find** the probability that it was type O.

3 This table shows the distribution of blood type in the US by gender.

	Probability		
Blood type	Male	Female	Total
O	0.21	0.21	0.42
A	0.215	0.215	0.43
B	0.055	0.055	0.11
AB	0.02	0.02	0.04
Total	0.50	0.50	1.0

a **Determine** if the events 'gender' and 'blood type' are mutually exclusive.

Let M be the event 'selecting a male'.

Let F be the event 'selecting a female'.

b **Find i** $P(M)$ **ii** $P(F)$

c Copy and complete the tree diagram.

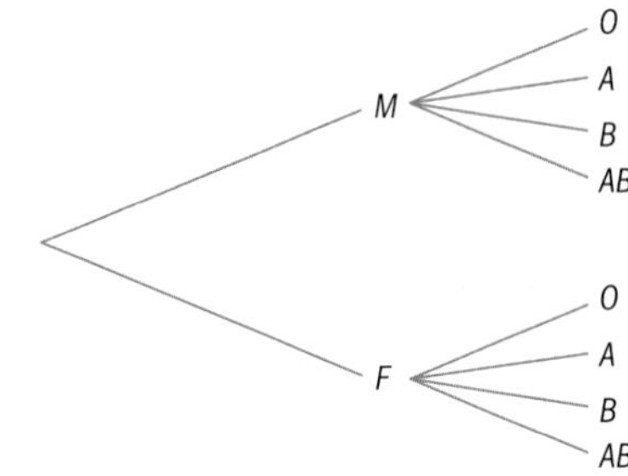

d Confirm that blood type and gender are independent events, using the multiplication rule for independent events.

Reflect and discuss

How have you explored the statement of inquiry? Give specific examples.

Statement of Inquiry:

Understanding health and making healthier choices result from using logical representations and systems.

Answers

Problem-solving

You should already know how to:

1 17.3 cm

2 a $P = 15.4$ cm, $A = 10$ cm^2 b $P = 18.8$ cm, $A = 28.3$ cm^2

3 a 120 cm^3 b 402 cm^3

4 $4x + 20 = 180$; $x = 40°$

Practice 1

1 1 500 000 000 = 1.5×10^9 times around the Earth

2 1 month and 1 day

3 14 km/h

4 A: 80 ml; B: 120 ml

5 23 pieces of jewellery

6 11.0 mph

Practice 2

1 48 students

2 a 92% b 91% c 4 more tests

3 20 minutes

4 8 minutes 20 seconds

5 a 4 hours 48 minutes
b 10 hours 30 minutes

6 176 steps

7 120 more days

8 a 170 diagonals b $\frac{n(n-3)}{2}$ diagonals

Practice 3

1 75 cm^3

2 a 1 : 6.29
b The ratio is about $1{:}2\pi$
c Student's own explanation.

Practice 4

1 25π

2 1000 cm^3

3 length = 3 cm, width = 12 cm, height = 18 cm

4 Its volume will decrease by $\frac{1}{16}$

5 40 minutes

6 358 cm^3

7 Use the 4L container to fill the 9L container, leaving 3L in the 4L container.
Empty the 9L container.
Pour the 3L from the 4L container into the 9L container.
Use the 4L container to fill the remaining 6L of the 9L container, leaving 2L in the 4L container.
Empty the 9L container.
Pour the 2L from the 4L container into the 9L container.
Fill the 4L container and pour it into the 9L container.
The 9L container will now contain the 6L you need.

Practice 5

1 You need the base and the perpendicular height.

2 The $4 per ticket is irrelevant. You need to know how much she paid.

3 You need to know the depth, length and width of the TV.

4 The time he gets up is irrelevant. You need to know his walking speed.

5 The bottle's weight is irrelevant.

6 The take-off time and speed are irrelevant. You need no more information, ignoring the curvature of the Earth.

Mixed practice

1 2

2 $x = \frac{360°}{7} = 51.4$

3 $\frac{7}{13}$

4 4 hours

5 3 hours

6 4 mg

7 47.5%

8 12 hours

9 $\frac{5}{6}$ of the catalogue

10 a after 5 months b 330 hamsters

11 36 apples

12 510 000 Euros

Review in context

1 a i 128 grains of rice
ii 32 768 grains of rice
iii 9.22×10^{18} grains of rice
b i 2.56 cm^2
ii 655.36 cm^2
iii 167 772.16 cm^2
c 16.78 m^2
d 18 446 744.07 km^2
e i 255 grains of rice
ii 65 535 grains of rice
iii 16 777 215 grains of rice
iv 1.845×10^{19} grains of rice

2 a i 7 moves
ii 15 moves
iii 31 moves
b i 7 seconds
ii 31 seconds
iii 63 seconds = 1 minute 3 seconds
c i 1023 seconds = 17 min 3 seconds
ii 1 048 575 seconds = 12 days 3 h 16 min 15 s

3 a Student's own answer
b $2\pi r$
c $2\pi(r + 1 \text{ inch}) = 2\pi r + 2\pi$ inches
d 2π inches
e No, the size of the sphere does not matter. The circumference increases by 2π inches independently of the radius r.

There may be more solutions to each question than provided here.

4 a $(7 + 1) \times (2 + 1) = 24$ b $(2 + 2 + 4) \times 3 = 24$
c $(2 + 3 \times 6) + 4 = 24$ or $(3 - 2) \times 4 \times 6 = 24$
d $(12 + 7 \times 2) - 2 = 24$

1.1

You should already know how to:

1 a Real, Rational, Integers, Natural
b Real, Rational, Integers
c Real, Rational, Integers, Natural
d Real, Rational
e Real, Irrational
f Real, Rational
g Real, Irrational
h Real, Rational, Integers

Practice 1

1 a {Monday, Tuesday, Wednesday, Thursday, Friday, Saturday, Sunday}; $n(A) = 7$
b {May, June, July, August}; $n(B) = 4$
c $\{1, 2, 3, 4, 6, 12\}$; $n(C) = 6$ OR $\{-1, -2, -3, -4, -6, -12, 1, 2, 3, 4, 6, 12\}$: $n(C) = 12$
d $\{4, 8, 12, 16, 20, 24, 28\}$; $n(\mathrm{D}) = 7$

2 a J is the set of odd positive integers less than 10.
b K is the set of types of triangle.
c L is the set of types of angle, not including straight line or full turn.
d M is the set of multiples of 4 less than or equal to 40.

3 a True **b** False: $7 \in C$
c False: $1 \notin A$ **d** False: $27 \notin A$
e False: $8 \notin D$ **f** False: $n(\mathrm{C}) \neq n(D)$

Practice 2

1 a $\{-1, 0, 1, 2, 3, 4\}$; Finite; 6
b $\{1, 2, 3, \ldots\}$; Infinite
c $\{5, 10, 15, \ldots\}$; Infinite
d $\{1, 2, 4, 7, 14, 28\}$; Finite; 6
e $\{1, 2, 3, 4\}$; Finite; 4
f {Red, Blue, Yellow}; Finite; 3

2 a $\{s \mid s = x^2, x \in \mathbb{N}$ and x is odd$\}$ **b** $\{t \mid t = 10x, x \in \mathbb{Z}\}$
c $\{u \mid u \in \mathbb{R}, 1 < u \leq 2\}$ **d** $\{v \mid v \in \mathbb{Q}, 0 < v < 1\}$

3 For this question, answers other than those below are possible.
a $\{x \mid x = 2n - 1, x \in \mathbb{N}\}$ **b** $\{x \mid x = 3n, x \in \mathbb{N}\}$
c $\{x \mid x = 2^n, n \in \mathbb{Z}, 0 \leq v \leq 5\}$

Practice 3

1 a Yes **b** No **c** No

2 a True **b** True
c False: 3 should not be bracketed
d False: 4 should be bracketed **e** True

3 a True
b False: answers will vary, e.g. $R = \{1, 2\}$, $S = \{1, 2, 3\}$
c True **d** True

4 $\{1, 2, 3, 4, 5\}$

5 24 is not a power of 2

Practice 4

1 a $\varnothing$ **b** U **c** $\{3, 5, 7, 11, 13, 17, 19\}$
d $\{1, 3, 4, 5, 6, 7, 8, 9, 10, 11, 12, 13, 14, 15, 16, 17, 18, 19, 20\}$
e A

2 a $\{1, 2, 3, 4\}$
b $\{1, 2, 3, 4, 5, 6, 7, 8, 9, 10, 11, 12, 13, 14, 15\}$
c $\{10, 11, 12, 13, 14, 15\}$
d $\{5, 6, 7, 8, 9, 10, 11, 12, 13, 14, 15\}$
e $\{5, 6, 7, 8, 9, 10, 11, 12, 13, 14, 15\}$

3 Answers will vary, e.g. $G = \{12, 14, 16, 18, 20\}$; $F = \{12, 14\}$; $H = \{13, 17, 19\}$

Practice 5

1 a $(A \cap B)'$

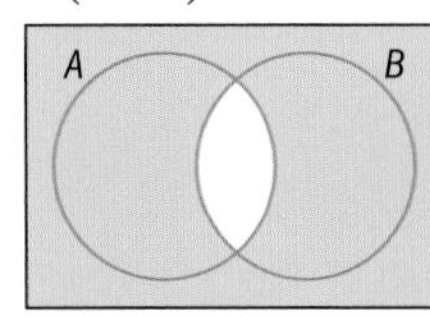

b $A \cup (A \cap B)$

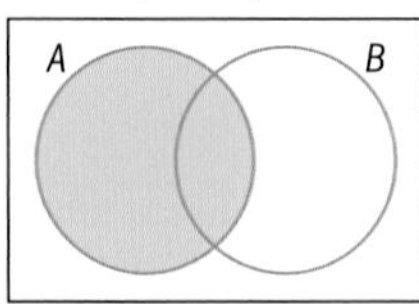

c $(A' \cap B)'$

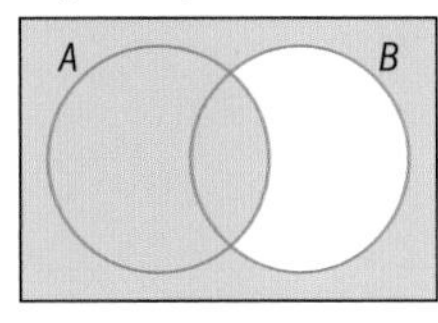

d $(A' \cup B')'$

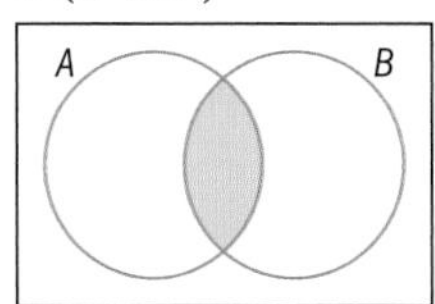

e $A \cup (B \cap C)$

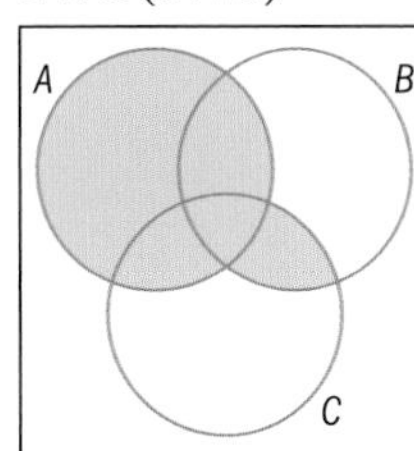

f $(A \cap B)' \cup C$

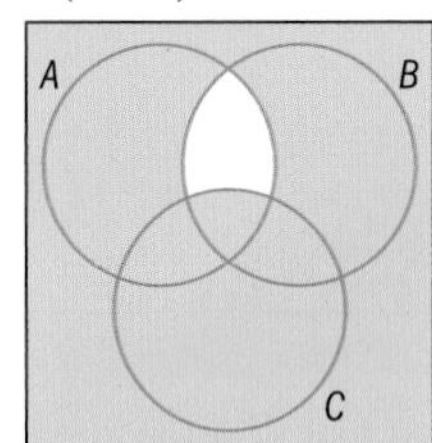

g $A \cap (B \cup C)'$

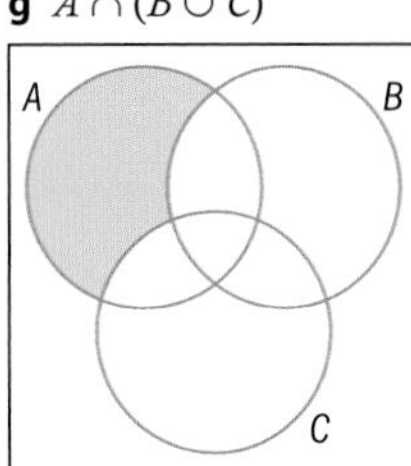

h $(A \cap B) \cup (A \cap C) \cup (B \cap C)$

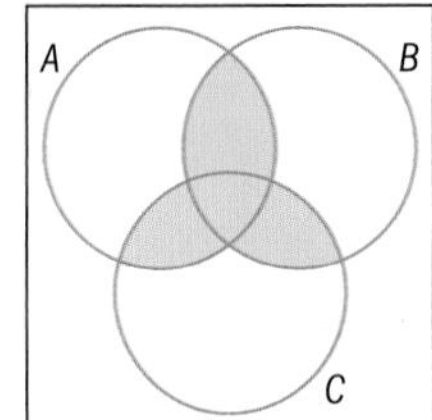

2 a A' **b** B
c $B \cap A'$ **d** $A \cap B \cap C$
e $(C \cap A') \cup B$ **f** $B \cap C \cap A'$
g $A \cup C \cup B'$ **h** $B' \cup A' \cap C'$

Practice 6

1 a i Live birth, breathe air; lay eggs, have scales, breathe water; live in water, have fins, can swim
ii Student's own answers
b i **Sea creatures**

W F S
1 2 3 5 6 7

Region 1: Breathes air, live births
Region 2: Fins
Region 3: Has scales
Region 5: Lives in water
Region 6: Lays eggs, breathes water
Region 7: Has legs

ii Student's own answers

2 8

3 218

4 115

5 a 35 **b** 20 **c** 45

6 a 2 **b** 19

7 17

8 a Students who do table tennis, basketball and squash
b Students who only do table tennis
c Students who do table tennis and basketball but not squash

Mixed practice

1 All answers correct at time of going to press.
a {Europe, Asia, Africa, North America, South America, Oceania, Antarctica}
b {Everest, K2, Kangchenjunga}
c {Burj Khalifa, Shanghai Tower, Abraj Al-Bait Clock Tower}
d {20, 25, 30, 35, 40, 45, 50}

2 a infinite **b** infinite **c** finite, 5 **d** finite, 6

3 a $\{x \mid x \in \mathbb{Z},\ x > 0\}$
b $\{x \mid x \in \mathbb{N}, x = 3n,\ 1 \le n \le 7\}$
c $\{x \mid x \in \mathbb{Z}$, x is a square number$\}$
d $\{x \mid x \in \mathbb{Z}^+$, x is a cube number $\le 1000\}$

4 a {7, 14}
b {11, 13, 17, 19}
c {even multiples of 7: 14, 28, 42, ...}
d $\{x \mid x \in \mathbb{N}, x \ne 7\}$

5 a

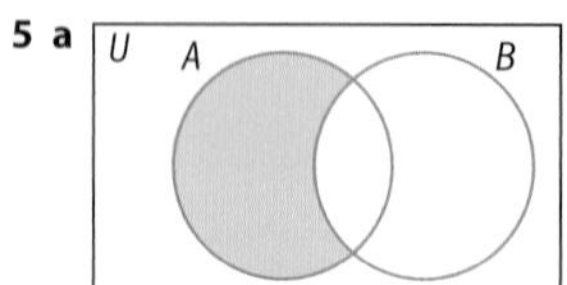

b

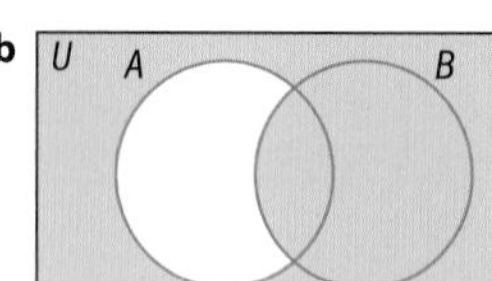

6

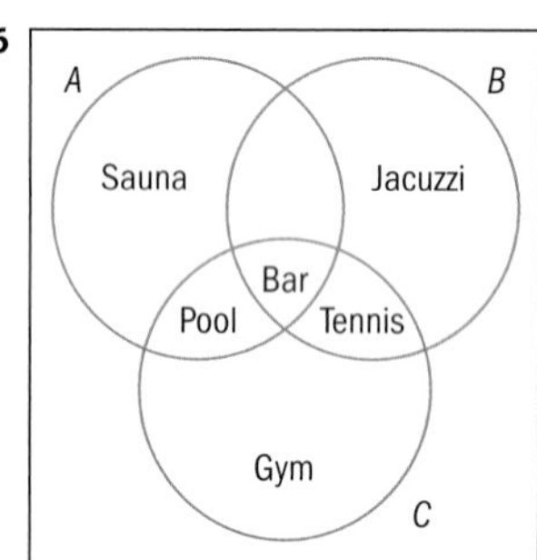

7 Real Numbers

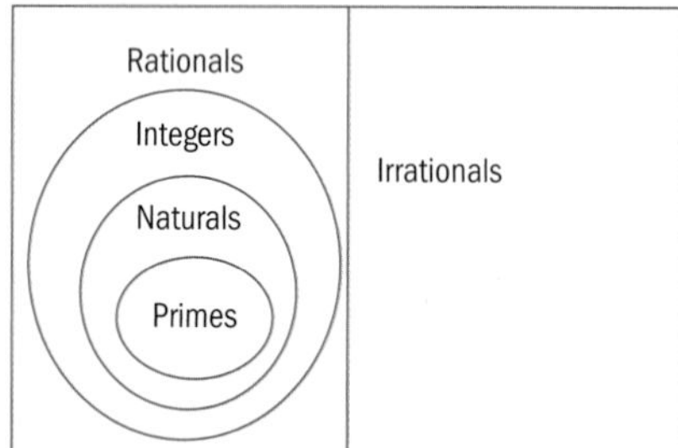

8

Quadrilaterals
Parallelograms
Squares
Rectangles
Trapezoids
Kites

9 a {0}
b {−10, 6, 7, 8, 9, 10}
c {−5, −4, −3, −2, −1, 0, 1, 2, 3, 4, 5}
d {−9, −8, −7, −6, −5, −4, −3, −2, −1}
e {0, 1, 2, 3, 4, 5}

10 a True **b** False: $2 \in$ {primes}
c True **d** True

11 13

12 a 6 **b** 3 **c** 5 **d** 21

Review in context

1 a 464, 1101, 3154 **b** 14
c Student's own answers

2 a

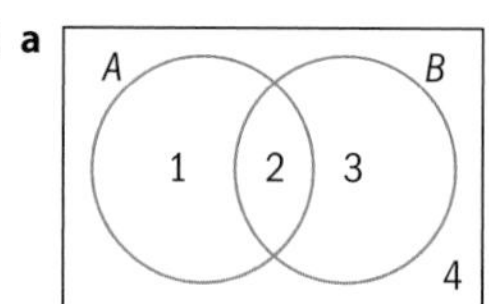

A: Truths; B: Beliefs
Region 2 defines Knowledge

b

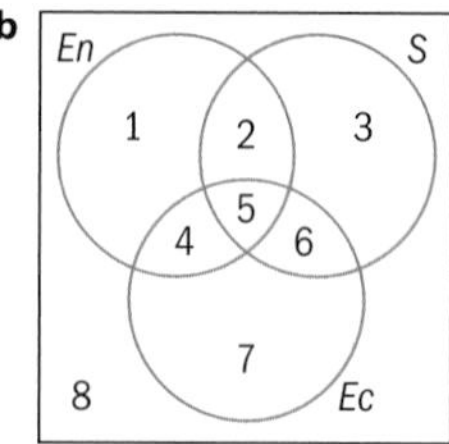

Region 5: Sustainable
Regions 5 and 2: Bearable
Regions 5 and 4: Viable
Regions 5 and 6: Equitable

3 a 10 660 **b** 18 402
c Mouse: 15 213, Chicken: 11 705, Zebrafish: 12 897
d Mouse: 82.7%, Chicken: 63.6%, Zebrafish: 70.1%
e Mouse

2.1

You should already know how to:

1 a 729 **b** 0.015625

2 a $\frac{1}{18}$ or $0.\dot{5}$ **b** 18

3 a $\{x \mid x \in \mathbb{R}\}$ **b** $\{x \mid x \in \mathbb{Z}, x \ge 0\}$

Practice 1

1 a many-to-one
b one-to-one (unless there are siblings in your class)
c one-to-many

2 a Student's own definitions
b Student's own relation

3 For $y = x + 2$, one-to-one
For $x^2 = y^2$, many-to-many

Practice 2

1 a Yes. Every input value has only one output value.
b Domain = $\{1, 2, 3, 4, 5, 6\}$, Range = $\{1, 2, 3\}$
c i 1 **ii** 2 **iii** 2

2 a i The country where a brand of chocolate comes from.
ii Because each brand of chocolate comes from only one country.
iii Domain = {Neuhaus, Cote d'Or, Leonidas, Toblerone, Lindt, Hershey's, Baci}
Range = {Belgium, Switzerland, USA, Italy}
iv Student's own choice.
b i y is the square of x.
ii Because every input value has only one output value.
iii Domain = $\{\pm 1, \pm 2, \pm 5\}$
Range = $\{1, 4, 25\}$
iv Student's own choice.
c i Whether a person is a girl or a boy.
ii Because no person can be both male and female.
iii Domain = {Joanne, James, Jessica, Jennifer, Joseph}
Range = {Girl, Boy}
iv Student's own choice.

Practice 3

1 a Domain = $\{2, 4, 5, 7, 8\}$, Range = $\{1, 2, 4, 6, 9\}$
b Domain = $\{3, 4, 6, 8\}$, Range = $\{-4, 2, 4, 7, 8\}$
c Domain = $\{-7, -3, -2, 3, 5, 8\}$, Range = $\{-1, 1\}$
d Domain = $\{a, j, k, p, s, t\}$, Range = $\{c, g, k, p, w\}$
e Domain = $\{a, b\}$, Range = $\{x, y, z\}$

2 a i $\{(-2, -2) (-1, -1) (0, 0) (1, 1) (2, 2)\}$
ii $B = \{-2, -1, 0, 1, 2\}$
iii

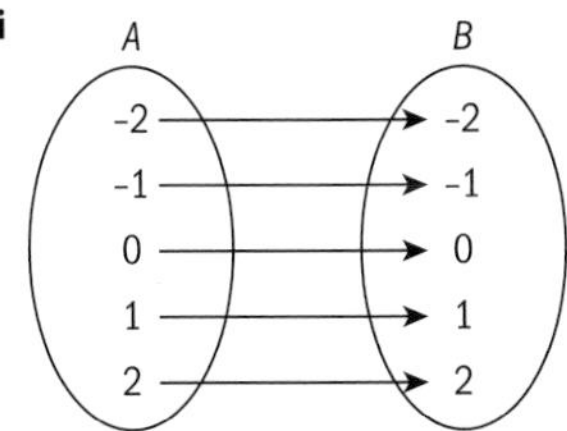

iv one-to-one
v Yes, it is a function
b i $\{(-2, -3) (-1, 0) (0, 3) (1, 6) (2, 9)\}$
ii $B = \{-3, 0, 3, 6, 9\}$
iii

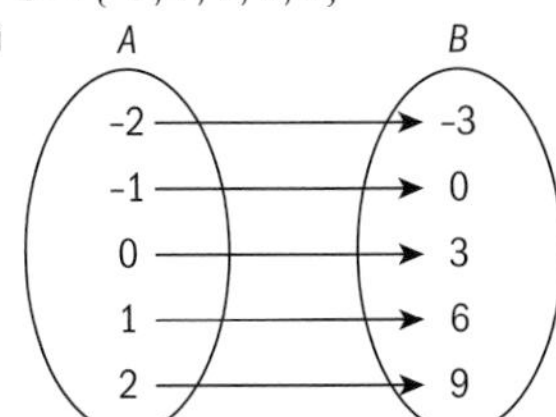

iv one-to-one
v Yes, it is a function
c i $\{(-2, 4) (-1, 1) (0, 0) (1, 1) (2, 4)\}$
ii $B = \{0, 1, 4\}$
iii

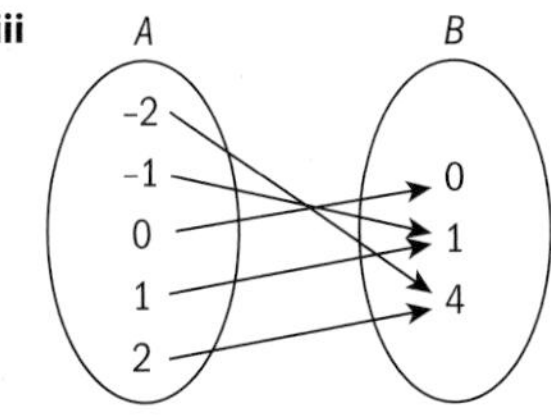

iv many-to-one
v Yes, it is a function

Practice 4

1 The domain is $x \in \mathbb{R}$.
The range is $y \in \mathbb{R}$.

2 The domain is $x \in \mathbb{R}$.
The range is $y \in \mathbb{R}$.

3 The domain is $x \in \mathbb{R}$.
The range is $y \in \mathbb{R}$.

4 The domain is $x \in \mathbb{R}$.
The range is $\{y \mid y \in \mathbb{R}, y \geq 0\}$.

5 The domain is $x \in \mathbb{R}$.
The range is $\{y \mid y \in \mathbb{R}, y \geq 5\}$.

6 The domain is $x \in \mathbb{R}$.
The range is $\{y \mid y \in \mathbb{R}, y \geq -3\}$.

7 The domain is $\{x \mid x \in \mathbb{R}, x \neq 0\}$.
The range is $\{y \mid y \in \mathbb{R}, y \neq 0\}$.

8 The domain is $\{x \mid x \in \mathbb{R}, x \neq 0\}$.
The range is $\{y \mid y \in \mathbb{R}, y \neq 0\}$.

9 The domain is $\{x \mid x \in \mathbb{R}, x \neq -1\}$.
The range is $\{y \mid y \in \mathbb{R}, y \neq 0\}$.

10 The domain is $\{x \mid x \in \mathbb{R}, x \neq 0\}$.
The range is $\{y \mid y \in \mathbb{R}, y \neq 1\}$.

Practice 5

1 a Yes **b** No **c** No **d** Yes

Practice 6

1 a Yes. Each input has a single output.
b No. –1 and 5 both have two different outputs.
c Yes. Each input has a single output.
d No, because 9 has two different output values.
e Yes. Each input has a single output. Also, the graph passes the vertical line test.
f No, because 2, for example, has two different output values. Also, the graph does not pass the vertical line test.
g No, because 4, for example, has two different output values. Also, the graph does not pass the vertical line test.
h Yes. Each input has a single output. Also, the graph passes the vertical line test.

Practice 7

1 a $f(x) = 30x + 40$
b Domain $x \geq 0$
Range $f(x) \geq 40$
c It is a function because each input has a unique output.

2 a $f(x) = 120 - 25x$
b Domain $0 \leq x \leq 4.8$
Range $0 \leq f(x) \leq 120$
c It is a function because each input has a unique output.

3 a $f(x) = 230x$
  b Domain $x \geq 0$
    Range $f(x) \geq 0$
  c It is a function because each input has a unique output.

Practice 8

1 a It multiplies it by 4, then subtracts 2.
  b Student's own table of values.

2 a It squares it, then adds 2.
  b Student's own table of values.

3 a It divides 2 by the input value.
  b Student's own table of values.

4 a 18 b −10 c −3 d 137

5 a quadrilateral b hexagon
  c octagon d decagon

Practice 9

1 a 2 b −2 c 6
  d −78 e $2 + 4a$ f $2 - 8x$

2 a −5 b 7 c −17
  d 31 e $9x - 5$ f $3x - 2$

3 a 13 b 13 c 13
  d 13 e 13 f 13

4 a 1 b 26 c 26
  d 82 e $x^2 + 2x + 2$ f $16x^2 + 1$

5 a 0 b 1 c 1
  d 36 e 121 f $9x^2 - 6x + 1$

6 a 29 b −1 c 0.5
  d 103 e −27 f $-4x^3 + 2x + 1$

7 a 20 b 35 c $6x + 5$
  d $9x + 5$ e $3x + 20$ f $9x - 1$

8 2 9 0 10 6

11 6 or −6 12 7 or −7 13 2

14 10 15 315 16 Student's mother

Mixed practice

1 a A function, because each input has a unique output.
  b A function, because each input has a unique output.
  c A relation, because some inputs have more than one output.
  d A relation, because the input 3 has more than one output.
  e A function, because each input has a unique output.
  f A relation, because some inputs have more than one output.
  g A function, because each input has a unique output.
  h Not a function, as e.g. $x = 0$ maps to 3 and −3. Also, the graph does not pass the vertical line test.
  i Not a function, as $x = 1$ maps to all values between 0 and 1. Also, the graph does not pass the vertical line test.
  j A function, as e.g. $x = 1$ maps to 5, not 10 as the open circle shows the value is not included. Also, the graph passes the vertical line test.

2 a Domain = {−2, 1, 2}, Range = {−4, 1, 4}
  b Domain = {3, 4, 5}, Range = {5, 6}
  c Domain = {−3, −2}, Range = {−3, −2, 5}
  d Domain = {1, 2, 5}, Range = {1, 3, 8}

3 a The domain is $x \in \mathbb{R}$.
    The range is $f(x) \in \mathbb{R}$.
  b The domain is $\{x \mid x \in \mathbb{R}, x \geq 0\}$
    The range is $\{f(x) \in \mathbb{R} \mid f(x) \geq 0\}$
  c The domain is $x \in \mathbb{R}$.
    The range is $f(x) \in \mathbb{R}$.
  d The domain is $\{x \mid x \in \mathbb{R}, x \neq 0\}$
    The range is $\{f(x) \in \mathbb{R} \mid f(x) \neq 0\}$
  e The domain is $\{x \mid x \in \mathbb{R}, x \neq 4\}$
    The range is $\{f(x) \in \mathbb{R} \mid f(x) \neq 0\}$
  f The domain is $x \in \mathbb{R}$.
    The range is $\{f(x) \in \mathbb{R} \mid f(x) \geq -1\}$

4 a i −2 ii 0 iii −2
  b i 28 ii 20 iii 4
  c i 0 ii 2
    iii undefined for real numbers
  d i 11 ii $8x + 3$ iii $7 - 2x$

5 a i $x = 5$ ii $x = 0.5$ iii $x = 1$
  b i $x = 5$ ii $x = -4$ iii $x = 0.5$
  c i $-\sqrt{5}$ ii 1 iii 3

6 a Because a word only ever has one amount of letters in it.
  b p(word) = number of letters
  c i 5 ii 11 iii 2
  d i Student's own 3-letter word
    ii Student's own 8-letter word
  e 5 f 4

7 a Each value n maps to one and only one value $T(n)$.
  b Domain = {−3, −2, −1, 0, 1, 2, 3, 4, 5, 6, 7, 8} (ground floor is zero, so the 9 floors above ground are 0 to 8)
  c Student's suggestions for values of $T(n) \geq 0$
  d Range $T(n) \geq 0$ with maximum value and suitable justification. E.g. $0 \leq T(n) < 1440$ because there are 1440 minutes in 24 hours and the lift will not return to a floor in less than one minute.

8 a, b Weight determines price

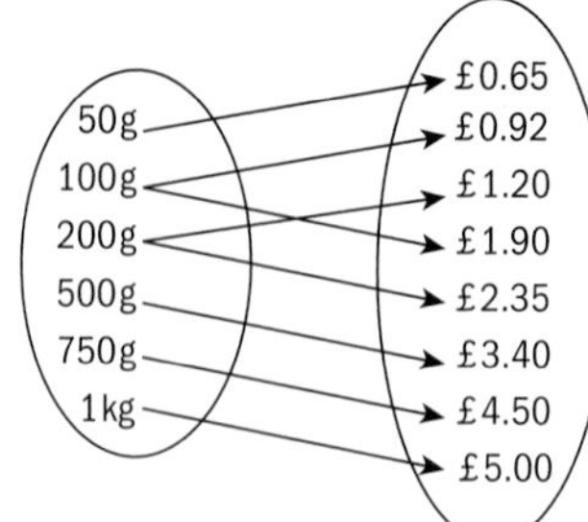

  c Not a function, because e.g. 100 g maps to two possible prices.

Review in context

1 a

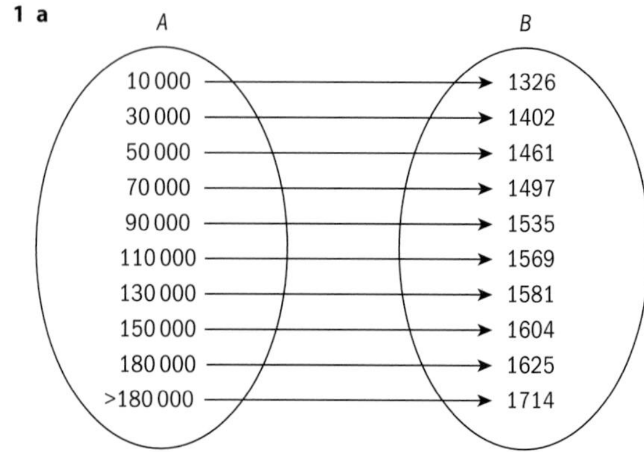

b Yes, it is a function.
c Higher income may enable the student to have access to tutoring or better books etc. or the higher income may be a result of higher intelligence.
d Student's own decision and explanation.

2 a Input = Percentage of world population,
Output = Percentage of world's resources used.
Domain = $\{x \mid x \in \mathbb{R}, 0 \le x \le 100\}$
Range = $\{y \mid y \in \mathbb{R}, 0 \le y \le 100\}$
b No, because the input value 3, for percentage of world population, has multiple output values.
c

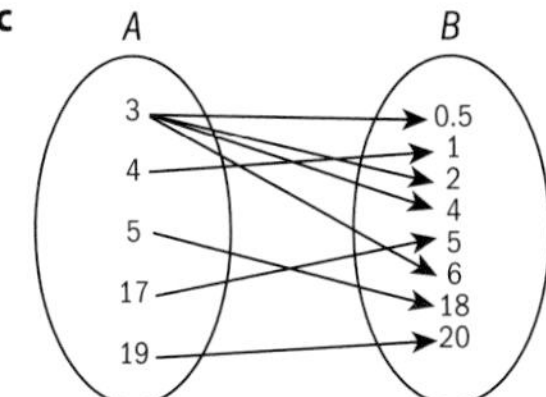

d Student's own opinion.
e Student's own opinion.

3 a $C = 24\,000 - 1000a$
b $C(18)$ represents the average cost of car insurance for an 18 year old male.
$C(18) = \$6000$
c Yes
d Student's own explanation.
e $C(21)$ represents the cost of car insurance for an 21 year old male.
$C(21) = \$3000$
f Because otherwise it would have a negative cost over the age of 24.

3.1

You should already know how to:

1 Trapezoid

2 10.3 cm

3 Student's own diagram

Practice 1

1 5

2 13

3 a $PQ = 25$, $QR = 25$, $RS = 25$, $SP = 25$
b Robin is not correct.
c The second premise is false; having four equal sides makes it a rhombus.

4 a Student's own diagram
b It is not equilateral

5 a Student's own diagram
b They are all $\sqrt{11\,482} = 107.2$ away from the origin
c Four lines of symmetry through centers of opposite pairs of sides on student's diagrams, and another four through opposite vertices.
d Order 4
e $AB = 82.02$, $BC = 82$
f It would be appropriate as it is almost regular

6 a All sides have length 5.
b Student's own explanation e.g. the triangle formed by three vertices satisfies Pythagoras' theorem.
c Student's own explanation e.g. the gradients of opposite sides are equal.

Practice 2

1 Student's own diagram, 3.40 (to 2 d.p.)

2 Line L_1 passes closer to point B.
Lengths are 1.9 units and 2.2 units.

3 (2, 9) is closer to L.
Distances are 3.58 and 4.02 units.

4 a 4.21
b The speed of plane A and the direction of flight of plane B.

Practice 3

1 (16, 13) **2** (7.5, −10.5)

3 a

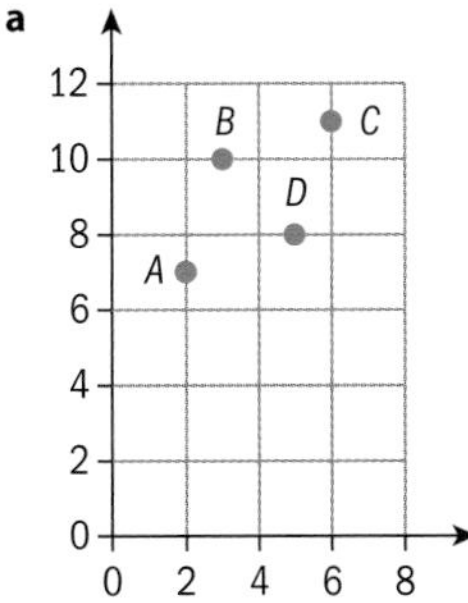

b (4, 9) and (4, 9)
c Diagonals bisect, therefore quadrilateral is a parallelogram; sides equal in length but not perpendicular, therefore rhombus.

4 a Segments PQ and QR both have length $2\sqrt{13}$ units.
b $L = (0, 0)$ $M = (4, 0)$ $N = (2, -3)$
c Segments PQ and QR both have length $2\sqrt{13}$ units.

5 Midpoint is (2.5, 6)
$2x + 1 = 5 + 1 = 6 = y$

6 a Gradients of diagonals are $\frac{1}{7}$ and −7, which multiply together to make −1. Therefore the diagonals are perpendicular.
b Midpoint of both diagonals is (−0.5, 1.5). Therefore they bisect each other.

Practice 4

1 a 5 **b** 9.43 **c** 232.4

2 They all equal $\sqrt{50}$ so the area is $50\pi = 157.1$

3 $FH = GH = \sqrt{17}$ therefore triangle FGH is isosceles.

Practice 5

1 a (10, 3) **b** (10, 4.5)

2 a $AB = BC = CD = DA = \sqrt{200}$
Therefore $ABCD$ is a rhombus.
b $E = (3, 12)$ $F = (9, 24)$ $G = (13, 22)$ $H = (7, 10)$
c gradient of EF = gradient of $GH = 2$
gradient of FG = gradient of $HE = -0.5$
Opposite sides are parallel so the quadrilateral is a parallelogram.

d $EG = \sqrt{200}$, $FH = \sqrt{200}$
Therefore $EFGH$ is a rectangle.

3 i $AB = \sqrt{160} = 12.6$
$BC = \sqrt{464} = 21.5$
$CD = 12$
$DA = \sqrt{1184} = 34.4$
Gradient of $AB = -3$
Gradient of $BC = 0.4$
Gradient of $CD = 0$
Gradient of $DA = 0.71$

ii $E = (-2, 14)$
$F = (6, 24)$
$G = (22, 28)$
$H = (14, 18)$

iii Gradient of EF = gradient of $GH = 1.25$
Gradient of FG = gradient of $HE = 0.25$
Opposite sides are parallel so the quadrilateral is a parallelogram.

b i $T = \left(\frac{x_1 + x_2}{2}, \frac{y_1 + y_2}{2}\right)$ $U = \left(\frac{x_2 + x_3}{2}, \frac{y_2 + y_3}{2}\right)$

$V = \left(\frac{x_3 + x_4}{2}, \frac{y_3 + y_4}{2}\right)$ $W = \left(\frac{x_4 + x_1}{2}, \frac{y_4 + y_1}{2}\right)$

ii Gradient of TU = gradient of $VW = \frac{y_1 - y_3}{x_3 - x_1}$

Gradient of UV = gradient of $WT = \frac{y_4 - y_2}{x_4 - x_2}$

Opposite sides are parallel so the quadrilateral is a parallelogram.

5 It is true. Two of the lengths are $\sqrt{2873}$. The perpendicular height (52) is exactly double the base (26).

6 Horizontal distance from A to P is $\frac{c - a}{3}$ and horizontal distance from P to B is $\frac{2(c - a)}{3}$.
Vertical distance from A to P is $\frac{d - b}{3}$ and vertical distance from P to B is $\frac{2(d - b)}{3}$.

7 (X, Y)
$C(x_C, y_C)$ and $D(x_D, y_D)$
$x_C = 2X - x_A$
$x_D = 2X - x_B$, $y_C = 2Y - y_A$ and $y_D = 2Y - y_B$
$\frac{y_B - y_A}{x_B - x_A}$
$\frac{y_D - y_C}{x_D - x_C} = \frac{(2Y - y_B) - (2Y - y_A)}{(2X - x_B) - (2X - x_A)} = \frac{y_B - y_A}{x_B - x_A}$
AB and CD are parallel
The length of AB
The length of CD is given by
$\sqrt{[(2X - x_B) - (2X - x_A)]^2 + [(2Y - y_B) - (2Y - y)]^2} =$
$\sqrt{(x_A - x_B)^2 + (y_A - y_B)^2}$ equal in length
AB and CD are parallel; AB and CD are equal in length; $ABCD$ is a parallelogram.

8 Gradient of OA = Gradient $CB = \frac{b}{a}$

Gradient of AB = Gradient $OC = \frac{d}{c}$

9 Gradient of BC = Gradient $AD = \frac{d - b}{c - a}$

So we have one pair of parallel lines.

Gradient $AB = \frac{b}{a}$ Gradient $DC = \frac{d}{c}$

If $\frac{b}{a} = \frac{d}{c}$ then it will be a parallelogram. If they are not equal then $ABCD$ is a trapezoid.

Mixed practice

1 a 13 **b** 5 **c** 6 **d** 25

2 a 7.81 **b** 11.4 **c** 6.45

3 a $\sqrt{80}$ **b** $\sqrt{185}$ **c** $\sqrt{117}$

4 a (5, 7) **b** (1, 3.5) **c** (−5.5, −9.5)

d (−0.65, 8.55) **e** $\left(\frac{7}{24}, \frac{5}{12}\right)$

5 Distance (0, 0) to (6, 0) = 6
Distance (0, 0) to (3, 4) = 5
Distance (3, 4) to (6, 0) = 5
Two sides are the same length, therefore it is an isosceles triangle.

6 Distance $AB = \sqrt{50}$

Distance $BC = \sqrt{40}$

Distance $AC = \sqrt{50}$

Two sides are the same length, therefore ABC is an isosceles triangle.

7 3.6 units (1 d.p.)

8 B

9 Midpoint of AB, $M = \left(\frac{a + c}{2}, \frac{b + d}{2}\right)$

Midpoint of AM divides AB in the ratio 1 : 3

Midpoint of $AM = \left(\frac{a + \frac{a + c}{2}}{2}, \frac{b + \frac{b + d}{2}}{2}\right) = \left(\frac{3a + c}{4}, \frac{3b + d}{4}\right)$

10 a $M = \left(\frac{a + c}{2}, \frac{b + d}{2}\right)$

b $AB = \sqrt{(c - a)^2 + (d - b)^2}$

c, d $AM = \frac{1}{2}\sqrt{(c - a)^2 + (d - b)^2}$

11 a $AC = 8$, $AB = 9.85 = BC$
b Midpoints: (5, 0), (3, 4.5), (7, 4.5)
c Distance (3, 4.5) to (7, 4.5) = 4 units
Distance (5, 0) to (3, 4.5) = 4.92 = distance (5, 0) to (7, 4.5)
d Triangle has base 4 and perpendicular height 4.5, area = 9

12 a The 4th column needs to contain a 2 and a 1, but only the 3rd row doesn't include 2 and 1 already.

b i The only number missing from the top row is 2.

ii Student's explanation

1	4	3	**2**
	2		4
		2	
2	1		

iii

1	4	3	**2**
3	**2**	**1**	4
4	**3**	**2**	**1**
2	1	**4**	**3**

13 (11, 3)

Review in context

1 (9, 9)
2 (9.5, 4)
3 (12, 12)
4 The crown is at Mermaid Pool.
This is because the distance from Falling Palms to Grey Cliffs is $\sqrt{7^2+1^2}=\sqrt{50}\approx 7.07$ leagues.
The distance from Kaspar's cave to Mermaid Pool is $\sqrt{5^2+5^2}=\sqrt{50}\approx 7.07$ leagues.
Drawing a circle centered on Kaspar's Cave and passing through Mermaid Pool only gives one location, so there are no others the same distance from the cave.
5 The midpoint between Kaspar's Cave and the Grey Cliffs is $\left(\frac{7+3}{2},\frac{7+10}{2}\right)=(5, 8.5)$.
The midpoint between (5, 8.5) and Pirates' Port is (12, 8.5).
6 a 5 leagues
b 5 leagues
c Since it has two sides of equal length, it is isosceles.
d Since isosceles triangles have a line of symmetry through the vertex, the point on the base that is closest to the vertex is the midpoint of the base.
e (2.5, 6.5)

4.1

You should already know how to:

1 a 3, 5, 4.82 (3 s.f.), 8 **b** 19, 19, 19.3 (3 s.f.), 8

Practice 1

1 Quantitative, discrete
2 Quantitative, continuous
3 Qualitative
4 Quantitative, discrete
5 Quantitative, continuous
6 Quantitative, continuous
7 Quantitative, discrete
8 Quantitative, continuous
9 Quantitative, discrete
10 Qualitative

Practice 2

1

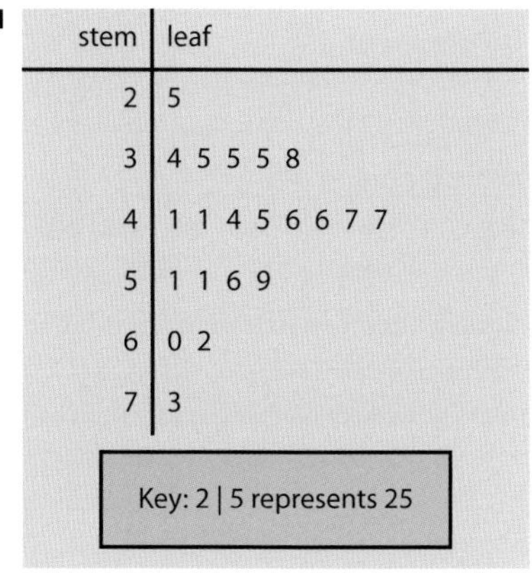

stem	leaf
2	5
3	4 5 5 5 8
4	1 1 4 5 6 6 7 7
5	1 1 6 9
6	0 2
7	3

Key: 2 | 5 represents 25

2 a

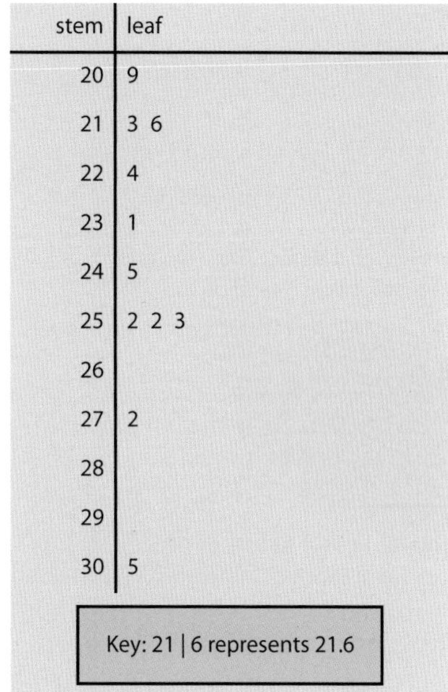

stem	leaf
20	9
21	3 6
22	4
23	1
24	5
25	2 2 3
26	
27	2
28	
29	
30	5

Key: 21 | 6 represents 21.6

3 a i 4 km **ii** 40 km **b** 18 **c** 3 **d** 50%

4

stem	leaf
16	1 1 6 6 7 9
17	3 3 5 7 7
18	0 3 4
19	2 5 5 7

Key: 16 | 9 represents 169

Practice 3

1 49, 43, 28
2 30, 29, 30
3 a 11 **b** 63 **c** 42
d 70 **e** 46, 35
f Mathematics: since the range is smaller, data is more closely bunched.

4 a

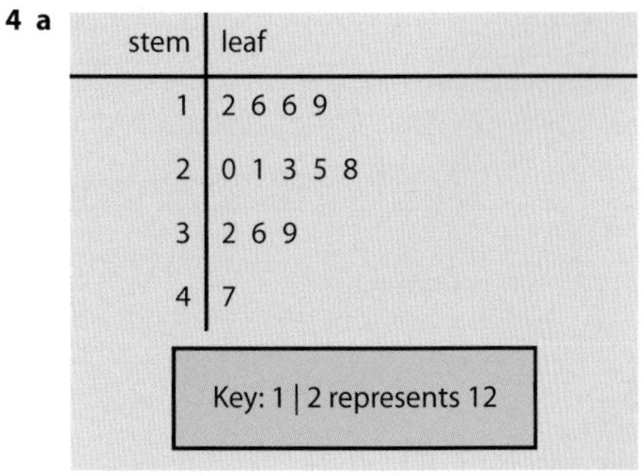

stem	leaf
1	2 6 6 9
2	0 1 3 5 8
3	2 6 9
4	7

Key: 1 | 2 represents 12

b 4 **c** 16, 23, 35

Practice 4

1 a 730, 670, 750 **b** 360, 80

2 a 45.5, 51, 41 **b** 39, 10

3 37.2, 21.5, 12.9 **4** 12, 48, 13

5 a £70 000 **b** £48 000
c £32 000, £18 000 **d** £23 500, £11 500
e Senior employees earn more on average and the salaries are more spread out.

6 $x = 34$, $y = 46$, $z = 54$ **7** Student's own answers

Practice 5

1 a 10 **b** 20 **c** 9

2 a 47, 52, 61, 66, 75 **b** 28, 14 **c** 25

3 a 1, 2, 3, 4, 4 **b** 10, 13, 15, 18, 24
c 10 000, 10 300, 11 750, 12 250, 64 000

4

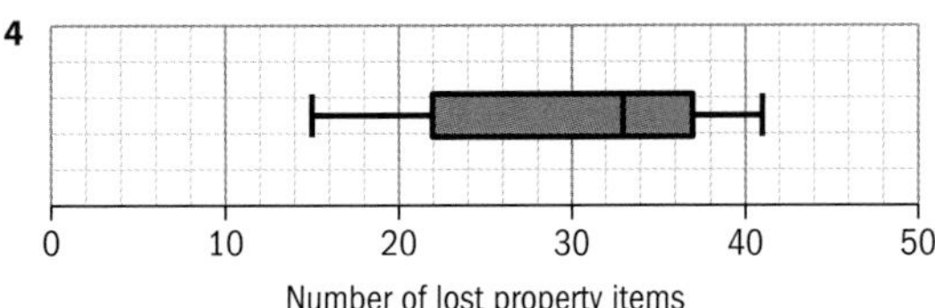

5 a 21 **b** 13, 27

c

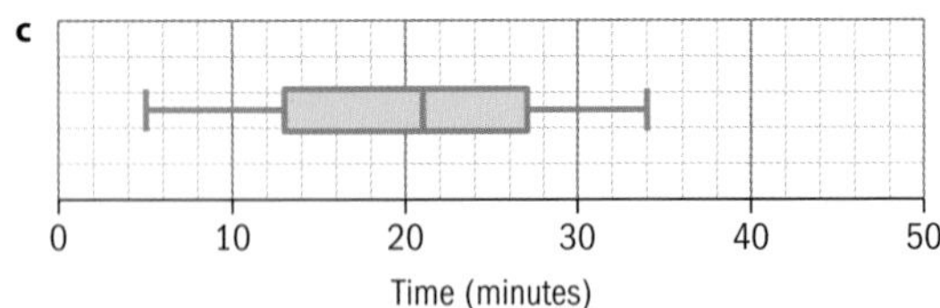

6

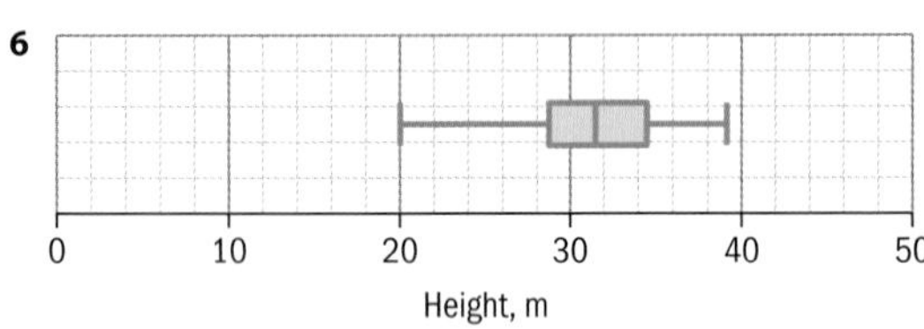

7 a 38 **b** 45.5 **c** 17

d

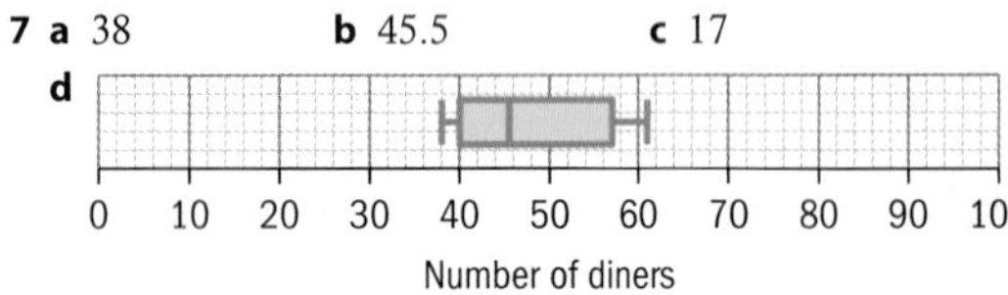

8

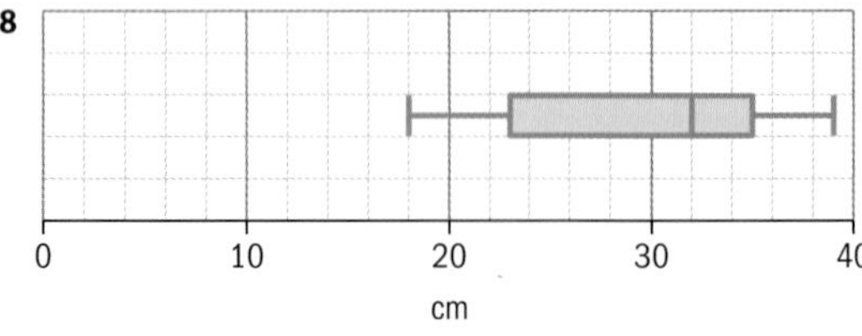

A = 8, B = 3, C = 2, D = 5, E = 9

Practice 6

1 a Med G: 38, Med B: 34; IQR G: 22.5, IQR B: 19.5
b Both distributions are fairly symmetrical and unimodal. Boys a little more centered on the lower values.
c On average, boys took less time and have less variation of times.

2 a

Santa Fe		Saint Paul
	2	6 9
	3	1
8 4 3	4	2 3
6 3	5	8 9
7 5	6	
8 4	7	1 3
6 3 3	8	0 2 5

Key: 3|4|2 represents 43 in Santa Fe and 42 in Saint Paul

b There is no modal value for either city. In Sante Fe the temperatures are evenly distributed between 43 and 86 degrees.
In Saint Paul, the temperatures are are evenly distributed between 26 and 85 degrees.
c The temperature in Santa Fe is on average higher but less varied.

3 a

apple trees		pear trees
9 7 6 6 1 1	16	0 0 5 5 6 8
7 7 5 3 3	17	0 1 2 4 7 8
4 3 0	18	2 3
7 5 5 2	19	0 2 3 4

Key: 1 | 16 | 0 represents 161 for apple trees and 160 for pear trees

b Both centered on lower values with a few taller trees. The apple trees are multi-modal, the pear trees are bimodal.
c Very similar. Slightly higher average for apple trees with a similar spread of heights.

4 Train B had a higher length if delay on average and a wider spread of delays.

5 a

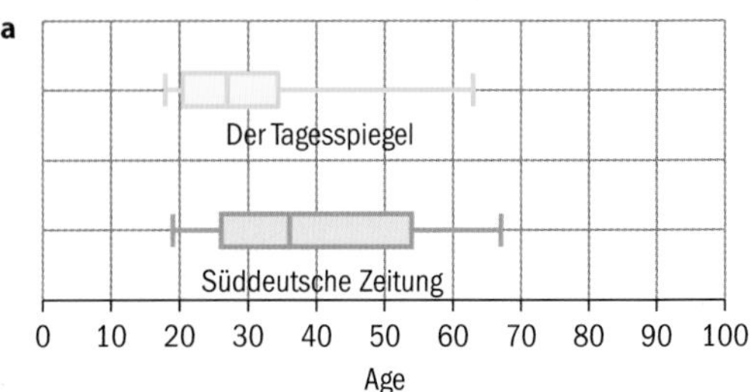

b Süddeutsche Zeittung has on average older users but the ages are more varied.

6 Ray's sales are on average higher and they are also more varied.

Practice 7

1 a 28.4, 28, Multi-modal: 25, 31, 32 all have 2
b Either mean or median – both fairly central. No clear mode.

2 a 28.9, 28 **b** 16.5 **c** 62

d 54, outlier included since could be legitimate value.

e Median – not affected by outliers.

3 a Boys: 60, 85

b Girls: 65, Boys: 85; Outliers included since could be legitimate values.

4 a Male: 100 kg

b Both sets symmetrical and unimodal

c With the outlier excluded, both data sets are almost the same.

5 a

stem	leaf
1	2 5 6
2	1 4 4 5 5 5 5 6 6 8 9
3	0 1 3 5 6 8
4	2
5	
6	4
7	4
8	4 6
9	
10	7

Key: 1 | 2 represents 1.2%

Outliers at 6.4, 7.4, 8.4, 8.6, 10.7

b These may be the cantons that have large urban centers.

c Unimodal and centered on lower values.

Mixed practice

1 a Qualitative **b** Quantitative

c Quantitative **d** Quantitative

2 a Continuous **b** Discrete

c Continuous **d** Continuous

3 a

stem	leaf
4	6 8
5	2 9
6	0 1 4
7	0 2 5
8	2

Key: 4 | 6 represents 46

b 4 **c** 61 kg **d** 36 kg

4 a 39 min **b** 16 min **c** 32 min

5 a $r = 10$, $s = 13$ **b** 18

6 a 3, 17, 26, 32, 90

b

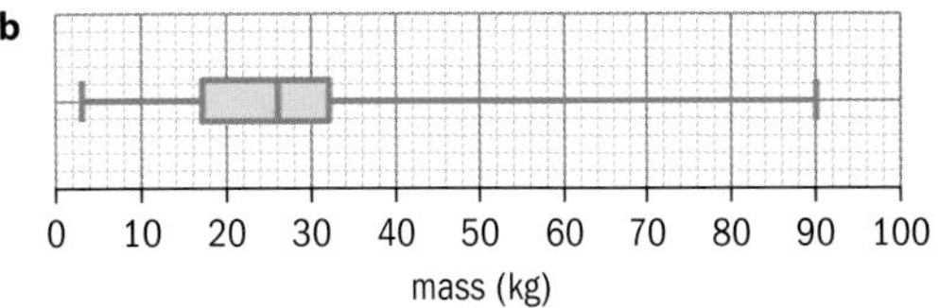

7 a 18 **b** $a = 10$, $b = 44$ **c** 40

8 a Film A had on average older people watching and a greater spread of ages.

b A: Young and old people attended.

9 a £132 000 **b** £560 000

c £559 000, included since a possible value

d Median: £85 000, not affected by outliers.

10 a

Farm A		Farm B
9	1	
	2	
	3	
9 9 8 8 5 3	4	9 9
7 6 5 5	5	0 4 4 6 7
5 4 4 3 0	6	2 4 6 8 8
	7	0 0 1 2

Key: 3 | 4 | 9 represents 43 for Farm A and 49 for Farm B

b Farm B had a higher average and greater spread.

c Both look like Average/good hill because the range of values fits that profile.

Review in context

1 a

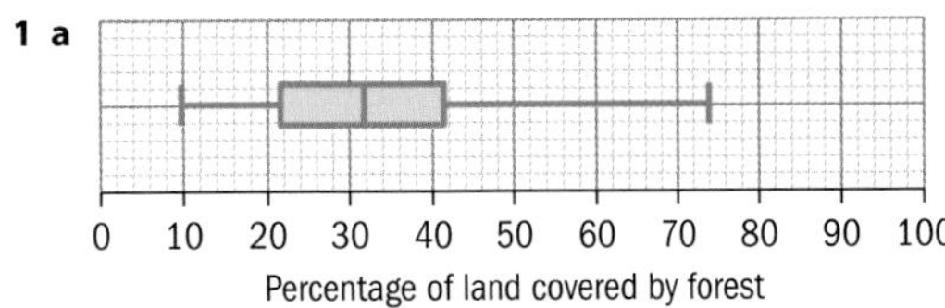

b Finland is a possible outlier.

2 2012 had a slightly lower average and values generally lower overall. Suggests forest areas reduced.

3 Student's own answers

4.2

You should already know how to:

1 4.2 and 6.0

2 a discrete

b continuous

c discrete

d continuous

3 Mean = 3.53 (3 s.f.)

4 a

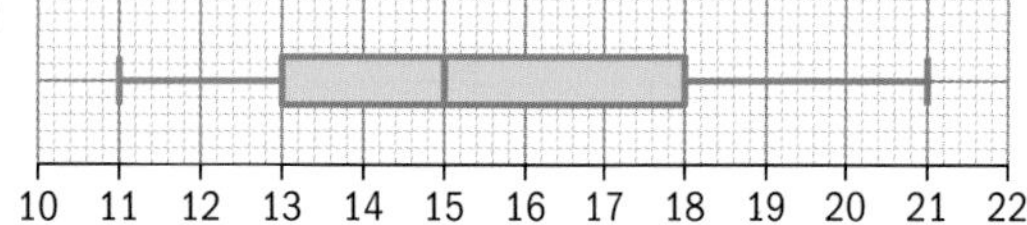

b IQR = 5

5 a $\frac{1}{4} = 25\%$ **b** $\frac{1}{2} = 50\%$ **c** $\frac{3}{4} = 75\%$

Practice 1

1 a $1.40 < x \le 1.50$ **b** $1.40 < x \le 1.50$

2 a

Time x (sec)	Frequency
$300 < x \le 350$	3
$350 < x \le 400$	4
$400 < x \le 450$	4
$450 < x \le 500$	2
$500 < x \le 550$	3
$550 < x \le 600$	2
$600 < x \le 650$	2

b $350 < x \le 400$ and $400 < x \le 450$

c $400 < x \le 450$

3 a Other ways of grouping the data are acceptable.

Height x (cm)	Frequency
$160 < x \le 170$	6
$170 < x \le 180$	6
$180 < x \le 190$	2
$190 < x \le 200$	4

b $160 < x \le 170$ and $170 < x \le 180$

c $170 < x \le 180$

4 a $a = 5, b = 17, c = 2, d = 27$

b The 15th and 16th apples are both in this interval.

Practice 2

1 a

Time (T hours)	Frequency	Mid-interval value	Mid-interval value × frequency
$0 < T \le 10$	8	5	40
$10 < T \le 20$	12	15	180
$20 < T \le 30$	16	25	400
$30 < T \le 40$	11	35	385
$40 < T \le 50$	3	45	135
Total	50		1140

b 22.8 hours

2 a 54 **b** $54 < w \le 57$ **c** 19 **d** $54 < w \le 57$

3 a 1.1 kg **b** $1.0 < w \le 1.2$ **c** 51 **d** $1.0 < w \le 1.2$

e 1 kg

f The estimate of the mean would change slightly but the median would stay the same.

Practice 3

1 a 1.96

b Upper Quartile = 2.06, Lower Quartile = 1.86

c 0.2

2 a The three table entries are 24, 32, 36

b

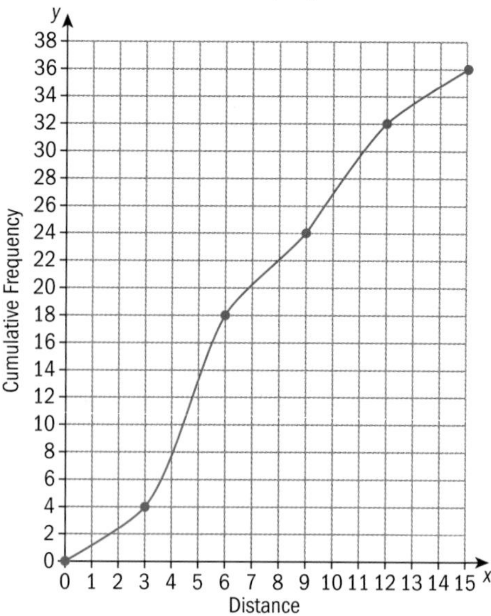

c Lower Quartile = 4.2, Median = 6, Upper Quartile = 10.2

d Range = 15, IQR = 6

3 a

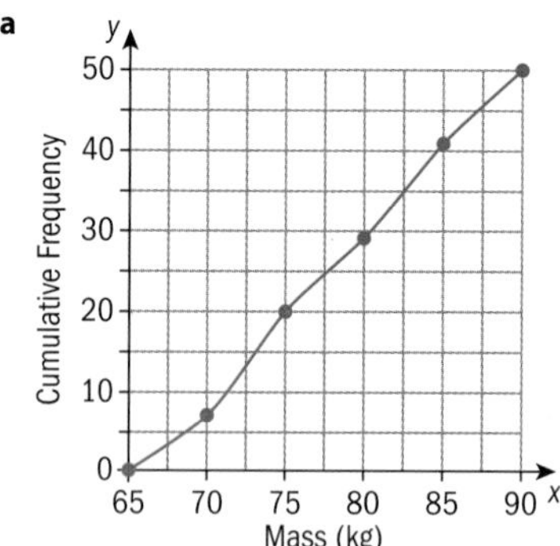

b 65, 72.25, 77.75, 83.5, 90

c

4 a

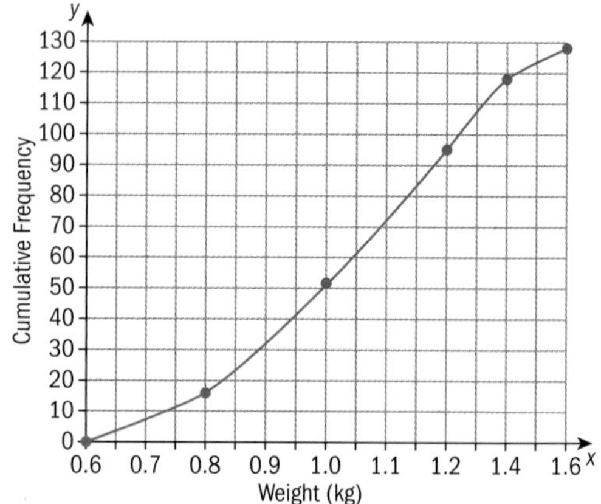

b

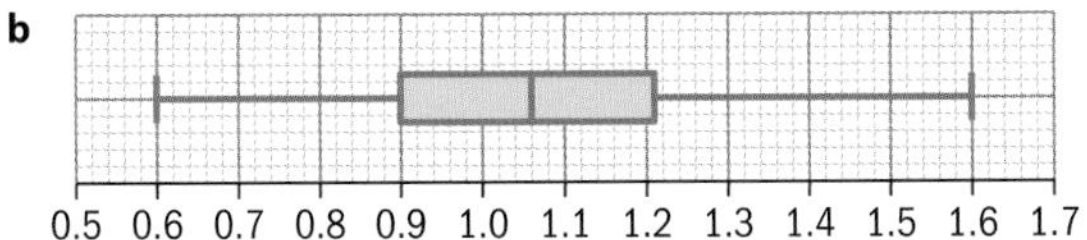

c i false **ii** false

d 1.2 kg (to the nearest 0.1kg)

5 a

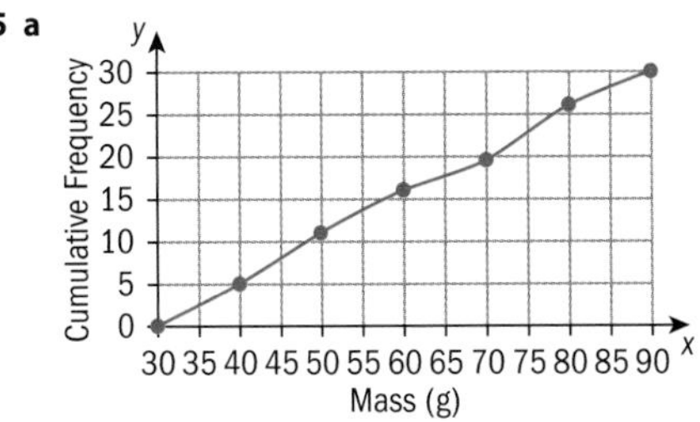

b 8 **c** 80

6

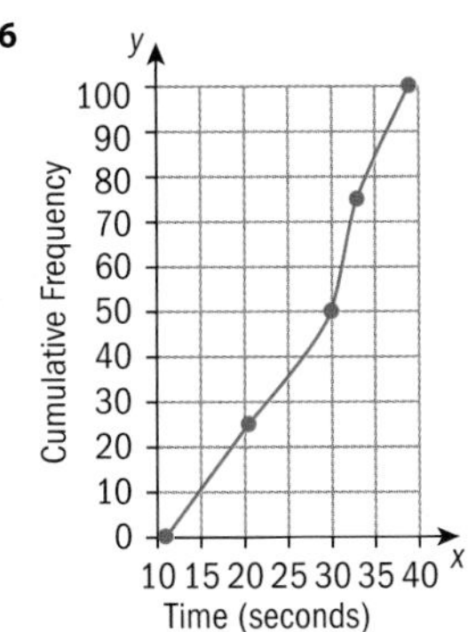

7 a 122 km/h

b About 16%

c

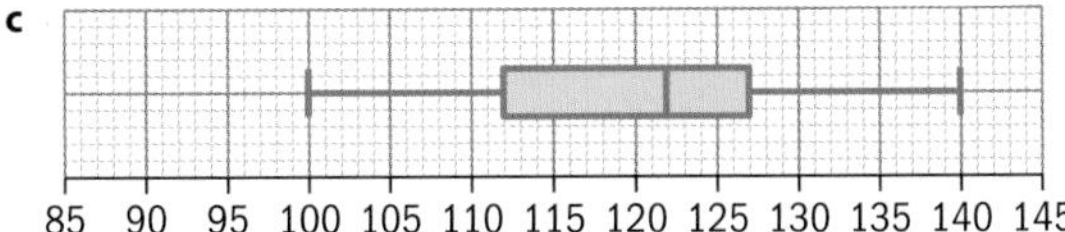

d On average, people drive faster on Motorway A. The range of speeds is greater on Motorway B.

Practice 4

1 i a continuous, overlapping boundary values with continuous intervals **b** 10

ii a discrete, no overlapping boundary values, gaps between intervals **b** 5

iii a continuous, overlapping boundary values with continuous intervals **b** 3

iv a discrete, no overlapping boundary values, gaps between intervals **b** 20, 25

2 a 11.8

It makes sense, although the number of words in a sentence must in reality be a whole number.

b $10 \le x \le 14$

c $10 \le x \le 14$

3 a 10.3

b There are two classes with a frequency of 6.

c $8 \le x \le 11$

d 19

4 a

Score	Frequency	Cumulative Frequency
$0.5 \le x < 5.5$	1	1
$5.5 \le x < 10.5$	9	10
$10.5 \le x < 15.5$	15	25
$15.5 \le x < 20.5$	15	40
$20.5 \le x < 25.5$	5	45
$25.5 \le x < 30.5$	5	50

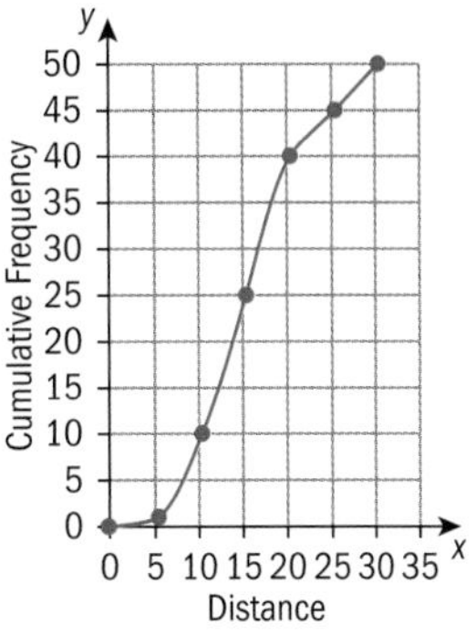

b 15.5

c Upper Quartile = 19.5

Lower Quartile = 11.5

d

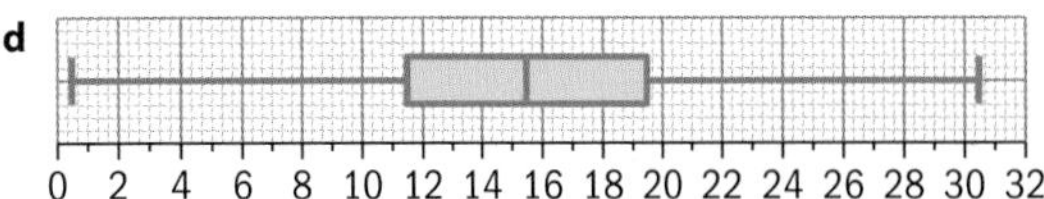

5 a $a = 20$, $b = 29$

b

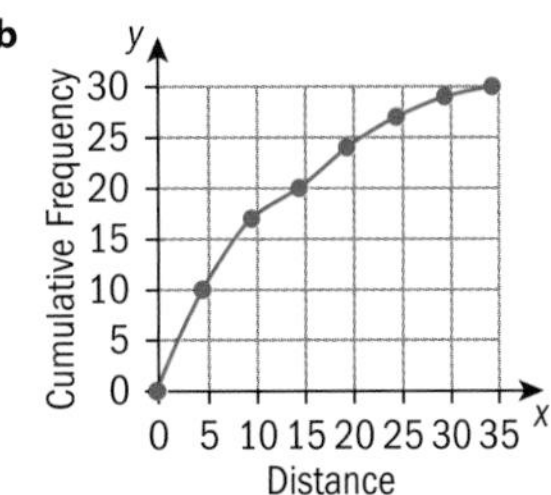

c Five point summary: 0, 3, 7.5, 17.5, 34

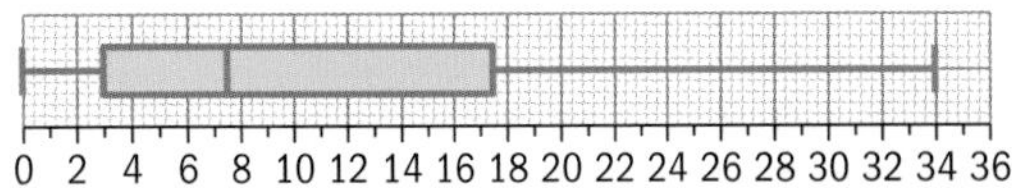

6 a

Age	Frequency	Cumulative Frequency
$0 \le x < 10$	1	1
$10 \le x < 20$	3	4
$20 \le x < 30$	8	12
$30 \le x < 40$	7	19
$40 \le x < 50$	7	26
$50 \le x < 60$	1	27
$60 \le x < 70$	3	30

Other intervals are acceptable.

b

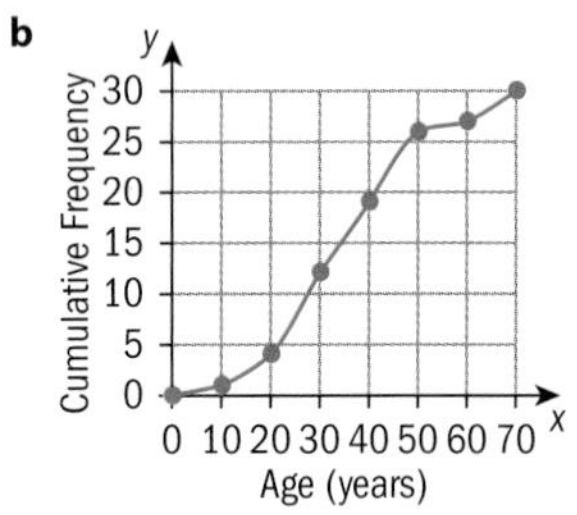

c 25–27% **d** 36 or 37%

7 a and
b on same diagram

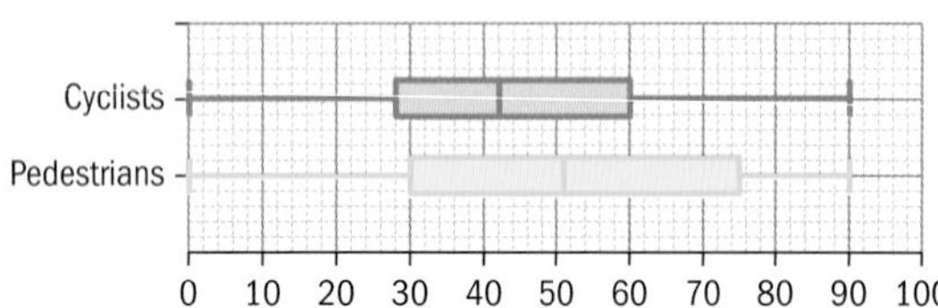

c On average, the pedestrians who were killed were older. There were also a wider range of ages of pedestrians who were killed.

Mixed practice

1 a

Mass (grams)	Frequency
$250 \le x < 300$	5
$300 \le x < 350$	7
$350 \le x < 400$	6
$400 \le x < 450$	4
$450 \le x < 500$	10
$500 \le x < 550$	3

Other intervals are acceptable.

b $450 \le x < 500$ **c** $350 \le x < 400$ **d** 398

2 a 60

b

Mark	Frequency	Cumulative Frequency
$16 \le x \le 27$	9	9
$28 \le x \le 39$	21	30
$40 \le x \le 51$	18	48
$52 \le x \le 63$	23	71
$64 \le x \le 75$	19	90

c

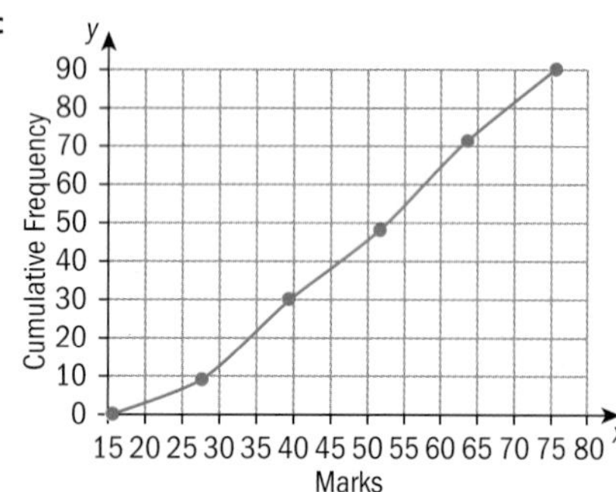

3 a 26 cm **b** About 5.75 cm

c

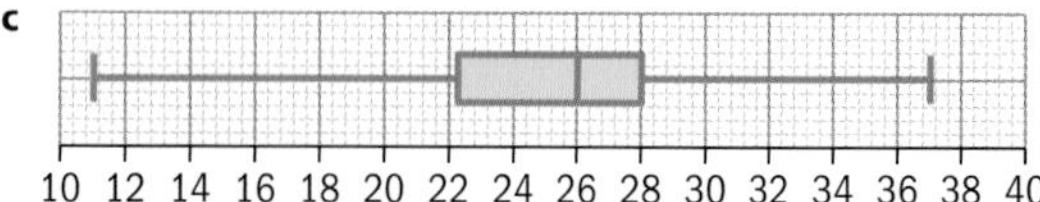

4

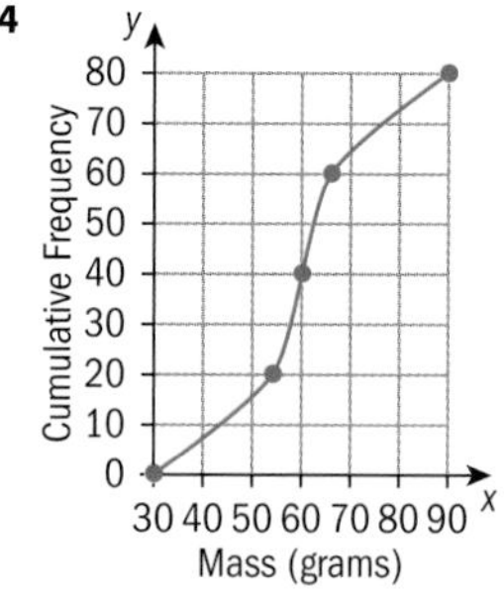

5 a 98.0 km/h to 1 d.p.
b $a = 167$ and $b = 277$
c

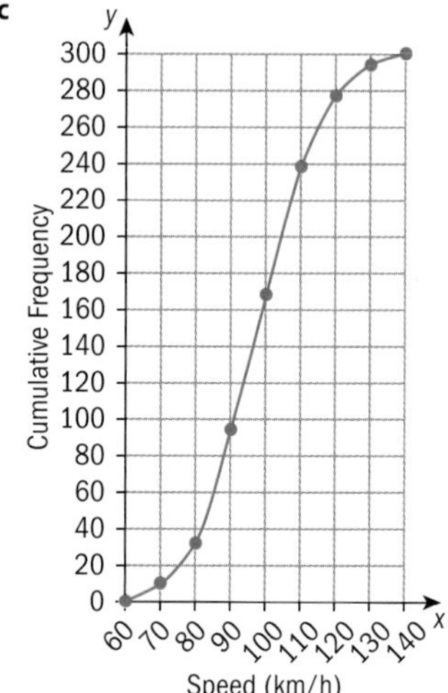

d 108 km/h

6 a i 72 grams **ii** 76 grams
b $x = 82$

7 a 24.2 cm
b About 11 cm
c 62 fish
d 76 small fish, 62 large fish
e About $1076

8 Any sensible comparisons.
e.g. the distributions are very similar.
The two box and whisker plots are the same.

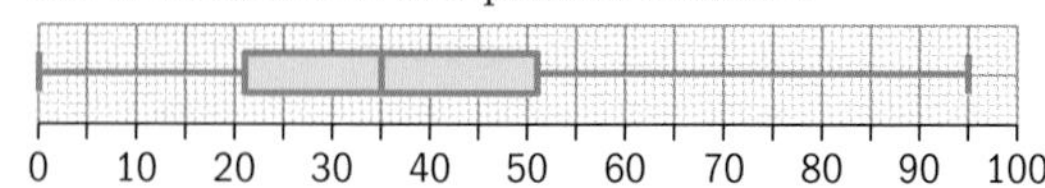

Review in context

1 a

Length L cm	Frequency	Cumulative Frequency
$25 < L \le 27$	1	1
$27 < L \le 29$	5	6
$29 < L \le 31$	9	15
$31 < L \le 33$	21	36
$33 < L \le 35$	29	65
$35 < L \le 37$	19	84
$37 < L \le 39$	16	100

b

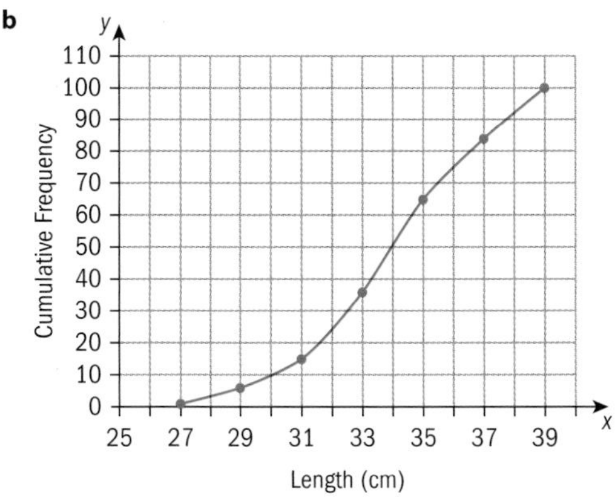

c 25, 32, 34, 36, 39 **d** 32 cm

e **i** About 70%

ii About 56%

iii About 63%

4.3

You should already know how to:

1 **a**

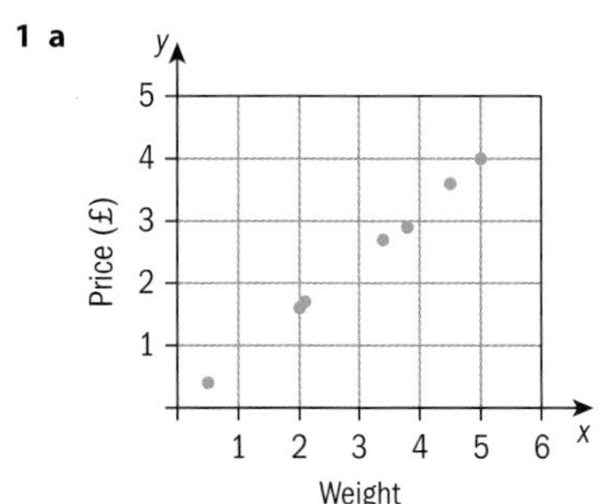

2 4.3

3 20: it is more than 1.5 times the IQR from the Upper Quartile

Practice 1

1 **a** No linear correlation – there is no correlation between scores in Mathematics and Art

b Weak positive correlation (linear) – there is a weak correlation between height and shoe size

c Strong positive correlation (linear) – there is a strong correlation between the amount of revision done and the score in the test

d Weak negative correlation (linear) – there is a weak negative correlation between the weight of the car and the gas mileage

e Strong negative correlation (linear) – the heating bill goes up as the temperature goes down.

f Moderate, non-linear association – the probability of having a short or tall height is low, however the probability of having a height in the middle is higher.

Practice 2

1 a b i

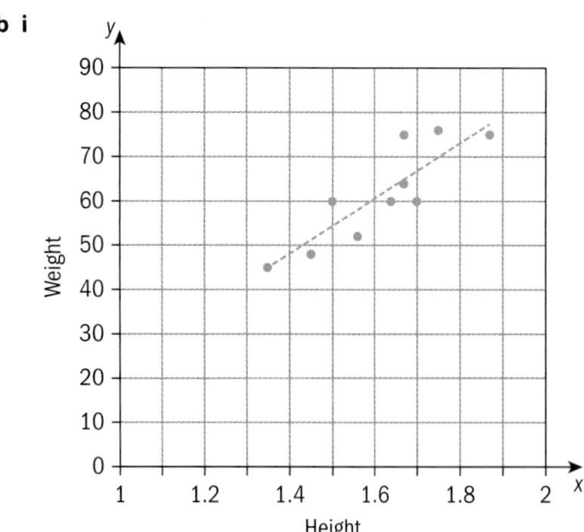

a b ii

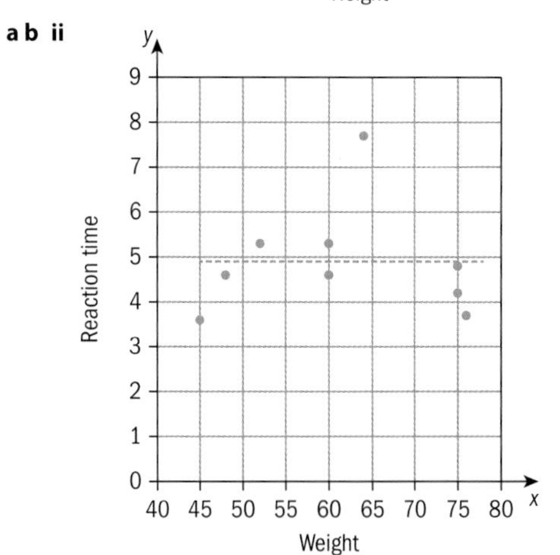

a b iii

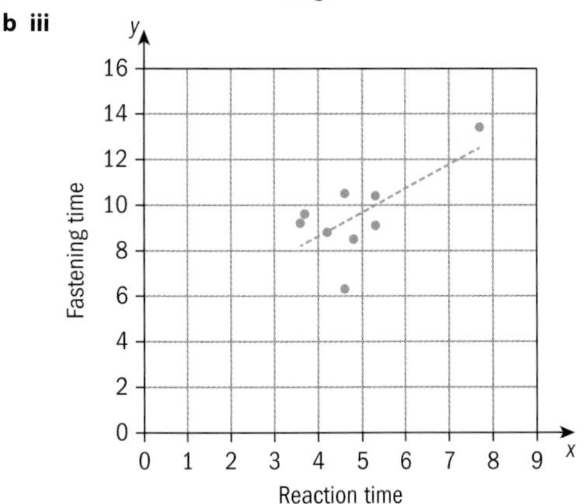

a b iv

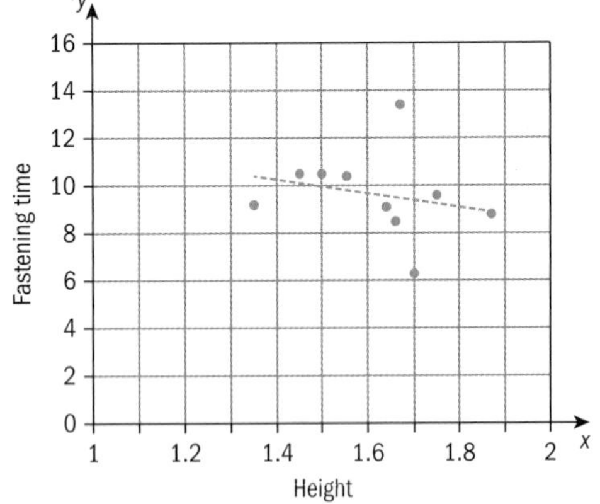

c i Strong positive linear correlation

ii No correlation

iii Weak positive correlation

iv Weak negative correlation

d 65 kg

e 7.1 seconds

f Graph **i** since it has the strongest correlation

2 a b Weak positive correlation

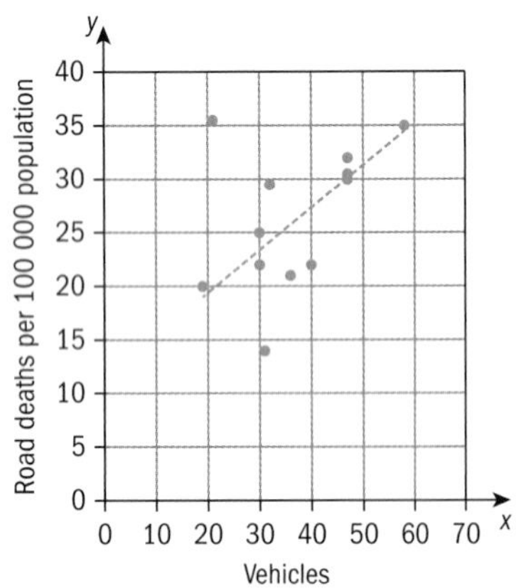

c 20–22

Practice 3

1 a b

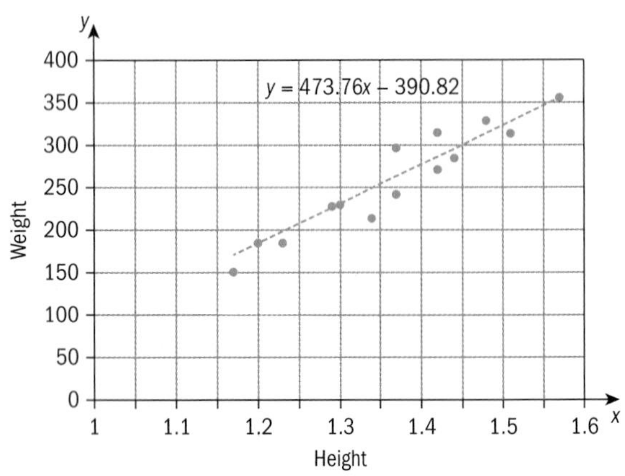

c Strong positive correlation

d 263 kg

2 a b

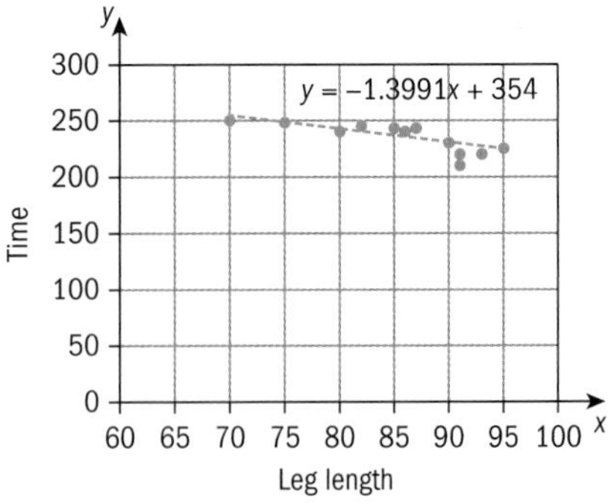

c Moderate negative correlation

d 236.5 seconds

3 a b c $511

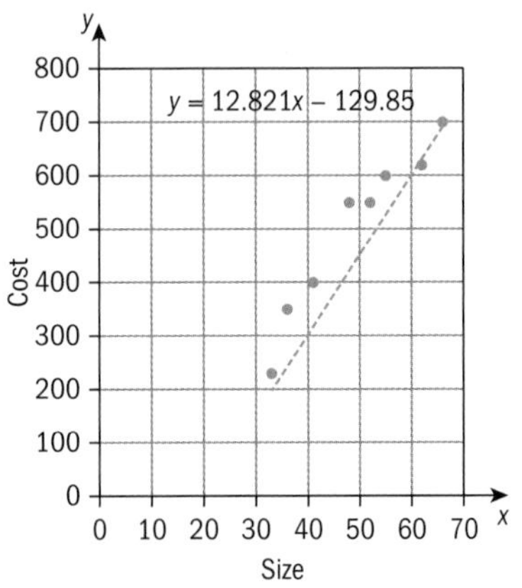

Practice 4

1 a

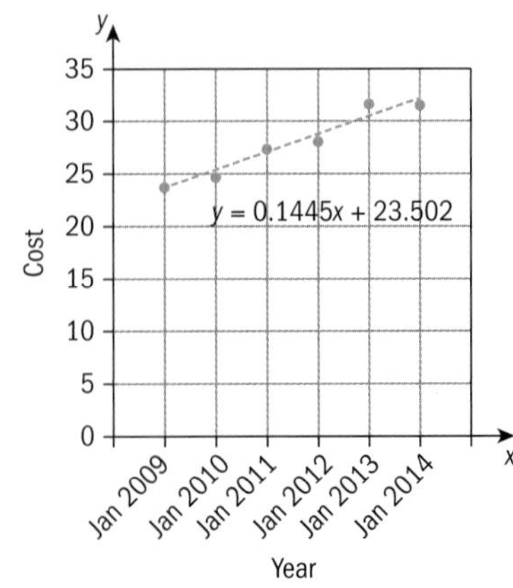

b i 24.37 **ii** 27.84 **iii** 33.91

c A very big difference from the value predicted – outside the scope of the data so prediction not reliable

2 a Spurious correlation – no relationship

b Strong positive linear correlation – suggests a very good relationship between the variables

c Spurious correlation – no relationship

3 a

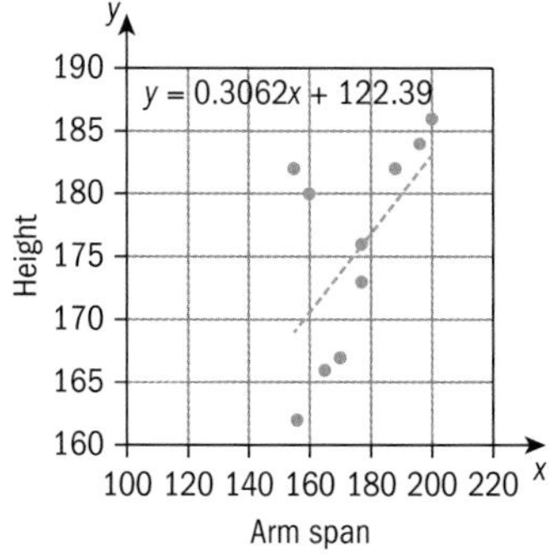

b 175 cm

c There are no outliers by the strictest definition, but the two points in the top left look 'odd'. New value 170 cm

d Part **c** is more reliable if the data does prove to be incorrect.

4 a 0.52 g

b

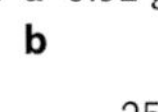

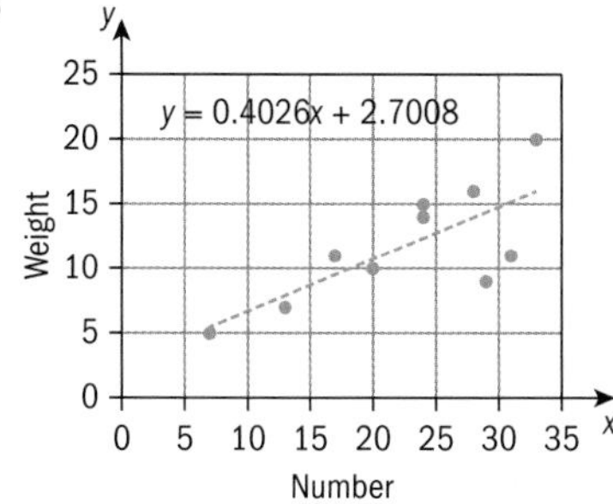

c Mean weight 0.58 g

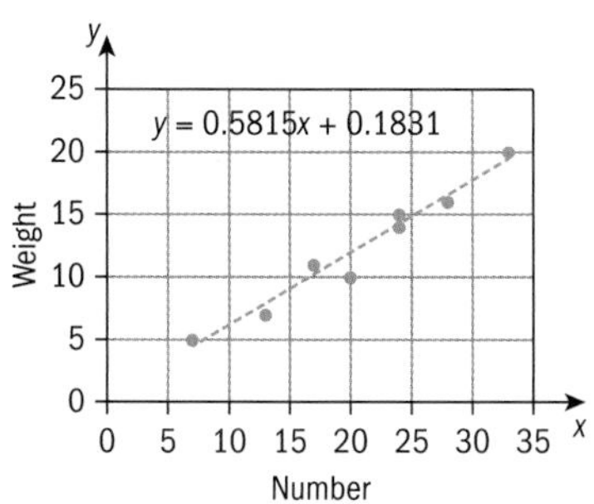

d Group 2 since they have removed the faulty data

Mixed practice

1 a c

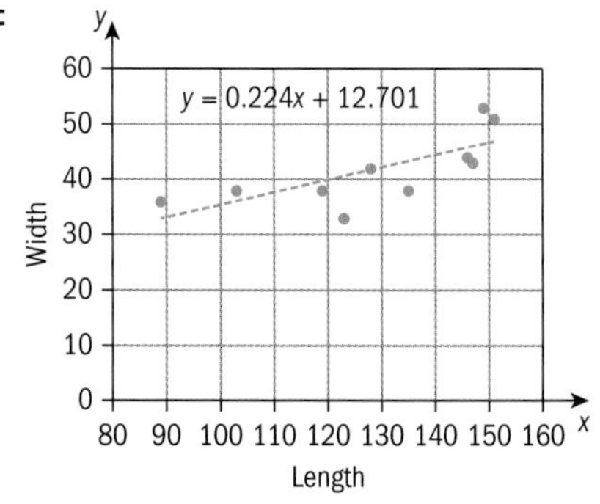

b Moderate positive correlation

2 a You cannot correlate non-numerical data – here 'correlation' is used informally at best.

b This could be correct since they both have risen – the correlation is, however, spurious.

c This should be *negative* correlation.

Review in context

1 a

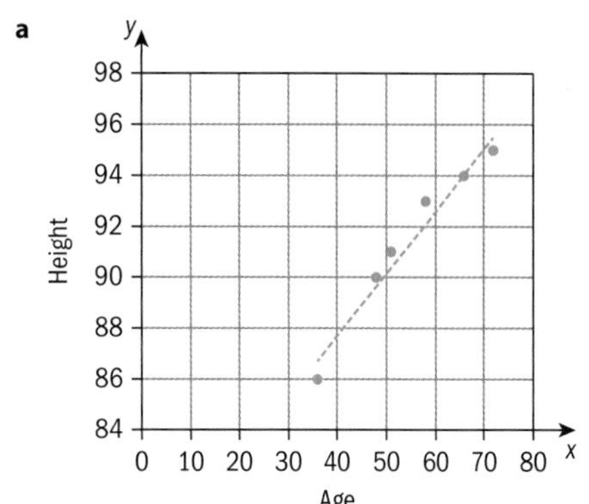

b 88.2 cm

c No – outside the range of data – relationship may not continue

2 a Moderate negative correlation **b** 48 min

c No – outside the scope of the data – relationship unlikely to continue that far out.

4.4

You should already know how to:

1 a $\frac{1}{3}$ **b** $\frac{1}{6}$ **c** $\frac{5}{6}$ **d** $\frac{1}{2}$ **e** $\frac{5}{6}$

2 a P = {2,3,5,7} **b** 1,4,6,8,9,10 **c** 4, 6

3

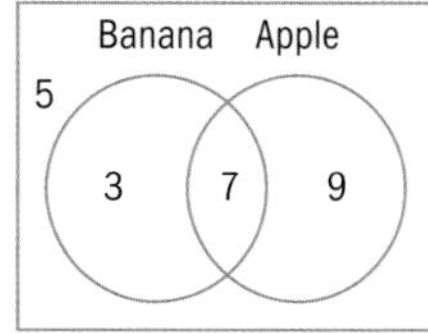

Practice 1

1 (I, I) (I, II) (I, III) (I, IV) (II, I) (II, II) (II, III) (II, IV) (III, I) (III, II) (III, III) (III, IV) (IV, I) (IV, II) (IV, III) (IV, IV)

	I	II	III	IV
I	I,I	I,II	I,III	I,IV
II	II,I	II,II	II,III	II,IV
III	III,I	III,II	III,III	III,IV
IV	IV,I	IV,II	IV,III	IV,IV

2 HHH, HHT, HTH, THH, HTT, THT, TTH, TTT

3

	1	2	3	4
1	2	3	4	5
2	3	4	5	6
3	4	5	6	7
4	5	6	7	8

4 (SAG, A) (SAG, O) (SAG, B) (GG, A) (GG, O) (GG, B)

5

	1	2	3	4	5	6
1	1,1	1,2	1,3	1,4	1,5	1,6
2	2,1	2,2	2,3	2,4	2,5	2,6
3	3,1	3,2	3,3	3,4	3,5	3,6
4	4,1	4,2	4,3	4,4	4,5	4,6

6

	1	2	3	4	5
1	1,1	1,2	1,3	1,4	1,5
2	2,1	2,2	2,3	2,4	2,5
3	3,1	3,2	3,3	3,4	3,5
4	4,1	4,2	4,3	4,4	4,5
5	5,1	5,2	5,3	5,4	5,5

7

	2	3	4	5	6
0	0	0	0	0	0
1	2	3	4	5	6
2	4	6	8	10	12
3	6	9	12	15	18
4	8	12	16	20	24

8 A tetrahedral dice and a coin are both thrown.

Practice 2

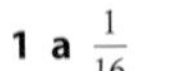

1 a $\frac{1}{16}$ b $\frac{15}{16}$

2 a $\frac{4}{16}=\frac{1}{4}$ b $\frac{7}{8}$

3 a $\frac{1}{8}$ b $\frac{3}{8}$ c $\frac{1}{8}$

4 $\frac{9}{16}$ 5 $\frac{11}{24}$

6 a $\frac{12}{52}=\frac{3}{13}$ b $\frac{13}{52}=\frac{1}{4}$ c $\frac{20}{52}=\frac{5}{13}$

d $\frac{10}{13}$ e $\frac{3}{4}$ f $\frac{8}{13}$

7 15

Practice 3

1 a

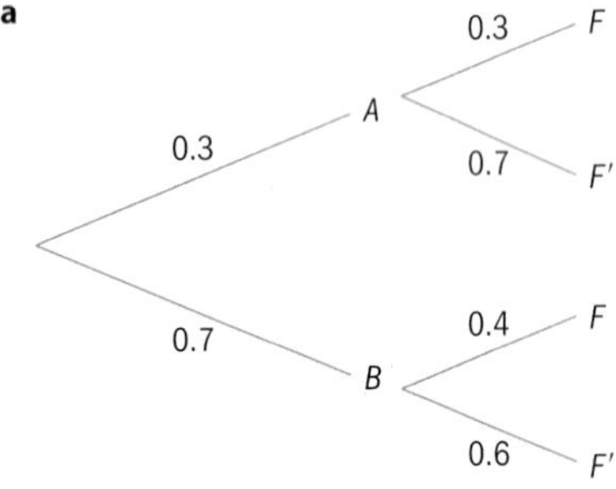

b 0.9 c 0.21

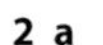

2 a

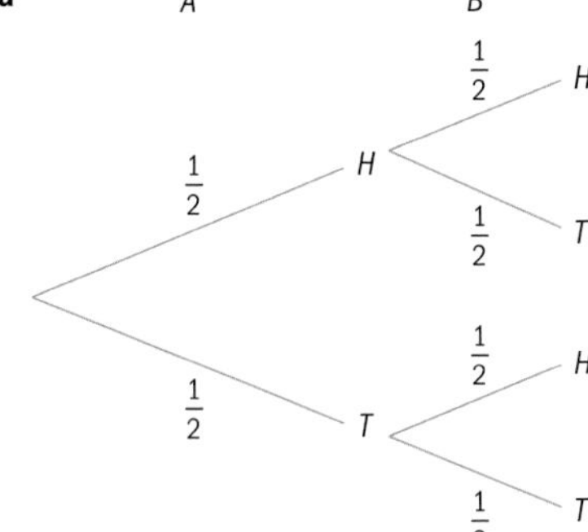

b $\frac{1}{4}$

3 $\frac{1}{36}$

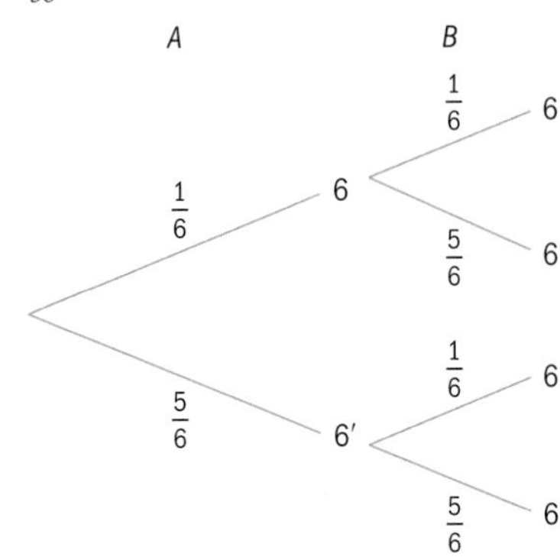

4

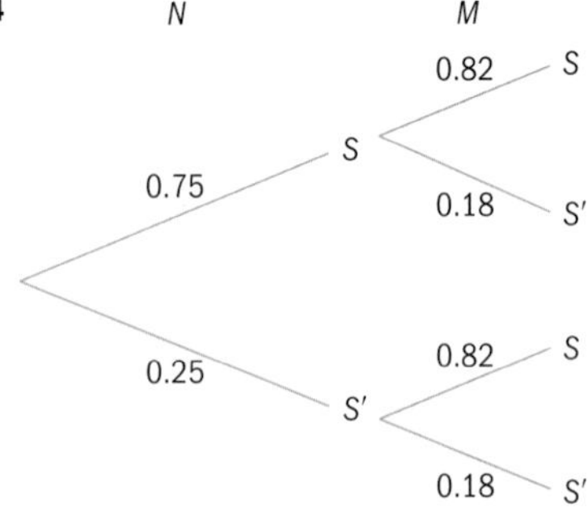

a 0.615 b 0.045

5 a $\frac{3}{5},\frac{2}{5}$

b

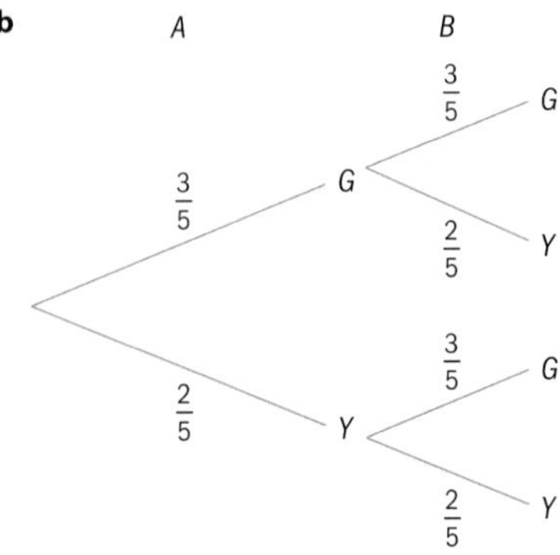

c $\frac{3}{5}\times\frac{3}{5}>\frac{2}{5}\times\frac{2}{5}$

Practice 4

1 a

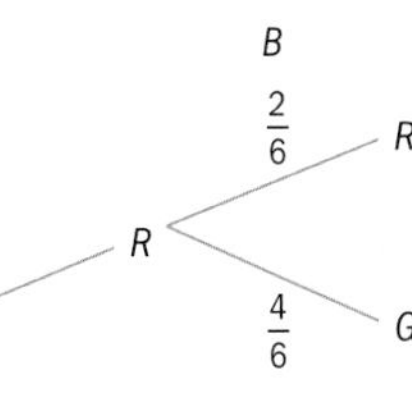

b $\frac{6}{42}$ **c** $\frac{12}{42}$ **d** $\frac{30}{42}$

2 a

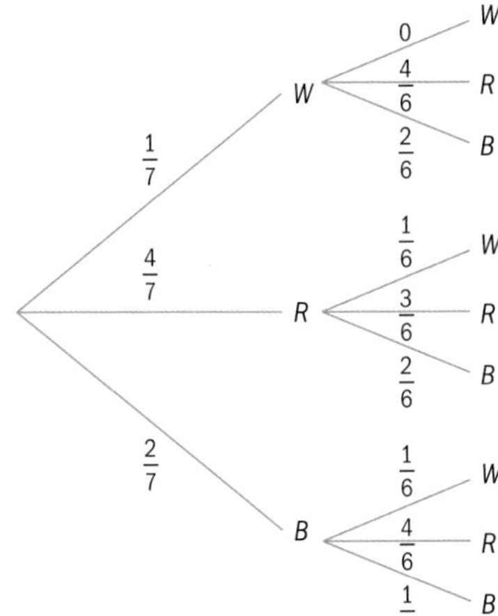

b $\frac{8}{42} = \frac{4}{21}$

3

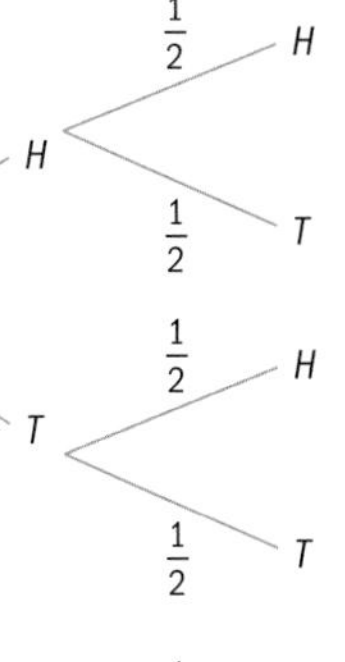

a $\frac{3}{8}$ **b** $\frac{7}{8}$ **c** $\frac{1}{8}$

d They are complementary probabilities – add to 1

4 a

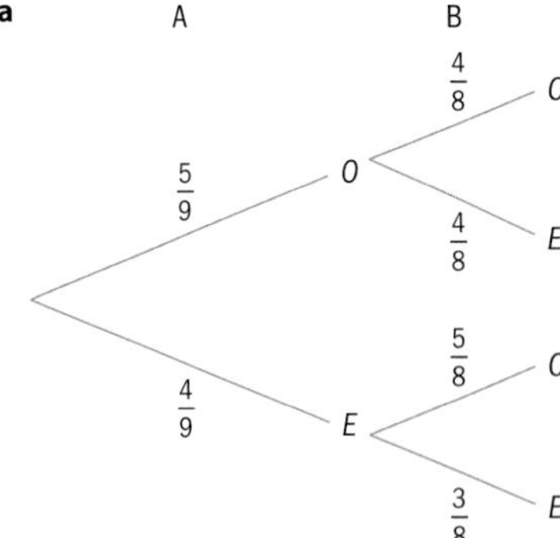

b $\frac{1}{6}$ **c** $\frac{5}{6}$

5 a 10

	J	M	R	H	P
J	–	JM	JR	JH	JP
M	MJ	–	MR	MH	MP
R	RJ	RM	–	RH	RP
H	HJ	HM	HR	–	HP
P	PJ	PM	PR	PH	–

b $\frac{1}{10}$

6 a 0.35 **b** 8

c

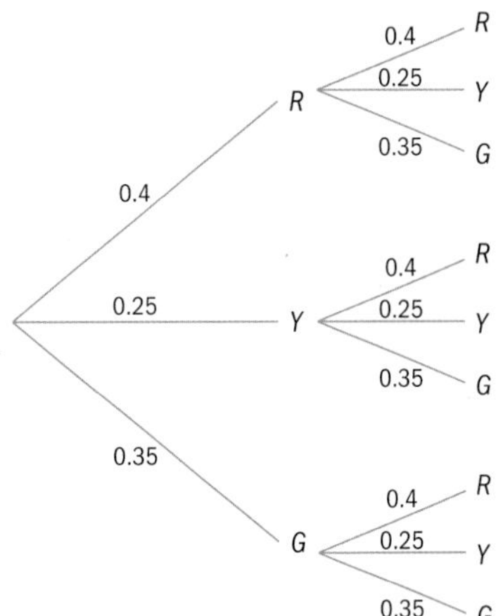

d 0.64

Practice 5

1 a

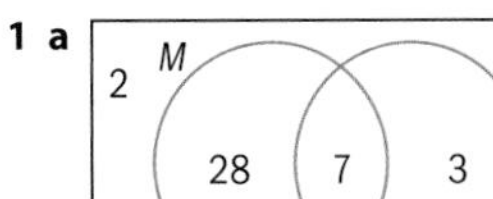

b $\frac{28}{40} = \frac{7}{10}$

2 a

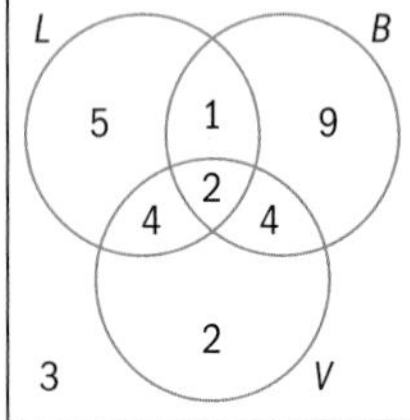

b i $\frac{4}{30}=\frac{2}{15}$ **ii** $\frac{5}{30}=\frac{1}{6}$ **iii** $\frac{2}{30}=\frac{1}{15}$

3 a

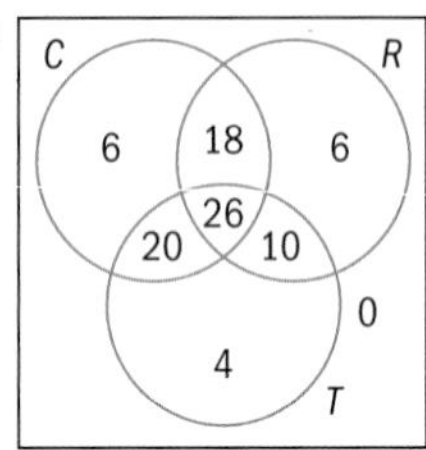

b $\frac{4}{90}=\frac{2}{45}$ **c** $\frac{6}{90}=\frac{1}{15}$

4

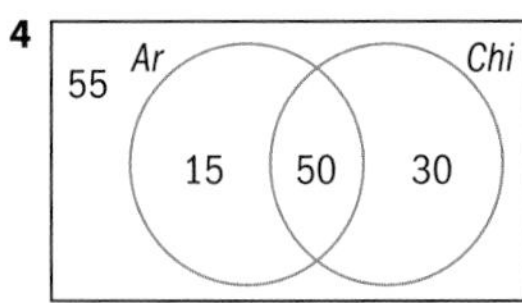

a $\frac{15}{150}=\frac{1}{10}$ **b** $\frac{30}{150}=\frac{1}{5}$ **c** $\frac{55}{150}=\frac{11}{30}$

Practice 6

1 a 29 **b** 26 **c** $\frac{14}{50}=\frac{7}{25}$

2 a

	A	I	B	Total
M	10	4	16	30
F	12	6	12	30
Total	22	10	28	60

b $\frac{10}{60}=\frac{1}{6}$

3 a 47.4% (3 s.f.) **b** $\frac{16}{57}$ **c** $\frac{4}{57}$

4

	15	15-20	20+	Total
M	4	6	7	17
F	8	15	11	34
Total	12	21	18	51

a i $\frac{34}{51}$ **ii** $\frac{6}{51}=\frac{2}{17}$ **iii** $\frac{39}{51}$

b $\frac{4}{17}$

Practice 7

1 Method c **2** Student's own answers

3 a No – not enough evidence with only 10 throws
b The coin is most likely biased, as this number is too high to occur by chance.

4 a Yes it is fair

	1	2	3	4	5	6
1	2	3	4	5	6	7
2	3	4	5	6	7	8
3	4	5	6	7	8	9
4	5	6	7	8	9	10
5	6	7	8	9	10	11
6	7	8	9	10	11	12

b i No – Hadley has more chance to win
ii In this one case, both win so 'roll again'

5 a

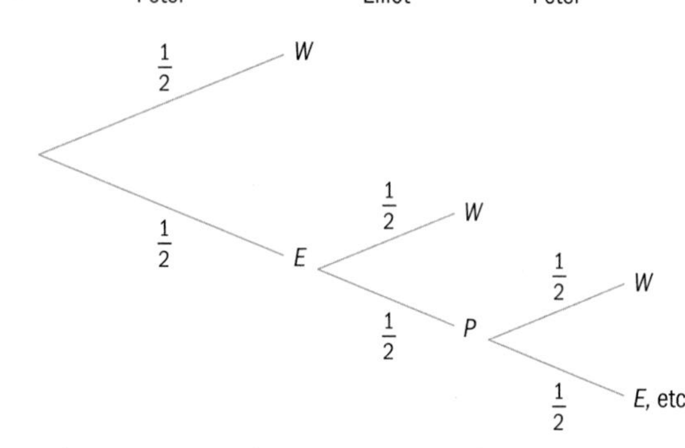

b i $\frac{1}{2}$ **ii** $\frac{1}{4}$ **iii** $\frac{1}{32}$

c Not fair: Peter has a greater chance of winning than Eliott.

Mixed practice

1 a (P, A) (P, B) (P, C) (P, F) (F, A) (F, B) (F, C) (F, F)

	A	B	C	F
P	P,A	P,B	P,C	P,F
F	F,A	F,B	F,C	F,F

b i $\frac{3}{8}$ **ii** $\frac{1}{2}$ **iii** $\frac{1}{8}$

2 a A+, A–, B+, B–, AB+, AB–, O+, O–

b $\frac{1}{8}$ **c** $\frac{2}{8}=\frac{1}{4}$ **d** $\frac{1}{2}$ **e** $\frac{1}{2}$

3

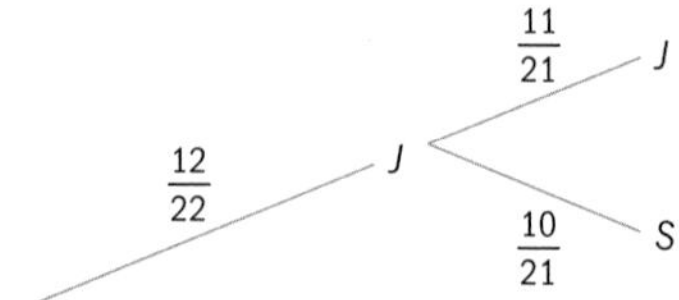
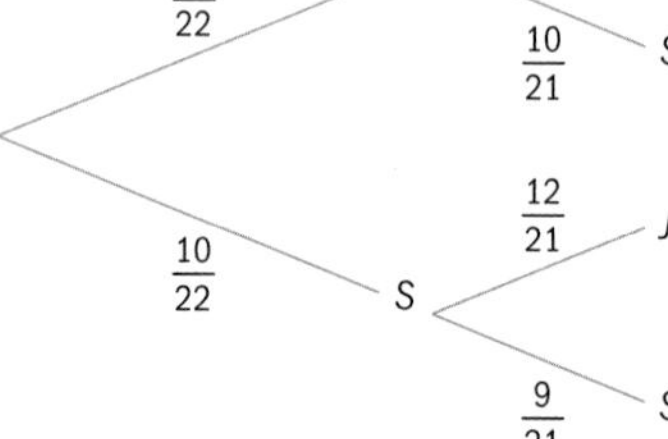

a $\frac{2}{7}$ **b** $\frac{37}{77}$

c $\frac{40}{77}$ **d** $\frac{15}{77}$

4 a 1 2 **b** $\frac{1}{4}$ **c** $\frac{1}{6}$ **d** $\frac{1}{6}$

5 a

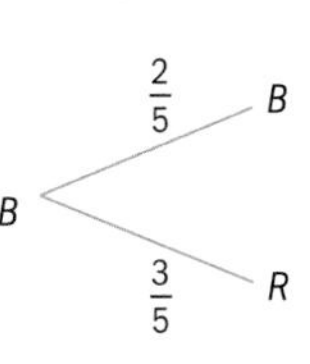
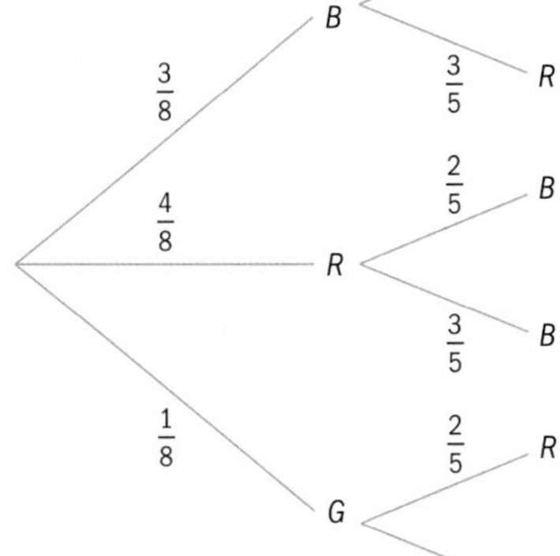

b $\frac{3}{20}$ **c** $\frac{11}{20}$

6 a

10 C S
2 6 6

b $\frac{6}{24}=\frac{1}{4}$ **c** $\frac{2}{24}=\frac{1}{12}$ **d** $\frac{10}{24}=\frac{5}{12}$

7 a

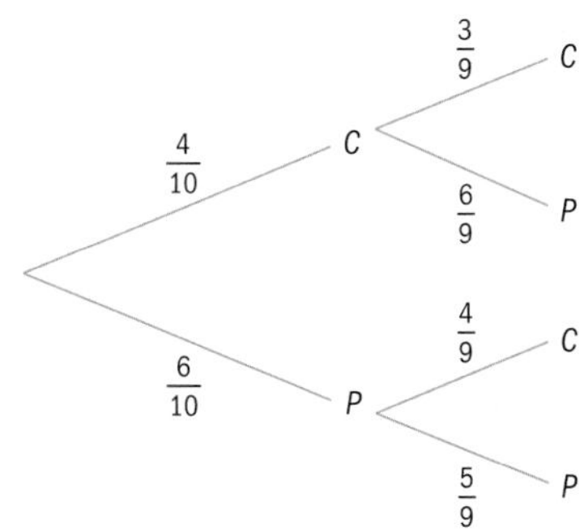

b $\frac{8}{15}$

8 a As above but second level probabilities are $\frac{4}{10}, \frac{6}{10}, \frac{4}{10}, \frac{6}{10}$

b $\frac{12}{25}\left(\text{down by } \frac{4}{75}\right)$

9 a

	More	Right	Less	
M	60	4	8	72
F	20	4	24	48
	80	8	32	120

b $\frac{8}{120}=\frac{1}{15}$ **c** $\frac{8}{120}=\frac{1}{15}$ **d** $\frac{20}{48}=\frac{5}{12}$

10 a $\frac{23}{50}$ **b** $\frac{11}{50}$ **c** $\frac{40}{50}=\frac{4}{5}$ **d** $\frac{17}{50}$

Review in context

1 a 15.8% **b** The risk increases
c About 5 times less **d** Student's own answers

2 a 0.003375 **b** 0.06075 **c** 0.614125

3 a

			Mother					
			MM		Mm		mm	
			M	M	M	m	m	m
Father	MM	M	MM	MM	MM	Mm	Mm	Mm
		M	MM	MM	MM	Mm	Mm	Mm
	Mm	M	MM	MM	MM	Mm	Mm	Mm
		m	mM	mM	mM	mm	mm	mm
	mm	m	mM	mM	mM	mm	mm	mm
		m	mM	mM	mM	mm	mm	mm

b $\frac{9}{36}=\frac{1}{4}$ **c** $\frac{27}{36}=\frac{3}{4}$ **d**

MM	Mm
mM	mm

¼

e $\frac{6}{16}$ **f** $\frac{9}{16}$ **g** Mm and mm

5.1

You should already know how to:

1 a 5 **b** 7 **c** 3
d $\sqrt{11}$ **e** 12 **f** $\frac{1}{3}$

Practice 1

1 a 11 **b** 8 **c** 3.8–3.9
d 3.4–3.5 **e** 5.1–5.2 **f** 7.1 (just over 7)

2 a 4.8–4.9 **b** 7.4–7.5 **c** 9.1
d About 9.5 **e** About 7.5 **f** 9.9

3 a $<$ **b** $<$ **c** $>$
d $>$ **e** $<$ **f** $<$

Practice 2

1 a $\sqrt{130}$ **b** $63\sqrt{105}$ **c** $\sqrt{77}$
d 3 **e** $\frac{\sqrt{15}}{5}$ **f** $\sqrt{7}$
g $\sqrt{21}$ **h** $\frac{1}{4}$

2 3

Practice 3

1 $2\sqrt{6}$ **2** $4\sqrt{2}$

3 $6\sqrt{2}$ **4** $5\sqrt{5}$

5 $3\sqrt{15}$ **6** $8\sqrt{15}$

7 $15\sqrt{3}$ **8** $12\sqrt{6}$

9 $\sqrt{991}$ **10** $4\sqrt{62}$

11 $3a^3\sqrt{11}$ **12** $10b^2c\sqrt{2}$

13 $12x^4y\sqrt{2y}$

Practice 4

1 $-2\sqrt{3}$
2 $17\sqrt{17}$
3 $3\sqrt{a+3}$
4 $6\sqrt{pq}$
5 $\sqrt{3}$
6 $8\sqrt{3}$
7 $-5\sqrt{3}$
8 $13\sqrt{2}+4\sqrt{3}$
9 $\sqrt{5}+\sqrt{6}$
10 $13\sqrt{2}+7\sqrt{3}$
11 $17\sqrt{3x}$
12 $(x-4)\sqrt{3}$
13 $\sqrt{3x}(3-x)$
14 $-11\sqrt{x+2y}$

Practice 5

1 2
2 3
3 $\frac{\sqrt{3}}{3}$
4 $\frac{\sqrt{2}}{4}$
5 $5\sqrt{3}$
6 $5\sqrt{5}$
7 $5\sqrt{2}$
8 $\frac{6\sqrt{10}}{5}$
9 $\sqrt{3}$
10 $\sqrt{7}$
11 $\frac{\sqrt{6}}{3}$
12 $\frac{\sqrt{30}}{10}$
13 $6x$
14 $\frac{\sqrt{5}}{5x}$
15 $\frac{\sqrt{10xy}}{2xy}$
16 $\frac{\sqrt{2x}}{x}$

Mixed practice

1 $>$
2 $>$
3 $<$
4 $>$
5 6.7
6 4.1
7 7.3 or 7.4
8 3.4 or 3.5
9 $\frac{\sqrt{2}}{2}$
10 $2\sqrt{3}$
11 $\frac{3\sqrt{2}}{16}$
12 $\sqrt{3}$
13 $2\sqrt{3}$
14 $3\sqrt{2}$
15 $5\sqrt{5}$
16 $11x\sqrt{x}$
17 60
18 6
19 $x^3y^2\sqrt{y}$
20 7
21 $\frac{\sqrt{5x}}{x}$
22 $\frac{\sqrt{6}}{2}$
23 $\frac{5\sqrt{3}}{3}$
24 $10\sqrt{13}$
25 $-5\sqrt{2}$
26 $21\sqrt{x}$
27 $16\sqrt{5t}$
28 $16\sqrt{2a}$

6.1

You should already know how to:

1 a 96.5 kg b 0.83 mm
c 15.6 Ω d 6 liters

2 22.9 %

Practice 1

1 675 EUR

2 a 1224 USD b 1700 CHF

3 1673.08 CAD

4 a 756 AUD b 396.83 USD

5 a 66 Yuan b 7125 Yen c 16 Peso

6 a 4 TRY b 1.25 CHF c 0.31 CHF

7 a 1935 BRL b 329 400 CRC c 241.3 USD

8

	AOA	BGN	CLP
1 AOA	1	0.0170	6.03
1 BGN	58.8	1	355
1 CLP	0.166	0.00282	1

Practice 2

1 a 756 EUR b 7.56 EUR c 748.44 EUR

2 3534.30 USD 3 16722.72 RND

4 a 169 CHF b 333.77 GBP c $S = 1.3(B-3)$

5 255 GBP

Practice 3

1 a 965.30 AUD b 702.51 USD

2 a 1326 MXN b 458.04 DKK
c 204.55 CAD

3 a i p = 1.54 ii q = 1.23
b 965.25 GBP
c Yes, he has enough money.

4 Bank 1

Mixed practice

1 a 1045 USD b 2500 EUR

2 a 242 USD b 2.42 USD c 239.58 USD

3 1128.96 GBP

4 a 148120 RUB b 13.23 GBP

5 $p = 0.79$, $q = 0.65$

6 a 3847 THB b 0.2990 USD
c 0.04090 NZD d 0.1368 NZD

7 a 560 EUR b 767.12 AUD c 32.88 AUD

8 a 274.40 CHF b 70 GBP

9 483.74 EUR

10 a 38 860 THB b 818.00 NZD c 1 NZD = 18.25 CZK

Review in context

1 a 2732.40 SGD b s = 2.07(b – 10) c The British bank

2 a 3960 CHF, 7500 BRL, 21750 SEK
b 4082.47 USD, 4934.21 USD, 3398.44 USD
c 4658.82 USD, 2279.64 USD, 2338.71 USD

6.2

You should already know how to:

1

Height	Frequency
$2.0 \le h < 3.0$	8
$3.0 \le h < 4.0$	5
$4.0 \le h < 5.0$	6
$5.0 \le h < 6.0$	5

2 Modal class is $2.0 \le h < 3.0$; Median class is $3.0 \le h < 4.0$; Mean estimate = 3.83 cm (3 s.f.); Range estimate = 4.0 cm

3 a Quantitative **b** Qualitative **c** Quantitative

Practice 1

1

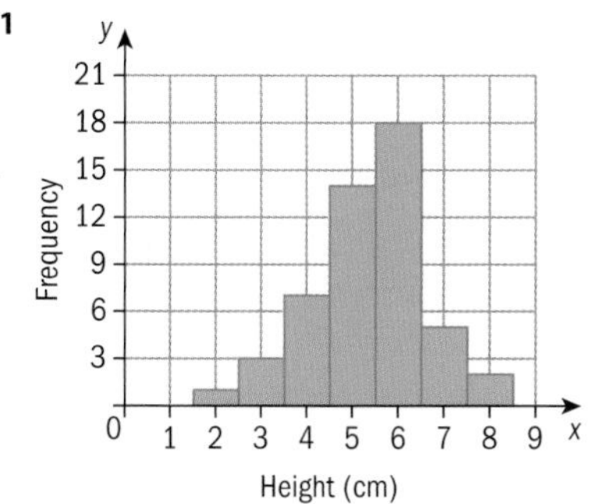

2

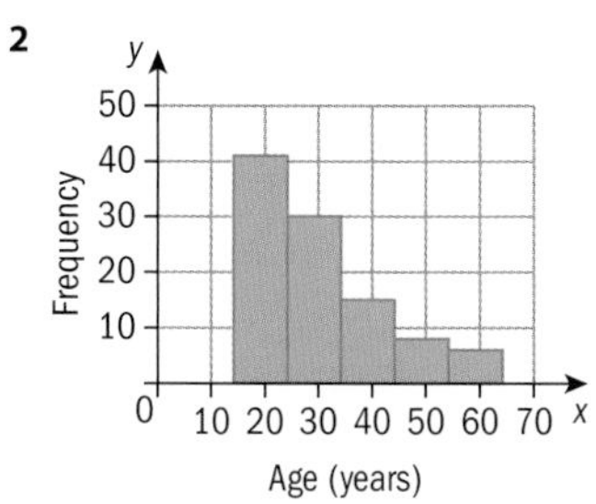

3

Class	Class boundaries	Frequency
80–119	79.5–119.5	15
120–159	119.5–159.5	6
160–199	159.5–199.5	9
200–239	199.5–239.5	8

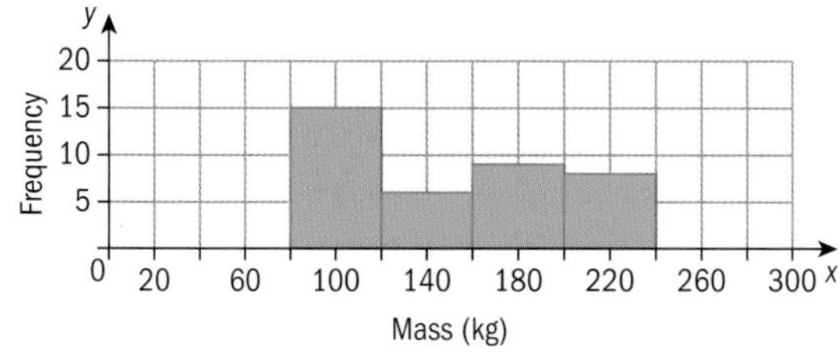

Practice 2

1 a Discrete, ungrouped

b

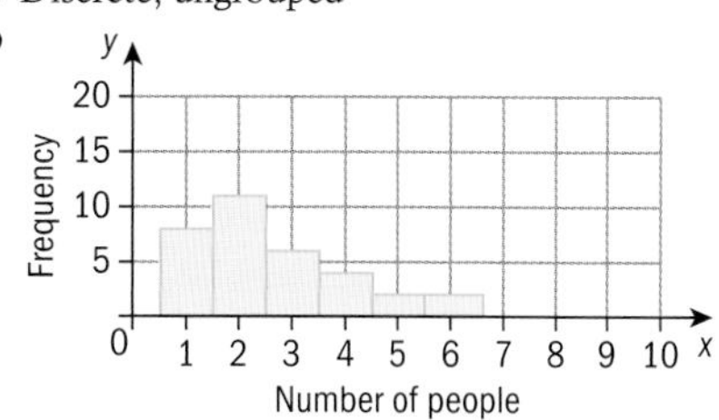

2 a

Number of eggs hatching	Frequency
1–4	2
5–8	2
9–12	6
13–16	9
17–20	3
21–24	3

b

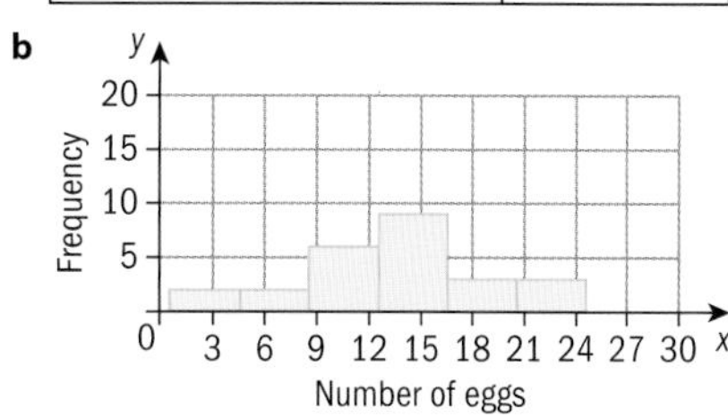

3 a Continuous

b

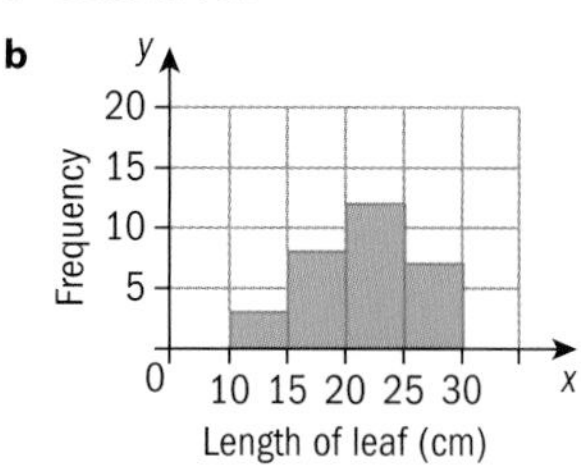

4 a 1–5

b 0.5–5.5, 5.5–10.5, 10.5–15.5, 15.5–20.5, 20.5–25.5

c

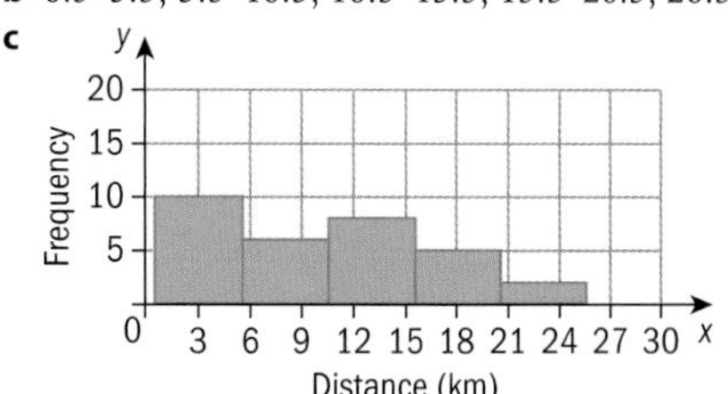

5 a Continuous

b Student's own answer with continuous variable

c No outliers, the data is evenly distributed within the range 9–35.

6 a

Time (h)	Frequency
$1.5 \le t < 2.5$	5
$2.5 \le t < 3.5$	6
$3.5 \le t < 4.5$	8
$4.5 \le t < 5.5$	7
$5.5 \le t < 6.5$	4

b 3 hours 58 minutes

Practice 3

1 i Not symmetrical, multi-modal, contains possible outlier
ii Not symmetrical, multi-modal, no outliers
iii Symmetrical, unimodal, no outliers

2 a 36–40 cm **b** 36–40 cm
c 20 cm **d** 38 cm
e Relatively symmetrical, unimodal, no outliers

Length (x)	Frequency
$28 \le x < 32$	4
$32 \le x < 36$	4
$36 \le x < 40$	9
$40 \le x < 44$	6
$44 \le x < 48$	3

3 a 40
b \$163.25
c Fairly symmetrical, unimodal, no outliers

4 a Discrete data
b Student's own answers (bar chart with uniform distribution would be most likely – bars 2–3 units high.)

5 a

Weight (x)	Frequency
$10 \le x < 20$	3
$20 \le x < 30$	8
$30 \le x < 40$	23
$40 \le x < 50$	10
$50 \le x < 60$	4
$60 \le x < 70$	2

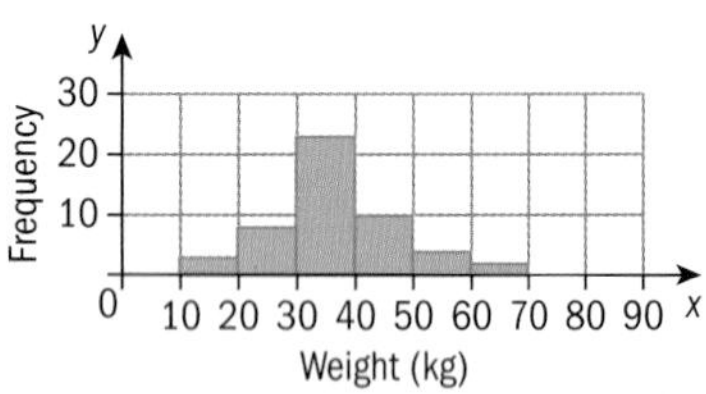

b Symmetrical, unimodal, no outliers

Practice 4

1 a Different number of data points in each group
b Men: $\frac{1}{6}, \frac{4}{15}, \frac{1}{3}, \frac{1}{6}, \frac{1}{15}$

Women: $\frac{2}{25}, \frac{1}{10}, \frac{7}{50}, \frac{7}{25}, \frac{2}{5}$

c

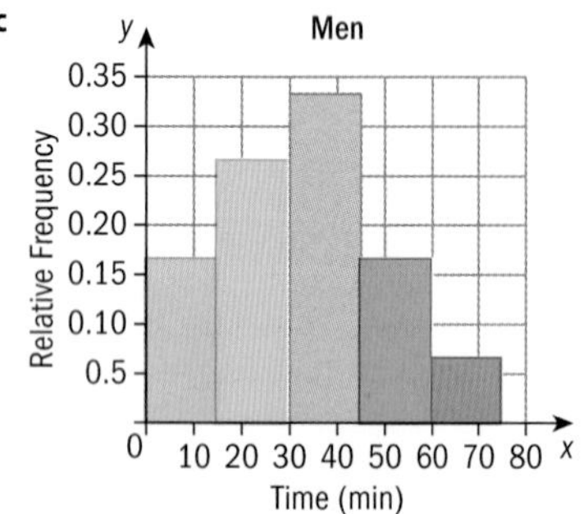

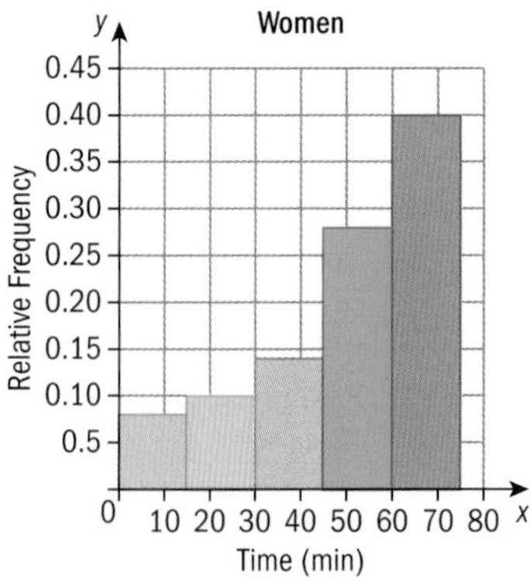

d Men: symmetrical, unimodal; Women: unimodal but centered mainly on the longer times.
e Women talked for longer on the phone than men. The spread of times is similar.

2 a Male: 9, 17, 17, 31, 26
Female: 26, 42, 21, 11, 0 (nearest percent throughout)

b

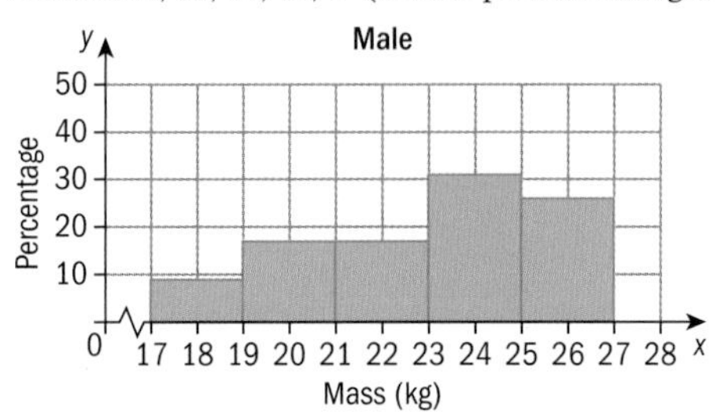

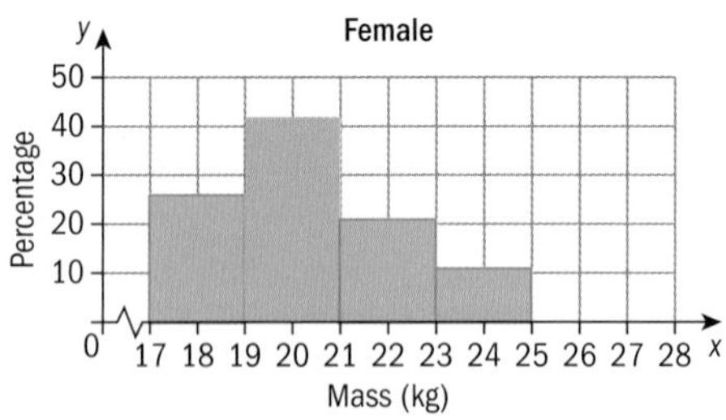

c Male huskies tend to weigh more than female huskies. The spread is narrower for the female huskies as well. The distribution for the females is fairly symmetrical.

3 18

Practice 5

1 a Missing values are 5,5,7 and 4,5.2,2.71 (3 s.f.)
b

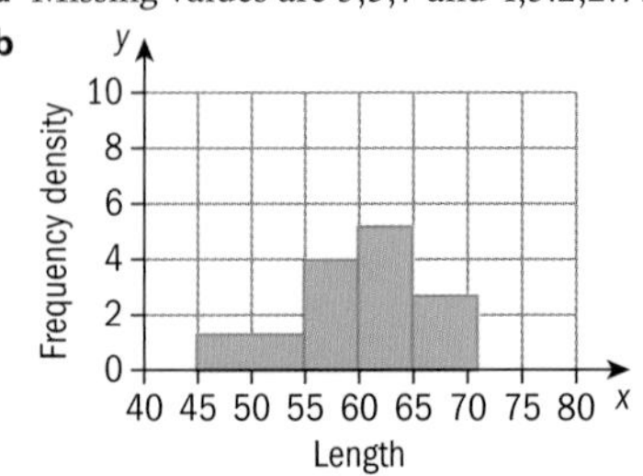

2 a

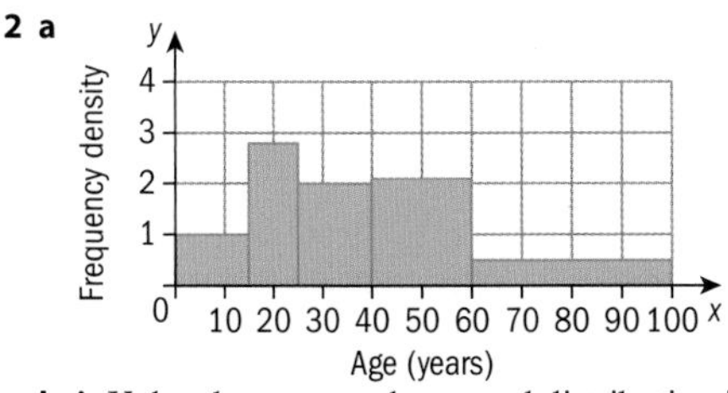

b i Voleta has assumed an equal distribution in the final class interval.

ii Not reasonable since, as can be seen from the graph, there may be more people at the lower end of the 60–100 class.

c 15–25

3 a

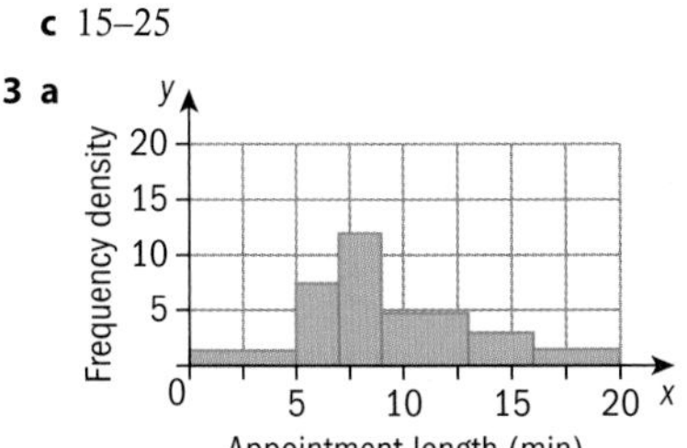

b 20

4 a

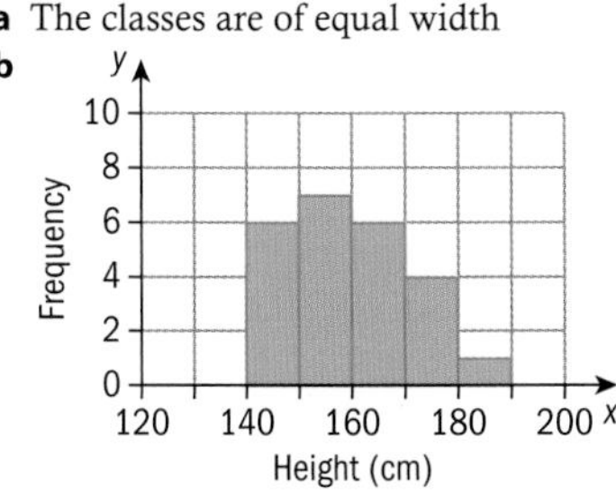

b 145–160 **c** 117

5 a

Distance (x)	Frequency
$0 \le x < 20$	60
$20 \le x < 35$	90
$35 \le x < 45$	100
$45 \le x < 60$	90
$60 \le x < 65$	10

b 350 **c** 35.5 meters (3 s.f.) **d** 35–45

6 115 **7** 2.5 cm

Mixed practice

1 a The classes are of equal width

b

2

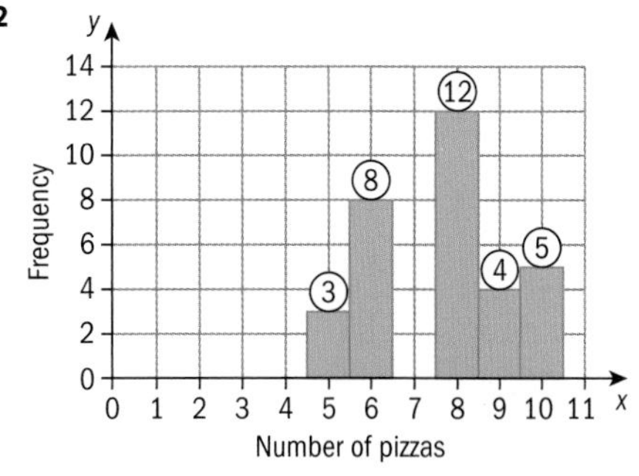

3 a

Weight (x)	Frequency
$80 \le x < 90$	7
$90 \le x < 100$	14
$100 \le x < 110$	11
$110 \le x < 120$	7
$120 \le x < 130$	9
$130 \le x < 140$	2

b

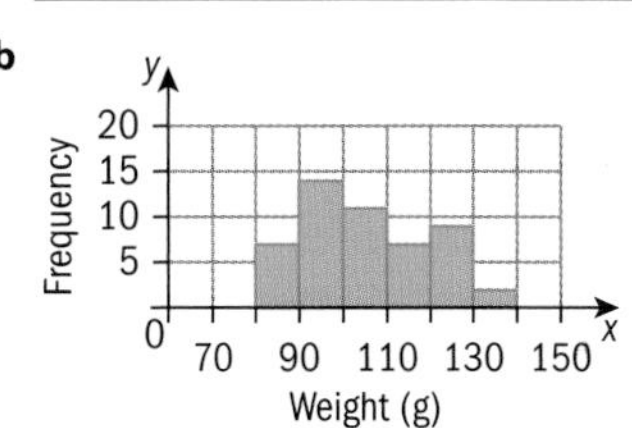

4 a

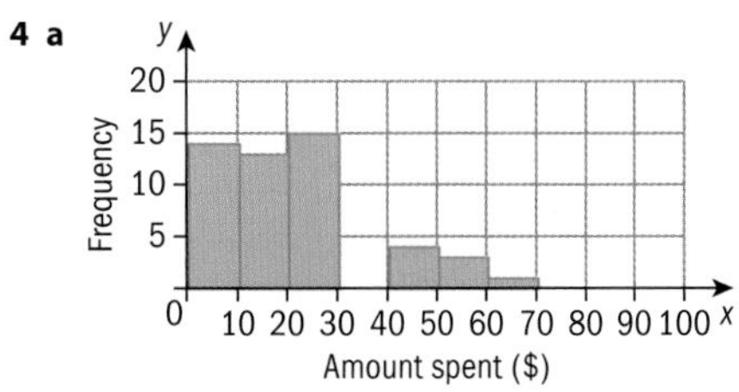

b 21–30

c 11–20

d Distribution is centered towards the lower amounts for eating out.

5 a 35.6 (3 s.f.)

b

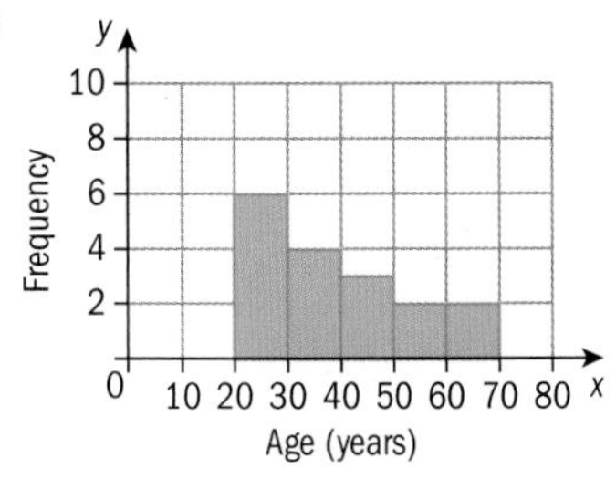

c Skewed to the left, unimodal

6 a Different number of students in each group

b Both are unimodal, Sri Lanka is more symmetrical, Peru is skewed to the left.

7 a 5.5–8.5, 8.5–11.5, 11.5–17.5, 17.5–20.5, 20.5–29.5
b Class widths are not equal
c Missing values are $\frac{4}{3}$, 2, $\frac{5}{3}$, 1, $\frac{4}{3}$
d

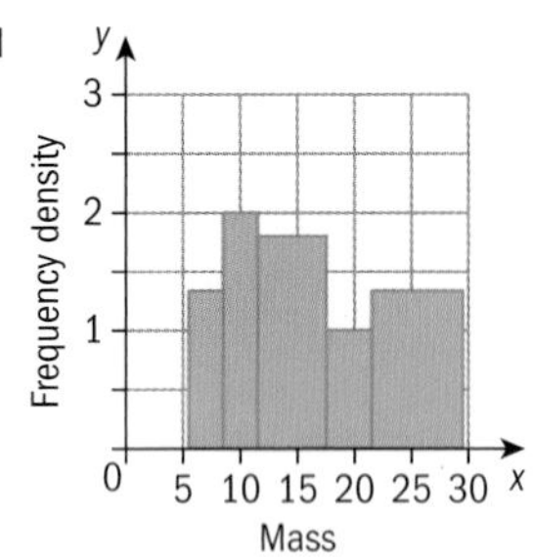

8 a Frequency density
b

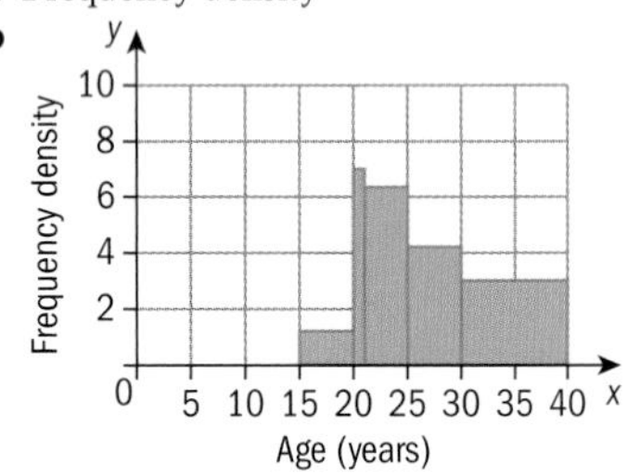

c Modal class 20–22, Mean ≈ 27.5 years (3 s.f.)

9 a 72–80 **b** 77.6 cm (3 s.f.)

10 a 60–70
b 106 seconds (3 s.f.)
c 140
d i 50 **ii** 70

Review in context

1 a $(0 \times 4) + (1 \times 10) + (2 \times 7) + (3 \times 3) + (4 \times 1) = 37; \frac{37}{25} = 1.48$
b More children on average in Australia; Bigger spread of children in Australia. Both distributions unimodal and skewed slightly to the left.

2 a

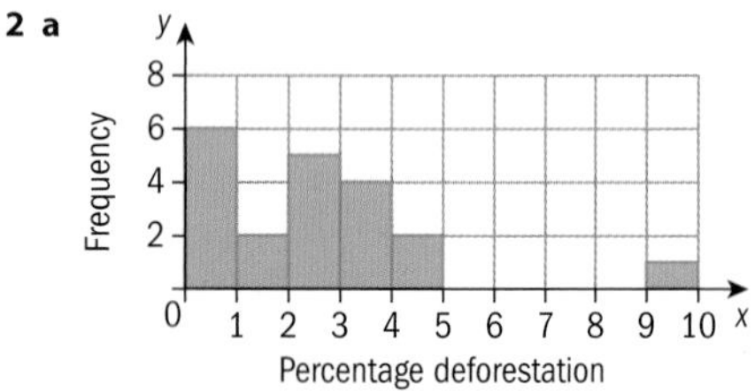

b Skewed to the left **c** 9.2%

7.1

You should already know how to:

1 a −2 **b** 4 **c** −15 **d** −9
2 a 11 **b** 7 **c** $\frac{1}{2}$ **d** 0.1
3 a 2 **b** 6 **c** 16
d 4 **e** 2 **f** 8

Practice 1

1 a 35 **b** −35 **c** 234 **d** −234
e 5.6 **f** 2.8 **g** 0 **h** $\frac{1}{2}$
i $\frac{2}{5}$ **j** 0.01 **k** 64 **l** 6
m −5 **n** $4\sqrt{2}$ **o** $-2\sqrt{6}$ **p** −1000

2 a 13 **b** 3 **c** 3 **d** 5
e 2.1 **f** 18 **g** 15 **h** 6
i 11 **j** −5 **k** 8 **l** 8

3 a 4, −4
b Student's own answer

4 a Yes, they are driving safely, since the difference between the cars' speeds is 16 km/h.
b Minimum = 90 km/h, Maximum = 130 km/h

Practice 2

1 a 1 **b** 15 **c** 18
d 10 **e** 33 **f** 64
2 a −3 **b** 10 **c** 6
d −20 **e** 6 **f** 625
3 a 2 **b** −2 **c** −1
d 1 **e** Undefined **f** 0
4 When $b < 5.5$

Practice 3

1 13%
2 a 240.46 ml **b** 0.15% **c** Yes. It's very small.
3 4.17%
4 \$23.31 or \$33.54
5 $56.8 \le x \le 59.2$
6 Less than 440 g or more than 560 g
7 500 kg measured with percentage error of 1%; it has an absolute error of 5 kg, whereas the 30 kg measured with a 10% error has an absolute error of 3 kg.

Mixed practice

1 a 33 **b** −2.5 **c** −5.75
d 4.5 **e** $\frac{2}{9}$ **f** 32
g $3\sqrt{3}$ **h** −9
2 a 20 **b** −36 **c** $-\frac{1}{2}$
d $\frac{9}{2}$ **e** $\frac{9}{20}$ **f** 2.4
g 70 **h** 5
3 3.15% **4** 4.76%
5 a 25 EUR **b** 10 EUR
c i 11.11 EUR
ii 9.99%. The true value is an hourly rate of 11.11 EUR, and her average hourly rate is 9.99% less.
d 28%. The true value is an hourly rate of 7.81 EUR, and her average hourly rate is 28% more.

Review in context

1 a 57, 3951057 **b** 0.00%, 0.287%
c Percentage errors are very small because the population is extremely large. However an absolute error of 3.95 million represents quite a large number of people, even though the percentage error is less than 0.5%.
d 24 768 881
e 1 351 280 062 to 1 400 817 824

f If the government claims the official figure may be inaccurate by 1.8% (which is 24.7 million people), then 1.4 billion is good estimate since the absolute error (24 million) is within the acceptable range, and gives an idea of the size of the total population.

g **United States:**
Absolute error: 5 785 956, Reliable Value: 320 million (320 million has a smaller percentage error than 1.8%, but 300 million has a larger percentage error than 1.8%)
Germany:
Absolute error: 1 452 394, Reliable Value: 80 million (rounded to the nearest 10 million; 80 million has a smaller percentage error than 1.8%)
Malaysia:
Absolute error: 545 958, Reliable Value: 30 million (30 million has a smaller percentage error than 1.8%)
Australia:
Absolute error: 431 442, Reliable Value: 24 million (24 million has a smaller percentage error than 1.8%, but 20 million has a larger percentage error than 1.8%)
Monaco:
Absolute error: 679, Reliable Value: 38 000 (38 000 has a smaller percentage error than 1.8%, but 40 000 has a larger percentage error than 1.8%)

2 a 403 597 **b** 23.9%
c 263 757 **d** an increase
e Student's own answer. Ideas include the aging of the population (increase in percentage of population over 65) and the increase in the total population.

3 a 4.45%
b To bias should let Felix know how the amount had been rounded and let Felix decide if it was reasonable.
c 13 cents; 4.45%
d The percentage errors are the same.

7.2

You should already know how to:

1 Length: 34 mm, 9 km, 13 miles
Mass: 2 kg, 15 g
Volume: 5 ml, 6 liters, 8.4 cm^3
24 s is the odd one out.

2 $10^3 = 1000$
$10^6 = 1\,000\,000$
$10^{-2} = 0.01$

3 0.28, 345

4 Area = 144 cm^2

5 Surface Area = 94 m^2
Volume = 60 m^3

Practice 1

1 a mm **b** mm **c** cm
d ml **e** m **f** m^2
g m

2 a tonnes **b** m **c** m^2
d km/h **e** m **f** m
g m/s

Practice 2

1 a 3 500 000 tonnes **b** 8 000 000 000 bytes
c 0.257 meters **d** 0.65 liters

2 a 1 000 000 000 **b** 1 000 000
c 1000 **d** 100

3 a 2 000 000 micrometers **b** 3200 mg
c 2 500 000 000 microwatts
d 500 000 000 nanoseconds

4 a 2.85 m **b** 0.9235 g **c** 4.358 km
d 4 358 000 mm **e** 726 300 cg **f** 1245 cl
g 18.655 l **h** 56 cm **i** 5000 cm
j 0.380506 kg

5 200 **6** 1 875 000 000

Practice 3

1 a 0.0254 m **b** 0.0353 ounces
c 0.454 kg (3 s.f.) **d** 0.222 UK gallons (3 s.f.)
e 0.568 liters (3 s.f.) **f** 0.0328 feet

2 a 80.45 km **b** 31.08 miles **c** 154 pounds
d 15.24 cm **e** 91.44 cm **f** 2.84 liters
g 16 pints **h** 160 900 cm

3 a 58 miles **b** 60 lbs **c** 1 pint
d 4 oz **e** 20 cm **f** 6 feet

4 a 45.72 cm, not effective **b** feet

5 a 20 089 liters should have been added to the 7682 liters already in the tank.
b 4917 liters
c 15 172 liters

Practice 4

1 a 2000 mm^2 **b** 6 cm^2 **c** 0.5 m^2
d 45 000 cm^2 **e** 290 mm^2 **f** 700 000 mm^2

2 a 40 000 000 cm^3 **b** 5 cm^3 **c** 2.4 m^3
d 3 ml **e** 2500 cm^3 **f** 10 000 liters

3 0.0005 m^3

4 a 1 dm^3 = 1000 cm^3 **b** 1000 dm^3 **c** 6000 dm^3

5 1 135 200 liters **6** Student's own answers

Practice 5

1 196 m^2 (3 s.f.)

2 8030 m^2, 0.00803 km^2, m^2 is easier to visualize

3 a 62.4 m^2 **b** 352 m^3 (3 s.f.)
4 36 ha

Practice 6

1 a 1.25 by 1.875 miles
b 2.34 square miles

2 a i 1966.45 cm^3 **ii** 954.84 cm^2
b i 1966.45 cm^3 **ii** 1019.35 cm^2
The boxes have the same volume but different surface areas.

3 81.94 cm^3

4 8.36 m^2

5 882.87 cubic feet

6 a 4 inches **b** 10.16 cm

7 40 meters

Practice 7

1 a 10 000 m/h b 54 000 m/h
c 4800 m/h d 280 000 m/h

2 a 1 m/s b 12 m/s
c 2.5 m/s d 0.6 m/s

3 18.06 m/s
4 7.2 km/h

5 2700 kg/m^3
6 14.3 mg/cl

7 11 340 kg
8 9660 kg

9 No, 30 mph = 48.57 km/h, so the speed limit is a bit lower in the US than in Europe.

10 The cat is faster (48 km/h = 13.3 m/s)

11 He is not correct. It is 1.716 grams.

12 He is not correct. It will take 1 hour 13 minutes.

13 Yes. It travelled at roughly 341 m/s.

Mixed practice

For question 1, other answers may be acceptable.

1 a m^3 or liters b km/h or mph
c cm^3 or liters d kg or pounds
e grams or ounces f km or miles

2 a 300 cm b 4800 g
c 7.6 m
d 9.845 km
e 1000 cm f 0.4006 kg

3 a 113.4 g b 38.1 cm
c 26.3 feet (3 s.f.) d 1.32 gallons
e 1.25 UK gallons f 1.76 pounds

4 a 5000 mm^2 b 0.95 m^2 c 500 000 mm^2
d 6.4 cm^3 e 10 ml f 100 000 l

5 a 0.95 km/h b 36 000 m/h c 36 000 m/h
d 2 m/s e 15 m/s

6 96 doses

7 a Person A (1.88 m) b Person B (Person A weighs 86.3 kg)

8 a Yes b 26 Euros

9 50 373 liters

10 0.549 cubic inches

11 5.017 m^2

12 14.69 m

13 8.9 g/cm^3

14 Platinum is denser.

15 Yes, the car is exceeding the speed limit.

16 The Formula 1 racing car is faster.

Review in context

1 a i 227.38 m ii 2331.72 m iii 128 720 km
b 7850 kg/m^3 c 278 377.48 m^3
d i 53.975 mm ii 32.725 mm

2 a 1435 mm
b i 1422 mm ii 13 mm iii 2.4% error
c 3 ft 3 in
d Student's own answers, such as 'some trains travel from Spain to France.'

7.3

You should already know how to:

1 a 15.1 cm b 18.1 cm

2 a 18.1 cm^2 b 26.0 cm^2

3 2.5 cm

Practice 1

1 18.5 cm
2 10.7 cm

3 12.1 cm
4 6.6 cm

5 106.26°

Practice 2

1 a Area = 9π cm^2
Circumference = 6π cm
b Area = 400π mm^2
Circumference = 40π mm
c Area = 0.16π m^2
Circumference = 0.8π m

2 a Area = 76.97 mm^2
Perimeter = 35.99 mm
b Area = 113.10 cm^2
Perimeter = 42.85 cm

3 a 7.4 cm b 23.4 cm c 34.8 cm d 55.3 cm^2

4 a $\frac{14\pi}{3}$ cm b $\frac{14\pi}{3}+14$ cm c $\frac{49\pi}{3}$ cm^2

5 a 86.64 m b 410.50 m^2

6 a 197.92 cm b 269.92 cm c 3562.57 cm^2

7 17.55 cm
8 163.86 cm

9 62.83 cm^3
10 706.86 cm^3

Practice 3

1 a 110° b 80°

2 a 24.0° b 3.15 cm

3 a 200° b 13.96 cm c 21.96 cm

4 a 9.17 cm b 6.94 m

5 347.86 mm
6 4.71 cm^2

7 radius = 7.9 cm
8 3 cm

9 5.91 cm

10 a

40°
120 m

b 323.78 m

c 5026.55 m^2 d 82.08 m

Mixed practice

1 a 2.09 mm b 8.78 mm c 5.17 mm^2
2 a 62.31 km b 436.16 km^2
3 a Perimeter = 25.07 feet, Area = 31.81 square feet.
b Perimeter = 67.48 feet, Area = 222.66 square feet.

4 a 3.93 m b 13.93 m c 68.72 m^2

5 a 120° b 16.76 cm
c 32.76 cm d 13.86 cm

6 a 110° b 39.2 km c 16.38 km

7 a Student's own diagram: a correctly labelled rectangle.
b Student's own diagram, with center of the circle at the center of the rectangle.
c 70.36 m^2
d 12 %

8 30 cm

Review in context

1 a 23.56 cm b 19.10°

2 a 190.46 m^2 b 0.347 m^2 c 190.1 m^2

3 a 12.86° b 16.16 m^2
c 23.64 m d 15.43 m

7.4

You should already know how to:

1 a 16 cm b 16.2 cm c 15.6 cm

2 $\sin A = a/c$; $\cos B = a/c$; $\tan A = a/b$; $\tan B = b/a$

3 a 5.77 m b 42.0° (3 s.f.)

Practice 1

1 a 0.731 b 0.848 c 0.306
d 0.0349 e 57.3 f 0.5
g 1.03 h 0.0138
i 1 j $\frac{\sqrt{3}}{3}$

2 a 27.8° b 45° c 45°
d 26.1° e 30° f 60°
g 82.9° h 87.9°

Practice 2

1 a 13.77 cm b 4.60 cm c 5.5 cm
d 11.92 cm e 2.05 cm f 7 cm

2 a 17.1° b 45°
c 50.0° d 66.4°

3 a $AB = 25.6$ cm, $BC = 12.0$ cm, $\angle A = 28°$
b $AC = 4.0$ cm, $BC = 3.0$ cm, $\angle A = 37°$
c $AC = 3.0$ cm, $BC = 1.9$ cm, $\angle B = 50°$
d $AC = 12.8$ cm, $BC = 5.7$ cm, $\angle B = 66°$
e $AC = 3.2$ cm, $AB = 6.8$ cm, $\angle A = 62°$

Practice 3

1 $a = 5\sqrt{3}$ m, $b = 10$ m

2 $c = 6\sqrt{2}$ cm, $d = 6\sqrt{2}$ cm

3 $e = 2$ m, $f = 2\sqrt{3}$ m

4 $g = 6$ cm, $h = 6\sqrt{2}$ cm

5 $x = 60°$

Practice 4

1 56 m

2 24.0° (Depends on the accuracy of measurement and the height of the person)

3 115 m (3 s.f.)

4 37.6° (3 s.f.)

5 3 m, 5.20 m (3 s.f.)

6 93.0 m (3 s.f.)

Practice 5

1 a 35.5 cm (3 s.f.) b 23.2 cm (3 s.f.)

2 a 15.7 mm (3 s.f.) b 14.6 mm (3 s.f.)

3 44.0° (3 s.f.)

4 5.20 cm (3 s.f.)

5 A: 9.85 m B: 14.0 m C: 7.21 m (all the 3 s.f.)

6 367 m (3 s.f.)

7 a 308 km b 073.7°

8 47.9 km (3 s.f.) 9 200 m

Mixed practice

1 17.0 m (3 s.f.)

2 a 145 m (3 s.f.) b 13.5 cm (3 s.f)

3 a 24.6° b $\sqrt{119}$ cm (10.9 cm to 3 s.f.)
c 3.9 cm d 5.9 cm
e 122.7° (1 d.p.)

4 a 4.24 m (3 s.f.) b 5.15 m (3 s.f.)

5 35.1° 6 73.9 m (3 s.f.)

Review in context

1 29000 feet (3 s.f.) 2 12.6° (3 s.f.)

8.1

You should already know how to:

1 a 10, 12 b 24, 29 c 20, 30

2 a 27 b −13
c 3 d 13

3 a $n = 4$ b $n = 3.2$ c $n = \pm 6$

4 a $u_n = 4n$ b $u_n = 12n$ c $u_n = -3n$

Practice 1

1 a 3, 7, 11, 15, 19 b 8, 26, 56, 98, 152
c 9.75, 9, 7.75, 6, 3.75 d 0, 2, 6, 14, 30
e −10, −6, 0, 8, 18 f −5, −1, 3, 7, 11
g 0, 1, 8, 6, 5

2 47

3 It is not; all terms of the sequence are odd.

4 The 22nd term

5 $9\frac{1}{3}$

6 18 (Note: the first term of the sequence is u_0.)

7 a The 11th term
b The 14th term $= 2 \times 14^2 + 4 = 396$

8 80

9 **a, iv b, v c, vi d, i e, iii f, ii**

Practice 2

1 a 1, 3, 11, 43, 171
b 2, 5, 11, 23, 47
c −4, 3, 10, 17, 24
d 3, 2.5, 2.25, 2.125, 2.0625
e 0.8. 0.32, 0.4352, 0.491602, 0.499859
f 0.8, 0.64, 0.9216, 0.28901376, 0.821939226
g 4, 7, 10, 13, 16

2 −3, 6, −3, 6

3 a 2, 0.5, −1, 2, 0.5, −1
b The numbers 2, 0.5, −1 keep repeating.
c 2, 0.5, −1, 2, 0.5, −1

4 The 9th term **5** 6562

Practice 3

1 Sequence
First difference
Second difference

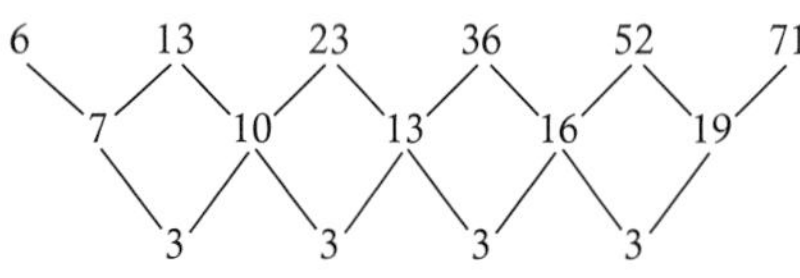

The second difference is constant, therefore it is a quadratic sequence.

2 a Sequence
First difference
Second difference

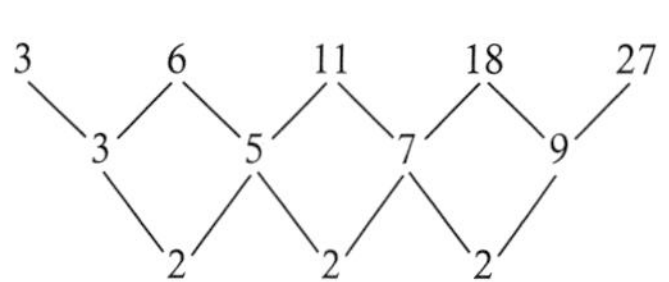

b Sequence
First difference
Second difference

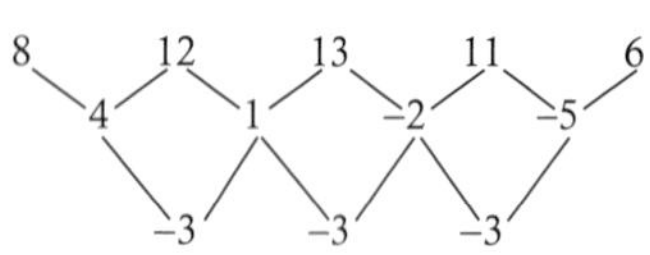

c Sequence
First difference
Second difference

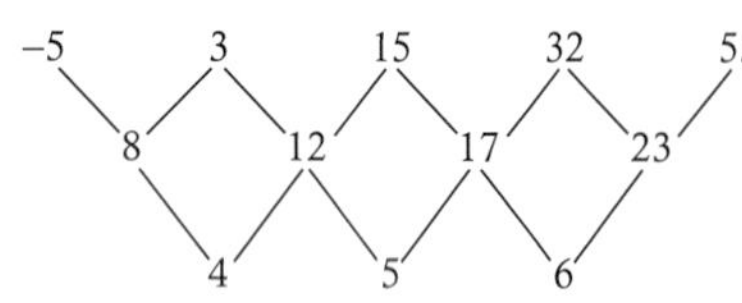

3 a linear **b** linear **c** quadratic
d quadratic **e** linear **f** neither

Practice 4

1 a $a_n = 8n$ **b** $b_n = 3n + 11$ **c** $c_n = -14n$
d $d_n = 60 - 11n$ **e** $e_n = 18n - 7$ **f** $f_n = \frac{1}{2}n + 16\frac{1}{2}$
g $g_n = 9 - \frac{2}{3}n$ **h** $h_n = 4.6n - 6$ **i** $i_n = 33 - 18n$

2 a $a_n = 2n - 1$ **b** $b_n = 50 - 3n$
c $c_n = \frac{1}{2}n + 9$ **d** $d_n = 4n - 13$

3 a Because the distance there and back from the plug to the first bulb is 2×250 cm = 500 cm.
However, between the bulbs is only 10 cm, adding 2×10 cm = 20 cm to each new bulb and back.
b $d_n = 500 + 20n$

4 a 6 + 14 + 14 = 34
b $6 + 28n$
c 14

5 a 690
b $7n - 10 = 102$
$7n = 112$
$n = 16$
The 16th term

6 $u_n = 5n - 6$

7 a 9 **b i** 16 **ii** −56
8 47 **9** −2

Practice 5

1 a $n^2 + 3$ **b** $n^2 - 8$ **c** $n^2 + 2.5$
d $n^2 + n$ **e** $n^2 - 2n$ **f** $n^2 + 5n$
g $n^2 - 1.5n$ **h** $n^2 - 2n + 2$ **i** $n^2 + 3n - 1$
j $n^2 + 6n + 6$ **k** $n^2 - 1.5n + \frac{1}{2}$ **l** $n^2 - 2n - 3$

2 a $n^2 - 5n - 14$ **b** 136

3 a $n^2 - 4n$ **b** 320 **c** 2300

4 a $u_n = n^2 - 16n - 10$
Therefore $u_{16} = 256 - 256 - 10 = -10$
b $u_{17} = 289 - 272 - 10 = 7$
c Since the second difference is positive and the sequence is already increasing, it will keep increasing by increasing amounts.

5 $u_{13} = -10$

Practice 6

1 a $2n^2$ **b** $0.5n^2$ **c** $3n^2$
d $2n^2 + 1$ **e** $3n^2 - 4$ **f** $1.5n^2 - 6$
g $5n^2 + 2n$ **h** $1.5n^2 + 0.5n$ **i** $1.25n^2 + 0.75n$
j $3n^2 + 2n + 1$ **k** $2n^2 - 4n + 1$ **l** $1.25n^2 + 1.25n + 0.5$

2 a 40, 58, 80 **b** $2n^2 + 8$

3 a 10 **b** 15 **c** $T_n = \frac{1}{2}n(n + 1)$
d $T_7 = 28$, $T_8 = 36$ **e** $T_{15} = 120$ **f** $T_{49} = 1225 = 35^2$

4 20, 34, 52 **5** 13 **6** $\frac{1}{2}n(n - 1)$

Mixed practice

1 a $u_n = 4n + 2$
b $u_n = 49 - 7n$
c $u_n = n^2 + 1$
d $u_n = n^2 + 2n$
e $u_n = 4.5n - 2$
f $u_n = 2.5n^2 - 3n$
g $u_n = -2n^2 + 10n + 1$
h $u_n = -0.5n^2 + 2.5n + 4$

2 311

3 a 12, 7, 2, −3, −8
b $u_n = 17 - 5n$

4 a 15, 22, 29, 36, 43
b $u_n = 7n + 15$
$u_{10} = 7 \times 10 + 15 = 85$

5 1008

6 a $u_n = n^2 + 3n$
b 70
c $u_{20} = 20^2 + 3 \times 20 = 460$

7 a $u_n = \frac{1}{3}n^2 + n + \frac{2}{3}$
b $u_{20} = 154$

8 a 7 b 34

9 7.5

10 19

11 a 1230 b 1860 c $5n^2 + 605n$ d 22650 e 15

12 a constant difference of 6, therefore a linear sequence
b $T_n = 6n + 2$
c 9

Review in context

1 a $u_n = 6n + 6$ b 186
2 a constant difference of 70, therefore a linear sequence
b $P_n = 70n - 30$
c 530
d e.g. Because a greater proportion of a one carriage train is taken up by the place where the driver sits, engine, or service areas.

3 a AB, BC, AC, BA, CB, CA
b AB, AC, AD, BC, BD, CD, BA, CA, DA, CB, DB, DC
12 journeys
c AB, AC, AD, AE, BC, BD, BE, CD, CE, DE, ED, EC, DC, EB, DB, CB, EA, DA, CA, BA
20 journeys
d $n^2 - n$ or $n(n - 1)$
e 2970
f 222

8.2

You should already know how to:

1 a $x^2 + 7x + 12$ b $x^2 - 4x - 5$
c $x^2 - 5x + 6$ d $x^2 - 4$

2 a $3(x + 4)$ b $x(x + 5)$
c $3(2x^2 + x + 4)$ d $(4x + 7)(3x - 5)$

Practice 1

1

p	*p*	*a*	*b*	*c*
2	−3	1	−1	−6
4	−2	1	2	−8
8	−8	1	0	−64
−3	−2	1	−5	6
−3	−7	1	−10	21
9	−9	1	0	−81
−6	4	1	−2	−24

2 $p = q = 2$ and $b = c = 4$

3 Yes, as $pq = 0$

Practice 2

1 a $(x + 3)(x + 4)$ b $(x + 3)(x + 5)$ c $(x + 6)(x - 3)$
d $(x + 9)(x - 2)$ e $(x - 9)(x + 2)$ f $(x + 7)(x - 2)$
g $(x - 4)(x - 9)$ h $(x - 3)(x - 8)$ i $x(x - 3)$

2 a $(x + 7)^2$ b $(x + 11)^2$ c $(x - 6)^2$

3 a $3(a + 1)^2$ b $2b(2b + 1)$
c $5(c + 3)(c - 5)$ d $3d(2d - 1)$
e $4(e + 9)(e - 4)$ f $3(f - 3)(f - 5)$

4 a $x + 5$ b $x + 3$

5 a $x^2 - 7x + 12 = (x - 3)(x - 4)$
b $x^2 - 10x + 16 = (x - 8)(x - 2)$
c $x^2 - 5x - 14 = (x + 2)(x - 7)$
d $x^2 + 9x + 20 = (x + 4)(x + 5)$
e $x^2 + 10x + 16 = (x + 2)(x + 8)$
f $x^2 - 5x - 24 = (x - 8)(x + 3)$

6 Two sides of length $(x - 7)$, and two sides of length $(x - 3)$

Practice 3

1 a $(x + 5)(x - 5)$ b $(x + 11)(x - 11)$
c $(2x + 3)(2x - 3)$ d $(x + y)(x - y)$
e $(3x + 1)(3x - 1)$ f $(x^2 + 1)(x + 1)(x - 1)$
g $(4x + 13)(4x - 13)$ h $9(3x + 1)(3x - 1)$
i $(5u + 4v)(5u - 4v)$ j $(4 + 7x)(4 - 7x)$
k $(4 + x^2)(2 + x)(2 - x)$ l $(1 + 9y^2)(1 + 3y)(1 - 3y)$

2 a $9x^2 - 100 = (3x - 10)(3x + 10)$
b $25y^2 - 16 = (5y - 4)(5y + 4)$
c $16a^2 - 49b^2 = (4a - 7b)(4a + 7b)$
d $9u^2 - 4 = (3u + 2)(3u - 2)$
e $36t^2 - 100 = (3t + 5)(12t - 20)$
f $18x^2 - 32 = (3x + 4)(6x - 8)$

Practice 4

1 $(2x + 3)(x + 1)$ 2 $(3x + 1)(x + 2)$

3 $(3x - 2)(x + 3)$

Practice 5

1 a 12, 20 b −12, 15 c −32, −3 d −125, 8

2 a $10x^2 - 31x - 14$
$= 10x^2 + 35x - 4x - 14$
$= 5x(2x + 7) - 2(2x + 7)$
$= (5x - 2)(2x + 7)$
b $4x^2 + 27x + 18$
$= 4x^2 + 3x + 24x + 18$
$= x(4x + 3) + 6(4x + 3)$
$= (x + 6)(4x + 3)$

3 a $(5x - 4)(x - 1)$ b $(2x + 1)(3x + 2)$
c $(2x + 1)(2x + 3)$ d $(2x - 1)(3x - 4)$
e $(4x + 3)(2x - 5)$ f $(3x - 2)(4x - 3)$
g $(3x + 1)(2x + 1)$ h $(3x + 2)(2x - 1)$
i $(6x + 1)(x - 1)$ j $(5x + 1)(3x - 2)$
k $(4x - 1)(2x + 3)$ l $(7x - 3)(3x - 2)$
m $(2x - 1)(x - 1)$ n $(2x - 5)(x - 2)$
o $(3x + 1)(5x - 15)$ p $(x + 3)(4x - 5)$
q $(2x - 3)(2x + 7)$ r $(6x + 5)(2x - 1)$
s $(5x + 3)(x - 1)$ t $(7x - 2)(x - 2)$
u $8(x + 1)(x + 2)$ v $(5x + 1)(x - 3)$
w $(9x + 2)(x + 1)$ x $(5x + 7)(2x - 3)$

4 a −7, −5, 5, 7 **b** 4, 7, 9, 10
c Any number of the form $10n - 6n^2$ where n is a positive integer

Mixed practice

1 a $(x + 2)^2$ **b** $(x - 4)(x - 9)$
c $(x + 7)(x - 2)$ **d** $(x + 9)^2$
e $(x - 8)(x + 1)$ **f** $(x - 3)(x - 8)$
g $(x - 10)^2$ **h** $(x - 20)(x + 5)$
i $(x + 1)(x + 3)$ **j** $(x + 14)(x - 3)$
k $(x + 12)(x - 8)$ **l** $(x + 1)(x + 6)$
m $(x - 7)(x - 8)$ **n** $(x - 15)(x + 4)$

2 a $(x + 7)(x - 7)$ **b** $(x + 13)(x - 13)$
c $(8 + x)(8 - x)$ **d** $(5x + 2)(5x - 2)$
e $(12x + 9)(12x - 9)$ **f** $(16x + 13)(16x - 13)$
g $(2x + y)(2x - y)$ **h** $(4x + 3y)(4x - 3y)$
i $(5x + 17y)(5x - 17y)$
j $(x^2 + 4)(x + 2)(x - 2)$
k $(4x^2 + 1)(2x + 1)(2x - 1)$
l $9(x^2 + 9y^2)(x + 3y)(x - 3y)$

3 a $2(x + 1)(x + 3)$ **b** $2(x^2 + 7x + 10)$
c $2(x - 1)(x - 5)$ **d** $3(x - 1)(x - 2)$
e $2(x + 6)(x - 2)$ **f** $3(x + 8)(x - 1)$
g $4(x + 2)(x - 4)$ **h** $2(x + 4)(x - 5)$

4 a $(2x + 1)(x + 2)$ **b** $(2x + 5)(x + 4)$
c $(2x - 3)(x - 4)$ **d** $(3x + 1)(x - 3)$
e $(5x - 2)(x - 2)$ **f** $(7x - 4)(x + 6)$
g $(6x + 1)(x + 5)$ **h** $(3x - 1)(2x - 5)$
i $(4x + 3)(2x + 3)$ **j** $(8x - 1)(x - 5)$
k $(3x - 4)(3x + 8)$ **l** $(5x + 6)(2x - 5)$

5 a $a(4a - 3)$ **b** $7b(b + 3)(b - 8)$

6 a $(a - 9)(a + 4)$ **b** $(4b + 3)(4b - 3)$
c $(c + 8)(c + 3)$ **d** $(4d + 3)(d - 5)$
e $e(4e - 3)$ **f** $(f + 8)(f - 6)$
g $(3g - 2)(g - 7)$ **h** $(4h + 1)(4h - 1)$

7 a $(n + 4)^2 - 9$
b It factorizes to
$(n + 4 + 3)(n + 4 - 3)$
$= (n + 7)(n + 1)$
which are two integers with a difference of 6.

8 a $x^2 - 8x + 15 = (x - 3)(x - 5)$
b $x^2 + 9x + 20 = (x + 4)(x + 5)$
c $x^2 - 8x - 33 = (x + 3)(x - 11)$
d $x^2 - 11x + 30 = (x - 5)(x - 6)$

9 $(2n)^2 - 1 = (2n + 1)(2n - 1)$
which is the product of two consecutive odd integers.

10 a $2n^2 + 13n + 15$
b Since it factorizes to give $(2n + 3)(n + 5)$, a rectangle with sides $(2n + 3)$ and $(n + 5)$ could be formed.

11 $4n^2 + 28n + 45 = (2n + 5)(2n + 9)$
The pitch is 5 m by 9 m.

12 a

Sequence	8		45		112		209		336
First difference		37		67		97		127	
Second difference			30		30		30		

2nd difference is constant.

b $u_n = 15n^2 - 8n + 1$
c $u_n = (5n - 1)(3n - 1)$
d An extra 5 columns and 3 rows.

9.1

You should already know how to:

1 $5\sqrt{2}$ m **2** 40 cm
3 36 cm^3, 84 cm^2 **4** 2410 m^3, 1010 m^2
5 56 mm^3, 92 mm^2 **6** 61.4 cm^2, 15.4 cm

Practice 1

1 52 m^2
2 80.9 cm^2
3 65.5 cm^2
4 a 7.60 m **b** 101 m^2
5 84.9 cm^2
6 a The triangles **b** 9140 mm^2
7 20 cm
8 144 cm^2

Practice 2

1 18.6 m^2
2 283 cm^2
3 170 cm^2 (There is no paper on the top circular face.)
4 a

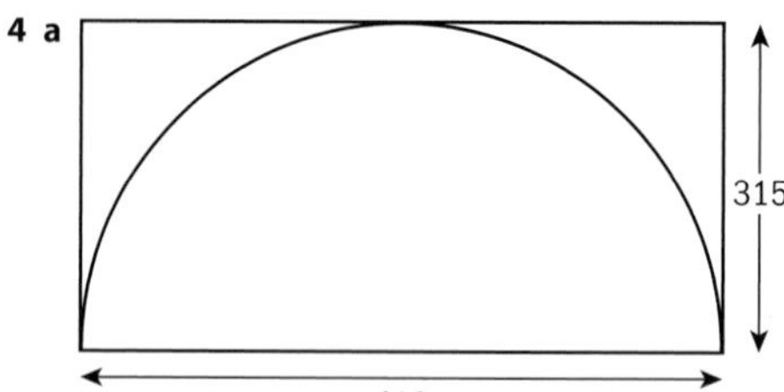

b $d = 410$ mm, $r = 205$ m **c** 66 000 mm^2
d $d = 644$ mm **e** 102.5 mm
5 a 75.4 cm^2 **b** 6.00 cm **c** 4.48 cm

Practice 3

1 50.3 cm^2
2 1810 cm^2
3 a 236 cm^2 **b** 214 cm^2
4 5.64 cm
5 The surface area of the cylinder is 1.5 times larger than the surface area of the sphere.

Practice 4

1 213 cm^3 **2** 112 cm^3
3 56 cm^3 **4** 3620 cm^3
5 314 cm^3 **6** 1700 cm^3
7 524 cm^3 **8** 452 cm^3
9 1530 cm^3

10 a 1 950 000 mm^3 or 1950 cm^3

b Because there is no air in between the French fries, so the potato does not fill the entire cone.

11 3456 cm^3

12 a They will always be enlargements of each other.

b i 2144 cm^3 **ii** 33.5 cm^3

13 10 mm

14 2.00 cm (3 s.f.)

15 5.00 cm (3 s.f.)

16 The water would reach $\frac{2}{3}$ of the height of the cylinder.

Mixed practice

1 a 466 cm^2, 576 cm^3 **b** 77.6 cm^2, 40 cm^3

c 118 cm^2, 84.8 cm^3 **d** 13.1 cm^2, 3.14 cm^3

e 845 cm^2, 2310 cm^3 **f** 452 cm^2, 905 cm^3

2 a 393 cm^2, 407 cm^3 **b** 221 cm^2, 172 cm^3

3 a 8.00 cm (3 s.f.) **b** 163 cm^3

4 Yes the soccer ball satisfies the regulations. It is 69.2 cm in circumference.

5 a 3.00 cm **b** 56.7 cm^3

6 200 cm^2

Review in context

1 a 1.58… – pretty close

b 85 500 m^2

c Ratio is 1.61… pretty close

d 2 570 000 m^3

e 921.6 m and 914.2 m – they are very close

2 a 3.90 cm^2, 49.7 cm^2

b 0.724 cm^3, 33.0 cm^3

3 Circumference of the ball is 94.3 cm so it was a pretty tight fit!

4 a The 215 cm tall tree has more area for decorations (4 720 000 cm^2; the 180 cm tree has area 3 520 000 cm^2).

b 8100 ft^2 is available for decorations.

9.2

You should already know how to:

1 m is the gradient, $(0, c)$ is the y-intercept

2 a 2 **b** 4

3 a negative **b** $y = -2x + 4$

4 a $(x + 2)(x + 3)$ **b** $(2x + 1)(x - 2)$

c $(x - 7)(x + 7)$ **d** $(3 + x)(1 - x)$

Practice 1

1 i (0.5, −6.25)

ii $x = 0.5$

iii concave up

iv −6

v −2 and 3

vi

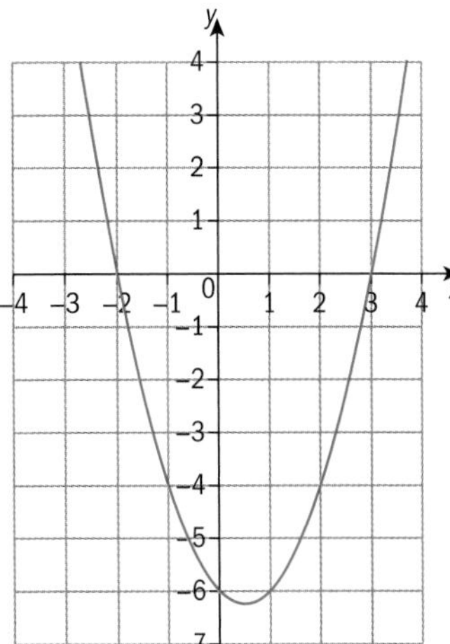

2 i (1, −3)

ii $x = 1$

iii concave down

iv −4

v not factorizable

vi

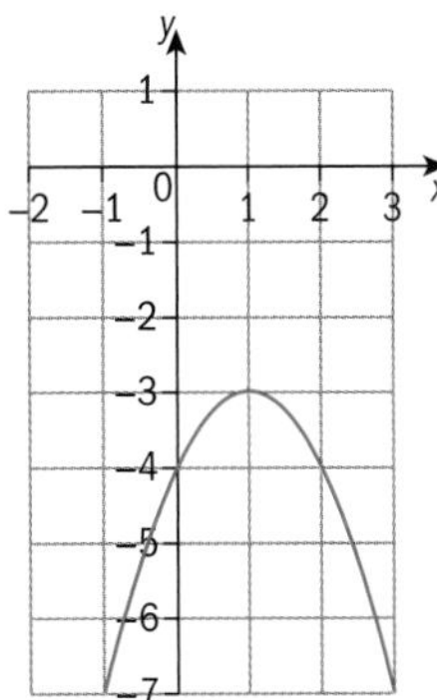

3 i (2, −6)

ii $x = 2$

iii concave up

iv −2

v not factorizable

vi

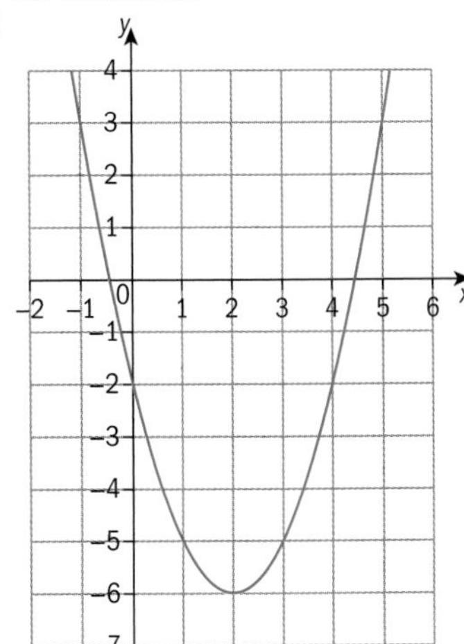

4 i (1, −3)
ii $x = 1$
iii concave down
iv −11
v not factorizable
vi 

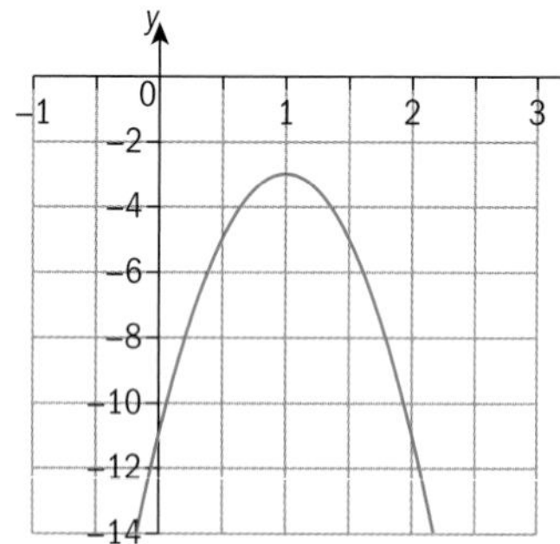

5 i (5, −1)
ii $x = 5$
iii concave down
iv −51
v not factorizable
vi 

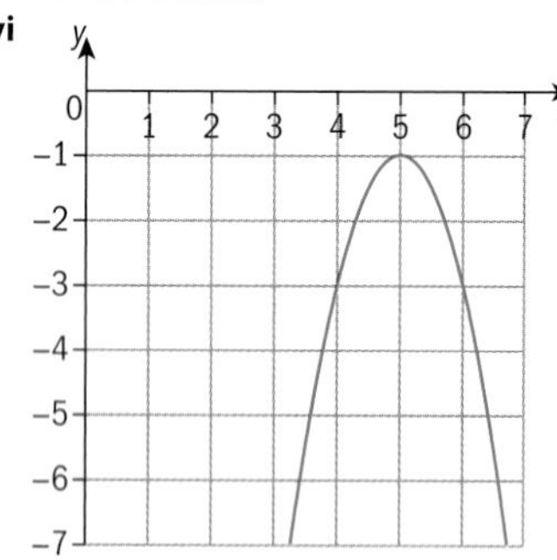

6 i (1, −2)
ii $x = 1$
iii concave up
iv 1
v not factorizable
vi 

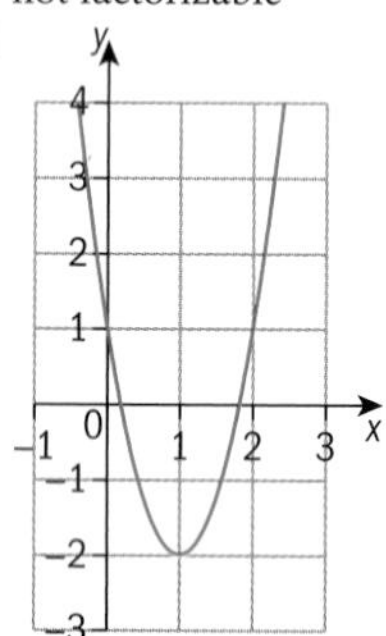

7 Student's own example, e.g. $y = x^2 - x - 6$
8 Student's own example, e.g. $y = x^2 - 8x + 16$

Practice 2

1 $T = 0.5\,°\text{C}$

2 a 45 m **b** 1.67 sec, 58.3 m **c** 5.15 seconds
3 44 balls

4 a 45 000 Euros **b** 9.375 months
5 40

Practice 3

1 (3, 1) concave up
2 (4, −3) concave down
3 (−1, −1) concave up
4 (−2, 1) concave down
5 (0, 2) concave down
6 (0, −1) concave up
7 $y = 2(x + 3)^2$

Practice 4

1 a i $y = (x + 1)^2$
ii

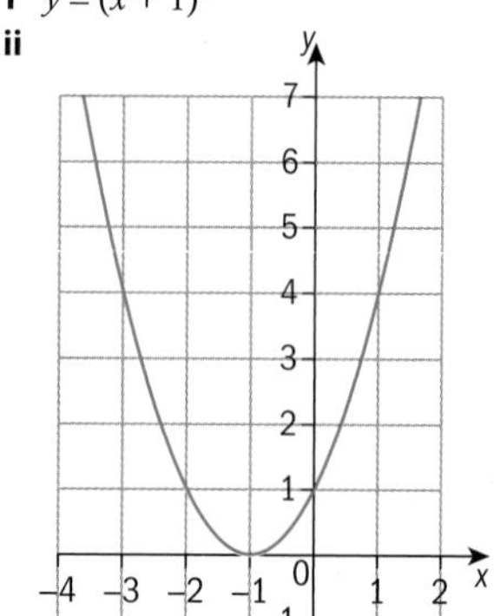

b i $y = (x - 2)^2 - 6$
ii

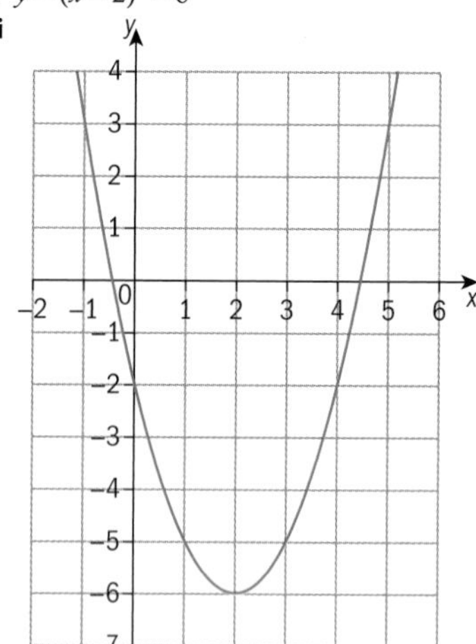

c i $y = (x + 3)^2 - 6$
ii

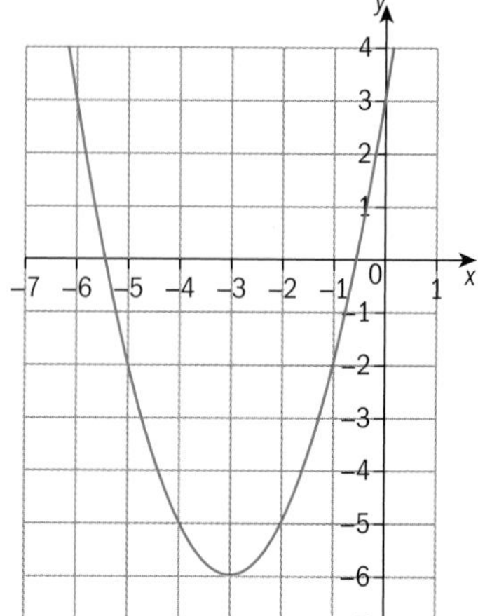

d i $y = -(x + 3)^2 + 10$

ii

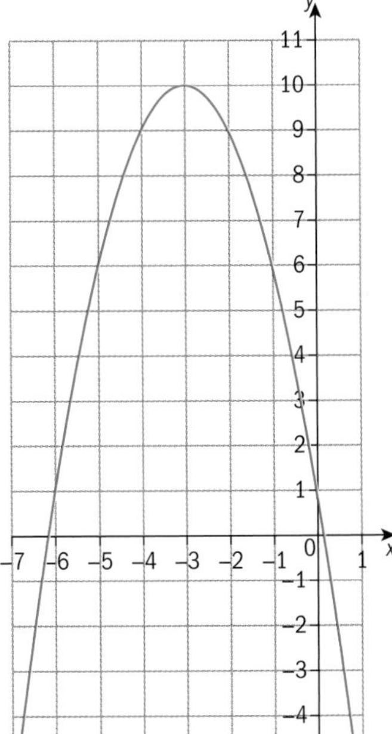

e i $y = (x - 1)^2 + 2$

ii

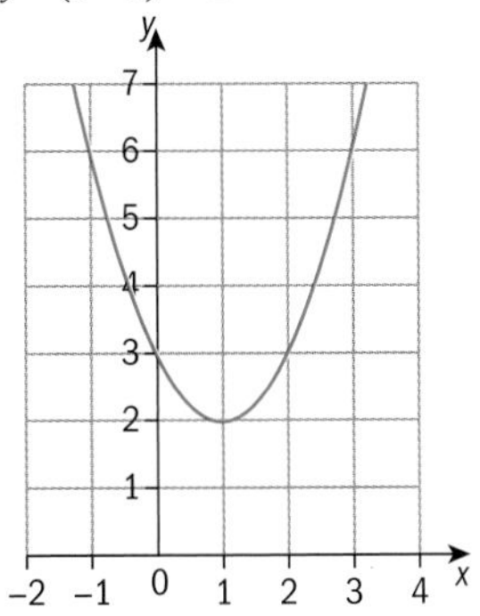

f i $y = (x - 1)^2$

ii

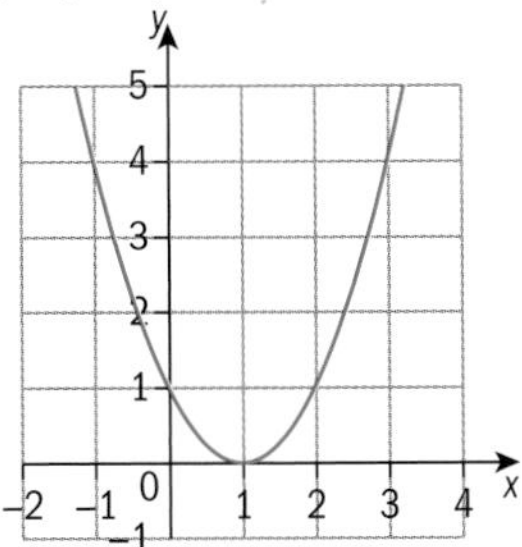

g i $y = (x + 0.5)^2 + 1.75$

ii

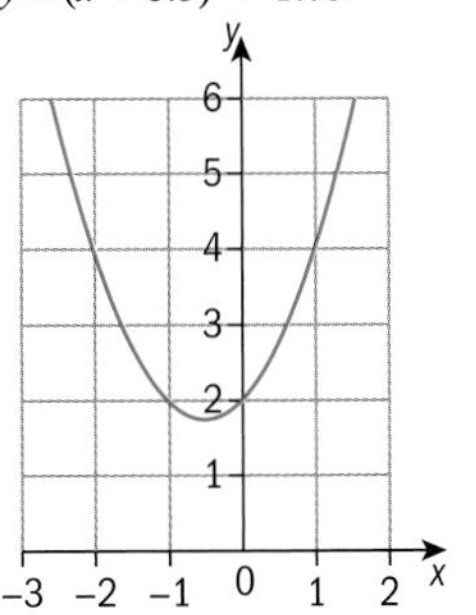

h i $y = (x - 0.5)^2 - 1.25$

ii

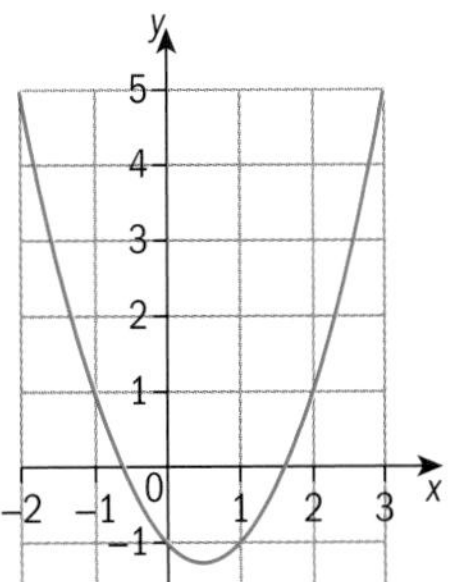

i i $y = -(x + 1.5)^2 + 3.25$

ii

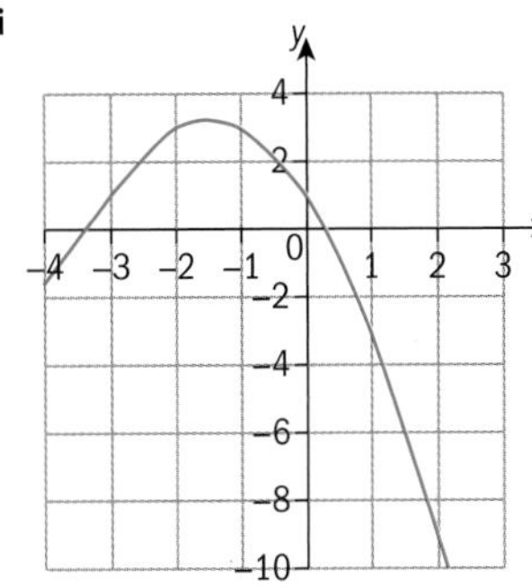

2 a i (−2, 5)

ii Student's own points

iii

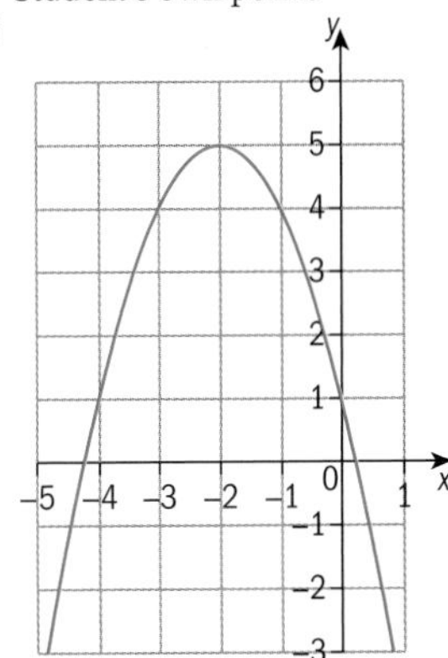

b i (1, −1)

ii Student's own points

iii

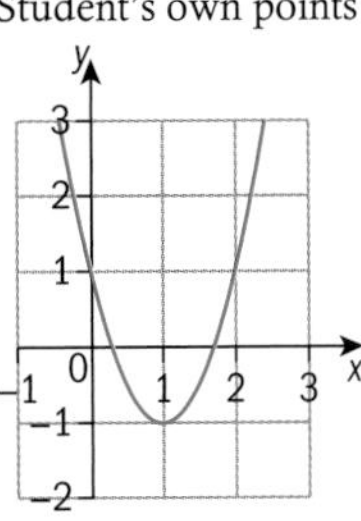

c i (−1, 4)

ii Student's own points

iii

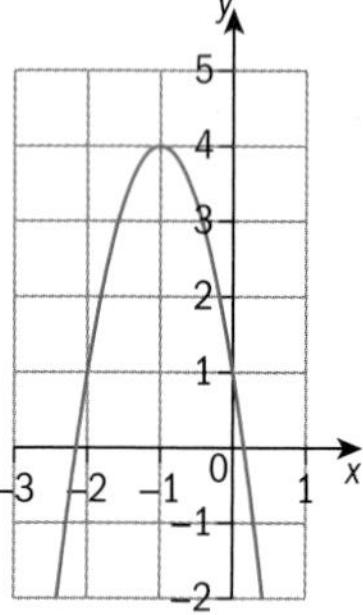

3 a C **b** E **c** F **d** B **e** A **f** D

Practice 5

1 Maximum area = 11 250 m^2
Dimensions are 150 m by 75 m

2 Dimensions are $33\frac{1}{3}$ m by 25 m

3 Maximum area = 12.5 m^2

4 $x = 0.1$ km, $r = \frac{0.1}{\pi}$ km; Area = 9549 m^2

Practice 6

1 Student's own work.

2 Student's own work.

Mixed practice

1 a i vertex is (–0.5, –12.25)
x-intercepts are –4 and 3
y-intercept is –12

ii axis of symmetry is $x = -0.5$
It is concave up.

iii

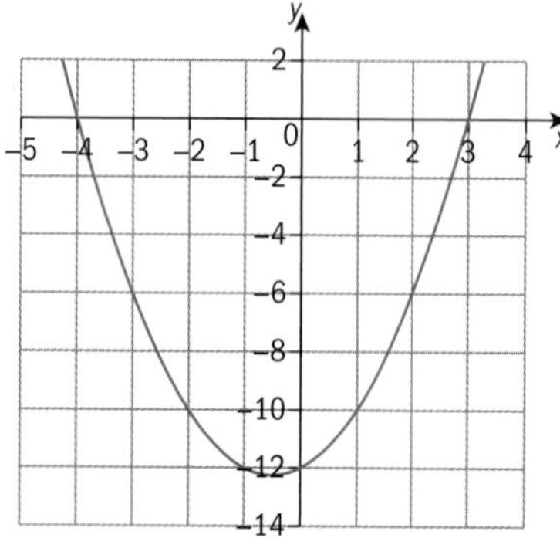

b i vertex is (–3.5, –0.25)
x-intercepts are –4 and –3
y-intercept is 12

ii axis of symmetry is $x = -3.5$
It is concave up.

iii

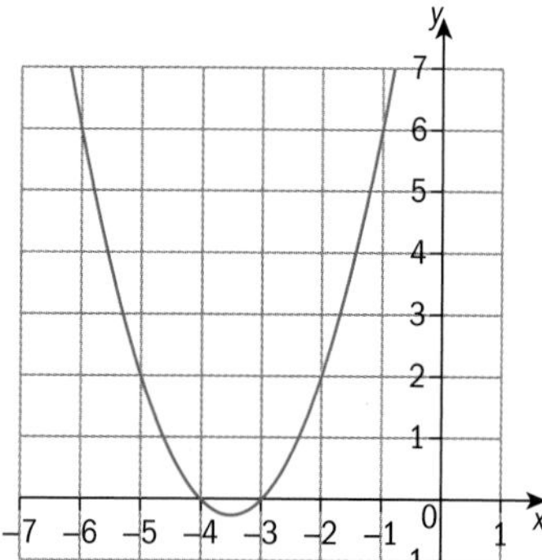

c i vertex is (0.25, –3.125)
x-intercepts are –1 and 1.5
y-intercept is –3

ii axis of symmetry is $x = 0.25$
It is concave up.

iii

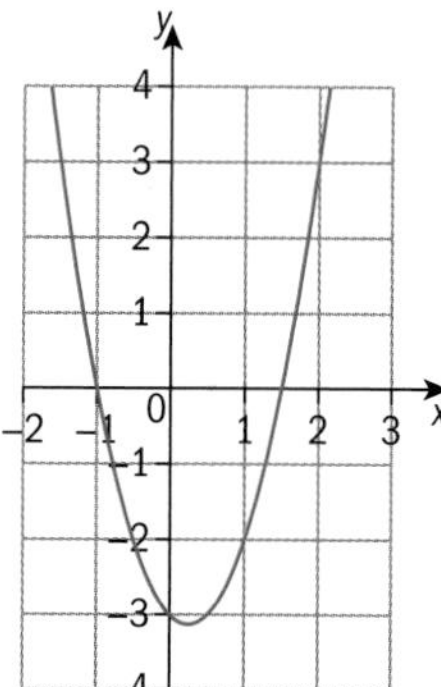

d i vertex is $\left(-\frac{1}{6}, 2\frac{1}{12}\right)$
x-intercepts are –1 and $\frac{2}{3}$
y-intercept is 2

ii axis of symmetry is $x = -\frac{1}{6}$
It is concave down.

iii

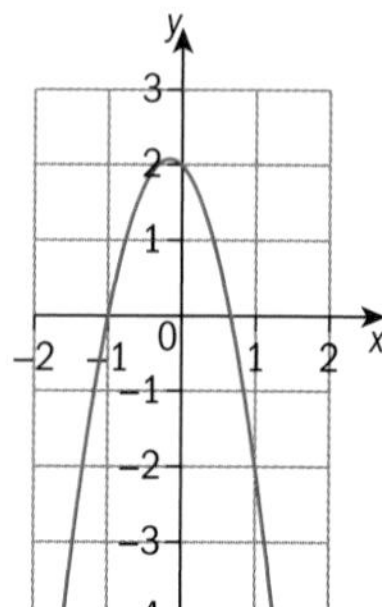

e i vertex is $\left(\frac{5}{12}, \frac{1}{24}\right)$
x-intercepts are $\frac{1}{3}$ and $\frac{1}{2}$
y-intercept is –1

ii axis of symmetry is $x = \frac{5}{12}$
It is concave down.

iii

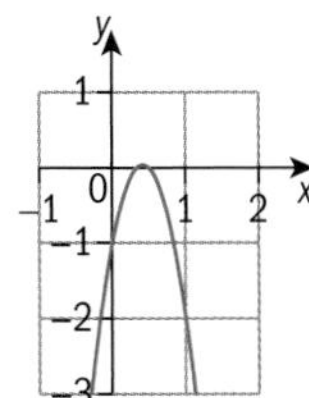

f i vertex is $(2.25, -15\frac{1}{8})$

x-intercepts are -0.5 and 5

y-intercept is -5

ii axis of symmetry is $x = 2.25$

It is concave up.

iii

2 a $y = (x - 2)^2 + 2$

$(2, 2)$

b $y = (x + 3)^2 - 1$

$(-3, -1)$

c $y = (x + 1)^2 - 10$

$(-1, -10)$

d $y = (x - 1)^2 + 6$

$(1, 6)$

e $y = (x + 0.5)^2 - 5.25$

$(-0.5, -5.25)$

f $y = (x - 0.5)^2 + 6.75$

$(0.5, 6.75)$

3 a Area $= 300x - 3x^2$ **b** 50 m **c** 7500 m^2

4 10 m by 10 m

5 a 176.32 months **b** 12913

c During the 361st month

6 $y = -(x - 2)^2 + 9$ or $y = 5 + 4x - x^2$

Review in context

1 Student's own work.

2 a Answers will vary, e.g. $y = -0.408(x - 4.9)^2 + 9.8$

b A rectangle.

c 8 meters at the very most.

d Student's own explanation.

e 3.13 m

f Student's own answers, e.g. dimensions of vehicles that are likely to be travelling through the tunnel.

10.1

You should already know how to:

1 a

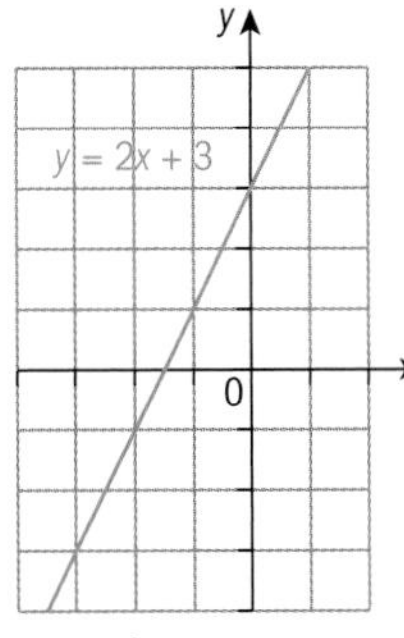

b

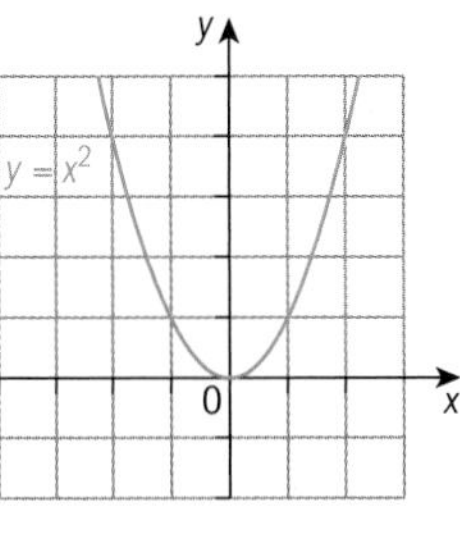

2 a $(x + 3)^2 - 1$ **b** $(-3, -1)$

Practice 1

1 a horizontal translation of 4 units

b horizontal dilation of scale factor $\frac{1}{3}$

c reflection in the x-axis

d reflection in the y-axis

e vertical dilation of scale factor 4

f horizontal translation of 3 units and vertical translation of 4 units

2 a vertical dilation of scale factor 2, vertical translation of -3 units

b horizontal dilation of scale factor $\frac{1}{2}$, reflection in the x-axis

c horizontal translation of 4 units, vertical dilation of scale factor $\frac{1}{2}$

d reflection in the x-axis and vertical translation of 4 units

3 Both are correct. The transformation from $f(x) = (x - 2)^2$ to $g(x) = (x + 2)^2$ can be seen as a horizontal translation of -4 units ($g(x) = f(x + 4)$) or a reflection in the y-axis ($g(x) = f(-x)$).

Practice 2

1 a A: $y = -3x - 3$ B: $y = -3x - 1$ C: $y = -3x + 4$

b D: $y = 0.5x$ E: $y = 0.5x - 2$ F: $y = 0.5x - 3$

c G: $y = x^2 - 3$ H: $y = (x - 4)^2$ J: $y = (x - 7)^2 - 4$

d K: $y = -0.5x^2 - 4$ L: $y = -0.5(x + 7)^2$ M: $y = -0.5(x + 2)^2 - 3$

2 a

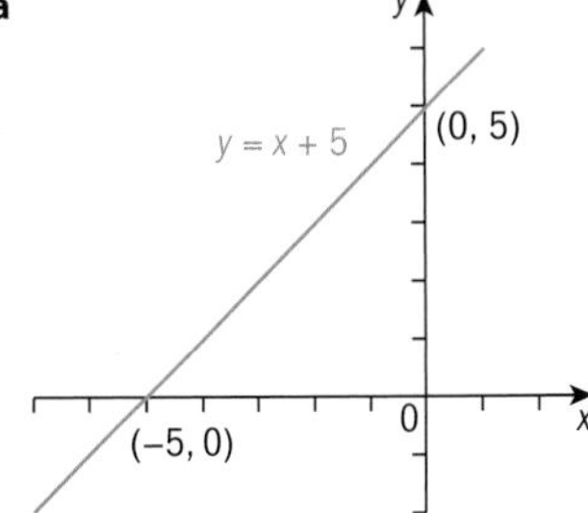

b

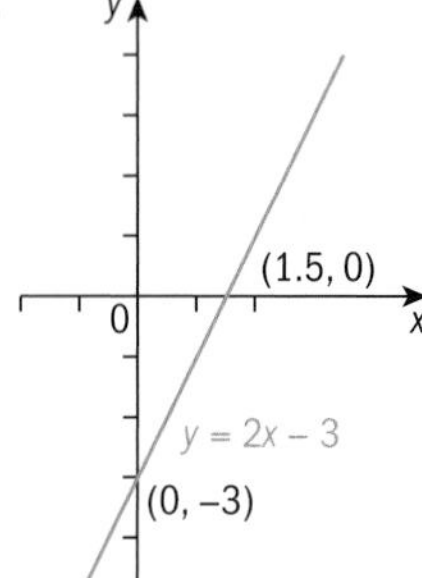

c

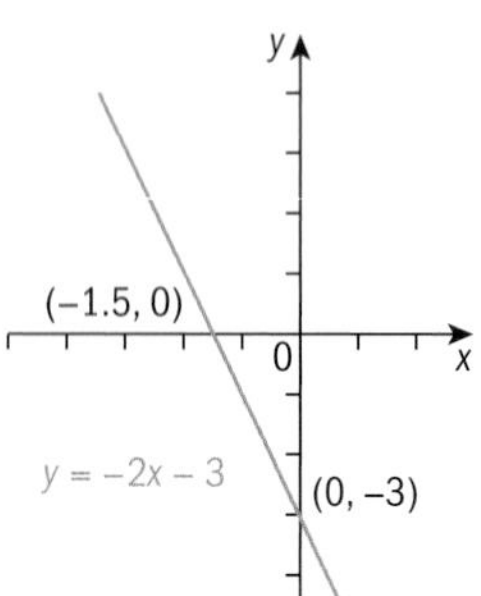

d

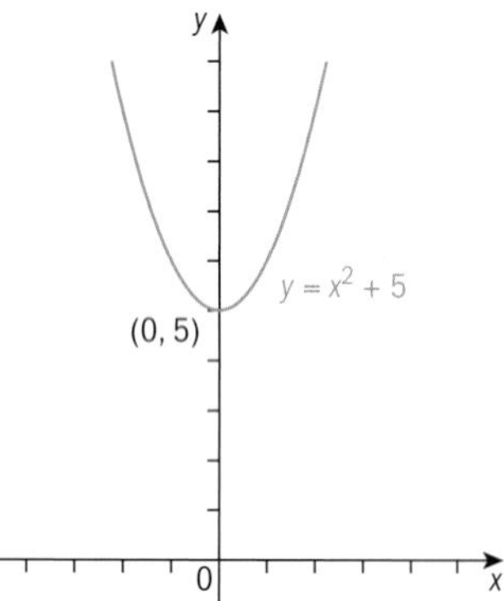

e

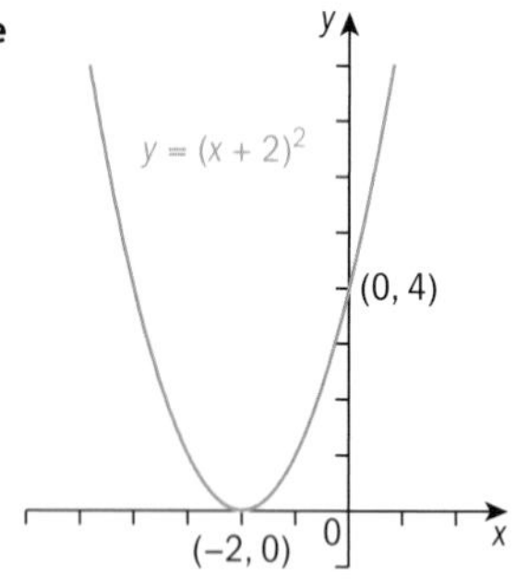

f

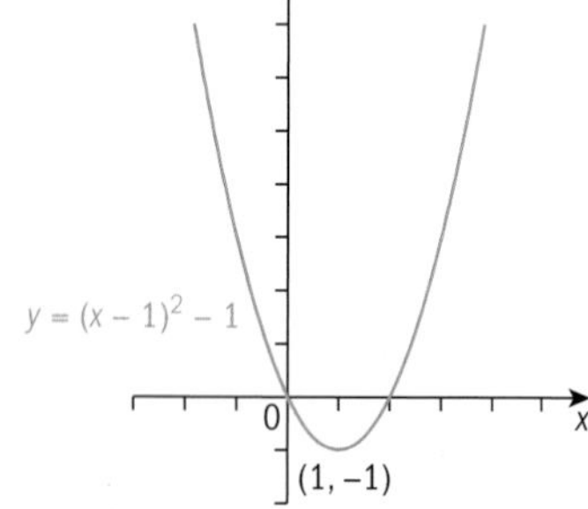

Practice 3

1

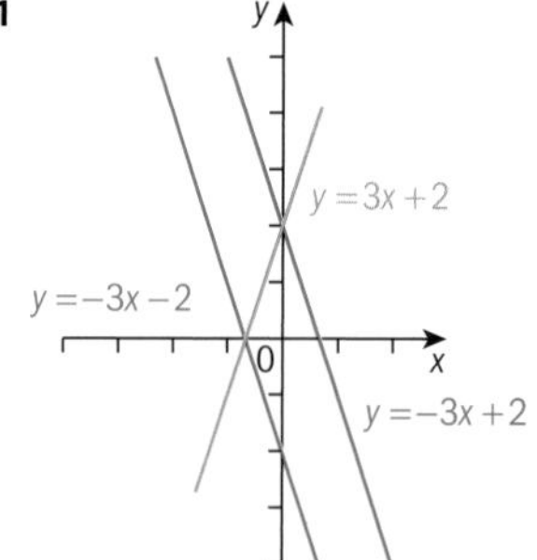

2

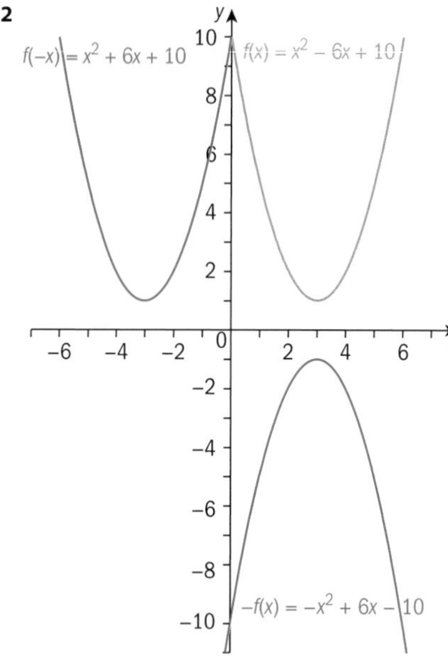

3 a $L_1: y = x + 1 \quad L_2: y = x - 1$

b $L_3: y = x^2 + 4x + 4 \quad L_4: y = -x^2 + 4x - 4$

c $L_5: y = (x + 3)^2 + 1 \quad L_6: y = -(x - 3)^2 - 1$

d $L_7: y = (x - 3)^2 - 2 \quad L_8: y = -(x + 3)^2 + 2$

4 $y = -x + 4$

5 Curve C: $y = -(x - 2)^2$

Practice 4

1

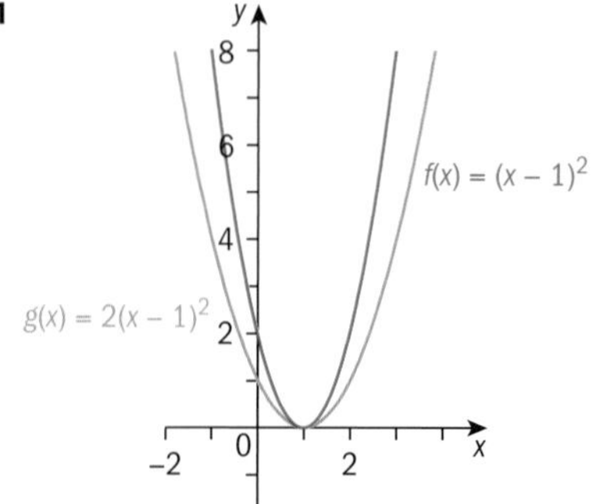

2

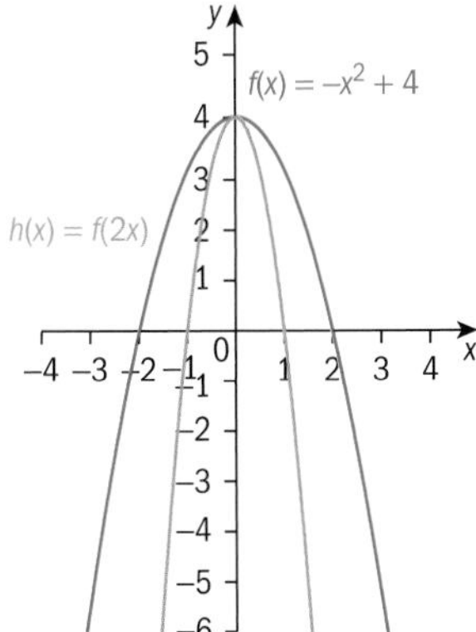

3

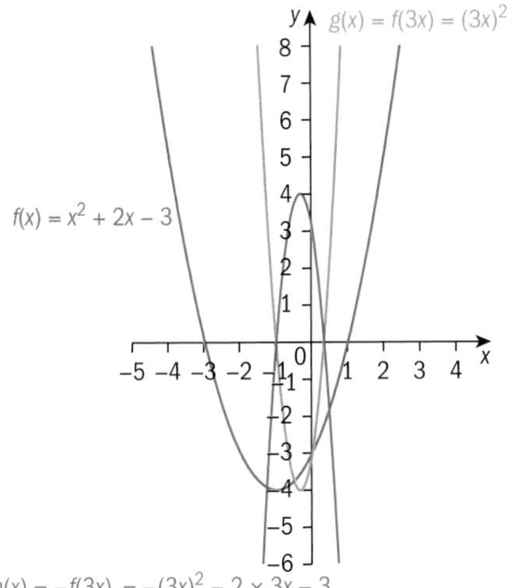

4

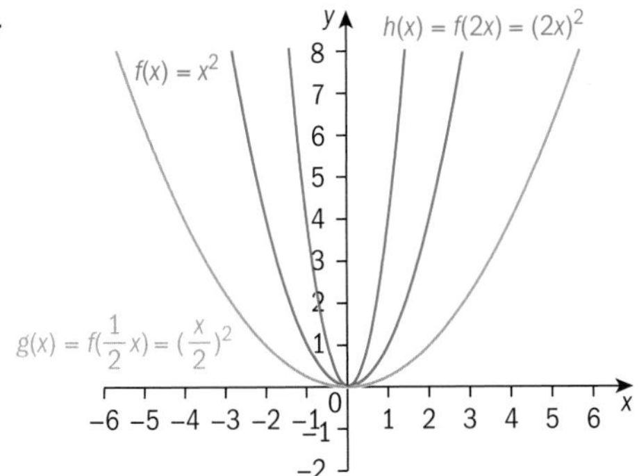

5 a Dilation scale factor 3 parallel to the y-axis

b Dilation scale factor $\frac{1}{2}$ parallel to the y-axis

c Dilation scale factor $\frac{1}{4}$ parallel to the x-axis

d Dilation scale factor 3 parallel to the x-axis

6 a $y = 5x^2$ **b** $y = \left(\frac{x}{4}\right)^2$ **c** $y = \frac{x^2}{3}$

7 a Dilation scale factor 3 parallel to the y-axis

b Dilation scale factor 2 parallel to the x-axis

Practice 5

1 a Dilation scale factor 4 parallel to the y-axis and reflection in x-axis

b Dilation scale factor 2 parallel to the y-axis and vertical translation of –9 units

c Dilation scale factor 0.5 parallel to the y-axis and vertical translation of 4 units

d Dilation scale factor 1.2 parallel to the y-axis, reflection in x-axis and vertical translation of –2.5 units

2 a Dilation scale factor 25 parallel to the y-axis

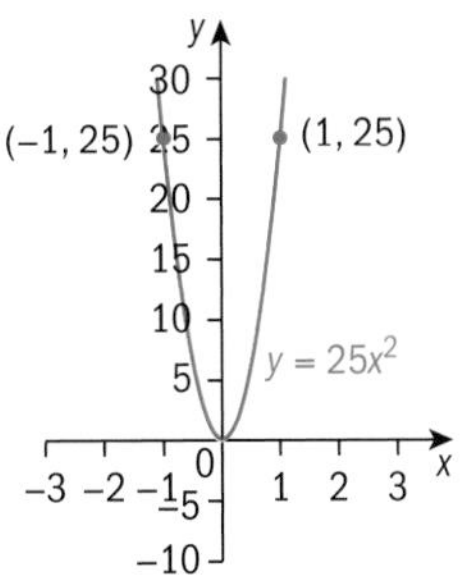

b Horizontal translation of 9 units then dilation scale factor 3 parallel to the y-axis

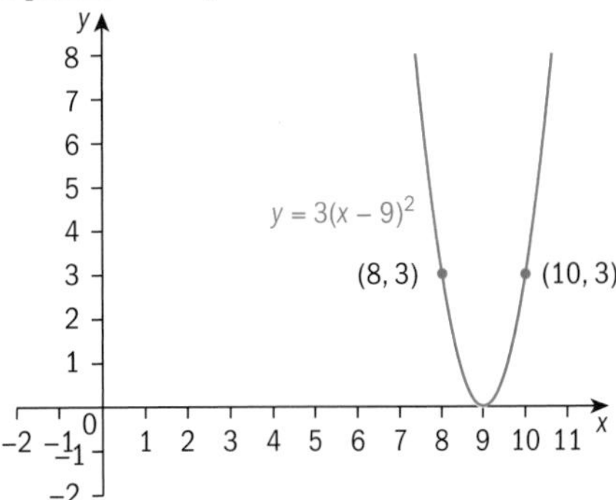

c Reflection in the x-axis and vertical translation of 4 units

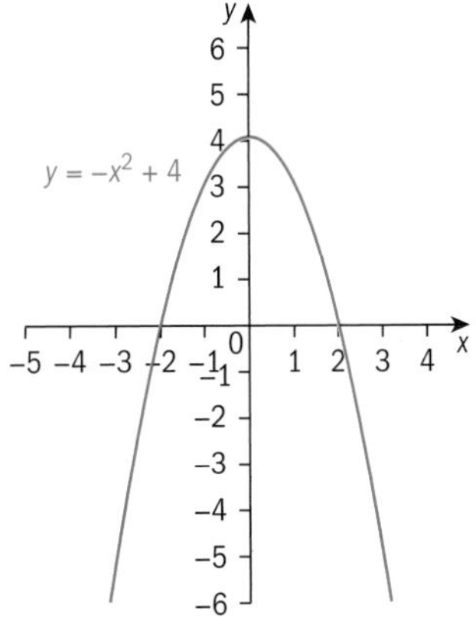

d Horizontal translation of –4 units

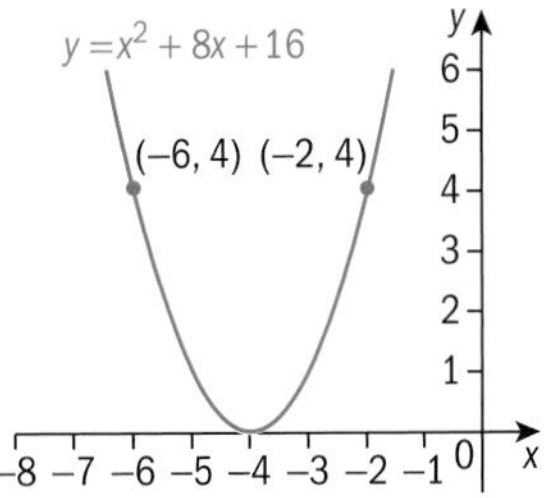

e Dilation scale factor 2 parallel to the y-axis and horizontal translation of 3 units, vertical translation of –9 units

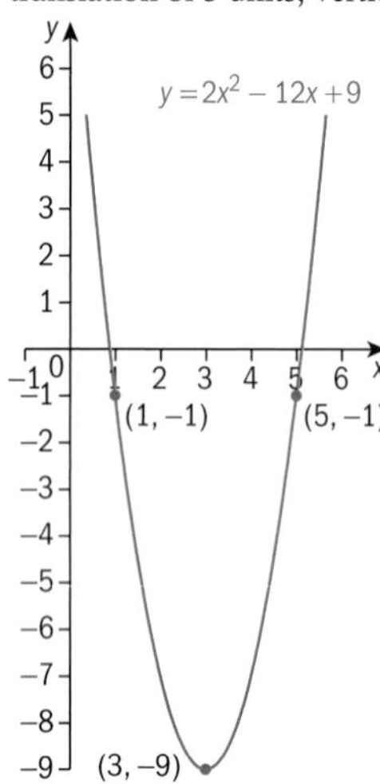

f Dilation scale factor $\frac{1}{2}$ parallel to the y-axis, reflection in the x-axis, horizontal translation of –4 units and vertical translation of 9 units.

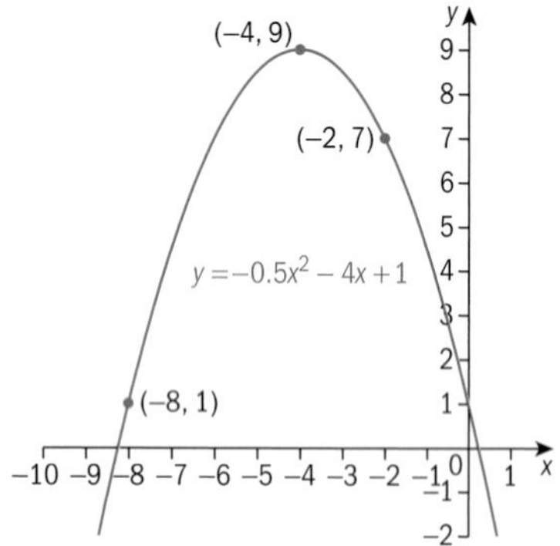

3 a $g(x) = 4f(x)$ vertical dilation scale factor 4

b $g(x) = f(3x)$ horizontal dilation scale factor $\frac{1}{3}$

c $g(x) = -f(x)$ reflection in the x-axis

d $g(x) = f(-x)$ reflection in the y-axis

e $g(x) = f(x) - 12$ vertical translation of –12 units

f $g(x) = f(-x)$ reflection in the y-axis

g $g(x) = f(x - 2)$ horizontal translation of 2 units

h $g(x) = \frac{1}{4}f(x)$ vertical dilation of scale factor $\frac{1}{4}$

4 a $g(x) = -f(x) - 5$ or $g(x) = f(-x) + 11$: reflection in the x-axis and vertical translation of –5 units

b $g(x) = \left(\frac{1}{2}\right)f(x) - 6$ or $g(x) = f\left(\left(\frac{1}{2}\right)x\right) - 9$: vertical dilation of scale factor $\frac{1}{2}$ and vertical translation of –6 units

c $g(x) = f(-x) + 4$: vertical dilation of scale factor $\frac{1}{2}$ and vertical translation of –6 units

d $g(x) = -\left(\frac{1}{2}\right)f(x)$: reflection in the x-axis and vertical dilation of scale factor $\frac{1}{2}$

e $g(x) = -f(x) + 9$: reflection in the x-axis and vertical translation of 9 units

f $g(x) = f(x - 5)$: horizontal translation of 5 units (only 1 transformation)

g $g(x) = f(x + 4) - 1$: horizontal translation of –4 units and vertical translation of –1 unit

h $g(x) = f(-3x)$: reflection in the y-axis and horizontal dilation of scale factor $\frac{1}{3}$

5 a $f_2(x) = -f_1(x)$: Reflection in the x-axis

b $f_2(x) = f_1(2x)$: Dilation scale factor $\frac{1}{2}$ parallel to the x-axis

c $f_2(x) = 2f_1(x)$: Dilation scale factor 2 parallel to the y-axis

d $f_2(x) = -f_1(x)$: Reflection in the x-axis

e $f_2(x) = f_1(x + 4)$: Horizontal translation of –4 units

f $f_2(x) = \frac{1}{2}f_1(x)$: Dilation scale factor $\frac{1}{2}$ parallel to the y-axis

6 a $f_2(x) = -f_1(x) + 3$. Reflect $f_1(x)$ in the x-axis, and vertical translation of 3 units OR $f_2(x) = f_1(-x) - 3$: Reflect $f_1(x)$ in the y-axis and vertical translation of –3 units.

b $f_2(x) = 2f_1(x - 1)$. Horizontal translation of $f_1(x)$ of 1 unit then a vertical dilation of scale factor 2.

c $f_2(x) = -f_1(x + 2)$. Reflect $f_1(x)$ in the x-axis then horizontal translation of –2 units.

Practice 6

1 a $f_2(x) = 2f_1(x + 3)$ **b** $f_2(x) = f_1(2x)$

c $f_2(x) = -f_1(x + 4)$ **d** $f_2(x) = 3f_2\left(\frac{x}{2}\right)$

2 a

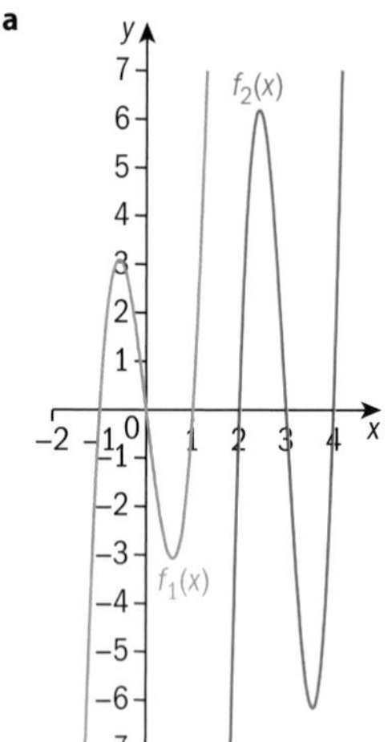

b

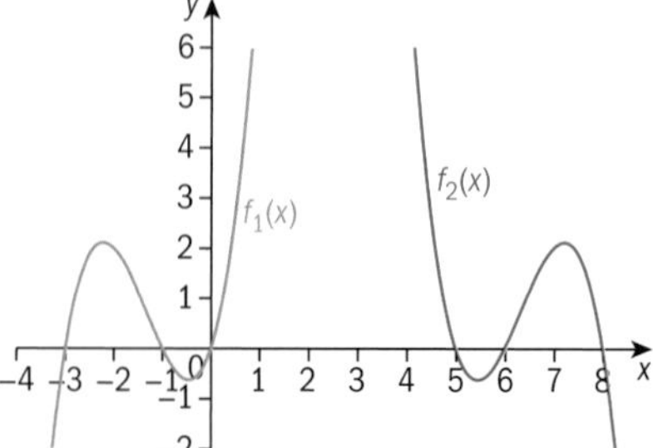

c

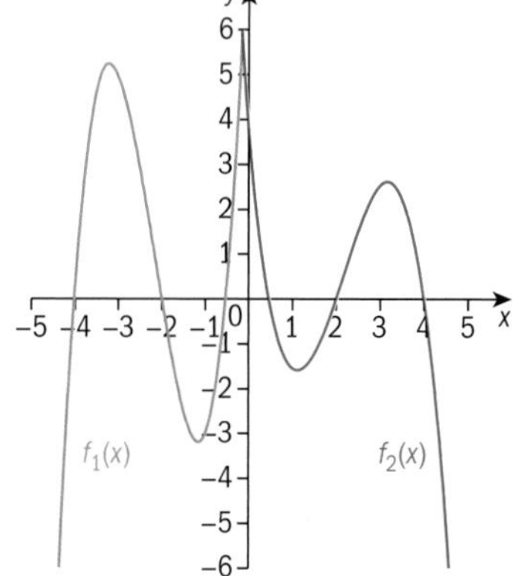

d

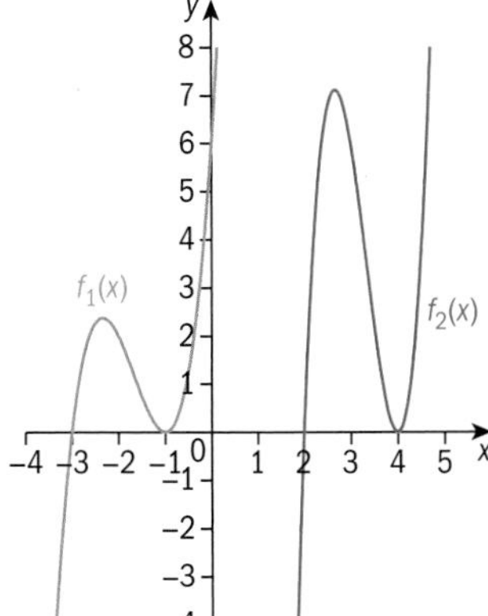

e

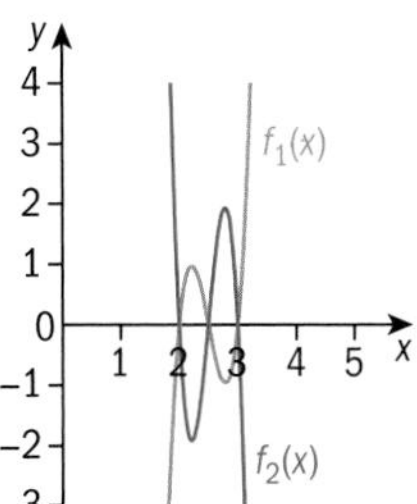

f

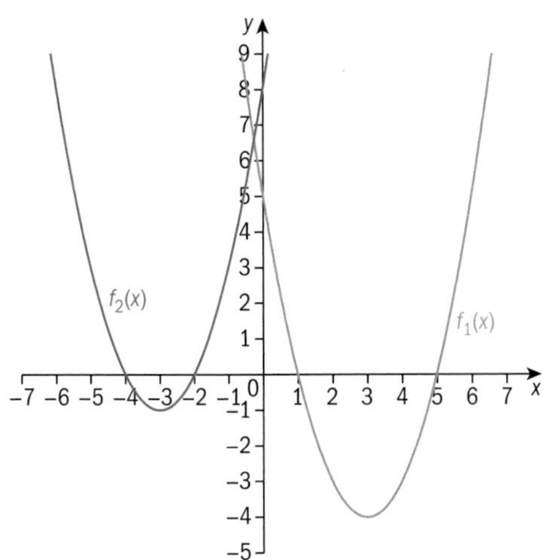

Mixed practice

1 a (−1, −2)

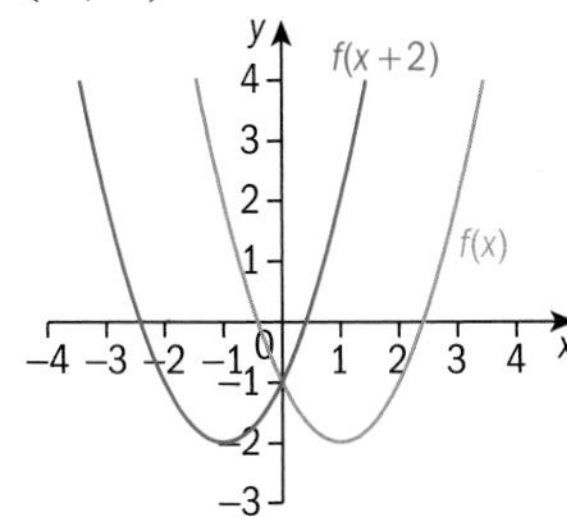

b (1, −6)

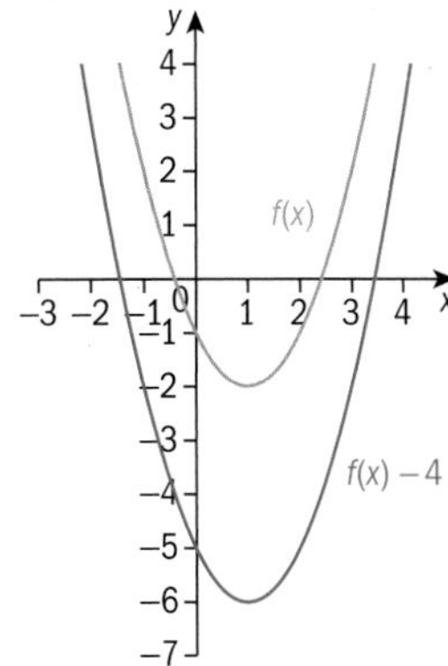

c (1, 2)

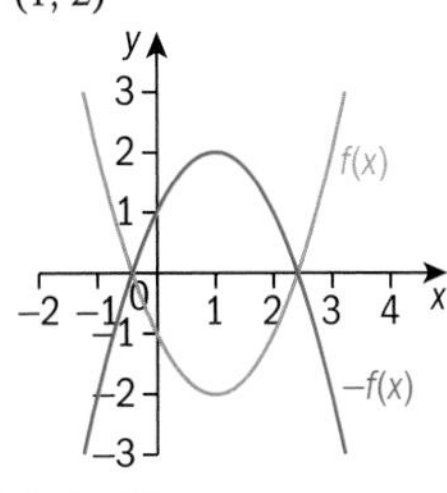

d (−1, −2)

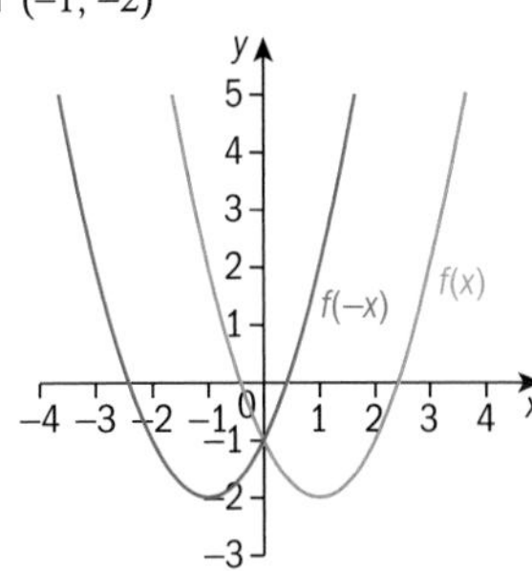

e (1, −4)

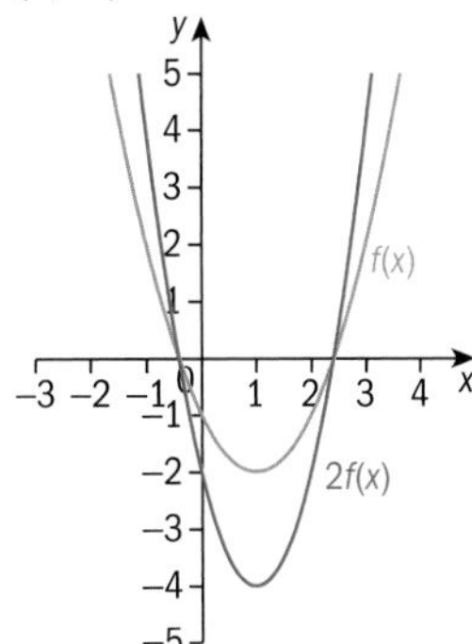

f $-\frac{1}{3}, -2$

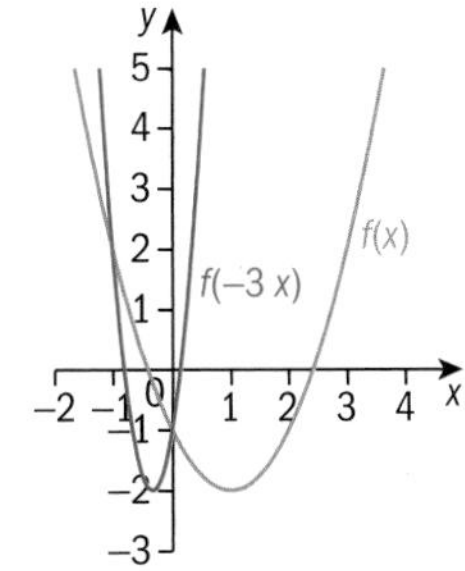

g (0, −5)

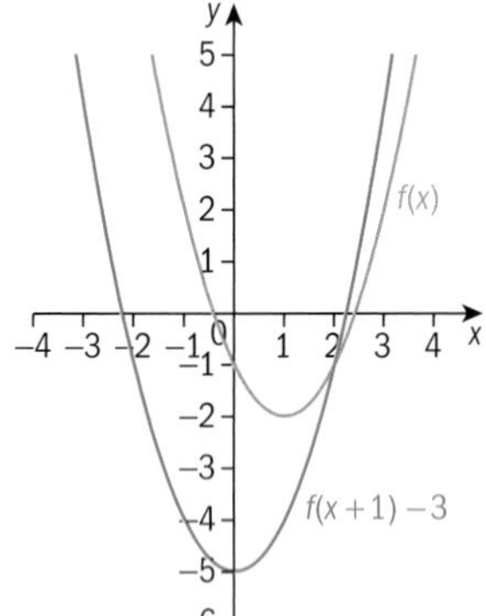

h (1, 5)

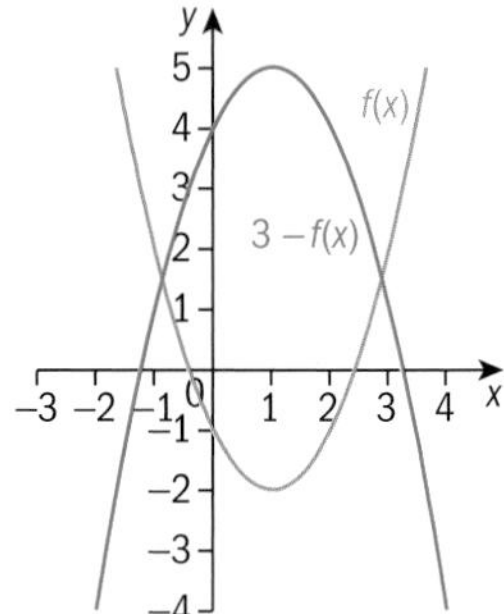

i (1, 3)

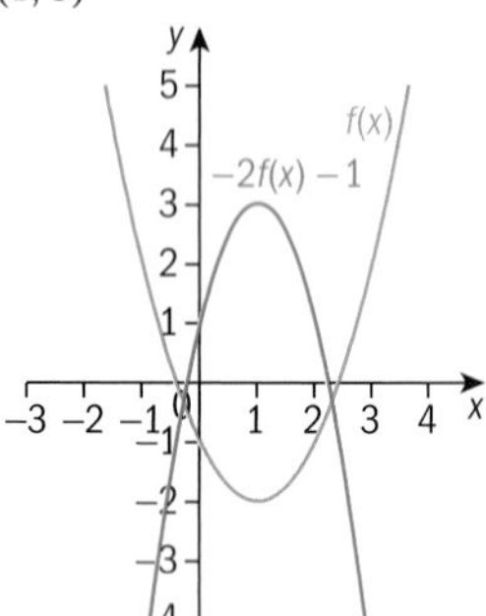

j (−2, −6)

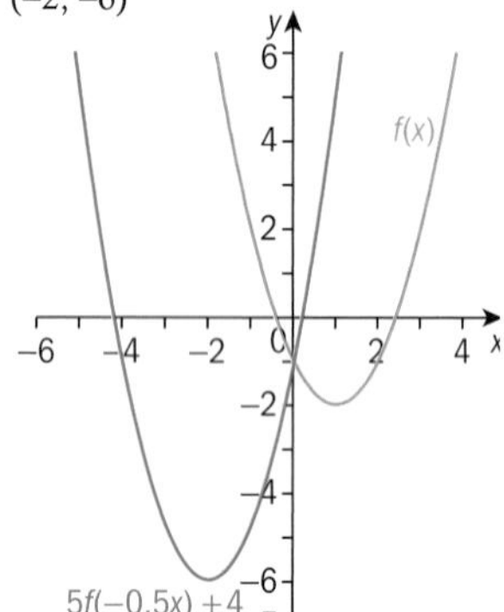

2 a Dilation scale factor $\frac{1}{2}$ parallel to the x-axis and vertical translation of −8 units OR dilation scale factor 2 parallel to the y-axis and vertical translation of −8 units

b Dilation scale factor 3 parallel to the x-axis, reflection in the y-axis and vertical translation of 8 units OR dilation scale factor $\frac{1}{3}$ parallel to the y-axis and reflection in the x-axis

c Dilation scale factor $\frac{1}{2}$ parallel to the y-axis and vertical translation of 3.5 units OR Dilation scale factor 2 parallel to the x-axis and vertical translation of 1 unit

d Horizontal translation of 2 units and vertical translation of −5 units

e Horizontal translation of −3 units and vertical translation of −3 units

f Dilation scale factor $\frac{1}{2}$ parallel to the y-axis and vertical translation of 2 units

3 a Reflection in the y-axis: $f_2(x) = f_1(-x)$

b Horizontal dilation scale factor 2: $f_2(x) = f_1\left(\frac{x}{2}\right)$

c Horizontal translation −3 units: $f_2(x) = f_1(x + 3)$

d Reflection in the x-axis: $f_2(x) = -f_1(x)$

e Vertical dilation scale factor $\frac{1}{2}$: $f_2(x) = 0.5f_1(x)$

f Horizontal dilation scale factor $\frac{1}{2}$: $f_2(x) = f_1(2x)$

4 a Reflection in the x-axis and vertical translation of 3 units $f_2(x) = -f_1(x) + 3$ or reflection in the y-axis and vertical translation of 5 units: $f_2(x) = f_1(-x) + 5$

b Reflection in the x-axis and vertical translation of −1 unit $f_2(x) = -f_1(x) - 1$

c Reflection in the x-axis and horizontal translation of −6 units $f_2(x) = -f_1(x + 6)$

d Vertical translation of 4 units and horizontal translation of 4 units $f_2(x) = f_1(x - 4) + 4$

e Vertical dilation of scale factor 2 and horizontal translation of 3 units $f_2(x) = 2f_1(x - 3)$

f Reflection in the x-axis and vertical translation of −2 units $f_2(x) = -f_1(x) - 2$

5 a

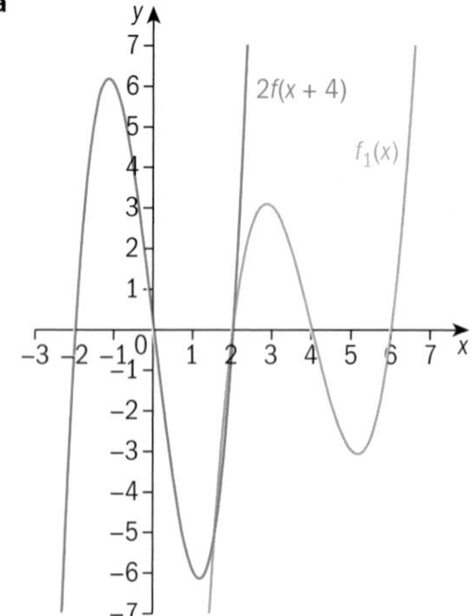

b

c

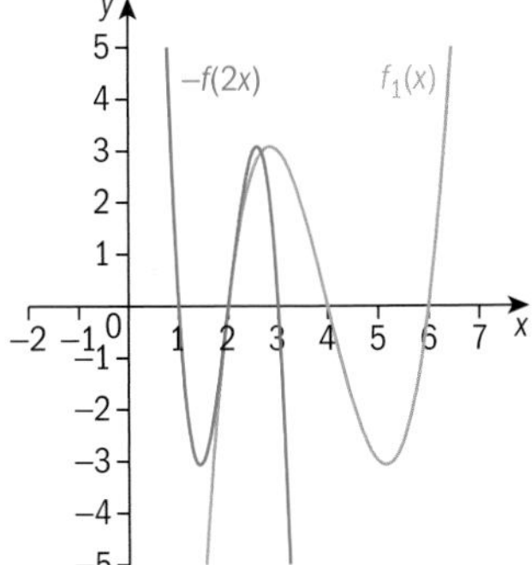

d

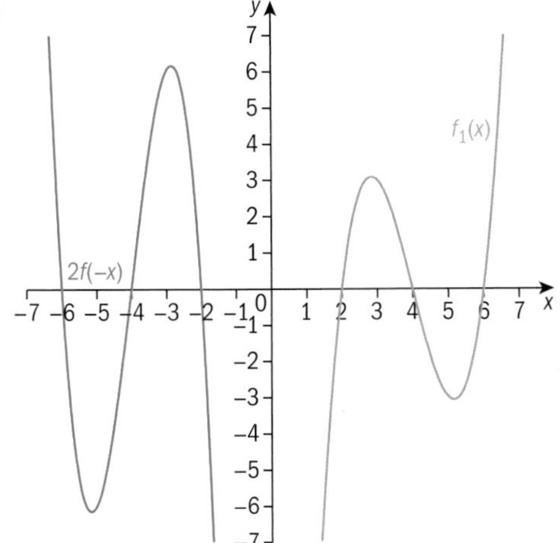

Review in context

1 a

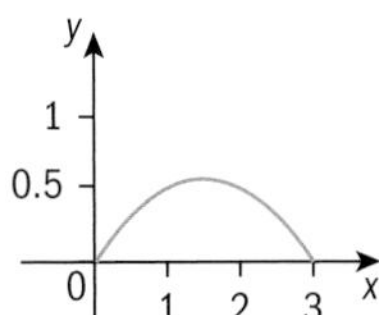

b 3 m

c i It would reach the same height but it would only travel half as far horizontally.

ii It would reach a greater maximum height, 0.675 m, and a greater overall distance, 4.29 m.

iii The frog has jumped to the left, not to the right.

d Because this would suggest that the frog has jumped downwards, then returned back upwards to its initial starting height.

2 a It translates the graph upwards – the ball would have been thrown from a greater height.

b It stretches the graph horizontally – the ball would have been thrown at a greater speed.

3 a i Horizontal translation of 10 units, reflect in the x-axis, dilate by a scale factor of $\frac{1}{20}$ in the y direction and translate by 5 units in the y direction.

ii $-\frac{1}{20}(0-10)^2+5=-\frac{1}{20}\times 100+5=-5+5=0$.

iii Minimum point starts at (0,0) and is mapped to a maximum point at (10,5).
It would hit the ceiling.

b The coordinates of the maximum point are (10, 2.5). With a maximum height of 2.5 m, the ball would not hit the ceiling.

4 a $b(t)=12\,000-150t$

b

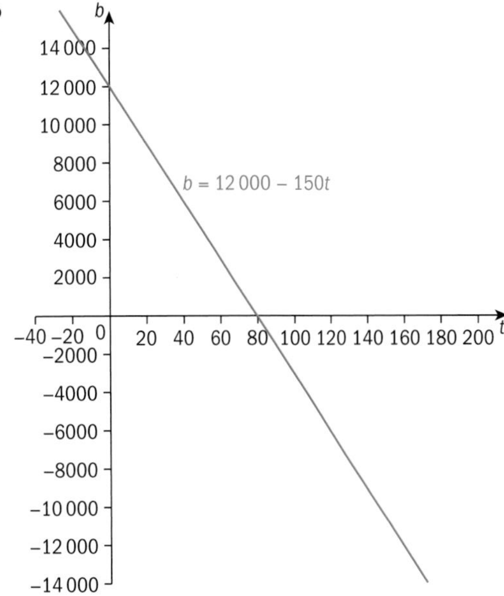

c 4500 – the loan balance after 50 months.

d 80 months

e A larger loan

f A smaller loan and smaller monthly repayments

g It will be steeper with the same y-intercept. $b(t)=12000-250t$

10.2

You should already know how to:

1 a $\frac{1}{4}$ **b** $\frac{2}{5}$ **c** $\frac{1}{x}$ **d** $\frac{1}{20}$

2 a $(x+3)(x+2)$ **b** $(x-5)(x+2)$ **c** $(x-3)(x+3)$

3 a $x=3$ **b** $x=3$ **c** $x=4$

Practice 1

1 a $x=\frac{y+4}{5}$ **b** $x=\pm\sqrt{3(y-3)}$

c $m=p^2-2$ **d** $w=\frac{5\pm\sqrt{V}}{2}$

e $V_i=\sqrt{V_f^2-2ad}$ **f** $a=\frac{V_f-V_i}{t}$

g $R=\frac{V}{I}-r$ **h** $r=\frac{mV^2}{F}$

i $V=\sqrt{\frac{Fr}{m}}$ **j** $m=\frac{F}{\frac{1}{2}V^2+gh}$

k $M=\frac{3RT}{2}$ **l** $a=\frac{1}{3}\left[\pm\sqrt{c+1}-2b\right]$

2 $A=\pi r^2$; $r=\sqrt{\frac{A}{\pi}}$ **3** $r=\sqrt{\frac{3V}{\pi h}}$

4 $t=\sqrt{\frac{2d}{a}}$ **5** $L=g\left(\frac{T}{2\pi}\right)^2$

6 $r=\sqrt{\frac{kq_1q_2}{F}}$

Practice 2

1 $5a^2$; $a \neq 0$

2 $\frac{5x+3}{2}$; $x \neq 0$

3 $\frac{6x^3}{5y^2}$; $y \neq 0$

4 $\frac{x-7}{x+7}$; $x \notin \{-8, -7\}$

5 $\frac{a+2b}{2b-a}$; $a \notin \{b, 2b\}$

6 $\frac{x+4}{x+5}$; $x \notin \{-5, -2\}$

7 $\frac{x+4}{x+2}$; $x \notin \{-2, 3\}$

8 $\frac{x+2}{x+7}$; $x \notin \{-7, 5\}$

9 $\frac{x(x-8)}{x-3}$; $x \notin \{3, 4\}$

10 Does not simplify; $x \notin \{-3, 6\}$

11 Any unsimplified expression where the numerator is a multiple of either $(x + 2)$ or $(x - 5)$ or both, and where the denominator is a multiple of $(x + 2)(x - 5)$.

12 $\frac{x^2-3x-18}{x^2-16x+60}$ or an equivalent algebraic fraction where the numerator and denominator have both been multiplied by the same factor(s).

Practice 3

1 $\frac{6p^2}{k^2}$; $k \neq 0$

2 $\frac{c}{4p^2}$; $c, p \neq 0$

3 $\frac{5}{2}$; $x \notin \{-2, 3\}$

4 $\frac{2(x-2y)(2x+9y)}{x+3y}$; $x \notin \{-\frac{9}{2}y, -3y\}$

5 $\frac{x-2}{x-1}$; $x \notin \{-1, 1\}$

6 $\frac{a-2b}{a}$; $a \notin \{0, 2b\}$

7 $\frac{x-5}{x-2}$; $x \notin \{-3, -2, -1, 2\}$

8 $\frac{(x-3)(x+6)}{(x-6)^2}$; $x \notin \{-8, -6, 3, 6\}$

9 $\frac{2}{3}$; $x \notin \{-3, -2, 0, 1, 3\}$

10 $\frac{6(x+2)^2}{(x-1)^2(x-2)}$; $x \notin \{-2, -1, 1, 2, 5\}$

Practice 4

1 $\frac{3y+4x}{xy}$; $x, y \neq 0$

2 $\frac{5}{4x}$; $x \neq 0$

3 $\frac{2x+1}{6x^2}$; $x \neq 0$

4 $\frac{2c+3b-5a}{abc}$; $a, b, c \neq 0$

5 $\frac{2x}{x^2-4}$; $x \notin \{-2, 2\}$

6 $\frac{(x-10)(x+2)}{(x+4)(x-3)} = \frac{x^2-8x-20}{x^2+x-12}$; $x \notin \{-4, 3\}$

7 $\frac{2x^2+3x+16}{(x+4)(x-2)}$; $x \notin \{-4, 2\}$

8 $\frac{x+1}{x-3}$; $x \neq 3$

9 $\frac{-(x+2)}{x^2+3x-10}$; $x \notin \{-5, 2\}$

10 $\frac{5x^2+15x+4}{(x-8)(x+3)} = \frac{5x^2+15x+4}{x^2-5x-24}$; $x \notin \{-3, 8\}$

11 $\frac{58+12x-6}{x^3-x^2}$; $x \notin \{-1, 0, 1\}$

12 $\frac{-x(x^2-2x-27)}{4(x+3)(x-5)} = \frac{-x^3+2x^2+47x}{4x^2-8x-60}$; $x \notin \{-3, 5\}$

Mixed practice

1 $r = \sqrt[3]{\frac{3V}{4\pi}}$

2 $T = \frac{Pv}{500}$

3 $r = \sqrt{\frac{Gm_1m_2}{F}}$

4 $C = \frac{2W}{U^2}$

5 $c = \sqrt{\frac{E}{m}}$

6 $c = \frac{v\left(1+\left(\frac{v}{v_0}\right)^2\right)}{\left(\frac{v}{v_0}\right)^2 - 1} = \frac{v\left(v_0^2+v^2\right)}{v_0^2-v^2}$

7 $\frac{x^3}{3}$; $x, y \neq 0$

8 $\frac{5x^4-6x^2}{4} = \frac{x^2\left(5x^2-6\right)}{4}$; $x \neq 0$

9 $\frac{x-7}{x-11}$; $x \notin \{-3, 11\}$

10 $\frac{1}{x-4}$; $x \notin \{-4, 4\}$

11 $\frac{10+xy}{5x}$; $x \neq 0$

12 $\frac{6x^2y}{z^2}$; $z \neq 0$

13 $\frac{x^3y^3}{x-y}$; $x \notin \{-y, 0, y\}$

14 $\frac{x^2-10xy-4}{4x}$; $x \neq 0$

15 $\frac{12x}{(x+5)(x-7)}$; $x \notin \{-5, 7\}$

16 $\frac{6-5x}{(x+9)(x-8)} = \frac{-5x+6}{x^2+x-72}$; $x \notin \{-9, 8\}$

17 $\frac{10x}{(x-1)(x+4)} = \frac{10x}{x^2+3x-4}$; $x \notin \{-4, 1\}$

18 $\frac{13}{x^2-1} = \frac{13}{(x+1)(x-1)}$; $x \notin \{-1, 1\}$

19 $\frac{12b+20a-ab(a+b)}{4ab}$; $a, b \neq 0$

20 $\frac{x^3-7x^2+10x-24}{6x(x-2)}$; $x \notin \{0, 2\}$

Review in context

1 **a** 830 km/h **b** 900 km/h **c** $(870 + x)$ km/h

d $(870 - y)$ km/h **e** $t = \frac{d}{V}$

f 6.39 hours = 6 hours 23 minutes 27 seconds

g 6.70 hours = 6 hours 42 minutes (6 hours 41 minutes 56 seconds)

h 6.18 hours = 6 hours 10 minutes (40 seconds)

i 70 km/h (headwind)

j 50.3 km/h (tailwind – 3 s.f.)

2 **a** $\frac{1}{x} + \frac{1}{x+4} + \frac{1}{2x+2} = \frac{5x^2+15x+8}{x(x+4)(2x+2)}$

b $\left(\frac{1}{x} + \frac{1}{x+4} + \frac{1}{2x+2}\right) + \frac{1}{2x+2} = \frac{3x^2+10x+4}{x(x+4)(2x+2)}$

3 **a** $I = \frac{V}{R+r}$ **b** $\frac{48r+72}{(6+r)(5+3r)}$

4 **a** 4 **b** Decreased 8 times

c Increased 96 times **d** It needs to double

10.3

You should already know how to:

1 **a** $\frac{5}{3}$ **b** $\frac{1}{2}$ **c** $\frac{45}{133}$ **d** $\frac{1}{6}$

2 $f(h) = 10h$; $c(m) = 4.5m + 2$

3

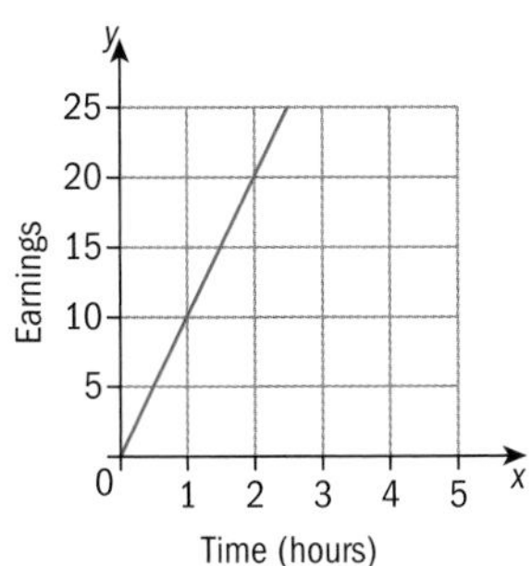

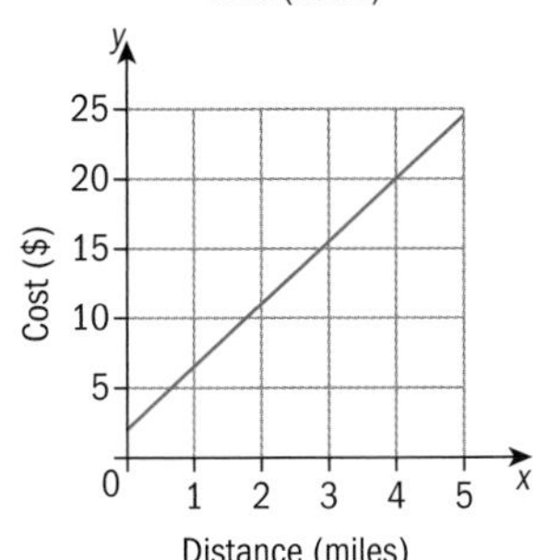

4

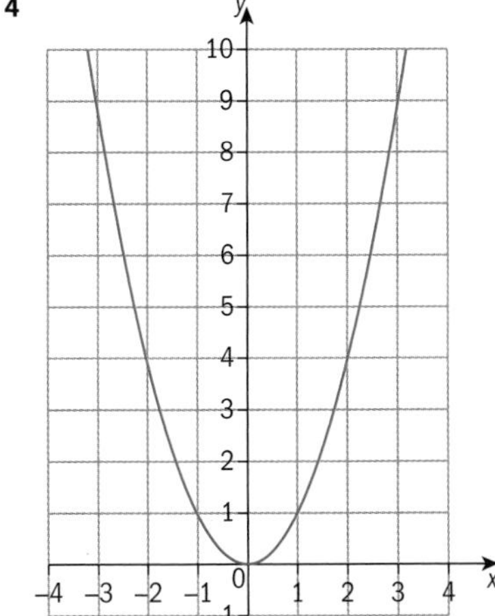

Practice 1

1 $c(b) = kb$
- **a** c is doubled
- **b** b is tripled
- **c** c is multiplied by $\frac{2}{3}$
- **d** $4k$ is added to c

2 **a** $v(t) = kt$ **b** $k = 8$ **c** $v = 200$

3 $V = 2$

4 0, 8.54, 9.8

5 **a** $k = 1.6 \div 2 = 2.4 \div 3 = 5.6 \div 7 = 7.2 \div 9$
- **b** $k = 0.8$
- **c** 4 kg

6 $a: k = 3,\ c: k = \frac{1}{2},\ d: k = \frac{4}{5},\ f: k = -4$

Practice 2

1 **a** **i** Direct proportion: the perimeter varies directly as the length of the sides of the triangle.
- **ii** $k = 3$

b **i** Direct proportion: the area that can be painted varies directly as the number of liters of paint.
- **ii** $k = 0.2$

c **i** Direct proportion: y varies directly as x.
- **ii** $k = 4$

d **i** Inverse proportion: y increases proportionally as x decreases, and vice versa; y varies inversely as x.
- **ii** $k = 3$

e **i** Direct proportion: y varies directly as x.
- **ii** $k = \frac{1}{5}$

2 **a** $f(v) = \frac{k}{v}$ **b** $k = 24$ **c** $f = 12$

3 $y = 6.5$

4 $T = 2.32$ (3 s.f.)

5 1.6, 0.667 (3 s.f.)

6 **a** $d(w) = \frac{k}{w}$ **b** $k = 30$ **c** 24 kg

7 **a** $k = 1 \times 300 = 4 \times 75 = 8 \times 37.5 = 10 \times 30 = 20 \times 15$
The cost per person halves when the number of people doubles, so this is an inverse relationship.
- **b** $k = 300$
- **c** \$12

8 **a** Yes. Direct proportion.
$k = 390 \div 1.5 = 624 \div 2.4 = 1274 \div 4.9 = 2158 \div 8.3 = 260$.
The increase in tiger population varies directly as the amount spent on conservation.
- **b** Increase in population = 260 × Amount spent
- **c** \$12.3 million (3 s.f.)

9 **a** **i** Yes, time (in seconds) to see the lightning varies directly as the distance (in km) $\left(k = \frac{1}{300000}\right)$.
- **ii** Yes, time (in seconds) to hear the thunder varies directly as the distance (in m) $\left(k = \frac{1}{340}\right)$
- **iii** Yes, the time to hear the thunder varies directly as the time to see the lightning $\left(k = \frac{300000}{340} = \frac{15000}{17}\right)$

b **i** 0.000 016 7 seconds (3 s.f.)
- **ii** 14.7 seconds (3 s.f.)
- **iii** $14.7 = \frac{15\,000}{17} \times 0.0000167$
The time to hear the thunder and the time to see the lightning are directly proportional.

Practice 3

1 **a** **i** direct variation, $y = kx^3$ and k is constant ($k = 1$)
- **ii** $y = x^3$

b **i** inverse variation, $y = \frac{k}{x}$, k is constant ($k = 24$)
- **ii** $y = \frac{24}{x}$

c **i** direct variation $y = kx^2$, k is constant ($k = 5$)
- **ii** $y = 5x^2$

d **i** inverse variation, $y = \frac{k}{x}$, k is constant ($k = 60$)
- **ii** $y = \frac{60}{x}$

e neither, the first three table entries follow the rule $y = \frac{1}{x}$ but the last one doesn't.

f neither, this sequence follows no recognisable pattern

2 a $y = 48$ **b** $y = 96$
c $y = 6$ **d** $y = 3$

3 a $x = \frac{5}{12}$
b $x = 2.18$ (3 s.f.)
c $x = 17.3$ (3 s.f.)
d $x = 11.4$ (3 s.f.)

4 a $16\pi\ \text{cm}^2$ **b** $A = 4\pi r^2$
c The surface area increases by a factor of 25
d The radius is multiplied by a factor of $\frac{1}{\sqrt{2}}$

5 691 N (3 s.f.)

6 3.0% to the nearest 0.1%

7 5.93 cm (3 s.f.)

8 2.58 cm (3 s.f.)

Practice 4

1 a $y = 6x$

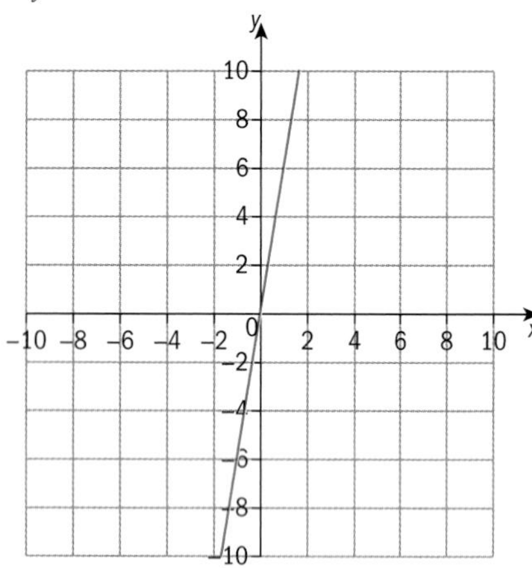

b $y = \frac{12}{x}$

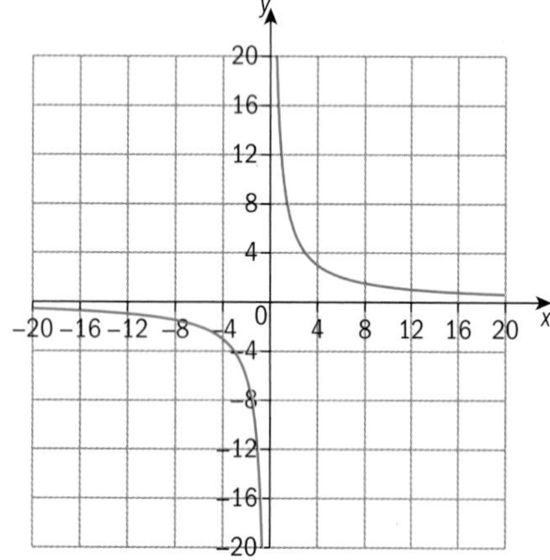

c $y = 7.5x^2$

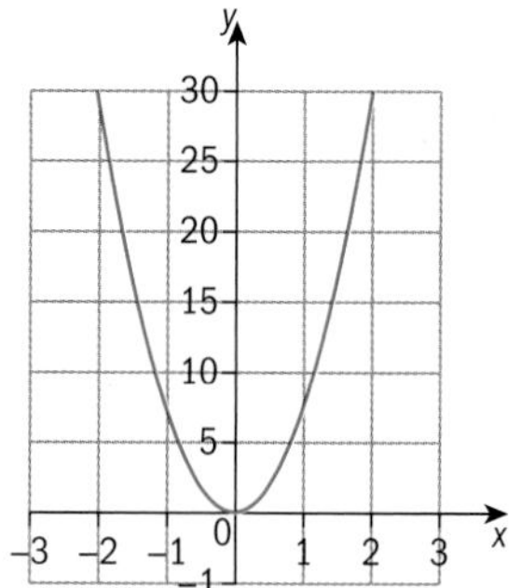

d $y = \frac{80}{x^2}$

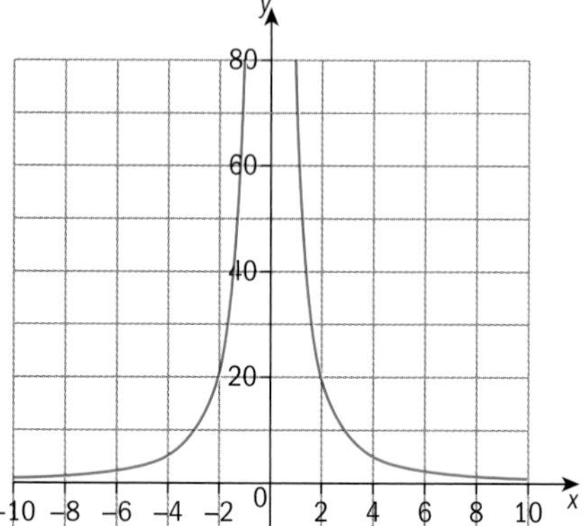

e $y = 7.8125x^3$

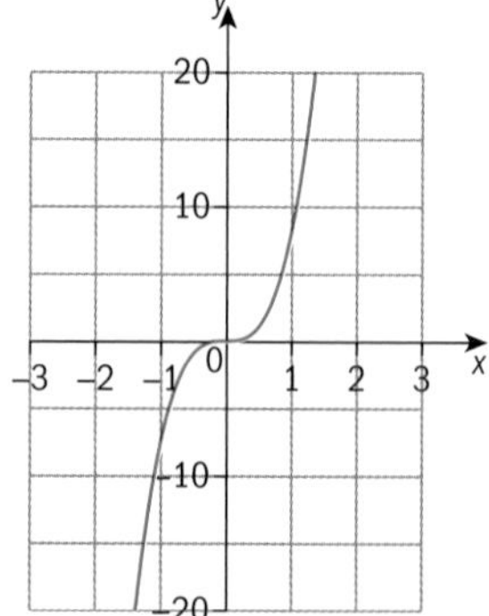

f $y = \frac{1}{2x^3}$

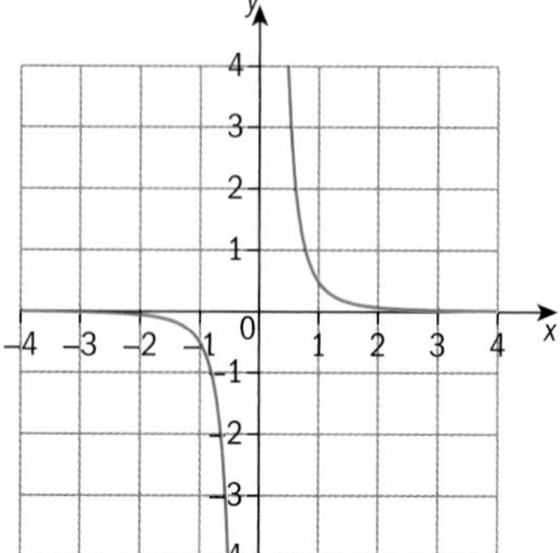

2 a Neither, because the line does not pass through the origin.

b Direct, as x increases y increases. y is directly proportional to a positive power of x.

c Neither, the graph crosses the x-axis but no power of x with a non-zero coefficient can give a y-value equal to 0.

d Direct, y varies directly as x^2, with a negative proportionality constant k.

e Direct, y varies directly as x, with a negative proportionality constant k.

f Direct, y varies directly as a higher power of x, with a negative proportionality constant k.

3

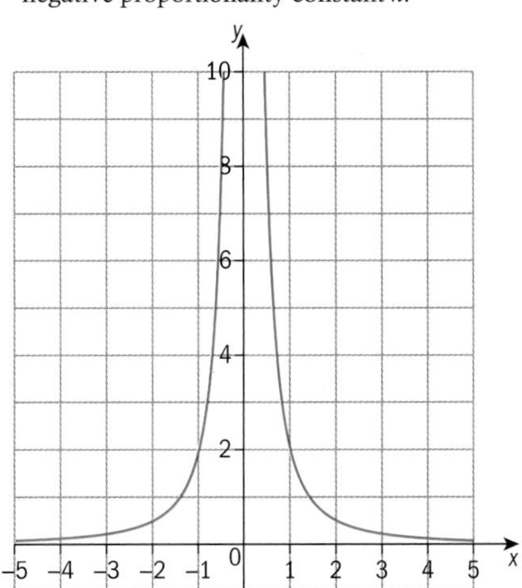

a y can't have a negative value because $2x^2$ can take only positive values.

b y approaches zero

c y becomes extremely large

d Because y becomes smaller as x becomes larger, and vice versa. x varies inversely as y.

Mixed practice

1 a direct relationship

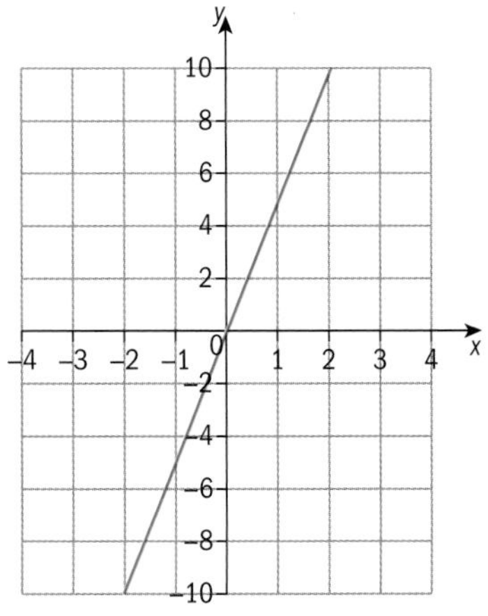

b direct relationship

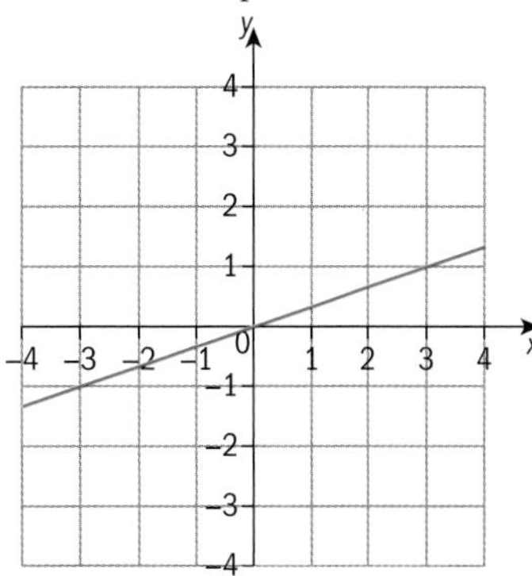

c direct relationship

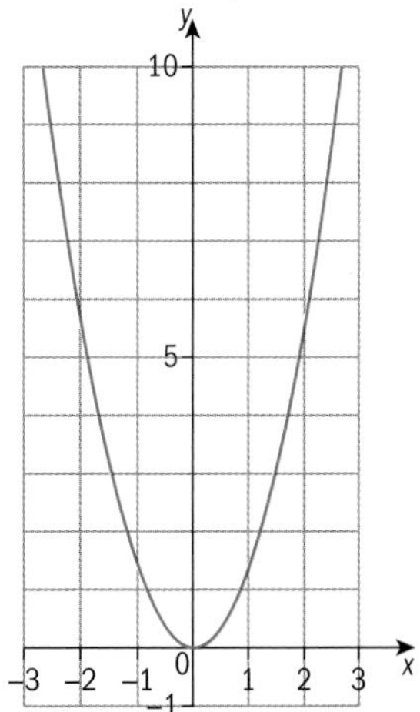

d inverse relationship

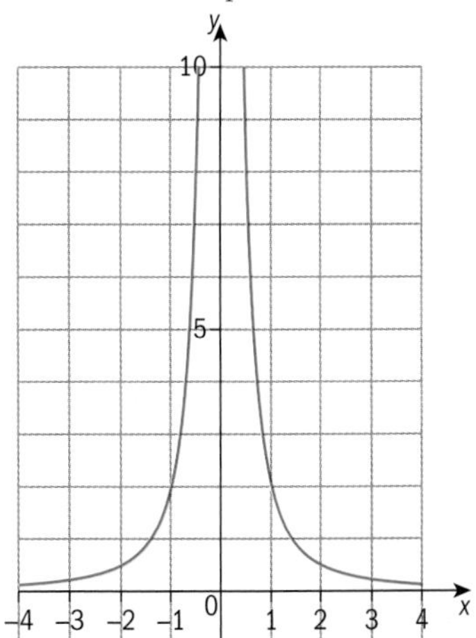

e inverse relationship

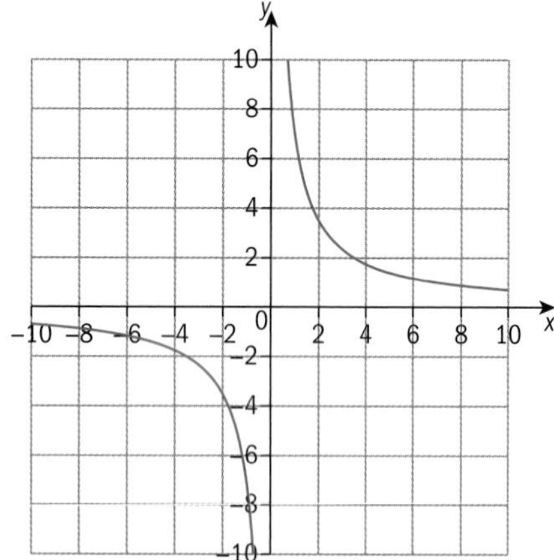

f inverse relationship

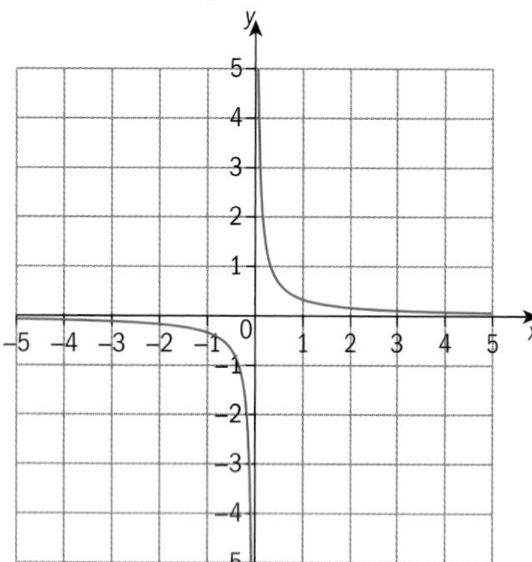

g direct relationship

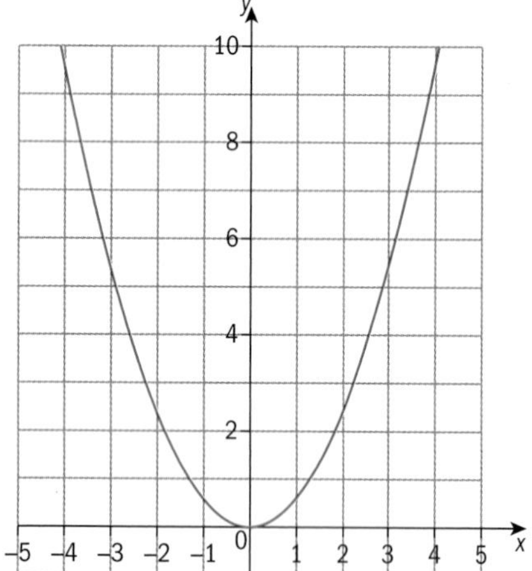

h inverse relationship

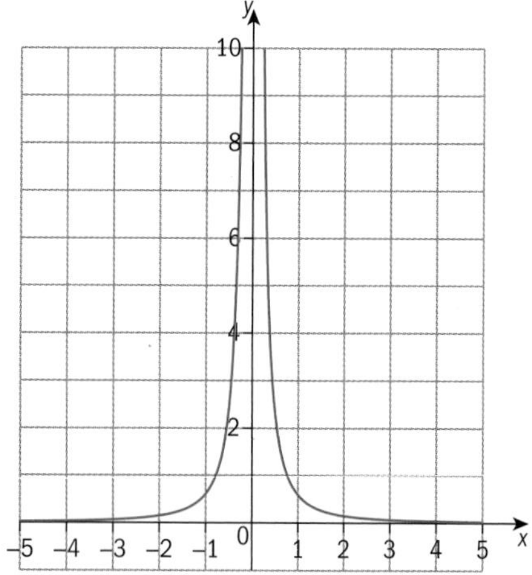

2 a $k = 0.75$ **b** $x = 0.75t$ **c** $x = 6.15$

3 $P = 2$ **4** $c = \pm 3.5$ **5** $p = 25$, $q = 0.02$

6 a $y = \frac{x^2}{4}$ **b** $y = -\frac{5}{x}$

7 a The more bottles are bought, the higher the cost. Cost = 1.3 × number of bottles, direct proportion

b No, because the cost for 2 bottles or 3 bottles is the same, £2.60. The cost does not increase proportionally to the number of bottles bought.

8 360 N

9 N = the whole number part of $\frac{3k}{4\pi r^3}$

10 70.4 kg **11** 367.2 m **12** 500

13 a $V = kr^2$, $k = \pi h$ **b** $V = kh$, $k = \pi r^2$

Review in context

1 Alphatown \$62 500
Boomcity \$125 000

2 a

Pipe diameter (cm)	10	20	30	40	50
Number of houses	50	200	450	800	1250

b Between 300 and 350 houses

c

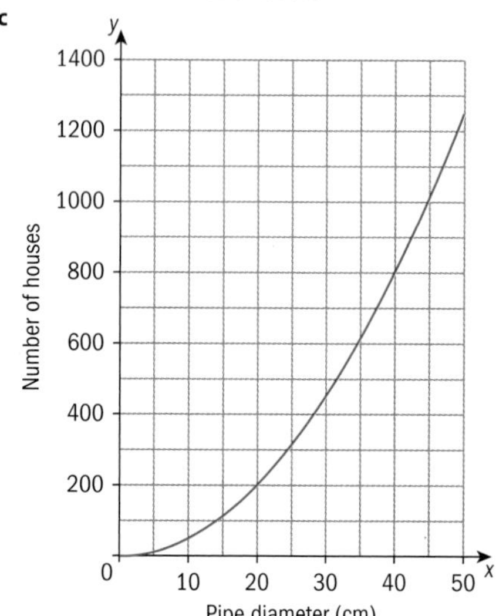

d Between 310 and 340 houses
e 312 houses
f Student's own answer
g Student's own answer. A possible argument could be to find ways to reduce household consumption so that a pipe can serve more houses without needing to be enlarged.

3 Moving the heater to B reduces the heat received by 36%.

4 a Temperature change –0.71°C
b Diameter of trees 2.24 m

5 a 1220 pairs of Classic jeans
b They should sell Classic jeans at $113 per pair to double their sales.

6 a $2.5 \div 3^3 \neq 5 \div 5^3$, there is no constant proportion between the cube of the radius and the price.
b One large scoop would cost $11.57
c Student's own suggestions. Possible arguments could include that the price of an ice-cream should cover other costs, such as the salary of the waiter, the cost of the equipment to make ice-cream, etc. These costs can be covered in the first scoop or two, which will therefore be more expensive than each additional scoop.

11.1

You should already know how to:

1 a $4x + 12$ **b** $10x - 5$ **c** $-24 + 18x$ **d** $28x + 4x^2 - x^3$

2 a $3(x + 2)$ **b** $5(x - 3)$ **c** $7(2 + 5x)$ **d** $17(2 - 5x)$

3 a 15 **b** 13 **c** 3 **d** 56

4 Red: $y = -x + 1$ Blue: $y = 2x + 3$

Practice 1

1 a –10 **b** $\frac{10}{3}$ **c** $-\frac{2}{3}$ **d** –1
e 36 **f** 3 **g** 1 **h** $\frac{1}{2}$

2 Student's own answers

3 The left-hand side equals the right-hand side – this is an example of an identity which is true for all x.

Practice 2

1 $x = 2, y = 3$ **2** $x = 4, y = 1$

3 $x = -2, y = -3$ **4** $x = -1, y = 3$

5 $x = \frac{87}{5}, y = -\frac{68}{5}$

6 Rupa: $x + 24 - 15x = 10$ $14 = 14x$ $x = 1$
$1 + 3y = 10$ $y = 3$
George: $50 - 15y + y = 8$ $42 = 14y$ $y = 3$
$x + 9 = 10$ $x = 1$

Practice 3

1 $x = 5, y = 2$ **2** $x = -1, y = 3$

3 $x = -2, y = 4$ **4** $x = 3, y = 1$

Practice 4

1 $x = 2, y = 4$ **2** $a = \frac{24}{5}, b = -\frac{31}{5}$

3 $x = 1, y = 2$ **4** $r = 4, s = 1$

5 $x = 5, y = -2$ **6** $x = 3, y = -1$

Practice 5

1 $x = 2, y = 3$ **2** $x = 1, y = 6$ **3** $x = 2, y = 5$

4 $x = 2, y = 5$ **5** $x = 1, y = -2$ **6** $x = 3, y = 5$

Practice 6

1 $R = 8800x$; $C = 90\,000 + 7800x$
$R = C$ when $x = 90$ weeks.
90 weeks

2 $x = -\frac{1580}{1223}$ (–1.29 to 2 d.p.) so working backwards makes it part way through 2003

3 A: $y = 31.45 + 0.155x$
B: $y = 5.20 + 0.37x$
Graphing both to see where they are equal, cost of both plans is $48.80 when $x = 112$ minutes. For $x < 112$, plan B is cheaper. For $x > 112$, plan A is cheaper.

4 $500 in the 2% one, $4000 in the 5% one

5 6 hours 40 minutes cycling and 3 hours 20 minutes walking

Mixed practice

1 a 5 **b** $\frac{7}{8}$ **c** –9 **d** $\frac{12}{7}$

2 a $x = -1, y = -1$ **b** $x = -9, y = -4$
c $x = \frac{18}{5}, y = \frac{3}{5}$ **d** $x = \frac{3}{8}, y = -\frac{3}{4}$
e $x = 2, y = 4$

Review in context

1 $38x + 5y = 550$; $56x + 60y = 1200$
13.5 ounces of milk and 7.4 ounces of orange juice.
This is realistic.

2 $x + y = 8$; $0.12x + 0.32y = 1$
4.8 liters of 12% acid and 3.2 liters of 32% acid

3 $x + y = 20$; $10.50x + 8.25y = 9(x + y)$
$6\frac{2}{3}$ kg of premium and $13\frac{1}{3}$ kg of standard

4 Laser: $y = 150 + 0.015x$
Inkjet: $y = 30 + 0.06x$
For $x < 2666$, the inkjet printer is cheaper.
For $x > 2667$ the laser printer is cheaper.

11.2

You should already know how to:

1 a $3x^2 + 5x - 2$ **b** $x^2 - 2x - 15$

2 a $(x + 3)(x - 2)$ **b** $(2x - 1)(x + 3)$
c $(2x + 9)(2x - 9)$

3 a $(x - 3)^2 + 5$ **b** $3(x - 1)^2 + 4$ **4** $\frac{11}{5}$

Practice 1

1 a 0, 8 **b** 2, 3 **c** $-2, \frac{4}{3}$ **d** 0, 9
e $\frac{1}{2}, -\frac{1}{2}$ **f** 4, –4 **g** –5, 2 **h** $-\frac{1}{2}, \frac{2}{3}$
i $0, \frac{10}{9}$ **j** 3, –3 **k** –2, 2 **l** $\frac{3}{2}, -6$

2 a (e.g.) $x^2 - 2x - 8 = 0$ **b** (e.g.) $2x^2 - 13x + 6 = 0$
c (e.g.) $5x^2 - 13x - 6 = 0$

3 a $x^2 + 4.25x - 3.75 = 0$ **b** $4x^2 + 17x - 15 = 0$

Practice 2

1 a -1 **b** 3 **c** 5
d $\frac{1}{3}$ **e** $\frac{3}{2}$ **f** $\frac{2}{3}$

2 a $x^2 + 4x + 4 = 0$
b $25x^2 - 10x + 1 = 0$
c $16x^2 + 24x + 9 = 0$

Practice 3

1 6cm, 3cm **2** 5 cm by 9 cm
3 15 m by 28 m **4** 22 cm, 5 cm
5 10 cm **6** 7, 24, 25
7 8 m

Practice 4

1 $2 \pm \sqrt{3}$ **2** $-1 \pm \sqrt{2}$
3 $1 \pm \sqrt{3}$ **4** No solutions
5 No solutions **6** $-3 \pm 2\sqrt{2}$
7 $-10 \pm 2\sqrt{15}$ **8** $\frac{-2 \pm \sqrt{6}}{2}$
9 $\frac{-3 \pm \sqrt{15}}{3}$ **10** $\frac{-10 \pm \sqrt{94}}{2}$
11 $\frac{1}{2}, -\frac{3}{2}$ **12** $\frac{-6 \pm 3\sqrt{2}}{2}$

Practice 5

1 a $\frac{-2 \pm \sqrt{5}}{2}$ **b** No solutions **c** No solutions
d $1 \pm \sqrt{6}$ **e** 1 **f** $\frac{7 \pm \sqrt{3}}{2}$
g $\frac{1 \pm 2\sqrt{3}}{3}$ **h** $\frac{-1 \pm \sqrt{10}}{2}$ **i** $\frac{1 \pm \sqrt{129}}{8}$

2 a 1.3, -2.3 **b** 3.2, -1.2 **c** 0.1, -2.1
d No solutions **e** No solutions **f** -1.3, 5.3

3 a $\frac{1 \pm \sqrt{21}}{2}$ **b** $\frac{-5 \pm \sqrt{29}}{2}$ **c** 1
d $\frac{7 \pm \sqrt{61}}{2}$ **e** $\frac{3 \pm \sqrt{199}}{2}$

4 a 1.34, -1.94 **b** 14.9, 0.101 **c** 0.869, -1.54
d 2.10, -1.27 **e** 1.66, -0.300

Practice 6

1 a i 0 **ii** No
b i 2 **ii** Yes
c i 2 **ii** No
d i 0 **ii** No
e i 2 **ii** Yes

2 a ± 8 **b** ± 12 **c** $k > -4$ **d** $k < \frac{9}{20}$

Practice 7

1 2.28 m by 4.39 m (both 3 s.f.)

2 3.49 m by 43.0 m or 21.5 m by 6.97 m (all 3 s.f.)

3 10.1 m (3 s.f.)

4 1.17 m (3 s.f.)

5 9 cm

6 8 km, 15 km

7 8.5 cm

8 a 8.5 cm **b** 6562.5 cm^2

9 1, 11 or -8, -11

10 11, 13 or -11, -13

11 12, -3

12 $2 + \sqrt{5}$ or $2 - \sqrt{5}$

Practice 8

1 Just over 14%
2 19 months
3 a

15	0	0	9500	142 500
14	1	1000	10500	147 000
13	2	2000	11500	149 500
12	3	3000	12500	150 000
p	$15 - p$	$1000(15 - p)$	$9500 + 1000(15 - p)$	$p(9500 + 1000(15 - p))$

b $s = p(9500 + 1000(15 - p))$
c $p(9500 + 1000(15 - p)) = 0$; \$24.50
d \$12.25; 12 250
e \$9.50
f Student's own answers

4 a $s = -3.07p + 178$ (both to 3 s.f.)
b $R = ps$
c \$29
d \$58

5 a $R = (50 + 5x)(36 - 2x)$, where x is the number of price increases.
b \$70 **c** \$140

6 a i 62 000 **ii** 26 000
iii 0 (function gives $-10\,000$)

b $R = sp = (80\,000 - 180p)p$
c $C = 800\,000 + 120(80\,000 - 180p)$
d $P = (80\,000 - 180p)p - (800\,000 + 120(80\,000 - 180p))$
e \$134.33 or \$430.11; \$282.22

Mixed practice

1 a 6, -4 **b** 9, -3 **c** $\frac{4}{3}, -\frac{1}{2}$
d $\frac{3}{2}$, 7 **e** $\frac{2}{5}$, 1 **f** 3, 1

2 a $-3 \pm 2\sqrt{17}$ **b** $-6 \pm \sqrt{13}$ **c** $5 \pm \sqrt{7}$
d $\frac{-3 \pm \sqrt{30}}{3}$ **e** $\frac{4 \pm \sqrt{10}}{2}$ **f** 3, $-\frac{1}{2}$

3 a −1.54, −8.46 **b** 6.14, −1.14 **c** 1.26, −0.26
d 1.59, −1.26 **e** 0.12, −2.12 **f** No solutions

4 a $1, \frac{1}{2}$ **b** No solutions **c** $2 \pm \sqrt{11}$

d $\frac{5}{3}, -1$ **e** $3 \pm \sqrt{3}$ **f** No solutions

5 9 cm by 12 cm **6** 5 cm by 17 cm

7 5 cm by 6 cm **8** 3.5 m

9 12 cm and 7 cm

10 15 cm, 8 cm and 17 cm

11 9 and 11 **12** 9 or −8

13 7 and 8

14 a 5.04 seconds **b** 0.96 seconds **c** 3.08 seconds

15 \$8.01 or \$25.99 **16** 18.9 cm

Review in context

1 10.1 seconds (3 s.f.)

2 a $h = -4.9t^2 + 24t + 30$
b i 59.4 meters (3 s.f.)
ii 2.45 seconds (3 s.f.)
iii 5.93 seconds (3 s.f.)

3 6 seconds

4 a 8.51 seconds (3 s.f.)
b 6 times as long
c Student's own answers

11.3

You should already know how to:

1 a 4 **b** 5.5 **c** $\frac{20}{3}$

2 a $\frac{13x}{12}$ **b** $\frac{2(x-5)}{x(x+2)}$ **c** $\frac{2}{5}$ **d** $\frac{4(x-7)(x+7)}{(x-3)(x-14)}$

3 a $(x-4)(x+4)$ **b** $3x(x-2)$
c $(x-6)(x+1)$ **d** $(3x+4)(x-3)$

4 a ±3 **b** −3, 1 **c** $-3, -\frac{1}{2}$ **d** $\frac{1}{3}, 1$

Practice 1

1 $\frac{40}{17}$ **2** $\frac{129}{47}$

3 $-\frac{14}{9}$ **4** $\frac{15}{58}$

5 $-\frac{19}{8}$ **6** 0

7 $\frac{73}{29}$ **8** $\frac{3}{26}$

9 1.426, −1.870 (3 d.p.)

Practice 2

1 a x **b** $6x$ **c** $3x^4$
d $x(x-2)$ **e** $2x$ **f** $(x-2)(x+2)$
g $x(x-2)(x+2)$ **h** $x(x^2+1)$ **i** $(x-1)(x+1)$
j $x(x-1)(x+1)$ **k** $2x+8$ **l** $(x+1)(x-1)$
m $(x-a)(x+1)(x-1)$ **n** $x(x-3)(x+2)$ **o** $2x(2x-3)(x+1)$
p $6x^2(x+2)(x-2)$

Practice 3

1 $-\frac{8}{5}$ **2** $\frac{4}{5}$ **3** $\frac{4}{5}$ **4** $\frac{4}{3}$

5 No solutions, $x = 2$ is an extraneous solution

6 $\frac{9}{7}$

7 No solutions, $x = 2$ is an extraneous solution

8 8, −9

9 3.303, −0.303 (3 d.p.)

Practice 4

1 1 **2** −1

3 No solutions, $x = 2$ is an extraneous solution

4 2 **5** −6

Practice 5

1 No solutions, $x = 2$ is an extraneous solution

2 8

3 No solutions, $x = 2$ is an extraneous solution

4 1, 1.5

5 $\frac{8}{5}$

6 ±4

Practice 6

1 a 30 **b** 20 cents (A is cheaper)

2 a 4 hours
b The other solution was −1 which does not make sense since you cannot have negative time
c 6 hours

3 4 mph

4 a If is equal to 20 when $d = 0$
b 99 m

5 80 km/h and 60 km/h

6 $8\frac{4}{7}$ min

7 1 m/s

Mixed practice

1 $-10 \pm 4\sqrt{7}$ **2** $\frac{11 \pm \sqrt{13}}{6}$

3 −4 **4** 18

5 $8 \pm 8\sqrt{3}$

6 a 2 and 3 **b** No **c** 2 and 4 **d** No

Review in context

1 a His speed upstream will be 2 km/h less due to the current; likewise his speed downstream will be 2 km/h greater.

b i $\frac{15}{x-2}$ **ii** $\frac{15}{x+2}$

c $\frac{15}{x-2} + \frac{15}{x+2} = 3$

d $(5 + \sqrt{29})$ km/h [10.39 km/h (2 d.p.)]

2 a 16 and 48 ohms **b** 3 and 6 ohms
c 3, 5 and 6 ohms **d** 4 and 12 ohms

3 3 km/h

4 a 22.5 cm **b** 2 cm **c** 1.83 cm (3 s.f.)

5 48.5 km/h (3 s.f.)

12.1

You should already know how to:

1 a $6x - 8x^2$ **b** $4 - 11x - 3x^2$

2 a $3xy$ **b** $2x^3 - 4x$

3 a 24 **b** 18

4 a 6 **b** 5

Practice 1

1 5, 25, 125, ...
Powers of 5 always end with a 5.
When the power is > 1, the last 2 digits are always 25.

2 The digit sum of a multiple of 9 will always be a multiple of 9.

3 The difference between the sums of the entries in the two diagonals will always be one.

4 The arithmetic mean will always be greater than or equal to the geometric mean.

5 a Student's own answers
b Student's own answers
c There will always be a remainder of 1.

6 a Student's own answers
b Student's own answers
c Student's own answers
d The product will always be equal to the LCM multiplied by the GCD.

7 a $14 = 7 + 7$, $16 = 13 + 3$, $18 = 11 + 7$, $20 = 17 + 3$
b i $30 = 23 + 7$
ii $98 = 61 + 37$
iii $128 = 97 + 31$
iv 2 can't be done.
c Every even number greater than 2 can be written as the sum of exactly two primes.

Practice 2

1 Their last digits repeat 4, 0, 6, 0 and they are always a multiple of 4

2 They are always a multiple of 3.

3 They are always a multiple of 24.

Practice 3

1 −1

2 1000

3 0.001

4 1000

5 1

Mixed practice

1 e.g. It is always a multiple of 6.

2 It is always a multiple of 5.

3 a $2^2 - 1 = 3$ which is prime.
$2^3 - 1 = 7$ which is prime.
$2^5 - 1 = 31$ which is prime.
$2^7 - 1 = 127$ which is prime.
b The evidence so far suggests that $2^n - 1$ will always be a prime whenever n is prime.
c $2^{11} - 1 = 2047 = 23 \times 89$

4 a 1000
b 1 000 000 000
c $\frac{5}{3}$
d 538

5 The total is 5 times the number in the center of the cross.

12.2

You should already know how to:

1 a 5, 6, 7 **b** 1, 3, 5 **c** 2, 4, 8

2 $u_{n+2} = 2u_{n+1} - u_n + 2$, $u_1 = 1$, $u_2 = 4$; $u_n = n^2$

3 a 3^7 **b** 6^2 **c** 2×5^n **d** 4^5

4 $x = 3$, $y = 9$

5 $173.64

Practice 1

1 a $u_n = 7n - 2$ **b** $u_n = 3n + 1$
c $u_n = 10 \times 4^{n-1}$ **d** $u_n = 3 \times \left(\frac{1}{3}\right)^{n-1}$
e $u_n = 20 \times \left(\frac{1}{4}\right)^{n-1}$

2 a 3, 5 **b** −2, 2 **c** 11, −4 **d** −13, −5

3 a 4, 5 **b** $\frac{1}{2}$, 7 **c** 15, 5
d $\frac{8}{3}, \frac{1}{3}$ **e** 4000, 10 **f** $7, \frac{1}{2}$

4 $l = 6.4 - 0.08n$; 4 km

5 a $t = 60 \times 2^n$ **b** 7 864 320

Practice 2

1 31

2 43

3 16, −6.5, −2.5

4 −3, −9

5 $\frac{1}{3}$, 27th

6 3, 648

7 2, 192

8 20, $u_n = 20 \times 1.5^{n-1}$, 11th

9 1st term = 4
1st sequence common difference = 3
2nd sequence common difference = 4

Practice 3

1 \$35 516.96

2 **a** Constant distance between rungs
b 20 cm, 15 cm
c 305 cm
d 22

3 **a** 12.5 cm
b 8.14 hours (3 s.f.)

4 **a** Each infected computer infects 5 more per day, meaning 6 per day are infected for each infected computer. These six then infect another 5 each meaning the total number of infected computers is 6 times more each day.
b 2
c On the 9th day

5 81 seats

6 **a** Deaths a month is 2 times the previous month.
b 63; $d = 63 \times 2^n$
c 2016
d e.g. The infection can double in size only for a short period of time because soon the population size will be exceeded.

Mixed practice

1 **a** 18 **b** $u_{n+1} = u_n + 18, u_1 = 7$ **c** $u_n = 18n - 11$
d 259 **e** 21st

2 **a** 38 **b** 139 **c** 531

3 **a** $31 = a + 2d$; $52 = a + 5d$
b 17, 7, 80

4 −13

5 **a** 30, 15 **b** No

6 **a** 6, 7 **b** $u_{n+1} = 7u_n, u_1 = 6$ **c** $u_n = 6 \times 7^{n-1}$
d 14 406 **e** 4 941 258 (8th term)

7 **a** 6, 0.5 **b** $u_{n+1} = 0.5u_n, u_1 = 6$
c $u_n = 6 \times 0.5^{n-1}$ **d** 0.0117

8 **a** $4a, a + 9$ **b** $4a = a + 9, a = 3$ **c** 5

9 **a** $r^2 = \frac{48}{12}$ hence $r^2 = 4$ **b** 2, −2 **c** 384, −384

Review in context

1 **a** 4, 6, 8 – sequence has a common difference
b $u_n = 2n + 2$
c 42
d 70

2 **a** Same scale factor means common ratio
b 232 831 (nearest person)
c 3 388 132 (nearest person)

3 \$183, \$58

4 **a** Constant number of components = common difference
b 2400
c 31 200

12.3

You should already know how to:

1

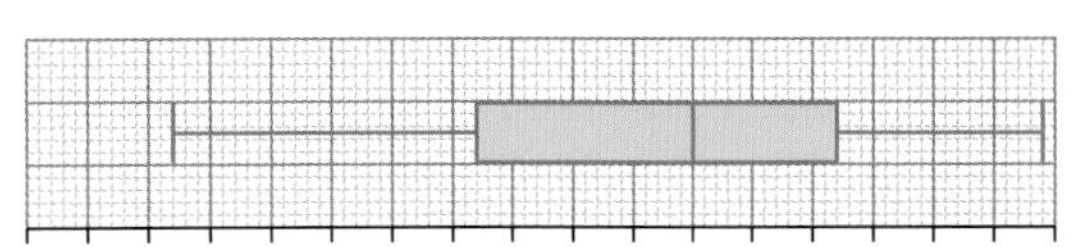

2

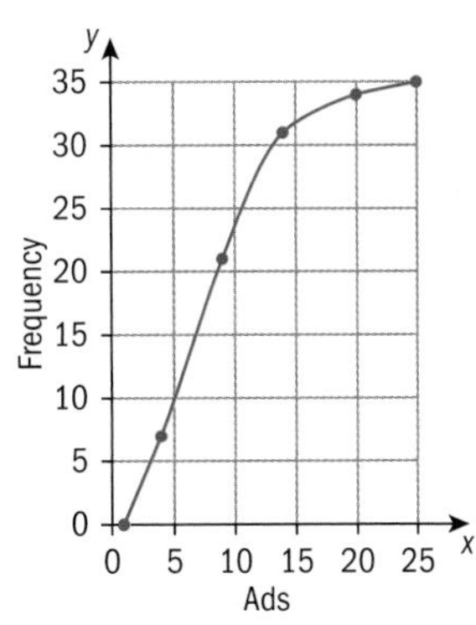

3 Mean = 1.667 to 3 d.p.
Median = 1.69
Range = 0.32
Lower Quartile = 1.58
Upper Quartile = 1.73
Interquartile Range = 0.15

Practice 1

1 **a** You don't want to break all the eggs because then there would be none left to eat.
b Either number the eggs and use a random number generator to choose which eggs to break open, or place the eggs in a large bag and take them without looking.

2 **a** Use a random number generator to generate 20 random numbers between 1 and 80.
Choose the corresponding apartments.
b Because the apartments with numbers that are 4 apart might all share some characteristic that would influence the number of people who live in it. For example, if there are four units per floor, the first apartment on each floor might happen to be the largest.
c The different types of apartment within the building.

3 A systematic sample. It may not be suitable, as he may not only have his friends on his contacts list.

4 **a** Number the employees from 1 to 900.
Use a random number generator to generate 40 random numbers between 1 and 900.
Choose the corresponding employees.
b Number the employees from 1 to 900.
Use a random number generator to pick a random starting point. Choose every 20th employee from that point.
c Take a random sample of 1 man and 2 women from pay grades A and B combined; 4 men and 3 women from pay grade C; 8 men and 32 women from pay grade D; 24 men and 16 women from pay grade E.
There are too few people on pay grade A for it to be considered separately.

5 a Yes. The school will have a register of all students.

b No. There is no complete list of the supporters of a team.

c Yes. The council keeps an electoral roll, which is a sampling frame for all voters.

d Yes, provided it was possible to access the government or city authority's employment registers.

Practice 2

1 In some cases alternative answers may be correct. Students must justify their answers.

a quota sampling, since it is not possible to form a sample frame of the leaves on the tree. The botanist could create categories by location, or tree size.

b stratified sampling divided into strata by age

c systematic sampling using the school register as a sampling frame.

d stratified sampling, taking 15 teachers, 2 administrative staff and 3 facilities staff.

e systematic sampling

2 Student's own list of advantages and disadvantages. They are not random samples, since in the first case everyone gets a feedback form, and in the second case only those that left their telephone number can be sampled.

3 a This sample could be very biased since height may affect athletic ability.

b This could be suitable, depending on the gender distribution within the group.

c Number the students from 1 to 150.
Use a random number generator to generate 20 random numbers between 1 and 150.
Choose the corresponding students.

d Divide the year group into natural subgroups, such as male/female or by month of birth. Take random samples from these subgroups in the same proportion as they appear in the population.

Practice 3

1 a Chemistry mean = 66.2, Physics mean = 66.8

b The tutor's claim is supported by the data but the difference between the two scores is very small. It would be better to say that the average scores in Physics and Chemistry are roughly the same.

c I have taken the scores in the last six tests to be representative of my ability.

2 a Quota sampling

b Pupils who are first to the canteen may not be representative of the whole population of pupils.

c Boys' mean spend: \$3.73
Girls' mean spend: \$3.14
This sample suggests the claim may be true.

d I have used the sample mean to infer things about the population mean.

Practice 4

1 a On average, the train is faster.

b The bus.

c He is more likely to be late using the train, due to the variability of the journey times.

2 a Programme A is faster.

b i Programme A. It never takes longer than 12.5 seconds.

ii Programme B. Programme A never takes less than 11.5 seconds but Programme B sometimes does.

iii Programme A. On average it is faster.

3 a Mean for Machine A = 1000.2, Mean for Machine B = 1005.3

b IQR for Machine A = 6.7, IQR for Machine B = 11.75

c On average, Machine A is more accurate and more consistent.

Mixed practice

1 Tom's marks are more consistent, but the IQR tells us nothing about who has done better. Tom may be consistently bad.

2 a On average, students in the chess club are taller, but they are also more variable.

b The IQR is not affected by extreme values.

c There is simply not enough information about the clubs to attempt to make these sorts of inferences; the ages and genders of the students within the clubs will have a large effect. Furthermore, while physical attributes are unlikely to have much effect on progression from youth chess clubs to adult chess playing, adult ballet dancers are selected on athletic and physical grounds.

3 a Quota sampling

b He will catch fish and measure them until he has 20 salmon and 30 trout.

4 e.g. The sample may be biased because the only people who will respond will be those who do not consider the survey 'unwanted'. People may have a similar attitude to surveys as they do to advertising. People who have strong feelings about unwanted advertising are likely to have asked to be removed from the directory.

Review in context

1 a

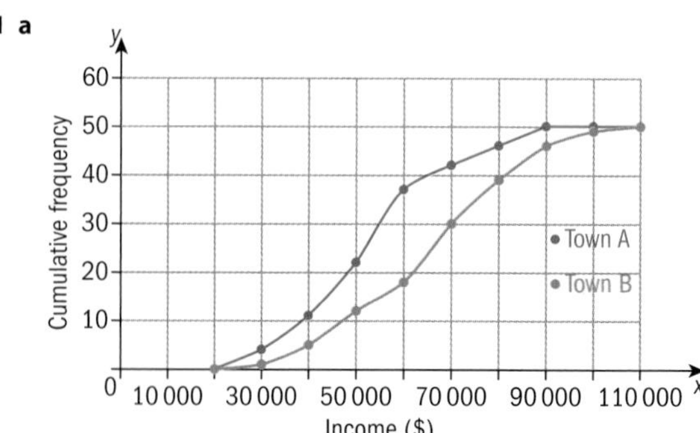

b On average, town B has higher household incomes. There is less variability though in town A.

c Different types of people were chosen for the sample in the same proportion as they appear in the population.

2 a

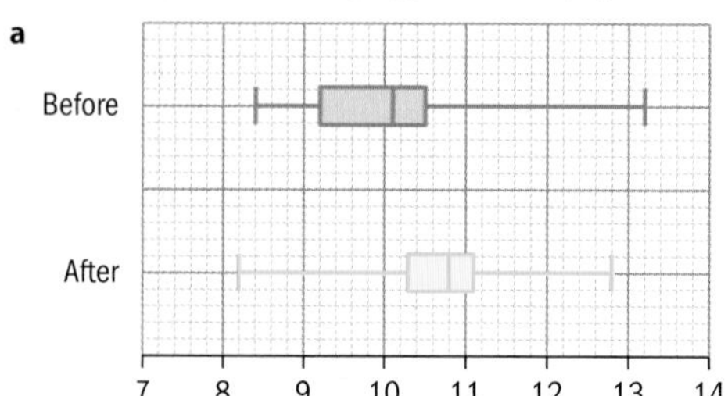

b Overall range similar; after data has smaller IQR so more consistent; after data has higher median.
c The intervention appears to have been effective.
d Small sample, limited age range covered.

3 a, b Student's own investigations.
c To take a census, you would use the data for every country in the world.

4 a The mean has increased from 5.6 to 6.4 hours, suggesting the scheme has been successful. Now, all residents report at least two hours of social contact, so everybody hits a minimum standard.
b Quota sampling.
c There are obvious flaws: those people using the bus already have a degree of mobility; just asking those who appear over 75 is not objective; asking if people would be interested is likely to bias the sample towards those who are likely to show improvement.
d e.g. To use a sampling frame from the electoral roll.

5 a e.g. People may not reliably remember what they donate to; people might be unwilling to share financial information.
b Create a stratified sample according to where the parents grew up.
c 40 local, 60 from elsewhere and 33 from different countries will make a stratified sample of 133.

13.1

You should already know how to:

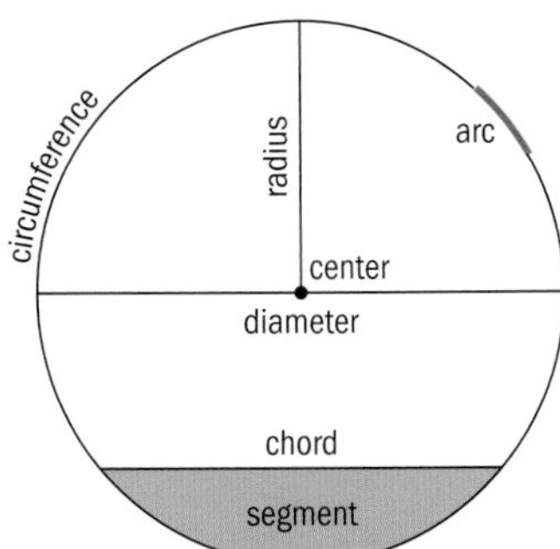

Practice 1

1 a $a = 41°$, $b = 49°$
b $c = 54°$, $d = 63°$
c $e = 130°$, $f = 50°$
d $g = 65°$, $h = 115°$
e $j = 58°$, $k = 64°$
f $m = 28°$, $n = 64°$

2 a $a = 50°$
$\angle ACB = 90°$ (Thales' Theorem)
$a = 180° - 90° - 40° = 50°$ (Angles in a triangle sum to 180°)
b $b = 45°$
$\angle DEF = 90°$ (Thales' Theorem)
$\angle EFD = 45°$ (Base angles in an isosceles triangle)
$b = 45°$ (Angles in the same segment)
c $c = 48°$
$\angle IKH$ = Angle $KHI = 76°$ (Base angles in an isosceles triangle)
$\angle IJK = 104°$ (Opposite angles in a cyclic quadrilateral are supplementary)
$\angle KIJ = 28°$ (Angles in the same segment subtended by equal chords are equal in size)
$c = 180° - 28° - 104° = 48°$ (Angles in a triangle sum to 180°)
d $d = 50°$
$\angle SRP = 70°$ (Alternate Segment Theorem)
$\angle PTR = 110°$ (Opposite angles in a cyclic quadrilateral are supplementary)
$\angle STR = 180° - 110° = 70°$ (Angles on a straight line)
$\angle SRT = 180° - 90° - 70° = 20°$ (Angles in a triangle sum to 180°)
d = Angle SRP – Angle $SRT = 70° - 20° = 50°$
e $e = 12°$
Construct the line WX.
$\angle UOV = 156°$ (The central angle is twice the inscribed angle)
$e = \angle OUV = \angle OVU = \frac{180 - 156}{2} = 12°$ (Base angles in an isosceles triangle)
f $f = 110°$
$\angle BAE = 180° - 70° - 20° = 90°$ (Angles in a triangle sum to 180°)
Therefore BE is a diameter of the circle and F is the center (Thales' Theorem)
Hence $AF = FB$ and $\angle FAB = \angle FBA = 70°$ (Base angles in an isosceles triangle)
$f = 180 - 70 = 110$ (Opposite angles in a cyclic quadrilateral are supplementary)

Practice 2

1 Draw a point D on the major arc between A and C.
$\angle ADC = \frac{1}{2} \times \angle AOC$ (The central angle is twice the inscribed angle)
$\angle ADC = 180° - \angle ABC$ (Opposite angles in a cyclic quadrilateral are supplementary)
Now equate the two expressions for Angle ADC.
$\frac{1}{2} \times \angle AOC = 180° - \angle ABC$
Therefore $\angle ABC = 180° - \frac{1}{2} \times \angle AOC$

2 Draw in the line OQ.
$\angle OYQ = \angle OXQ = 90°$
Consider the quadrilateral $OXQY$.
$\angle XOY = 2 \times \angle XPY$ (The central angle is twice the inscribed angle)
$\angle XOY = 360° - 90° - 90° - \angle XQY$ (Angles in a quadrilateral sum to 360°)
Equating the two expressions for $\angle XOY$ gives:
$2 \times \angle XPY = 180° - \angle XQY$
$\angle XPY = \frac{1}{2}(180° - \angle XQY)$

3 $\angle BDE = \angle CBE$ (Alternate Segment Theorem)
$\angle BED = \angle CBE$ (Alternate angles)
Therefore $\angle BDE = \angle BED$ and triangle BDE is isosceles.

4 $\angle BAC = \angle BDC$ (Angles in the same segment)
Both triangles share the angle at X.
Therefore $\angle DBX = \angle ACX$
(If the two triangles share two common angles, their other angles will be equal)
Triangles ACX and DBX are therefore similar.

5 $\angle XEY = \angle YEZ$ (Angles in the same segment subtended by equal chords are equal in size)
$\angle EYZ = \angle EYX = 90°$ (lines BE and DF are perpendicular)
Therefore triangles EXY and EZY are both right-angled triangles with $\angle XEY = \angle ZEY$.
Since $\angle XEY$ is complementary to $\angle EXY$ and $\angle ZEY$ is complementary to $\angle EZY$
$\angle EXY = \angle EZY$.

Mixed practice

1 a $a = 25°, b = 40°$ **b** $c = 84°$

2 a $a = 23°$ **b** $b = 132°$

3 a Converse: If $3a - 2 = 19$, then $a = 7$. Both statement and converse are true.
b Converse: If $a - b = 0$, then $a = b$. Both statement and converse are true.
c Statement is true (as wingless Moa bird is now extinct.) Converse: If it has wings, then it is a bird is false. (e.g. it could be an insect.)
d Converse: If a polygon is a square, then it has four sides is true. The statement is false (e.g. it could be a rectangle) but the converse is true.
e Converse: If $c^2 = 81$, then $c = 9$. The statement is true but the converse is false (c could be -9).
f Converse: If a right-angled triangle's hypotenuse is the diameter of a circle, then its vertices will be on the circumference of the circle. Both statement and converse are true.

4 $\angle ADC = 180° - \angle ABC$ (Opposite angles in a cyclic quadrilateral are supplementary)
Since $\angle ABC = \angle DAB$, $\angle DAB + \angle ADC = 180°$
Therefore AB and DC are parallel.
$\angle BAC = \angle ACD$ (alternate angles)
Therefore $AD = BC$ (Equal angles are subtended by equal chords)

5 Draw in lines OA and OC.
$\angle OCB = \angle OBC$ (Base angles in an isosceles triangle)
$\angle OAB = \angle OBA$ (Base angles in an isosceles triangle)
Since $\angle OBA = \angle OBC$, all four of these angles are equal.
$\angle OAC = \angle OCA$ (Base angles in an isosceles triangle)
So therefore the two base angles in triangle ABC are equal, since $\angle OAC + \angle OAB = \angle OCA + \angle OCB$.
Therefore triangle ABC is isosceles.

Review in context

1 44°

2 4.94 m

13.2

You should already know how to:

1 $(2x - 3)(x + 1) \Rightarrow x = \frac{3}{2}$ or $x = -1$

2 12 cm

Practice 1

1 a $x = 8$ **b** $x = 6$ (6, 12)
c $x = 7$ (6, 8) **d** $x = 8$ (5, 8)
e $x = 9$ (6, 8, 9, 12) **f** $x = 3$ (1, 2, 3) or $x = 4$ (2, 3, 4)

Practice 2

1 a $x = 3$
b 5 (2, 3, 4)
c 2 (2, 3, 4, 12)

Mixed practice

1 a Student's own sketch
b AX and BX take values 1 and 10 or vice versa. CX and DX take values 2 and 5 or vice versa.

2 a $x = 5$ **b** $x = 9$
c $x = 2$ **d** $x = 1, y = 4$
e $x = 18$ **f** $x = 5$

3 $x = 8, y = 4.5, z = 7.5$

4 $x = 3.6, y = 6.4$

5 $x = 9, y = 12$

6 a Student's own demonstration. **b** $x = 4, y = 3$

Review in context

1 a Student's own sketch.
b Student's own demonstration.
c 25.25 cm

2 a 370 m **b** 381 m
They are very similar.

14.1

You should already know how to:

1 a 1 **b** 16 **c** $\frac{1}{4}$ **d** $\frac{1}{9}$

2 a

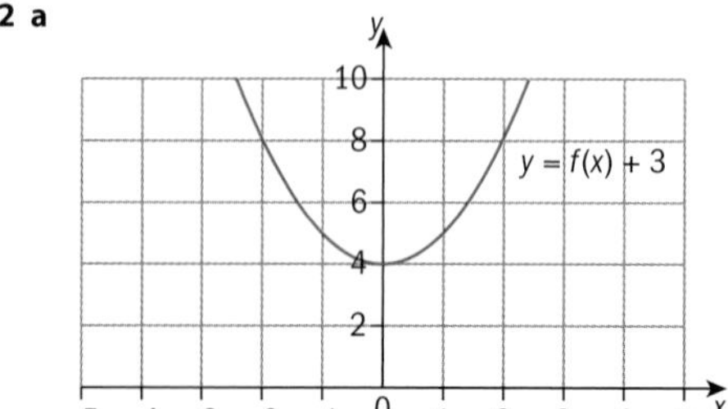

b

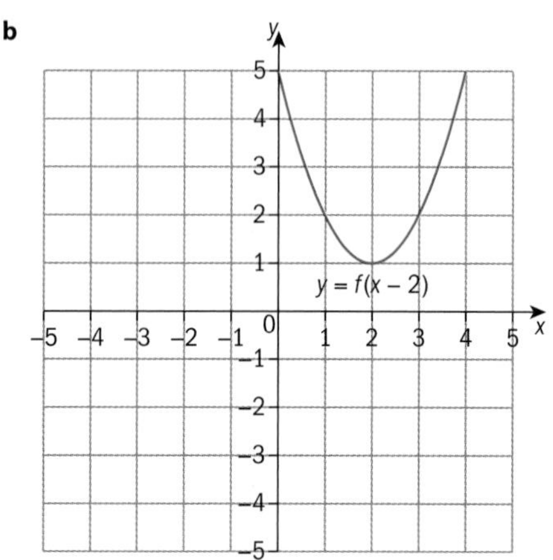

c

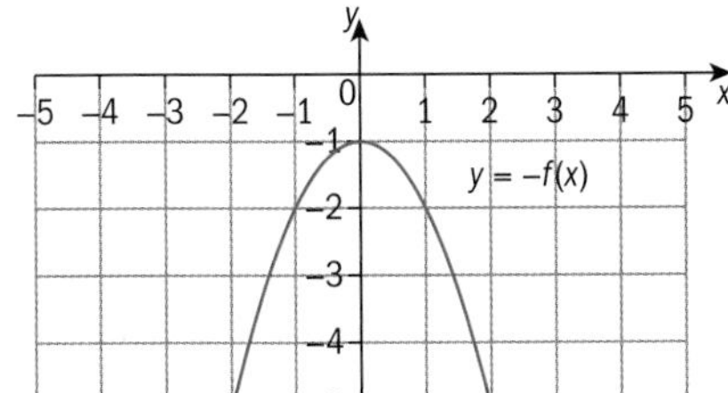

d

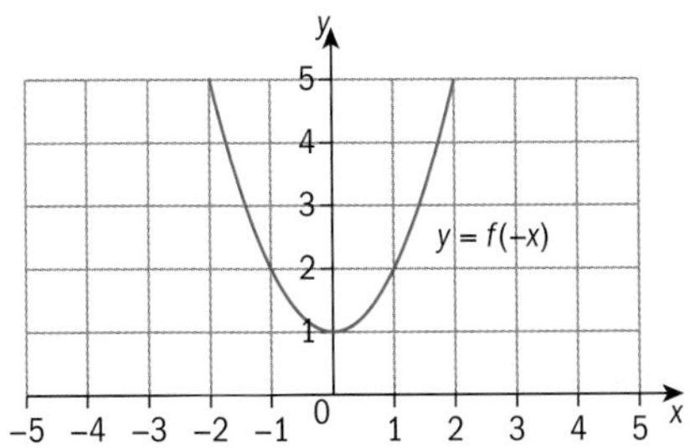

e

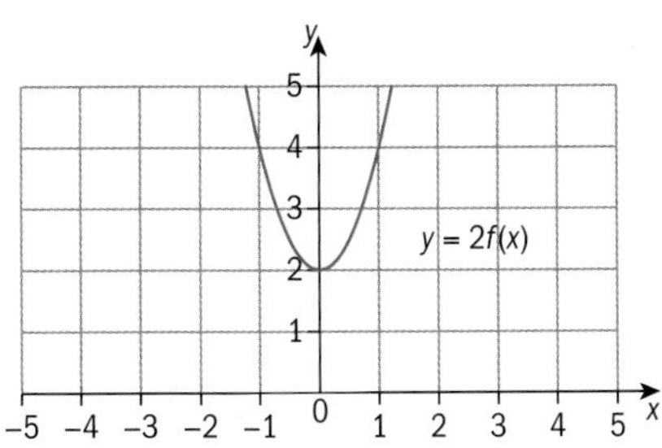

f

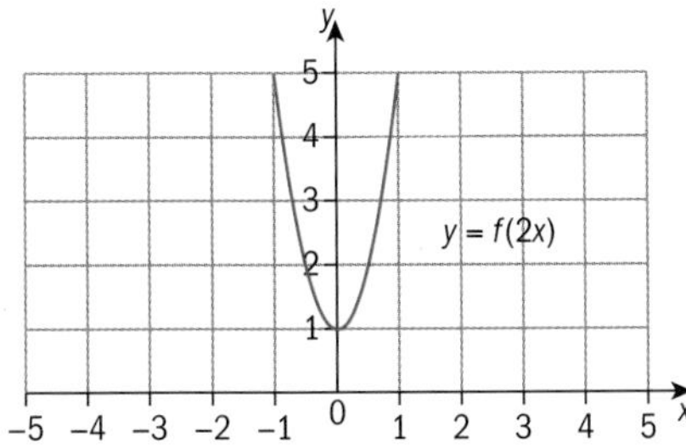

Practice 1

1 a $f(x) = 2^x$
b 32 768
c i 1 073 741 824 **ii** 20 min

2 a $f(x) = 10\,000\left(\frac{1}{2}\right)^x$

0	1	2	3	4	5
10000	5000	2500	1250	625	313

b 20
c > 66.4 minutes

3 2.4 hours

Practice 2

1 a i 1.43 **ii** 0.43 **iii** 1
b i 0.62 **ii** 0.38 **iii** 3
c i 0.667 (3 s.f.) **ii** 0.333 (3 s.f.) **iii** 4
d i 1.2 **ii** 0.2 **iii** 0.8

2 a 1.12 **b** $y = 300 \times 1.12^x$ **c** £1642.07

3 a $y = 470 \times 0.78^x$ **b** 64.40

4 100 468

5 a i 1.12 **ii** $y = 600 \times 1.12^x$ **iii** 3284, 17976

b

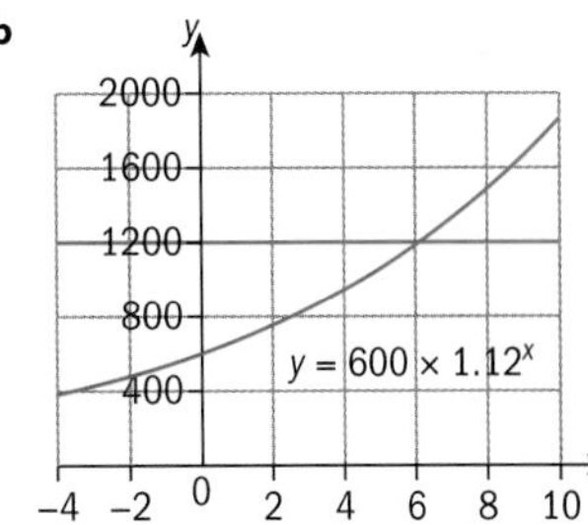

Just over 6 weeks

6 $y = 10000 \times 0.91^x$
a i 7536 **ii** 3894
b 7.35 years

7 18 years

8 20 years

9 38 hours

Practice 3

1 a i Horizontal translation 2 units in the negative direction.
ii $g(x) = 3^{x+2}$
iii

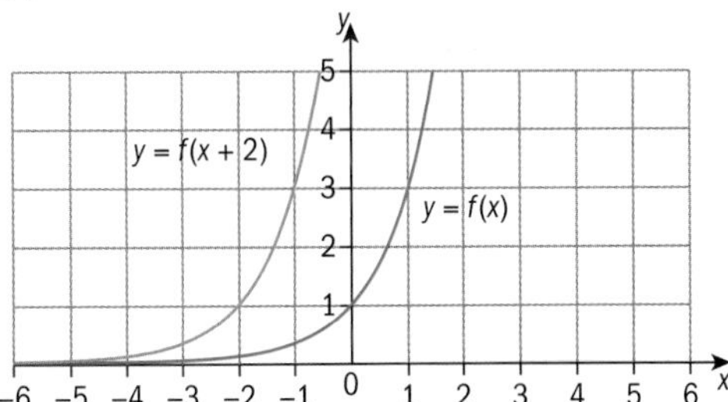

b i Reflection in the x–axis
ii $g(x) = -3^x$
iii

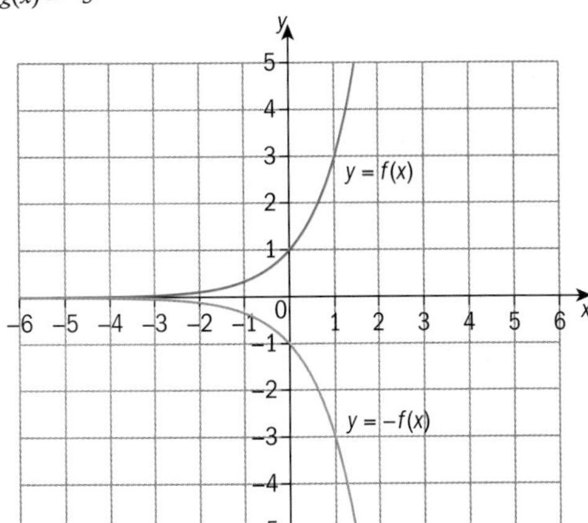

c i Vertical translation 2 units in the negative direction

ii $g(x) = 3^x - 2$

iii

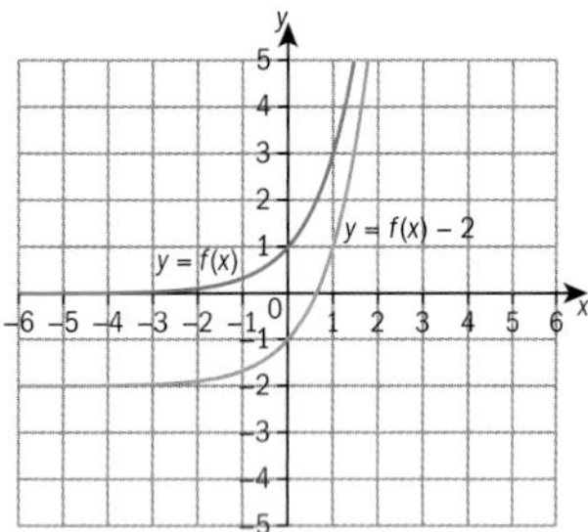

d i Reflection in the y-axis

ii $g(x) = 3^{-x}$

iii

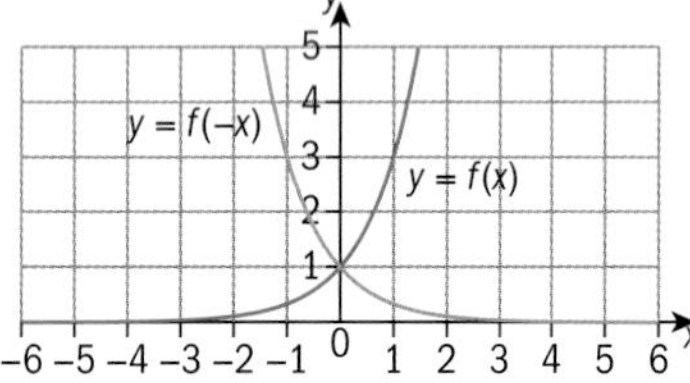

e i Horizontal translation 1 unit in the positive direction and vertical translation 3 units in the positive direction.

ii $g(x) = 3 + 3^{x-1}$

iii

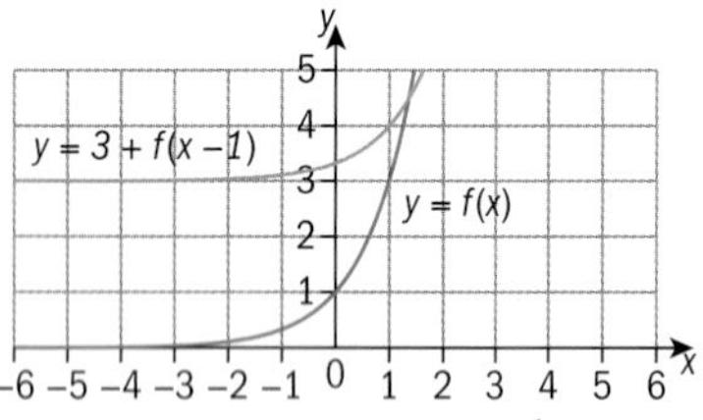

f i Horizontal dilation scale factor $\frac{1}{2}$

ii $g(x) = 3^{2x}$

iii

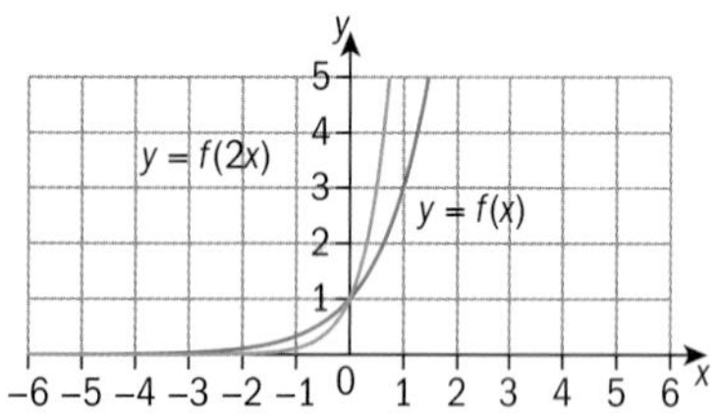

g i Vertical dilation scale factor 2

ii $g(x) = 2 \times 3^x$

iii

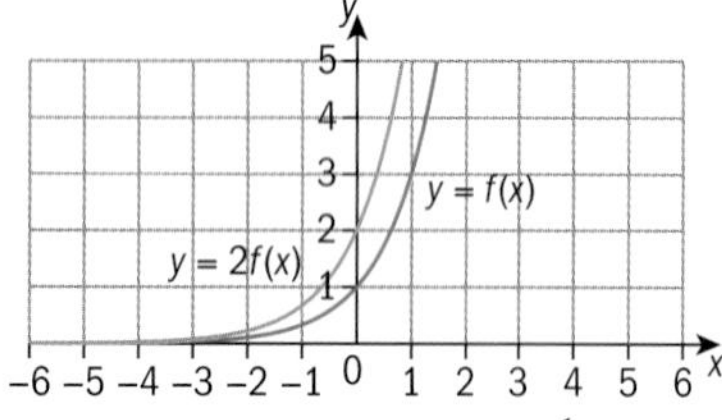

h i Horizontal dilation scale factor $\frac{1}{2}$

ii $g(x) = 3^{0.5x}$

iii

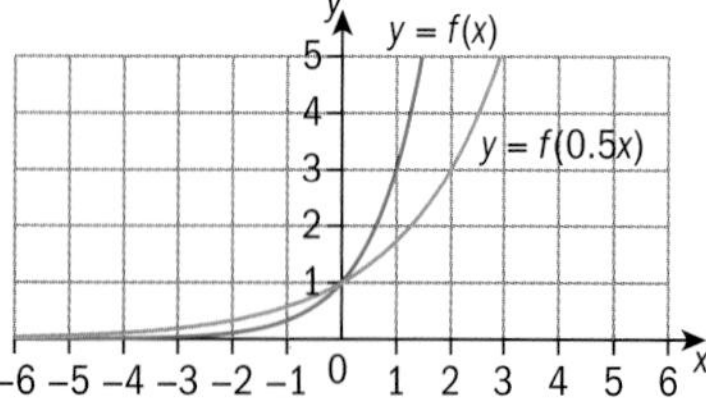

i i Vertical dilation scale factor 2 and reflection in the x-axis

ii $g(x) = -2 \times 3^x$

iii

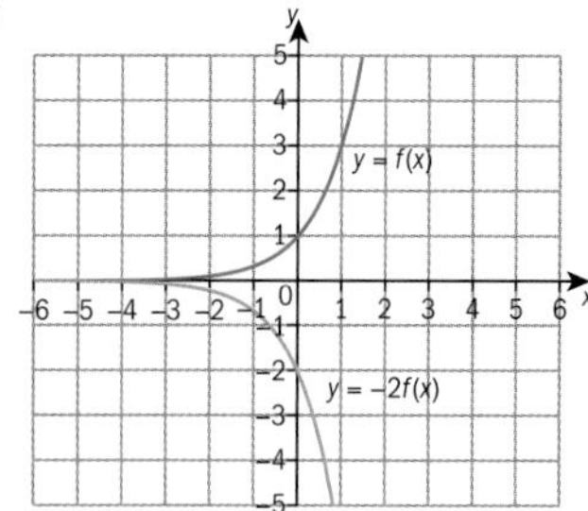

j i Reflection in the x-axis, and horizontal dilation scale factor $\frac{1}{2}$

ii $g(x) = 3^{-2x}$

iii

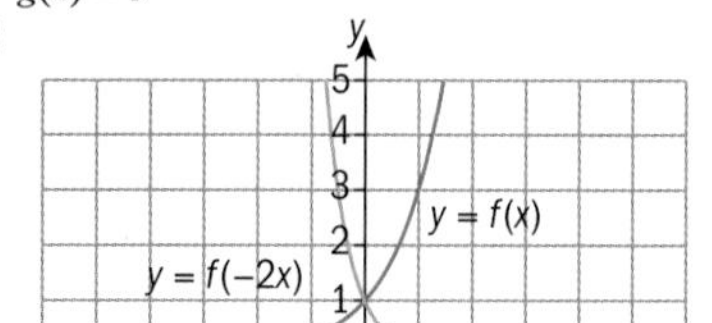

2 a i Reflection in the x–axis

ii $g(x) = -f(x)$

iii

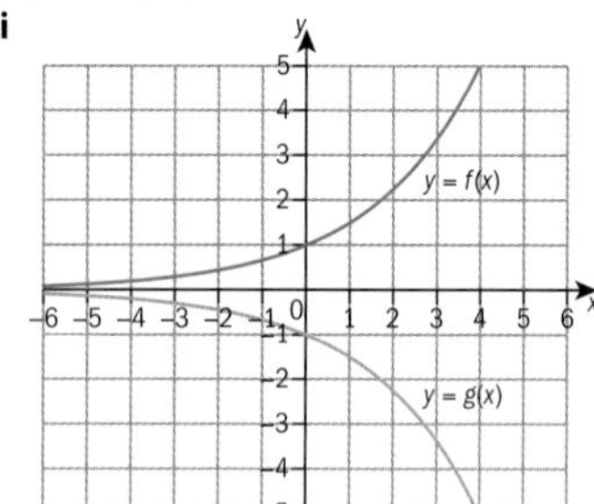

b i Reflection in the y–axis

ii $g(x) = f(-x)$

iii

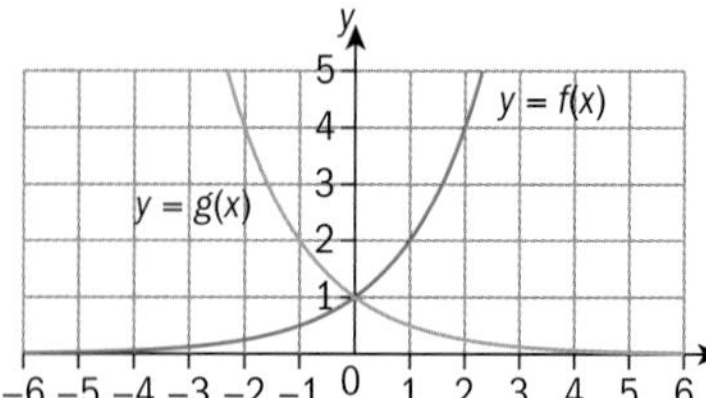

c i Horizontal translation 2 units in negative direction and vertical translation 3 units in negative direction.

ii $g(x) = f(x+2) - 3$

iii

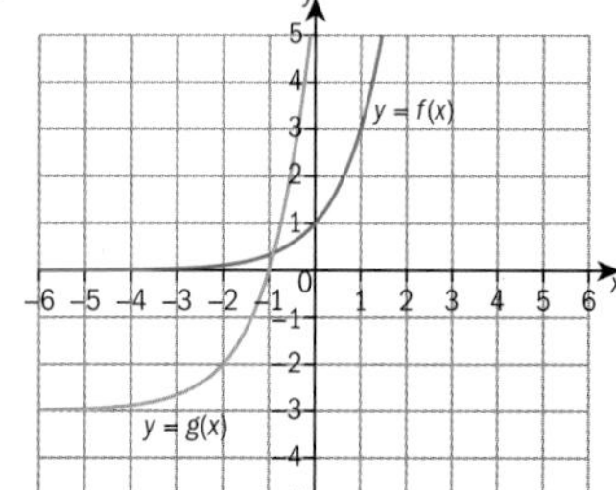

d i Horizontal dilation scale factor $\frac{1}{2}$

ii $g(x) = f(2x)$

iii

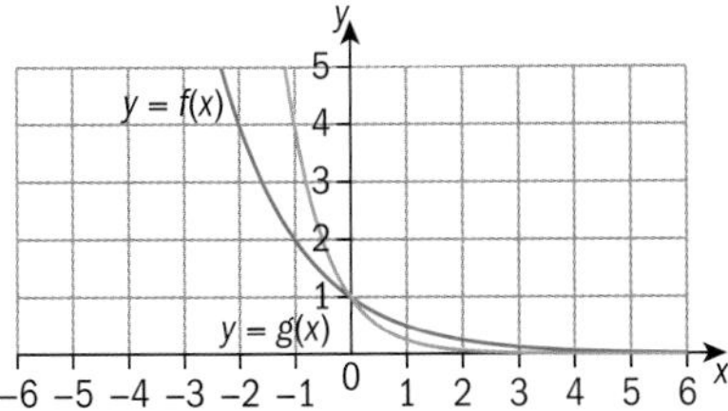

e i Horizontal dilation scale factor 2.

ii $g(x) = f(0.5x)$

iii

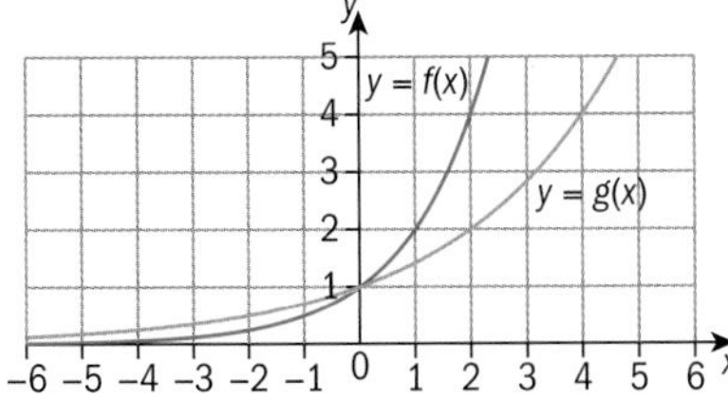

f i Vertical dilation scale factor 3

ii $g(x) = 3f(x)$

iii

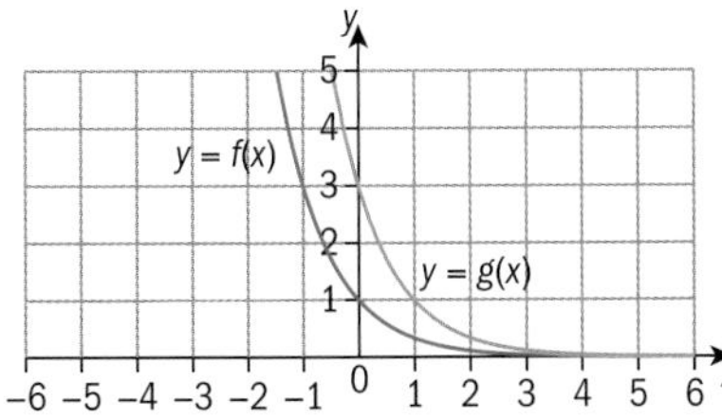

g i Reflection in the y-axis, and then horizontal dilation scale factor $\frac{1}{2}$

ii $g(x) = f(-2x)$

iii

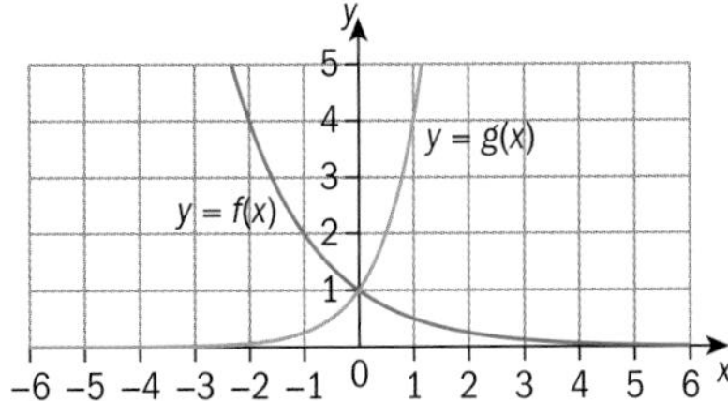

3

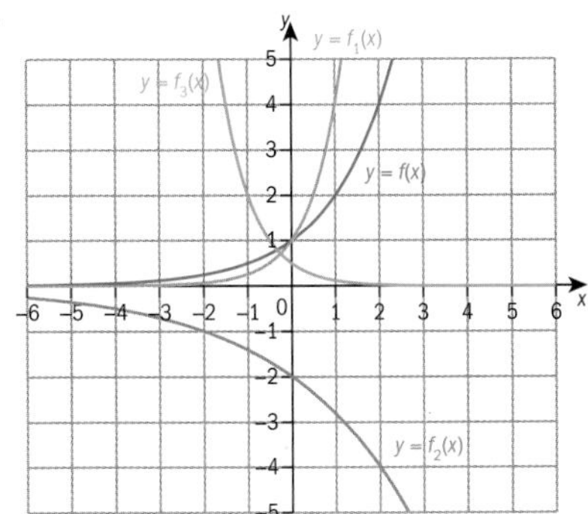

Mixed practice

1 a 0.85 **b** $y = 18\,000 \times 0.85^x$ **c** \$8000

2 81 920 **3** 9371 **4** 4 grams **5** 57 600

6 a 1.014 **b** $y = 400 \times 1.014^x$ **c** 50

7 a 2^{-x} **b** $2^{x+3} - 4$ **c** -2^x **d** $2^{\frac{x}{3}}$ **e** 4×2^x

8

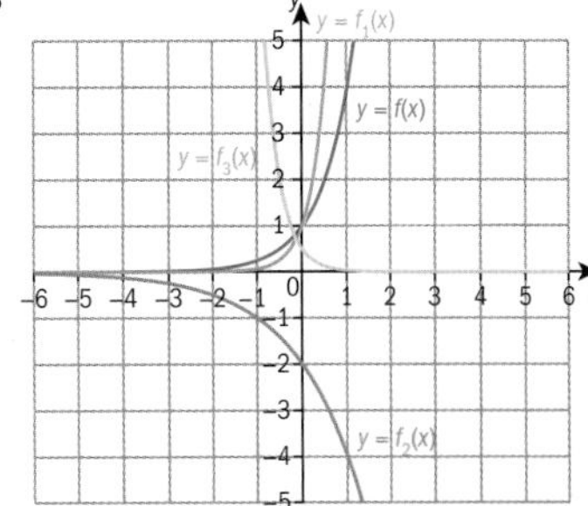

9 a $f(x) = 3^{2x}$ Horizontal dilation scale factor $\frac{1}{2}$

b $f(x) = 81^{\frac{x}{2}}$ Horizontal dilation scale factor 2

Review in context

1 $N = N_0 \times 0.5^x$

2 7500

3 11 400

4 13.4%

5 23 136

6 1.42%

7 18 935 to 24 635

8 The amount of carbon left is approximately 0.23%, a very small amount

14.2

You should already know how to:

1 a $\frac{5}{13}$ **b** $\frac{12}{13}$

2 a 0.707 **b** 0.799 (3 s.f.)

3 a

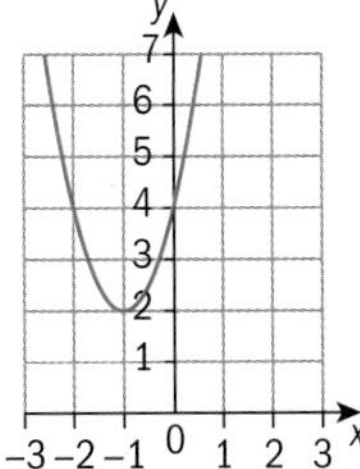

b

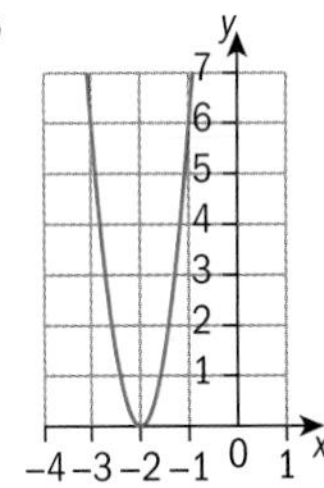

c

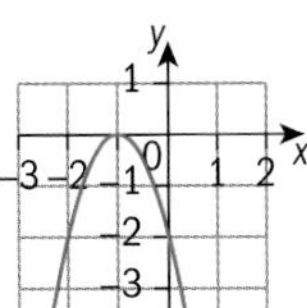

d

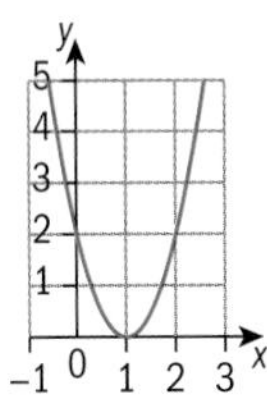

e

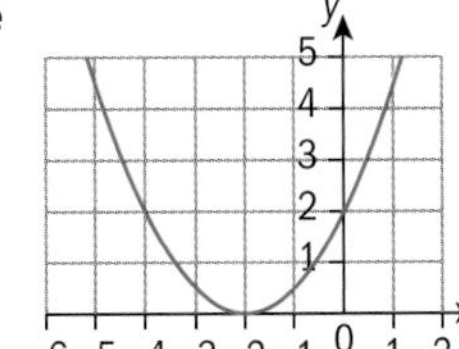

f

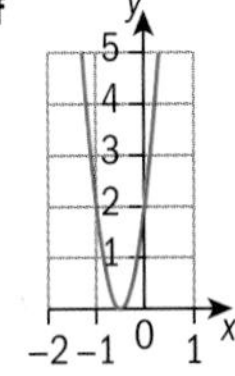

Practice 1

1 a

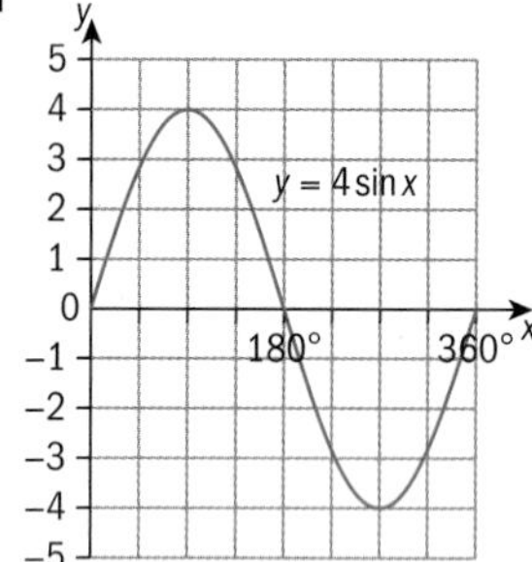

b

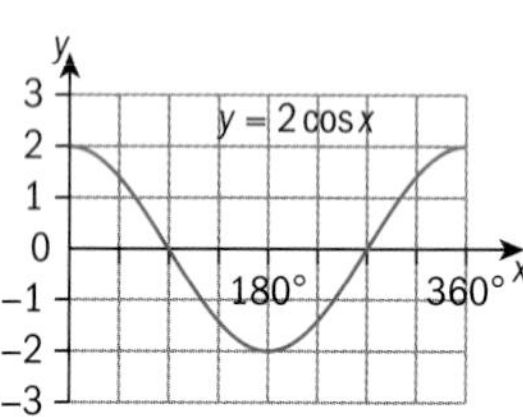

2 a $y = 3\sin x$ **b** $y = 5\cos x$
c $y = 0.25\cos x$ **d** $y = 1.5\sin x$

3 a $y = 2\cos x$
b $y = 0.5\sin x$ and $y = 2\sin x$
c $y = 2\cos x$ and $y = 4\cos x$
d $y = 0.5\cos x$

Practice 2

1 a Period 120°, Frequency 3

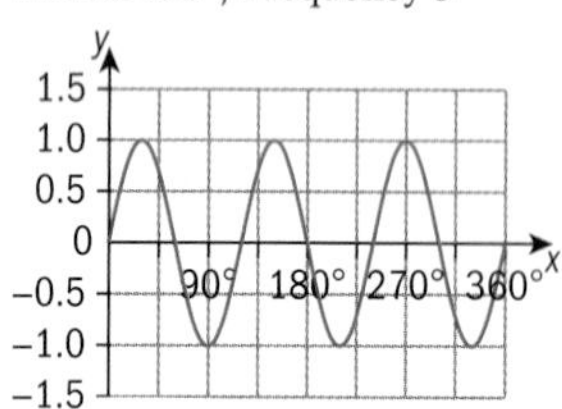

b Period 90°, Frequency 4

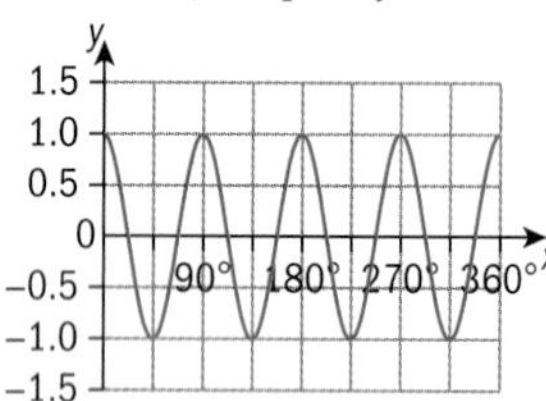

2 a Period 720°, Frequency $\frac{1}{2}$

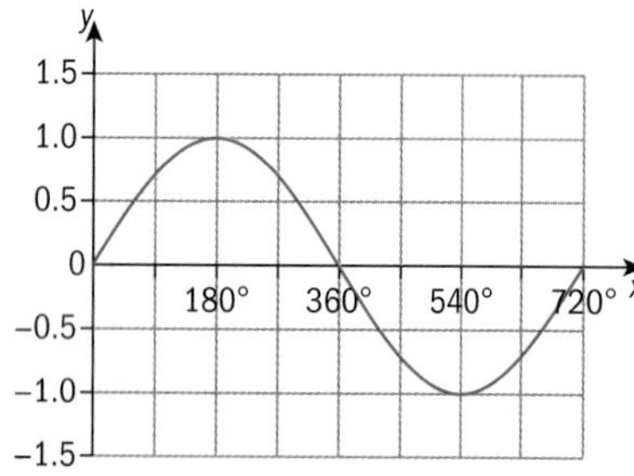

b Period 1080°, Frequency $\frac{1}{3}$

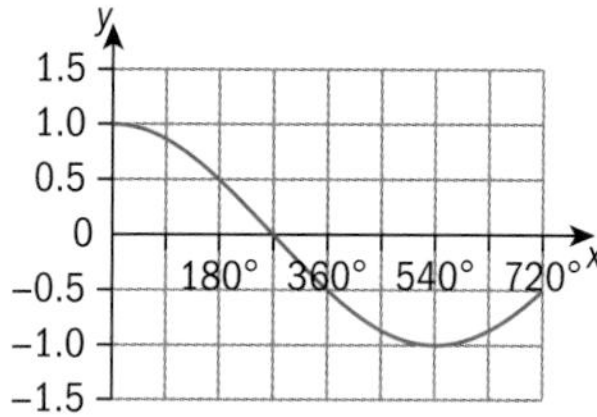

3 a $y = \sin\left(\frac{x}{3}\right)$ **b** $y = \cos(3x)$
c $y = \cos\left(\frac{x}{4}\right)$ **d** $y = \sin(4x)$

4 a $y = 3\sin\left(\frac{x}{2}\right)$ **b** $y = 0.5\cos(3x)$
c $y = 0.5\cos(0.5x)$ **d** $y = 3\sin(2x)$

Practice 3

1 a Amplitude 2, Period 90°, Frequency 4

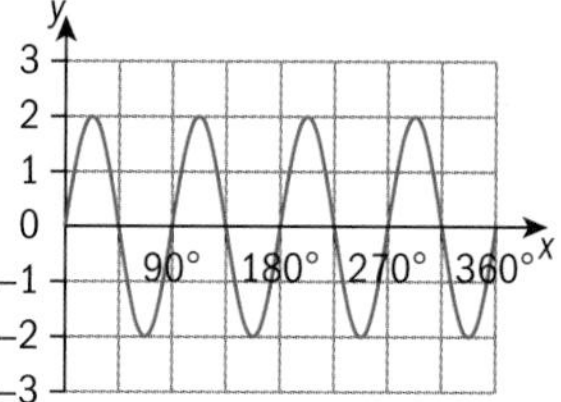

b Amplitude 4, Period 120°, Frequency 3

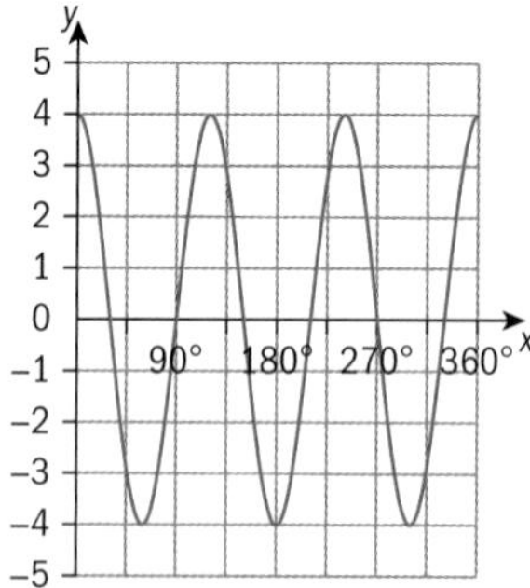

c Amplitude $\frac{1}{3}$, Period 1080°, Frequency $\frac{1}{3}$

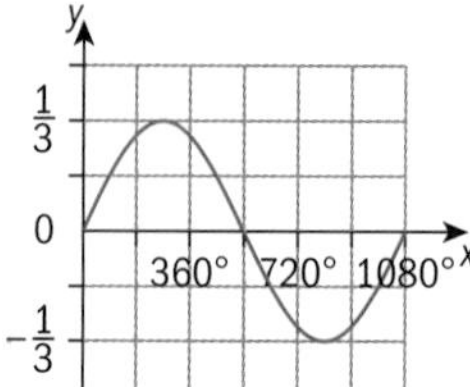

d Amplitude 0.7, Period 1440°, Frequency $\frac{1}{4}$

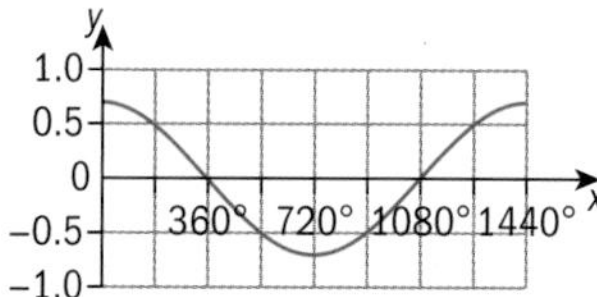

Practice 4

1 a Vertical translation 7 units in the positive direction
  b Reflection in the x-axis
  c Vertical translation 2 units in the negative direction
  d Reflection in the y-axis

2 a $y = \sin x + 2$ b $y = -\sin x$
  c $y = -\cos x$ d $y = \sin x - 4$

3 a Amplitude 1, Period 360°

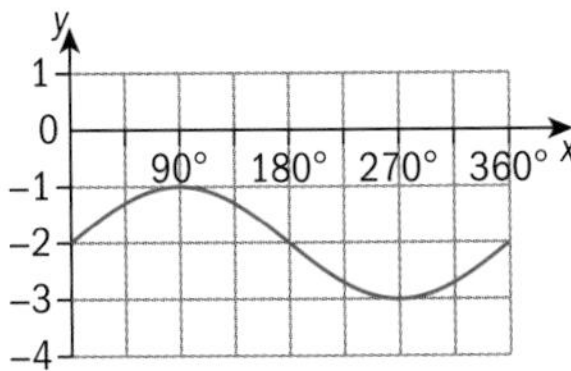

  b Amplitude 1, Period 360°

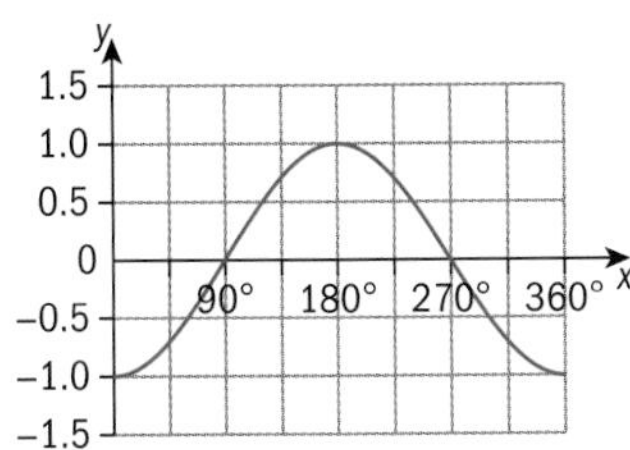

  c Amplitude 1, Period 360°

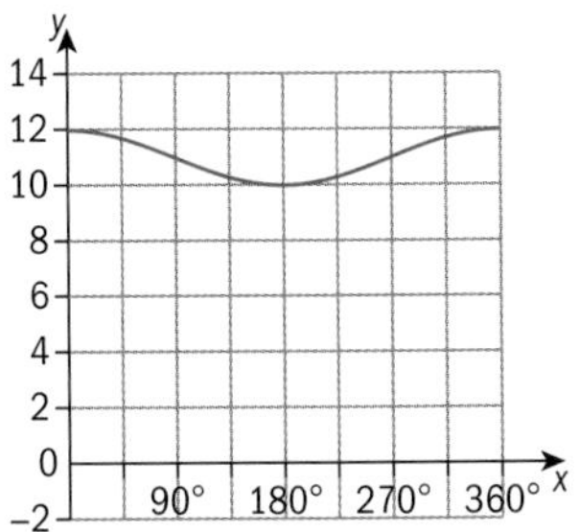

  d Amplitude 1, Period 360°

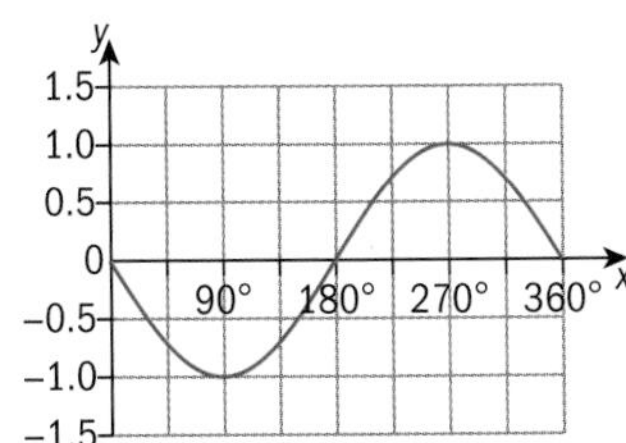

  e Amplitude 1, Period 360°

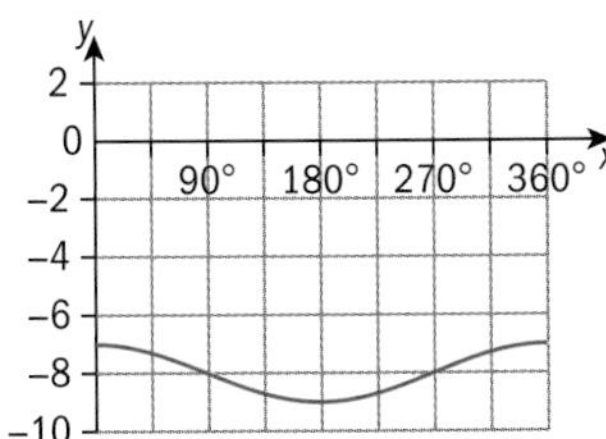

  f Amplitude 1, Period 360°

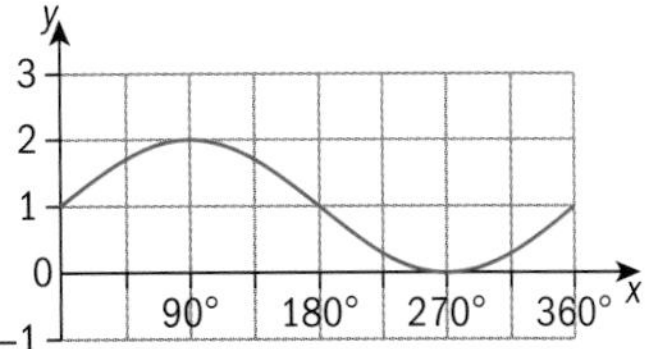

4 a Amplitude 1, Period 360°, Frequency 1
  b $y = \sin x + 4$
  c Move down 3 units
  d $y = \sin x + 1$

5 Both could be correct – both transformations on the graph of the given function are equivalent.

Practice 5

1 a Vertical dilation scale factor 4 and reflection in the x-axis
  b Vertical dilation scale factor 2 and vertical translation 9 units in the negative direction
  c Vertical dilation scale factor 0.5 and horizontal translation 4 units in the positive direction
  d Horizontal dilation scale factor 2 and vertical translation 3 units in the positive direction
  e Horizontal dilation scale factor 3 and reflection in the x-axis
  f Horizontal dilation scale factor $\frac{1}{2}$ and reflection in the y-axis

2 a Vertical dilation scale factor 3 and reflection in the x-axis
  b Vertical dilation scale factor 2 and vertical translation 1 unit in the positive direction
  c Horizontal dilation scale factor 2 and vertical translation 2 units in the positive direction
  d Vertical dilation scale factor 0.5 and vertical translation 2 units in the negative direction
  e Horizontal dilation scale factor $\frac{1}{3}$ and reflection in the x-axis
  f Horizontal dilation scale factor 2, reflection in the x-axis and vertical translation 3 units in the negative direction

3 a $y = 3\sin(2x)$ b $y = -3\sin\left(\frac{x}{2}\right)$
  c $y = -\cos x + 3$ d $y = -3\sin\left(\frac{x}{2}\right) - 2$
  e $y = 0.5\cos(2x) + 1$ f $y = -2\cos(3x) + 2$
  g $y = -2\sin\left(\frac{x}{2}\right) + 1$ h $y = -1.5\cos x + 3$

Practice 6

1 a i 1 liters/s b 2.5 liters/s c 0 liters/s
  b Exhaling
  c Period is 4 seconds – represents your breathing rate
  d Amplitude = 2, Max velocity = 2 liters/s
  e

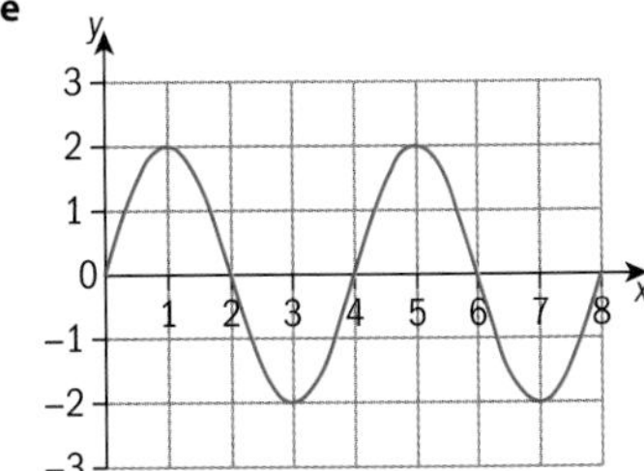

2 $y = 0.5\sin(90°t) + 2$

3 a $y = 2.5\cos(2x) - 0.5$ **b** $y = 3\sin(18°x) + 1$
c $y = 2\cos(6x) - 3$ **d** $y = -3\sin 12°x + 4$

4 a 2 m
b

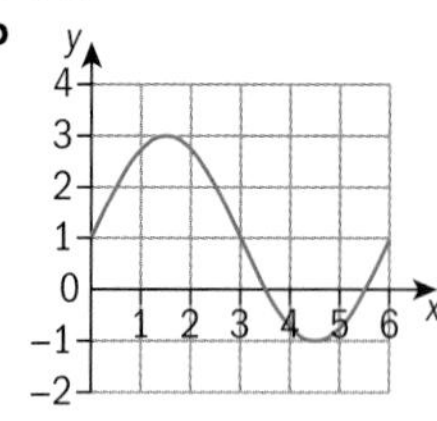

c No – max height is 3 m

5 a 1 m **b** 7 m **c** 2 seconds
d 1 m **e** 8 seconds **f** 3 m
g $y = -3\cos(45°t) + 4$

6 a 0, 0, 0, 0, 0
b

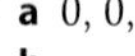

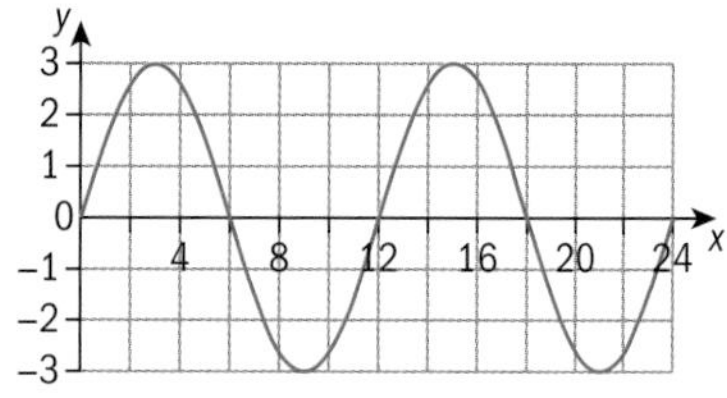

c 3 am and 3 pm, both 3 meters
d 2.60 m
e From 1:24 to 4:36 (both am and pm)

7 a $y = 1.6\cos(30°t) + 1.7$
b

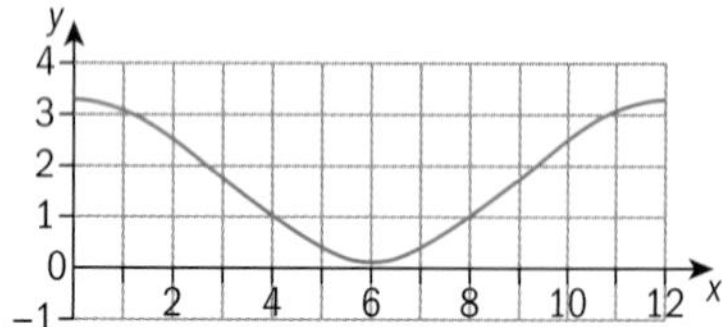

Mixed practice

1 a Amplitude 4, Period 360°, Frequency 1, $y = 4\cos x$
b Amplitude 2, Period 90°, Frequency 4, $y = 2\cos(4x)$
c Amplitude 0.5, Period 120°, Frequency 3, $y = 0.5\cos(3x)$

2 a Amplitude 0.4, period 180°

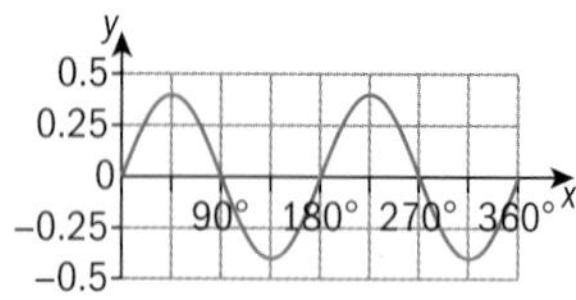

b Amplitude 7, period 72°

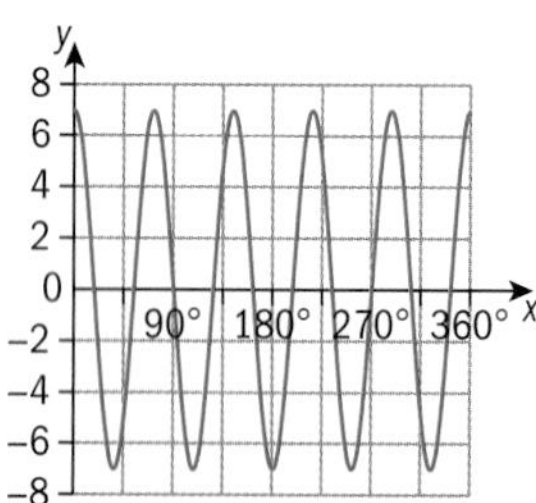

c Amplitude $\frac{2}{3}$, period 45°

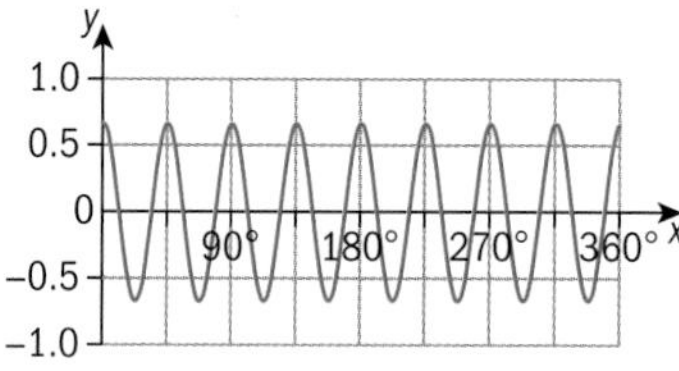

d Amplitude $\frac{5}{2}$, period 120°

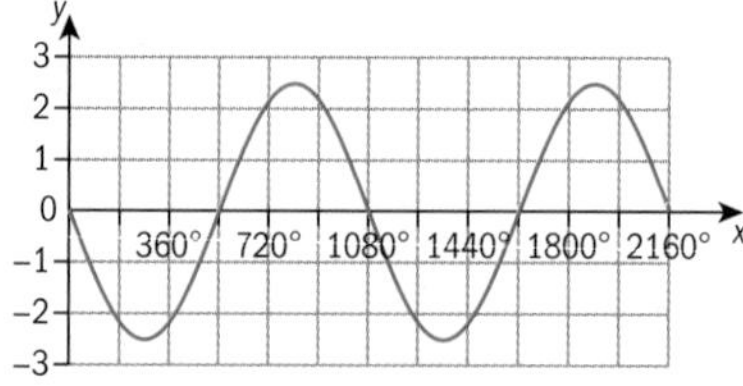

3 a Amplitude 4, Period 120°, Frequency 3

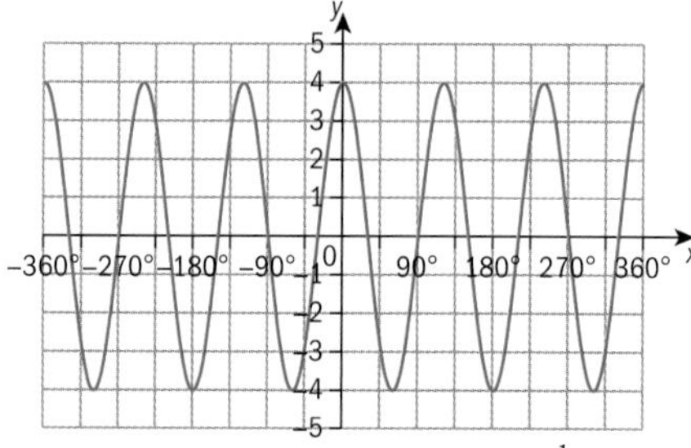

b Amplitude 1, Period 720°, Frequency $\frac{1}{2}$

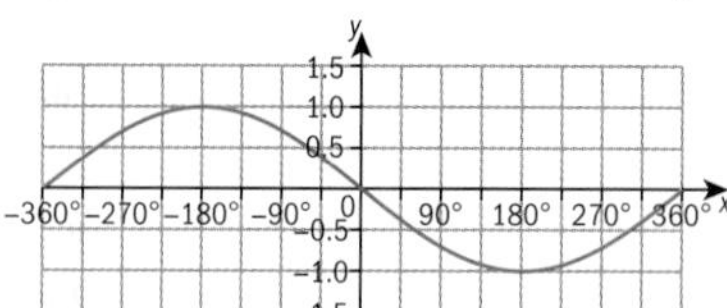

c Amplitude 0.5, Period 180°, Frequency 2

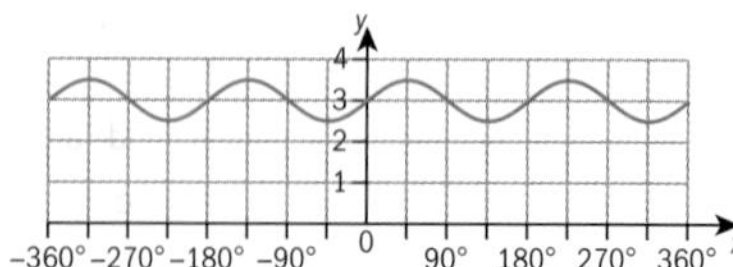

d Amplitude 3, Period 720°, Frequency $\frac{1}{2}$

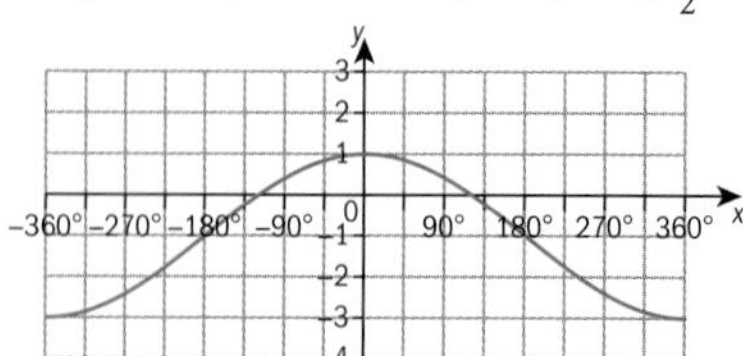

4 $y = -2\sin\left(\frac{x}{2}\right)$

5 a Amplitude 1, period 360°

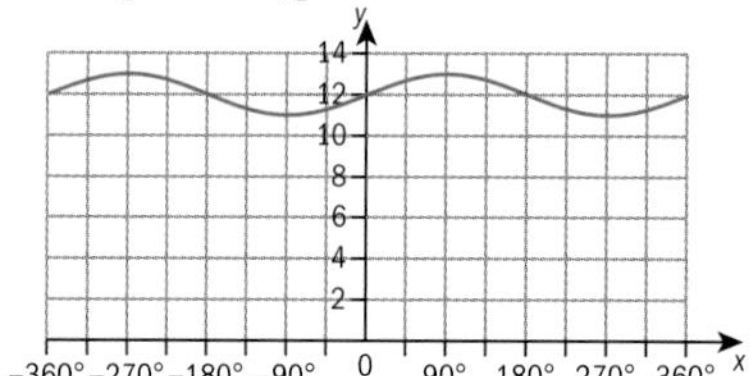

b Amplitude 1, period 360°

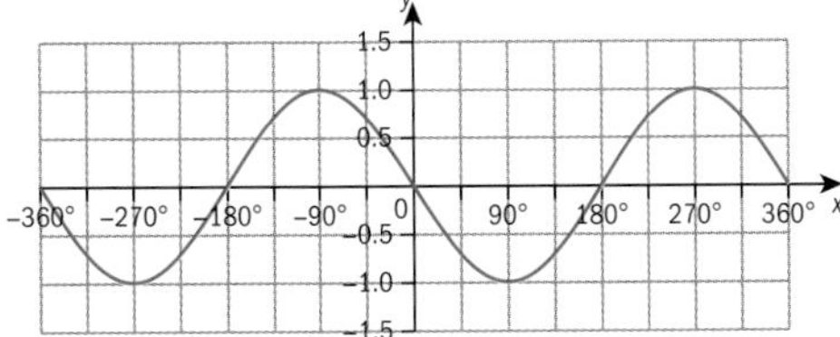

c Amplitude 1, period 360°

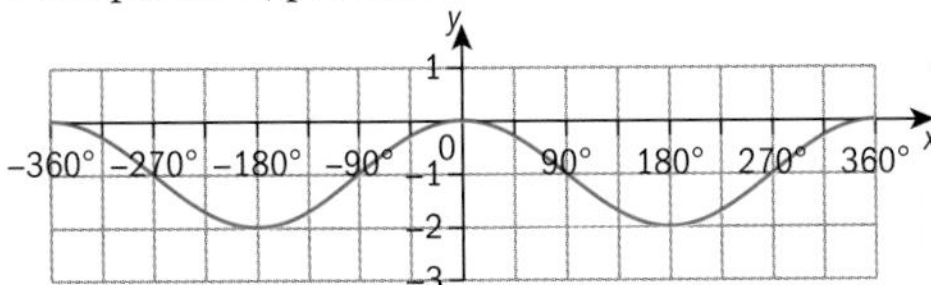

d Amplitude 1, period 360°

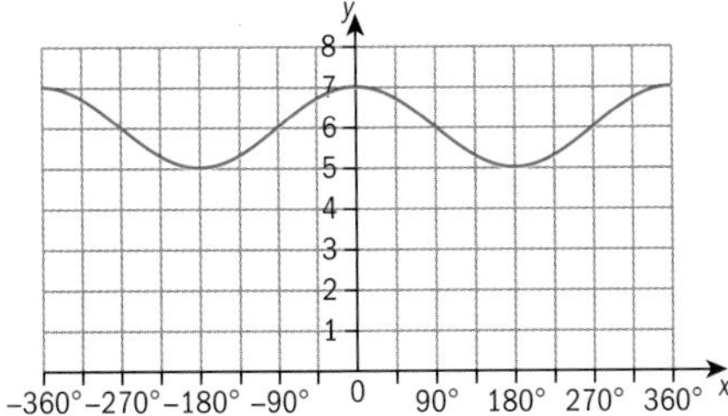

6 a Amplitude 1, Period 360° $y = \sin x + 3$

b Amplitude 1, Period 360° $y = \cos x - 3$

c Amplitude 1, Period 360° $y = -\cos x$

d Amplitude 1, Period 360° $y = \sin x - 1$

7 a Vertical dilation scale factor 3 and vertical translation 2 units in the positive direction

b Vertical dilation scale factor 2, reflection in the x-axis, and horizontal dilation scale factor $\frac{1}{4}$

c Horizontal dilation scale factor $\frac{1}{2}$, reflection in the x-axis, vertical translation 3 units in the negative direction

d Horizontal dilation scale factor $\frac{1}{2}$, vertical dilation scale factor $\frac{1}{2}$, and vertical translation 4 units in the negative direction

e Horizontal dilation scale factor $\frac{1}{3}$, and vertical dilation scale factor 3, reflection in the x-axis, and vertical translation 3 units in the positive direction

f Horizontal dilation scale factor $\frac{4}{3}$, vertical dilation scale factor $\frac{3}{4}$.

8 a $y = 2\sin x - 1$ **b** $y = -\cos(2x) - 2$

c $y = -\sin\left(\frac{x}{2}\right) + 1$ **d** $y = \sin(4x) + 1$

e $y = -2\cos(2x) - 2$

9 a i $y = -10\cos(3.6°t) + 12$

ii

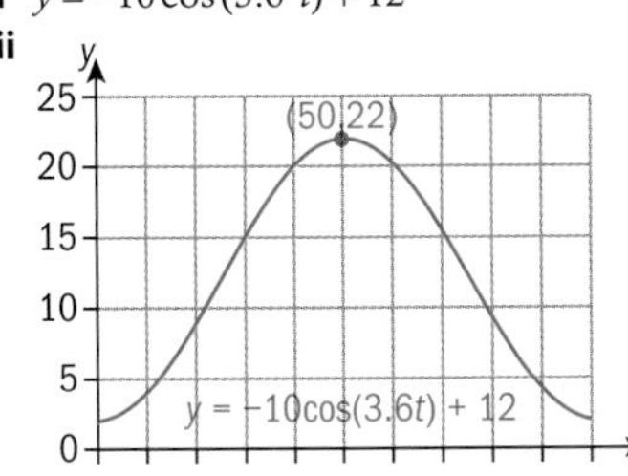

b i $y = -5\cos(3°t) + 6$

ii

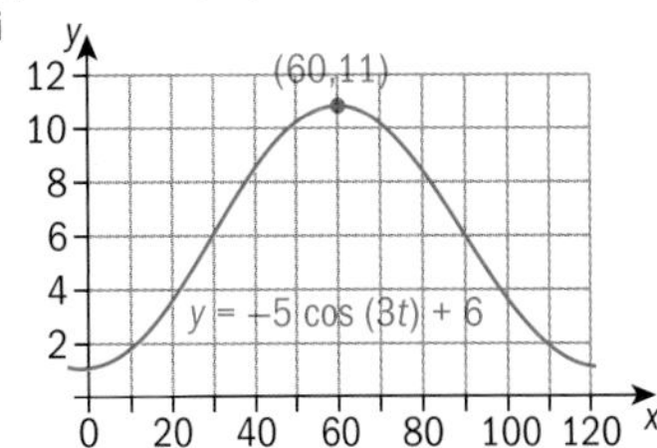

iii Children

10 a The period is 10 seconds – it is the time taken for the building to sway back to its original position.

b 20 cm

c $y = 20\sin(36t)$

11 a 180 cm – represents the distance travelled in one revolution

b 30 cm – represents the radius of the wheel

c 30 cm

d $y = -30\cos(2x) + 30$

12 a

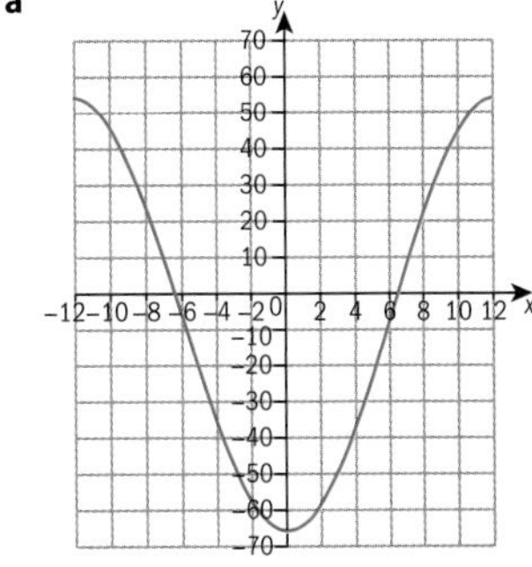

b Sunrise and sunset

c The sun is below the horizon – it is dark

d 37.4°, 47°

e 6:19 am

Review in context

1 a $d = -10\sin\left(\frac{360x}{15}\right) + 9$

b i 4.9 m **ii** 0.40 m **iii** 9.3 m

c –1 m

d The shore waters would recede leaving the ocean floor exposed.

2 a

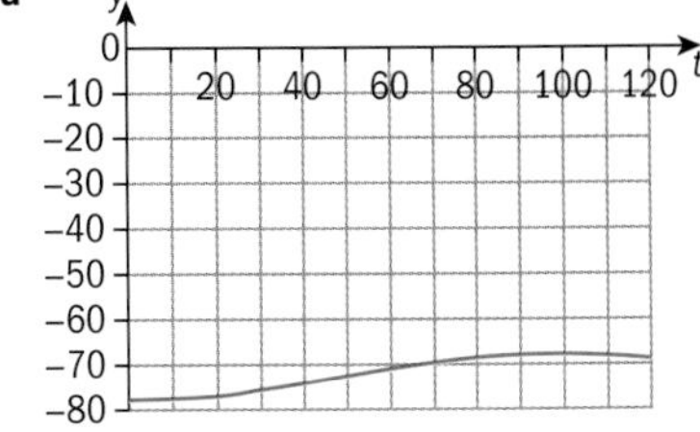

b $y = -4.5\cos(3x) - 73.5$

c –69.6°

d Horizontal dilation scale factor $\frac{1}{3}$, vertical dilation scale factor 4.5, reflection in the x-axis, and vertical translation 73.5 units in the negative direction

e No – it would travel linearly down the latitude

3 a

b $y = -8.5\cos(15°t) + 10$ **c** 4 m

d Midnight and midday

e Tides provide constant energy, whereas wind varies.

4 a

b 1.5 million, 20 days, freq 18

c More than half – is at 3.5 million (> 2.5 million)

d Before – the next low would happen sooner and be over quicker.

5 a

b frequency $= \frac{1}{50}$ seconds/cycle, hence period is $\frac{360°}{\left(\frac{1}{50}\right)} = 18\,000°$

c $y = 220\cos(18\,000°t)$

d

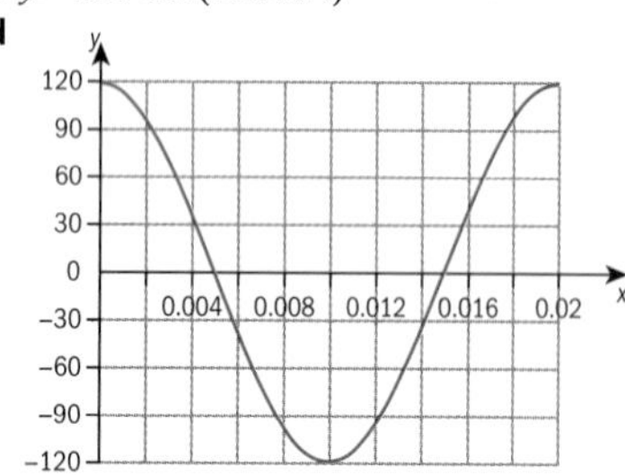

e $\frac{1}{60}$, 120, $y = 120\cos(21\,600°t)$

f Vertical dilation scale factor $\frac{6}{11}$ and horizontal dilation scale factor $\frac{5}{6}$

14.3

You should already know how to:

1 a $6 < 8$, valid **b** $2 < 4$, valid

c $6 < 12$, valid **d** $-3 < -6$, not valid

2 a $x = 2$ **b** $x = 12$

3 a $x > 3$ **b** $x < 2$

c $x \geq 3$ **d** $-2 < x < 2$

4 a

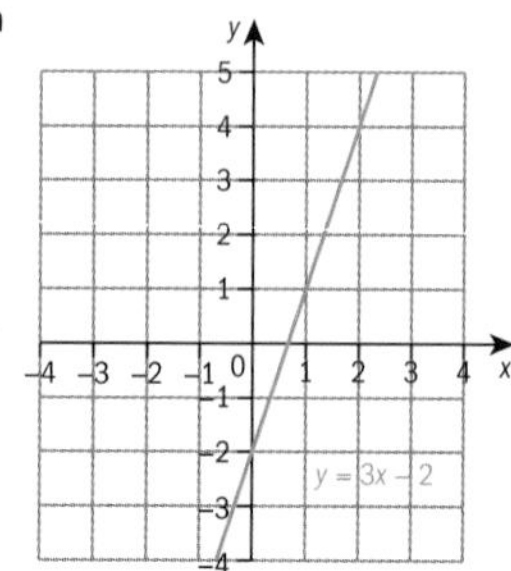

b

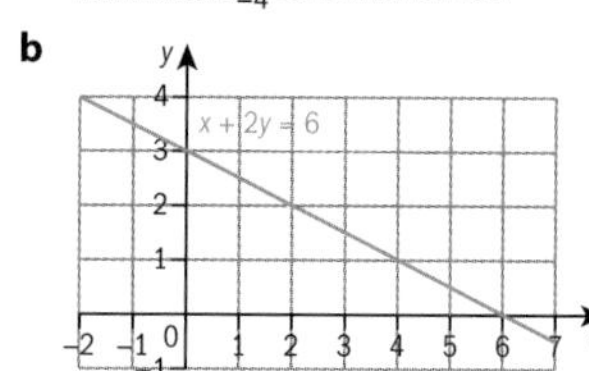

Practice 1

1 $x < 4$ **2** $x > 54$ **3** $m < 1$

4 $x \geq 8$ **5** $x \geq 2$ **6** $x < -\frac{7}{17}$

7 $a > 4$ **8** $b \leq 1.25$ **9** $k \leq 6.4$

Practice 2

1 Solutions the same as in **Practice 1**

2 Answers will vary, e.g. $y \leq -x + 6$; $y < 4$

3 $3x + 2 < 9x + 6$

$-6x < 4$

$x > \frac{-2}{3}$

which is an infinite set of solutions.

Practice 3

1 The unshaded area is the solution set.

a

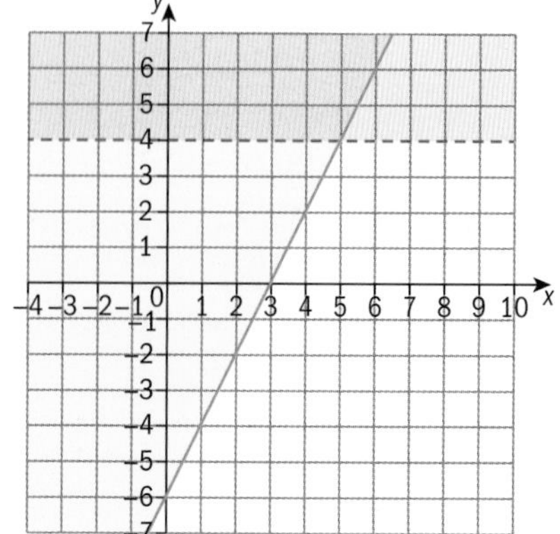

b

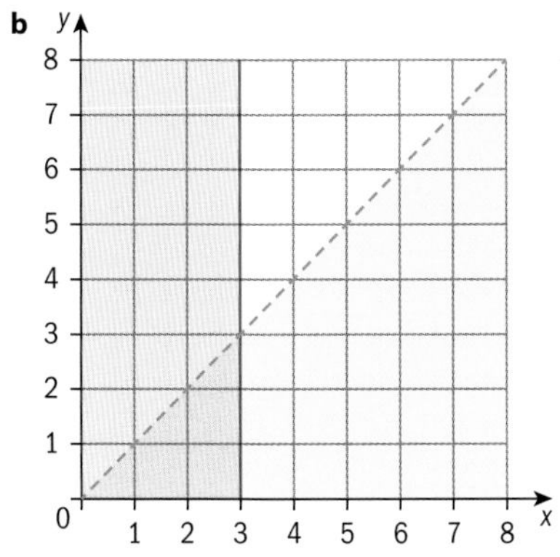

c

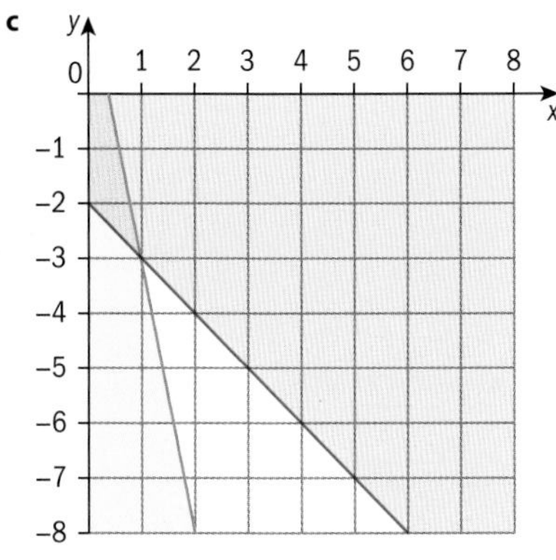

d

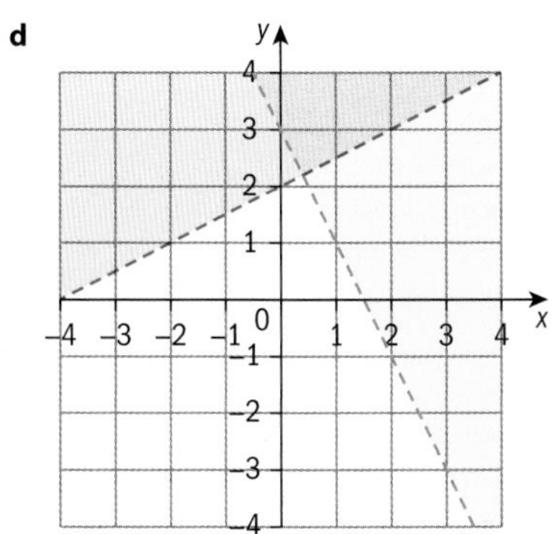

e

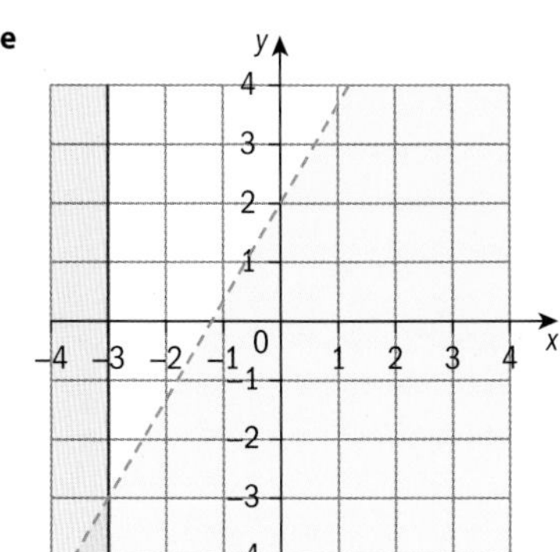

f

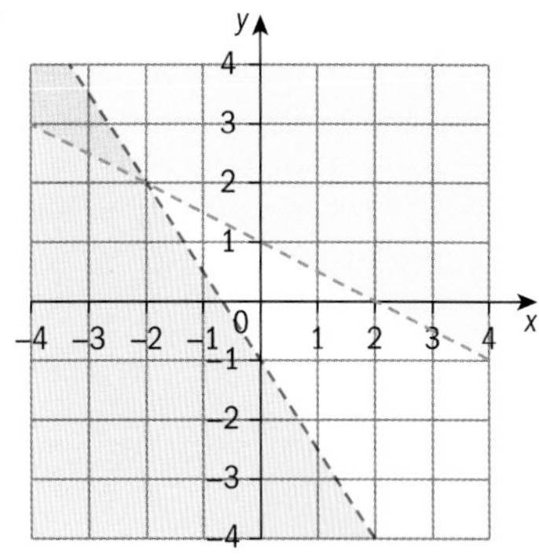

g

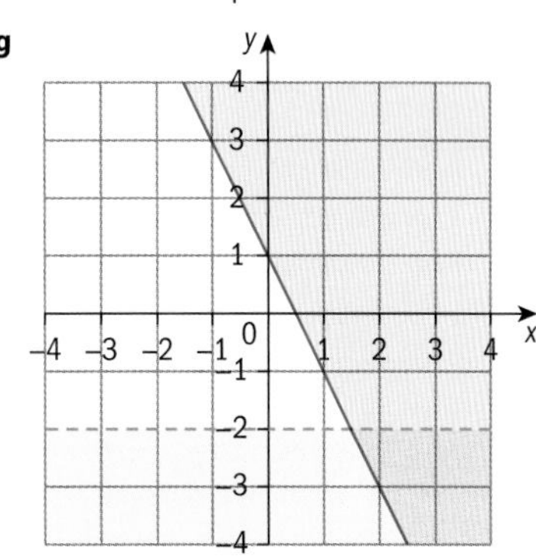

h

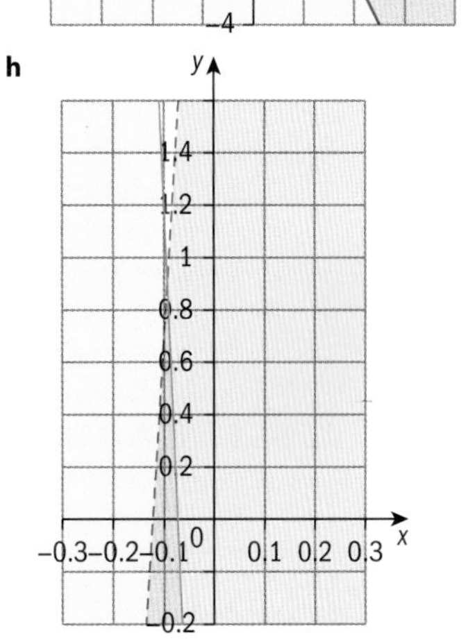

2 $x < 6$; $y \le x$; $y \ge 6 - 2x$

3 The unshaded area is the solution set.

a

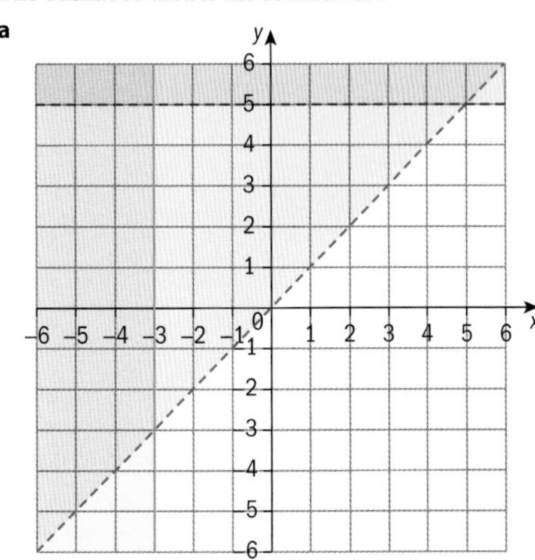

b

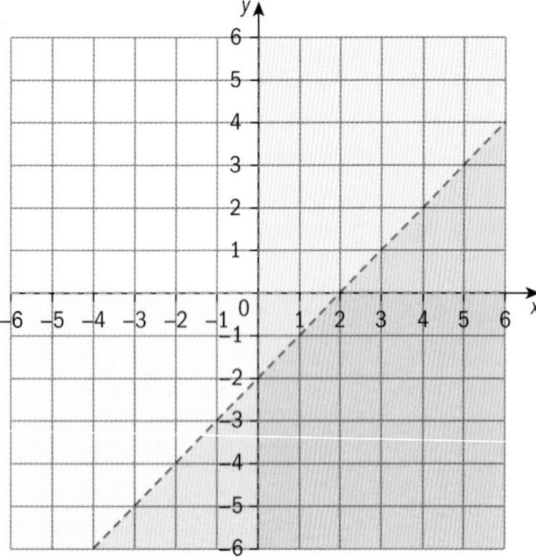

4 The unshaded area is the solution set.

a

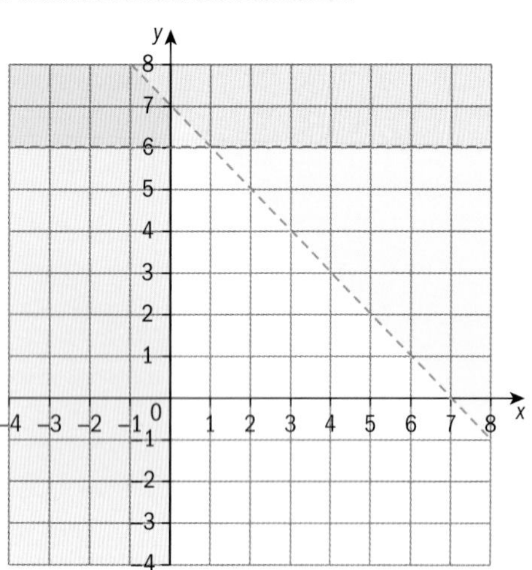

b

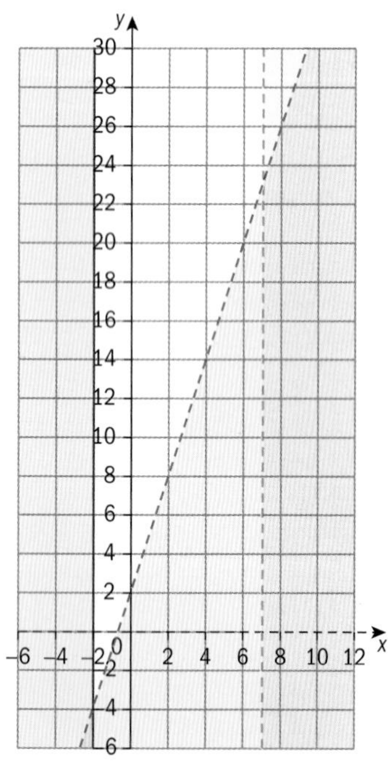

c

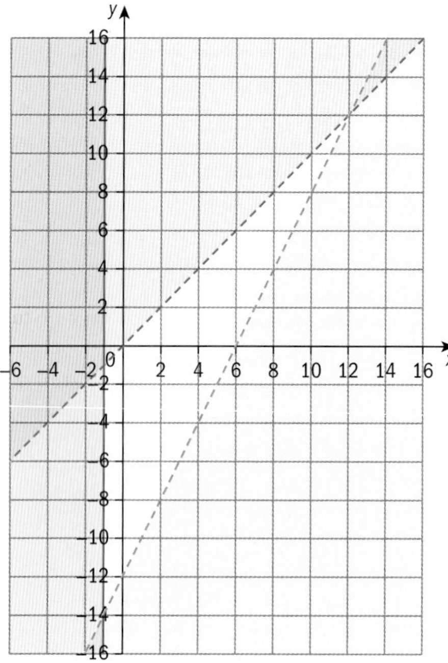

d

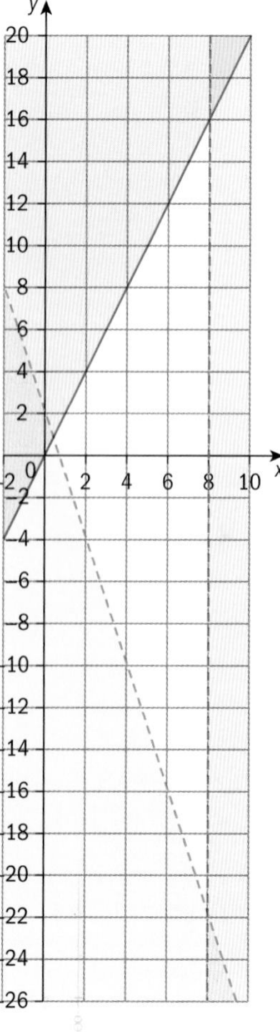

Practice 4

1 Maximum = 480 at (15, 1.5)

2 Maximum = 19.6 at (1, 6)

3 Maximum = 707 at (99, 181)

4 Maximum = 2085 at (105, 45)

5 4 small and 7 large minibuses

6 100 algebraic solvers and 170 graphing programs

7 40 downhill skis and 30 cross country skis

8 $75 000 in municipal bonds and $25 000 in bank mutual fund

Mixed practice

1 **a** $x \geq 1$
b $x < 18$
c $x > -9$
d $x < -5$
e $x \geq 4$
f $5 \leq x \leq 9$
g $\frac{1}{2} < x < 8$

2 Solutions same as **1 a – d**

3 The unshaded area is the solution set.

a

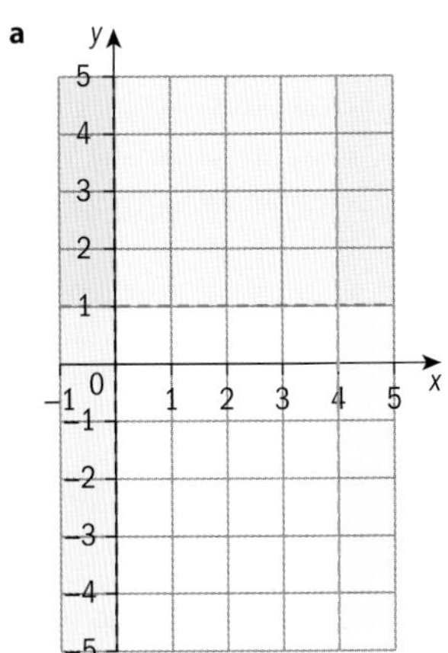

b

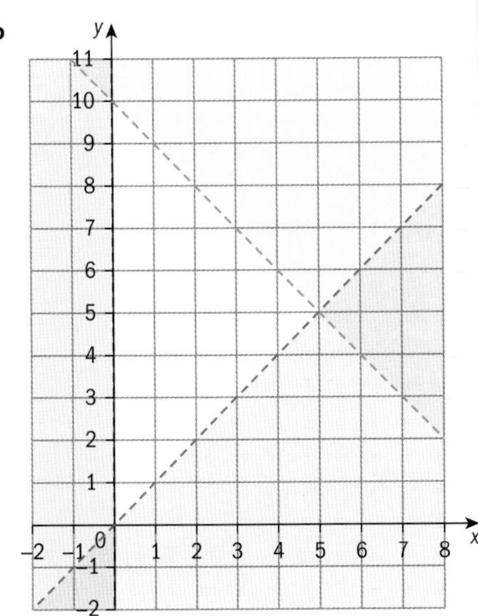

c

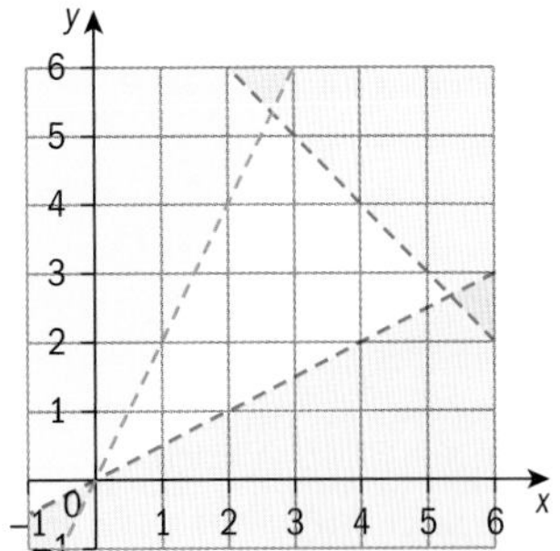

d

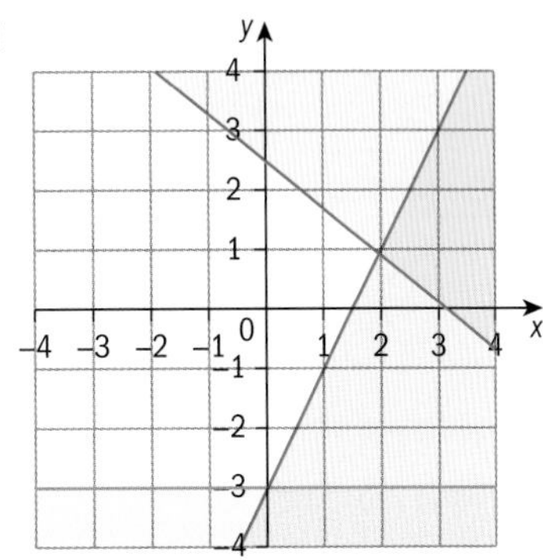

4 Three 40-seater and no 24-seater buses

5 The factory can meet its target.
It can buy 8 Machine A and 0 Machine B, or 6 Machine A and 3 Machine B.

6 12 weekday and 8 weekend advertisements

Review in context

1 6 large trees and 20 small trees

2 21 type A and 8 type B costing £83

15.1

You should already know how to:

1 **a** 1,2,3,4,5,6 **b i** $\frac{1}{6}$ **ii** $\frac{1}{2}$

2 $\frac{12}{22} = \frac{6}{11}, \frac{19}{22}$ 3 0.2, 0.6, 0.08 4 $\frac{15}{50} = \frac{3}{10}$

Practice 1

1 **a** Yes **b** No **c** Yes

2 **a** No **b** No **c** No

3 **a** 0.2
b Intersection is not zero, it is possible to be stung by both.

4 **a**

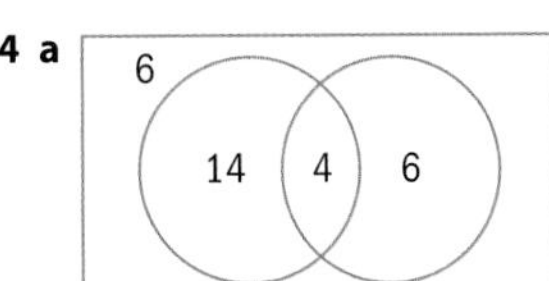

b 4 **c** No – intersection is not zero

5 a $p = 4, q = 1$

b $P(\text{boy}) = \frac{18}{30} = \frac{3}{5}$, $P(\text{girl}) = \frac{12}{30} = \frac{2}{5}$, Yes

c $P(S) = \frac{24}{30} = \frac{4}{5}$, $P(B) = \frac{19}{30}$, No

6 a Pick a heart, pick a diamond (for example)

b Pick a red card, pick a 10 (for example)

Practice 2

1 a $\frac{1}{24}$

b Yes since $P(A) \times P(B) = P(A \cap B)$

c $\frac{1}{4} \times \frac{1}{6} = \frac{1}{24}$

2 a $\frac{1}{18}$

b Yes since $P(A) \times P(B) = P(A \cap B)$

c $\frac{1}{6} \times \frac{1}{3} = \frac{1}{8}$

3 a 28

b Yes since the first domino chosen is replaced

c $\frac{7}{28} \times \frac{7}{28} = \frac{1}{16}$

4 $\frac{9}{20} \times \frac{10}{20} \neq \frac{6}{20}$ so no, not independent

5 $\frac{6}{24} \times \frac{8}{24} = \frac{2}{24}$ so yes, independent

6 a $\frac{39}{90} = \frac{13}{30}$ **b** $\frac{60}{90} = \frac{1}{3}$ **c** $\frac{35}{90} = \frac{7}{18}$

d No since $P(\text{trampolining}) \times P(\text{table tennis}) \neq 0$

Practice 3

1 a

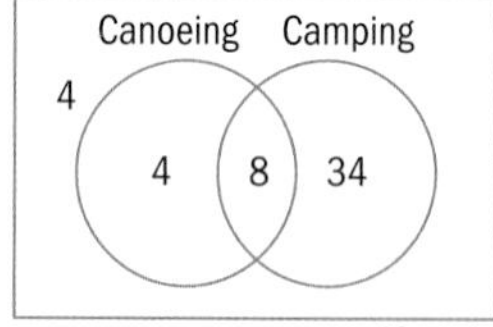

b 4 **c** $\frac{12}{50} = \frac{6}{25}$ **d** $\frac{12}{50} \times \frac{42}{50} \neq \frac{8}{50}$ so no

e Axiom 2, Axiom 3, Theorem 1, Theorem 2

2 a

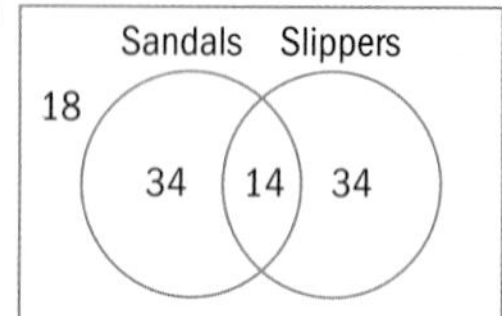

b 48 **c** $\frac{14}{48} = \frac{7}{24}$

d i No **ii** No

e Axiom 2, Axiom 3, Proposition 1, Theorem 1, Theorem 2

3

	R	L	
M	58	12	70
F	27	3	30
	85	15	100

a $\frac{3}{30} = \frac{1}{10}$

b Yes – a person is left or right handed, not both.

c No since $P(L) \times P(R) \neq 0$

c Axiom 2, Axiom 3, Proposition 1, Theorem1, Theorem 2

Practice 4

1 a i 20 **ii** 12 **iii** 15

iv 18 **v** $\frac{12}{20} = \frac{3}{5}, \frac{15}{20} = \frac{3}{4}$ **vi** $\frac{18}{20} = \frac{9}{10}$

b i $\frac{9}{20}$ **ii**

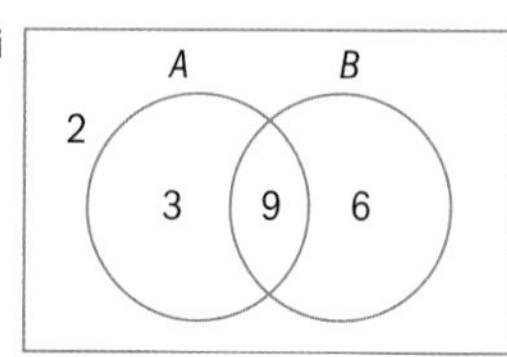

c i $\frac{9}{12} = \frac{3}{4}$ **ii** $\frac{3}{12} = \frac{1}{4}$

d i $\frac{6}{8} = \frac{3}{4}$ **ii** $\frac{2}{8} = \frac{1}{4}$

e

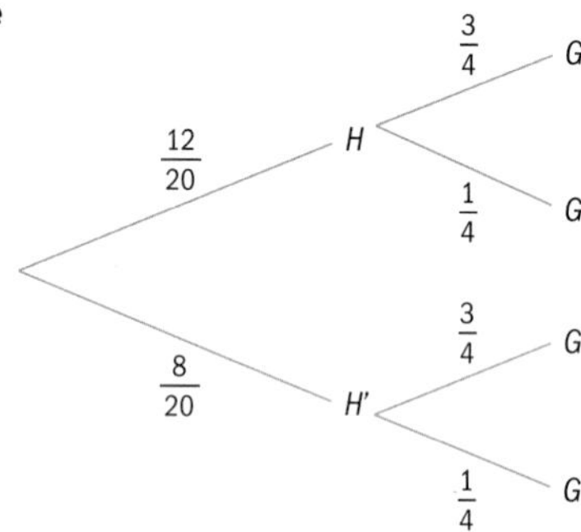

2 a i 0.7 **ii** 0.12

b

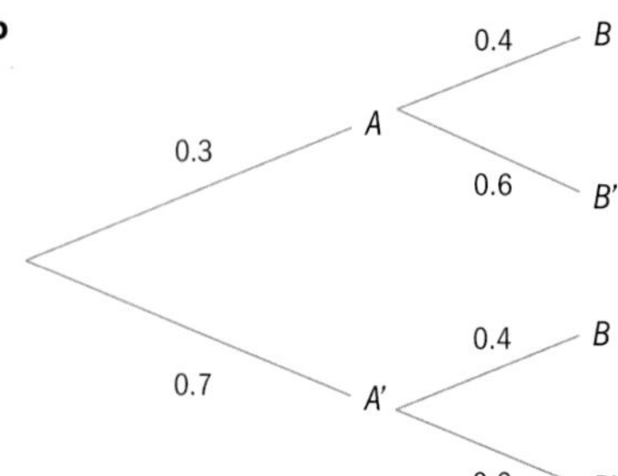

Mixed practice

1 a i $\frac{18}{30}=\frac{3}{5}$ **ii** $\frac{24}{30}=\frac{4}{5}$ **iii** $\frac{10}{30}=\frac{1}{3}$ **iv** $\frac{26}{30}=\frac{13}{15}$

b $\frac{4}{30}=\frac{2}{15}$

c

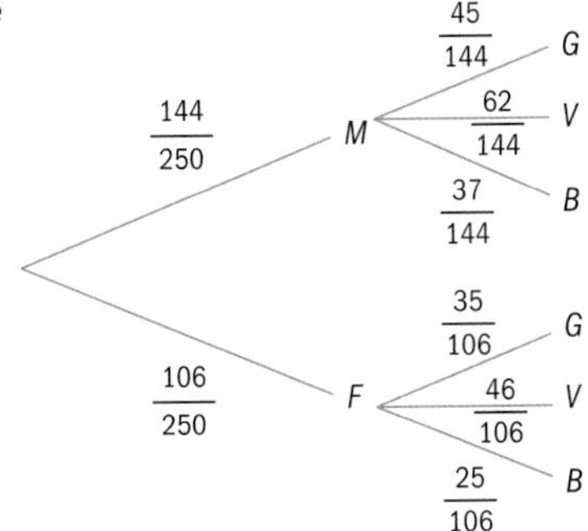

d $\frac{4}{18}=\frac{2}{9}$

2 a i Yes **ii** Yes

b $\frac{108}{250}=\frac{54}{125}$ **c** $\frac{35}{106}$ **d** $\frac{60}{142}=\frac{30}{71}$

e

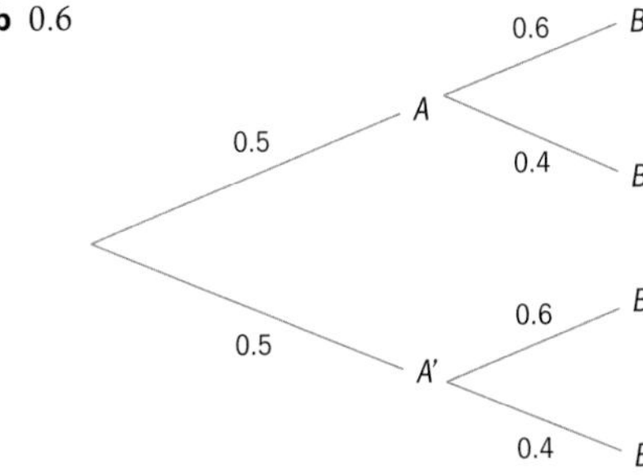

Yes

3 a 0.3

b 0.6

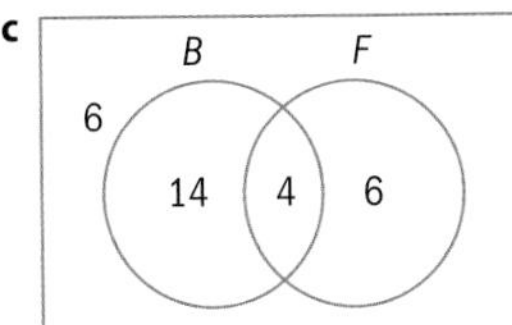

c Yes: $0.5 \times 0.6 = 0.3$

4 a

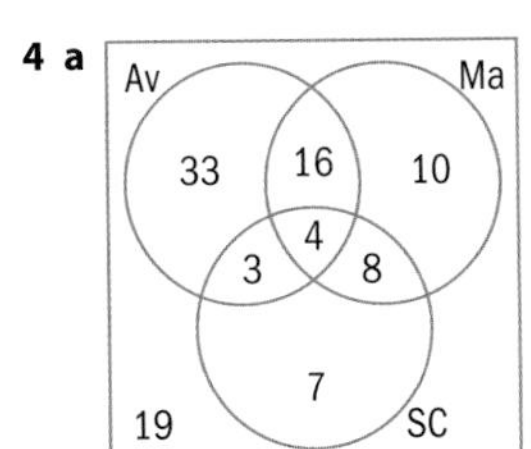

b 19 **c** No **d** $\frac{12}{22}=\frac{6}{11}$

5 a 0.2 **b** No **c** No

6 a

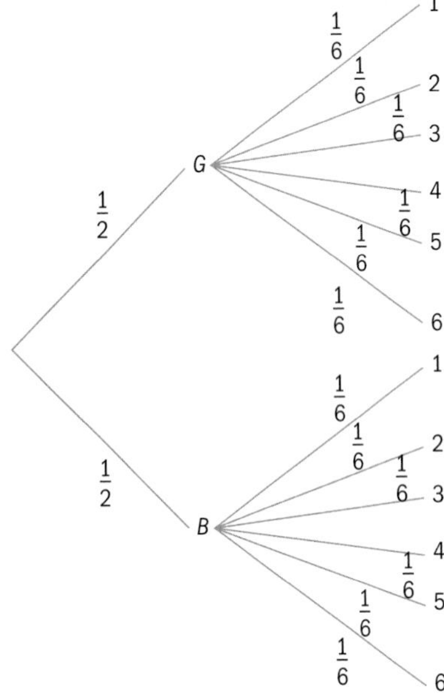

b $\frac{1}{6}$ **c** $\frac{2}{6}=\frac{1}{3}$ **d** $\frac{1}{4}$

e Yes $P(E) \times P(G) = P(E \cap G)$

7 a i $\frac{3}{50}$ **ii** $\frac{45}{50}=\frac{9}{10}$ **iii** $\frac{16}{18}=\frac{8}{9}$

b No, No

Review in context

1 a i $\frac{32}{60}=\frac{8}{15}$ **ii** $\frac{28}{60}=\frac{7}{15}$ **iii** $\frac{8}{17}$

iv $\frac{49}{60}$ **v** $\frac{3}{25}$

b i $\frac{14}{885}$ **ii** $\frac{253}{1770}$

2 a i 0.04 **ii** 0.96

b Yes **c** 0.54 **d** 0.53 **e** $\frac{42}{53}$

3 a No **b i** 0.5 **ii** 0.5

c

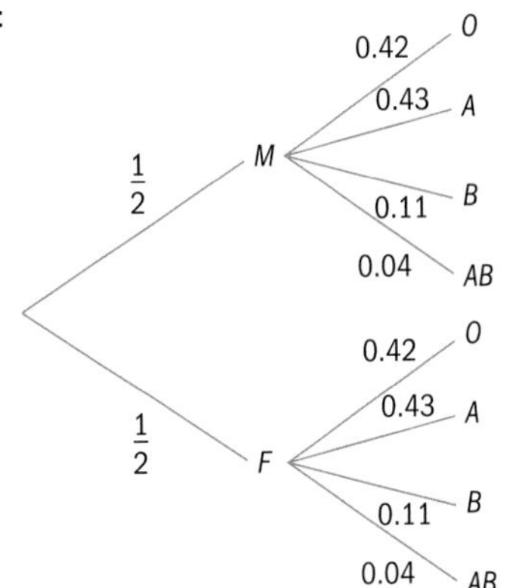

d E.g. $P(M) \times P(O) = P(M \cap O)$

Index